Extended Linear Chain Compounds

Volume 3

Extended Linear Chain Compounds

Volume 3

Edited by

Joel S. Miller

Occidental Research Corporation
Irvine, California

PLENUM PRESS • NEW YORK AND LONDON

Library of Congress Cataloging in Publication Data

Main entry under title:

Extended linear chain compounds.

Includes bibliographical references and index. 1. Polymers and polymerization. I. Miller, Joel S.

QD381.E96 547.7 81-17762

ISBN 0-306-40941-0 AACR2

A Division of Plenum Publishing Corporation
233 Spring Street, New York, N.Y. 10013

Printed in the United States of America

Contributors

L. Alcácer, Serviço de Química-Física Estado Sólido, LNETI, Sacavém, Portugal

Stephen R. Bondeson, Department of Chemistry, Princeton University, Princeton, New Jersey 08544

Jill C. Bonner, Department of Physics, University of Rhode Island, Kingston, Rhode Island 02881

James W. Bray, General Electric Corporate Research and Development, P.O. Box 8, Schenectady, New York 12301

I. D. Brown, Department of Physics, McMaster University, Hamilton, Ontario, Canada

W. R. Datars, Department of Physics, McMaster University, Hamilton, Ontario, Canada

A. F. Diaz, IBM Research Laboratory, San José, California 95193

Helmut Endres, Anorganisch-Chemisches Institut der Universität, Im Neuenheimer Feld 270, 6900 Heidelberg 1, West Germany

William E. Estes, Department of Chemistry, University of North Carolina, Chapel Hill, North Carolina 27514

R. M. Gaura, Program in Chemical Physics, Department of Chemistry, Washington State University, Pullman, Washington 99164

R. J. Gillespie, Department of Chemistry, McMaster University, Hamilton, Ontario, Canada

William E. Hatfield, Department of Chemistry, University of North Carolina, Chapel Hill, North Carolina 27514

Brian M. Hoffman, Department of Chemistry and Materials Research Center, Northwestern University, Evanston, Illinois 60201

James A. Ibers, Department of Chemistry and Materials Research Center, Northwestern University, Evanston, Illinois 60201

Leonard V. Interrante, General Electric Corporate Research and Development, P.O. Box 8, Schenectady, New York 12301

Israel S. Jacobs, General Electric Corporate Research and Development, P.O. Box 8, Schenectady, New York 12301

K. K. Kanazawa, IBM Research Laboratory, San José, California 95193

C. P. Landee, Program in Chemical Physics, Department of Chemistry, Washington State University, Pullman, Washington 99164

Wayne E. Marsh, Department of Chemistry, University of North Carolina, Chapel Hill, North Carolina 27514

Jens Martinsen, Department of Chemistry and Materials Research Center, Northwestern University, Evanston, Illinois 60201

Paul J. Nigrey, Laboratory for Research on the Structure of Matter, University of Pennsylvania, Philadelphia, Pennsylvania 19104

H. Novais, Serviço de Química-Física Estado Sólido, LNETI, Sacavém, Portugal

Laurel J. Pace, Department of Chemistry and Materials Research Center, Northwestern University, Evanston, Illinois 60201

M. Wayne Pickens, Department of Chemistry, University of North Carolina, Chapel Hill, North Carolina 27514

Zoltán G. Soos, Department of Chemistry, Princeton University, Princeton, New Jersey 08544

Leonard W. ter Haar, Department of Chemistry, University of North Carolina, Chapel Hill, North Carolina 27514

Robert R. Weller, Department of Chemistry, University of North Carolina, Chapel Hill, North Carolina 27514

R. D. Willett, Program in Chemical Physics, Department of Chemistry, Washington State University, Pullman, Washington 99164

Preface

Linear chain substances span a large cross section of contemporary chemistry ranging from covalent polymers, organic charge transfer complexes to nonstoichiometric transition metal coordination complexes. Their commonality, which coalesced intense interest in the theoretical and experimental solid-state-physics/chemistry communities, was based on the observation that these inorganic and organic polymeric substrates exhibit striking metal-like electrical and optical properties. Exploitation and extension of these systems has led to the systematic study of both the chemistry and physics of highly and poorly conducting linear chain substances. To gain a salient understanding of these complex materials rich in anomalous anisotropic electrical, optical, magnetic, and mechanical properties, the convergence of diverse skills and talents was required. The constructive blending of traditionally segregated disciplines such as synthetic and physical organic, inorganic, and polymer chemistry, crystallography, and theoretical and experimental solid state physics has led to the timely development of a truly interdisciplinary science. This is evidenced in the contributions of this monograph series. Within the theme of *Extended Linear Chain Compounds*, experts in important, but varied, facets of the discipline have reflected upon the progress that has been made and have cogently summarized their field of specialty. Consequently, up-to-date reviews of numerous and varied aspects of "extended linear chain compounds" has developed. Within these volumes, numerous incisive contributions covering all aspects of the diverse linear chain substances have been summarized.

I am confident that assimilation of the state-of-the-art and clairvoyance will be rewarded with extraordinary developments in the near future. Clearly, commercially viable applications of this class of materials is imminent and we look forward to them.

I wish to thank all of the contributors and their families for relinquishing the time necessary to consummate this endeavor.

Irvine, California Joel S. Miller

Contents

2. The Synthesis and Static Magnetic Properties of First-Row Transition-Metal Compounds with Chain Structures

William E. Hatfield, William E. Estes, Wayne E. Marsh, M. Wayne Pickens, Leonard W. ter Haar, and Robert R. Weller

3. Ferromagnetism in Linear Chains

R. D. Willett, R. M. Gaura, and C. P. Landee

4. Magnetic Resonance in Ion-Radical Organic Solids

Zoltán G. Soos and Stephen R. Bondeson

5. Salts of 7,7,8,8-Tetracyano-*p*-quinodimethane with Simple and Complex Metal Cations

Helmut Endres

6. Linear Chain 1,2-Dithiolene Complexes

L. Alcácer and H. Novais

7. The Spin–Peierls Transition

James W. Bray, Leonard V. Interrante, Israel S. Jacobs, and Jill C. Bonner

8. Polypyrrole: An Electrochemical Approach to Conducting Polymers

A. F. Diaz and K. K. Kanazawa

9. Compendium of Synthetic Procedures for One-Dimensional Substances

Paul J. Nigrey

10. Structural, Magnetic, and Charge Transport Properties of Stacked Metal Chelate Complexes

Brian M. Hoffman, Jens Martinsen, Laurel J. Pace, and James A. Ibers

Contents of Other Volumes

Volume 1

Volume 2

1

The Infinite Linear Chain Compounds $Hg_{3-\delta}AsF_6$ and $Hg_{3-\delta}SbF_6$

I. D. Brown, W. R. Datars, and R. J. Gillespie

1. History and Introduction

The compound $Hg_{3-\delta}AsF_6$ was first prepared in 1971 by Gillespie and Ummat[1] during the course of experiments which had as their aim the preparation of polyatomic cations of mercury having the general formula Hg_n^{2+}. The mercurous cation Hg_2^{2+}, **1**, has been known for a very long time and it was the unique example of a metal cation containing two covalently bonded metal atoms.

$$\overset{+}{Hg}\text{—}\overset{+}{Hg} \quad \overset{+}{Hg}\text{—}Hg\text{—}\overset{+}{Hg} \quad \overset{+}{Hg}\text{—}Hg\text{—}Hg\text{—}\overset{+}{Hg}$$

1 **2** **3**

It seemed reasonable to suppose that, if two mercury atoms could be bonded together in this way, it might be possible to prepare cations containing three and even more mercury atoms covalently bonded together. In view of the very strong tendency of mercury to form two colinear covalent bonds these ions were expected to have linear structures such as **2** and **3**.

At that time it had been found that solutions of AsF_5 and SbF_5 in liquid sulfur dioxide are very effective reagents for oxidizing nonmetals and metalloids to previously unknown polyatomic cations containing the element in unusual low oxidation states. For example, it had been found that sulfur, selenium, and tellurium could be oxidized to S_4^{2+}, Se_4^{2+}, and Te_4^{2+}, respectively, as well as to other cationic species such as S_8^{2+} and Te_6^{2+}.[2] Therefore, solutions of AsF_5 and SbF_5 in SO_2 were reacted with mercury in order to attempt the preparation of cations such as **2** and **3**. The first

I. D. Brown and W. R. Datars • Department of Physics, McMaster University, Hamilton, Ontario, Canada. *R. J. Gillespie* • Department of Chemistry, McMaster University, Hamilton, Ontario, Canada.

experiments showed immediately that previously unknown mercury species were produced when a solution of AsF_5 in liquid SO_2 reacted with liquid mercury to give a golden crystalline solid in a deep red solution which then changed to orange with the disappearance of the golden yellow solid to give finally a colorless solution. It appeared that these color changes corresponded to different and previously unknown oxidation states of mercury because both mercuric Hg(II) and mercurous Hg(I) salts are in general colorless.

From the orange solution an orange crystalline material having the composition $Hg_3(AsF_6)_2$ was isolated. The overall reaction may be represented by the equation

$$3Hg + 3AsF_5 \xrightarrow{SO_2} Hg_3(AsF_6)_2 + AsF_3.$$

An x-ray crystallographic determination of the structure of $Hg_3(AsF_6)_2$ showed that it did indeed contain the linear Hg_3^{2+} cation[3,4] with an Hg–Hg bond length of 2.55 Å.

At about the same time, Torsi *et al.*[5] found that the Hg_3^{2+} ion could also be prepared by the reaction of mercury with mercuric chloride, $HgCl_2$, in molten aluminum trichloride. The tetrachloraluminate $Hg_3(AlCl_4)_2$ was isolated from this reaction and a crystal structure determination[6] showed that the Hg_3^{2+} ion has an almost linear structure with a bond angle of 174.4° and a bond length of 2.56 Å.

With a sufficient quantity of AsF_5 the final product of the reaction with mercury is a colorless solution that was shown to contain mercurous hexafluorarsenate $Hg_2(AsF_6)_2$. It seemed reasonable to suppose that the deep red solution, preceding the formation of the orange solution containing $Hg_3(AsF_6)_2$, might contain mercury in a still lower oxidation state. By using smaller amounts of AsF_5, the reaction can be stopped at the deep red stage and crystallization of this solution gave a mixture of orange crystals of $Hg_3(AsF_6)_2$ and deep red-black needles.[7]

An x-ray crystallographic examination of this latter product showed it to be the compound $Hg_4(AsF_6)_2$ and to contain Hg_4^{2+} cations and octahedral AsF_6^- anions.[7] The Hg_4^{2+} cation has an almost linear structure with bond angles of 176° and a *trans* configuration **4**.

$$Hg \xrightarrow{2.57} Hg_{176°} \xrightarrow{2.70} Hg^{176°} \xrightarrow{2.57} Hg$$

4

The formation of the Hg_4^{2+} cation can be represented by the equation

$$4Hg + 3AsF_5 \longrightarrow Hg_4(AsF_6)_2 + AsF_3.$$

In the first stages of the reaction of mercury with a solution of AsF_5 in SO_2, the liquid mercury is rapidly transformed into a golden yellow crystalline mass that persists in equilibrium with the deep red solution. It disappears only when sufficient AsF_5 is added to give the orange $Hg_3(AsF_6)_2$. The crystalline compound, which had a distinct metallic luster, appeared to contain mercury in an even lower oxidation state than in Hg_4^{2+}. The chemical analysis of the crystals confirmed this supposition by showing that they had the composition Hg_3AsF_6.[1] In the light of the structure of Hg_3^{2+} and Hg_4^{2+} it was proposed that these crystals contained Hg_6^{2+} cations, i.e., that the compound was $Hg_6(AsF_6)_2$. However, an x-ray crystal structure determination subsequently showed this conclusion to be incorrect.[8] Precession photographs (Figure 1) showed the rather unusual feature of prominent sheets of diffuse intensity perpendicular to the *a* and *b* axes of the tetragonal crystals. A complete determination of the structure showed these diffuse intensity sheets to arise from chains of mercury atoms running through the crystal lattice in two mutually perpendicular directions parallel to the *a* and *b* axes (see Section 3). The mercury–mercury distance in the chains was found to be about 2.64 Å, and, since this distance is not commensurate with the dimensions of the tetragonal lattice ($a = 7.54$ Å), the compound appeared to have the nonstoichiometric composition $Hg_{2.86}AsF_6$. Each mercury atom has a formal nonintegral charge of $+0.35$ and the conduction band arising from the overlap of the mercury orbitals was expected to be partially filled. It was therefore predicted that the compound would be highly conducting along the directions of the chains, and, as the interchain distance of 3.09 Å is substantially greater than the Hg–Hg distance within

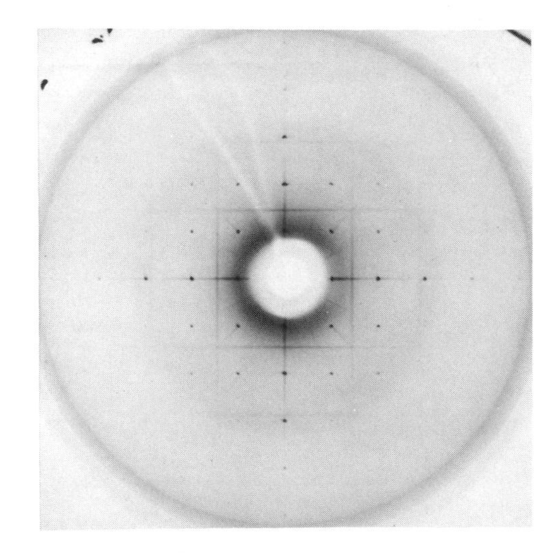

FIGURE 1. X-ray precession photograph of the (*hk*0) plane of $Hg_{3-\delta}SbF_6$. The lattice of Bragg peaks is associated with the SbF_6^- lattice, the lines with the Hg chains. Photographs of $Hg_{3-\delta}AsF_6$ are similar.

the chains, it was expected that the conductivity would exhibit a considerable anisotropy (see Section 4).

It was subsequently found that a solution of SbF_5 in SO_2 also reacted with liquid mercury in a quite analogous fashion and the same series of color changes were observed as with AsF_5. An orange compound $Hg_3(Sb_2F_{11})_2$ was isolated from the solution. The formation of this compound can be represented by the equation

$$3Hg + 5SbF_5 \longrightarrow Hg_3(Sb_2F_{11})_2 + SbF_3.$$

With smaller amounts of SbF_5 a deep red solution together with golden yellow "metallic" crystals were obtained. The crystals have a structure essentially identical to that of $Hg_{2.86}AsF_6$ (see Section 3).

It should be stressed that the composition $Hg_{2.86}AsF_6$ is an ideal composition determined from the crystal structure and that chemical analysis gives compositions closer to Hg_3AsF_6.[1,9] In view of the fact that there is some uncertainty about the exact composition and because it appears that the composition may vary with temperature, the formulas are written here in the more general form as $Hg_{3-\delta}AsF_6$ and $Hg_{3-\delta}SbF_6$. The question of the composition of these compounds is discussed in detail in Section 4.1.

2. Chemistry

Several different procedures have been developed for the preparation of the compounds $Hg_{3-\delta}AsF_6$ and $Hg_{3-\delta}SbF_6$. They all involve the oxidation of liquid mercury with a solution of a suitable oxidizing agent in liquid sulfur dioxide. Instead of using AsF_5 or SbF_5 as oxidizing agents, as in the original preparations described above, the salts $Hg_3(AsF_6)_2$ and $Hg_3(Sb_2F_{11})_2$ may be used.[9]

$$Hg_3(AsF_6)_2 + 3Hg \longrightarrow 2Hg_3AsF_6\dagger$$

$$Hg_3(Sb_2F_{11})_2 + 3Hg \longrightarrow 2Hg_3SbF_6\dagger + 2SbF_5.$$

These salts have the advantage of being somewhat more easily handled than the simple fluorides, and solutions of known concentration are more easily prepared.

Relatively large crystals are needed for some of the physical measurements, such as electrical conductivity, and these can be grown if steps are taken to prevent the nucleation of a large number of crystals. This can be accomplished by reducing the rate of reaction and by keeping the surface area of the mercury very small. The reactions proceed at a convenient rate

† The approximate formulas Hg_3AsF_6 and Hg_3SbF_6 are used here in order to simplify the equations.

at room temperature but large crystals are obtained when the rate is slowed down by reducing the temperature to about −20°C. A more convenient method of reducing the reaction rate is to keep the concentration of the oxidizing agent rather low throughout the course of the reaction by adding the oxidizing agent in small successive amounts. If, in addition, the area of the surface between the mercury and the liquid sulfur dioxide is kept small by arranging that it is in a capillary tube, very frequently only one crystal forms, which in the case of $Hg_{3-\delta}AsF_6$ may grow as large as $10 \times 10 \times 2$ mm.[9,10] The crystal grows on the surface of the mercury and it is evident that mercury is transported through the crystal to the growing crystal faces in contact with the SO_2–AsF_5 solution. The crystals of $Hg_{3-\delta}SbF_6$ obtained in this way are generally appreciably smaller. The growth of large crystals seems to be inhibited by the simultaneous formation of a precipitate of insoluble SbF_3.

It has been of considerable interest to attempt to prepare compounds analogous to $Hg_{3-\delta}AsF_6$ and $Hg_{3-\delta}SbF_6$ but containing anions other than AsF_6^- and SbF_6^-. An obvious extension of the preparations described above was to study the reaction of mercury with other pentafluorides.[11] It was found that mercury does not react with PF_5 even under pressure at 60°C in liquid SO_2. Bismuth pentafluoride BiF_5 was found to be insoluble in SO_2 and no reaction with mercury was observed. Bismuth pentafluoride has some solubility in HF, but in this solvent the only products of the reaction with mercury were mercurous and mercuric salts. Other fluorides were also investigated including WF_6 and UF_6 and either there was no reaction or the products were the well-known mercurous and mercuric salts. Other types of oxidizing agents were therefore studied. In particular, it was found that $O_2^+AsF_6^-$, $O_2^+SbF_6^-$, $NO^+AsF_6^-$, and $NO^+SbF_6^-$ could be used to oxidize mercury to $Hg_{3-\delta}AsF_6$ and $Hg_{3-\delta}SbF_6$. However, when other O_2^+ salts such as $O_2^+BiF_6^-$ and $O_2^+PtF_6^-$ were used, no evidence for the formation of new long-chain mercury compounds was obtained. In the cases of $O_2^+BiF_6^-$ and $O_2^+PtF_6^-$ the only products obtained from the reactions were $Hg_2(BiF_6)_2$ and HgO, respectively.

Still other types of oxidizing agents such as $F_5TeOOTeF_5$ were studied. This compound is reduced to $OTeF_5^-$ which is a pseudo-octahedral ion that is expected to resemble closely SbF_6^-.

$$F_5TeOOTeF_5 + 2e^- \longrightarrow 2F_5TeO^-$$

However, the only product of the reaction of mercury with $F_5TeOOTeF_5$ in liquid sulfur dioxide was found to be $Hg_2(OTeF_5)_2$.

It is disappointing that an intensive search for long-chain mercury compounds containing anions other than SbF_6^- and AsF_6^- has been unsuccessful. It is difficult to understand why only these two anions should have the ability to form a lattice that stabilizes these mercury cations.

It is a common feature of polyatomic cations containing an element in a low and unusual oxidation state that, in the presence of a suitably strong base, they undergo disproportionation to more stable oxidation states. For example, although the cation I_2^+ is relatively stable in the presence of SO_3F^- it is completely disproportionated in the presence of the more-basic anion HSO_4^-

$$I_2^+ \xrightarrow{HSO_4^-} I_3^+ + I(SO_4H)_3$$

and, in the presence of the still more-basic water molecule, I_3^+ is disproportionated to free iodine I_2 and iodate IO_3^-.[12] It is noteworthy that AsF_5 and SbF_5 are the strongest known Lewis acid pentafluorides and hence the corresponding anions AsF_6^- and SbF_6^- are the weakest known hexafluoride bases. Thus, it is possible that in the presence of more-basic hexafluoride anions, compounds that contain mercury in the approximate +1/3 oxidation state are completely disproportionated to elemental mercury and a mercurous salt. The polymeric anions $Sb_2F_{11}^-$ and $Sb_3F_{16}^-$ are certainly more weakly basic than SbF_6^- but their elongated shape may not be suitable for the formation of a lattice containing long straight tunnels capable of accommodating the infinite cation chains. Alternatively, it may be that because these anions produce SbF_5 by dissociation

$$Sb_2F_{11}^- \rightleftharpoons SbF_6^- + SbF_5$$

$$Sb_3F_{16}^- \rightleftharpoons Sb_2F_{11}^- + SbF_5$$

they oxidize the mercury chain cations further to give Hg_4^{2+} and Hg_3^{2+}.

We need to improve our understanding of the factors responsible for the stabilization of the infinite-chain mercury cations so that the search for new compounds containing such cations can be pursued further.

3. *Room-Temperature Structure* (*D Phase*)

The structure of $Hg_{3-\delta}AsF_6$ was originally determined by Brown *et al.*[8] using x-ray diffraction and confirmed using neutron diffraction by Schultz *et al.*[13] The structure is unusual in that it contains two components, AsF_6^- anions and cationic Hg chains, whose lattice parameters are incommensurate—a unique feature that gives rise to most of the interesting properties reviewed here. In order to understand the structure, it is convenient to consider the two components separately, the host AsF_6^- lattice and the two mutually perpendicular sets of parallel chains of mercury atoms. The structure is shown in Figure 2.

The arrangement of the chains which lie in channels within the host lattice is shown in Figure 2b. The host lattice itself is tetragonal, space

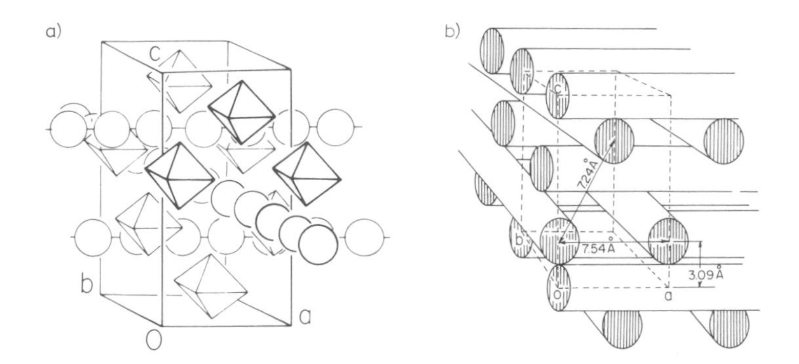

FIGURE 2. (a) Structure of $Hg_{3-\delta}AsF_6$. The octahedra represent AsF_6^- ions. (b) Arrangement of the Hg chains in $Hg_{3-\delta}AsF_6$. The AsF_6^- ions (not shown) lie in the cavities between the chains.

group $I4_1/amd$ with four AsF_6^- ions in the unit cell, and has the lattice parameters and coordinates given in Table 1. It consists of discrete almost octahedral AsF_6^- ions (D_{2d} crystal symmetry) with two axial As–F distances of 1.702(5) Å and four slightly longer equatorial distances of 1.720(5) Å. The difference in the lengths of these bonds can be attributed to the different cationic environments of the F atoms, the axial F atoms have a single neighboring Hg chain at 2.95 Å while the equatorial F atoms have two such chains at 3.00 Å.

The host lattice gives rise to a perfectly normal x-ray diffraction pattern with well-defined Bragg peaks. The unusual feature found in the x-ray and neutron diffraction patterns is a series of sheets of diffuse scattering arranged perpendicular to the a^* and b^* axes at positions† $h = (3 - \delta)n$ and $k = (3 - \delta)n$, where n is an integer and $\delta = 0.18$. These sheets can be seen as lines in the $hk0$ precession photograph of the analogous $Hg_{3-\delta}SbF_6$ compound shown in Figure 1. Such sheets represent the scattering that one would expect from a one-dimensional array of atoms. In this case, the array can be identified with the infinite chains of Hg atoms that thread the host lattice along the a and b directions.

From the position and structure of the diffuse sheets, one can obtain the following information about the mercury chains:

(i) The positions of the sheets correspond to a spacing‡ between the Hg atoms of 2.670(5) Å.

† Here, and throughout this chapter, all coordinates (in reciprocal and real space) are given in terms of the spacings of the host lattice, i.e., integral values of h, k, and l correspond always to the reciprocal lattice points of the host lattice. The unsubscripted lengths a, b, c and a^*, b^*, c^* also refer to the host lattice.

‡ This spacing will be referred to as d or d_{Hg} throughout this chapter.

TABLE 1. Selected Structural Data for $Hg_{3-\delta}AsF_6$ and $Hg_{3-\delta}SbF_6$ [a]

	$Hg_{3-\delta}AsF_6$				$Hg_{3-\delta}SbF_6$
	Ref. 8	Ref. 13	Ref. 14		Ref. 15
	(293 K)	(293 K)	(293 K)	(0 K)	(293 K)
AsF_6 lattice					
Space group	$I4_1/amd$	$I4_1/amd$	$I4_1/amd$	— [b]	$I4_1/amd$
$a = b$ (Å)	7.538(4)	7.549(5)	7.534(7)	7.44	7.711(2)
c (Å)	12.339(5)	12.390(9)	12.395(8)	12.248	12.641(2)
Unit cell volume (Å^3)	701.1	706.1	703.6	678.0	751.6
As–F(1) × 2	1.61(7)	1.702(2)			1.84(1)
As–F(2) × 4	1.71(4)	1.720(2)			1.88(1)
Hg chains					
d_{Hg}(Hg–Hg intrachain)	2.64(1)	2.64(2)	2.670(5)	2.670(5)	2.66(2)
Interchain Hg–Hg distances					
Perpendicular chains					
if chains are straight	3.085(2)	3.098(2)	3.099(2)	3.06	3.160(2)
if chains undulate		3.24(2)			3.23(1)
Parallel chains					
along [100]	7.538(4)	7.549(5)	7.534(7)	7.44	7.711(2)
along [101]	7.227(3)	7.254(5)	7.253(7)	7.17	7.404(2)
Other					
Hg–F(1)	3.02(6)	2.87(1)			2.86(1)
Hg–F(2)	2.99(4)	2.981(5)			2.98(1)
Crystallographic value of δ	0.14(1)	0.14(2)	0.178(6)	0.210(4) (at 80 K)	0.10(2)
Density (measured), g/cm^3		7.06(2)			

[a] Numbers given in parentheses are the standard errors in the last digit quoted.
[b] Monoclinic, $\gamma = 90.1°$.

(ii) Both Schultz *et al.*[13] and Pouget *et al.*[14] have examined the thickness of the sheet and both have found that their measurements were limited by instrumental resolution. The thinness of the sheets indicates that the chains have a large coherence length so that the 2.67 Å spacing between Hg atoms is precisely maintained over long distances.

(iii) At room temperature, the sheets, at least for values of $n \neq 0$, are without structure, indicating that there is no coherence between adjacent parallel Hg chains. The positions of Hg atoms in one chain are therefore unaffected by the position of Hg atoms in adjacent chains. This is not surprising since the incommensurability of the chains and lattice makes it difficult for them to couple to each other, and adjacent parallel chains are too far away from each other (over 7 Å) to couple directly.

There are no diffuse sheets observed for $n = 0$, since in these planes the scattering from the chains contributes only to the Bragg reflections of the host lattice. This can be understood by considering just one of the two perpendicular arrays of chains, say that parallel to *a*. The arrangement of atoms along a chain does not influence the scattering in the zero layer. For this layer, each chain can therefore be considered as a structureless rod of scattering matter. The diffraction pattern observed will then be the diffraction pattern of a two-dimensional array of rods, namely, a two-dimensional array of Bragg peaks lying in the plane perpendicular to the rods. Since the chains (or rods) are embedded in channels within the host lattice, the spacings between the chains will be the spacings of the host lattice and the Bragg peaks arising from the chains will be superimposed on the Bragg peaks of the host lattice. The chains are arranged in a centered array belonging to the plane group *amm* (Figure 2) and contribute to all the Bragg reflections with $h = 0$, $k + l = 2n$. However, the host lattice, because its atoms are on special positions, only contributes to the $h = 0$ reflections with $k = 2n$ and $l = 2n$. Thus, half the $0kl$ and $h0l$ Bragg reflections have contributions only from the chains while the other half have contributions from both the chains and the host lattice.

The symmetry of the host lattice does not require that the chains be perfectly straight and it is possible for them to undulate with a period equal to the lattice spacing of the host lattice. In this case, there would be a small contribution from the chains to those Bragg reflections of the host lattice that do not lie in the zero layer. Schultz *et al.*[13] included this degree of freedom in their refinement and found that the chains are undulating in such a way as to increase slightly the contact distance between perpendicular chains at the point where they cross from 3.10 to 3.24 Å.

4. Structure-Related Properties

4.1. Composition

The exact composition of $Hg_{3-\delta}AsF_6$ has been the subject of some debate. Chemical analyses[1, 10, 11] give a composition close to the stoichiometry Hg_3AsF_6, but the dimensions of the host lattice and the distance between Hg atoms within the chains would suggest that the contents of the unit cell are $4x(Hg_{3-\delta}AsF_6)$, where $\delta = (3 - a/d_{Hg})$. Using the observed values of a and d, a value of 0.18 is found for δ in the crystallographic studies at room temperature. This differs significantly from the value of $\delta = 0$ found in the chemical analyses. In order to resolve this discrepancy, Miro *et al.*[10] made careful measurement of the density of the crystals. By comparing the observed value [7.06(2) g/cm^3] with those calculated for a number of models—Hg_3AsF_6 ($Hg_{2.82}AsF_6$ + 0.14 interstitial Hg, 7.49 g/cm^3) $Hg_{2.82}AsF_6$ (7.16 g/cm^3), and $Hg_{2.82}(AsF_6)_{0.94}$ ($Hg_{2.82}AsF_6$ with anion vacancies, 7.05 g/cm^3)—they concluded that the compound contains Hg and AsF_6 in the ratio of 3 : 1 and that this stoichiometry is maintained by the introduction of anion vacancies. Such a formulation suggests that each Hg atom carries a formal positive charge of precisely 1/3 and that the electronic structure of the chains does not permit this value to vary enough to allow all anion sites to be occupied. Against this it must be noted that there are formidable problems in making accurate density and chemical composition measurements on these very reactive crystals.

4.2. Extrusion of Mercury

A most interesting feature of these compounds is that, on cooling, the lattice contracts but the mercury–mercury distance in the chains remains constant. This implies that in an ideal crystal the composition should change continuously with temperature. The observed 0.5% contraction in the lattice between room temperature and 160 K implies that, if the crystal retains its ideal structure, 0.5% of the mercury should be lost from the crystal over this temperature range.

In order to investigate possible phase transitions, particularly any involving a loss of mercury, differential thermal analysis experiments were carried out by Datars *et al.*[16] In these experiments the heat changes associated with melting or freezing or other phase changes in the sample are observed. It was found that when a sample of $Hg_{3-\delta}AsF_6$ or $Hg_{3-\delta}SbF_6$ was cooled to 160 K and then warmed again to room temperature, a pronounced endotherm was observed during the warming cycle corresponding to the melting of mercury at 235 K. The cooling curves showed no corresponding exotherm in this range although a broad and rather ill-defined exotherm was observed over the temperature range 185–205 K, which varied somewhat in shape and size from sample to sample.

Another very important observation made during these experiments was that the endotherm due to the melting of mercury was only observed during warming if the sample had previously been cooled below a certain threshold temperature which varied somewhat from sample to sample, but was normally in the range 200–210 K. The area of the mercury melting endotherm depends on the lowest temperature to which the sample has been cooled. As this temperature is lowered from the threshold to 15° below the threshold the area increases. For lower temperatures the area then remains constant. This 15° temperature range corresponds approximately to that of the broad exotherm observed on cooling the sample, which, it seems reasonable to conclude, corresponds to a phase transition in which solid mercury is liberated from the sample. This solid mercury does not reenter the crystal on raising the temperature but subsequently melts when the temperature is raised to 235 K. The liquid mercury then appears to be reabsorbed although no separate isotherm corresponding to the uptake of mercury is normally observed on warming. The heat associated with this phase change is either combined with the endotherm due to melting or is too small and occurs over too great a temperature range to be observed.

Assuming that the endotherm observed on warming is due only to the melting of mercury, its area corresponds to the liberation on cooling of 1.3 ± 0.4 wt. % of the total mercury in the sample. From the 0.4% contraction of the lattice between room temperature and 200 K, it would be expected that only 0.4% of the mercury would have been liberated.

The exact nature of the process by which mercury is extruded from the crystal is not certain, but it does not occur continuously as the crystal is cooled from room temperature. This implies that down to the threshold temperature the average mercury–mercury distance in the chains is able to remain constant while the AsF_6^- or SbF_6^- lattice contracts without the crystal losing mercury. How this is accomplished is not clear. Possibly there are vacancies in the mercury chain that become filled as the lattice contracts, but this view is not consistent with the chemical analysis discussed in Section 4.1. Buckling of the chains within the channels can also be ruled out since this would reduce the apparent Hg–Hg distance as recorded in the diffraction sheets.

The extrusion may be related to the ordering of Hg and AsF_6^- ions (see Sections 6.2.1 and 9.4) and changes in resistance (Section 5) and thermopower (Section 9.5) that also occur at this temperature.

4.3. Thermal and Hydrolytic Decomposition

Crystals of both compounds can be kept in sealed tubes for over a year without losing their characteristic golden metallic luster. However, on exposing the crystals to the atmosphere they immediately become grey and within 30 s distinct droplets of mercury become visible on the surface.

Microscopic observation of crystals whose rate of decomposition was decreased by covering them with Saranwrap showed that almost none of the mercury produced by hydrolytic decomposition appears on the large flat (001) face of the crystal. This observation is consistent with the structure, since this face is the only one on which no mercury chains terminate.[11]

A similar anisotropic decomposition was observed upon heating crystals to approximately 420 K. At temperatures above 485 K there is a rapid loss of a large quantity of mercury from the crystals and finally only a yellow-white powder of $Hg_2(AsF_6)_2$ remains.[11]

5. Electrical Resistivity

The electrical resistivity was studied by Cutforth *et al.*[17] and Chiang *et al.*,[18] who found that it is isotropic in the (001) plane, in agreement with the tetragonal symmetry of the lattice and the equivalence of the two sets of perpendicular chains aligned along *a* and *b*. The anisotropy ρ_a/ρ_c (where ρ_a and ρ_c are resistivities along the *a* and *c* directions, respectively) is 100 for $Hg_{3-\delta}AsF_6$ and 40 for $Hg_{3-\delta}SbF_6$ and is essentially independent of temperature between 300 K and 4.2 K. This anisotropy is also seen in the reflectivity of polarized light.[19–21] Light polarized parallel to the mercury chains exhibits a strong metallic-like plasma edge and is strongly reflected in the infrared, whereas the reflectance for light polarized perpendicular to the mercury chains is an order-of-magnitude weaker in both compounds.

The anisotropy results from the localization of the conduction electrons on the Hg chains. Because of the short metal–metal distance within the chains, each chain acts as a one-dimensional metal. Conduction in the *c* direction occurs only by electrons moving across the 3.1-Å gap between chains and is therefore much lower.

A better understanding of the electrical properties of $Hg_{3-\delta}AsF_6$ can be gained by considering the structure of the Fermi surface in momentum (reciprocal) space which is presented in Section 9. The Fermi surface of an isolated chain of Hg atoms consists of two parallel planes perpendicular to the chain direction and equidistant from the origin. The Fermi surface of two perpendicular sets of chains will therefore consist of two such pairs of planes arranged at right angles to each other as in Figure 3a. If there is no coupling between the electrons in different chains these planes will remain independent, but if electrons can move from one chain to an adjacent perpendicular one the sheets will form connected surfaces as shown in Figure 3b. The electrons in states on the Fermi surface move along the chains since the velocity is perpendicular to the Fermi surface. Only for states on the corners of the Fermi surface does the electron wave function have sufficient amplitude on both sets of chains to allow scattering between chains and hence conduction along the *c* direction.

The temperature dependence of the resistivity and conductivity, shown in Figure 4, exhibits several interesting properties. The resistivity at room

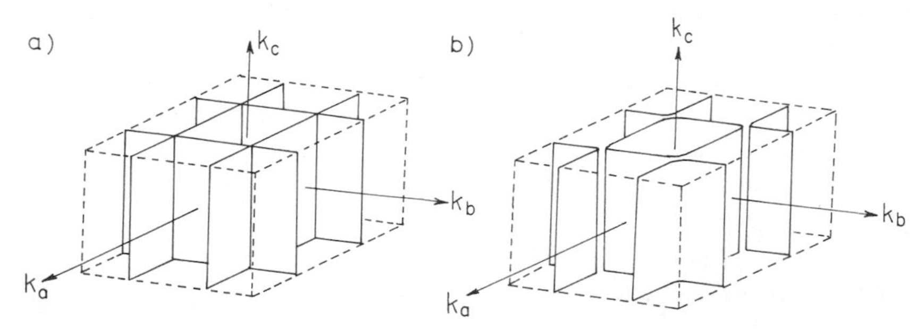

FIGURE 3. (a) Fermi surface of two perpendicular sets of noninteracting chains. The Brillouin zone is shown with broken lines. (b) Same As (a) but with interaction between the chains.

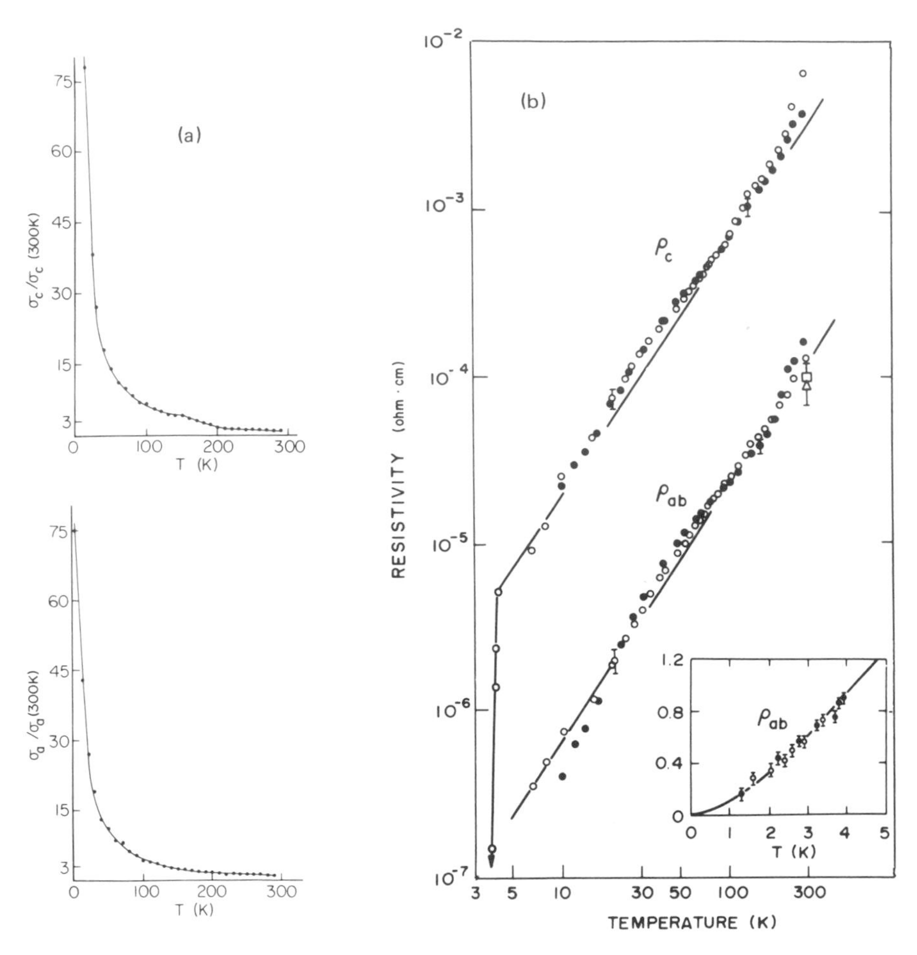

FIGURE 4. (a) Temperature dependence of the conductivity σ of $Hg_{3-\delta}SbF_6$ along the *c* direction and in the *ab* plane normalized to the room temperature conductivity σ_{300K}. (from Reference 17.) (b) Temperature dependance of the resistivity of $Hg_{3-\delta}AsF_6$. (From Reference 18.)

temperature is small with a value of 1×10^{-4} Ω cm for both compounds. It then decreases by factors of 300 and 75 in $Hg_{3-\delta}AsF_6$ and $Hg_{3-\delta}SbF_6$, respectively, between room temperature and 4.2 K. This change is comparable to that for simple, pure metals and is much larger than that of most compound metals. The resistivity follows the power law $\rho \sim T^m$ over a wide temperature (T) range. The value of the exponent m is approximately 1.5, but can vary between 1.3 and 1.8 in different samples so that it is difficult to attribute physical significance to the value 1.5. It is also interesting that down to at least 1.2 K there is no metal–insulator transition, such as is

intensity of the sheets along the line $(3 - \delta, k, 0)$, that is, along the intersection of one of the sheets with the $(hk0)$ plane (Figure 5). They found three different phases. At room temperature, the sheets are essentially structureless, indicating no correlation between the positions of the Hg atoms in one chain and those in the next (disordered or D phase). Between 120 K and about 200 K, they found broad peaks at $k \sim \pm 0.4$ which they associated with a short-range order between parallel chains (S phase). Cooling below 120 K, they found that the broad peaks rapidly vanished and a new set of narrow, intense peaks appeared at $k = 1 \pm \delta$. The intensity of these peaks increased as the temperature was lowered to 10 K, the lowest temperature at which measurements were made. They associated this behavior with a long-range, three-dimensional ordering involving both sets of chains (L phase).

6.2.1. Short-Range-Order Phase (S Phase, 120–200 K)

The diffuse peaks that appear at $(3 - \delta, \pm 0.4, 0)$ have a half-width of $\Delta k = \pm 0.2$ consistent with only a partial ordering of the Hg atoms. An examination of other parts of reciprocal space showed that similar peaks occur at $(3 - \delta, \pm 0.6, 1)$ but that the line $(2(3 - \delta), k, 0)$ remains structureless. Pouget *et al.*[14] showed that these results are consistent with a poorly ordered reciprocal lattice that can be indexed by $(h(3 - \delta), k \pm 0.4, l)$ with $k + l = 2n$. This corresponds to a face-centered monoclinic cell with

$$a_s = d_{\mathrm{Hg}}, \qquad b_s = \frac{a}{\cos \gamma_s}, \qquad c_s = c, \qquad \gamma_s = 94\text{–}102^\circ\dagger$$

The range of values observed for γ_s indicates that the ordering of adjacent chains is not perfect and that their relative longitudinal displacement may be anywhere in the range from zero to 1.6 Å.

Because of this variation in displacement, long-range order is not possible. Figure 6 shows the average arrangement of neighboring chains and the influence of the ordering on the Bragg scattering in reciprocal space. Since the broadening of the reflections at $(3 - \delta, \pm 0.4, 0)$ is attributable to the variablity of γ_s, the reflections in the second layer at $(2(3 - \delta), \pm 0.8, 0)$ should be twice as broad. This is sufficient to account for the absence of any observed structure along $(2(3 - \delta), k, 0)$.

Two types of domain are possible for each set of chains, depending on whether γ_s is greater or less than 90°. The diffraction pattern from the

† Pouget *et al.*[14] use a body-centered cell. We have chosen to use the more conventional face-centered setting. The less common c-axis setting is chosen to preserve the relative orientations of a, b, and c in the different phases. The subscript s refers to the parameters of the S phase.

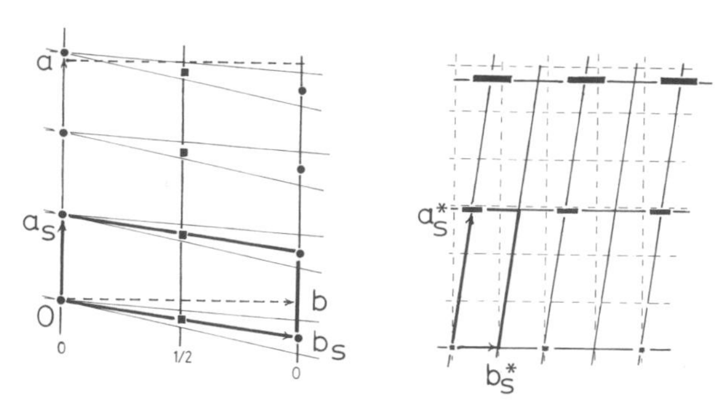

FIGURE 6. (a) Arrangement of the Hg atoms in neighboring parallel chains in the *S* phase. The AsF_6^- cell is outlined with a broken line, the Hg cell with a solid line. The thin lines represent the range of γ values observed. (b) Reciprocal lattice corresponding to (a) showing the neutron scattering from the short-range ordering.

second type of domain is identical to that of the first except that the sign of k is reversed. The presence of both domains in a crystal further serves to smear out the structure in the second sheet.

The diffuse peaks first appear on cooling at around 200 K. On further cooling they become more intense (Figure 7) but remain broad. The width

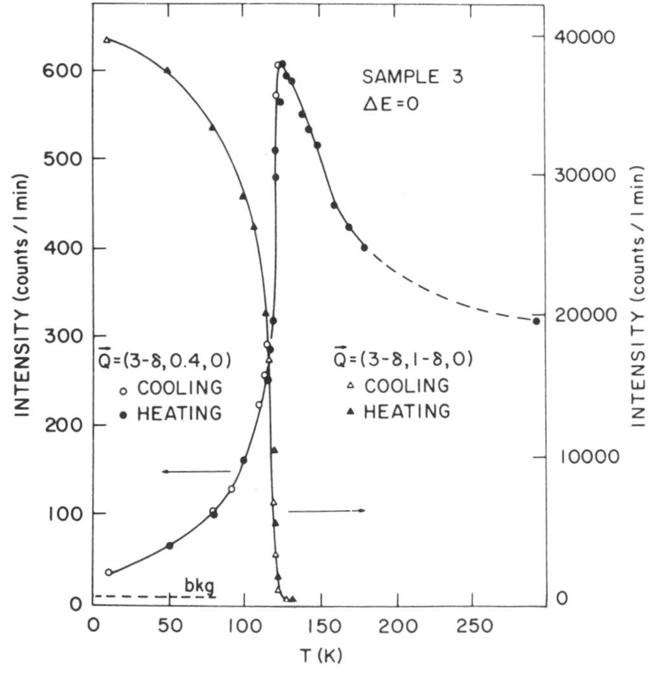

FIGURE 7. Temperature dependence of the intensity of neutron scattering for a reflection of the *S* phase (120–200 K) and for a reflection of the *L* phase (<120 K). (From Reference 14.)

of the peaks indicates that the order remains poor throughout the temperature range but the increase in intensity indicates an increase in the proportion of the crystal that is ordered. Below the 120-K phase transition, the intensity of the broad peaks drops rapidly, but some short-range order remains to the lowest temperature. With rapid cooling through 120 K, much of the short-range order can be frozen in.

6.2.2. Long-Range-Order Phase (L Phase, < 120 K)

Below 120 K, a new set of intense, narrow peaks appears at $(3 - \delta, 1 \pm \delta', 0)$. Pouget *et al.*[14] found that the value of δ' was identical (within the experimental error of 2%) with δ, suggesting that in this new phase both sets of chains ordered relative to each other. A survey of other points in reciprocal space indicated that the new lattice also consisted of two domains that could be indexed by $(h(3 - \delta), k \pm h\delta, l)$ with $h + k + l = 2n$. In the domain with $k - h\delta$, the reflections with $h = 1$ and $k = 3$ will occur at the intersection of two perpendicular sheets $(3 - \delta, 3 - \delta, l)$ which, as indicated above, is the place where a Bragg reflection must occur if the perpendicular sets of chains are to be coupled. The sharpness of the peaks indicates that the order is long range, that is to say, the Hg atoms form their own completely ordered lattice which is, however, still incommensurate with the host lattice. The Hg atoms in the [100] chains are arranged on a body-centered monoclinic lattice with $a_l = d$, $b_l = a/\cos \gamma_l$, $c_l = c$, and $\gamma_l = 93.5°$† as shown in Figure 8. The Hg atoms in the [010] chains lie on a similar lattice with a_l and b_l interchanged.

Both lattices have a common reciprocal lattice point at the sheet intersection $(3 - \delta, 3 - \delta, 0)$ corresponding to the 130 reflection of the [100] chain lattice and 310 reflection of the [010] chain lattice as shown in Figure 9. Therefore, both lattices must contain planes of atoms perpendicular to

† The subscript l indicates parameters of the L phase.

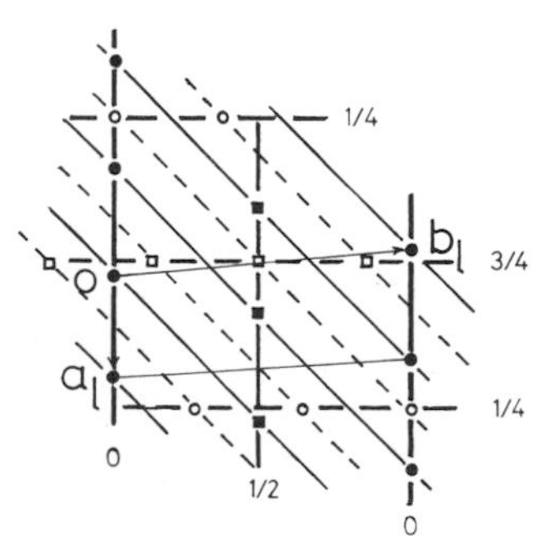

FIGURE 8. Arrangement of the Hg atoms in perpendicular chains in the L phase. [100] Hg lattice, solid; [010] Hg lattice, broken. The planes that correspond to the common reciprocal lattice point are shown.

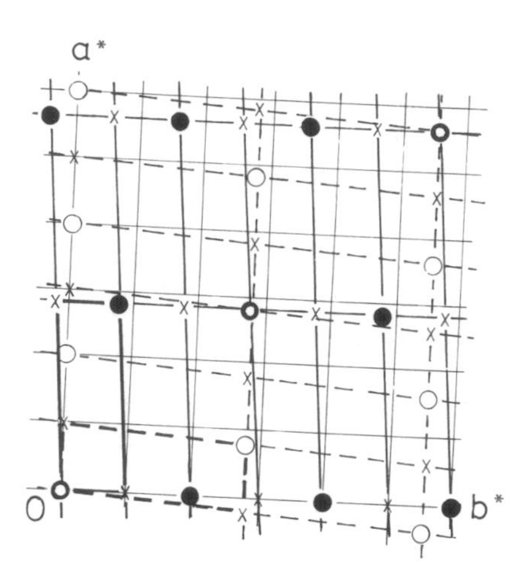

FIGURE 9. Superposition of the reciprocal lattices of the two perpendicular sets of chains in the L phase. The AsF_6^- reciprocal lattice is shown with light lines and the two Hg lattices with heavy ([100], solid; [010], broken) lines. The solid circles are observed reflections from the [100] Hg lattice, the open circles those from the [010] Hg lattice. Crosses mark positions of systematically absent reflections.

this reciprocal lattice vector. The relative spacing between the planes is given by the intensities of the reflections at $(n(3-\delta), n(3-\delta), 0)$ which is zero for $n = 1$ but large for $n = 2$, indicating that the two lattices are scattering out of phase. That is to say the atoms on the (130) planes of the [100] lattice lie between the atoms on the (310) planes of the [010] lattice as shown in Figure 8.

As in the case of the S phase, each set of chains has two possible domains corresponding to $\gamma_l = 90^\circ + 3.5^\circ$ and $\gamma_l = 90^\circ - 3.5^\circ$. Of the four possible combinations of domains, two cannot exist since they do not share a common reciprocal point. In the other two combinations, the domains share the common reciprocal lattice points $(m(3-\delta), m(3-\delta), 0)$ in one case and $(m(3-\delta), -m(3-\delta), 0)$ in the other.

The fully ordered phase consists of two incommensurate interpenetrating lattices—the tetragonal host lattice and the monoclinic Hg chain lattice. Even though the coupling between them is small, one would expect the strict tetragonal symmetry of the host lattice to be destroyed. This in fact was observed by Pouget *et al.*[14] but the effect is small amounting to a change in γ of only 0.1°.

In searching for the influence of the host lattice on the chains at low temperature, Pouget *et al.*[14] observed satellites that can be indexed as $h \pm 2\delta$, 0, l (and 0, $k \pm 2\delta$, l) associated with the $h0l$ reflections from the chains. They represent a lateral modulation of the chains with a propagation vector of $2\delta a^*$. No explanation has been proposed for this modulation which is only found in the L phase.

One of the predictions of the theory of Emery and Axe[25] is that if the perpendicular interactions had not been available to produce the L phase below 120 K (T_c), the interaction between parallel chains would have produced its own ordering at a slightly lower temperature (T_1). They suggest that interesting effects might be observed if the order of these two transitions can be changed (i.e., $T_1 > T_c$), perhaps by a change of pressure. In particular, a tricritical point would exist when $T_1 = T_c$. So far these possibilities have not yet been examined.

7. Dynamical Properties

7.1. Lattice Dynamics

Sacco and Sokoloff[28] have developed a theory for the motion of a chain passing through an incommensurate lattice. They point out that an infinitely long rigid chain will be perfectly free to slide through the lattice. The reason for this is that the Hg atoms will all lie at different positions along the periodic potential of the host lattice. An infinitesimal shift of the chain will move some Hg atoms to positions of higher potential but others will move to positions of lower potential with the net result that the potential energy of the chain remains unchanged. It therefore requires only an infinitesimal energy to start the chain sliding through the lattice. This is referred to as the zero-frequency phonon mode since it represents a translation of the whole chain, i.e., a wave of infinite wavelength and zero frequency.

In practice a number of factors might prevent such free sliding. If the chain is finite in length, Sacco and Sokoloff[28] show that the chain will be pinned to the lattice by the end atoms. Impurities and defects are irregularities in the host lattice that might also provide particularly favorable sites for pinning, and, if the chain were not rigid—i.e., the bonds between the Hg atoms could stretch—the chain might be able to distort itself sufficiently to accommodate to the host lattice spacings. Sacco and Sokoloff show that this latter effect will only occur when the chain–lattice coupling is sufficiently strong relative to the Hg–Hg force. In $Hg_{3-\delta}AsF_6$ the chain–lattice coupling appears to be too weak to provide any resistance to the sliding of the chains.

In their theory, Emery and Axe[25] ignore the chain–host lattice coupling but include the coupling between adjacent chains, both parallel and perpendicular. Like Sacco and Sokoloff they only include intrachain interactions between nearest-neighbor Hg atoms. They assume that small harmonic displacements of neighboring atoms are allowed, and therefore since the force is short range, there is no requirement for long-range order. That is, although each individual Hg–Hg bond has a length very close to $d_{Hg} = 2.670$ Å, there is no requirement that the distance to the mth atom down the chain will be exactly md_{Hg} Å; it will in general differ by roughly $m^{1/2}\sigma$ Å from this value, where σ is the rms deviation of the length of an individual bond from the value d_{Hg}. Thus, the further down the chain one looks, the further from exact registry the atoms will be.

Such a chain will be able to support one-dimensional longitudinal vibrations (acoustic phonons) which will be quite independent of the phonons of the host lattice. Using inelastic neutron scattering, such phonons have been observed at room temperature by Heilmann *et al.*,[29] who found that the one-dimensional longitudinal acoustic (LA) phonons of the chains have a much higher velocity (3600 ± 150 m/s) than the corresponding [100] LA phonons of the three-dimensional host lattice (2130 m/s). The chains are thus more rigid than the lattice and there is essentially no coupling between them. As expected, the acoustic-mode dispersion curves for the host lattice go to zero frequency at the host-lattice Bragg reflections while those for the chains go to zero frequency at the sheet positions, $n(3 - \delta)a^*$ and $n(3 - \delta)b^*$.

As discussed in Section 6, Emery and Axe[25] show that in the *L* phase below 120 K the chains become pinned by the interaction between neighboring perpendicular chains. This ordering results in the replacement of the sheets by a new set of Bragg reflections associated with the ordered chains. Heilmann *et al.*[30] measured the phonon spectrum of this phase close to the point $(3 - \delta, 0, 0)$, i.e., at a position close to where a sheet occurs at higher temperatures but away from the Bragg reflections associated with the ordered chains. In this case, they found that the dispersion curves did not go to zero frequency but showed a well-defined gap of about 0.13 meV. In addition, the velocity of these acoustic waves had increased to 4400 m/s, showing that the chains had not only become pinned (i.e., they required 0.13 meV to set them sliding) but were more rigid than at room temperature.

According to the theory of Emery and Axe[25] the sheets of scattering at room temperature do not result from Bragg scattering since this only occurs when each atom moves about an equilibrium position on a fixed lattice. Within the Hg chains each atom is relatively tightly fixed with respect to its two neighbors in the chain, but its position within the host lattice can vary with time. The atom does not have a fixed equilibrium position but will move back and forth along the chain. In this sense, the chains can be considered to be an ordered one-dimensional liquid. The sheets in the diffraction pattern then arise from the limiting case of inelastic scattering in which the energy loss on scattering is zero. On the basis of this model Emery and Axe[25] show that the shape and width of the sheets will be different from that expected from Bragg scattering; in particular, the width of the nth sheet will be proportional to n^2. Heilmann *et al.*[29] have made careful measurements of the neutron scattering (both elastic and inelastic) in the neighborhood of the sheets and have confirmed the predictions of Emery and Axe.

From these theoretical and experimental results we conclude that the Hg chains are relatively stiff and have a negligible interaction with the host lattice. At room temperature they can slide freely through the channels in the lattice and so behave as essentially ordered one-dimensional liquids. It is this fluidity that is used during crystal growth (see Section 2) to transport the Hg atom from the Hg metal to the crystal surface in contact with the AsF_5/SO_2 solution. Below 120 K the chains become pinned, not to the host lattice but to each other, and scatter neutrons and x-rays like a normal crystal.

7.2. Elastic Constants

Inelastic neutron scattering from the host lattice can be used to measure the speed of sound of various types of acoustic waves in the crystal, and these measurements in turn can be used to determine the elastic constants. Heilmann *et al.*[26] have measured the transverse acoustic (TA) and longitudinal acoustic (LA) modes of the host lattice along the [100] and [001] directions at room temperature. The results are summarized in Table 2.

According to elasticity theory, for a tetragonal crystal the wave propagating along [100] polarized along [001] must have the same velocity as the wave propogating along [001] polarized along [100]. The difference between the observed values has been attributed by Heilmann *et al.*[30] to the fact that the chains are strongly coupled to the lattice in the [001] direction (since they are confined to the channels within the lattice) but are uncoupled along their length in the [100] or [010] direction. The chains must necessarily vibrate with the lattice for waves polarized perpendicular to the chains, but not for waves polarized along the chains. The TA waves polarized along [001] involve both the lattice and all the chains, whereas the TA waves polarized along [100] involve only the lattice and half the chains. As expected, the observed ratio of the velocities is equal to $(m_1/m_2)^{1/2}$, where m_1 and m_2 are the masses of the components of the crystal vibrating in each of the two modes.

TABLE 2. Velocity of Sound in the Host Lattice of $Hg_{3-\delta}AsF_6$ in the D Phase [a]

v (TA, 100) =	912 (35) m/s
v (TA, 001) =	1140 (75) m/s
v (LA, 100) =	2130(300) m/s
v (LA, 001) =	1800(150) m/s

[a] The transverse modes are polarized along [001] and [100], respectively.

7.3. Specific Heat

Wei *et al.*[31] report that the specific heat of $Hg_{3-\delta}AsF_6$ between 1.8 and 5 K can be represented by

$$C = \gamma T + \beta T^3$$

where γ is 46.3 mJ K^{-2} (g-at. Hg)$^{-1}$ and $\beta = 6.09$ mJ K^{-4} (g-at. Hg)$^{-1}$. The value of γ represents an unusually large linear term and the value of β corresponds to a Debye temperature of 78 K. The data are discussed by Wei *et al.*[31] in terms of several contributions to the linear term γ, including phonon excitations of one-dimensional and three-dimensional lattices and excitations arising from disorder within the mercury chains.

Further measurements by Moses *et al.*[32] down to 0.35 K confirm the unusually large linear term and show a dramatic change in the functional form of the temperature dependence below 1.4 K as shown in Figure 10. Comparison of the experimental data with Tarasov's[33] phenomenological theory for the specific heat of a coupled chain system indicates that the unusual specific heat of $Hg_{3-\delta}AsF_6$ results from the excitations of the coupled incommensurate linear chains. According to this theory a T^3 behavior of the data at the lowest temperatures changing monotonically to T at the higher temperatures arises from a crossover from three-dimensional to one-dimensional behavior of the incommensurate Hg chain structure. The crossover is claimed to result from the small coupling between mercury chains. Moses *et al.*[32] also show that the large linear term is not caused by disorder within mercury chains as suggested by Wei *et al.*[31]

Some care is necessary in interpreting these low-temperature specific-heat data. Usually, a T^3 dependence does not hold accurately above one-

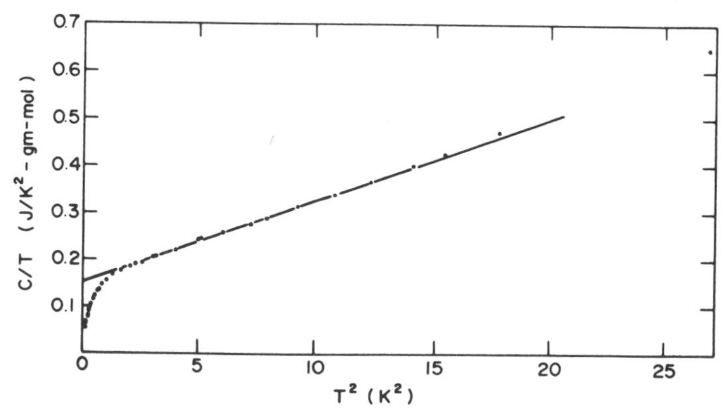

FIGURE 10. Specific heat (C) shown as a plot of C/T versus T^2 for $Hg_{3-\delta}AsF_6$. (From Reference 32.)

hundredth of the Debye temperature θ_D, which means that it may not be reliable above approximately 1 K for $Hg_{3-\delta}AsF_6$. Also, it is difficult to know whether the data of Figure 10 go to low enough temperatures to extract the form of the specific heat at very low temperatures. Thus, more measurements are required to establish this interpretation of the specific heat.

8. Superconducting Properties

Chiang *et al.*[18] report that the *c*-axis resistivity drops to zero at 4.1 K and remains zero for lower temperatures, indicating superconductivity along the *c* axis as shown in Figure 4b. The critical magnetic field of this superconductivity is 0.038 T at a temperature of 1.4 K. At the same time, resistivity in the (001) plane changes smoothly and continuously with temperature down to 1.4 K and shows no superconducting properties. Additional resistivity measurements were made by Spal *et al.*[34] with two current electrodes on one (001) crystal face and two voltage probes in contact with the opposite face. The measured ratio of voltage/current increased at 4.1 K by a factor of more than 8000. This indicates again that the *c*-axis resistivity becomes zero at 4.1 K. However, similar experiments by Batalla *et al.*,[35] using many electrode configurations including those described above, did not show any drop in the *c*-axis resistivity nor an increase in the ratio of voltage/current when voltage and current probes were placed on opposite (001) faces. Thus, the *c*-axis superconductivity is not reproducible in all resistivity measurements.

Magnetization measurements by Spal *et al.*[22] show that there is partial flux exclusion between 4.1 and 1.4 K consistent with the Meissner effect of superconductivity. The magnetization is anisotropic, with the ratio of the magnetization in the *ab* plane to that along the *c*-axis as large as 17. This ratio is reduced to about 6 in a magnetic field of 10^{-4} T and to 1 in a field of 10^{-2} T as shown in Figure 11. Another unusual feature of the magnetization is that it varies nonlinearly with field (Figure 11) from 10^{-2} T down to fields as small as 1.5×10^{-6} T. With this magnetic field dependence, it is difficult to relate with confidence the effects observed in the magnetization to those observed in the resistivity.

The differential ac susceptibility measured by Batalla *et al.*[35] shows unusual effects as a function of temperature and magnetic field. In powders, there is a large change in susceptibility with magnetic field for fields less than 0.005 T. There is also an indication of a superconducting transition with a critical temperature and magnetic field of 4.15 K and 0.038 T, respectively. The change in dM/dB at this transition is increased after partial decomposition of the sample. In a single crystal studied by Batalla *et*

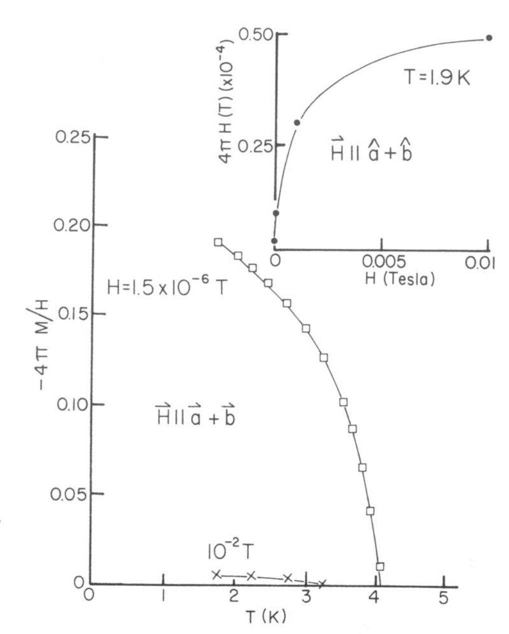

FIGURE 11. Magnetization of $Hg_{3-\delta}AsF_6$ as a function of temperature and magnetic field. (From References 22 and 34.)

al.[35] the low field susceptibility is very anisotropic and consistent with the magnetization reported by Spal *et al.*[22] However, the single-crystal susceptibility appears to be sample dependent.

The results of Batalla *et al.*[35] show that there are two processes causing changes in the susceptibility. The transition at 4.15 K is attributed to the superconductivity of elemental mercury since the observed critical temperature and the observed variation of the critical field with temperature are similar to those of mercury. Some of this mercury could come from mercury extrusion as described in Section 4.2. However, it is likely that a larger portion of the mercury is formed as a result of decomposition by reducing agents in the atmosphere surrounding the sample. Since the critical temperature and magnetic field observed in the resistivity observed by Chiang *et al.*[18] are also close to the corresponding values for mercury, it is likely that this transition is also due to mercury. Mercury in cracks and on crystal surfaces with a *c*-axis component could account for the anisotropy of the transition.

A second process is evident at low fields. The field dependence of the susceptibility at low field values suggests that the compound enters the intermediate state of a type-II superconductor at a very low value of magnetic field. Thus, $Hg_{3-\delta}AsF_6$ may be an anisotropic type-II superconductor at low fields. However, it is also possible that superconducting electron pairs in elemental mercury on the surface and in the interior of the sample are leaking to normal regions of the compound making a larger volume participate in the Meissner effect. However, neither of these explanations is completely satisfactory and more work should be done to explain the unusual effects in magnetization and susceptibility measurements.

Measurements of the ac susceptibility below 1 K by Spal *et al.*[34] indicate that there is a state at low temperatures in which there is complete flux exclusion. The critical temperature and magnetic field of this effect are 0.43 K and 0.002 T, respectively.

9. Electron Properties

9.1. De Haas–van Alphen Effect and Fermi Surface

The de Haas–van Alphen (dHvA) effect is a variation in the magnetic susceptibility of a metal with magnetic field and is periodic in reciprocal magnetic field.[36] It is useful in the study of metal Fermi surfaces—the surfaces of constant energy at 0 K that surround all occupied electron states in reciprocal space (k space). A dHvA frequency, measured usually in tesla, is proportional to an extremal (maximum or minimum) cross-sectional area of the Fermi surface perpendicular to the magnetic-field direction. For a complex Fermi surface, there can be more than one extremal cross section giving rise to more than one dHvA frequency for a given direction of the magnetic field. By measuring the dHvA frequencies in different magnetic-field directions, it is often possible to develop a model of the Fermi surface. In an ideal one-dimensional conductor, no dHvA effect is expected because the Fermi surface consists of two infinite parallel planes that enclose a nonfinite area. However, if the conductor has two-dimensional or three-dimensional character, the dHvA effect may occur. However, it has not been observed in other quasi-one-dimensional materials because it can only occur when there is a long electron mean free path at low temperatures.

Razavi *et al.*[37] observed the dHvA effect in $Hg_{3-\delta}AsF_6$ at 1.1 K in a magnetic-field range between 3 and 5.5 T. The dHvA frequencies for various magnetic-field directions are shown in Figure 12. Six frequency branches are found, each with a minimum along the c axis. Five branches, α, β, γ, δ, and ε, can be observed up to $64^\circ \pm 10^\circ$ from the c axis. Branch μ is only observed over a smaller angular range lying within 20° of the c axis and is not observed below a magnetic field of 4.6 T. A low-frequency branch with a value of 48 T, which is not shown in Figure 12, exists over a very small angular range about the c axis. Table 3 shows the values of the dHvA frequencies and the corresponding cross-sectional areas of the Fermi surface with magnetic field along the c axis. No dHvA frequencies were found in a careful search with magnetic-field directions in the (001) plane indicating that no closed Fermi-surface cross sections exist perpendicular to the a and b axes.

The dHvA frequencies in Figure 12 follow a sec θ dependence, where θ is the angle between the magnetic-field direction and the *c* axis. This shows that the Fermi surface consists of a set of cylinders with their axes along *c*, because for a straight cylinder the extremal cross-sectional area, which is proportional to the dHvA frequency, varies as A_0 sec θ, where A_0 is the cross-sectional area perpendicular to the axis of the cylinder. The sec θ behavior is independent of the shape of the area A_0, so that no information about the shape of the cross section A_0 is obtained from the fit in Figure 12.

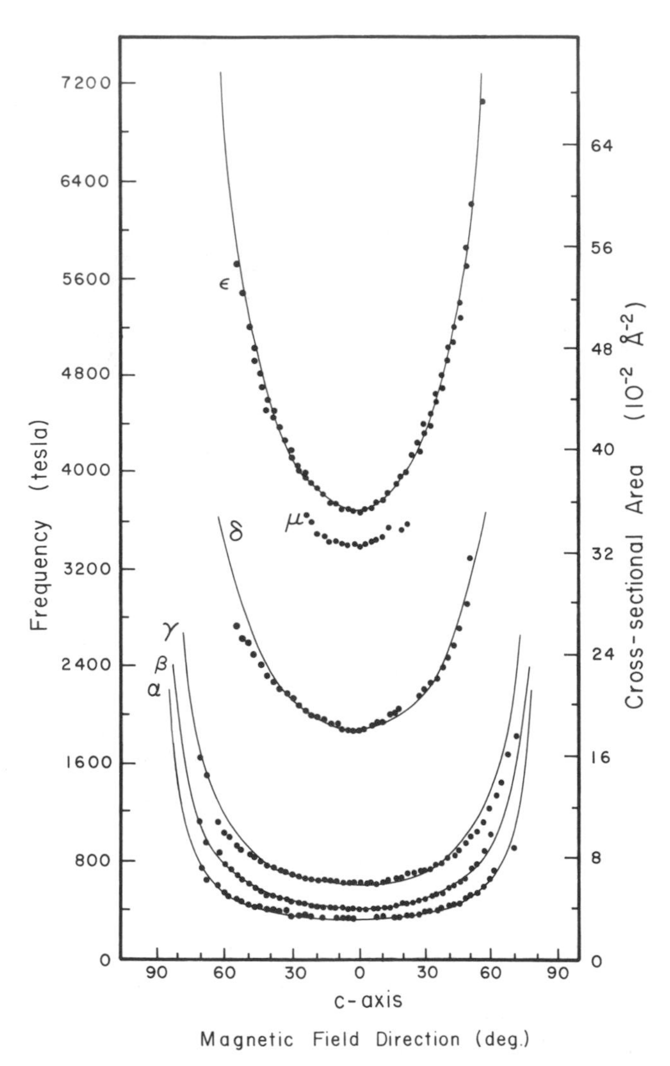

FIGURE 12. De Haas–van Alphen frequencies of $Hg_{3-\gamma}AsF_6$ as a function of magnetic-field directions from the *c* axis. (From Reference 37.)

TABLE 3. Experimental and Theoretical dHvA Frequencies and Corresponding Fermi-Surface Cross-Sectional Areas for $Hg_{3-\delta}AsF_6$ Using $a = 7.443$ A and $\delta = 0.21$

Orbit	Experiment		Theory	
	Frequency (T)	Area ($Å^{-2}$)	Frequency (T)	Area ($Å^{-2}$)
α^*	48	0.0046	48	0.0046
α	340	0.0325	298	0.0285
β	413	0.0395	—	—
γ	626	0.0598	627	0.0599
δ	1860	0.1780	1860	0.1780
μ	3410	0.3260	4330	0.3290
ε	3680	0.3520	3760	0.3590

A model of the Fermi surface is constructed by first assuming that conduction electrons move in a cylindrically symmetric uniform potential along the direction of the chains. Each mercury atom contributes two valence electrons, but one electron for every $(3 - \delta)$ mercury atom is used in bonding the mercury chains to the octahedral anions. If the chain's length is L, then the number of mercury atoms along a chain is L/d_{Hg} where d_{Hg} is the Hg–Hg distance. The total number of electrons N is

$$N = \frac{L}{d_{\text{Hg}}}\left(2 - \frac{1}{3 - \delta}\right)N_c$$

where N_c is the number of chains along one direction. Equating the total number of electrons to the number of k states occupied between $\pm k_F$, where k_F is the Fermi wave vector, gives

$$2k_F \frac{L}{2\pi} 2N_c = \frac{L}{d_{\text{Hg}}}\left(2 - \frac{1}{3 - \delta}\right)N_c$$

where $L/2\pi$ is the density of state for a wave vector along the chain and $2N_c$ is the number of filled states, including spin, for each value of the wave vector along the chain. Solving this equation gives

$$k_F = \frac{\pi(2.5 - \delta)}{a}$$

This indicates that the first two zones are filled and the reduced wave vector in the third zone is $\pi(0.5 - \delta)/a$. The Fermi surface for one set of chains consists of two planes extending to the Brillouin-zone boundary perpendicular to the chain wave vector at $\pm\pi(0.5 - \delta)/a$ in the third zone.

The Fermi surface for the two perpendicular sets of chains consists of four planes at $\pm(\pi/a)(0.5 - \delta)$ along the k_a and k_b directions for each reciprocal lattice point. Interaction between perpendicular chains removes the

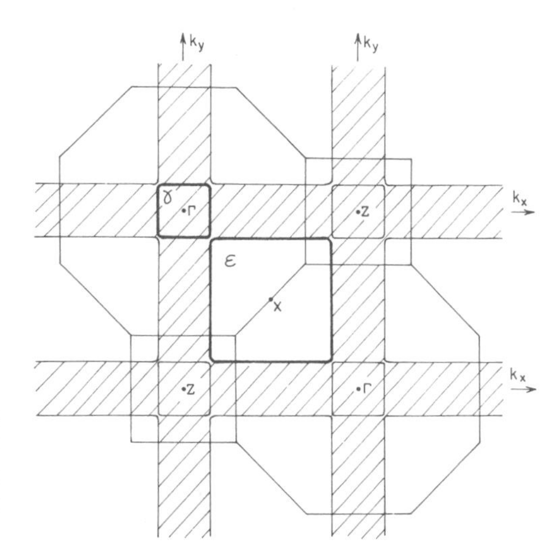

FIGURE 13. Cross section of the Fermi surface of $Hg_{3-\delta}AsF_6$ at $k_c = 0$ in the extended-zone scheme showing the electron orbit γ and the hole orbit ε.

degeneracies at the crossings of the Fermi surfaces of the two sets of chains as shown in Figure 3. The $k_c = 0$ cross section of this Fermi surface for $Hg_{3-\delta}AsF_6$ is shown in Figure 13. This Fermi surface consists of electron and hole square cylinders with axes along k_c.

$$\text{Area (electron)} = [(\pi/a)(1 - 2\delta)]^2$$

$$\text{Area (hole)} = [(\pi/a)(1 + 2\delta)]^2$$

These areas with $\delta = 0.21$ and $a = 7.443$ Å (see final column of Table 1) agree well with those derived from the γ and ε branches of the dHvA effect of $Hg_{3-\delta}AsF_6$ as shown in Table 3.

Ehrenfreund *et al.*[38] and Berlinsky,[39] assuming a uniform density of electrons with free-electron wave functions along a chain, have calculated the effect of the overlap of nearest-neighbor perpendicular chains on the band structure. The electron and hole Fermi surfaces have flat sides and corners that are only slightly rounded even with quite large chain interaction. The electron states that are part of the hole and electron Fermi surfaces correspond to electrons localized on chains running along either *a* or *b* except for the electron states at the corners of the Fermi surface. Thus, the Fermi surface closely approximates that expected for coupled one-dimensional conductors even in the presence of quite large interchain coupling.

The nonzero overlap between the electron wave functions of different chains makes the Fermi surface depend on k_c within the Brillouin zone. The Fermi surface undulates along k_c and up to six extremal orbits are possible on electron and hole surfaces for a magnetic-field direction along the *c* axis.[39] However, these orbits cannot be identified with the dHvA data because the angular dependence of the dHvA frequencies does not deviate sufficiently from that for a straight cylinder (see Figure 12) to be caused by undulating cylinders. Also, there is no evidence of two frequency branches merging at large values of θ as is required of frequencies that come from the minimum and maximum cross sections of an undulating cylinder. Thus, any undulations of the Fermi surface are too small to be detected with the present dHvA data.

The effect of the Hg–Hg intrachain spacing which is incommensurate with the tetragonal lattice could be included with the theory of Azbel[40] but is introduced here more simply by using the difference **q** between the tetragonal reciprocal-lattice vector **G**(lattice) and the chain reciprocal-lattice vector **G**(chains).

$$\mathbf{q} = \mathbf{G}(\text{lattice}) - \mathbf{G}(\text{chains})$$

The smallest nonzero **q** is given by

$$\mathbf{q} = \frac{2\pi}{a}(\delta, \ \pm\delta, 0)$$

where the choice of signs corresponds to two possible domains in the crystal in the L phase (see Figure 9). Gaps will be created at the intersections of the original Fermi surface and the surface displaced by **q**. This makes additional orbits possible when magnetic breakdown is allowed across some of the energy gaps. There are electron orbits α and α^* and hole orbits δ and μ as shown in Figure 14. The β frequency cannot be assigned to a particular orbit unambiguously. The excellent agreement between predicted and observed dHvA frequencies and areas for $Hg_{3-\delta}AsF_6$ shown in Table 3 is good support for this Fermi-surface model.

The area of the δ orbit is $(\pi/a)^2$ and the ratios of the areas of other orbits to that of the δ orbit depend only on the incommensurability parameter (δ) at low temperatures. Differences in δ produce large differences in the areas of possible orbits but do not change the geometry of the Fermi surface except for some of the more complicated orbits.

This method has been used by Batalla *et al.*[41] to determine the incomensurability parameter of $Hg_{3-\delta}SbF_6$ at 1.25 K. The dHvA effect in this compound also has six frequency branches arising from cylindrical Fermi-surface sections. Their minimum frequencies and corresponding areas are shown in Table 4. In the calculation of cross-sectional areas, the lattice parameter a was chosen to be 7.843 Å in order to give agreement between the calculated and measured area of the δ orbit. This value of a is somewhat larger than that found by x-ray diffraction (Table 1). The calculated values of all the frequencies are in excellent agreement with the measured values when δ is chosen to be 0.135 which is in good agreement with the value of δ

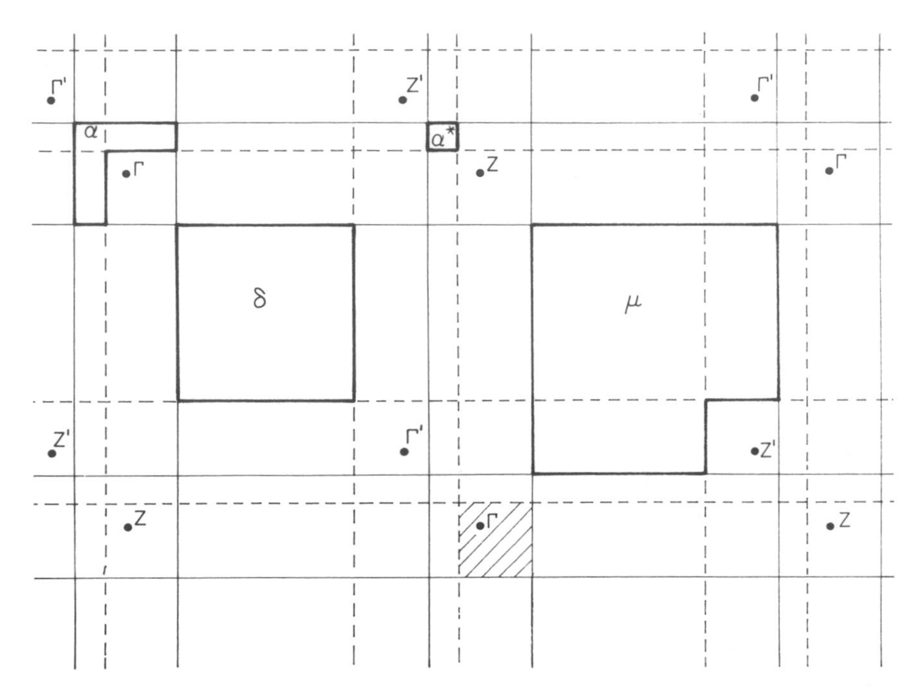

FIGURE 14. The $k_c = 0$ cross-section Fermi surface of $Hg_{3-\delta}AsF_6$ and the Fermi surface displaced by **q** showing electron orbits α and α^* and hole orbits δ and μ.

derived from a 1% contraction of the SbF_6^- lattice from room temperature to 1.2 K. It is 35% smaller than the value of δ in $Hg_{3-\delta}AsF_6$ at 10 K so that the dHvA frequencies associated with the same orbits in both compounds differ by as much as a factor of 7 for the α^* orbit and by 10% for the δ orbit. Still, the two sets of frequencies are correctly accounted for by the same model using the two different values of the incommensurability parameter δ.

TABLE 4. Experimental and Theoretical dHvA Frequencies and Corresponding Fermi-Surface Cross-Sectional Areas for $Hg_{3-\delta}SbF_6$ Using $a = 7.843$ A and $\delta = 0.135$

	Experiment		Theory	
Orbit	Frequency (T)	Area (Å^{-2})	Frequency (T)	Area (Å^{-2})
α^*	345	0.0329	355	0.0339
α	770	0.0735	773	0.0738
γ	975	0.0931	895	0.0855
δ	1680	0.1610	1680	0.1610
μ	2580	0.2460	2590	0.2470
ε	2700	0.2580	2710	0.2590

Cyclotron masses were determined from the temperature dependence of the amplitude of the dHvA oscillations. For the magnetic field along the c axis, the cyclotron masses of the α, δ, and ε branches for $Hg_{3-\delta}SbF_6$ are 0.15 ± 0.02, 0.18 ± 0.03, and 0.3 ± 0.01, respectively, and of the γ, δ, and ε branches for $Hg_{3-\delta}AsF_6$ are 0.23 ± 0.02, 0.27 ± 0.08, and 0.33 ± 0.01, respectively, all being expressed in terms of the free-electron mass.

The dHvA effect of $Hg_{3-\delta}AsF_6$ was studied as a function of pressure up to 5 kbars.[41] The pressure dependence of the dHvA frequencies shows that hydrostatic pressure of 5 kbars causes a lattice contraction of 0.7%, which is of the same order as the 1% contraction observed from room temperature to 10 K, without changing the Hg–Hg distance in the chains significantly. At 1.25 K, this causes δ to change from 0.21 at zero pressure to 0.23 at 5 kbars. Excellent agreement between theoretical and experimental frequencies allows one to conclude that all changes in the dHvA frequencies are a result of a contraction of the arsenic hexafluorides lattice and the consequent change in the incommensurability parameter δ.

9.2. *Induced Torque*

Induced-torque experiments are done by rotating a sample in a magnetic field (or rotating a magnetic field with a stationary sample) and measuring the torque about the axis of rotation.[42] Currents induced by the changing magnetic field in the sample interact with the magnetic field to produce the induced torque. At high magnetic fields, the induced torque is small and almost independent of field for closed orbits but is large with a quadratic magnetic-field dependence for open orbits.[43] This makes the method an ideal open-orbit detector: large induced torque appears whenever the magnetic field is perpendicular to an open orbit. This torque is a measure of the magnetoresistance of the sample.

Dinser *et al.*[44] have shown that in $Hg_{3-\delta}AsF_6$, there is a large induced-torque peak for magnetic-field directions in the (001) plane. This means that there are open orbits normal to this plane, along the k_c direction. These orbits run along the electron and hole cylinders in our Fermi-surface model.

The induced torque is small and practically independent of magnetic field at high fields when the magnetic field is in the c direction. This shows that there are only closed orbits with the magnetic field in the c direction and that the Fermi surface is not connected in the $k_a k_b$ plane in the extended-zone scheme. This result agrees with the Fermi-surface model given in Figure 13.

The induced-torque experiments also show a set of secondary peaks at magnetic fields larger than 5 T. The field dependence of the amplitude of these secondary peaks is much larger than quadratic, indicating that they arise from open orbits produced by magnetic breakdown across small energy gaps. The open orbits responsible for the secondary peaks run along host-lattice directions of the type [01l] where $l \geq 2$. This can occur when there is a small energy gap at certain points between the electron and hole cylinders. Such a gap could be produced by the spin–orbit interaction and indicates a small undulation along the Fermi-surface cylinders which is not detected in the dHvA effect.

9.3. *Magnetoresistance*

The magnetic-field, H, dependence of the resistivity of $Hg_{3-\delta}AsF_6$ was measured by Chakraborty *et al.*[23,45] in magnetic fields up to 18 T. They found that, for temperatures less than 25 K, the resistivity in the (001) plane could be divided into two parts

$$\rho(H, T) = \rho_0(H) + \rho_1(T)$$

where $\rho_0(H)$ is a function of magnetic field only and $\rho_1(T)$ is a function of temperature only. $\rho_1(T)$ changes by a factor of 30 between 4.2 and 25 K, and $\rho_0(H)$ appears as a magnetic-field-dependent residual resistivity. The change in resistivity with magnetic field is quadratic only at very low fields ($H < 10^{-4}$ T). It increases at a rate less than quadratic at higher fields, but shows no sign of saturation up to 18 T when the magnetic field is along the c axis.

Chakraborty *et al.*[45] give an explanation for this magnetoresistance using a model of electron scattering on the cylindrical Fermi surface shown in Figure 3. They assume a long relaxation time for electron scattering on flat sections of the Fermi surface and a short relaxation scattering time for electrons at the corners. Since electrons move to the corners during their cyclotron motion in a magnetic field, the number of electrons scattered by the corners increases and the resistivity becomes larger with increasing magnetic field.

Resistivity is expected to become independent of magnetic field at high fields when there are only closed orbits in a metal. There are only closed orbits on the cylindrical Fermi-surface model with the magnetic field along the c axis if there is no magnetic breakdown between cylinders. The induced-torque measurements by Dinser *et al.*[44] show no evidence of open orbits with the magnetic field along the c axis. This indicates that, contrary to the results of Chakraborty, the magnetoresistance determined from the induced-torque experiment saturates with the magnetic field along the c direction.

This lack of agreement between the results of the induced-torque and magnetoresistance experiments for the magnetic-field dependence of the resistance for the c direction may lie with the use of the inductive technique employed for the magnetoresistance measurements. The technique was developed by Rosenthal and Maxfield[46] in order to measure the resistivity of a metal. A sample is placed inside two coils and its resistance is determined by measuring the impedance of the secondary coil when an ac current in the primary coil induces currents in the sample. It was developed for zero magnetic field and Chakraborty *et al.*[45] did not consider the effect of a magnetic field on the measured impedance. Since the induced-current pattern is influenced profoundly by a magnetic field its effect must be considered. It is likely that the result depends on a number of effects including the Hall resistivity which has a linear field dependence at high fields.

9.4. Nuclear Magnetic Resonance

The Knight shift and the spin–lattice relaxation of the magnetic resonance of nuclei in a metal depend on the amplitude of the conduction electron wave function at nuclear sites, and thereby provide information about conduction electrons in the metal. For this reason, a nuclear magnetic resonance (nmr) study of $Hg_{3-\delta}AsF_6$ was carried out by Ehrenfreund *et al.*[38]

The Knight shift of ^{199}Hg in $Hg_{3-\delta}AsF_6$ is 1.9% compared with 2.7% for metallic mercury and is independent of magnetic field and temperature in the range 1.5–300 K. It corresponds to a magnetic susceptibility of the conduction electrons of 1.5×10^{-5} cm^3 per mole of $Hg_{3-\delta}AsF_6$. This susceptibility gives a density of states at the Fermi energy of 8×10^{-2} per eV per Hg atom for single-spin direction, assuming no enhancement of the Pauli susceptibility.

The nuclear spin–lattice relaxation rate, τ_1^{-1}, shows metallic behavior and is proportional to temperature for the nmr of ^{199}Hg in the temperature range between 1.5 and 300 K and for the nmr of ^{19}F for $T < 100$ K. For these data, $\tau_1 T = 340$ s K for ^{19}F and $\tau_1 T = 7.5 \times 10^{-3}$ s K for ^{199}Hg. If the spin–lattice relaxation is due to conduction electrons at the nucleus, the ratio between the relaxation rates is

$$\frac{[\tau_1(^{19}\mathrm{F})]^{-1}}{[\tau_1(^{199}\mathrm{Hg})]^{-1}} = \left(\frac{\gamma(^{19}\mathrm{F})\langle\psi_\mathrm{F}^2(0)\rangle}{\gamma(^{199}\mathrm{Hg})\langle\psi_\mathrm{Hg}^2(0)\rangle}\right)^2$$

where $\langle\psi_i^2(0)\rangle$ (i = F or Hg) is the mean square of the conduction electron wave function at the Fermi energy. The relaxation measurements give $\langle\psi_\mathrm{F}^2(0)\rangle/\langle\psi_\mathrm{Hg}^2(0)\rangle = 10^{-3}$, confirming that the conduction electrons are primarily on the mercury chains.

Above 100 K, the fluorine relaxation rate is no longer proportional to temperature but increases rapidly to a maximum at 220 K. This behavior is believed to arise from the mechanical rotation of the AsF_6^- octahedra which would affect the dipolar coupling between ^{19}F spins. It is likely that the

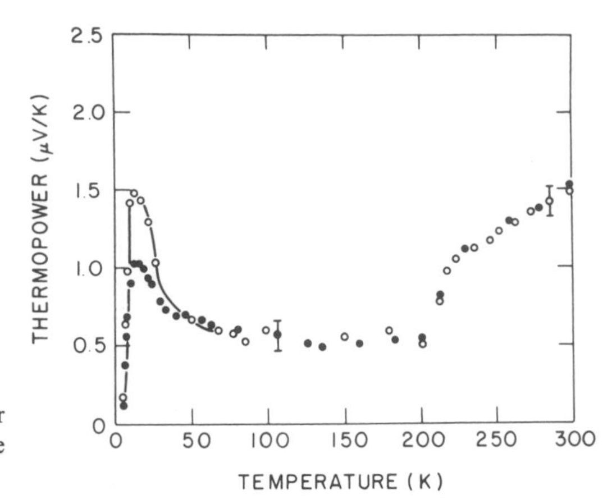

FIGURE 15. Thermoelectric power vs temperature of $Hg_{3-\delta}AsF_6$ in the *ab* plane. (From Reference 47.)

effect is related to the changes in resistivity, heat capacity, and thermopower that occur at the same temperature.

9.5. *Thermoelectric Power*

Both Chiang *et al.*[47] and Scholz *et al.*[48] report measurements of the thermoelectric power of $Hg_{3-\delta}AsF_6$. At room temperature, both agree that S_c, the thermopower along the *c* direction is positive, but they differ as to the sign of S_{ab}, the thermopower in the *ab* plane. They also differ on the temperature dependence of S_c and S_{ab} between 300 and 4.2 K.

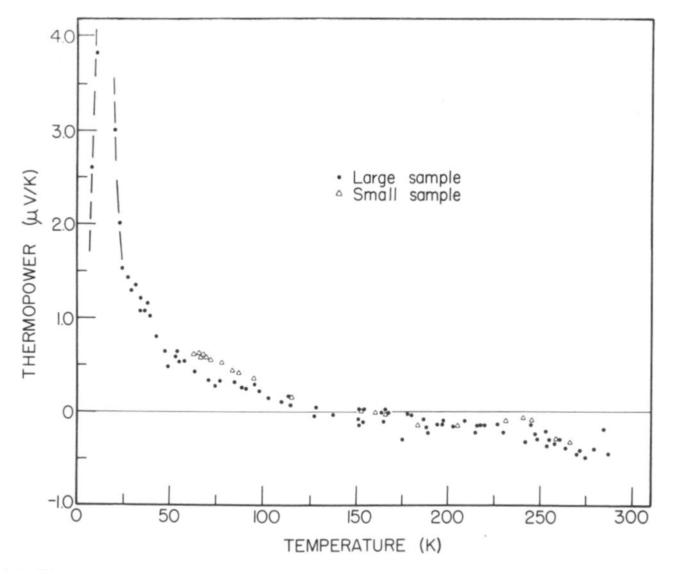

FIGURE 16. Thermoelectric power vs temperature of $Hg_{3-\delta}AsF_6$ in the *ab* plane. (From Reference 48.)

In the measurement of S_{ab} by Chiang *et al.*,[47] shown in Figure 15, a sharp increase between 200 and 225 K indicates the existence of a phase transition that may be related to mercury extrusion. However, this transition is not observed by Scholz *et al.*[48] (Figure 16). They find that S_{ab} decreases linearly with increasing temperature above 120 K as would be expected for the diffusion term of a degenerate electron gas. The negative slope shows that conduction is by electrons in the (001) plane as expected by the Fermi-surface model. The magnitude of the slope corresponds to an electron Fermi energy of 4.3 ± 0.5 eV. The peak found by both groups in S_{ab} at 13 ± 3 K is typical of phonon-drag thermopower enchancement and indicates a Debye temperature in the range 65–130 K. In the measurement by Chiang *et al.*[47] the thermopower peak is smaller in S_{ab} than that observed by Scholz *et al.*[48]

The anisotropic nature of the compound is evident in the results of Scholz *et al.*[48] The slope of the thermopower versus temperature is positive for S_c and negative for S_{ab} in the temperature range between 75 and 300 K.

9.6. *Optical Reflectance*

Polarized reflectance studies were carried out by Peebles *et al.*[19] and Batalla[20] on single-crystal faces of $Hg_{3-\delta}AsF_6$ in the spectral range 0.5–4.2 eV. The reflectance of the (001) crystal face illustrated in Figure 17 shows metal-like values below 2.7 eV but falls rapidly to zero above this energy. The frequency dependence of the reflectance is consistent with a plasma

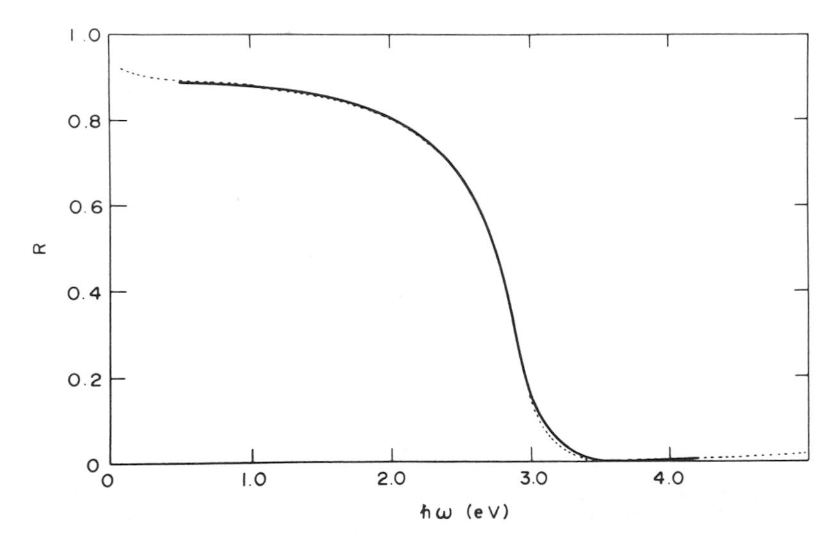

FIGURE 17. Reflectance spectra (solid line) and the Drude fit (dashed line) of the *ab* face of $Hg_{3-\delta}AsF_6$ at room temperature. (From Reference 19.)

edge occurring when the frequency changes from $\omega > \omega_p/\varepsilon_\infty^{1/2}$ to $\omega < \omega_p/\varepsilon_\infty^{1/2}$, where ω_p is the plasma frequency and ε_∞ is the core dielectric constant in the Drude model. The derived plasma energy depends somewhat on the parameters used in the model and has values of 4.8 and 5.2 eV in the two experiments. The plasma energy has very little temperature dependence between 300 and 25 K.

The reflectance of a crystal face perpendicular to the (001) plane in the far-infrared region ($\hbar\omega < 1$ eV) is 10% with light polarized along the *c* direction and 90% with the direction of polarization in the *ab* plane. There is no plasma edge observed between 0.5 and 4.2 eV in the frequency dependence of the reflectance with the *c*-axis polarization. These results are consistent with the quasi-one-dimensional metallic character of $Hg_{3-\delta}AsF_6$.

The reflectance of the (001) face of $Hg_{3-\delta}SbF_6$, measured by Koteles *et al.*,[21] also increases at 2.7 eV as the incident energy is decreased. However, between 1 and 2 eV there are two shallow minima in the reflectance which are attributed to intraband absorption. The reflectance of a crystal face normal to the (001) plane with optical polarization along the *c* axis is small and shows no plasma edge. Thus, the optical measurements show that $Hg_{3-\delta}SbF_6$ is also an anistropic conductor.

10. Summary

Crystals of $Hg_{3-\delta}AsF_6$ (and the virtually identical $Hg_{3-\delta}SbF_6$) have very unusual properties that result from their unique structure. This consists of a lattice of AsF_6^- ions through which pass chains of metallically bonded Hg atoms running along the *a* and *b* crystallographic directions. Much of the interest in these structures arises because the lattice and chains have incommensurate spacing. At room temperature the chains behave like ordered one-dimensional metals and are perfectly free to slide through the lattice, a property that is exploited in the preparation of large crystals. As the temperature is decreased the chains order, at first below 200 K through interactions between parallel chains (*S* phase) and then (below 120 K) through interactions between perpendicular chains (*L* phase). The structural and dynamical properties have been the subject of careful experimental and theoretical studies and are reasonably well understood. The high electrical conductivity of $Hg_{3-\delta}AsF_6$ and its marked anistropy result from the metallic character of the Hg chains and the fact that these chains run along the *a* and *b* directions but not along *c*. This behavior can be explained by a Fermi-surface model formed from one-dimensional bands by the interaction of mutually perpendicular mercury chains and by the incommensurability of the chains with respect to the host lattice.

The electrical behavior at low temperatures is less well understood and the evidence is conflicting. It has been established that the conductivity increases as the temperature is lowered and that there is no evidence for a metal–insulator transition, but some studies suggest that the crystals become superconducting below 4.1 K. The results are not reproducible and the similarity with the superconductivity of metallic Hg suggests that the effect may be attributable to Hg impurities, which appear on the surface as a result of extrusion during cooling or decomposition in the presence of traces of moisture.

Surprisingly, the composition is not precisely known. Crystallographic measurements suggest that in the formula $Hg_{3-\delta}AsF_6$, $\delta = 0.18$ at room temperature changing to 0.21 at 0 K. Chemical analyses, however, suggest a value of δ close to zero. Density measurements favor the latter value and a structure containing an appreciable concentration of anion defects.

Indications of a phase transition have been observed at around 200 K in a variety of measurements. The character of the transition is not clear although there is a partial ordering of the Hg chains, a change in the dynamic state of the AsF_6^- ions, and evidence that metallic Hg is extruded from the crystal as it cools through this range.

Solid-state theorists have not been slow to exploit the opportunities that $Hg_{3-\delta}AsF_6$ offers. As a result $Hg_{3-\delta}AsF_6$ is in some respects among the best understood materials. However, there are many properties for which a satisfactory explanation does not exist and indeed for which there are conflicting experimental reports. Work on these two highly reactive compounds is continuing and there are good reasons to suppose that future studies will resolve some of the difficulties outlined in this chapter.

Notation

Latin

A_0	Cross sectional area
a, b, c	Crystallographic axes
a^*, b^*, c	Reciprocal space crystallographic axes
ac	Alternating current
C	Specific heat
d, d_{Hg}	Intrachain Hg–Hg separation
dHvA	De Haas–van Alphen effect
D, L, S	Phases
$\mathbf{G}$	Reciprocal lattice vector
H	Magnetic field
h, k, l	Miller indices
k	Wave vector
k_F	Fermi wave vector

l	L phase
L	Chain length
LA	Longitudinal acoustic phonons
m	Power law exponent
m_1, m_2	Mass
N	Number of electrons
N_c	Number of chains in one direction
S	Thermoelectric power
s	s Phase
TA	Transverse acoustic phonon

Greek

α, β, γ, σ, ε, μ	Frequency branches
α, α^*	Electron orbits
β	Frequency
β, γ	Specific-heat coefficients
δ	Deviation from stoichiometry (incommensurability parameter)
θ	Angle between the magnetic field and c axis
θ_D	Debye temperature
δ, μ	Hole orbits
ε_∞	Core dielectric constant
γ	Monoclinic unit cell angle
ρ	Resistivity
σ	Root mean square of Hg–Hg bond length
σ	Conductivity
τ^{-1}	Nuclear spin–lattice relaxation rate
ψ	Conduction electron wave function
ω	Frequency
ω_p	Plasma frequency

References

1. Gillespie, R. J., and Ummat, P. K., *J. Chem. Soc. Chem. Commun.* 1168 (1971).
2. Gillespie, R. J., and Passmore, J., *Acct. Chem. Res.*, **4**, 413 (1971).
3. Davies, C. G., Dean, P. A. W., Gillespie, R. J., and Ummat, P. K., *J. Chem. Soc. Chem. Commun.*, 782 (1971).
4. Cutforth, B. D., Davies, C. G., Dean, P. A. W., Gillespie, R. J., Ireland, P. R., and Ummat, P. K., *Inorg. Chem.* **12**, 1343 (1973).
5. Torsi, G., Fung, K. W., Begun; G. H., and Mamantov, G. *Inorg. Chem.* **10**, 2285 (1975).
6. Ellison, R. D., Levy, H. A., and Fung, K. W. *Inorg. Chem.* **11**, 833 (1972).
7. Cutforth, B. D., Gillespie, R. J., and Ireland, P. R., *J. Chem. Soc. Chem. Commun.*, 723 (1973). Cutforth, B. D., Gillespie, R J., Ireland, P. R., Sawyer, J F., and Ummat, P. K., *Inorg. Chem.* (in press).
8. Brown, I. D., Cutforth, B. D., Davies, C. G., Gillespie, R. J., Ireland, P. R., and Vekris, J. E., *Can. J. Chem.* **52**, 791 (1974).
9. Cutforth, B. D., and Gillespie, R. J., *Inorg. Syntheses* **19**, 22 (1979).
10. Miro, N. D., MacDiarmid, A. G., Heeger, A. J., Garito, A. F., Chiang, C. K., Schultz, A. J., and Williams, J. M., *J. Inorg. Nucl. Chem.* **40**, 1351 (1978).

11. Chartier, D., Ph.D. thesis, McMaster University, 1982.
12. Gillespie, R. J., and Morton, M. J., *Q. Rev. Chem. Soc.* **25**, 53 (1971).
13. Schultz, A. J., Williams, J. M., Miro, N. D., MacDiarmid, A. G., and Heeger, A. J., *Inorg. Chem.* **17**, 646 (1978).
14. Pouget, J. P., Shirane, G., Hasting, J. M., Heeger, A. J., Miro, N. D., and MacDiarmid, A. G., *Phys. Rev. B* **18**, 3645 (1978).
15. Tun, Z., and Brown, I. D., *Acta. Crystallogr. B.* (in press).
16. Datars, W. R., Van Schyndel, A., Lass, J. S., Chartier, D., and Gillespie, R. J., *Phys. Rev. Lett.* **40**, 1184 (1978).
17. Cutforth, B. D., Datars, W. R., Van Schyndel, A., and Gillespie, R. J., *Solid State Commun.* **21**, 377 (1977).
18. Chiang, C. K., Spal, R., Denenstein, A. M., Heeger, A. J., Miro, N. D., and MacDiarmid, A. G., *Solid State Commun* **22**, 293 (1977).
19. Peebles, D. L., Chiang, C. K., Cohen, M. J., Heeger, A. J., Miro, N. D., and MacDiarmid, A. G., *Phys. Rev. B* **15**, 4607 (1977).
20. Batalla, E., M.Sc. thesis, McMaster University, 1976.
21. Koteles, E. S., Datars, W. R., Cutforth, B. D., and Gillespie, R. J., *Solid State Commun.* **20**, 1129 (1976).
22. Spal, R., Chiang, C. K., Denenstein, A. M., Heeger, A. J., Miro, N. D., and MacDiarmid, A. G., *Phys. Rev. Lett.* **39**, 650 (1977).
23. Chakraborty, D. P., Spal, R., Chiang, C. K., Denenstein, A. M., Heeger, A. J., and MacDiarmid, A. G., *Solid State Commun.* **27**, 849 (1978).
24. Kaveh, M., and Ehrenfreund, E., *Solid State Commun.* **31**, 709 (1979).
25. Emery, V. J., and Axe, J. D., *Phys. Rev. Lett.* **40**, 1507 (1978).
26. Janner, A., and Janssen, T. *Acta Crystallogr. A* **36**, 408 (1980).
27. Hastings, J. M., Pouget, J. P., Shirane, G., Heeger, A. J., Miro, N. D., and MacDiarmid, A. G., *Phys. Rev. Lett.* **39**, 1484 (1977).
28. Sacco, J. E., and Sokoloff, J. B., *Phys. Rev. B* **18**, 6549 (1978).
29. Heilmann, I. U., Axe, J. D., Hastings, J. M., Shirane, G., Heeger, A. J., and MacDiarmid, A. G., *Phys. Rev. B* **20**, 752 (1979).
30. Heilmann, I. U., Hastings, J. M., Shirane, G., Heeger, A. J., and MacDiarmid, A. G., *Solid State Commun.* **29**, 469 (1979).
31. Wei, T., Garito, A. F., Chiang, C. K., and Miro, N. D., *Phys. Rev. B* **16**, 3373 (1977).
32. Moses, D., Denenstein, A., Heeger, A. J., Nigrey, P. J., and MacDiarmid, A. G., *Phys. Rev. Lett.* **16**, 3373 (1979).
33. Tarasov, V. V., in *New Problems in the Physics of Glass*, Oldbourne, London (1963).
34. Spal, R., Chen, C. E., Denenstein, A. M., McGhil, A. A., Heeger, A. J., and MacDiarmid, A. G., *Solid State Commun.* **32**, 641 (1979).
35. Batalla, E., Datars, W. R., Chartier, D., and Gillespie, R. J., to be published.
36. Gold, A. V., in *Solid State Physics. I. Electrons in Metals*, J. F. Cochrane and R. R. Haering, Eds., Gordon and Breach, New York (1967), pp. 39–126.
37. Razavi, F. S., Datars, W. R., Chartier, D., and Gillespie, R. J., *Phys. Rev. Lett.* **42**, 1182 (1979); Batalla, E., Razavi, F. S., and Datars, W. R., *Phys. Rev. B* **25**, 2109 (1982).
38. Ehrenfreund, E., Newman, P. R., Heeger, A. J., Miro, N. D., and MacDiarmid, A. G., *Phys. Rev. B* **16**, 1781 (1977).
39. Berlinsky, J., private communication.
40. Azbel, M. Ya., *Phys. Rev. Lett.* **43**, 1954 (1979).
41. Batalla, E., Datars, W. R., Chartier, D., and Gillespie, R. J., *Solid State Commun.*, **40**, 465 (1982).
42. Cook, J. R., and Datars, W. R., *Phys. Rev. B* **1**, 1415 (1970).
43. Visscher, P. B., and Falicov, L. M., *Phys. Rev. B* **2**, 1518 (1970).

44. Dinser, R. J., Datars, W. R., Chartier, D., and Gillespie, R. J., *Solid State Commun.* **32**, 1041 (1979).
45. Chakraborty, D. P., Spal, R., Denenstein, A. M., Lee, K-B., Heeger, A. J., and Azbel, M. Ya., *Phys. Rev. Lett.* **43**, 1832 (1979).
46. Rosenthal, M. D., and Maxfield, B. W., *Rev. Sci. Instrum.* **46**, 398 (1975).
47. Chiang, C. K., Spal, R., Denenstein, A., Heeger, A. J., Miro, N. D., and MacDiarmid, A. G., in *Thermoelectricity of Metallic Conductors*, F. J. Blatt and P. A. Schroder, Eds., Plenum Press, New York (1978), p. 393.
48. Scholz, G. A., Datars, W. R., Chartier, D., and Gillespie, R. J., *Phys. Rev. B* **16**, 4209 (1977).

2

The Synthesis and Static Magnetic Properties of First-Row Transition-Metal Compounds with Chain Structures

William E. Hatfield, William E. Estes, Wayne E. Marsh, M. Wayne Pickens, Leonard W. ter Haar, and Robert R. Weller

1. Introduction

The most authoritative reviews of exchange coupling in one-dimensional linear chain systems of transition-metal ions are now several years old.[1–4] Recent advances are largely a result of systematic studies of magnetic properties in conjunction with structural determinations. Since a large number of systems have been investigated, we will discuss only one example of each type of behavior that has been found for each of the first-row transition-metal ions, and tabulate data for other systems which have figured prominently in the development of the field. For the most part, only those chain compounds for which rather complete structural and magnetic data are available will be considered here.

1.1. General Magnetic Interaction Hamiltonian

The Hamiltonian $\hat{H}$ describing the magnetic properties and behavior of first-row transition-metal ions is given by

$$\hat{H} = \hat{H}_{ex} + \hat{H}_z + \hat{H}_a \tag{1}$$

William E. Hatfield, William E. Estes, Wayne E. Marsh, M. Wayne Pickins, Leonard W. ter Haar, and Robert R. Weller • Department of Chemistry, University of North Carolina, Chapel Hill, North Carolina 27514.

where individual terms contributing to $\hat{H}$ may be concisely summarized as follows:

1. The nearest-neighbor exchange coupling $\hat{H}_{ex}$,

$$\hat{H}_{ex} = -2J \sum_{i \leq j}^{N} \alpha \hat{S}_i^z \hat{S}_j^z + \beta \hat{S}_i^x \hat{S}_j^x + \gamma \hat{S}_i^y \hat{S}_j^y \tag{2}$$

2. The Zeeman energy due to an applied magnetic field, $\hat{H}_z$,

$$\hat{H}_z = -g\mu_B \sum_{i=1}^{N} \mathbf{H} \cdot \hat{S}_i \tag{3}$$

3. The single-ion anisotropy, $\hat{H}_a$, often referred to as the zero-field splitting,

$$\hat{H}_a = D(\hat{S}_z^2 - \hat{S}^2/3) + E(\hat{S}_x^2 + \hat{S}_y^2) \tag{4}$$

The operators and symbols have their usual meanings, and the index j can be taken to be $i + 1$ for linear chain interactions. The exchange coupling constant, J, can be either positive or negative. As defined in Eq. (2), a negative J value corresponds to the antiparallel coupling of unpaired spins and the interaction is designated antiferromagnetic. Alternately, a positive coupling constant indicates ferromagnetic behavior where the unpaired spins are coupled parallel to each other.

The values of α, β, and γ in Eq. (2) define the nature of the exchange coupling. When $\alpha = \beta = \gamma = 1$, the isotropic Heisenberg exchange Hamiltonian results. In the extreme situation of $\alpha = 1$ and $\beta = \gamma = 0$, the Ising model is described. Similarly, the case when $\alpha = 0$ and $\beta = \gamma = 1$, leads to the other extreme and is labeled the XY model. The totally anisotropic case arises when α, β, and γ take on arbitrary values. Symbols which are used in this review to identify the various theoretical models are listed in the Notation section. Measurements on powders are often sufficient for isotropic or weakly anisotropic materials. In situations where the limiting cases of Eq. (2) apply, magnetic measurements along the crystal axes are necessary to identify the nature of the anisotropy and to allow comparison with theory.

1.2. Isotropic Heisenberg Interaction

The isotropic ($\alpha = \beta = \gamma$) Heisenberg Hamiltonian has been extensively examined with $\hat{H}_z$ and $\hat{H}_a$ taken to be zero. No exact solutions are presently known, but results of many approximate methods exist. Analytical results for one-dimensional Heisenberg systems have recently been surveyed by Johnson[5] who notes that the theoretical literature dates from an early paper by Bethe.[6] A key work remains the classical paper by Bonner and Fisher[7] on short chains and rings of $S = 1/2$ ions with antiferromagnetic interaction. The maximum magnetic susceptibility, and the temperature at which it occurs, are uniquely defined by

$$k_B T_{max}/|J| = 1.282 \tag{5}$$

and

$$\chi_{max}|J|/Ng^2\mu_B^2 = 0.07346 \tag{6}$$

Hall[8] noted that the shape of the reduced coordinate magnetic susceptibility versus temperature curve given by Bonner[7b] could be fit by the function

$$f(y) = (Ay^2 + By + C)/(y^3 + Dy^2 + Ey + F) \tag{7}$$

Using the theoretical results that have been tabulated by Bonner,[7b] the best-fit values for the parameters are $A = 0.25$, $B = 0.14995$, $C = 0.30094$, $D = 1.9862$, $E = 0.68854$, and $F = 6.0626$. Consequently, the expression for the magnetic susceptibility of $S = 1/2$ ions isotropically coupled in antiferromagnetic linear chains becomes

$$\chi_M = \left(\frac{Ng^2\mu_B^2}{k_B T}\right)\frac{A + Bx + Cx^2}{1 + Dx + Ex^2 + Fx^3} \tag{8}$$

where

$$x = |J|/k_B T$$

An alternate set of equally convenient procedures has been developed by Jotham[9] for antiferromagnetic systems. For ferromagnetic systems with $S = 1/2$, the Padé approximation technique has been used by Baker *et al.*[10] to derive the expression in Eq. (9) for the reduced magnetic susceptibility. The ferromagnetic case has also been described by a high-temperature series expansion.[11]

$$\begin{aligned}\chi_M(K) = [(1.0 &+ 5.7979916K + 16.902653K^2 + 29.376885K^3 \\ &+ 29.832959K^4 + 14.036918K^5)/(1.0 + 2.7979916K \\ &+ 7.0086780K^2 + 8.6538644K^3 + 4.5743114K^4)]^{2/3}\end{aligned} \tag{9}$$

where

$$K = J/2k_B T$$

Other results, most of which are also not exact solutions, exist for the general case of spin S greater than 1/2 for both ferromagnetic and antiferromagnetic exchange. Fisher[12] solved Eq. (2) for the classical infinite-spin case for the linear chain. This solution is applicable to both antiferromagnetic and ferromagnetic exchange coupling. Friedberg and

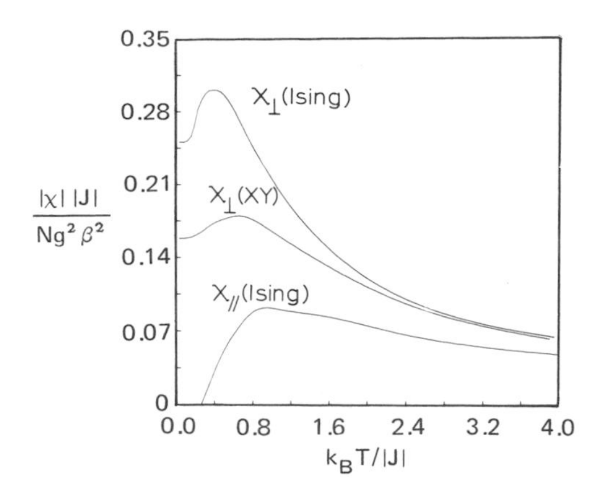

FIGURE 2. Theoretical curves in reduced coordinates for the Ising and XY models of $S = 1/2$ systems. (After References 17–19.)

1.5. The Ising Interaction

When $\alpha = 1$ and $\beta = \gamma = 0$ in Eq. (2), the Ising model Hamiltonian is described. Closed-form solutions exist for both the parallel[19] and perpendicular[17, 19] magnetic susceptibilities of $S = 1/2$ chains. These expressions are given in Eqs. (16) and (17).

$$\chi_{\parallel} = \frac{Ng^2\mu_B^2}{4k_B T} \exp\left(\frac{2|J|}{k_B T}\right) \tag{16}$$

$$\chi_{\perp} = \frac{Ng^2\mu_B^2}{8|J|}\left[\tanh\left(\frac{|J|}{k_B T}\right) + \frac{|J|}{k_B T}\operatorname{sech}^2\left(\frac{|J|}{k_B T}\right)\right] \tag{17}$$

Plots of these functions in reduced coordinates are shown in Figure 2. Various exact solutions also exist for the parallel susceptibility of the $S \geq 1$ case.[20–22] The general results for the Heisenberg, XY, and Ising models which relate χ, J, and $T_{\max}$ are tabulated in Table 1.[1] The detailed information required for the application of these results for the analyses of experimental data are available in the original works.

1.6. Effects of Zero-Field Splitting

Low-symmetry ligand fields can produce single-ion effects, implied by Eq. (4), that are nonnegligible in linear chain compounds, when the single-ion spin is 1 or greater. In axial fields, the zero-field splitting parameter (D) describes the removal of spin degeneracy, while crystal fields of orthorhombic or lower symmetry require that the rhombic term (E) of Eq. (4) be nonzero. The E term consequently removes any remaining degeneracy. It must be pointed out that $\hat{H}_a$ is written in many forms in the literature and, as written in Eq. (4), the energy barricenter has been preserved.[23, 24] As an

TABLE 1. Relationships between Exchange Coupling Constants and the Maximum in the Magnetic Susceptibility of Antiferromagnetically Coupled Linear Chains[a]

Exchange coupling model	Spin	Model[b] abbreviation	$\frac{k_B T_{(\chi_{max})}}{\lvert J \rvert}$	$\frac{\chi_{max} \lvert J \rvert}{N g^2 \mu_B^2}$
Heisenberg	1/2	H–BF	1.282	0.07346
Heisenberg	1	H–W	2.70	0.088
Heisenberg	3/2	H–W	4.75	0.091
Heisenberg	2	H–W	7.1	0.094
Heisenberg	5/2	H–W	10.6	0.095
Heisenberg	3	H–W	13.1	0.096
XY	1/2	*XY*	0.64 ($\chi_\perp$)	0.174 ($\chi_\perp$)
Ising	1/2	I–F	1.0 ($\chi_\parallel$)	0.09197 ($\chi_\parallel$)
			0.4168 ($\chi_\perp$)	0.2999 ($\chi_\perp$)
Ising	1	I–GS	2.55 ($\chi_\parallel$)	0.098 ($\chi_\parallel$)
Ising	3/2	I–GS	4.70 ($\chi_\parallel$)	0.10 ($\chi_\parallel$)
Ising	2	I–GS	7.46 ($\chi_\parallel$)	0.101 ($\chi_\parallel$)
Ising	5/2	I–GS	10.8 ($\chi_\parallel$)	0.1015 ($\chi_\parallel$)
Ising	3	I–GS	14.8 ($\chi_\parallel$)	0.102 ($\chi_\parallel$)

[a] After Reference 1.
[b] See p. 131 for Notation.

example, the effect of an axial field ($E = 0$) on the energy levels of an $S = 1$ ion is plotted in Figure 3 as a function of the applied magnetic field. The example shows $D > 0$; $D < 0$ reverses the energy-level scheme.[25]

Although the observed magnitudes, signs, and experimental anisotropies are spin and metal-ion dependent, some general features are common to all linear chains in which single-ion effects are important. Experimentally, the observed anisotropies in the principal magnetic susceptibilities, $\Delta\chi = \chi_\perp - \chi_\parallel$, are positive for positive D. The reverse situation results when $D < 0$ and $\Delta\chi$ becomes negative. For very small values of $D/|J|$, the theoretical analyses can be carried out by treating $\hat{H}_a$ as a perturbation to the exchange interaction.[13] Generally, a positive D value

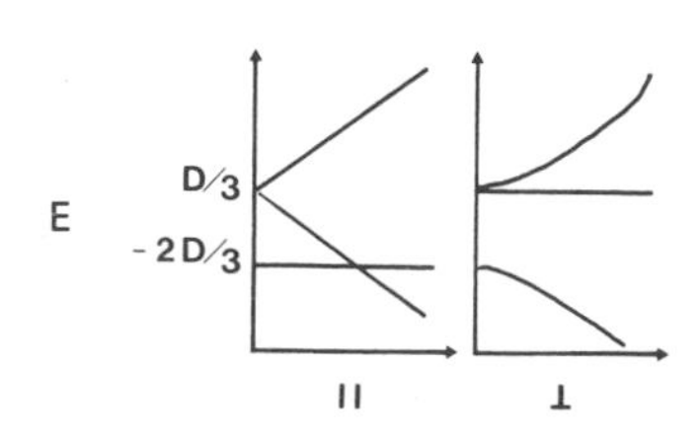

FIGURE 3. Splitting diagram of an axial $S = 1$ ion in an applied magnetic field where $D > 0$.

tends to force the exchange-coupled spins to lie in the plane perpendicular to the unique z axis, but a negative D value tends to favor the spins lying along the z direction. Larger values of either sign of $D/|J|$ present much more difficult theoretical problems, and they are very dependent on whether the total spin of the ground state is even or odd. At this point it is sufficient to state that large zero-field splittings can induce anisotropies in the exchange interaction itself and causes the choice of the appropriate model to be difficult. Further discussion of these effects and their consequences is taken up on an individual ion basis.

2. Chain Compounds of Titanium

The only well-known and widely studied one-dimensional system of titanium is that of β-$TiCl_3$. The compound may be prepared by photochemical reduction of $TiCl_4$ vapor in a stream of ultrapure hydrogen at 25°C, and any adsorbed species can be removed by heating and evacuation.[26] This preparation results in microscopic crystals; attempts to grow macroscopic crystals have failed. The crystal structure consists of chains of titanium(III) ions formed by face-sharing $TiCl_6$ octahedra, with an intrachain Ti–Ti distance of 2.91 Å and an interchain distance of 6.1 Å.[27] (See Table 2.)

The $TiCl_6$ octahedra are slightly elongated along the c axis, resulting in a stabilization of the d_{xy} and $d_{x^2-y^2}$ orbitals. If this were the only factor to be considered, the susceptibility would be expected to obey the Curie–Weiss law. On the contrary, the observed susceptibility shows an unusual temperature dependence, and this has been explained in terms of interactions between titanium ions along the chain as a result of the short interchain Ti–Ti distance.[28] Based on a calculation similar to those of Fletcher and Wohlfarth,[29] it was concluded that d_{z^2}–d_{z^2} interactions exceed the ligand field stabilization and are at least 0.25 eV. This narrow d band accounts for

TABLE 2. Structural and Magnetic Properties of Titanium Chain Compounds[a]

Compound	Structure			Intrachain	Interchain	Experimental			
	M–M (Å)	M–B–M (deg)	M–M′ (Å)	J (cm^{-1})	T_c (K)	Temp. (K)	Technique	Model	Ref.
β-$TiCl_3$	2.91		6.1			77–400	χ_p	B–HH	26
$RbTiI_3$	3.398		8.097	−23	<4.2	4.2–293	χ_p		67

[a] See p. 131 for Notation.

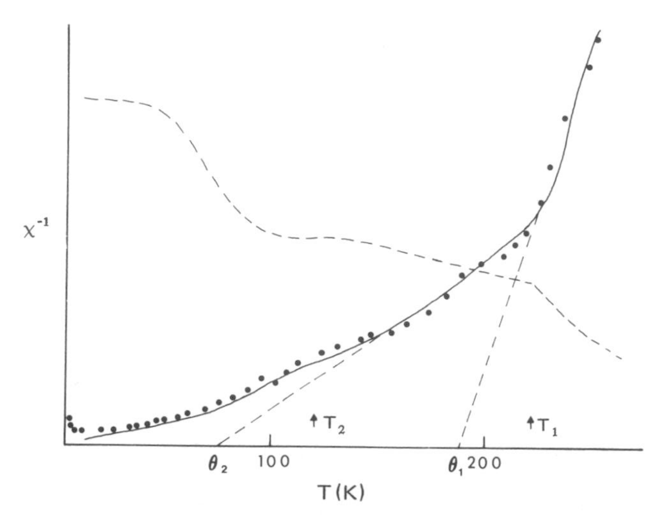

FIGURE 4. Inverse dynamic susceptibility of β-$TiCl_3$ as a function of temperature. Dots are experimental points. Solid line is calculated with Berlin model. Upper dashed line is χT plot. Lower dashed lines are Curie–Weiss plots for the different temperature ranges. (From Reference 28, with permission.)

the antiferromagnetism of the system, and Berlin *et al.*[30] have shown that the Hubbard Hamiltonian, Eq. (18), is applicable.

$$\hat{H} = \sum_{m,n,\sigma} T_{mn} a^{\dagger}_{m\sigma} a_{n\sigma} + \frac{\lambda}{2} \sum_{n,\sigma} a^{\dagger}_{n-\sigma} a_{n-\sigma} a^{\dagger}_{n\sigma} a_{n\sigma} \tag{18}$$

In this equation $T_{m,n} = 0$, except $T_{n+1,n} = T_{n-1,n} = -\beta$ ($\beta > 0$), β is the resonance integral, and γ is the Coulomb integral of two electrons on one center. The indices for the atoms are m and n, and σ is the spin index. In this treatment, a dense set of low-lying singlet and triplet excitations can exist. It is thought that mobile triplet excitations account for the temperature dependence of the susceptibility which is shown in Figure 4.

3. *Chain Compounds of Vanadium*

The vanadium systems best characterized magnetically and structurally are those of the ABX_3 type [with A = Cs^+, Rb^+, $(CH_3)_4N^+$ and X = Cl^-, Br^-, I^-) and some oxovanadium or vanadyl complexes. A noteworthy exception to the above generalizations is $BaVS_3$. The preparation and handling of many of these compounds is well documented, and several systems are available for experimental study. However, much of the work with vanadium chains (or suspected chains) has been on systems that have been incompletely documented and these will not be covered extensively in this chapter.

The vanadium ions in these compounds are either V(II) or V(IV) and thus have d^3 or d^1 electronic configurations, with spins of 3/2 or 1/2, respectively. The Ising model has been applied often for the description of the magnetic properties of vanadium chains because ground-state properties can be calculated,[31–33] and the model gives good limiting values of χ_m as the number of spins in the chains tends to infinity for the antiferromagnetic case. In the case of strong exchange coupling, the Ising model exhibits some defects and, as a result, there has been some criticism of the application of the Ising model to these problems.[9, 34] Jotham[9] has demonstrated that the Heisenberg model is to be preferred in many instances.

3.1. Vanadium Chain Compounds of the AVX_3 Type

The structures of the AVX_3 compounds consist of octahedral units of VX_6 sharing opposite faces, thus forming infinite chains that run parallel to the hexagonal c axis. Each VX_3^- unit has associated with it an A^+ ion, and as shown in Figure 5, these A^+ cations separate the chains from each other.

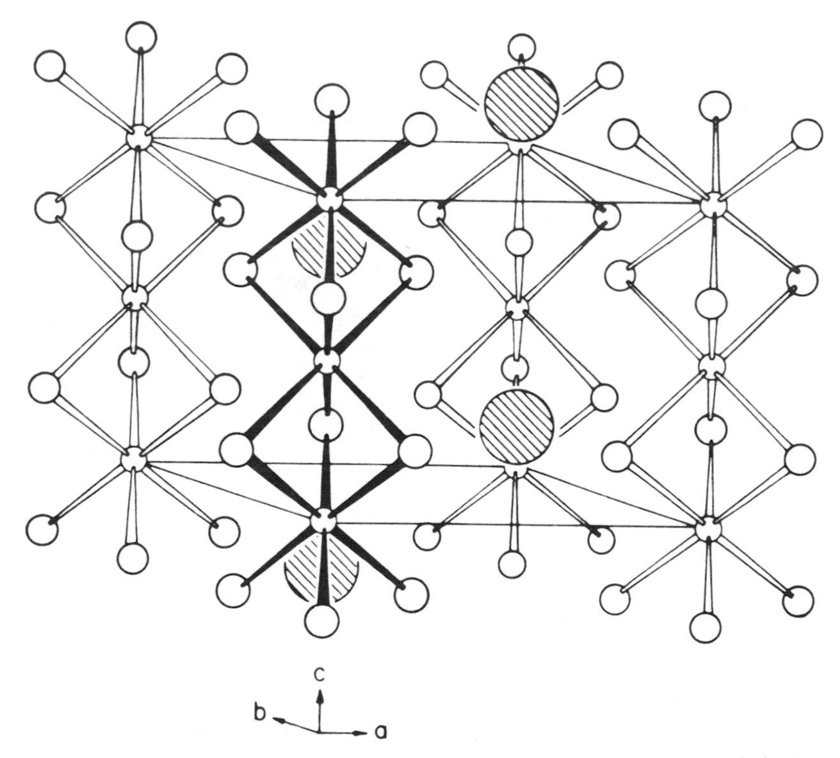

FIGURE 5. The general AMX_3 structure. (From Reference 138, with permission.)

3.1.1. The $CsVX_3$ Compounds

The $CsVX_3$ compounds may be prepared by solid-state and other simple reactions[35–37] and single crystals can be grown by the Bridgeman technique or sublimation methods. All three compounds (with $X = Cl^-$, Br^-, I^-) crystallize with the one-dimensional $CsNiCl_3$ structure, although only the crystal structure of $CsVCl_3$ has been determined by single-crystal x-ray techniques.[38] The cell parameters and intrachain vanadium–vanadium distances of these compounds have been collected[36,37] and are summarized in Table 3.

The magnetic susceptibilities, χ, of the $CsVX_3$ compounds have been measured for temperatures ranging from liquid helium to well above room temperature. According to the calculations of Figgis[39] for single ions, the measured susceptibility is related to the spin-only value of the $^4A_{2g}$ term by the following expression:

$$\chi_M^S = \frac{\chi_M^{\text{meas}} - 8k_1^2 N\mu_B^2/\Delta}{1 - 8k_1^2\lambda_0/\Delta} \tag{19}$$

where N is Avogadro's number, μ_B the Bohr magneton, g the Landé factor, Δ the octahedral crystal field splitting, and λ the spin–orbit coupling parameter. Values for the constants are derived from a fit of experimental data.

Intrachain exchange-coupling constants have been obtained by application of the scaled Heisenberg chain model with $S = 3/2$, with J for $CsVCl_3$ being $-79.93\ \text{cm}^{-1}$. Slight variations of the J value for $CsVBr_3$ and $CsVI_3$ with temperature are probably due to thermal expansion of the cell itself. The thermal variation of the corrected susceptibility of $CsVI_3$ is shown in Figure 6.

The cation–cation exchange interactions and the cation–anion–cation superexchange interactions according to the Goodenough–Kanamori rules for a d^3 ion are

(1) direct exchange: $t_{2g} - t_{2g}$ antiparallel spin

(2) superexchange:

$t_{2g} - t_{2g} - t_{2g}$	antiparallel spin
$t_{2g} - p - e_g$	antiparallel spin
$e_g - s - e_g$	antiparallel spin
$t_{2g} - p - e_g$	parallel spin
$e_g - p_\sigma - p_\sigma - e_g$	parallel spin
$t_{2g} - p_\pi - p_\pi - e_g$	parallel spin

The fact that the overall interaction in the $CsVX_3$ compounds is strongly antiferromagnetic indicates that the exchange is due primarily to the first four interactions listed above.

Charlot *et al.*[40] have recently presented a molecular orbital approach to explain antiferromagnetic coupling in $CsVCl_3$ and other compounds. In

TABLE 3. Structural and Magnetic Properties of Vanadium Chain Compounds

	Structure			Intrachain	Interchain	Experimental			
Compound	M–M (Å)	M–B–M (deg)	M–M′ (Å)	J (cm^{-1})	T_c (K)	Temperature (K)	Technique	Model	Reference
$BaVS_3$	2.805		6.7			20–130	χ_i	SM	52
$(CH_{34}NVCl_3$	3.11		9.146			80–289	χ_p		2
$CsVCl_3$	3.015	74.95	7.23	−80		4.2–800	χ_i	H–S	2, 36
				−160				MO–K	40
$CsVBr_3$	3.16		7.56	−63 to −56		4.2–800	χ_i	H–S	36
$CsVI_3$	3.41		8.21	−47 to −38	32	4.2–800	χ_i	H–S	2, 36
$RbVCl_3$	3.00		7.04	−105	350	77–500	χ_p	F–L	2, 41
$RbVI_3$	3.404		8.004	−38	25	1.2	N		67
$(sal_2\,pn)VO$		158.2				1.7–68	χ_p	I–F	50

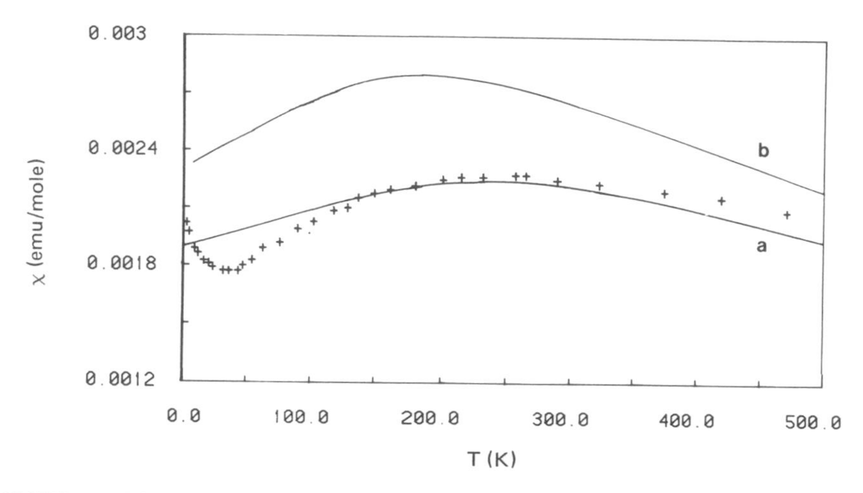

FIGURE 6. The corrected magnetic susceptibility of $CsVI_3$ as a function of temperature, pluses are experimental points. Curves (a) and (b) are calculated with the two J values obtained. (After Reference 36.)

this work the exchange interaction is expressed as the sum of antiferromagnetic (J_{AF}) and ferromagnetic (J_F) components where each is expressed as orbital integrals related to the mechanism of the exchange interaction. The expression for J is

$$J = J_{AF} + J_F \tag{20}$$

$$J_{AF} = -\frac{1}{n^2} \sum_{\mu=1}^{n} \Delta_\mu s_{\mu\mu}, \qquad J_F = \frac{2}{n^2} \sum_{\mu,\,\nu=1}^{n} J_{\mu\nu} \tag{21}$$

$S_{\mu\mu}$ is the overlap integral between two magnetic orbitals of the same symmetry, Γ_μ, centered on nearest-neighbor transition-metal ions, Δ_μ is the width of the molecular-orbital band built from the magnetic orbitals previously mentioned for the state of highest spin multiplicity, and $J_{\mu\nu}$ is the exchange integral between nearest-neighbor magnetic orbitals of differing symmetries. A magnetic orbital is defined as a partially occupied orbital centered on a transition ion and delocalized toward the ligands surrounding the ion. The extended Hückel method with charge iteration and Madelung correction for potential was used to obtain Δ_μ and $S_{\mu\mu}$ and to evaluate the terms in Eq. (21). The Wolfsberg–Helmholtz approximation was used for the magnetic orbitals. This method gives reasonable values for the exchange-coupling constants for the sequence of compounds studied; however, the calculated value for $|J_{AF}|$ is considerably smaller than the measured value of $|J|$. This is probably due to the quality of the d atomic orbitals used and probably could be improved by use of more elegant representations of the orbitals involved.

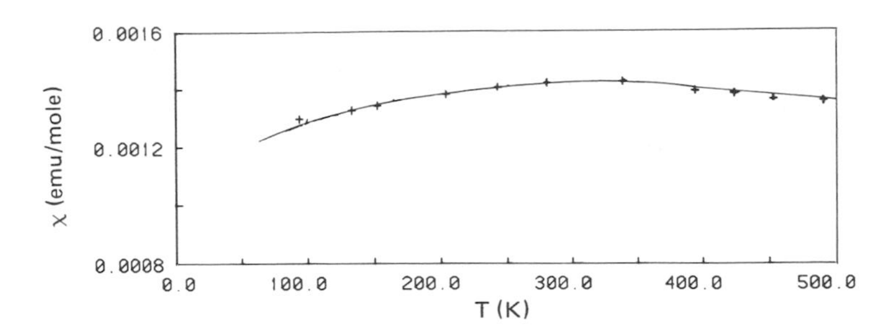

FIGURE 7. Magnetic susceptibility of $RbVCl_3$ as a function of temperature. Pluses are experimental points. The line is a fit to the Van Vleck equation for an exchange-coupled pair; the equation has been modified to include a large $(T + \theta)$ term. (After Reference 41).

3.1.2. $RbVCl_3$

$RbVCl_3$ can be prepared by careful dehydration of $RbVCl_3 \cdot 6H_2O$ at elevated temperatures under reduced pressure. The compound is formally isostructural with $CsNiCl_3$.[2]

Grey and Smith[41] report that the magnetic and spectral properties of $RbVCl_3$ display unusual behavior not consistent with linear chain properties.[42] In their analysis, the magnetic behavior was explained in terms of a model in which the primary interaction was between pairs of metal ions along the chain with weaker interactions between adjacent pairs.[43] To accomplish this, the Van Vleck equation for an exchange-coupled pair of vanadium(II) ions was modified to include a $(T + \theta)$ term. The results of the analysis are shown in Figure 7, but it should be recognized that the data must be analyzed in terms of more refined theory before the exact nature of the interactions can be known.

3.2. Vanadyl Chains

Precise structural determinations do not exist for many vanadyl chain compounds, and magnetic properties have been used as evidence for linear chain structures.[44–46] Therefore, any conclusions from this work must be considered to be tentative until such time as exact structural determinations are available. One compound that has been completely characterized is *N,N'*-propylenebis(salicylaldiminato)oxovanadium(IV).

3.2.1. N,N'-propylenebis(salicylaldiminato)oxovanadium(IV)

The complex *N,N'*-propylenebis(salicylaldiminato)oxovanadium, $(sal_2pn)VO$, has been known for many years. Its preparation was reported

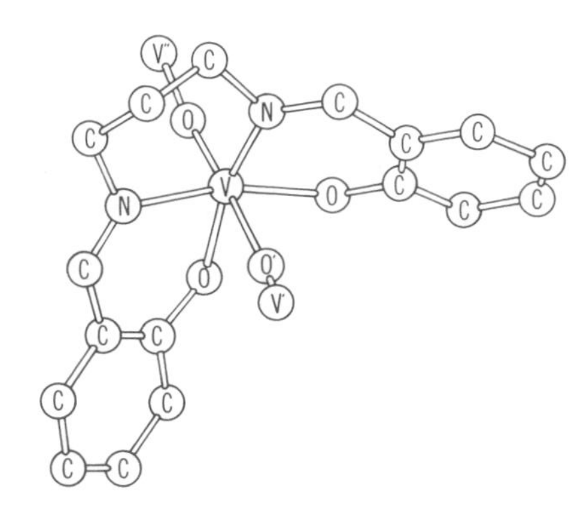

FIGURE 8. The structure of $(sal_2pn)VO$. (After Reference 50.)

as early as 1937,[47] and the structure was determined by Mathew and coworkers in 1970.[48] The molecules are packed such that the vanadyl oxygen occupies the sixth coordination site of the vanadium atom in the neighboring molecule, with the result being infinite chains of ··· V–O–V ··· bonds running along the *c* direction. The structure is shown in Figure 8.

The physical properties of this compound are rather atypical of most vanadyl(IV) complexes, with the difference between $(sal_2pn)VO$ and most other vanadyl complexes being attributed to the more planar nature of the $(sal_2pn)V$ moiety and to the interaction of the vanadyl oxygen of one molecular unit with the neighboring vanadium atom.[49]

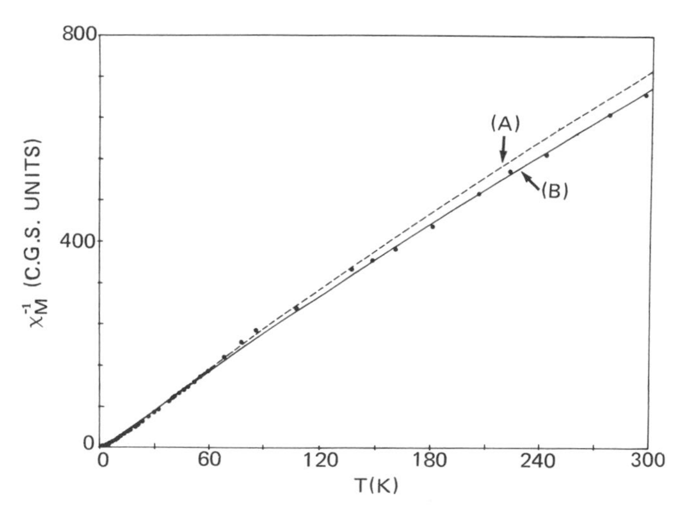

FIGURE 9. Inverse susceptibility plotted as a function of temperature for the complex $(sal_2pn)VO$. Line (A) represents the best fit for the temperature range 1.73–297 K, with $J = 2.3$ cm^{-1}. Line (B) represents the best fit for the temperature range 1.73–78 K, with $J = 4.1$ cm^{-1}. (After Reference 50.)

The inverse magnetic susceptibility of $(sal_2pn)VO$ is shown in Figure 9, where a small ferromagnetic interaction is indicated by the data.[50, 51] This is unusual behavior since there are large antiferromagnetic interactions in other compounds suspected to be oxovanadium chains. Accurate crystal-structure determinations of other vanadyl chains will provide useful information for comparison and will lead to a better understanding of the properties of these chain compounds.

3.2.2. *Oxovanadium Carboxylates*

A large number of oxovanadium carboxylates exhibit similar magnetic properties, that being substantially reduced paramagnetism. Reports[44–46] indicate that crystal growth is either extremely difficult or impossible for oxovanadium carboxylates, and consequently there are few structural determinations for these compounds and none for chains. The basis given for the polymeric nature of these compounds is their insolubility in all solvents with which they do not react. Casey and Thackeray[46] report that the magnetic data collected for oxovanadium acetate are in agreement with the Ising model for one-dimensional systems, but conflict with the Heisenberg model. This is taken to suggest that exchange interactions involve the z component of spin, but not the x and y components. Here again, it should be pointed out that the basis for assuming that oxovanadium acetate is a one-dimensional chain is the fact that the magnetic data fits the Ising model.

A total of 26 mono and dicarboxylates of oxovanadium have been prepared and their magnetic properties studied by Casey *et al.*[45] and Kalinnikov *et al.*[44] A large volume of indirect evidence and the fact that they fit one-dimensional models were used to infer that 22 of the compounds were indeed one-dimensional infinite chains.

3.3. $BaVS_3$

The preparation and study of the magnetic and electrical properties of $BaVS_3$ have been given by Massenet *et al.*,[52] and the crystal structure is shown in Figure 10. The structure consists of linear chains of vanadium(IV) ions which run parallel to the c axis. The chains are formed by face-sharing octahedra of sulfur atoms that surround each vanadium atom with an intrachain vanadium–vanadium distance of 2.805 Å, and an interchain vanadium–vanadium distance of 6.7 Å. The cell dimensions and magnetic susceptibility of $BaVS_3$ as a function of temperature are shown in Figure 11. The magnetic susceptibility shows a peak around 70 K, and Massenet and co-workers[53] have attributed this peak to the onset of short-range antiferromagnetic ordering along the chains. However, the sharpness of the peak may be indicative of longer-range interactions or magnetic ordering.

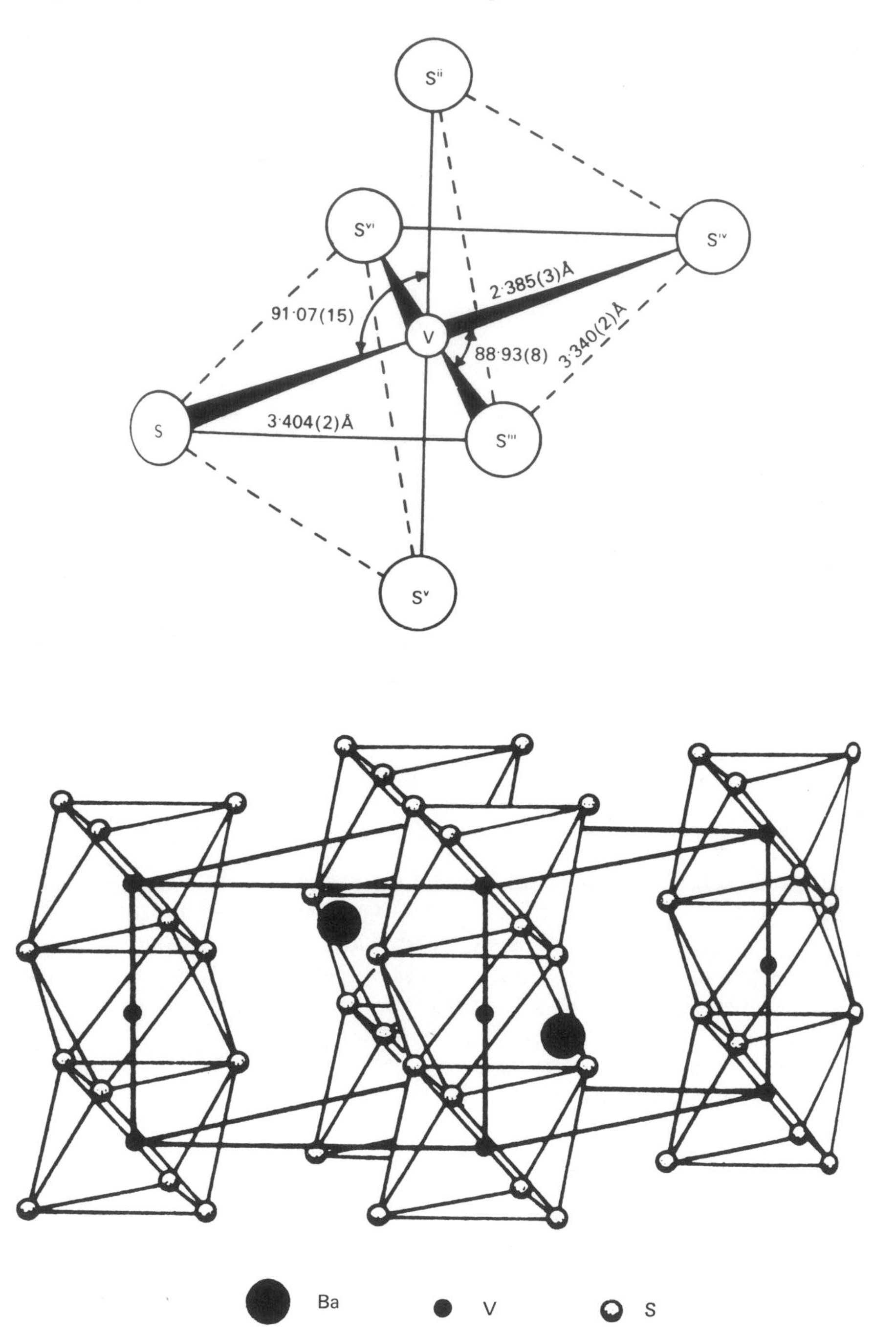

FIGURE 10. Structure of $BaVS_3$. (After Reference 54.)

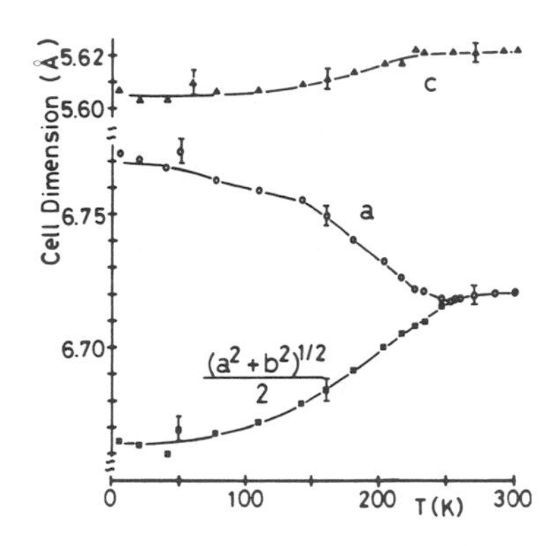

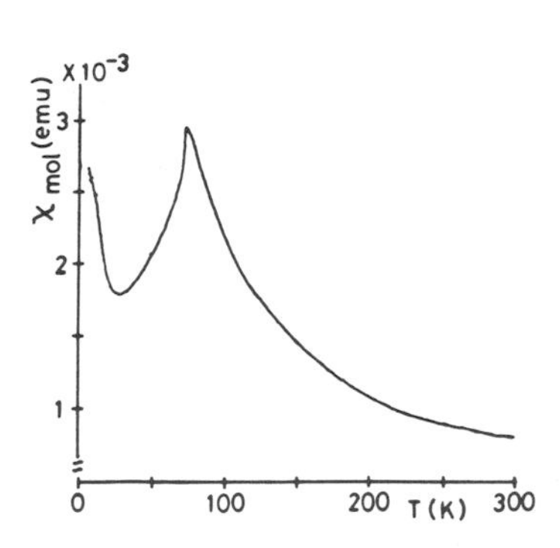

FIGURE 11. Cell dimensions and magnetic susceptibility plotted as a function of temperature for $BaVS_3$. (After Reference 54.)

4. Chain Compounds of Chromium

The study of cooperative magnetic phenomena in one-dimensional arrays of ligand bridged chromium ions has been limited to three broad classes of compounds. They can generally be summarized as hexagonal pervoskites, fluoride salts, and metal phosphinate polymers, and only compounds of $S = 3/2$ and $S = 2$ spin systems have been studied. In all cases, the magnetic behavior has been observed to be antiferromagnetic. There are more chain systems of Cr(II) ions than there are of Cr(III), even though the synthesis and characterization of the former compounds are more difficult.

4.1. Crystal-Field Background for the Chromium Ion

In a cubic crystal field, the ground-state free-ion ^4F term of Cr(III) is split such that there is a nondegenerate $^4A_{2g}$ ground state and two excited states, $^4T_{1g}$ and $^4T_{2g}$. There is no orbital angular momentum associated with the ground-state term, and, consequently, spin–orbit coupling cannot lift the degeneracy. Spin-only magnetic moments are expected, and isotropic-exchange behavior should occur. Since some mixing of the excited states into the ground state can occur,[55] the magnetic susceptibility of octahedrally coordinated chromium(III) ions becomes

$$\chi_M = \frac{N\mu_B^2}{3k_B T}(\mu_{\text{eff}}^2) + \frac{8N\mu_B^2}{10Dq} \tag{22}$$

where

$$\mu_{\text{eff}} = \mu_{so}\left(1 - \frac{4\lambda}{10Dq}\right) \tag{23}$$

where the symbols have their usual meanings. Since the spin–orbit coupling constant is small and $10Dq$ is large for chromium(III), near spin-only magnetic moments are often observed.

In the case of Cr(II) with $S = 2$, the ^5D term splits in a cubic field to yield a ^5E_g ground state and a $^5T_{2g}$ excited state, and the magnetic susceptibility and magnetic moment are given by

$$\chi_M = \frac{N\mu_B^2}{3k_B T}(\mu_{\text{eff}}^2) + \frac{4N\mu_B^2}{10Dq} \tag{24}$$

$$\mu_{\text{eff}} = \mu_{so}\left(1 - \frac{2\lambda}{10Dq}\right) \tag{25}$$

When the cubic symmetry of the ligand field is reduced to axial symmetry, the orbital degeneracy of the ground state is removed. This is especially important for the ^5E term of Cr(II). To a first approximation the orbital contribution to either the A_2 or E_2 term is small, and distortions and subsequent orbital contributions will not affect the susceptibility. As mixing of the ground and excited states becomes important, the susceptibility values differ from the spin-only prediction and become anisotropic. The spin degeneracy of the energy levels may be lifted through crystal-field and spin–orbit coupling effects in the absence of an external magnetic field, but this zero-field splitting is apparently small in chromium chain compounds since it has rarely been found.

4.2. Additional Theoretical Considerations for Chromium Chains

A theoretical technique that has been used for the analysis of the magnetic data for some chromium systems, and which was not discussed in Section 1, requires some additional comment here. The method treats all three contributions to the Hamiltonian given in Eq. (1) simultaneously. Pesch and Selke[56] solved the problem with all three terms present using an expansion in $1/n$, where n is the spin dimensionality, while Stahlbush and Scott[57] performed the same calculation, perhaps more conveniently, for all three terms using the transfer integral method.[58] In this latter approach, the magnetic field was assumed to be in the direction parallel to the single-ion anisotropy axis, and the spin–spin interaction was assumed to be isotropic. By rewriting Eq. (1) as

$$\hat{H} = -2J\sum_{i,j}^{N}\hat{S}_i\cdot\hat{S}_j - DS^2\sum_{i}^{N}(\hat{S}_i^z)^2 - g\mu_B[S(S+1)]^{1/2}H\sum_{i}^{N}\hat{S}_i^z \tag{26}$$

for the system with N classical spins, the transfer integral becomes

$$\int \exp\left(\frac{2J}{k_B T}\hat{S}_i\cdot\hat{S}_j + \frac{DS^2}{k_B T}(\hat{S}_i^z)^2 + g\mu_B[S(S+1)]^{1/2}\frac{H}{k_B T}\hat{S}_i^z\right)\Psi_n(\Omega_i)\frac{d\Omega_i}{4\pi}$$
$$= \lambda_n\Psi_n(\Omega_i) \tag{27}$$

where the eigenfunctions $\Psi_n(\Omega_i)$ are the spherical harmonics Y_{1m}, and λ_n is the corresponding eigenvalue λ_{1m}. From the thermodynamic limit of the N-spin partition function, the susceptibility is given as

$$\chi_M = Nk_B T \frac{\delta}{\delta H}\left(\frac{1}{\lambda_{\max}} \frac{\delta \lambda_{\max}}{\delta H}\right) \tag{28}$$

Therefore, magnetic susceptibilities may be calculated from the maximum eigenvalue for a sequential series of magnetic fields, and the subsequent numerical determination of the derivative.

4.3. Chromium Chain Compounds with Halide Bridges of the AMX_3 Type

4.3.1. General Aspects

Most of the AMX_3 compounds have the hexagonal perovskite structure in which the MX_6 octahedra share faces (Figure 5) and form infinite chains along the crystallographic c axis.[59] Deviations from this structure have been observed, and these deviations have given rise to considerable discussion for chain compounds containing the Jahn–Teller active chromium(II) ion. Briefly, the high-temperature phases have hexagonal symmetry in most of these linear chains, but as the temperature is lowered transitions to lower symmetry usually occur. This reduction of symmetry is reflected by the large thermal anisotropies for the X^- ions along the principal chain axis. The elegant evidence that surrounds the development of this topic can not be presented here, but the interested reader will find a starting base from which to begin.

4.3.2. $CsCrCl_3$, An Antiferromagnetic Hexagonal Perovskite

By far, $CsCrCl_3$ is the most extensively studied chromium chain in the $ACrX_3$ series, and the compound has been synthesized by several methods: The first of which includes fusing together CsCl and $CrCl_2$ in the appropriate 1 : 1 ratio.[60] A convenient alternate method consists of mixing chromium(II) acetate with an appropriate portion of cesium acetate in a solution of acetyl chloride and acetic acid.[61] In another method, $CsCrCl_3 \cdot 2H_2O$ may be isolated from concentrated hydrochloric acid, and subsequently dehydrated in vacuum at 120°C.[62] Single crystals of the compounds that have been used for optical and structural work have usually been obtained by the Bridgeman method.[64]

The structure of $CsCrCl_3$ was first suggested to be analogous to that of $CsCuCl_3$ by Iberson *et al.*[65] and later, Klatyk and Seifert,[60] using the results of zero-level Weissenberg photographs, reported the structure to be that of $CsNiCl_3$. One might suspect the structure to resemble the Cu(II) complex, since the ground state is an orbitally degenerate E level in a cubic field for both Cu(II) and Cr(II) ions. McPherson *et al.*[63] carried out a complete x-ray diffraction study and observed that the structure was not identical with either $CsNiCl_3$ or $CsMgCl_3$. The structure also differs significantly from that of $CsCuCl_3$. Anisotropic motion of the chloride ions along the crystallographic c axis result in large thermal parameters, unlike the case for $CsMgCl_3$ in which the motion is nearly isotropic. The major difference between $CsCrCl_3$ and $CsCuCl_3$ arises from the fact that the Cu(II) ions do not lie on the sixfold screw axis, presumably because of a static Jahn–Teller distortion. The Cr(II) ions lie on C_{3v} symmetry sites in $CsCrCl_3$, and the coordination does not show the distinct static distortions found for Cu(II) in $CsCuCl_3$. Results of additional spectral studies are in agreement with the x-ray structure, thus indicating that there is little or no Jahn–Teller distortion; i.e., the spectra do not show the characteristic splitting of distorted Cr(II) complexes. It has been suggested[63] that the Jahn–Teller active E_{1g} vibrations of the anionic chain (the motion of the chloride ions parallel to the c axis) could explain the large thermal anisotropies in the c direction. Day and co-workers[66] confirmed the structure determined by McPherson *et al.*[63] by powder neutron diffraction. Powder profiles at 4.2 K indicated three-dimensional antiferromagnetic ordering, while single-crystal neutron diffraction studies indicated a first-order structural phase transition at 171.5 K. The structural changes are minor and have little influence on the magnetic properties. The structure of $CsCrCl_3$ was investigated again by x-ray diffraction by Zandbergen.[67] Hexagonal symmetry with slight anisotropic distortions of the Cl^- ion in the c direction was found.

Magnetic susceptibility data for $CsCrCl_3$ have been interpreted using several of the models discussed in Section 1. Earnshaw *et al.*[75] found that the Van Vleck model for an exchange-coupled pair of chromium(II) ions gave a good fit to the experimental results using $J = -35$ cm^{-1} and $g = 2.0$.[76] Larkworthy and Trigg[62] also found good agreement between theory and experiment. Having assumed $CsCrCl_3$ to be isostructural with $CsNiCl_3$, Leech and Machin[68,69] determined the temperature dependence

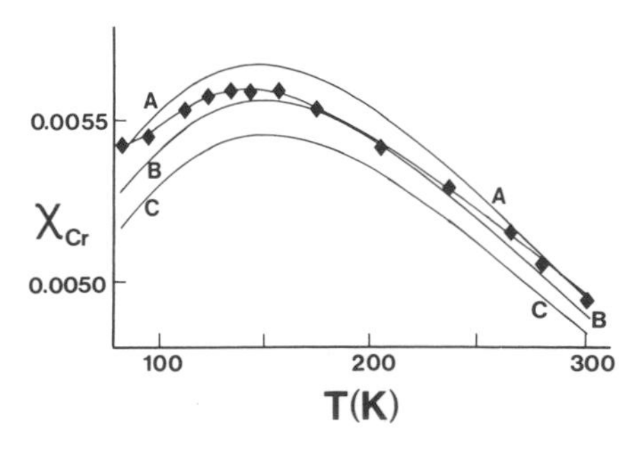

FIGURE 12. Magnetic susceptibility data for $CsCrCl_3$ fitted with the scaled Fisher model using $S = 2$, $g = 1.96$; and $J = -25.5$, 26.0, and 26.5 cm^{-1} for curves A, B, and C, respectively. (After Reference 69.)

TABLE 4. Structural and Magnetic Properties of Chromium Chain Compounds

	Structure			Intrachain	Single ion	Interchain	Experimental			
Compound	M–M (Å)	M–B–M (deg)	M–M′ (Å)	J (cm^{-1})	D (cm^{-1})	T_c (K)	Temperature (K)	Technique	Model	Reference
				Chain Compounds with Chloride Bridges						
$CsCrCl_3$	3.112	76.20	7.256				RT	x ray		63
				−35.0			80–300	χ_p	F–L	62
				−33.4		16	4.2–300	N	H–I	66
				−25.0			80–300	χ_p	H–S	68
$(CH_3)_4NCrCl_3$	3.27		9.05				RT	x ray		61
				−12.5		104	80–300	χ_p	H–S	77
$RbCrCl_3$	3.04		7.03				RT	x ray		60
$KCrCl_3$				−5.0			80–300	χ_p	H–S	68
				Chain Compounds with Iodide Bridges						
$CsCrI_3$	3.42		8.12				RT	x ray		59
		73.84		−9.7		27	1.0–300	N	H–WD	67
β-$RbCrI_3$	3.53		8.06	−7.6			1.0–300	N	H–WD	67
				Chain Compounds with Flouride Bridges						
$NaCrF_4$		90,150				8	4.2–300	χ_p, x ray	C–W	70
$KCrF_4$						15	4.2–300	χ_p	C–W	70, 72
$RbCrF_4$						20	4.2–300	χ_p	C–W	70
$CsCrF_4$	3.741	149,178				40	4.2–300	χ_p, x ray	C–W	70, 71
$CaCrF_5$ [a]				−2.75		<4.2	4.2–350	χ_p	H–I	79
				Polychromium phosphinates						
$Cr(O_2PMePh)_2(OH)$				−9.0			4.2–80	amor.	H–S	74
$Cr(O_2PMePh)_2(O_2P(C_8H_{17})_2)$				−17.4			4.2–80	χ_p	H–S	74
	4.52		7.8, 13.4				1.0–400	x ray		57
					0.007–0.004		1.0–400	χ_{film}	H–TI[b]	57

[a] A monofluoride bridge, not a triple chain structure.
[b] Transfer integral method of Section 4.2.

of the susceptibility and ascribed the broad maximum in the susceptibility near 140 K to short-range ordering in a one-dimensional array. For the analysis of their data, Leech and Machin applied the isotropic form of the Smith and Friedberg scaled model of the Fisher $S = \infty$ case. With $S = 2$ and g in the range 1.94–1.98, good theoretical fits to the data were obtained (Figure 12) with J values from -25.0 to $-26.5\ \text{cm}^{-1}$.

Day and co-workers[66] have observed an abrupt three-dimensional ordering transition in $CsCrCl_3$ at $T_c = 16$ K. In the ordered phase, the spins are antiferromagnetically aligned in the parent cell along the b and c directions, and ferromagnetically aligned along the a direction. Powder susceptibility measurements from 4.2 K to room temperature indicated a coupling constant of $\sim -34\ \text{cm}^{-1}$ with $S = 2$ and $g = 2$. Day *et al.*[66] also gave interpretations of spin-wave phenomena in the crystal and details of the single-ion anisotropy along the c axis.

4.4. $CsCrF_4$: *An Infinite Antiferromagnetic Triple Chain*

Several $ACrF_4$ systems have been synthesized[70] and these are listed in Table 4. $CsCrF_4$ has been closely examined by Babel and Knoke.[71] The compound may be prepared from equimolar mixtures of either CsF, $CsHF_2$, or $CsBF_4$ with CrF_3 in a vacuum (or under argon) at 600–700°C. Single crystals have been grown in Bridgeman platinum ampoules using a CsF/CrF_3 mixture. The structure consists of three linear chains of octahedra that share *trans* corners along the chain axis. The chains are linked together by sharing *cis* corners to form the triple linkage system shown in Figure 13. The magnetic susceptibility of this interesting system is displayed in Figure 14 where a maximum in χ_M at 55 K may be seen. Three-dimensional ordering occurs at $T_c = 40$ K, and the antiferromagnetic nature of the exchange interaction is indicated by the θ value of -104 K from the

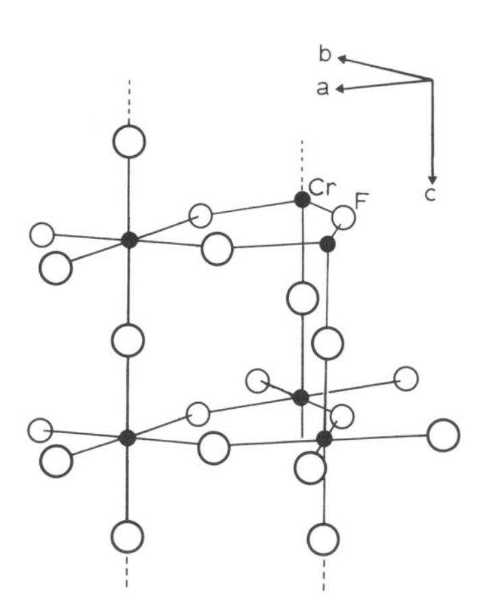

FIGURE 13. View of the $CsCrF_4$ triple chain structure showing the almost planar Cr_3F_9 units. (After Reference 71.)

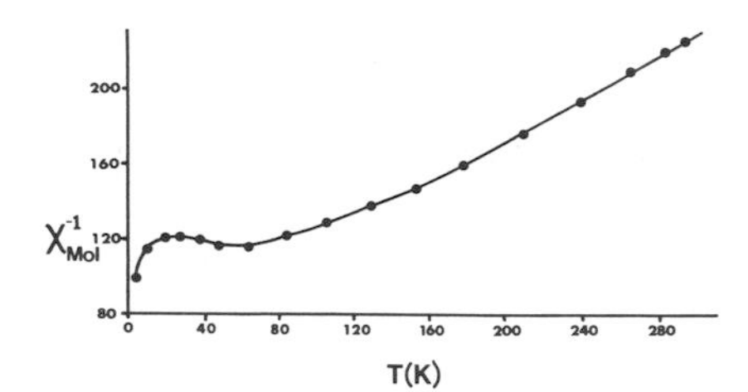

FIGURE 14. Magnetic behavior of $CsCrF_4$ which shows it to be an antiferromagnetic chain compound. (After Reference 70.)

Curie–Weiss fit. A theoretical analysis of this magnetic susceptibility data for this unusual compound would be worthwhile since several such systems are known.

4.5. Poly(chromiumphosphinates)

The syntheses of various metal dioctylphosphinates has been discussed by Gillman.[73] In the case of Cr(II) systems, the products are air sensitive and the syntheses are carried out under nitrogen. The preparation usually involves reaction between chromium(II) halide and the potassium phosphinate salt. Infrared and visible spectra have also been described. Despite the fact that the crystal structures are not known for any of these compounds, the compounds are of interest with respect to this chapter. The polymeric nature of these compounds arises from the phosphinate anion bridges as shown in Figure 15. The substitution of different R groups in the bridges determines the structures that are adopted and affects physical properties such as the exchange interaction. For example, the phosphinate polymer poly[bis(methylphenylphosphinate) dioctylphosphinatechromium] consists of trivalent chromium ions in sites of octahedral symmetry. The exchange is expected to be isotropic because of the $^4A_2$ ground state.

Sheets of these polymers have been pressed from the reaction products, and x-ray diffraction[57] studies indicate preferential alignment of the chains in the plane of the sheet. The x-ray results also reveal a uniform azimuthal distribution. The intrachain Cr–Cr separation is 4.52 Å, while transverse

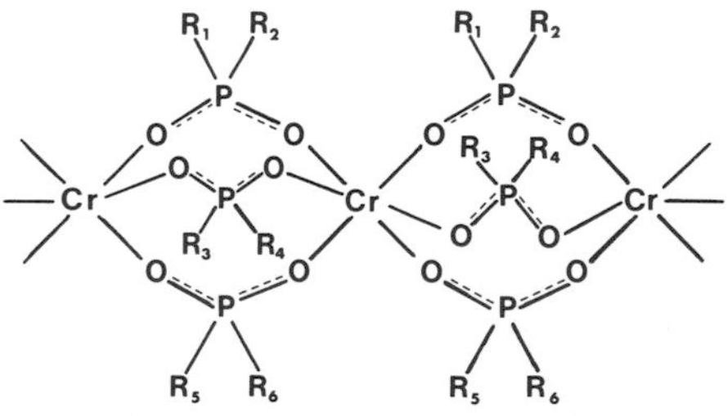

FIGURE 15. Bridging network in the chromium phosphinate polymer. The R substituents may be varied to achieve different physical properties.

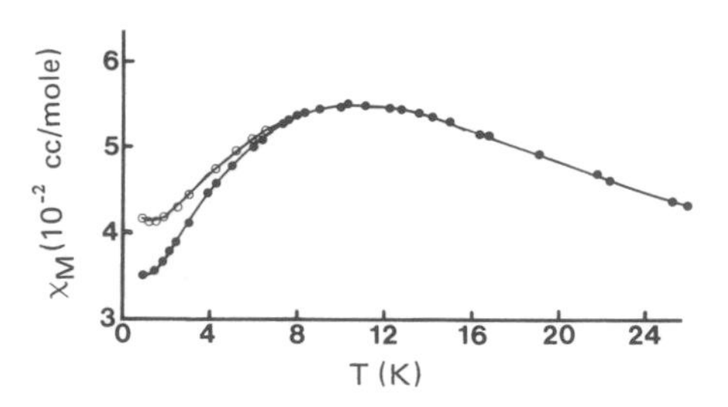

FIGURE 16. Magnetic susceptibility data for the polychromiumphosphinate of Section 4.5, indicating chainlike behavior and anisotropy below 6 K. The magnetic field is parallel to the sheet in the upper curve and perpendicular to the sheet in the lower curve. (After Reference 57.)

separations are 7.81 and 13.4 Å. Magnetic data in the region 1–300 K, which are shown in Figure 16, reflect the linear chain structure. No anisotropy is observed above approximately 6 K, and the data for $T > 6$ K may be fit by the zero-field model of Fisher with $g = 2.010(5)$ and $J = -1.93(14)$ cm^{-1}. Below 6 K anisotropic effects become important, and field dependence is also observed. Stahlbush and Scott[57] applied the transfer integral method summarized above to their experimental data and found $D = 0.02$ cm^{-1}. For the most part, the theoretical plots were in excellent agreement with experimental data. The discrepancy between theory and experiment may be interpreted as being due to the limitations inherent in a classical calculation on a quantum system. From arguments based on the competition between anisotropy and thermal energies, Stahlbush and Scott[57] suggested that the limits on the single-ion anisotropy parameter were $0.007 < D(\text{cm}^{-1}) < 0.04$.

5. *Chain Compounds of Manganese*

Although a large number of chain compounds of manganese exist (see Table 5), basically only three different structural or magnetic cases are known. All of the bridged manganese chain compounds that have been studied exhibit antiferromagnetic intrachain interactions, and most of the compounds undergo long-range ordering at low temperatures. Typically, the manganese(II) ion occupies an octahedral site with a $^6A_1$ ground state. As a result, the magnetic properties are expected and found to be largely isotropic with a g tensor that is close to the free-electron g value.[24] The analysis of the experimental data has involved only two models, those being the scaled infinite-spin chain of Smith and Friedberg[13] and the interpolation method of Weng.[15] Anisotropy in the magnetic susceptibility is sometimes observed at low temperatures; the anisotropy is usually very small and arises from single-ion effects. In many studies the three-dimensionally ordered phase has received more attention than the chain interactions, but the focus here is on the magnetic behavior above the ordering temperature.

TABLE 5. Structural and Magnetic Properties of Manganese Chain Compounds

Compound	Structure			Intrachain	Single ion	Interchain		Experimental		Model	Reference
	M–M (Å)	M–B–M (deg)	M–M′ (Å)	J (cm^{-1})	D (cm^{-1})	T_c (K)	J'/J	Temperature (K)	Technique		
Chain Compounds with μ-Fluoro Bridges											
$(NH_4)_2Mn(III)F_5$	3.87	143.4		−6.95		7.5		4.2–250	χ_i	H–S	95, 96
Chain Compounds with μ-Chloro Bridges											
$CsMnCl_3 \cdot 2H_2O$	3.1			−2.16	+0.08	4.89		0.35–80	χ_i	H–S	13, 94, 103
				−2.08	+0.08	4.89		0.35–80	χ_i	H–I	13, 97, 103
Chain Compounds with Di-μ-Chloro Bridges											
$MnCl_2 \cdot 2H_2O$	3.7	~90	>5.7	−0.31		6.9	~1	1.5–20	χ_i	H–S	104, 105
$[(CH_3)_3NH]MnCl_3 \cdot 2H_2O$	3.717	85.28		−0.25		0.98	$\frac{0.38}{0.25z}$	1.5–20	χ_i	H–S	106–108
$PyHMnCl_3 \cdot 2H_2O$	~3.7			−0.48				4–300	χ_p	H–RW, H–S	109
$Mn(py)_2Cl_2$	3.76			−0.48				2–80	χ_p	H–S, H–I	109, 111, 113, 115
$Mn(pz)_2Cl_2$	3.761			−0.63		1.3		2–80	χ_p	H–S	110–114
				−0.60		1.3		2–80	χ_p	H–I	110–114

				Chain Compounds with Tri-μ-Chloro Bridges						
$[(CH_3)_4N]MnCl_3$	3.25	84.09	9.15	−4.38	0.84		0.3–300	χ_i	H–S	80–86
$[(CH_3)_2NH_2][MnCl_3]$	3.232	84.28		−4.87			1.6–20	χ_p	H–S	117
				−4.80		0.072	1.6–20	χ_p	H–S	116
				Chain Compounds with Di-μ-Bromo Bridges						
$Mnpy_2Br_2$	~3.7			−0.6			2–80	χ_p	H–S	109, 111, 113–115
				−0.59			2–80	χ_p	H–I	109–111, 113–115
$Mnpz_2Br_2$	~3.7			−0.69			2–80	χ_p	H–S	110–114
				−0.68			2–80	χ_p	H–I	110–114
				Chain Compounds with Tri-μ-Bromo Bridges						
$CsMnBr_3$	3.2		7.7	−6.67	8.3		4.2–300	χ_p	H–S	117–119
				Chain Compounds with Chalcogenide Bridges						
$Mn(S_2CN(CH_3CH_2)_2)$	3.76	88.7		<0			84–298			120–121
				Chain Compounds with Salts of Oxyacids Bridges						
Mn(III)(Salen)(AcO)	6.536		~8.6	−1.4			80–300	χ_p	H–S	122–125
$Mn(N_2H_5)_2(SO_4)_2$	~5.3			−0.41			2–80	χ_p	H–I	126–129
$MnC_2O_4 \cdot 2H_2O$				−0.42		0.03	2–80	χ_p	H–S	126–129
				−0.69			80–300	χ_p	I–GS	130–131
				Chain Compounds with No Ligand Bridges						
MnPc	4.75				8.6					87–94

5.1. *An Isotropic Antiferromagnetic Linear Chain*, $[(CH_3)_4N][MnCl_3]$

Tetramethylammonium manganese chloride $[N(CH_3)_4]MnCl_3$, TMMC, may be prepared by mixing stoichiometric amounts of tetramethylammonium chloride and manganous chloride in aqueous solution. Two compounds crystallize from solution, these being light green prisms of $[(CH_3)_4N]_2MnCl_4$ and rose pink rods of TMMC. Morosin and Graeber[80] have reported the structure of TMMC to be the typical AMX_3 structure as shown in Figure 5 with the chain running parallel to the *c* axis. The intrachain Mn–Mn separation is 3.25 Å, while the interchain Mn–Mn distance is 9.15 Å. Single-crystal magnetic susceptibility studies[81] show a broad maximum in the region of 55 K. Below this maximum, as shown in Figure 17, the magnetic susceptibility falls, and then rises sharply, possibly due to impurities. The data have been fit to the scaled model of Friedberg with $g = 2.00$ yielding an exchange coupling constant J of $-4.38\ \text{cm}^{-1}$.[82–85] The exchange coupling constant may also be determined from the position of the maximum in the magnetic susceptibility data, using the relationships tabulated in Table 1. Values for J are -4.0 and $-3.61\ \text{cm}^{-1}$ based upon the magnitude of χ_{max} and the temperature associated with the maximum in the magnetic susceptibility, respectively. Hutchings *et al.*[82] have also determined the exchange coupling constant from the magnetic susceptibility data

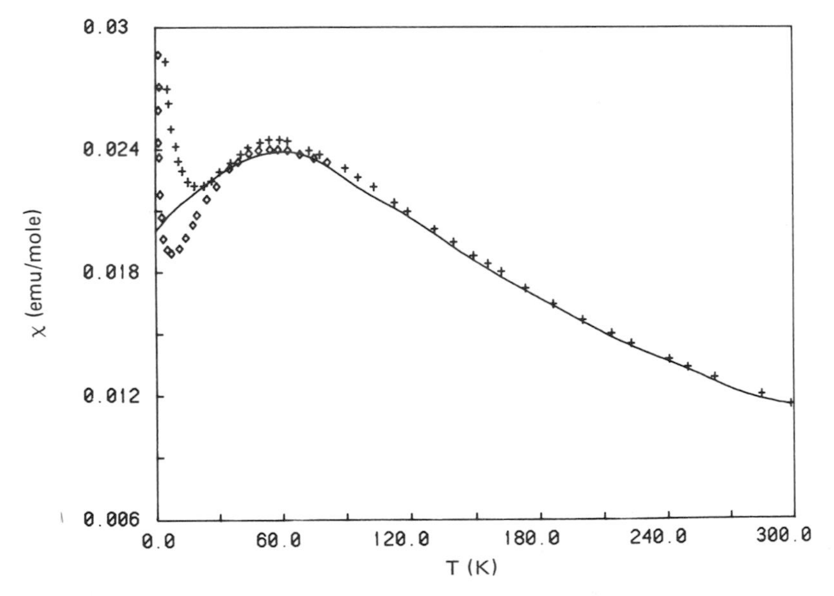

FIGURE 17. Experimental magnetic susceptibility vs temperature for TMMC with H ∥ *c* (+) and H ⊥ *c* (◇). Solid line is the best fit to the scaled linear chain with $S = 5/2$, $g = 2.00$, and $J = -4.38\ \text{cm}^{-1}$. (After Reference 81.)

between 60 and 170 K by using the high-temperature series expansion of Rushbrooke and Wood[11] and find $J = -4.52 \text{ cm}^{-1}$, while neutron experiments give a J value of -4.59 cm^{-1}. These values agree well with that reported by Dingle *et al.*[81] and with those from specific-heat measurements (-4.66 cm^{-1}).[83] De Jongh and Miedema[1] feel that the fit by Dingle may be fortuitous and that the interpolation model of Weng[15] should be more accurate. However, in most studies on manganese chains, the scaled model produces a better fit. The discrepancy between the values for the exchange coupling constants determined from the maximum in the magnetic susceptibility versus those obtained from the best fits of the data to the various models may be understood in terms of the presence of monomeric impurities. Paramagnetic impurities will raise the magnitude of the maximum and shift it to lower temperatures. Hutchings *et al.*[82] have shown that there is a very small amount of anisotropy ($\sim 1\%$) in the exchange and several reports[82,84–89] have determined that long-range order sets in at very low temperatures with $T_c = 0.84$ K.

5.2. *A Manganese Linear Chain without Bridging Ligands*

Manganese(II) phthalocyanine is a planar molecule with the manganese ion in a tetragonal environment which, along with the absence of ligand bridges, makes it unique among the structurally characterized manganese chains. A great deal of structural work[87–90] has been done on phthalocyanine compounds, and it has been shown that the Mn^{2+}, Co^{2+}, and Cu^{2+} compounds are isomorphous with the Ni^{2+} compound. The molecules stack parallel to the b axis with each molecular plane inclined by an angle[91] of 45.8° from the normal to the b axis (see Figure 18). The magnetic properties have been studied by Barraclough and co-workers[92,93]

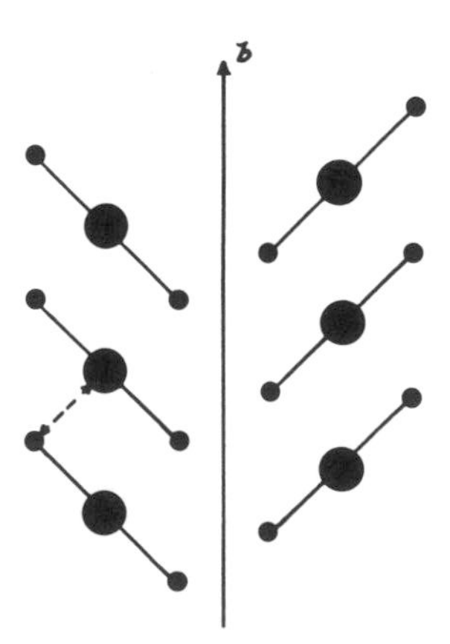

FIGURE 18. Stacking of the planar MnPc molecule. Large circles represent manganese atoms while the smaller circles are the azamethine nitrogen atoms which are the closest intercluster contact (3.4 Å). (After Reference 93.)

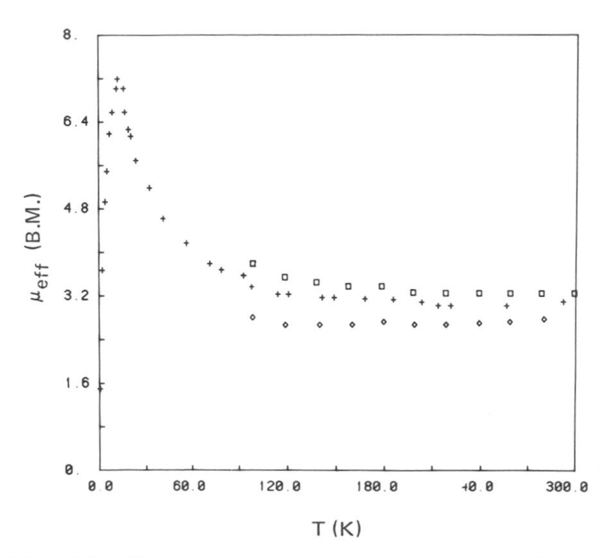

FIGURE 19. Plot of the effective magnetic moment vs temperature for MnPc. $\mu_{avg}(+)$, $\mu_{\perp}(\square)$, $\mu_{\parallel}(\diamond)$. (After Reference 92.)

and by Miyoshi.[91, 94] Anisotropy measurements show that the compound is highly anisotropic with $\mu_{\parallel}$ remaining constant at 4.0 B.M. while $\mu_{\perp}$ increases from 4.4 B.M. at 300 K to 5.0 B.M. at 80 K.[92] Below 80 K the moment rises sharply, reaches a maximum of 7.4 B.M. in the region of 14 K and then drops sharply as the temperature is lowered (see Figure 19). Below the maximum, the data are field dependent. Barraclough and co-workers[92, 93] originally explained the high-temperature anisotropy as resulting from a $^4A_{2g}$ ground state with a ^4E_g excited state being mixed in by spin–orbit coupling. This interpretation was later modified to take into account the ferromagnetic interactions that are present ($T_c = 8.6$ K).[94] The results are consistent with a ground state that is mainly $^4A_2(\pm 1/2)$ with a $^4A_2(\pm 3/2)$ more than 40 cm^{-1} above it. The $^6A_1(\pm 1/2)$ state lies approximately 600 cm^{-1} above the ground state and apparently is also mixed in. The magnetic susceptibility data were fit to the ligand field calculation modified by Heisenberg or Ising exchange interactions, with the Ising modified model giving the better results.[93] The low-temperature data have been analyzed by Miyoshi[91, 94] in terms of a canted ferromagnet.

5.3. An Anisotropic Linear Chain Compound, $CsMnCl_3 \cdot 2H_2O$

Cesium μ-chloro-dichloro manganese(II) dihydrate is a single chloro-bridged chain, unlike the related compound $CsMnCl_3$ which has the ABX_3 type structure. The compound may be prepared from aqueous solutions of stoichiometric amounts of CsCl and $MnCl_2 \cdot 4H_2O$ and crystals may be grown by slow evaporation. The manganese ion is octahedrally coordinated with the water molecules lying *cis* to one another. The two bridging chlorine ions are *trans* to one another and lie at slightly longer distances[97] (2.57 vs 2.50 Å) than those in the equatorial plane, thus yielding chains of apex sharing octahedra that run parallel to the *a* axis. The counterions lie between the chains.

Many magnetic susceptibility and related studies[13,97–103] have been carried out on $CsMnCl_3 \cdot 2H_2O$. The first was by Smith and Friedberg,[13] where the single-crystal magnetic susceptibilities were measured. As may be observed in Figure 20, the sample is highly anisotropic below 4.2 K, and moderate, though diminishing anisotropy, continues to about 20 K, above which the compound is largely isotropic. Long-range antiferromagnetic ordering sets in at low temperatures with $T_N = 4.98$ K. The broad maximum in the magnetic susceptibility in the region of 30 K has been attributed by Friedberg and co-workers[13,98] to antiferromagnetic linear chain interactions. The data has been fit to the Fisher model scaled to $S = 5/2$ with an isotropic g of 2.00, yielding an exchange coupling constant

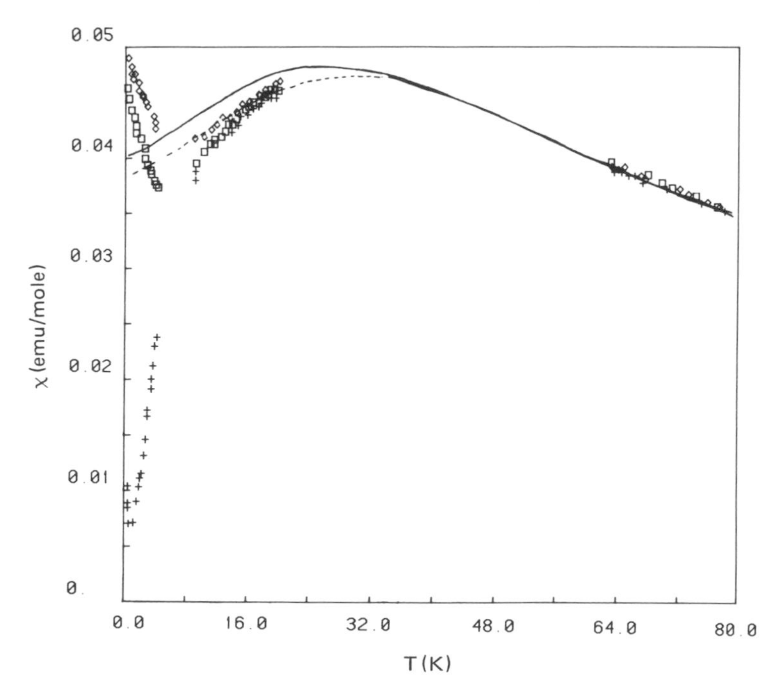

FIGURE 20. Single-crystal magnetic susceptibility of $CsMnCl_3 \cdot 2H_2O$. (+) *a* axis; (◇) *b* axis; (□) *c* axis; solid line, the scaled linear chain model with $S = 5/2$, $g = 2.00$, and $J = -2.16$ cm^{-1}; dashed line, the interpolation model with $J = -2.08$ cm^{-1}. (After Reference 13.)

J -2.16 cm^{-1}. The long-range antiferromagnetic ordering below T_N is apparently caused by interchain interactions with the preferred direction along the a axis. The axial anisotropy between T_N and T_{max} has been attributed by Smith and Friedberg[13] to single-ion, zero-field-splitting effects which arise from the $\hat{H}_a$ term in Eq. (1). Smith and Friedberg[13] have derived the expressions for the parallel and perpendicular magnetic susceptibilities for general spin S, in terms of D, the zero-field-splitting parameter is

$$\chi_{\parallel} = \chi_a - \frac{4Ng^2\mu_B^2[S(S+1)]^2 D}{45(k_B T)^2}\left(\frac{(1+u)(1+v)}{(1-u)(1-v)} + \frac{2u}{(1-u)^2}\right) \tag{29}$$

$$\chi_{\perp} = \chi_a + \frac{4Ng^2\mu_B^2[S(S+1)]^2 D}{45(k_B T)^2}\left(\frac{(1+u)(1+v)}{(1-u)(1-v)} + \frac{2u}{(1-u)^2}\right) \tag{30}$$

where u is defined in Eq. (11), χ_a is the average magnetic susceptibility, and

$$v = 1 - [3uk_B T/2JS(S+1)] \tag{31}$$

It is evident from Figure 21 that the crystallographic b axis must correspond to the parallel direction, and it is interesting to note that the easy

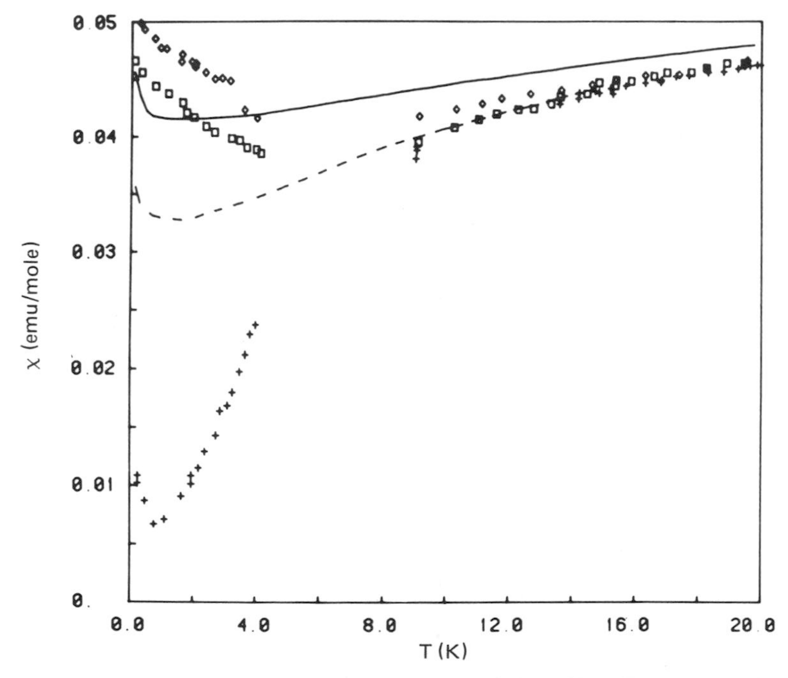

FIGURE 21. Low-temperature magnetic susceptibility of $CsMnCl_3 \cdot 2H_2O$. (+) a axis; (◇) b axis; (□) c axis; dashed line, the average magnetic susceptibility; solid line, the scaled model with $J = -2.16$ cm^{-1}. (After Reference 13.)

axis does not coincide with the chain axis. Smith and Friedberg[13] determined the zero-field-splitting parameter D to be 0.08 cm^{-1} and observed an unusual temperature dependence of unknown origin in the region of 0.35 K, where χ_b and χ_c rise sharply instead of remaining constant as expected. This may be due to the presence of paramagnetic impurities, probably Mn^{2+}, which in spite of the long-range ordering, causes the observed anomaly.

The number of linear chain compounds of Mn(II) and the few Mn(III) compounds which are known warrant further study. The bulk of the studies have been carried out on the ordered phase with little emphasis placed on intrachain interactions. Further structural and magnetic studies need to be undertaken so that any structure–function relationships that may exist in the manganese linear chain compounds can be identified and understood.

6. *Chain Compounds of Iron*

Although iron commonly forms compounds with formal oxidation states ranging from 0 to +4, and with high-spin, low-spin, and intermediate-spin electronic configurations, extensive magnetic and structural data exist only for chain compounds of formal oxidation states +2 and +3. All of the iron (II) compounds that have been studied in detail have six coordinate structures based on the octahedron and are high spin—that is, have $S = 2$ single-ion states. There are examples involving single ligand bridges, two ligand bridges, and three ligand bridges, and both ferromagnetic and antiferromagnetic exchange interactions have been characterized. In addition to the uniformly spaced chains, there are examples of unusual structures. One of these is a double-strand chain, while a second contains mixed ligand bridges and is alternatingly spaced.

A variety of chain compounds of iron in formal oxidation state +3 have been studied and unusual properties have been observed for some of them. As will be seen, exchange coupling in high-spin iron(III) compounds can be described in terms of isotropic Heisenberg theory, and comparable theory has been successfully used for organometallic systems. However, the situation is not settled for sulfur-bridged iron(III) chains. Available data suggest low-spin behavior, but the iron coordination environment is distorted tetrahedral and a low-spin electronic configuration is not expected. Representative results illustrating these specific cases are reviewed here.

6.1. *Theoretical Models Applicable for Iron Chains*

6.1.1. *High-Spin Iron(III) Chain Compounds*

The theory necessary for the description of the magnetic properties of high-spin iron(III) compounds with $S = 5/2$ ground states was outlined in

TABLE 6. Structural and Magnetic Properties of Iron Chain Compounds

Compound	Structure			Intrachain	Interchain		Experimental			
	M–M (Å)	M–B–M (deg)	M–M′ (Å)	J (cm^{-1})	T_c (K)	J'/J ($\times 10^2$)	Temperature (K)	Technique	Model	Reference
Chain Compounds with Fluoride Bridges										
K_2FeF_5		162–173						x ray		149
				−6.5(3)		0.1	4–600	χ_p	H–S, H–I	150
					11.2(5)	0.5	4.2–300	GR		151
Rb_2FeF_5					9.3(5)		4.2–300	GR		161
$CaFeF_5$				−7.5	21.0	1.0	1.5–300	χ_p, GR	H–I	162
N_2H_6FeF				−7.0	9.0	0.2		GR	H–I	163
				−7.2					H–S	164
Chain Compounds with Chloride Bridges										
$RbFeCl_3 \cdot 2H_2O$				27.0	11.96	1.8		C_m	I–R	165
										166
$Fe(pyrazole)_2Cl_2$					10		4.2–80	χ_p		167
					10.5		1.7–35	GR		167
$FeCl_2 \cdot 2H_2O$					23			M_i		168
										169
	3.637(2)							x ray		170
$TlFeCl_3$	3.00	73.46	6.976							2
				−5.5				χ_p	MF	142
				−1.9						142
$RbFeCl_3$	3.53		7.060							2
				−4.0			2–250	χ_p	MF	142
				−1.1 (⊥)			2–250	χ_i	MF	142
				+5 (‖)			1–70	χ_i		141
				+11 (⊥)			1–70	χ_i		141
					2.55			GR		144
					2.45			GR		140
	3.01	74.7				6	1–300	χ_i	CEF	135

$CsFeCl_3$	3.03	74.75	7.238					x ray		2
				−4.8	10		2–250	χ_p	MF	142
				−0.5 (⊥)						142
					< 1.24			MR		138
				2.5 (‖)			4–300	GR		140
				5 (⊥)			4–300	GR		140
NH_4FeCl_3	3.03	73.57	7.019					x ray		2
				−5.5			2–250	χ_p	MF	142
				−1.0						142
$KFeCl_3$	3.84		8.20		16		4–250	N, χ_p		143
				Chain Compounds with Bromide Bridges						
$RbFeBr_3$	3.14	73.3								146
				−1.7(3)	5.5	3	1–300	χ_i	CEF	134
				Chain Compounds with Sulfur Bridges						
$KFeS_2$	2.7		6.43	–	250			x ray, N		152
$Ba_{1.09}Fe_2S_4$				−70				χ_p	H	156
				Chain Compounds with Oxo-anion Bridges						
$Fe(ox)\cdot 2H_2O$	5.60							x ray		171
										172
				−4.4	< 30		4–65	χ_p	H–S	173
$(N_2H_5)_2Fe(SO_4)_2$							2–80	χ_p		126
				Chain Compound with Mixed Liquid Bridges						
$Fe_2(PO_4)Cl$	3.082(1)				< 9		9–78	x ray, χ_p		148
	3.044(1)									148
	3.499(1)									148
				Chain Compound without Ligand Bridges						
[DMeFc][TCNQ]	10.840							x ray		159
						2.55	1.5–50	χ_p		160

the theoretical section. Basically, the isotropic Heisenberg model applies, and theoretical results of Rushbrooke and Wood,[11] Bonner and Fisher,[7] Fisher,[12] Smith and Friedberg,[13] and Weng[15] are available for the analyses of experimental data. Experimental magnetic and structural data for iron chains are collected in Table 6.

At low temperatures, the effects of interchain interactions often become very important even for systems with antiferromagnetic intrachain exchange coupling. Oguchi[132] has examined this problem for the Heisenberg case and has derived an expression for the ratio η of the interchain J' and intrachain J exchange coupling constants. The expression is

$$(4 + z\eta)/z\eta = \exp(2|J|/k_B T_c) \tag{32}$$

where z is the number of magnetic neighbors.

6.1.2. Low-Spin Iron(III) Chains

Little experimental data have been collected on low-spin iron(III) chains. The limited amount of data that have been collected have been analyzed in terms of Heisenberg theory for $S = 1/2$.

6.1.3. High-Spin Iron(II) Chains

Several approaches have been used to describe the magnetic properties of exchange-coupled iron(II) chains. The isotropic Heisenberg models include the scaled form[13] of Fisher's exact solution[12] for the linear chain in

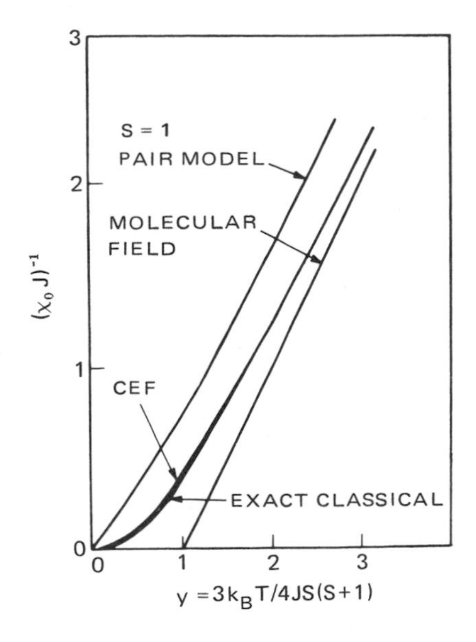

FIGURE 22. Reciprocal reduced magnetic susceptibility vs temperature for linear chains of ions with $S = 2$, exhibiting Heisenberg ferromagnetic interactions. (From Reference 134, with permission.)

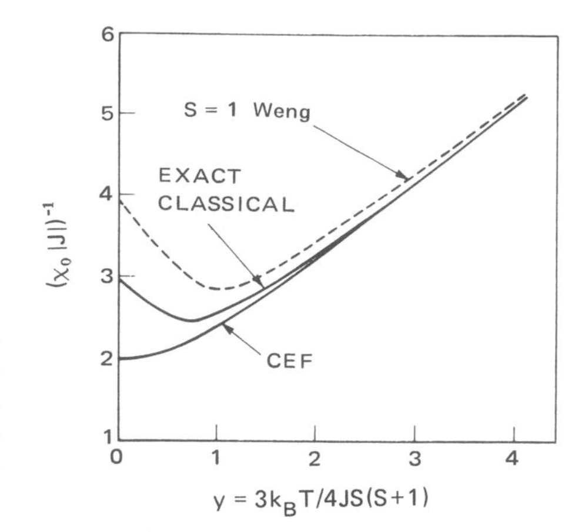

FIGURE 23. Reciprocal reduced magnetic susceptibility vs temperature for linear chains of ions with $S = 2$, exhibiting Heisenberg ferromagnetic interactions. (From Reference 134, with permission.)

the limit of infinite spin, the isolated pair model with $S = 1$, the molecular field approximation, Weng's results for $S = 1$,[15] and correlated effective field theory which was developed by Lines[133] and has the form[(34)]

$$(\chi J)^{-1} = \{1 + [3k_B T/4JS(S + 1)]^2\}^{1/2} - 1 \tag{33}$$

The inverse susceptibilities predicted by these models are compared in Figures 22 and 23. These models have radically different temperature dependencies, and the conclusions from the analysis of experimental data will depend on the model selected for the analysis. Therefore, care must be exercised when comparing the results of studies reported from different laboratories.

Magnetic and structural data exist only for high-spin iron(II) chains with quasi-octahedrally coordinated iron. In an octahedral crystal field the orbital degeneracy of the ^5D free-ion state is removed, and in O_h the ground state is $^5T_{2g}$ while the excited crystal-field state is ^5E_g. Only the ground state need be considered in the exchange problem, and calculations[134] have made use of the isomorphism of the T_2 and P symmetry groups. In the $AFeX_3$ compounds, the major distortion is trigonal, and if the threefold axis is taken as the z axis, the Hamiltonian including crystal-field distortion and spin–orbit interaction becomes

$$\hat{H} = \Delta(\hat{L}_z^2 - 2) + \lambda\hat{L}\cdot\hat{S} \tag{34}$$

Since the matrix elements of real orbital angular momentum L are the negative of the equivalent elements of L' in the isomorphic P state, and the matrix elements of $\hat{L}_z^2$ are three times those of $\hat{L}_z'^2$, the Hamiltonian in terms of the fictitious angular momentum becomes

$$\hat{H} = \Delta'(\hat{L}_z'^2 - 2/3) + |\lambda|\hat{L}'\cdot\hat{S} \tag{35}$$

where $\Delta' = 3\Delta$. Using the basis set $|M_{L'}, M_S\rangle$, the elements of the Hamiltonian matrix may be calculated. For positive values of $\Delta'/|\lambda|$, Lines and Eibschutz[134] obtained the following wave functions for the lowest three levels—one singlet and one doublet,

$$\begin{aligned} \Psi_0 &= a'|1, -1\rangle + b'|0, 0\rangle + a'|-1, 1\rangle \\ \Psi_\pm &= a|\pm 1, 0\rangle + b|0, \pm 1\rangle + c|\mp 1, \pm 2\rangle \end{aligned} \tag{36}$$

The other levels are expected to lie at higher energies and are not populated at low temperatures. The splitting of the levels within this fictitious triplet is determined by the parameter $\Delta'/|\lambda|$.

To account for exchange in the correlated effective-field approximation developed by Lines, the following term is added to the spin Hamiltonian

$$2J[\alpha(\text{per}) - \alpha(\text{par})]\hat{S}_{iz}^2 \tag{37}$$

and a summation is taken over all the spins. Here, α(per) and α(par) are measures of the nearest-neighbor perpendicular- and parallel-spin correlations along the chain. In the usual dot product $\hat{S}_i \cdot \hat{S}_j$ formulation of the exchange term the $\hat{S}_{jz}$ operator is approximated as $\alpha(\text{par})\hat{S}_{iz}$, and the perpendicular components, say $\hat{S}_{jx}$ are approximated by $\alpha(\text{per})\hat{S}_{ix}$. Lines and Eibschutz[134] give expressions for the static magnetic susceptibility in terms of only three parameters, the spin–orbit coupling constant, the trigonal-distortion parameter, and the exchange coupling constant. The application of the theory for the analysis of experimental magnetic susceptibility data for $RbFeBr_3$ is discussed below.

In a second paper, Eibschutz *et al.*[135] applied the correlated effective-field approximation in the analysis of the magnetic data for ferromagnetic exchange in the linear chain compound $RbFeCl_3$. The exchange Hamiltonian was formulated as

$$\hat{H}_{\text{ex}} = \sum_{i>j} -2J(\text{per})[\hat{S}'_{ix}\hat{S}'_{jx} + \hat{S}'_{iy}\hat{S}'_{jy}] - 2J(\text{par})\hat{S}'_{iz}\hat{S}'_{jz} \tag{38}$$

With the splitting between the ground singlet and the first excited doublet taken as $D\hat{S}_z'^2$, the effective-spin Hamiltonian becomes

$$\hat{H} = \sum_i [D\hat{S}_{iz}'^2 - \mu_B H \cdot g \cdot \hat{S}'_i] + \hat{H}_{\text{ex}} \tag{39}$$

This is transformed in correlated effective-field theory to become

$$\hat{H}_i(\text{eff}) = D\hat{S}_{iz}'^2 - \mu_B H \cdot g \cdot S_i + 2\hat{S}_{iz}'^2[J(\text{per})\alpha(\text{per}) - J(\text{par})\alpha(\text{par})] \tag{40}$$

Expressions for the magnetic susceptibility result from the application of the Hamiltonian and parallel and perpendicular correlation parameters. The general form of the expression is

$$\chi_k = N g_k^2 \mu_B^2 \langle S_k : S_k\rangle / [k_B T - 4J_k(1 - \alpha_k)\langle S_k : S_k\rangle] \tag{41}$$

where the subscript k represents the parallel (par) magnetic susceptibility or the perpendicular (per) magnetic susceptibility, and the colon products are

$$\langle S_{\text{par}} : S_{\text{par}} \rangle = 2 \exp(-E/k_B T)/[1 + 2 \exp(-E/k_B T)] \tag{42a}$$

and

$$\langle S_{\text{per}} : S_{\text{per}} \rangle = 2(k_B T/E)[1 - \exp(-E/k_B T)]/[1 + 2 \exp(-E/k_B T)] \tag{42b}$$

The singlet–doublet splitting in the correlated approximation is

$$E = D + 2J(\text{per})\alpha(\text{per}) - 2J(\text{par})\alpha(\text{par}) \tag{43}$$

These expressions do not include the temperature-independent higher-order contributions which become extremely important. These terms prevent the perpendicular magnetic susceptibility from going to zero at low temperature and account for as much as one-fourth of the parallel susceptibility at relatively high temperatures.

6.2. *Chain Compounds of Iron(II)*

6.2.1. *Rubidium Iron(II) Chloride Dihydrate, $RbFeCl_3 \cdot 2H_2O$: A Canted Ising Antiferromagnet*

Pale violet-brown single crystals of $RbFeCl_3 \cdot 2H_2O$ may be grown by slow evaporation at 37°C from a 0.03 *N* HCl solution of $FeCl_2 \cdot 4H_2O$ and RbCl in a molar ratio of 3.2 : 1.[136] The structure of the compound consists of *cis* octahedra coupled along the *a* axis by a shared chloride ion. The resulting chains are separated in the *b* direction by layers of rubidium ions, and are coupled in the *c* direction by hydrogen bonds.[137] Magnetization data for single crystals have been analyzed in terms of a psuedo-one-dimensional canted Ising antiferromagnet with very small coupling constants.

6.2.2. *$AFeCl_3$ ($A = NH_4^+$, Rb^+, Cs^+, Tl^+), Ferromagnetic Chains of Iron(II)*

The magnetic, optical, and structural properties of this series of compounds has attracted considerable attention.[134, 138–144] Single crystals of the rubidium and cesium salts may be grown in silica ampoules from stoichiometric melts of the alkali metal chloride and $FeCl_2$ using the Bridgeman technique, while a polycrystalline sample of $TlFeCl_3$ may be prepared by melting a stoichiometric amount of TlCl and $FeCl_2$ in an evacuated and sealed silica ampoule and holding the temperature at 600°C for one day followed by heating for several days at 350°C. The ammonium salt may be prepared by mixing in an agate mortar stoichiometric amounts of NH_4Cl and $FeCl_2$ and heating the mixture in a sealed gold tube for several days at 500°C under a pressure of about 1 kbar.[142]

The structure adopted by these compounds is shown schematically in Figure 5. Essentially the structure consists of chains formed by face-sharing octahedra, such that the iron(II) ions are triply bridged by chloride ions. The distance between nearest-neighbor iron ions in the chain is about 3.0 Å, while the nearest-neighbor distances between chains is about 7.0 Å. The Fe–Cl–Fe bridge angles differ widely, being 74° in $RbFeCl_3$ and 87° in $CsFeCl_3$. If perfect octahedra shared faces, the angle would be close to 70.5°. Inspection of the schematic structure of these compounds shown in Figure 5 leads to the conclusion that large cations will lead to large bridging angles, since the cations occupy crystal positions that have an influence on the trigonal distortion.

Detailed magnetic studies of $RbFeCl_3$ have been carried out by Eibschutz *et al.*,[135] and based on their persuasive arguments, earlier studies of the magnetic properties of these high-spin iron(II) chains should be viewed in historical perspective. Anisotropic magnetic susceptibility data for $RbFeCl_3$, collected with an applied field of 15.3 kOe, are displayed in Figure 24, where it may be seen that the reciprocal parallel and perpendicular magnetic susceptibilites are linear above about 100 K. There is a distinct maximum in the parallel susceptibility near 10 K, and long-range order sets in at $T_c = 2.55(5)$ K. Although correlated effective-field theory, including an interchain Heisenberg exchange interaction, can provide a very good fit of the anisotropic magnetic susceptibility data, using the parameters $\Delta'/|\lambda| = 1.3$, $J/|\lambda| = 0.015$, $2z'J'/|\lambda| = -0.038$, Eibschutz *et al.*[135] note that there are reasons for skepticism concerning the fit. First, the largest percentage deviations between theory and experiment occur at higher temperatures, and this is the region in which the theory should be most accurate, since the theory is expected to overestimate the magnetic susceptibility at lower temperatures. Next, the ratio of the interchain/intrachain exchange coupling constants is a factor of 10 larger than that found using $J/|\lambda| = 0.015$ and the observed Curie temperature in the Oguchi[132] relationship. In summary, the discrepancies that develop at low temperature are a result of the overestimation of susceptibility by correlated effective-field theory at low temperatures, the presence of interchain interactions, demagnetization corrections, and the presence of non-Heisenberg contributions to the exchange.

The Hamiltonian for exchange between orbitally degenerate paramagnetic ions may be written

$$\hat{H}_{ex} = \sum_{i>j} -2(J_S + J_D \hat{L}_{iz}^2 \hat{L}_{jz}^2)\hat{S}_i \cdot \hat{S}_j \tag{44}$$

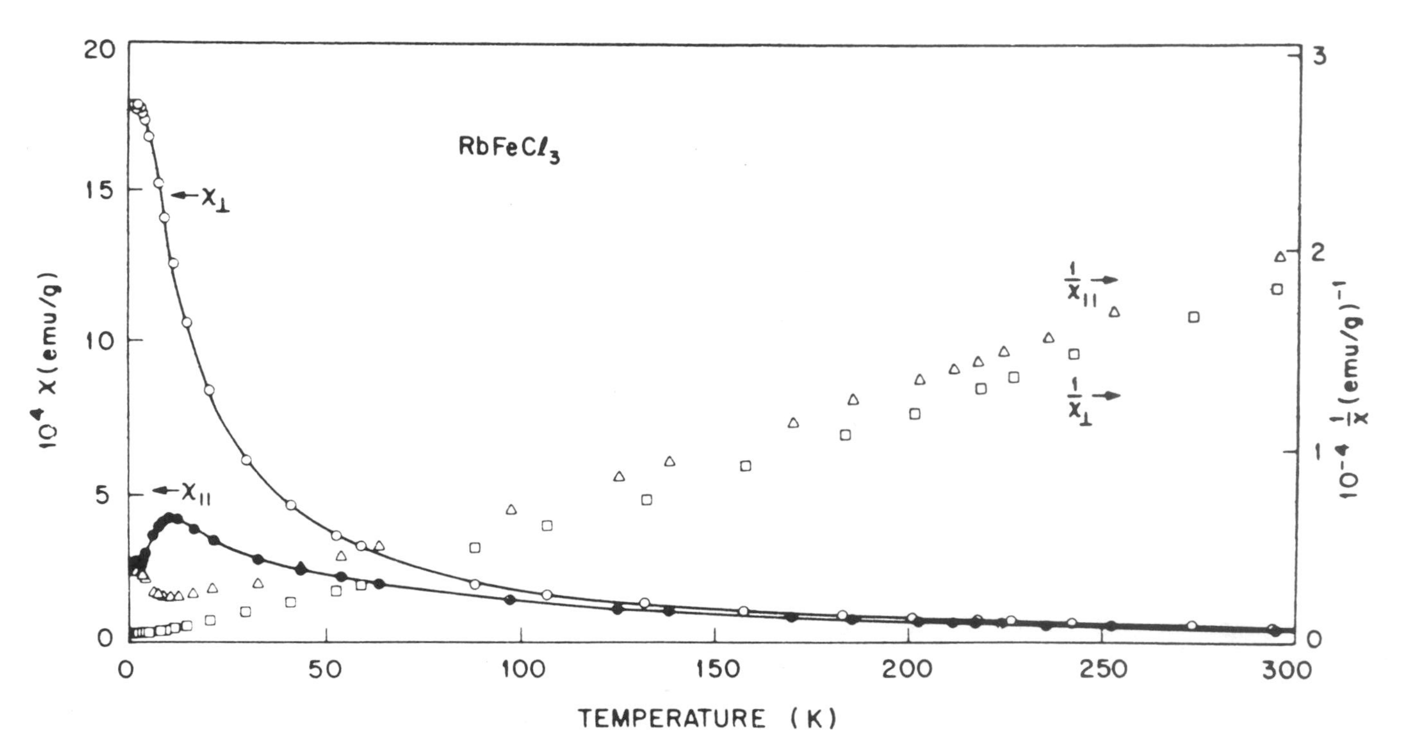

FIGURE 24. Anisotropic magnetic susceptibility collected on a single crystal of $RbFeCl_3$. (From Reference 135, with permission.)

where J_S denotes superexchange contributions and J_D is a measure of direct exchange, which is always negative. Since only the three lowest states are thermally populated, a fictitious spin-1 Hamiltonian may be used, and the splitting between the ground singlet and the excited doublet may be described by $DS_z'^2$. The effective spin-1 Hamiltonian becomes

$$\hat{H} = \sum [D\hat{S}_{iz}'^2 - \mu_B H \cdot g \cdot \hat{S}_i'] + \hat{H}_{ex} \tag{45}$$

The expression for the anisotropic magnetic susceptibility within this approximation becomes

$$\chi_k = Ng^2\mu_B^2\langle S_k : S_k\rangle/[k_B T - 4J_k(1 - \alpha_k)\langle S_k : S_k\rangle] \tag{46}$$

where the subscript k designates the parallel and perpendicular components of magnetic susceptibility, and the other terms in the expression have been defined earlier. The temperature dependence of the magnetic susceptibility computed from these expressions are shown in Figure 25 for $\Delta'/|\lambda| = 1.0$, $g(\text{par}) = 3.00$, $g(\text{per}) = 3.62$, $J(\text{par})/|\lambda| = 0.03$, and $J(\text{par})/|\lambda| = 0.02$, 0.01, and 0.0. It is important to note in Figure 25 that the reduction of J(par) from its isotropic value results in closer agreement between theory and experiment. It is unfortunate that the magnetic properties of the other members of this chloro-bridged series have not been analyzed in terms of

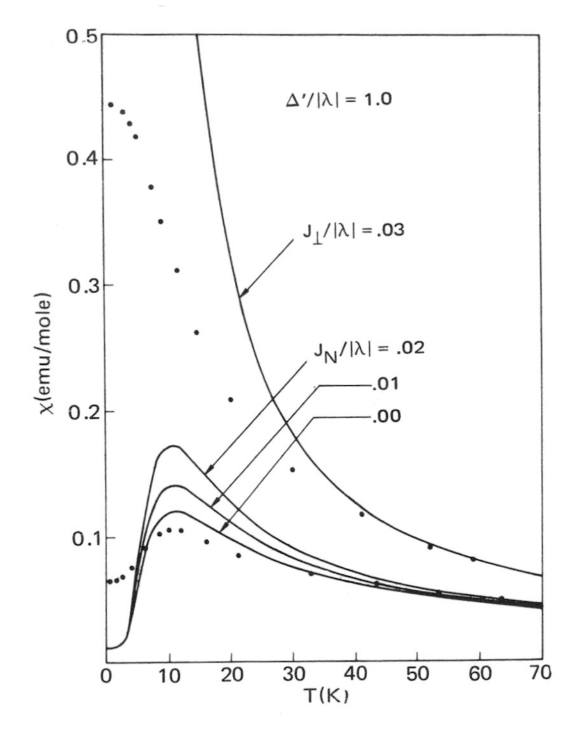

FIGURE 25. Magnetic susceptibility calculated for correlated effective-field theory for exchange in orbitally degenerate states with $S = 1$. (From Reference 135, with permission.)

this theory. More experimental results should guide theoretical refinements and provide support for the approximations that have been used.

6.2.3. *Antiferromagnetic Exchange Coupling in* $RbFeBr_3$

Single crystals of $RbFeBr_3$ may be grown by the Bridgeman technique.[145] The compound is prepared by careful dehydration and reduction of a solution of Rb_2CO_3 and Fe_2O_3 in dilute hydrobromic acid. The structure of $RbFeBr_3$ is schematically depicted in Figure 5. The nearest-neighbor iron–iron separation in the chain is 3.14 Å, and the Fe–Br–Fe angle at the bridging bromide ligand is 73.3°.[144] Below about 120 K, there is a crystallographic transition to a lower-symmetry structure,[146] and the iron ions occupy two nonequivalent sites. However, the site inequivalency is not great enough to affect nuclear quadrupole splitting in Mössbauer measurements.[134]

The magnetic susceptibility results may be interpreted in terms of the correlated effective-field theory of Lines[133] with $\Delta = 35(6)$ cm^{-1}, $\lambda = -80(7)$cm^{-1}, and $J = -1.9(3)$ cm^{-1}. When interchain exchange coupling is taken into consideration to explain the long-range ordering at $T_c = 5.5$ K, the intrachain exchange coupling constant is adjusted to be $-1.7(3)$ cm^{-1}. The interchain coupling constant is estimated to be about -0.07 cm^{-1}.[134] Although the intrachain exchange contains a direct component that is not rigorously of the Heisenberg form, the contribution of this term to the total exchange, which also includes the Heisenberg superexchange through the bridging bromide ligands, is not great enough to influence the magnetic susceptibility of $RbFeBr_3$.

6.2.4. $KFeCl_3$: *A Compound with a Double-Strand Chain Structure*

The structure of $KFeCl_3$ is isomorphous with that of $KCdCl_3$.[143] Double chains of distorted $FeCl_6$ octahedra that share an edge run parallel to the *b* axis. The nearest-neighbor intrachain iron–iron separation is 3.84 Å, while the nearest-neighbor interchain separation is 8.2 Å.[143] One consequence of the structure of $KFeCl_3$ as compared to $RbFeCl_3$ or $CsFeCl_3$ is the color. The latter two compounds are dark golden brown, but $KFeCl_3$ is a very pale brown color.[139] Good single crystals of $KFeCl_3$ may be grown from a stoichiometric melt of KCl and $FeCl_2$ in a silica ampoule by the Bridgeman technique. The rate of lowering of the ampoule through the temperature gradient must be very slow (about 1 mm/h).[139]

$KFeCl_3$ exhibits simple paramagnetic behavior between room temperature and 30 K, and at $T_c = 16$ K undergoes a transition to an antiferromagnetically ordered state.[143] The antiferromagnetic state may be described as consisting of ferromagnetic chains coupled antiferromagnetically.

The strong intrachain ferromagnetic coupling gives rise to unusual neutron diffraction results in the range $0.8T_c < T < 2T_c$. No attempts have been made to extract exchange coupling constants from magnetic data, and this may be understood in terms of the difficulty of the theoretical problem.

6.2.5. *An Iron(II) Chain Compound with Mixed Ligand Bridges*

The iron(II) chain compound with mixed ligand bridges $Fe_2(PO_4)Cl$ may be prepared by heating to 1025°C for several hours a 60 : 40 mixture of $FeCl_2$ and $Fe_3(PO_4)_2$, which is contained in a gold tube and sealed in a vitreous silica tube. Pale orange single crystals may be grown by slowly cooling the reaction mixture.[147]

The crystal structure of $Fe_2(PO_4)Cl$ consists of chains of face-sharing octahedra, with the three bridging ligands including two phosphate oxygens and a chloride.[148] There are three crystallographically unique iron (II) ions in the chain and each is coordinated by four phosphate oxygen atoms with two chloride ligands in *trans* positions. Preliminary magnetic susceptibility measurements taken in the range 9–78 K indicate antiferromagnetic behavior with $\theta = -29(2)$ K and a magnetic moment of 4.83 B.M. Additional measurements are required since the short iron–iron distances and the large θ value suggest extensive exchange coupling interactions. The iron–iron distances are Fe(1)–Fe(2) = 3.082(1) Å, Fe(1)–Fe(3) = 3.044(1) Å, and Fe(2)–Fe(3) = 3.499(1) Å. The alternating spacings in the iron–iron distances should also be reflected in the magnetic behavior, but there has not been any attempts to address the problem of exchange in alternating chains of exchange-coupled ions with large spin quantum numbers.

6.3. *Chain Compounds of Iron (III)*

6.3.1. *High-Spin, Antiferromagnetically Exchange-Coupled Chains of Iron (III)*

A number of chain compounds of the formula A_nFeF_5 (with $n = 2$ or 1, and A being an alkali metal, alkaline earth, or protonated ion) have been structurally characterized, and careful magnetic studies on this series of compounds should lead to an understanding of the effects of chemical and structural variations on the exchange processes in the high-spin monofluoride-bridged iron(III) chain compounds. Among the most carefully studied compounds is K_2FeF_5.[149, 150] Colorless, transparent crystals of K_2FeF_5 with a maximum size of 4 × 3 × 3 mm have been grown by the flux method in a fluorinating atmosphere[149] by slowly cooling (1.5°C/h) from 650°C a mixture of $PbCl_2$, KF, and FeF_3. The structure[149] is made up of infinite chains of FeF_6 octahedra that share *cis* corners. The chains

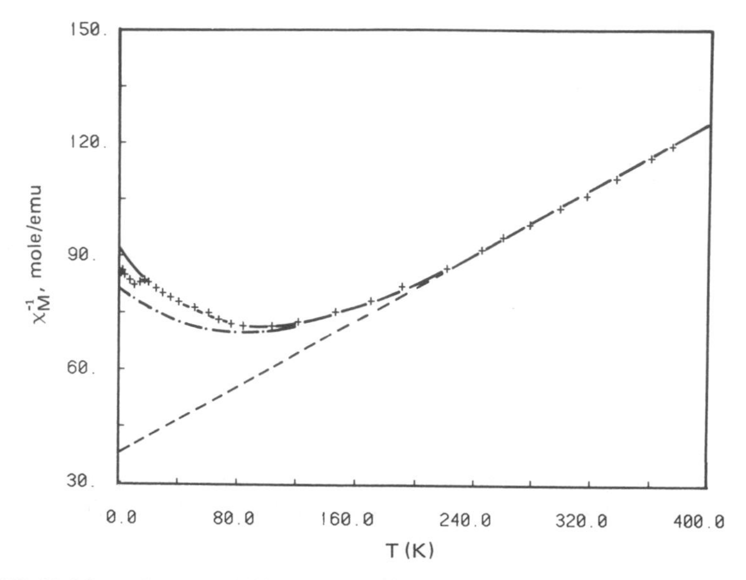

FIGURE 26. Magnetic susceptibility data for $(N_2H_6)FeF_5$ fit by the Curie–Weiss law (- - -); the scaled model of Smith and Friedberg (–·–·–) with $J = -7.2\ cm^{-1}$; and the results obtained by Weng (——) with $J = -7.0\ cm^{-1}$. (After Reference 164.)

run along the *a* axis with the average Fe–F(terminal) bond distance of 1.882 Å being clearly shorter than the average of the Fe–F(bridging) bond distance of 2.020 Å. The Fe–F–Fe bridge angles range from 162.0° to 173.4°.

The magnetic susceptibility versus temperature plot[150] exhibits a broad maximum at approximately 100 K and can be closely described in the temperature range 50–700 K by the theoretical results obtained by Weng[15] for $S = 5/2$ with $J = -6.5(3)\ cm^{-1}$. The theory fails at low temperatures because of a finite interchain coupling $[J(\text{inter})/J(\text{intra}) = 10^{-3}]$ which leads to three-dimensional antiferromagnetic ordering at 11.2(5) K.[151] The interchain exchange interaction is presumably transmitted by the Fe–F–K–F–Fe pathway. The scaled model for $S = 5/2$ developed by Smith and Friedberg[13] does not fit the magnetic susceptibility data except in the high-temperature region.[150] As shown in Figure 26, comparable results obtain for $(N_2H_6)^{2+}FeF_5^{2-}$,[164] with the interpolation model developed by Weng[15] providing a good fit of the data above T_c.

6.3.2. *Chain Compounds of Four Coordinate Iron(III)*

Potassium dithioferrate $KFeS_2$, as well as the rubidium and cesium analogues, may be prepared by a fusion reaction under a nitrogen atmosphere. The crystal structure consists of linear chains of edge-sharing FeS_4

tetrahedra lying along the crystallographic *c* direction,[152] with the potassium ions occupying positions between the chains. The iron–iron distance in the chain is 2.70 Å, while the closest iron–iron interchain distance is 6.43 Å. Single-crystal neutron diffraction results[153] show that the spins are coupled antiferromagnetically along the *c* axis, and these yield ferromagnetic sheets parallel with the *ab* plane. The transition temperature was determined from the temperature dependence of the (001) magnetic reflection to be 250.0(5) K, and the magnetic moments were found to have no component along the *b* axis. The magnetic moment values of 2.43(3)μ_B at 4.2 K and 2.37μ_B at 77 K are considerably reduced from that which is expected for high-spin iron(III), and approaches the value expected for low-spin iron(III). However, low-spin iron(III) in tetrahedral coordination has not been reported.

The electronic structures of $KFeS_2$ and its analogues are incompletely understood and provide the opportunity for additional experimental and theoretical investigations. A series of compounds that also have a bearing on this problem are the so-called infinitely adaptive barium iron sulfides. These are compounds of the general formula $Ba_{1+x}Fe_2S_4$ which were first prepared by Hong and Grey.[154] "Infinitely adaptive" compounds are those compounds for which every possible composition can attain a fully ordered structure with no defects or coexistence of successive structures.[155] The preparation and physical properties of these compounds have been reviewed recently.[156] In the preparations, BaS and iron powders are mixed to give specific ratios of barium and iron, and the mixture is packed in a loosely capped graphite tube positioned in one end of a Vycor tube. Excess sulfur is placed in the other end of the tube, and the tube is sealed under a vacuum of 10^{-3}–10^{-4} mm Hg. The sealed Vycor tube is put in a horizontal furnace with a temperature gradient such that the end containing the graphite vessel is kept at a constant temperature which is higher than the temperature at the end with the excess sulfur. After 2 days, the reaction is terminated by quenching the samples in water.

Infinitely adaptive phases are characterized by structures formed from sublattices having incommensurate repeat distances. In the $Ba_{1+x}Fe_2S_4$ series, one of these sublattices is a rigid chain of edge-sharing FeS_4 tetrahedra parallel to the *c* axis, while the second sublattice is formed by the barium ions situated in the channels formed by the chains.[156] Magnetic susceptibilities of $Ba_{1+x}Fe_2S_4$ have been measured and the results yield low effective moments similar to those observed for $KFeS_2$.[153] For example, above 125 K, magnetic susceptibility data for $Ba_{1.09}Fe_2S_4$ may be described by Heisenberg theory for $S = 1/2$ with $J = -70$ cm^{-1}. However, the theory fails completely below 125 K. Additional work is required since the Mössbauer spectra show no evidence of magnetic ordering at the temperatures of the magnetic measurements,[156] and as noted above, low-spin iron(III) is not expected in tetrahedral coordination.

6.3.3. An Iron(III) Chain Compound without Ligand Bridges

Decamethylferrocene (DMeFc) reacts with 7,7,8,8-tetracyano-*p*-quinodimethane (TCNQ) to form at least three compounds: one with the formula $(DMeFc)(TCNQ)_2$,[157] and two compounds with the formula (DMeFc)(TCNQ). Of these latter 1 : 1 compounds, one has an interesting dimer structure $(DMeFc)_2(TCNQ)_2$,[158] while the second one has a chain structure comprised of alternating $DMeFc^{+\cdot}$ radical cations and radical $TCNQ^{-\cdot}$ anions.[159] The distances between the cation and anion within a chain are 3.67(2) and 3.66(3) Å, and the distance between $S = 1/2$ iron(III) sites is 10.840(5) Å. The $TCNQ^{-\cdot}$ radical ion is reactive and forms α,α-dicyano-*p*-toluoylcyanide, $DCTC^{-\cdot}$. These latter ions are randomly substituted for $TCNQ^{-\cdot}$ in the chain, and as a result there is considerable disorder in the positions of the cyano groups and the oxygen of $DCTC^{-\cdot}$.[159]

One-dimensional (DMeFc)(TCNQ) exhibits metamagnetic behavior with T_c of 2.55,[160] and the data are consistent with an exchange model involving $S = 1/2$ iron(III) and $TCNQ^{-\cdot}$ ions. The low-spin formulation of the iron(III) ion is confirmed by Mössbauer spectral results. These results demonstrate that organometallic substrates offer a wide range of opportunities for the production of low-dimensional systems with new or unusual properties.

7. Chain Compounds of Cobalt

The magnetic properties of linear chain complexes of cobalt provide an excellent opportunity for investigating systems displaying large anisotropy effects. Although many compounds of first-row transition-metal ions have been structurally characterized as linear chains, only a few of these chain compounds show low spin-dimensionality in their magnetic behavior, and these are frequently cobalt compounds. Polymeric cobalt complexes are most frequently described as Ising spin-1/2 systems at low temperatures, but XY-type behavior is observed in at least one case. While only a few cobalt chain complexes have been thoroughly studied, the compounds that have been characterized have allowed the experimental testing of theoretical models and have aided in the understanding of anisotropic exchange interactions.

7.1. Theoretical Aspects

For an accurate theoretical assessment of a magnetic system, the determination of the ground state and the relationship of the excited states to the ground state is essential. For Co(II) ions, two cases must be considered: those being high-spin cobalt(II) in an octahedral crystal field, and high-spin cobalt(II) in a tetrahedral crystal field. The ground state of the free Co(II) ion is ^4F, but the orbital degeneracy is removed in an octahedral crystal field giving one ^4A and two ^4T levels with the lowest-lying state being a $^4T_{1g}$. The degeneracy of the $^4T_{1g}$ is removed by the action of axial and rhombic distortions of the crystal field, as well as by spin–orbit coupling. The overall effect of low-symmetry crystal-field components and spin–orbit coupling produces six Kramers doublets and results in a doublet ground state. Since the same doublet energy level remains lowest in energy for all values of the applied field strength, and since the energy difference between the two lowest-lying doublets is relatively large with respect to thermal energy present at low temperatures (<30 K), cobalt(II) systems may be described as having an effective spin of 1/2. At higher temperatures, thermal population of the higher-energy Kramers doublets becomes significant, and the magnetic problem becomes more difficult. This energy spectrum obtains frequently for linear chain Co(II) complexes, and for this reason most experimental observations are made at low temperatures to avoid the complexity of including the upper energy levels.

A similar situation arises for cobalt(II) ions in tetrahedral crystal fields. As mentioned above, the ^4F state splits into three levels, but in the presence of a tetrahedral crystal field, the lowest level is the $^4A_2$. Due to the significant spin–orbit coupling effects of Co(II), this level is split by axial, $D(\hat{S}_z^2 - 1/3\hat{S}^2)$, and nonaxial, $E(\hat{S}_x^2 - \hat{S}_y^2)$, distortions which generate two Kramers doublets. Depending upon the magnitude of these terms in comparison to the exchange coupling constant $|J|$, the spin value may be either 3/2 or 1/2. When $|D| \ll |J|$, the system is a spin-3/2 exchange-coupled case for temperatures greater than $|D/k_B|$. If $|J| \ll |D|$, then the system is dominated by single-ion effects at high temperatures, with exchange effects becoming significant only at lower temperatures. In addition, the upper Kramers level becomes depopulated and the system can be described as a spin-1/2 system.

7.2. Selected Examples

Most of the structurally characterized linear chain Co(II) complexes investigated up to this time (see Table 7) fall into the first category described above, that is, high-spin octahedral cobalt(II). These complexes display magnetic susceptibility reflective of spin-1/2 systems, and the data may be analyzed according to this formulation in the low-temperature region. The Ising model applies, and theoretical fits to experimental data are made by using the equations derived from this model[7,12] [see Section 1.5, Eqs. (16) and (17)]. The compound, $((CH_3)_3NH)CoCl_3 \cdot 2H_2O$, is a good example of an Ising-like complex that displays ferromagnetic behavior, while $CsCoCl_3$ is an example of a system that exhibits antiferromagnetic behavior. These compounds will be discussed below. Cs_2CoCl_4 provides a good example of spin-exchanged cobalt(II) ions in a tetrahedral crystal field, and it has been proposed that the exchange is *XY* like.

7.3. *An Ising-like Ferromagnetic Chain Compound,* $[(CH_3)_3NH]CoCl_3 \cdot 2H_2O$

Trimethylammonium catena-di-μ-chlorodiaquocobalt(II) chloride may be prepared by evaporating an aqueous solution containing equimolar quantities of $CoCl_2 \cdot 6H_2O$ and $[(CH_3)_3NH]Cl$.[174] Single crystals of appropriate size for a structural analysis using x-ray techniques and for magnetic susceptibility measurements are easily obtained, and these crystals are relatively stable. The empirical formula was confirmed by chemical analysis and by a crystal-structure determination. The structure consists of chains of edge-sharing *trans*-$[CoCl_4(H_2O)_2]$ octahedra which run parallel to the *b* axis (see Figure 27).[174] The trimethylammonium cations and the remaining chloride ions are present as interstitial ions. Each $[CoCl_4(H_2O)_2]$ molecular unit is tilted with respect to an adjacent unit, resulting in a dihedral angle between $CoCl_4$ planes of 15.58(6)°. This feature is the main structural difference between the chains in this compound and those observed in the compounds $CoCl_2 \cdot 2H_2O$ and $Co(py)_2Cl_2$. Except for the γ form of $Co(py)_2$-Cl_2, each of these complexes displays a symmetric bridging arrangement with Co–Cl bond distances between 2.45 and 2.50 Å and bridging angles ranging from 92.3° to 95.5°.

The magnetic susceptibilities along the principal axes of $[CH_3)_3NH]$ $CoCl_3 \cdot 2H_2O$ are shown in Figure 28. These display an anisotropy which dramatically emphasizes the difference in behavior along the axes.[174, 204–206] At 4.3 K, the susceptibility measured along the *c* axis is about 500 times greater than that measured parallel to the chain (*b* axis). Just below this temperature, the axial magnetic susceptibility undergoes an abrupt change indicating the onset of long-range ordering. The transition temperature of 4.14 K, determined from specific-heat data,[174, 204] agrees well with the value obtained from the magnetic susceptibility measurements. The heat-capacity data also indicate that the exchange between the polymeric chains, which are cross-linked by hydrogen bonding in the crystallographic *c* direction, is relatively small. These results support the analysis of the data in terms of noninteracting linear chains of Ising coupled paramagnetic ions. The fit of Eq. (16) to the susceptibility data measured parallel to the chemical chain, χ_b, produces values for J and g of 1.54 cm^{-1} and 4.08, respectively. However, the application of Eq. (17) with the value of $|J|$ determined from the heat-capacity data (5.4 cm^{-1}) yields an equally good fit with a g value of 3.90. The solid line in the top plot of Figure 28 represents the fit using Eq. (17) and is particularly satisfying because the

TABLE 7. Structural and Magnetic Properties of Cobalt(II) Chain Compounds

Compound	Structure			Intrachain		Interchain	Experimental			
	M–M (Å)	M–B–M (deg)	M–M′ (Å)	J (cm^{-1})	%S	T_c (K)	Temperature (K)	Technique	Model	References
Chain Compounds with tri-μ-Chloro Bridges										
$CsCoCl_3$	3.0158	74.8	7.2019				RT	x ray		2, 185
				‖−59		21.5	~0–300	χ_i	I–F	2, 186, 187
				⊥−12						
				‖−80						1
Chain Compounds with di-μ-Chloro Bridges										
$CoCl_2 \cdot 2H_2O$	3.55	92.3	5.6				RT	x ray		174, 188, 189, 192
				7.53[a]	40	17.2	2–95	χ_i	I–M	174, 188–196
				6.5						
α-$Co(py)_2Cl_2$	3.66	94.5	8.7				RT	x ray		174, 197–199
γ-$Co(py)_2Cl_2$	3.59	98.3	8.4				89	x ray		198
				6.6[a]	85	3.17	2–20	χ_p	I–F	174, 199–203
				7.6						
$[(CH_3)_3NH]CoCl_3 \cdot 2H_2O$	3.64	99.5	8.1				RT	x ray		174, 204, 206–207
				5.4[b]	90	4.14	2–20	χ_i	I–F	174, 204–207
Chain Compounds with μ-Chloro Bridges										
$CsCoCl_3 \cdot 2H_2O$		128					RT	x ray		208, 209, 211, 220
				−9[c]	84	3.4	1–3.4			209
				−36		3.4	5–240	χ_i	I–F	210
				−25→	84	3.38–3.4	1.3; 100	N		211
				−35						

$RbCoCl_3 \cdot 2H_2O$	4.372						x ray		212–214
				~0[d]	2.94	1.5–30	χ_i	D–M	212–214
Chain Compounds with Chloride–Chloride Bridges									
Cs_2CoCl_4	7.392					RT	x ray		178, 179
				−0.94	0.22	1.5–20	χ_i	SI	179–183
Chain Compounds with tri-μ-Bromo Bridges									
$CsCoBr_3$					28	~0–300	χ_i		187, 216
Chain Compounds with di-μ-Bromo Bridges									
$CoBr_2 \cdot 2H_2O$					9.5	10–50	χ_p	I–M	189, 190
Chain Compounds without Ligand Bridges									
PcCo				−3.40		1.8–12	χ_p	I–F	94, 217
TPPCo				1.27		2–300	χ_p	I–F	218
Chain Compounds with One Oxygen Bridges									
$Co(hipp)_2(H_2O)_3 \cdot 2H_2O$	3.966	128.3	6.903			RT	x ray		219, 220
						2.2–80	χ_p	C–W	219, 220
Chain Compounds with Carboxylic Acid Bridges									
$Co(ox)(H_2O)_2$				−9.3		4–30	χ_p	I–F	221, 222
Chain Compounds Bridge by Salts of Oxo-anions									
$Co(N_2H_5)_2(SO_4)_2$				−4.90[a]	1.57	2–80	χ_p	*XY*	126, 223
				−6.2					
				−3.1		2–80	χ_p	H–BF	126
				−8.0		2–80	χ_p	I–F	126

[a] From specific-heat measurements.
[b] This value of J was obtained from specific-heat measurements and was used in the fitting process of the magnetic susceptibility data.
[c] From antiferromagnetic resonance results.
[d] Dzaloshinsky–Moriya antisymmetric exchange term was used resulting in a value for $|D|$ of $\sim 41\ cm^{-1}$.

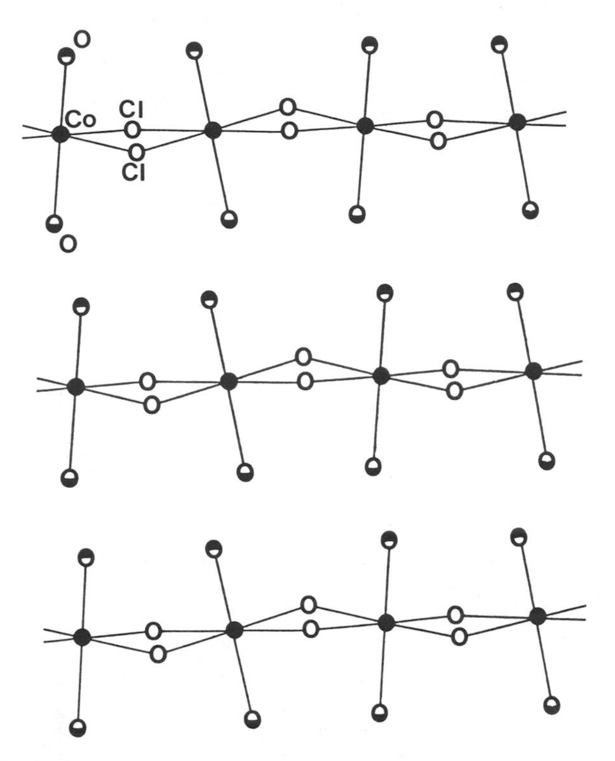

FIGURE 27. Projection of a portion of the crystal structure of $[(CH_3)_3NH]CoCl_3 \cdot 2H_2O$. The chains run parallel to the b axis; ●, Co atoms; ○, Cl atoms; ◓, H_2O molecules. (After Reference 174.)

exchange coupling constant was obtained independently and the values of the magnetic parameters predict the behavior of χ_b on either side of the transition temperature. On the other hand, the fitted values using the Ising parallel solution give neither quantitative nor qualitative agreement with the experimental data below the transition.

A similar approach for analyzing the susceptibility data measured along the c-axis direction showed that only a positive exchange coupling constant could generate values as large as the observed data. By including a term to account for the effects of a molecular exchange field[175, 176] along the c axis and using the heat-capacity value for J of 5.4 cm^{-1}, a theoretical curve that is at least qualitatively correct is obtained using a g value of 6.54. This curve is shown in the bottom plot of Figure 28 in conjunction with the χ_c data (□).

Although the a axis is also perpendicular to the chains, the attempts to explain the experimental data using the Ising theory were unsuccessful. It was suggested that the behavior of the magnetic susceptibility data below

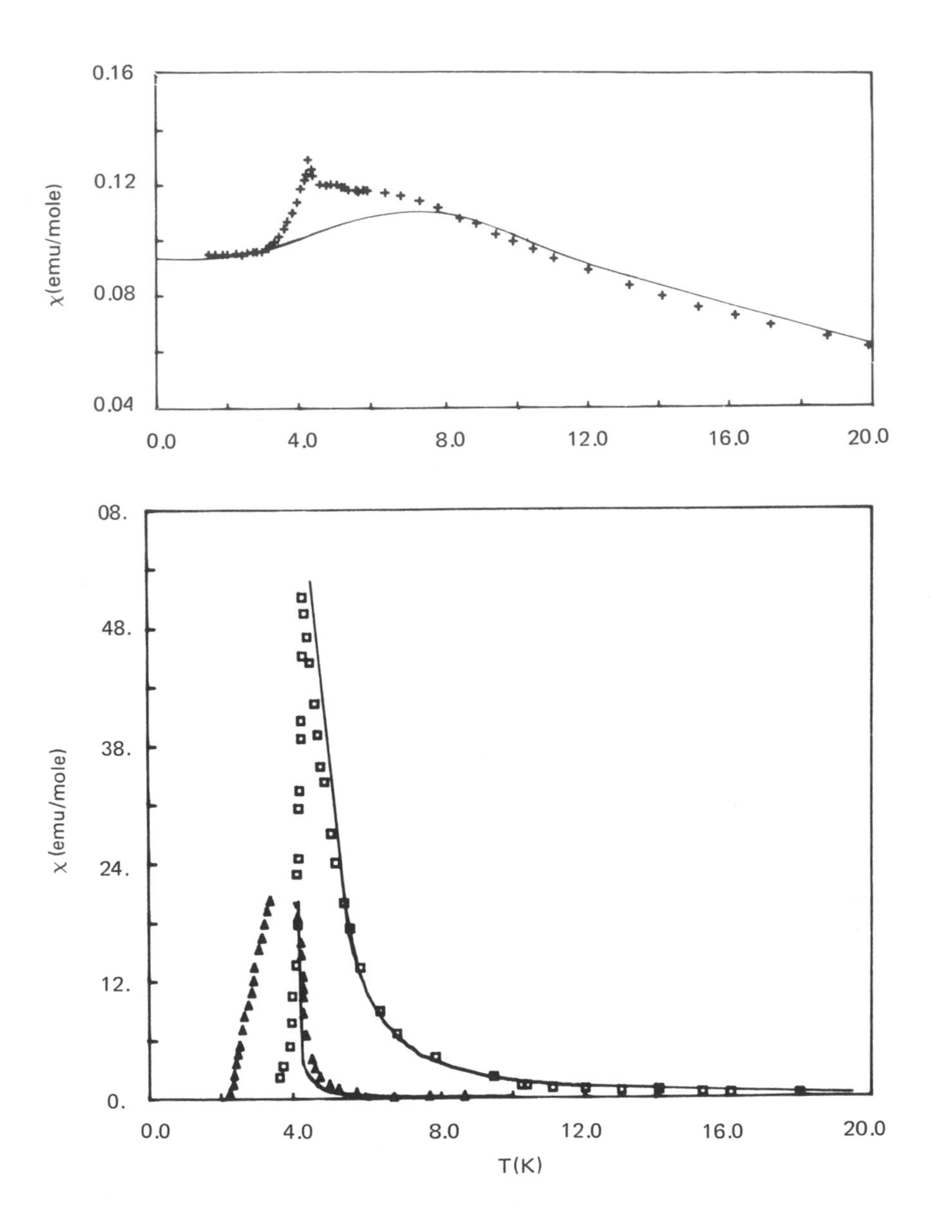

FIGURE 28. Plots of principal-axes magnetic susceptibilities vs temperature for $[(CH_3)_3NH]CoCl_3 \cdot 2H_2O$. Top: (+) results of measurements made along the *b* axis. Bottom: (□) measurements made along the *c* axis; (△) measurements made along the *a* axis. Solid lines are the theoretical curves mentioned in the text. (After Reference 174.)

4.3 K might be explained by a canted spin system that was weakly ferromagnetic and that the addition of an antisymmetric term,[177] $D \sum_{i,j}(\hat{S}_i^x\hat{S}_j^y - \hat{S}_i^y\hat{S}_j^x)$, to the general Hamiltonian was appropriate. The following equations, derived by Moriya,[177] were used to fit the χ_a data with the resultant curve and corresponding data (Δ) being shown in Figure 28:

$$\chi = Ng^2\mu_B^2(T - T_0)/4k_B(T^2 - T_C^2) \tag{47}$$

$$T_C = -(J/2k_B)[1 + (D/J)^2]^{1/2} \tag{48}$$

$$T_0 = -J/2k_B \tag{49}$$

$$T_0 = T_N/[1 + (D/J)^2]^{1/2} \tag{50}$$

If T_c is set equal to the actual transition temperature of 4.3 K, then the fitting procedure produces values of g and D/J of 2.95 and 0.64, respectively.

The experimental observations and theoretical analyses lead to a spin arrangement in the ordered state in which the spins lie nearly along the c axis, with some canting towards the a axis. The spins lie perpendicular to the b axis with the result that all of the spins in each bc plane of the cobalt(II) ions are aligned parallel, producing the ferromagnetic coupling along and between chains in this plane. The spins of the neighboring planes are then arranged in a manner that produces a net moment along the a axis but no net moment along the c axis.

7.4. *A Chain Compound with Ising-like Antiferromagnetic Coupling, $CsCoCl_3$*

The compound $CsCoCl_3$ may be prepared by evaporation at 80°C of an aqueous solution containing 5.0 g of CsCl and 34.0 g of $CoCl_2$.[185] With the onset of crystallization, the solution is cooled to 50°C in an open container and maintained at this temperature for approximately 1 h. The bright-blue crystals that form are filtered, washed with warm toluene, and dried at 80°C.

The crystal structure of $CsCoCl_3$ has been solved using intregrated Weissenberg and precession photographs.[185] The coordination polyhedron about the cobalt(II) ion is a trigonally distorted octahedron with Co–Cl bond distances of 2.447(3) Å. $CsCoCl_3$ adopts the typical ABX_3 structure involving the tri-μ-chloro bridging arrangement shown in Figure 5.

The magnetic susceptibility of $CsCoCl_3$ in the temperature range 80–300 K has been reported by Soling.[185] The plot of magnetic moment versus the absolute temperature shows that the magnetic moment decreases rapidly as the temperature decreases. Soling did not attempt an analysis of the experimental data. In a more detailed study by Achiwa,[186] magnetic susceptibility measurements were made on a single crystal. The parallel and perpendicular susceptibilities were analyzed with the Ising model and estimations for $J_{\parallel}$ and $J_{\perp}$ were given as -59 and -12 cm^{-1}, respectively. It has been suggested[1] that the reported value for $J_{\parallel}$ is too small in magnitude since the susceptibility data of Achiwa were not corrected for a second-order contribution, and a value of -80 cm^{-1} was thought to be a better estimate. While these studies have confirmed the presence of antiferromagnetic exchange of relatively large magnitude, the exact value is in question and additional studies on this interesting chain system are necessary.

7.5. Cs_2CoCl_4

The compound Cs_2CoCl_4 may be prepared by mixing aqueous solutions containing stoichiometric amounts of CsCl and $CoCl_2 \cdot 6H_2O$. Crystals of good size and quality may be grown at room temperature. Such crystals have been used for a structural investigation,[178] magnetic susceptibility measurements,[179–181] and heat-capacity measurements.[179, 182–183]

The crystal structure of Cs_2CoCl_4 was first determined by Porai-Koshits[178] using x-ray diffraction photographs; but more recent investigations on the zinc complex,[184] which is isomorphous with the cobalt complex, have provided a more accurate assessment of possible exchange pathways present in the structure. One exchange pathway, which joins the metal ions together into chains parallel to the *c* axis, is represented by the

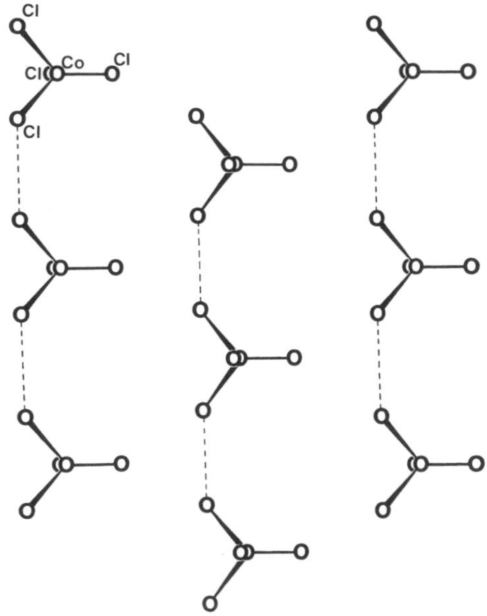

FIGURE 29. Projection of a portion of the crystal structure of Cs_2CoCl_4 onto the *bc* plane. Dashed lines represent the closest Cl–Cl contacts between tetrahedra. In this view, the Co atoms and one of the Cl atoms are almost on top of one another and alternate in nearly obscuring each other from chain to chain. The chain runs parallel to the *c* axis. (After Reference 179.)

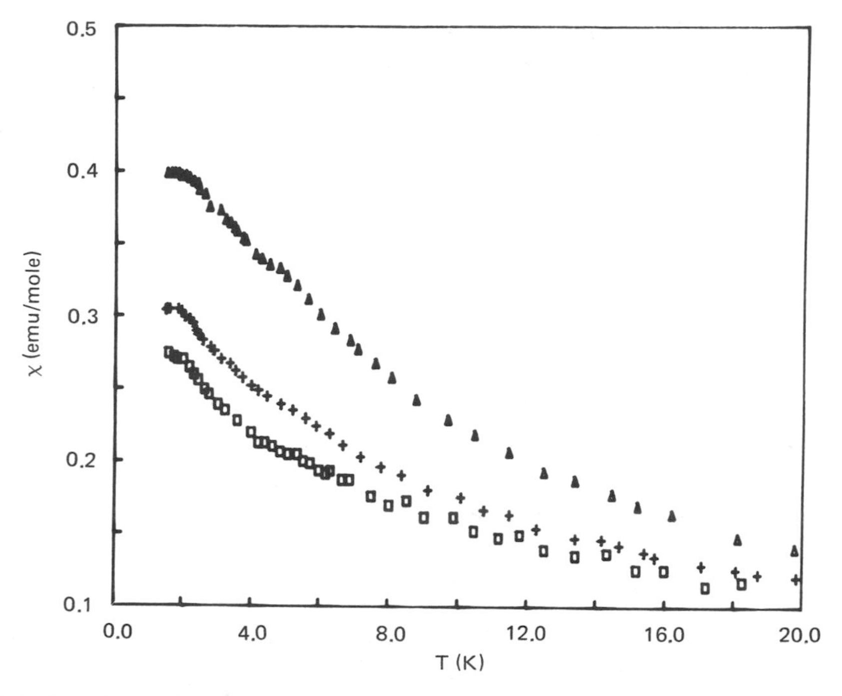

FIGURE 30. Principal-axes magnetic susceptibilities for Cs_2CoCl_4; (+) measurements made along the *a* axis; (□) measurements made along the *b* axis; (△) measurements made along the *c* axis. (After Reference 179.)

dashed lines in Figure 29. The Cl–Cl distance, which is the closest contact between the tetrahedra, is 3.733 Å and this corresponds to a metal–metal distance of 7.400 Å.

The results of the magnetic susceptibility measurements taken along the crystalline principal axes are shown in Figure 30. The presence of magnetic exchange is indicated by the apparent leveling off of the magnetic susceptibility data for all three directions near 1.5 K. The data were analyzed by McElearney *et al.*[179] using the theory for single-ion susceptibilities modified to account for exchange interactions. Satisfactory theoretical fits could be obtained only by allowing the molecular principal axes to vary in the fitting procedure. The resultant theoretical curves were characterized by large, positive D values (4.82 cm^{-1}) and nonnegligible E values (0.51 cm^{-1}). Significant antiferromagnetic exchange was indicated by the value of $J = -0.94$ cm^{-1}. The g values that were obtained from the analyses of the data were centered about $g_x = 2.62$, $g_y = 2.70$, and $g_z = 2.53$.

Analysis of the heat-capacity data suggest that the exchange coupling could be described by the XY model.[179] Unfortunately, since there are two magnetically inequivalent molecules in the unit cell, application of the solution of Katsura for the magnetic susceptibility of an XY linear chain[17] is not possible. There are few systems that display XY behavior, and a development of the appropriate theoretical framework for the exact description of the exchange interactions in Cs_2CoCl_4 would be especially valuable. The results may give clues that would be useful in the identification of additional XY systems.

8. *Chain Compounds of Nickel*

Chain complexes of nickel ions are restricted to the d^8 nickel(II) ions of roughly octahedral symmetry. The resulting ${}^3A_{2g}$ ($S = 1$) ground state is separated by about 8000–10,000 cm^{-1} from the lowest excited state, and the properties of this state are usually isotropic with g values in the range 2.15–2.3.

Since a first-order treatment of the spin–orbit interaction cannot mix properties of the excited states into the ${}^3A_{2g}$ ground state, nickel(II) ions in cubic octahedral ligand fields must show isotropic magnetism.[25] Thus, the observed g values are fully isotropic, and the isotropic $S = 1$ Heisenberg model is appropriate to the treatment of the thermal and magnetic properties. Various approximations to the unperturbed $S = 1$ Heisenberg model are known from the works of Weng,[15] Rushbrooke and Wood,[11] and Smith and Friedberg[13] (see Section 1 and Table 1). In practice, the fully isotropic model rarely applies to real chemical chains of nickel(II) ions because octahedral site symmetry is uncommon.

8.1. *Theory for Nickel(II) Chains In Distorted Ligand Fields*

The combined effects of ligand field distortions and spin–orbit coupling can lift the degeneracy of the ${}^3A_{2g}$ state and give rise to zero-field splittings. When single-ion effects are small, they may be treated as perturbations to the Heisenberg approximation.[13] Both signs of D and E are known in nickel(II) monomers and chains.[224]

A far more complicated situation is observed when the magnitudes of the zero-field splitting and exchange interaction are comparable. Watanabe[225] has used a molecular field approach to treat these situations through the use of the following Hamilitonian

$$\hat{H}_{MF} = D\hat{S}_z^2 + E(\hat{S}_x^2 - \hat{S}_y^2) + \mu_B(g_x \mathbf{H}_x \hat{S}_x + g_y \mathbf{H}_y \hat{S}_y + g_z \mathbf{H}_z \hat{S}_z) + A\hat{S}\cdot\langle\hat{S}\rangle \tag{51}$$

where the last term represents the thermal expectation value of the exchange interaction and is related to J by

$$A\hat{S}\cdot\langle\hat{S}\rangle = -2zJ\hat{S}\cdot\langle\hat{S}\rangle$$

where z is the number of near neighbors and is two for a linear array. This model essentially treats the exchange interaction as a correction to a form of the single-ion Hamiltonian. Anisotropic high-temperature magnetic susceptibilites are given by Eqs. (53a) and (53b).

$$\chi_{\parallel} = \chi_z = \frac{2Ng_{\parallel}^2\mu_B^2}{3k_BT}\left(1 - \frac{D - 4zJ}{3k_BT}\right) \tag{53a}$$

$$\chi_{\perp} = \chi_{x,y} = \frac{2Ng_{\perp}^2\mu_B^2}{3k_BT}\left(1 + \frac{D + 8zJ}{6k_BT}\right) \tag{53b}$$

Although the model is easy to use, it suffers from the fact that the molecular field approximation is known to be a poor model for linear chains.[1]

Independent computer calculations of finite rings (to $n = 8$) and chains (to $n = 7$) of $S = 1$ spins have been published by de Neef[226] and Blote[227] for Eq. (4). The published work of these authors deals exclusively with the specific heat of the various permutations of the signs of D and J. Their results indicate that for small values of $D/|J|$, the exchange interaction displays anisotropy such that $D < 0$ resembles an anisotropy of the Ising type while $D > 0$ shows XY behavior. When $D/|J|$ is large and negative, the situation resembles a single-ion term followed by Ising exchange interactions. For $D/|J|$ large and positive, the system approaches the behavior of a single ion, since the nonmagnetic $M_s = 0$ level lies lowest.

In an unpublished thesis de Neef[228] has calculated the parallel magnetic susceptibility of similar finite rings and chains of $S = 1$ ions which obey Eq. (4). Due to the complexity of the theoretical problem, this calculation has been carried out for the limited range of $D/|J| = -4, -2, -1, 0, 2, 4$. The low-temperature region, $1 \leq k_BT/|J| \leq 8$, is covered by an extrapolation procedure similar to that used by Bonner and Fisher.[7] Higher temperatures, $2 \leq k_BT/|J| \leq \infty$, were approximated by a series-expansion method which expresses $\chi_{\parallel}(J, D)$ in powers of $1/T$ up to the sixth order. From the above discussion it is obvious that not all ratios of $D/|J|$ are considered, but Dupas and Renard[229] have devised an interpolation scheme that should make the work of de Neef easier to apply.

At the present time no theoretical work is available to treat the perpendicular magnetic susceptibility of an infinite $S = 1$ chain, where exchange interactions and zero-field splittings are of comparable magnitude. This will prove to be a very formidable problem for theoreticians. Such a calculation would be of considerable utility in analyzing the magnetic properties of many $S = 1$ chains, and it should allow new insight into the effects of the ligand field on exchange interaction.

8.2. Selected Examples of Chain Compounds of Nickel(II)

The recent theoretical advances cited in Section 8.1 suggest that Ni(II) chains are best categorized by the ratio of the zero-field splitting to the exchange interaction, $|D|/|J|$. As long as this quantity is smaller than or comparable to J, a perturbed Heisenberg model is the best choice. Large ratios with D of either sign represent the limiting cases. Experimental examples of each of these cases will be discussed in the following sections.

8.3 $CsNiF_3$: A Compound with $J > 0, D > 0, D/J = 0.9$

Crystals of $CsNiF_3$ are readily obtained by the reaction of CsF or $CsBF_4$ with NiF_2 either under vacuum or inert atmospheres at 800–1000°C. Babel[230] has shown that the structure consists of face-sharing chains of NiF_6 octahedra running along the c axis in the familiar $2L$-pervoskite structure (see Figure 5, Table 8). Nickel ions are situated 2.61 Å apart within the chains, and the shortest Ni–Ni contacts between chains are 6.23 Å. Neutron diffraction work[231] has elucidated the magnetic structure and proven that no structural phase transition occurs above 2.6 K.

The intrachain magnetism in $CsNiF_3$ has been studied more thoroughly than that of any other nickel chain. Early susceptibility work revealed that the intrachain exchange was ferromagnetic and that a large planar anisotropy was present, i.e., $(\chi_\perp - \chi_\parallel) > 0$, at temperatures below 30 K.[232] From specific heat and powder magnetic susceptibility, Lebesque and co-workers[233] were able to locate the ordering temperature of 2.613 K and estimate an intrachain exchange of $+5.5 \text{ cm}^{-1}$. Roughly 87% of the expected entropy was gained above T_c, and the qualitative features of the specific heat were intermediate between the Heisenberg and Ising models. Neutron scattering experiments then revealed that a large positive zero-field splitting was present in addition to the ferromagnetic exchange.[234, 235] Later, single-crystal magnetic susceptibility and magnetization experiments provided values for the intrachain coupling and zero-field splitting parameters.[236–238] Dupas and Renard[229] carefully reexamined the zero-field susceptibilities and magnetization using the theory of de Neef[228] and presented convincing evidence that the parameters of interest are $J = +6.95 \text{ cm}^{-1}$ and $D = +5.9 \text{ cm}^{-1}$ (Figure 31).

8.4. $NiCl_2 \cdot 2H_2O$: A Compound with $J > 0, D < 0, |D|/J = 0.08$

Large crystals of $NiCl_2 \cdot 2H_2O$ are obtained from solutions maintained above 65°C which contain $NiCl_2 \cdot 6H_2O$ and LiCl in a 3 : 1 molar ratio.[239] The room-temperature structure[240] consists of chains of $NiCl_4(H_2O)_2$ octahedra that are tilted by 12° with respect to each other as in $[(CH_3)_3NHCoCl_3 \cdot 2H_2O]$ (Figure 27). A structural phase transition occurs at 230 ± 20 K, which shifts the chains with respect to each other, and moves the nickel(II) ions from centers of symmetry to general positions within the unit cell. Changes in the site symmetry of individual nickel ions are not known with any certainty but are thought to be small.[241]

TABLE 8. Structural and Magnetic Properties of Nickel(II) Chain Compounds

Compound	Structure: M–M (Å)	Structure: M–B–M (deg)	Structure: M–M′ (Å)	Intrachain: J (cm^{-1})	Intrachain: %S	Single ion: D (cm^{-1})	Interchain: T_c (K)	Interchain: J'/J ($\times 10^2$)	Experimental: Temperature (K)	Experimental: Technique	Model	Reference
Chain Compounds with Tri-μ-Fluoro Bridges												
$CsNiF_3$	2.61		6.23						300	x ray		230
									1–30	N		231
				+6.95		+5.9		1	1–77	χ_i	H–dN	229
				+7.99		+6.2				INS	H–SW	235
				+6.88		+5.2			15–170	χ_i	H–dN	238
				+5.5		+5.7			63–300	χ_i	H–MF	237
				+8.3						M_i	H–SW	236
				+8.2		+3.1				NS	H–SW	234
				+5.8				1	70–300	χ_p	H–MF	233
					87			1	1–30	C_m		233
Chain Compounds with Di-μ-Chloro Bridges												
$NiCl_2 \cdot 2H_2O$									300	x ray		240
									4.2–300	N		241
				+9.2		−0.76			1–150	χ_i	H–MF	239
						−0.75		8.27	1.1–7.3	M_i	H–MF	239
					50		6.3, 7.3		1–25	C_m		239
$Ni(py)_2Cl_2$				+3.5		−16.7		1.4	2–150	M_p, χ_p	H–MF	255
				+1.87	75	−22.9	6.4, 6.7	3–6	1–80	C_m	H–B	256
$Ni(pz)_2Cl_2$				+4.2		−18.1		1.3	2–150	M_p, χ_p	H–MF	255
				+5.07	92	−20.1	6.05	0.1–2	1–80	C_m	H–B	256

				Chain Compounds with Tri-μ-Chloro Bridges								
$(CH_3)_4NNiCl_3$	3.58	84.05	9.02						300	x ray		2
				+1.18		+2.3	1.2		1–77	χ_i	H–dN	229
				+1.18	81	+2.3	1.2		1–15	C_m	H–dN	257
				+0.76		+1.5	1.19	3	0.4–20	χ_i	H–MF	258
				−1.87	81		1.2		2–25	C_m	H–W	259
				+1.25					1–80	χ_p	H–S	260
$TlNiCl_3$	2.94	73.2	6.86						300	x ray		2
				−15.3			13	2	2–150	χ_p	H–W	245
NH_4NiCl_3	2.96		6.93	−11.1			9	2	2–150	χ_p	H–W	245
$CsNiCl_3$	2.97	74.3	7.18						300	x ray		2
									1.6–5.0	N		251
								0.7	1.1–5	M_i	H–MF	243
				−9.4		+0.46	4.3, 4.7		3.8–40	C_m	H–B	244
				−9			4.5	0.6	2–150	χ_p	H–W	245
					94		4.3, 4.8		1–20	C_m		246
				−11.1		−0.25			1–80	χ_i	H–SW	247
				−9.9					1–150	χ_p	H–S	248
				−7.6			4.5	1	1–300	χ_i	H–IS	242
				−8.4			4.5	1	1–300	χ_i	H–IS	242
$RbNiCl_3$	2.95	74.5	6.95						300	x ray		2
									4.5–11	N		250
				−12.8			11.1	2	2–150	χ_p	H–W	245
				−13.2		−1.1			2–80	χ_i	H–SW	247
				−9.8			11	1–10	1–300	χ_i	H–IS	242
				−11.2			11	1–10	1–300	χ_i	H–IS	242

(continued overleaf)

TABLE 8. (continued)

Compound	Structure			Intrachain		Single ion	Interchain		Experimental			
	M–M (Å)	M–B–M (deg)	M–M′ (Å)	J (cm^{-1})	%S	D (cm^{-1})	T_c (K)	J'/J ($\times 10^2$)	Temperature (K)	Technique	Model	Reference
Chain Compounds with Di-μ-Bromo Bridges												
$Ni(py)_2Br_2$				+1.7		−15.9		7	2–150	M_p, χ_p	H–MF	255
				+1.9	75	−20.8	2.85	4–9	1–80	C_m	H–B	256
$Ni(pz)_2Br_2$				+2.2		−18.7		1	2–150	M_p, χ_p	H–MF	255
				+1.9	87	−22.9	3.35	1–5	1–80	C_m	H–B	256
Chain Compounds with Tri-μ-Bromo Bridges												
$(CH_3)_4NNiBr_3$	3.17	76.45	9.35						300	x ray		2
				+3.3		+2.4	2.7	5	1–77	χ_i	H–dN	229
				+2.1			2.7	5	0.4–20	χ_p	H–S	258
$CsNiBr_3$	3.12		7.50						300	x ray		2
				−11.8		−1.0		2.7	1–150	χ_i	H–S	262
				−12.2	94		11.7, 14.2		1–30	C_m	H–B	262
				−12.8			14	4	2–150	χ_p	H–W	245
$RbNiBr_3$	3.1		7.27						300	x ray		245
				−17.7			23	4	2–150	χ_p	H–W	245
Chain Compounds with Tri-μ-Iodo Bridges												
	6.707		8.007						300	x ray		263
				−25.7			32			χ(?), N		264

				Chain Compounds with μ-Oxo Bridges								
$Ni(hipp)_2(H_2O)_3 2H_2O$	3.94	137.2	6.93						300	x ray		265
				−12.9					2.5–300	χ_p	H–S	265
				Chain Compounds with Di-μ- Sulfur Bridges								
$Ni(tu)_2(SCN)_2$									300	x ray		266
				+13.2			11(?)		2–300	χ_p	I	267
				Chain Compounds with Di-μ-Thiocyanato Bridges								
$Ni(tam)_2(SCN)_2$									300	x ray		268
				+3.1				3	2–300	χ_p	I	267
$Ni(tim)_2(SCN)_2$									300	x ray		268
				+2.1			8(?)	20	2–300	χ_p	I	267
				+2.5			8(?)	20	2–300	χ_p	I	267
				Chain Compounds with Carboxylic Acid Bridges								
$Ni(ox)(H_2O)_2$				−8		−8			2–150		H–dN	270
				−29					80–300		I	269
$Ni(sq)(im)_2(H_2O)_2$			7.1						300	x ray		254
				0(?)		+5.8			2–150	χ_p	SI	254
				Chain Compounds with Oxo-Anion Bridges								
$[Ni(en)_2(NO_3)]ClO_4$									300	x ray		271
				−18					80–300	χ_p	H–RW	272
$[Ni(bpy)(H_2O)_2]SO_4$									300	x ray		273
				−2.1					1.7–300	χ_p		274
$Ni(N_2H_5)_2(SO_4)_2$	5.3		5.8	−2.3					2–150	M_p, χ_p	H–W	252
				−1.1	93	−10.8			1–30	C_m	H–B	253
				−4.4	93	−10.6			1–30	C_m	SI + I	253

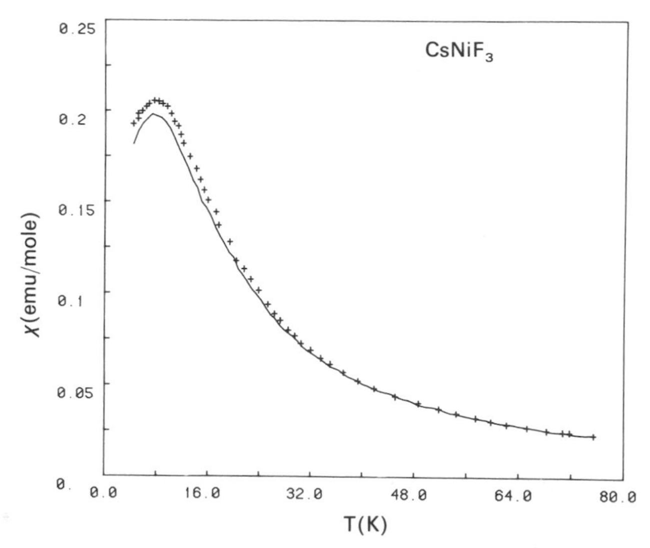

FIGURE 31. Comparison of the experimental magnetic susceptibility (+) parallel to the c axis of $CsNiF_3$ with the theory of de Neef. The solid line has been calculated for $J = +6.95$ and $D = +6.2\ cm^{-1}$. (After Reference 229.)

The Eindhoven group[239] has reported an extensive survey of the magnetic and thermal properties of $NiCl_2 \cdot 2H_2O$. They have found that this unusual material has ferromagnetic intrachain exchange, a negative zero-field splitting ($M_s = \pm 1$ lowest), two anomalies in the specific heat at 7.2 and 6.3 K, and metamagnetic behavior in applied fields. An analysis of the high-temperature magnetic susceptibilities along the a^*, b, and c axes using the theory of Watanabe yielded an intrachain exchange $J = +9.2\ cm^{-1}$ and $D = -0.76\ cm^{-1}$ (Figure 32). A very detailed study of the magnetic phase diagram shown in Figure 33 allowed additional analyses of the magnetic structure and the various interchain exchange terms that cause it. The magnetic structure consists of ferromagnetic chains along the b axis with neighboring chains along the a axis whose moments also point in the same

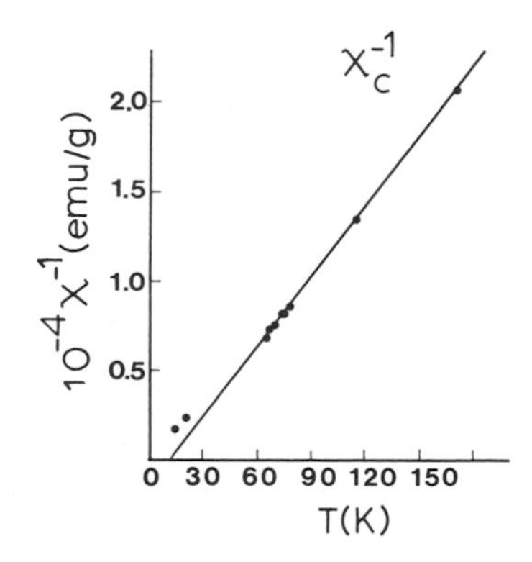

FIGURE 32. Inverse susceptibility along the c axis of $NiCl_2 \cdot 2H_2O$. The solid line is the best fit to the Curie–Weiss law with $g_c = 2.23$ and $\theta_c = +12.5$. (After Reference 239.)

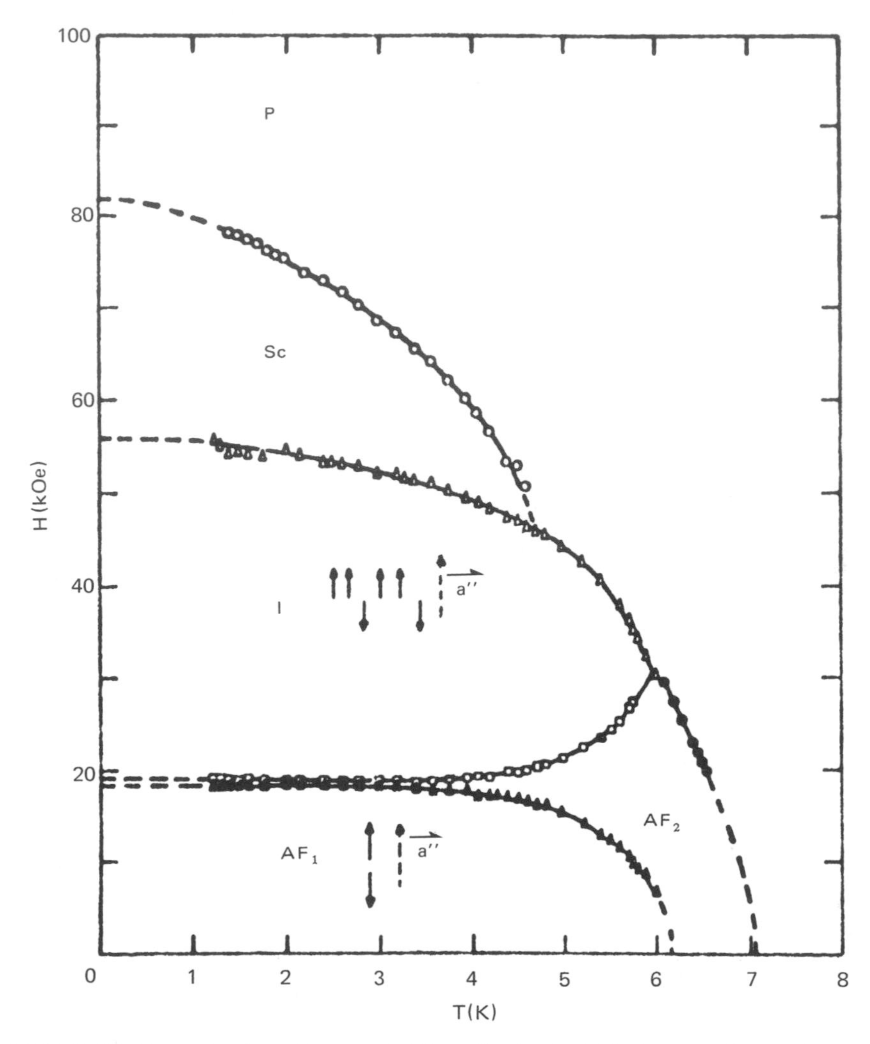

FIGURE 33. Magnetic phase diagram of $NiCl_2 \cdot 2H_2O$ with $H \parallel a^*$ axis. AF_1 and AF_2 are different antiferromagnetic phases. I is the metamagnetic phase, S_c the screw phase, and P the paramagnetic phase, S_c the screw phase, and P the paramagnetic state. (Data from Reference 239, with permission.)

direction. The ferromagnetic layers in the *ab* plane are coupled antiferromagnetically along the *c* axis. A study of the angular dependence of the phase boundaries in Figure 33 allowed the following parameters to be determined:

$$D = -0.75 \text{ cm}^{-1}$$

$$z_1J_1 = -2.5 \text{ cm}^{-1} \qquad (z_1 = 4)$$

$$z_2J_2 = -0.75 \text{ cm}^{-1} \qquad (z_2 = 2)$$

where J_1 is the interchain interaction between neighboring *ab* layers, J_2 is the interchain exchange between next-nearest neighbors, and the z_i are the number of neighbors.

8.5 $CsNiCl_3$: A Compound with $J < 0, D > 0, D/|J| = 0.05$

Achiwa[242] first reported the magnetic properties and preparative details for $CsNiCl_3$. The structure is again the familiar 2*L*-pervoskite with chains running along the *c* axis. Intrachain metal–metal distances are 2.97 Å with Ni–Cl–Ni angles being 74.3°. The chains are separated from each other by 7.18 Å. Unlike $CsNiF_3$, the intrachain interaction is antiferromagnetic with a broad maximum in the susceptibility near 35 K. Numerous authors[243–248] have determined the value of J using a variety of Heisenberg models, and this value is approximately $-9.4\ cm^{-1}$. Torque experiments performed by Achiwa[242] showed that $(\chi_\perp - \chi_\parallel)$ was positive (see Figure 34), but an initial analysis[247] of this data erroneously stated that the single-ion term was negative, i.e., $D = -0.25\ cm^{-1}$. Later specific-heat experiments,[244] which were analyzed after Blote,[227] corrected this error with $J = -9.4\ cm^{-1}$ and $D = +0.46\ cm^{-1}$. These values are in accord with the qualitative behavior of the torque measurements; i.e., $(\chi_\perp - \chi_\parallel)$ is positive.

8.6. $RbNiCl_3$: A Compound with $J < 0, D < 0, D/J = 0.08$

The single-crystal susceptibilities[242] along [0001] and [11$\bar{2}$0] in $RbNiCl_3$ are shown in Figure 35. The structure is isomorphous with $CsNiCl_3$, and the intrachain coupling is also antiferromagnetic and about $-13\ cm^{-1}$.[1, 242, 245] However, inspection of the experimental data immediately shows that $D < 0$ since $(\chi_\perp - \chi_\parallel) < 0$. A spin-wave analysis of the original data of Achiwa supported this observation and yielded $J = -13.2\ cm^{-1}$ and $D = -1.1\ cm^{-1}$.[247]

It is important to note that the original single-crystal magnetic susceptibility data for $CsNiCl_3$ and $RbNiCl_3$ were observed with the measuring field along the [0001] and [11$\bar{2}$0] directions.[242] The former is along the *c* axis and is a diagonal element of the χ tensor, but the latter is not and lies at an angle of 60° from the *a* axis. Thus, $\chi[11\bar{2}0]$ contains contributions from both of the unique directions. In view of the isomorphous nature of the room-temperature crystallography and the similarity of the low-temperature magnetic structures of these compounds, the fact that the single-ion terms are of different signs is surprising. Further comparative studies of the single-crystal magnetic susceptibilities along the diagonal elements of the χ tensor, i.e., [0001] and [1000], should help in resolving the remaining problems in these systems.[2, 249–251]

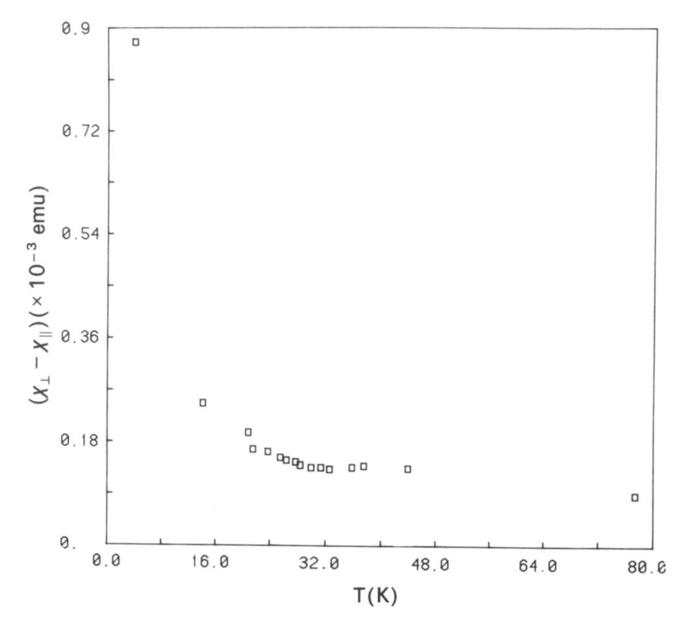

FIGURE 34. Anisotropy in the susceptibility of $CsNiCl_3$ obtained from torque measurements. The magnetic field was rotated in the $(11\bar{2}0)$ plane which contains the [0001] and $[11\bar{2}0]$ axes. Note that $\chi_\parallel = \chi[0001]$ is a diagonal element of the χ tensor, but $\chi_\perp = \chi[11\bar{2}0]$ lies 60° from the *a* axis. (After Reference 242.)

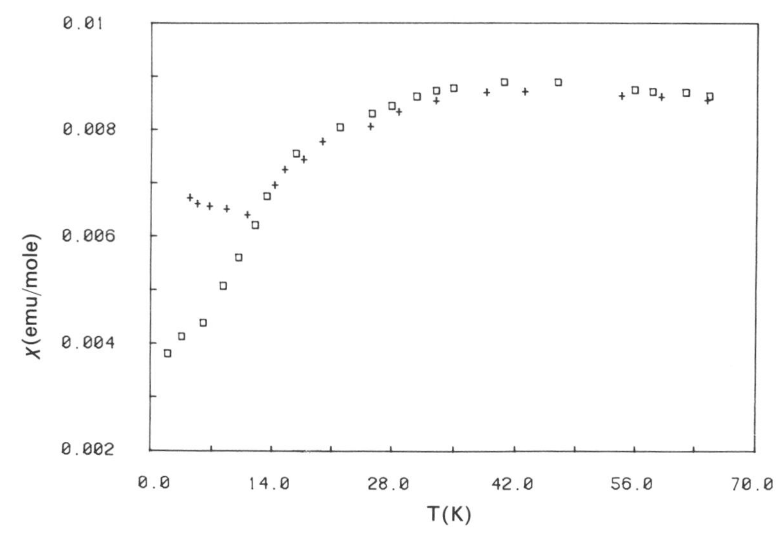

FIGURE 35. Anisotropic susceptibilities of $RbNiCl_3$ in the region of the maximum: (+), $H\parallel [11\bar{2}0]$ and (□) $H\parallel$ [0001]. Note that the sign of the anisitropy is different from that of $CsNiCl_3$. (After Reference 242.)

8.7. $Ni(N_2H_5)_2(SO_4)_2$: A Compound with $J < 0, D < 0, |D|/|J| = 9.2$

Treatment of hot aqueous solutions of $NiSO_4 \cdot 6H_2O$ with hydrazinium sulfate solutions yields the anhydrous $Ni(N_2H_5)_2(SO_4)_2$ complex. This material has been shown to be isomorphous with the zinc analog whose complete structure is known to consist of chains of di-sulfato bridged Zn(II) ions running along the *b* axis[252] (see Figure 36). The metal-ion environment is essentially a tetragonally compressed octahedron. The Ni–Ni separations in the chain are 5.3 Å while those between chains are 5.8 Å. Powder susceptibility work over the range 2–80 K revealed the presence of a broad maximum centered near 9 K. These data could be fit by the $S = 1$ Heisenberg model of Weng with $J = -2.33$ cm^{-1}. Another observation in this work was an unusual upward bend in the isothermal magnetization at 2.2 K. Later specific-heat experiments were not consistent with the susceptibility parameters and could only be understood by including a large, negative, single-ion term ($M_s = \pm 1$ lowest).[253] In fact, the above work was consistent with either $J = -1.1$ cm^{-1} and $D = -10.8$ cm^{-1} from the theory of Blote[227] or a combination of $D = -10.6$ cm^{-1} from a single-ion model and $J = -4.43$ cm^{-1} from an effective $S' = \frac{1}{2}$ Ising model. The unusually large value of D was corroborated by relaxation experiments. These authors[253] chose the Heisenberg chain with large anisotropy as more likely for a Ni(II) species. Also, the specific-heat data revealed that 93% of the expected entropy was removed by intrachain and single-ion effects. This result suggests that, even though intra and interchain distances are comparable, the chains are very well isolated.

Single-crystal susceptibility experiments would be of considerable interest for $Ni(N_2H_5)_2(SO_4)_2$. From an experimental point of view, this will be an extremely difficult task since the crystal symmetry is triclinic. However, this material is one of the best examples of an isolated nickel(II) chain with no long-range magnetic ordering down to 1.5 K, and the unusual synergism between the single-ion and exchange interactions should be easier to understand.

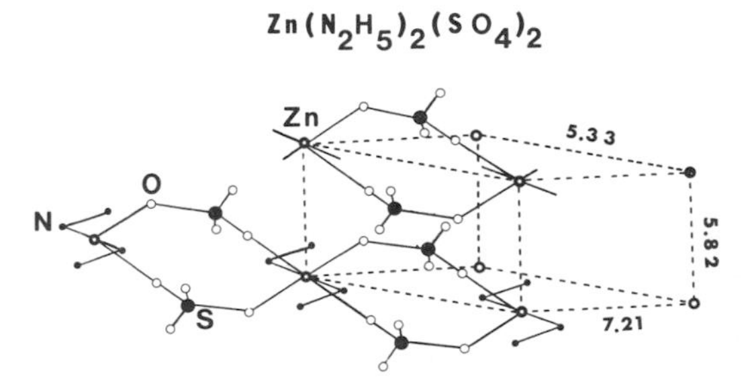

FIGURE 36. Structure of bis(hydrazinium)zinc sulfate. The nickel analog is known to be isomorphous with this structure. (After Reference 253.)

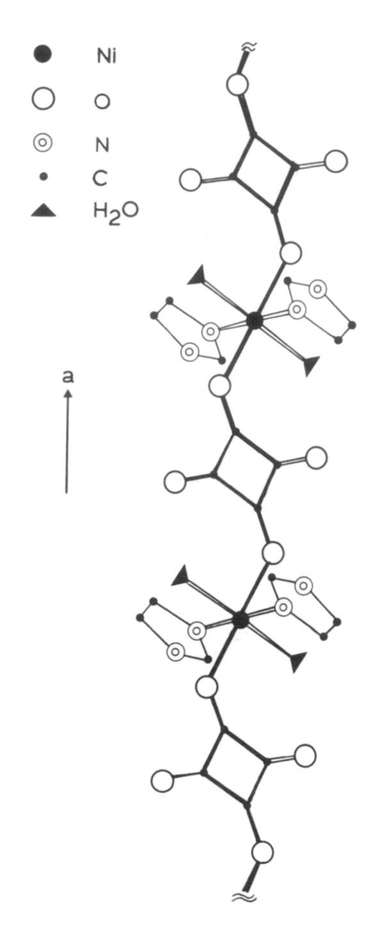

FIGURE 37. Structure of diaquobis(imidazole)-*catena*-μ-(squarato-1,3)nickel(II). Only one chain is shown, and the bonds that propogate it are blackened for clarity. (After Reference 254.)

8.8 $Ni(C_4O_4)(C_3N_2H_4)_2(H_2O)_2$: A Compound with $D > 0$, $D > 4|J|$

Blue-green crystals of diaquobis(imidazole)-catena-μ-(squarato-1, 3)-nickel(II)[254] can be obtained by mixing aqueous solutions of $NiCl_2 \cdot 6H_2O$, squaric acid ($H_2C_4O_4$), and imidazole ($C_3N_2H_4$) in a mole ratio of 1 : 1 : 2. The structure of this compound revealed chains of nickel(II) ions bridged by 1, 3-squarato ligands running parallel to the crystal *a* axis (see Figure 37). The powder susceptibility data (1.2–100 K) reach an almost constant value below about 2 K (see Figure 38). An analysis of this data yielded a best fit of $D = +5.8\ \text{cm}^{-1}$ and $g_\perp = g_\parallel = 2.28$ using a single-ion model only, Eqs. (54a) and (54b):

$$\chi_M = \frac{4Ng_\perp^2\mu_B^2}{3D}, \qquad k_BT \ll D \tag{54a}$$

$$\chi_M = \frac{2N\langle g\rangle^2\mu_B^2}{3k_BT}, \qquad k_BT \gg D \tag{54b}$$

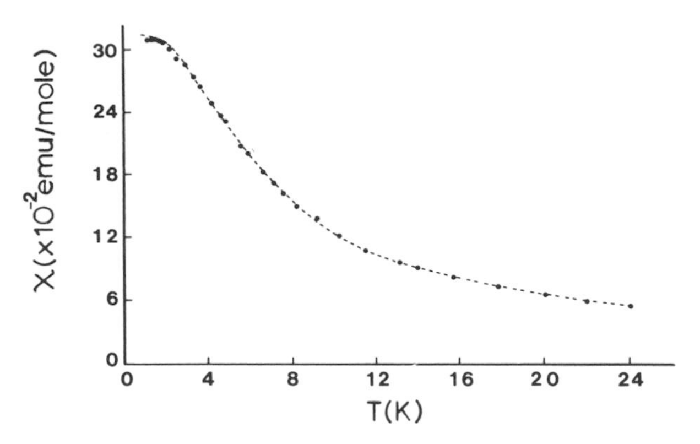

FIGURE 38. Powder magnetic susceptibility of $Ni(sq)(im)_2(H_2O)_2$ at low temperatures. The dashed curve is the best fit to the single-ion model with $D = +5.8\ cm^{-1}$ and $\langle g \rangle = 2.28$. (After Reference 254.)

Although a weak exchange interaction could not be eliminated by fitting procedures, the next best fit to the model of de Neef was obtained for $D/|J| = 4$ (the upper limit of the model). It was concluded that the curious tilting of the squarato ligand resulted in the poor π(ligand)–d(metal) orbital overlap and hence very weak exchange along the chain. This unusual compound is apparently one of the few examples of a chain structure whose magnetism is almost wholly dominated by single-ion effects. Additional heat-capacity and single-crystal susceptibility experiments are needed to corroborate the unusual magnetic properties of this interesting system.

9. Chain Compounds of Copper

Chains of ligand-bridged copper(II) ions provide a rich area for research since large numbers of compounds may be synthesized conveniently, the compounds are usually stable in the laboratory atmosphere, and it is frequently possible to obtain large single crystals for detailed studies. Exchange coupling in copper(II) systems is largely an isotropic phenomenon, and, consequently, the Heisenberg exchange theory is appropriate in the theoretical treatment of experimental data. Theoreticians have focused much attention on $S = 1/2$ systems and, as a result, a large body of theoretical results are readily available for a number of Heisenberg models.

9.1. Theoretical Models Applicable for Copper Chains

9.1.1. Uniformly Spaced Linear Chains

The available analytical results for one-dimensional Heisenberg magnets have recently been surveyed by Johnson,[5] and these results are summarized in the theoretical section along with the results for XY and Ising systems. Anistropic models are rarely necessary for the description of the static magnetic properties of exchange-coupled copper(II) systems, and before these equations are applied to linear copper(II) chains, the comments by Jotham must be taken into consideration.[275]

9.1.2. *The Alternating Heisenberg Chain*

The Hamiltonian for the Heisenberg alternating linear chain may be written

$$\hat{H} = -2J \sum [\hat{S}_{2i} \cdot \hat{S}_{2i-1} + \alpha \hat{S}_{2i} \cdot \hat{S}_{2i+1}] \tag{55}$$

where the summation runs from $i = 1$ to $n/2$, J is the exchange integral between a spin and one of its nearest neighbors, and αJ is the exchange integral between the same spin and the other nearest neighbor in the chain. The model of immediate interest is for antiferromagnetic exchange ($J < 0$) and for $0 \leq \alpha \leq 1$. At the extremes, when $\alpha = 0$, the model reduces to the dimer model with pairwise interactions, and when $\alpha = 1$, the model reduces to the uniform linear chain model. This problem has been studied in detail by Duffy and Barr,[276] Bonner and Friedberg,[227, 278] Diederix *et al.*[279] and more recently by Bonner *et al.*[280] and Hatfield and co-workers.[281, 282]

The reduced magnetic susceptibilities of short alternating rings of up to ten $S = \frac{1}{2}$ spins for $\alpha = 0$, 0.1, 0.2, 0.3, 0.4, 0.6, 0.6, 0.8, and 1.0 have been recalculated using the cluster approach[285] and the results are displayed in Figure 39. These results were used to generate the following expression for

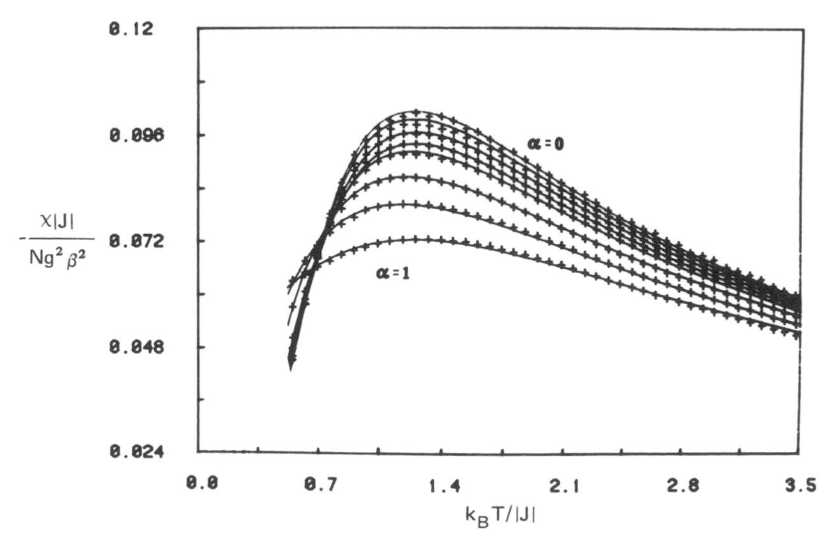

FIGURE 39. Plots of the expression for the alternating chain model with $\alpha = 0.0$, 0.1, 0.2, 0.3, 0.4, 0.6, 0.8, and 1.0. The points (+) were calculated for 10-membered rings with $H = 100$ G and $g = 2.1$. (From Reference 281, with permission.)

magnetic susceptibility in terms of the exchange integral and the alternation parameter:

$$\chi_M = [Ng^2\mu_B^2/k_BT][(A + Bx + Cx^2)/(1 + Dx + Ex^2 + Fx^3)] \quad (56)$$

where $x = |J|/k_BT$ and the values for the parameters A through F for $0 \leq \alpha \leq 0.4$ are

$$A = 0.25$$
$$B = -0.12587 + 0.22752\alpha$$
$$C = 0.019111 - 0.13307\alpha + 0.509\alpha^2 - 1.3167\alpha^3 + 1.0081\alpha^4$$
$$D = 0.10772 + 1.4192\alpha$$
$$E = 0.0028521 - 0.423462\alpha + 2.1953\alpha^2 - 0.82412\alpha^3$$
$$F = 0.37754 - 0.067022\alpha + 5.9805\alpha^2 - 2.1678\alpha^3 + 15.838\alpha^4 \quad (57a)$$

The values for the parameters A through F for $0.4 < \alpha \leq 1.0$ are

$$A = 0.25$$
$$B = -0.13695 + 0.26387\alpha$$
$$C = 0.017025 - 0.12668\alpha + 0.49113\alpha^2 - 1.1977\alpha^3 + 0.87257\alpha^4$$
$$D = 0.070509 + 1.3042\alpha$$
$$E = -0.0035767 - 0.40837\alpha + 3.4862\alpha^2 - 0.73888\alpha^3$$
$$F = 0.36184 - 0.065528\alpha + 6.65875\alpha^2 - 20.945\alpha^3 + 15.425\alpha^4 \quad (57b)$$

The expression with the two sets of parameters A–F is valid for $k_BT/|J| > 0.5$ and for negative J values, and reproduces the calculated magnetic susceptibilities of the ten membered rings, which should be good approximations for infinite systems for the various values of α much better than 1% in moderate magnetic fields.

9.1.3. *Next-Nearest-Neighbour Exchange Interactions*

Next-nearest-neighbor exchange interactions which are possible in certain structures are depicted schematically in Figure 40, where it may be seen that each spin interacts with its nearest neighbor as well as the next-nearest neighbor. Klein[283] and Ananthakrishna *et al.*[284] have calculated the magnetic susceptibilities of some of these systems as a function of the ratio of the exchange coupling constants using a Hamiltonian that is formally similiar to that for the alternating chain given above.

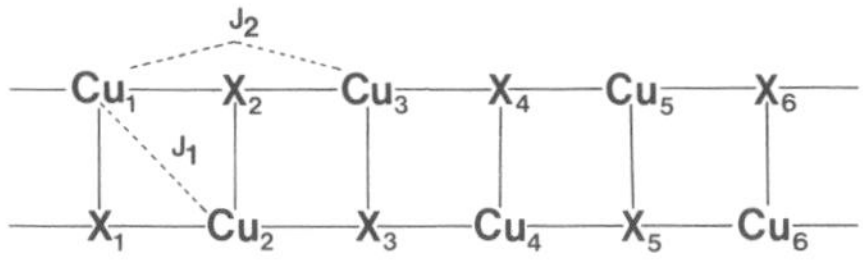

FIGURE 40. Schematic representation of a structure exhibiting next-nearest-neighbor exchange interactions.

At the present time, closed-form expressions for the magnetic susceptibilities in terms of the magnetic parameters are not available, but Weller[285] has carried out calculations for several of these systems using a versatile computer program, and results of this work are discussed below.

9.2. Antiferromagnetic Exchange in a Uniformly Spaced Linear Chain, Catena-di-μ-chloro-bis(pyridine)copper(II)

Although many copper(II) compounds with halide bridges have been synthesized, relatively few have been structurally and magnetically characterized. Among these are a few fluoro-bridged compounds, no iodo-bridged compounds, several bromo-bridged compounds, and a number of chloro-bridged compounds. The research with these compounds has been extremely fruitful in that a variety of bridges have been found, that is,

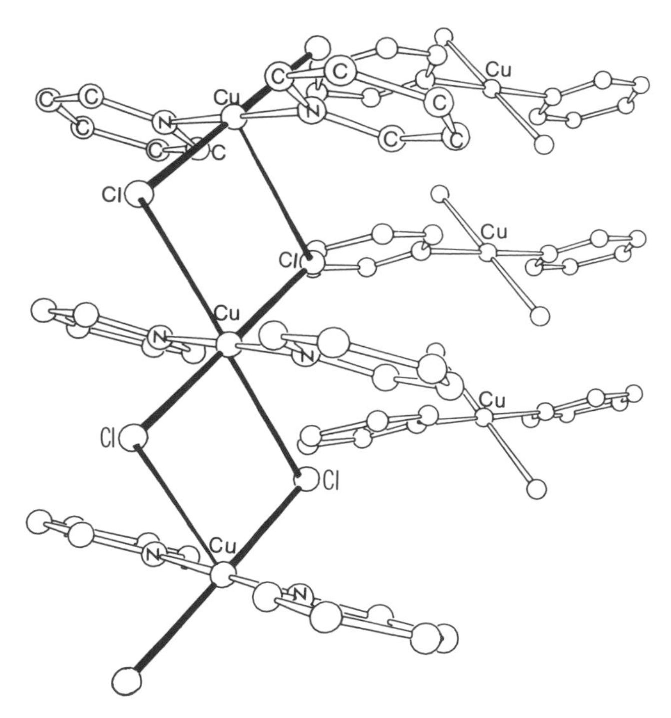

FIGURE 41. Structure of the linear chain compound $[(Cu(pyridine)_2Cl_2]$.

TABLE 9. Structural and Magnetic Properties of Copper Chain Compounds

Compound	Structure			Intrachain	Interchain		Experimental			
	M–M (Å)	M–B–M (deg)	M–M′ (Å)	J (cm^{-1})	T_c (K)	J'/J ($\times 10^2$)	Temperature (K)	Technique	Model	Reference
Chain Compounds with Fluoride Bridges										
$KCuF_3$ [a]								x ray		310
					39(1)	10				311
				−136(4)	39(1)	1.6		N		312
				−132	39	1.8		x ray		313
						2.7				314
				−129			39–580	OBF[c]	H–BF	315
$KCuF_3$ [b]					22(4)					316
Mono-μ-Chloro Bridged Chains										
$[Cu(DMSO)_2Cl_2]$	4.757	144.6						x ray		317
				−6.1			4–45	χ_p	H–BF	318
$[Cu(imH)_2Cl_2]$	4.37	117						x ray		319
				−2.1	7(?)		2–60	χ_p	H–BF	318
$[Cu(caf)(H_2O)Cl_2]$[d]	4.597	128.1						x ray		319
				+0.48			4–60	χ_p	H–BF	318
$[Cu(MAEP)Cl_2]$[e]	4.263	113.58						x ray		320
				+1.58			2–50	χ_p	H–BF	318
Di-μ-Chloro Bridged Chains										
$[(CH_3)_3NH]CuCl_3 \cdot H_2O$	3.739	91.0						x ray		321
		95.8						x ray		321
				+0.14	<1.0			χ_i	I–F	321
					0.157(3)			χ_i		322
				+0.59	0.165			χ_i	H–BRG	323

[Cu(4-Me-py)$_2$Cl$_2$]	3.932	89.0					x ray		292
				−9.6(2)	(alt. par. = 0.67)	1.8–120	χ_p	H–alt	281
[(CH$_3$)$_2$NH$_2$]CuCl$_3$	3.444		5.596			2–90	χ_p		324
HzHCuCl$_3$	3.751	92.79					x ray		298
				−5.46 to −7.5			χ_p	H–nnn	285
Cuen$_2$Cl$_2$	3.68		8.27				x ray		325
				neg.			χ_p		326
[Cu(py)$_2$Cl$_2$]	3.87	88.48					x ray		286
				−9.2		8–297	χ_p	H–BF	288
				−9.2	1.130(5)	1.5–300	χ_p	H–BF	289
							χ_p		287
[Cu(4-v-py)$_2$Cl$_2$] [f]	3.91	91.4					x ray		327
				−9.1			χ_p	H–BF	328
[Cu(4-Et-py)$_2$Cl$_2$] [g]	4.00	91.9					x ray		329
				−6.7			χ_p	H–BF	328
[Cu(tz)$_2$Cl$_2$] [h]	3.853	91.89					x ray		330
				−3.8			χ_p	H–BF	330
Tri-μ-Chloro Bridged Chains									
CsCuCl$_3$	3.062	81.12	7.216				x ray		331
		73.80					x ray		331
				positive					1
[(CH$_3$)$_4$N]CuCl$_3$	3.21–3.33 (five nonequivalent copper ions in the chain)						x ray		332
				+20.0	0.1	2–55	χ_p	H–BRG	333
				+27.0		85–280	χ_p	H–BRG	333
[(CH$_3$)$_4$NH]Cu$_3$Cl$_7$	3.15	77.6					x ray		334
		77.9					x ray		334
		82.0					x ray		334
				+34.5		2–280	χ_p	H–BRG	333
				−7		2–280	χ_p	H–BRG	333

(continued overleaf)

TABLE 9 (continued)

Compound	Structure			Intrachain	Interchain		Experimental			
	M–M (Å)	M–B–M (deg)	M–M′ (Å)	J (cm^{-1})	T_c (K)	J'/J ($\times 10^2$)	Temperature (K)	Technique	Model	Reference
				di-μ-Bromo Bridged Chains						
[Cu(py)$_2$Br$_2$]	4.050	89.64						x ray		286
				−18.9			8–297	χ_p	H–BF	288
				−16.0				χ_p	H–BF	335
[Cu(NMim)$_2$Br$_2$][i]	4.130	90.03						x ray		336
				−5.8			4–34	χ_p		337
				−7.2			1.2–300	χ_p	H–alt	338
[Cu(dmpy)$_2$Br$_2$][j]	4.097	90.0						x ray		339
				−21				χ_p	H–BF	335
				Chain Compounds with Water Bridges						
Cu(benzoate)$_2$·3H$_2$O	3.15		6.98					x ray		340
				−5.9	<16		1.5–50	χ_p	H–BF	341
Cu(NH$_3$)$_4$SO$_4$·H$_2$O				−2.2	0.37	6				1
Cu(NH$_3$)$_4$SeO$_4$·H$_2$O				−1.6	<1.2					1
				Chain Compounds with Sulfur Bridges						
Cu(dmdtc)$_2$[k]		94.2						x ray		342
				−1.22			2–60	χ_p	H–BF	297
CuKTS								x ray		296

Chain Compounds with Mixed Ligand Bridges										
$[Cu(DMSO)Cl_2]$	3.238	87.57						x ray		291
		80.92						x ray		291
		81.04						x ray		291
				+31(4)	5.4(?)	4		χ_p	H–BRG	291
$[Cu(TMSO)Cl_2]$	3.209	86.7						x ray		291
		81.0						x ray		291
		79.7						x ray		291
				+27(4)	3.9(?)	2		χ_p	H–BRG	291
Chain Compounds with Alkoxo-Bridges										
$[Cu(sal{=}gly)H_2O]0.5H_2O$[l]	5.33		5.003					x ray		343
				−1.5		4	1.6–160	χ_p	H–D	344
$[Cu(sal{=}aiba)H_2O]$[m]	4.85		5.00					x ray		345
				−0.6			1.6–160	χ_p	H–D	344
Chain Compounds with Oxo-Anion Bridges										
$Cu(NO_3)_2 \cdot 2.5H_2O$	4.94 (uniform chain							x ray		293
	5.5, 6.2 (alternating chain)							x ray		293
				−1.8 (alt. parameter = 0.27)			0.05–1.4	χ_i	H–alt	279
										346
										277
										278
$Cu(NH_3)_4(NO_3)_2$								x ray		347
				−2.7	0.15		1.5–20	χ_p	H–BF	348
$Cu(H_2O)_5SO_4$				−1.0	<0.03	0.01				1
$Cu(H_2O)_5SeO_4$				−0.6	0.045	0.13				1
$Cu(N_2H_5)_2(SO_4)_2$							2.1–55	χ_p	H–BF	126
Chain Compound with Single Cyanide Bridge										
$[Cu(terpy)CN]NO_3 \cdot H_2O$	168							x ray		349
				−0.44			4–100	χ_p	H–BRG	350

(continued overleaf)

TABLE 9 (continued)

Compound	Structure			Intrachain	Interchain		Experimental			
	M–M (Å)	M–B–M (deg)	M–M′ (Å)	J (cm^{-1})	T_c (K)	J'/J ($\times 10^2$)	Temperature (K)	Technique	Model	Reference
Chain Compound with an Extended Pi-System Ligand Bridge										
$[Cu(pyz)(NO_3)_2]$	6.712		5.142					x ray		299
				−3.7			1.7–60	χ_i	H–BF	301
				−3.8	<0.18	<4	0.18–25	χ_p	H–BF	302
				−3.7			0.18–2.0	C_m		302
				−3.7			4.8–100	χ_i	H–BF	303
Chain Compounds without Logand Bridges										
$[Cu(AdH)_2Cl_2]Cl_2^n$	8.21							x ray		351
				−7.6			4–45	χ_p	H–BF	351
$[Cu(AdH)_2Cl_2]Cl_2$	8.51							x ray		352
				−36.5			4–200	χ_p	H–BF	351

Chain Compounds without Bridging Ligands

$[Cu(NH_3)_4][PtCl_4]$	6.43	9.06					x ray		353
			0.21		10		MR		354
$(TTF)[Cu(S_2C_2(CF_3)_2)_2]$	7.80						x ray		306
			−53 to −76			12–250	χ_i	H–BF	307
						2–12	χ_i	Br[o]	307
$[N(bu)_4]_2[Cu(mnt)_2]$	9.403						x ray		355
			0.0107			4.2	MR		355
$Cu(NH_3)_2 \cdot Ni(CN)_4 \cdot 2C_6H_6$	4.12	7.39					x ray		356
			−1.6			0.2–20	χ_p	H–RW	357
$Cu(NSal)_2$ [p]	3.33						x ray		358
			−2.2			2–300	χ_p	H–BF	359
			−2.2			0.07–3.5	C_m	H–BF	360
				0.044					361
$CuL_6(ClO_4)_2$			−0.70	0.142			χ_p	H–BF	362

[a] Structure type (a), D_{4h}^{18}.
[b] Structure type (d), D_{4h}^{15}.
[c] Optical birefringence.
[d] caf is caffeine.
[e] MAEP is 2-(2-methylaminoethyl)pyridine.
[f] 4-v-py is 4-vinylpyridine.
[g] 4-Et-py is 4-ethylpyridine.
[h] tz is thiazole.
[i] *N*-methylimidazole.
[j] dimethylpyridine.
[k] dmdtc is dimethyldithiocarbamate.
[l] sal = gly is *N*-salicylideneglycinato.
[m] sal = aiba is *N*-salicylidene-aminoisobutyrato.
[n] AdH is protonated adenine.
[o] Bulaevskii results for dimerized chain (Reference 308).
[p] NSal is *N*-methylsalicylaldiminato.
[q] L is pyridine *N*-oxide.

examples of structurally symmetrical and antisymmetrical mono-μ-, di-μ-, and tri-μ-halo-bridged chains are known. The research has provided predominately examples of linear chains with antiferromagnetic exchange coupling, but recently ferromagnetic exchange coupling in copper(II) chain compounds has been observed. Antiferromagnetic interactions, especially if $|J|$ is greater than about 1.0 cm^{-1}, are more easily identified than ferromagnetic interactions, and it is for this reason that chains with positive J values have only recently been identified.

$Cu(pyridine)_2Cl_2$, $Cu(py)_2Cl_2$, may be easily prepared by mixing ethanolic solutions of pyridine and copper(II) chloride in a molar ratio of 2 : 1. Nice single crystals can be obtained from the pasty substance that precipitates by recrystallization from dimethylformamide. The structure[286] of the monoclinic crystals (space group $P2_1/n$) consists of chains of uniformly spaced $Cu(py)_2Cl_2$ units as shown in Figure 41. The in-plane Cu–Cl bond distances are 2.28 Å, and the out-of-plane bond distances are 3.05 Å. The nearest-neighbor copper–copper separation in a given chain is 3.87 Å.

Magnetic susceptibility measurements by Takeda *et al.*[287] in the high-temperature range and by Jeter and Hatfield[288] in the range 8–297 K are

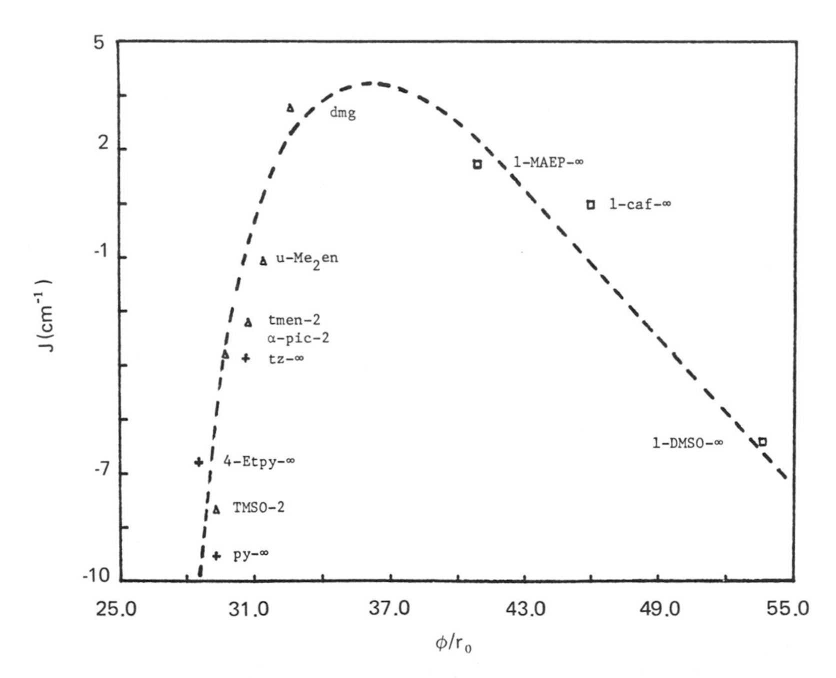

FIGURE 42. Correlation of the exchange coupling constant with the function ϕ/r_0, where ϕ is the angle at the bridging chloride ligand and r_0 is the long out-of-plane copper chloride bond distance.

consistent with antiferromagnetically coupled Heisenberg chains with $J = -9.15\ \text{cm}^{-1}$ and $g = 2.05$. Additional measurements have been carried out and very good agreement with the work of Jeter and Hatfield[288] reported.[289] In this latter work, J was found to be $-9.2(14)\ \text{cm}^{-1}$ and $g = 2.07(4)$. The heat capacity exhibits a sharp peak at 1.130(5) K which signals long-range ordering to an antiferromagnetic state.

Available data for other compounds of the same general formula and structure are collected in Table 9. All of these compounds have uniform chain structures and are antiferromagnetically coupled. There is an extremely interesting feature in these data. The magnitude of the exchange coupling constant is a smooth function of ϕ/r_0, where ϕ is the angle at the bridging chloride ligand, and r_0 is the long out-of-plane bond distance. This trend has been discussed in detail[290] and it has been pointed out that the magnetic and structural data for the chloro-bridged dimeric cluster and chains are highly correlated. The correlation displayed in Figure 42 may be understood in terms of molecular orbital theory and the contributing factors in the expression for the exchange integral.

9.3. Ferromagnetic Exchange Coupling in the Uniformly Spaced Copper(II) Chain Compounds, $Cu(DMSO)Cl_2$ and $Cu(TMSO)Cl_2$

The compound $Cu(TMSO)Cl_2$ may be prepared by dissolving tetramethylene sulfoxide (TMSO) in *n*-propanol at 70°C and adding a slight excess of anhydrous copper chloride. Thin orange needles crystallize from solution upon cooling. Crystals of $Cu(DMSO)Cl_2$ may be grown from a solution prepared by adding excess $CuCl_2$ to a mixture of dimethylsulfoxide (DMSO) in ethanol until all of the green $Cu(DMSO)_2Cl_2$ dissolves. The structure of $Cu(DMSO)Cl_2$ and $Cu(TMSO)Cl_2$ consist of triply bridged copper(II) ions, with the two chlorides and an oxygen atom from the sulfoxide acting as the bridging ligands. The structures of these chains are unusual. There is a symmetrical Cu–Cl–Cu bridge in each compound, with short Cu–Cl distances of about ~2.35 Å, and there are two antisymmetric bridges. In $Cu(DMSO)Cl_2$ the antisymmetric bridge involving the chloride ligand has Cu–Cl distances of 2.352 and 2.714 Å, and the related bond distances in $Cu(TMSO)Cl_2$ are 2.255 and 2.665 Å. The antisymmetric Cu–O–Cu distances are 1.972 and 2.909 Å in $Cu(TMSO)Cl_2$ and 1.959 and 2.901 Å in $Cu(DMSO)Cl_2$.[291]

The superexchange pathway that dominates the intrachain interaction is that one provided by the short symmetrical bridge. Since the Cu–Cl–Cu angle is 87.57° in $Cu(DMSO)Cl_2$ and 86.7° in $Cu(TMSO)Cl_2$, and since the chloride bridges the σ^* orbitals in which the unpaired electrons are located, then relatively strong ferromagnetic intrachain coupling is expected[290] and observed. Fits of the expression developed by Baker *et al.*[10] for ferromagne-

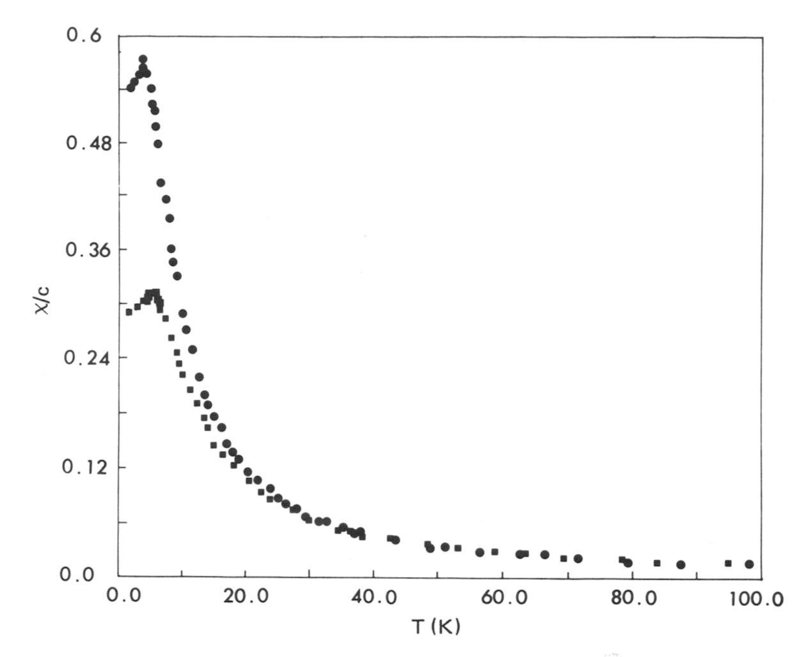

FIGURE 43. Reduced magnetic susceptibility (X_M/C) vs temperature for the ferromagnetically exchange-coupled chains: [Cu(DMSO)Cl_2], (■); [Cu(TMSO)Cl_2], (●). (After Reference 291.)

tic $S = +\frac{1}{2}$ chains (modified with a mean-field term for interchain interactions) to the data yielded the parameters summarized in Table 9, and the modified expression is successful in fitting the magnetic data shown in Figure 43 at higher temperatures. These are the largest ferromagnetic intrachain exchange coupling constants that have been reported for copper chain compounds, and it has been suggested that the compounds are good candidates for studying the spin-correlation function of one-dimensional ferromagnetic systems. Studies of the ordered state would also be of interest.

9.4. Alternately Spaced Chain Compounds

9.4.1. Cu(4-methylpyridine)$_2$Cl$_2$

Structural data are summarized in Table 9 for several di-μ-chloro-bridged copper(II) chain compounds. All of these compounds have uniformly spaced chains of CuL_2Cl_2 units with short in-plane bond Cu–Cl distances of approximately 2.3Å and longer out-of-plane Cu–Cl bond distances of 3.0–3.2 Å, and their magnetic properties in the low temperature region can be described accurately with Heisenberg exchange theory for uniformly spaced chains of $S = \frac{1}{2}$ ions. The compound Cu(4-methylpyridine)$_2$Cl$_2$ also has a uniformly spaced chain structure at room temperature,[292] but the magnetic data in the temperature range 4–70 K can not be described with the Bonner–Fisher results for Heisenberg linear chains.[281] In the initial work on this compound it was found that the data were consistent with the graphical results of Duffy and Barr[276] for an alternatingly spaced linear chain using the magnetic parameters $J = -8.5$ cm^{-1}, $g = 2.13$, and $\alpha = 0.6$. A subsequent, more accurate analysis of the data, using the results summarized in Section 9.1.2, yielded $J = -9.6(2)$ cm^{-1}, $g = 2.17(2)$, and $\alpha = 0.67(2)$.[281]

The dichotomy in the structural and magnetic results is removed by the observation of an anomaly in the dielectric constant of a pressed pellet sample at about 50 K. The anomaly signals a structural phase transition from the uniform chain structure presumably to an alternatingly spaced chain structure.[281]

9.4.2. Copper(II) Nitrate·2.5 Hydrate

An x-ray crystal-structure determination[293] on $Cu(NO_3)_2 \cdot 2.5H_2O$ has revealed that the copper(II) ions are coordinated by four oxygen atoms from two water molecules (Cu–O = 1.959 Å) and two nitrate oxygen atoms (Cu–O = 1.99 Å) in a near-square planar array. There are three additional oxygen atoms coordinated to copper with long copper–oxygen distances of 2.4–2.7 Å. Three superexchange pathways are set up by the packing, ligand bridging, and hydrogen bonding in this substance.[293] Two of these superexchange pathways lead to pairwise exchange interactions, and these combine to form an alternatingly spaced chain, with the shorter copper–copper distance being 5.5 Å, and the longer one being 6.2 Å. The third superexchange pathway, which also involves the copper ions in the alternating chain, leads to the formation of a uniformly spaced chain of singly bridged copper ions with the copper–oxygen bond distances in the superexchange pathway being 2.4 and 2.7 Å. Exchange interactions along this latter pathway must be less important than those in the alternating chain because of the dominant pairwise behavior of the magnetic susceptibility.[277–280, 294]

The dominant intrapair interaction gives rise to a ground-state singlet with ajlow-lying triplet state. In high external magnetic fields the $|1, -1\rangle$ level crosses the ground state $|0, 0\rangle$ at the "level-crossing magnetic field" $H_{lc} = 2|J|/g\mu_B$. This phenomenon permits additional exchange coupling as a result of the interpair interactions. When the external field is applied parallel to the b axis, a transition to an antiferromagnetically ordered state occurs at 175 mK at 36.0 kOe. Bonner *et al.*[277] proposed that the interaction between the pairs would exhibit one-dimensional magnetic characteristics of a ladderlike or alternating chain nature. Diederix *et al.*[295] concluded that it was not possible to distinguish between the ladderlike model and the alternating chain model based on measurements made in the paramagnetic state, but when data for the long-range ordered state were included,[279] it became clear that the alternating chain model provided an explanation of the magnetic properties of $Cu(NO_3)_2 \cdot 2.5H_2O$.

9.4.3. Bis(thiosemicarbazonato)copper(II) Chains

The structure[296] of the cancinostatic agent (3-ethoxy-2-oxobutraldehyde)bis(thiosemicarbazonato)copper(II), CuKTS, consists of stacks of planar units with alternating interplanar distances. The out-of-plane copper–sulfur distances in the alternating chain are 3.101(2) and 3.312(2) Å, with the Cu–S_b–Cu bridging angle in the more tightly bound fragment being 89.9°. The corresponding angle in the unit with the longer bridging bond distance is 86.5°.

Magnetic data have not been reported for CuKTS, but magnetic susceptibility data do exist for two analogous compounds, those being *catena*-hexanedionebis(thiosemicarbazonato)copper(II), CuHTS, and *catena*-octanedionebis(thiosemicarbazonato)copper(II), CuOTS.[297] The magnetic susceptibility data exhibit chainlike behavior, but can not be described by uniform chain Heisenberg results. In view of the structure of CuKTS, it is likely that CuHTS and CuOTS have alternatingly spaced chain structures. The magnetic susceptibility data for CuHTS and CuOTS may be understood in terms of the Heisenberg theory for exchange in alternatingly spaced chains.[281]

9.5. A Chain Compound with a Ladderlike Structure, Catena-trichlorohydraziniumcopper(II)

The structure and magnetic properties of *Catena*-trichlorohydraziniumcopper(II), $[(HzH)CuCl_3]_\infty$, have been reported by Brown *et al.*[298] The ladderlike structure of the compound is shown in Figure 44. The coordination about the copper ion is the familiar 4 + 2 form, with the equatorial plane being formed by three chloride ions and the monodentate hydrazinium ion. The apical positions in the coordination sphere are occupied by chloride ions from adjacent units. The Cu–Cl in-plane bridging distance is 2.297(1) Å and the out-of-plane bridging distance is 2.856 Å.

The magnetic susceptibility data reflect an antiferromagnetic interaction with a maximum in the magnetic susceptibility at about 12 K. The data above 25 K may be fit by the Curie–Weiss law to yield a g value of 2.196, while the entire data set may be fit by calculations on an eight-membered chain to yield the same g value, J_{12} (the 90° interaction) of -7.5 cm^{-1}, and J_{13} (the 180° interaction) of $-7.0\ cm^{-1}$.

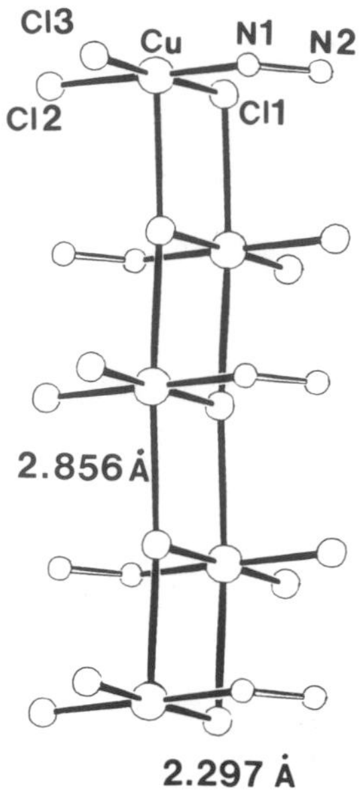

FIGURE 44. Structure of the ladderlike compound (Hydrazinium)$CuCl_3$.

A similar calculation was carried out for an eight-membered ring to yield $J_{12} = -5.4\ \text{cm}^{-1}$, $J_{13} = -5.3\ \text{cm}^{-1}$, and $g = 2.14$. The calculations firmly support the conclusion that the superexchange interactions are antiferromagnetic by both exchange pathways, but additional studies must be carried out before the magnitudes of the exchange coupling constants can be known with certainty.

9.6. A Copper Chain Compound Bridged by an Extended Pi System Ligand, Copper Pyrazine Nitrate

As shown in Figure 45, the crystal structure of copper pyrazine nitrate consists of linear chains of copper(II) ions bridged by pyrazine ligands.[299] The chains run parallel to the *a* axis, and the coordination environment of the copper ions is completed by two short bonds and two long bonds to nitrate oxygens which lie in a plane perpendicular to the *a* axis. Magnetic susceptibility measurements of a powdered sample of the compound revealed that the copper ions were antiferromagnetically exchange coupled,[300] and subsequent measurements revealed Heisenberg exchange with $J = -3.7\ \text{cm}^{-1}$.[301] These measurements were made on a bundle of the small needlelike crystals, since a single crystal large enough for measurement with a vibrating sample magnetometer could not be obtained. The small crystals are readily obtained by slow evaporation of an aqueous solution of copper nitrate and pyrazine mixed in a 1 : 1 molar quantity.

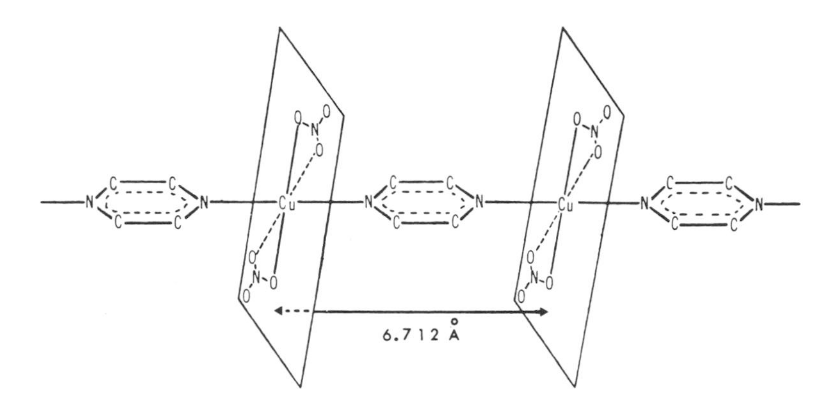

FIGURE 45. Structure of the linear chain compound [Cu(pyrazine)(NO_3)].

The low-temperature limit of the measurements was extended to 0.18 K by Koyama *et al.*[302] and no evidence for long-range order was seen in either the magnetic susceptibility or specific heat. The maximum in the magnetic susceptibility versus temperature plot, the magnitude of the susceptibilities, and the average g value were in excellent agreement with the earlier work.[301] Boyd and Mitra[303] subsequently obtained the principal crystal susceptibilities using the Krishan critical torque method in the range 4.8–100 K and confirmed the earlier conclusions concerning the nature of the exchange interaction.

From a study of the magnetic and spectral properties of several similar chain compounds with substituted pyrazines, Richardson and Hatfield[304] concluded that the superexchange mechanism involved the π orbitals of the bridging heterocyclic ligand. This conclusion was later confirmed by a broad line ^1H nmr study.[305]

9.7. Chain Compounds without Ligand Bridges

There are a number of chain compounds formed by stacking planar molecules or ions without obvious coordination by a bridging ligand. These chains are of two basic types. In one type, all the molecules in a given stack are identical, while the identity of the molecules alternate in the other type. Generally, exchange interactions in these chains without ligand bridges do not lead to large exchange coupling constants. The small exchange coupling constants may be understood in terms of single-ion electronic structure, since it is known that the unpaired electrons reside in σ^* orbitals, and these orbitals lie in the plane of the molecule. Effective overlap of the orbitals containing the unpaired electrons is not possible in this geometrical arrangement.

9.7.1. Tetrathiafulvalinium Bis(1,2-perfluoromethyl-1,2-dithiolato)cuprate(I)

The compound tetrathiafulvalinium bis-*cis*-(1,2-perfluoromethylethylene-1,2-dithiolato)cuprate(I), (TTF)$[Cu(S_2C_2(CF_3)_2)_2]$, may be obtained as needlelike crystals by slow cooling of an acetonitrile solution. The structure of the compound consists of stacks formed by alternating planar paramagnetic TTF^+ radical cations and diamagnetic $[Cu(S_2C_2(CF_3)_2)_2]^-$ anions.[306]

The paramagnetic ions are antiferromagnetically coupled and, as shown in Figure 46, the magnetic susceptibility data may be fit by Heisenberg exchange theory in the temperature range 12–300 K, with a temperature-dependent exchange coupling constant which varies from about 50 to 75 cm^{-1}. At 12 K, there is a spin Peierls transition with spin–lattice dimerization resulting in a singlet ground state;[307] however, the pairing does not occur along the molecular stacking axis. The dimerization

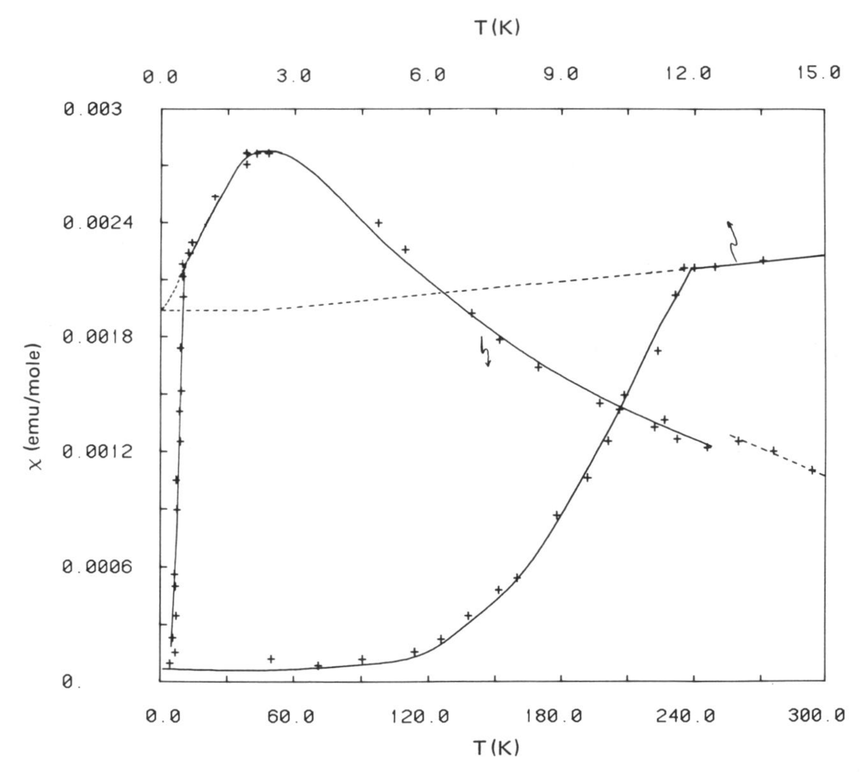

FIGURE 46. Magnetic susceptibility data for $(TTF)^{\dot{+}}[Cu(S_2C_2(CF_3)_2)_2]^-$ which show the antiferromagnetic intrachain coupling and the spin Peierls transition at 12 K. (After Reference 307.)

occurs along a diagonal axis and results from a first-order structural phase transition at 240 K.[308] The magnetic susceptibility of the alternating chain which exists below 12 K may be fit by the results of Bulaevskii theory.[309] Interchain magnetic coupling may be dominant in other such linear chain compounds and this leads to three-dimensional ordering rather than to spin Peierls transitions.

10. Summary

It is apparent from the contents of this chapter that there has been intense activity in the synthesis and characterization of linear chain systems involving paramagnetic first-row transition-metal ions with theoretical expositions and experimental programs being synergistic. However, only the foundation has been laid for research in this area, and new, important advances in the understanding of the chemical and structural effects that govern exchange coupling interactions are to be expected as results from systematic studies of related compounds become available. For the most part, these effects are more clearly revealed by results of studies in the paramagnetic region, and it is for this reason that we have concentrated our attention on the paramagnetic region. Additional questions dominate research devoted to the crossover and critical regions and to the ordered state, and consideration of these questions, along with considerations of dynamic behavior, must be deferred to other articles.

A number of directions exist for promising research activity. The absence of many examples of chain compounds of titanium, vanadium, and chromium is striking. For titanium, vanadium, and chromium (II) this absence of chain compounds may be partially explained by the chemical reactivity of compounds and their instability in the laboratory atmosphere, with the result that linear chain behavior is probably frequently masked by the properties of impurity products. However, studies on linear chain chromium(III) systems are especially inviting, since the single-ion electronic structure is relatively simple, and the theory of the exchange problem is not especially difficult.

The identification of structure–function relationships has largely involved copper(II) systems. This unfortunate situation will surely be corrected as attention is paid to systematic research of other transition-metal chain compounds. Initially, high-spin manganese(II) and iron(III) systems may be good candidates, but there are some difficulties with the detailed analyses of the magnetic susceptibilities of exchange-coupled linear chain compounds of these ions. When anisotropy is observed in these *A*-state ions, although small in most cases, it is rarely taken into account. Often, when single-crystal measurements have been made and anisotropy is observed, the data along one crystal direction has been chosen and analyzed in terms of some kind of average. Results from such studies will be adequate for structure relationships, but a full understanding of the problem will require attention to all the details.

Linear chain iron(II), cobalt(II), and nickel(II) compounds provide a rich opportunity for the investigation of spin exchange in low-dimensional systems. A wide range of phenomena arise, and ligand field effects, zero-field splitting, and Dzyaloshinsky–Moriya exchange terms become important with all permutations of the magnitudes of these effects being seen. In fact, recent advances in the theory of exchange as modified or dominated by ligand field effects should allow a better physical picture to be drawn. The presence of a ligand field may produce anisotropy and force *XY* or Ising behavior. Once these phenomena are understood, then structure–magnetism relationships should emerge from systematic studies. Initial results will stimulate more structural studies by x-ray diffraction and eventually permit the identification of chain structures from magnetic experiments alone. In the limit, it may be expected that the behavior of complex ordered states may be more fully understood as a result of modeling experiments involving chains. In any event, substantial advances in the designed synthesis, experimental characterization, and theoretical descriptions of the properties of linear chain magnetic systems are to be expected.

Notation

1. Key to the Tables

M–M	Intrachain metal distance (Å)
M–B–M	Intrachain metal–bridge atom–metal angle (degrees)
M–M′	Interchain metal–metal distance (Å)
J	Intrachain exchange coupling constant (cm^{-1})
D	Single-ion or zero-field-splitting parameter (cm^{-1})
$\%S$	Percent entropy gained from intrachain interactions and/or single-ion effects
T_c	Temperature below which long-range magnetic order is observed
J'/J	Ratio of interchain to intrachain exchange
Temperature	Experimental temperature range over which properties were reported
Techniques	Experimental technique employed to characterize the system
Model	Theoretical model used to analyze the experimental data

2. Abbreviations for Ligands

bpy	2,2′-bipyridine
en	ethylenediamine
hipp	hippurate

im	imidazole
ox	oxalate
Pc	phthalocyanine
Ph	phenyl
py	pyridine
pyz	pyrazine
pz	pyrazole
salen	*N*,*N*′-ethylenebis(salicylaldiminate)
salpn	*N*,*N*′-propylenebis(salicylaldiminate)
sq	1,3-squarate
tam	thioacetamide
terpy	2,2′,2″-terpyridine
tim	thioimidazolidinone
TPP	tetraphenylporphyrin
tu	thiourea

3. *Abbreviations for Experimental Techniques*

C_m	Magnetic portion of the specific heat
GR	Gamma resonance fluorescence spectroscopy (Mössbauer)
INS	Inelastic neutron scattering
M_p	Powder magnetization
M_i	Single-crystal magnetization
MR	Magnetic resonance
N	Neutron diffraction
NS	Neutron scattering
SCR	Spin-cluster resonance
χ_p	Powder magnetic susceptibility
χ_i	Single-crystal magnetic susceptibility
x ray	X-ray diffraction

4. *Abbreviations for Theoretical Models*

B–HH	Hubbard Hamiltonian of Berlin
CEF	Correlated effective-field theory of Lines
C–W	Curie–Weiss Law
D–M	Antisymmetric exchange of Dzyaloshinsky and Moriya
F–L	General spin Heisenberg chain model of Figgis and Lewis
H–ALT	$S = 1/2$ antiferromagnetic alternating Heisenberg chain of Duffy and Barr
H–B	$S \geq 1$ antiferromagnetic Heisenberg chain with single-ion anisotropy of Blote
H–BF	$S = 1/2$ antiferromagnetic Heisenberg chain model of Bonner and Fisher
H–BRG	$S = 1/2$ ferromagnetic Heisenberg chain from Padé approximation method of Baker, Rushbrooke, and Gilbert
H–D	$S = 1/2$ Heisenberg dimer
H–dN	$S \geq 1$ antiferromagnetic Heisenberg chain with single-ion anisotropy of de Neef
H–I	$S \geq 3/2$ antiferromagnetic Heisenberg chain interpolation scheme of Weng

H–IS	$S = 1$ antiferromagnetic Heisenberg chain scaled from $S = 1/2$ Heisenberg and $S = 1/2$, $S = 1$ Ising results of Achiwa
H–MF	Molecular field corrections applied to $S = 1$ ions with large single-ion anisotropy by Watanabe
H–RW	High-temperature series expansion model for $1/2 \leq S \leq 3$ Heisenberg chains of Rushbrooke and Wood
H–S	Scaled infinite-spin Heisenberg chain $1 \leq S \leq 3$
H–W	$S = 1$ antiferromagnetic Heisenberg chain model of Weng
H–WD	High-temperature series-expansion model for general spin anisotropic Heisenberg chains of Wood and Dalton
I–F	$S = 1/2$ Ising chain of Fisher
I–GS	$S \geq 1$ Ising chain
I–M	Modified Ising chain with fictitious $S' = 1/2$ of Oguchi
I–R	$S = 1/2$ quadratic Ising layer
MO–K	Molecular orbital calculation of Kahn
J–MC	Monte Carlo solution for antiferromagnetic $S = 1/2$ Heisenberg chain of Jotham
SI + I	Single-ion anisotropy followed by fictitious $S' = 1/2$ Ising behavior
SI	Single-ion anisotropy
SM	Model of Slater based on Hartree–Fock formalism
XY	$S = 1/2$ XY-chain model of Katsura

References

1. De Jongh, L. J., and Miedema, A. R., *Adv. Phys.* **23**, 1–260 (1974).
2. Ackerman, J. F., Cole, G. M., and Holt, S. L., *Inorg. Chim. Acta* **8**, 323–343 (1974).
3. Steiner, M., Villian, J., and Windsor, C. G., *Adv. Phys.* **24**, 87–209 (1976).
4. Stryjewski, E., and Giordano, N., *Adv. Phys.* **26**, 487–650 (1977).
5. Johnson, J. D., *J. Appl. Phys.*, **52**, 1991 (1981).
6. Bethe, H. A., *Z. Phys.* **71**, 205 (1931).
7. (a) Bonner, J. C., and Fisher, M. E., *Phys. Rev. A* **135**, 640 (1964). (b) Bonner, J. C., Ph.D. Dissertation, University of London, 1968.
8. Hall, J. W., Ph.D. Dissertation, University of North Carolina, Chapel Hill, 1977.
9. Jotham, R. W., *J. Chem. Soc. Dalton Trans.*, 266 (1977).
10. Baker, G. H. Jr., Rushbrooke, G. S., and Gilbert, H. E., *Phys. Rev. A* **135**, 1272 (1964).
11. Rushbrooke, G. S., and Wood, P. J., *Mol. Phys.* **1**, 257 (1958).
12. Fisher, M. E., *Am. J. Phys.* **32**, 343 (1964).
13. Smith, T., and Friedberg, S. A., *Phys. Rev.* **176**, 660 (1968).
14. Wagner, G. R., and Friedberg, S. A., *Phys. Lett.* **9**, 11 (1964).
15. Weng, C. H., Dissertation, Carnegie–Mellon University, 1968.
16. Wood, D. W., and Dalton, N. H., *J. Phys. C* **5**, 1675 (1972).
17. Katsura, S., *Phys. Rev.* **127**, 1508–1518 (1962).
18. Katsura, S., *Phys. Rev.* **129**, 2835 (1963).
19. Fisher, M. E., *J. Math Phys.* **4**, 124–135 (1963), and references therein.
20. Suziki, M., Tsuijiyama, B., and Katsura, S., *J. Math Phys.* **8**, 124 (1967).
21. Obokata, T., and Oguchi, T., *J. Phys. Soc. Jap.* **25**, 322 (1968).
22. Smith, J., Gerstine, B. C., Liu, S. H., and Stucky, G., *J. Chem. Phys.* **53**, 418 (1970).
23. Wertz, J. E., and Boulton, J. R., *Electron Spin Resonance*, McGraw-Hill, New York (1969), Chapter 11.

24. Abragam, A., and Bleaney, B., *Electron Paramagnetic Resonance of Transition Metal Ions*, Oxford University Press, London (1970), Chapters 3 and 7.
25. Casey, A. T., in *Theory and Applications of Molecular Paramagnetism*, L. N. Mulay, and F. A. Boudreaux, Eds., John Wiley, New York (1976), Chapter 2.
26. Drent, E., Emeis, C. A., and Kortbeek, A. G. T. G., *Solid State Commun.* **16**, 1351 (1975).
27. Natta, P., Gorradini, P., and Allegra, G., *J. Polym. Sci.* **51**, 399 (1961).
28. Drent, E., Emeis, C. A., and Kortbeek, A. G. T. G., *Chem. Phys.* **10**, 313 (1975).
29. Fletcher, G. C., and Wohlforth, E. P., *Philos. Mag.* **42**, 106 (1951).
30. Berlin, A. A., Vinogradow, G. A., and Ovchinnikow, A. A., *Int. J. Quant. Chem.* **6**, 263 (1972).
31. Orbach, R. L., *Phys. Rev.* **112**, 309 (1958).
32. Walker, L. R., *Phys. Rev.* **116**, 1089 (1959).
33. Griffiths, R. B., *Phys. Rev.* **133**, A768 (1964).
34. Emori, S., Inoue, M., and Kubo, M., *Bull. Chem. Soc. Jap.* **45**, 2259 (1972).
35. Sanchez, J. P., Friedt, J. M., and Djermouni, B., *J. Phys, Chem. Solids* **40**, 585 (1979).
36. Niel, M., Cros, C., LeFlem, G., Pouchard, M., and Hagenmuller, P., *Physica* (*Utr.*) **86–88B**, 702 (1977).
37. Li, T., Stucky, G. D., and McPherson, G. L. *Acta Crystallogr. B* **29**, 1330 (1973).
38. Niel, M., Cros, C., Vlasse, M., Pouchard, M., and Hagenmuller, P., *Mater. Res. Bull.* **11**, 827 (1976).
39. Figgis, B. N., *Introduction to Ligand Fields*, Interscience, New York (1961).
40. Charlot, M.-F., Girerd, J.-J., and Kahn, O., *Phys. Status Solidi B* **86**, 497 (1978).
41. Grey, I. E., and Smith, P. W., *J. Chem. Soc. Chem. Commun.* 1525 (1968).
42. Earnshaw, A., Figgis, B. N., and Lewis, J., *J. Chem. Soc. A*, 1656 (1966).
43. Earnshaw, A., and Lewis, J., *J. Chem. Soc.*, 396 (1961).
44. Kallinikov, V. T., Zelentsov, V. V., Kuzmicheva, O. N., and Aminov, T. G., *Russ. J. Inorg. Chem.* **15(3)**, 341 (1970).
45. Casey, A. T., Morris, B. S., Sinn, E., and Thackeray, J. R., *Aust. J. Chem.* **25**, 1195 (1972).
46. Casey, A. T., and Thackery, J. R., *Aust. J. Chem.* **22**, 2549 (1969).
47. Pfeiffer, P., Hesse, T., Pfitzner, H., Scholl, W., and Theileit, H., *J. Prakt. Chem.* **149**, 217 (1937).
48. Mathew, M., Carty, A. J., and Palenik, G. J., *J. Am. Chem. Soc.* **92**, 3198 (1970).
49. Ginsberg, A. P., Koubek, E., and Williams, H. J., *Inorg. Chem.* **5**, 1656 (1966).
50. Drake, R. F., Crawford, V. H., Hatfield, W. E., Simpson, G. D., and Carlisle, G. O., *J. Inorg. Nucl. Chem.* **37**, 291 (1975), and references therein.
51. Goodenough, J. B., *Magnetism and the Chemical Bond*, Interscience, New York (1963).
52. Massenet, O., Since, J. J., Mercier, J., Avignon, M., Buder, R., and Nguyen, V. D., *J. Phys. Chem. Solids* **40**, 573 (1979).
53. Massenet, O., Buder, R., Since, J. J., Schlenker, C., Mercier, J., Kelber, J., and Stucky, D. G., *Mater. Res. Bull.* **13**, 187 (1978), and references therein.
54. Takano, M., Kosugi, H., Nakanishi, N., Shimada, M., Wada T., and Koizumi, M., *J. Phys. Soc. Jap.* **43**, 1101 (1977).
55. Mabbs, F. E., and Machin, D. J., *Magnetism and Transition Metal Complexes*, Chapman and Hall, London (1973).
56. (a) Pesch, W., and Selke, W., *Z. Phys. B* **23**, 271 (1976). (b) Pesch, W., and Selke, W., *Z. Phys. B* **24**, 203 (1976).
57. Stahlbush, R. E., and Scott, J. C., *Sol. State Commun.* **33**, 707–711 (1980).
58. Domb, C., *Adv. Phys.* **9**, 164 (1960).
59. Li, T. J., Stucky, G. D., and McPherson, G. L., *Acta Crystallogr. B* **29**, 1330 (1973).
60. Klatyk, K., and Seifert, H. J., *Z. Anorg. Allg. Chem.* **334**, 113–224 (1964).
61. Hardt, H. D., and Streit, G., *Z. Anorg. Allg. Chem.* **373**, 97–208 (1970).
62. Larkworthy, L. F., and Trigg, J. K., *J. Chem. Soc. Chem. Commun.*, 1221 (1970).

63. McPherson, G. L., Kistenmacher, T. J., Flokers, J. B., and Stucky, G. D., *J. Chem. Phys.* **57**, 3771–3780 (1972).
64. Alcock, N. W., Putnik, C. F., and Holt, S. L., *Inorg. Chem.* **15**, 3175 (1976).
65. Iberson, E., Gut, R., and Gruen. O. M., *J. Chem. Phys.* **66**, 65 (1962).
66. Day, P., Gregson, A. K., Leech, D. W., Hutchings, M. T., and Rainford, B. D., *J. Magn. Magn. Mater.* **14**, 166 (1979).
67. Zandbergen, H. W., Dissertation, University of Leiden, 1981.
68. Leech, D. H., and Machin, D. W., *J. Chem. Soc. Chem. Commun.*, 866 (1974).
69. Leech, D. H., and Machin, D. W., *J. Chem. Soc. Dalton Trans.*, 1609 (1975).
70. Knoke, G., and Babel, D., *Z. Naturforsch.* **30b**, 454 (1975).
71. Babel, D., and Knoke, G., *Z. Anorg. Allg. Chem.* **442**, 533 (1978).
72. Dewan, J. C., and Edwards, A. J., *J. Chem. Soc. Chem. Commun.* 533 (1977).
73. Gillman, H. D., *Inorg. Chem.* **13**, 1921 (1974).
74. Scott, J. C., Garito, A. F., Heeger, A. J., Nannelli, P., and Gillman, H. D., *Phys. Rev. B* **12**, 356 (1975).
75. Earnshaw, A., Figgis, B. N., and Lewis, J., *J. Chem. Soc. A*, 1656–1663 (1966).
76. Figgis, B. N., and Lewis, J., *Progr. Inorg. Chem.* **6**, 91 (1964).
77. Larkworthy, L. F., Trigg, J. K., and Yauari, A., *J. Chem. Soc. Dalton Trans.*, 1879 (1975).
78. McPherson, G. L., Sindel, L. J., Quarls, H. F., Friedberg, C. B., and Dounit, C. J., *Inorg. Chem.* **14**, 1831 (1975).
79. Dance, J., Soubeyroux, J., Fournes, L., and Tressaud, A., *C. R. Acad. Sci. C* **288**, 37 (1979).
80. Morosin, B., and Graeber, E. J., *Acta Crystallogr. B* **23**, 766 (1967).
81. Dingle, R., Lines, M. E., and Holt, S. L., *Phys. Rev.* **187**, 643 (1969).
82. Hutchings, M. T., Shirane, G., Birgeneau, R. J., and Holt, S. L., *Phys. Rev. B* **5**, 1999 (1972).
83. de Jongh, W. M. J., Swuste, C. H. W., Kopinga, K., and Takeda, K., *Phys. Rev. B* **12**, 5858 (1975).
84. Walker, L. R., Dietz, R. E., Andres, K., and Dorack, S., *Solid State Commun.* **11**, 593 (1972).
85. Dupas, C., and Renard, J. P., *Phys. Lett. A* **34**, 119 (1973).
86. Birgeneau, R. J., Shirane, G., and Kitchings, T. A., *Proceedings of the 14th Low Temperature Conference*, Boulder, Colorado (1972).
87. Robertson, J. M., *J. Chem. Soc* 615 (1935).
88. Robertson, J. M., *J. Chem. Soc.*, 219 (1937).
89. Linstead, R. P., and Robertson, J. M., *J. Chem. Soc.*, 1736 (1936).
90. Brown, C. J., *J. Chem. Soc. A*, 2488 (1968).
91. Miyoshi, H., *J. Phys. Soc. Jap.* **37**, 50 (1974).
92. Barraclough, C. G., Martin, R. L., Mitra, S., and Sherwood, R. C., *J. Chem. Phys.* **53**, 1638 (1970).
93. Barraclough, C. G., Gregson, A. K., and Mitra, S., *J. Chem. Phys.* **60**, 962 (1974).
94. Miyoshi, H., *Bull. Chem. Soc. Jap.* **47**, 561 (1974).
95. Sears, D. R., and Huard, J. L., *J. Chem. Phys.* **50**, 1066 (1969).
96. Kida, J., *J. Phys. Soc. Jap.* **30**, 290 (1971).
97. Jensen, S. J., Anderson, P., and Rasmussen, S. E., *Acta Chem. Scand.* **16**, 1890 (1962).
98. Kobayashi, H., Tsujikawa, I., and Friedberg, S. A., *J. Low Temp. Phys.* **10**, 621 (1973).
99. Iwashita, T., and Uryu, N., *J. Phys. Soc. Jap.* **39**, 1226 (1975).
100. Nayata, K., Tazuke, Y., and Tsushima, K., *J. Phys. Soc. Jap.* **32**, 1486 (1972).
101. Skalyo, J., Shirane, G., Friedberg, S. A., and Kobayashi, N., *Phys. Rev. B* **2**, 4632 (1970).
102. Iwashita, T., and Uryu, N., *J. Chem. Phys.* **65**, 2794 (1976).
103. Nishihara, H., de Jonge, W. J. M., and de Neef, T., *Phys. Rev. B* **12**, 5325 (1975).
104. McElearney, J. N., Merchant, S., and Carlin, R. L., *Inorg. Chem.* **13**, 906 (1973).

105. Goto, T., Hirai, A., and Haseda, T., *Phys. Lett. A* **33**, 185 (1970).
106. Caputo, R. E., Willett, R. D., and Muir, J. A., *Acta Crystallogr. B* **32**, 2639 (1976).
107. Merchant, S., McElearney, J. N., Shankle, G. E., and Carlin, R. L., *Physica (The Hague)* **78**, 308 (1974).
108. Moriya, T., *Phys. Rev.* **117**, 635 (1960).
109. Richards, P. M., Quinn, R. K., and Morosin, B., *J. Chem. Phys.* **59**, 4474 (1973).
110. Cox, E. G., Sharratt, E., Wardlaw, W., and Webster, K. C., *J. Chem. Soc. Lond.*, 129 (1936).
111. Anderson, P. W., *Phys. Rev.* **115**, 2 (1959).
112. Gortner, S., van Ingen Schenau, A. D., and Verschour, G. C., *Acta Crystallogr. B* **30**, 1867 (1974).
113. Witteveen, H. T., Nieuwenhuijse, B., and Reedijk J., *J. Inorg. Nucl. Chem.* **36**, 1535 (1974).
114. Klaaijen, F. W., Blote, H. W. J., and Dokoupil, Z., *Physica (Amsterdam)* **81B**, 1 (1976).
115. Zanetti, R., and Serra R., *Gazz. Chim. Ital.* **90**, 328 (1960).
116. Caputo, R. E., and Willett, R. D., *Phys. Rev. B* **13**, 3956 (1976).
117. Eibschutz, E., Sherwood, R. C., Hsu, F. S. L., and Cox, D. E. *AIP Proc. Magn. Magn. Mater.*, 684 (1972).
118. Cox, D. E., and Merchant, F. C., *J. Cryst. Growth* **13/14**, 282 (1972).
119. Walker, R. L., Dietz, R. E., Anders, K., and Darack, S. *Solid State Commun.* **11**, 593 (1972).
120. Hill, D. M., Larkworthy, L. F., and O'Donoghue, M. W., *J. Chem. Soc. Dalton Trans.* 1726 (1975).
121. Ciampolini, M., Menguzzi, C., and Oridli, P., *J. Chem. Soc. Dalton Trans.*, 2051 (1975).
122. van den Bergen, A., Murray, K. S., O'Connor, M. J., and West, B. O., *Aust. J. Chem.* **22**, 39 (1969).
123. Lewis, J., Mabbs, F. E., and Weigold, H., *J. Chem. Soc. A* 1699 (1968).
124. Earnshaw, A., King, E. A., and Lackworthy, L. F., *J. Chem. Soc. A*, 1699 (1968).
125. Davies, J. E., Gatehouse, B. M., and Murray, K. S., *J. Chem. Soc. Dalton Trans.*, 2523 (1973).
126. Witteveen, W. T., and Reedijk, J., *J. Solid State Chem.* **10**, 151 (1974).
127. Prout, C. K., and Powell, H. M., *J. Chem. Soc.* 4177 (1961).
128. Hand, D. W., and Prout, C. K., *J. Chem. Soc. A*, 168 (1966).
129. Nieuwport, A., and Reedijk, J., *Inorg. Chim. Acta* **7**, 323 (1973).
130. Pezerat, H., Dubernat, J., and Lasier, J. P., *C.R. Acad. Sci. Paris C* **266**, 1357 (1968).
131. Sharov, V. A., *Zh. Neorg. Khim.* **22**, 2762 (1977).
132. Oguchi, T., *Phys. Rev.* **133**, A1098 (1964).
133. Lines, M. E., *Phys. Rev. B* **9**, 3927 (1974).
134. Lines, M. E., and Eibschutz, M., *Phys. Rev. B* **11**, 4583–4594 (1975).
135. Eibschutz, M., Lines, M. E., and Sherwood, R. C., *Phys. Rev. B* **11**, 4595–4605 (1975).
136. van Vlimmeren, Q. A. G., and de Jonge, W. J. M., *Phys. Rev. B* **19**, 1503–1514 (1979).
137. Basten, A. J., van Vlimmeren, Q. A. G., and de Jonge, W. J. M., *Phys. Rev. B* **10**, 2179 (1970).
138. Euler, W. B., Long, C., Moulton, W. G., and Garrett, B. B., *J. Magn. Res.* **32**, 23–32 (1978).
139. Krausz, E., Viney, S., and Day, P., *J. Phys. C: Solid State Phys.* **10**, 2685–2699 (1977).
140. Montano, P. A., Shechter, H., Cohen, E., and Makovsky, J., *Phys. Rev. B* **9**, 1066–1070 (1974).
141. Montano, P. A., Cohen, E., Shechter, H., and Makovsky, J., *Phys. Rev. B* **7**, 1180–1187 (1973).
142. Witteveen, H. T., and van Veen, J. A. R., *J. Chem. Phys.* **58**, 186–191 (1973).
143. Gurewitz, E., Makovsky, J., and Shaked, H., *Phys. Rev. B* **9**, 1071 (1974).

144. Davidson, G. R., Eibschutz, M., Cox, D. E., and Minkiewicz, V. J., *AIP Conf. Proc.* 436 (1971).
145. Cox, D. E., and Merkert, F. C., *J. Cryst. Growth* **13/14**, 282 (1972).
146. Eibschutz, M., Davidson, G. R., and Cox, D. E., *AIP Conf. Proc.* **18**, 386 (1973).
147. Kostiner, E., and Rea, J. R., *Inorg. Chem.* **13**, 2876–2880 (1974).
148. Anderson, J. B., Rea, J. R., and Kostiner, E., *Acta Crystallogr. B* **32**, 2427–2431 (1976).
149. Vlasse, M., Matejka, G., Tressaud, A., and Wanklyn, B. M., *Acta Crystallogr. B* **33**, 3377–3380 (1977).
150. Sabatier, R., Soubeyroux, J. L., Dance, J. M., and Tressaud, A., *Solid State Commun.* **29**, 383–387 (1979).
151. Gupta, F. P., Dickson, D. P. E., Johnson, C. E., and Wanklyn, B. M., *J. Phys. C: Solid State Phys.* **10**, L459 (1977).
152. Bronger, W., *Z. Anorg. Allg. Chem.* **359**, 225–233 (1968).
153. Nishi, M., and Ito, Y., *Solid State Commun.* **30**, 571–574 (1979).
154. Hong, H., and Grey, I. E., unpublished work, cited by Ref. 156.
155. Anderson, J. S., *J. Chem. Soc. Dalton Trans.*, 1107 (1973).
156. Swinnea, J. S., and Steinfink, H., *J. Chem. Educ.* **57**, 580–582 (1980).
157. Ritsko, J. J., Nielsen, P., and Miller, J. S., *J. Chem. Phys.* **67**, 687–690 (1977).
158. Reis, Jr., A. H., Preston, L. D., Williams, J. M., Peterson, S. W., Candela, G. A., Swartzendruber, L. J., and Miller, J. S., *J. Am. Chem. Soc.* **101**, 2756–2758 (1979).
159. Miller, J. S., Reis, Jr., A. H., Gebert, E., Ritsko, J. J., Salaneck, W. R., Kovnat, L., Cape, T. W., and Van Duyne, R. P., *J. Am. Chem. Soc.* **101**, 7111–7113 (1979).
160. Candela, G. A., Swartzendruber, L. J., Miller, J. S., and Rice, M. J., *J. Am. Chem. Soc.* **101**, 2755–2756 (1979).
161. Gupta, G. P., Dickson, D. P. E., Johnson, C. E., and Wanklyn, B. M., *J. Phys. C: Solid State Phys.* **11**, 3889–3909 (1978).
162. Dance, J.-M., Soubeyroux, J.-L., Fournes, L., and Tressaud, A., *C. R. Acad. Sci. Paris C* **288**, 37–40 (1979).
163. Dance, J. M., Menil, F., Hanzel, D., Sabatier, R., Tressaud, A., Le Flem, G., and Hagenmuller, P., *Physica* (*utr.*) **86–88B**, 699–701 (1977).
164. Hanzel, D., Tressaud, A., Dance, J.-M., and Hagenmuller, P., *Solid State Commun.* **22**, 215–218 (1977).
165. Kopinga, K., van Vlimmeren, Q. A. G., Bongaarts, A. L. M., and de Jonge, W. J. M., *Physica* (*utr.*) **86–88B**, 671–672 (1977).
166. Basten, J. A., van Vlimmeren, Q. A. G., and de Jonge, W. J. M., *Phys. Rev. B* **18**, 2179 (1978).
167. Gubbens, P. C. M., van der Kraan, A. M., van Ooijen, J. A. C., and Reedijk, J., *J. Magn. Magn. Mater.* **15–18**, 635–636 (1980).
168. Narath, A., *Phys. Rev.* **139**, A1221–A1227 (1965).
169. Katsumata, K., *J. Phys. Soc. Jap.* **39**, 42–49 (1975).
170. Morosin, B., and Graber, E. J., *J. Chem. Phys.* **42**, 898 (1965).
171. Mazzi, C., and Garavelli, F., *Periodico Mineral.* (*Rome*) **26**, No. 2–3 (1957).
172. Cavid, S., *Bull. Soc. Franc. Mineral. Crist.* **82**, 50 (1959).
173. Wrobleski, J. T., and Brown, D. B., *Inorg. Chem.* **18**, 2738–2749 (1979).
174. Losee, D. B., McElearney, J. N., Shankle, G. E., Carlin, R. L., Cresswell, P. J., and Robinson, W. T., *Phys. Rev. B* **8**, 2185 (1973).
175. Watanabe, T., *J. Phys. Soc. Jap.* **17**, 1856 (1962).
176. McElearney, J. W., Losee, D. B., Merchant, S., and Carlin, R. L., *Phys. Rev. B* **7**, 3314 (1973).
177. Moriya, T., *Phys. Rev.* **120**, 91 (1960).
178. Porai-Koshits, M. A., *Kristallografiya* **1**, 291 (1956).

179. McElearney, J. N., Merchant, S., Shankle, G. E., and Carlin, R. L., *J. Chem. Phys.* **66**, 450 (1977).
180. Figgis, B. N., Gerloch, M., and Mason, R., *Proc. R. Soc. Lond. A* **279**, 210 (1964).
181. Smit, J. J., and de Jongh, L. J., *Physica B + C* **97B**, 224 (1979).
182. Melia, T. P., and Merrifield, R., *J. Chem. Soc. A*, 1258 (1971).
183. Bartolome, F. G., Algra, H. A., de Jongh, L. J., and Huiskamp, W. J., *Physica (Amsterdam)* **86–88B**, 707 (1977).
184. McGinnety, J. A., *Inorg. Chem.* **13**, 1057 (1974).
185. Soling, H., *Acta Chem. Scand.* **22**, 2793 (1968).
186. Achiwa, N., *J. Phys. Soc. Jap.* **27**, 561 (1969).
187. Melamud, M., Pinto, H., Makovsky, J., and Shaked, H., *Phys. Status Solidi B* **63**, 699 (1974).
188. Narath, A., *Phys. Rev. A* **140**, 552 (1965).
189. Narath, A., *J. Phys. Soc. Jap.* **19**, 2244 (1964).
190. Oguchi, T., *J. Phys. Soc. Jap.* **20**, 2236 (1965).
191. Shinoda, T., Chihara, H., and Seki, S., *J. Phys. Soc. Jap.* **19**, 1637 (1974).
192. Morosin, B., and Graeber, E. J., *Acta Crystallogr.* **16**, 1176 (1963).
193. Narath, A., *Phys. Rev. A* **136**, 766 (1964).
194. Kobayashi, H., and Haseda, T., *J. Phys. Soc. Jap.* **19**, 765 (1964).
195. Motokawa, M., *J. Phys. Soc. Jap.* **35**, 1315 (1973).
196. Shinoda, T., Chihara, H., and Seki, S., *J. Phys. Soc. Jap.* **19**, 1088 (1964).
197. Dunitz, J. D., *Acta Crystallogr.* **10**, 307 (1957).
198. Clarke. P. J., and Milledge, H. J., *Acta Crystallogr. B* **31**, 1543 (1975).
199. Takeda, K., Matsukawa, S., and Haseda, T., *J. Phys. Soc. Jap.* **30**, 1330 (1971).
200. Foner, S., Frankel, R. B., Reiff, W. M., Little, B. F., and Long, G. J., *Solid State Commun.* **16**, 159 (1975).
201. Pires, A., and Hone, D., *J. Phys. Soc. Jap.* **44**, 43 (1978).
202. Foner, S., Frankel, R. B., Reiff, W. M., Wong, H., and Long, G. J., *J. Chem. Phys.* **68**, 4781 (1978).
203. Swuste, C. H. W., Phaff, A. C., and de Jonge, W. J. M., *J. Chem. Phys.* **73**, 4646 (1980).
204. Carlin, R. L., O'Connor, C. J., and Bhatia, S. N., *Phys. Lett. A* **50**, 433 (1975).
205. Bhatia, S. N., O'Connor, C. J., and Carlin, R. L., *Inorg. Chem.* **15**, 2900 (1976).
206. Interrante, L. V., *Extended Interactions Between Metal Ions in Transition Metal Complexes*, ACS Symposium Series, Washington D.C. (1974).
207. Spence, R. D., and Botterman, A. C., *Phys. Rev. B* **9**, 2993 (1974).
208. Thorup, N., and Soling, H., *Acta. Chem. Scand.* **23**, 2933 (1969).
209. Herweijer, A, de Jonge, W. J. M., Botterman, A. C., Bongaarts, A. L. M., and Cowen, J. A., *Phys. Rev. B* **5**, 4618 (1972).
210. McElearney, J. N., *Solid State Commun.* **24**, 863 (1977).
211. Bongaarts, A. L. M., and van Laar, B., *Phys. Rev. B* **6**, 2669 (1972).
212. Harkema, S., and van der Graaf, W., *Inorg. Nucl. Chem. Lett.* **11**, 813 (1975).
213. McElearney, J. N., and Merchant, S., *Phys. Rev. B* **18**, 3612 (1978).
214. Flokstra, J., Gerritsma, G. J., van den Brandt, B., and van der Marel, L. C., *Phys. Lett. A* **53**, 159 (1975).
215. Algra, H. A., de Jongh, L. J., Blote, H. W. J., Huiskamp, W. J., and Carlin, R. L., *Physica (Amsterdam)* **82B**, 239 (1976).
216. Yelon, W. B., Cox, D. E., Eibschutz, M., *Phys. Rev. B* **12**, 5007 (1975).
217. Barrett, P. A., Dent, C. E., and Linstead, R. P., *J. Chem. Soc.*, 1719 (1936).
218. Sato, M., Kon, H., Akoh, H., Tasaki, A., Kabuto, C., and Silverton, J. V., *Chem. Phys.* **16**, 405 (1976).
219. Eichelberger, H., Majeste, R., Surgi, R., Trefonas, L., and Good, M. L., *J. Am. Chem. Soc.* **99**, 616 (1977).

220. Morelock, M. M., Good, M. L., Trefonas, L. M., Karraker, D., Maleki, L., Eichelberger, H. R., Majeste, R., and Dodge, J., *J. Am. Chem. Soc.* **101**, 4858 (1979).
221. Sharov, V. A., *Russ. J. Inorg. Chem.* **22**, 1500 (1977).
222. van Kralingen, C. G., van Ooijen, J. A. C., and Reedijk, J., *Transition Met. Chem.* **3**, 90 (1978).
223. Klaaijsen, F. W., den Adel, H., Dokoupil, Z., and Huiskamp, W. J., *Physica (Amsterdam)* **B79**, 113 (1975).
224. Al'Shuler, S. A., and Kozyrev, B. M., *Electron Paramagnetic Resonance in Compounds of Transition Elements*, John Wiley, New York (1974), Appendix I.
225. Watanabe, T., *J. Phys. Soc. Jap.* **17**, 1856 (1962).
226. de Neef, T., and de Jonge, W. J. M., *Phys. Rev. B* **11**, 4402 (1975).
227. Blote, H. W. J., *Physica (Amsterdam)* **79B**, 427 (1975).
228. de Neef, T., *Dissertation*, Eindhoven University of Technology, 1975.
229. Dupas, C., and Renard, J. P., *J. Phys. C* **10**, 5057 (1977).
230. Babel, D., *Z. Anorg. Allgem. Chem.* **369**, 117 (1969).
231. Steiner, M., *Solid State Commun.* **11**, 73 (1972).
232. Steiner, M., *Z. Angew. Phys.* **32**, 116 (1971).
233. Lebesque, J. V., Snel, J., and Smit, J. J., *Solid State Commun.* **13**, 371 (1973).
234. Steiner, M., and Dorner, B., *Solid State Commun.* **12**, 537 (1973).
235. Steiner, M., and Kjems, J. K., *J. Phys. C* **10**, 2665 (1977).
236. McGurn, A. R., Montano, P. A., and Scalapino, D. J., *Solid State Commun.* **15**, 1463 (1974).
237. Dupas, C., and Renard, J. P., *Proceedings of the 14th International Conference on Low Temperature Physics*, North Holland, Amsterdam (1975), Volume 5.
238. Lebesque, J. V., and Huyboom, N. F., *Commun. Phys.* **1**, 38 (1976).
239. Swuste, C. H. W., Bottermann, A. C., de Jonge, W. J. M., and Millenaar, J., *J. Chem. Phys.* **66**, 5021 (1977), and references therein.
240. Morosin, B., *Acta Crystallogr. B* **23**, 630 (1967).
241. Bongaarts, A. L. M., van Laar, B., Botterman, A. C., and de Jonge, W. J. M., *Phys. Lett. A* **41**, 411 (1972).
242. Achiwa, N., *J. Phys. Soc. Jap.* **27**, 561 (1969).
243. Johnson, P. B., Rayne, J. A., and Friedberg, S. A., *J. Appl. Phys.* **50**, 1853 (1979), and references therein.
244. Moses, D., Shechter, H., Ehrenfreund, E., and Makovsky, J., *J. Phys. C* **10**, 433 (1977).
245. Witteveen, H. T., and van Veen, J. A. R., *J. Phys. Chem. Solids* **35**, 337 (1974).
246. Adachi, K., and Meketa, J., *J. Phys. Soc. Jap.* **34**, 269 (1973).
247. Montano, D. A., Cohen, E., and Shechter, H., *Phys. Rev. B* **6**, 1053 (1972).
248. Smith, J., Gerstein, B. C., Liu, S. H., and Stucky, G., *J. Chem. Phys.* **53**, 418 (1970).
249. Garrett, B. B., and Euler, W. B., *Solid State Commun.* **28**, 505 (1978).
250. Yelon, W. B., and Cox, D. E., *Phys. Rev. B* **6**, 204 (1972).
251. Yelon, W. B., and Cox, D. E., *Phys. Rev. B* **7**, 2024 (1973).
252. Witteveen, H. T., and Reedijk, J., *J. Solid State Chem.* **10**, 151 (1974), and references therein.
253. Klaaijsen, F. W., den Adel, H., Dokoupil, Z., and Huiskamp, W. J., *Physica (Amsterdam)* **79B**, 113 (1975).
254. van Ooijen, J. A. C., Reedijk, J., and Spek, A., *Inorg. Chem.* **18**, 1184 (1979).
255. Witteveen, H. T., Rutten, W. L. C., and Reedijk, J., *J. Inorg. Nucl. Chem.* **37**, 913 (1975).
256. Klaaijsen, F. W., Dokoupil, Z., and Huiskamp, W. J., *Physica (Amsterdam)* **79B**, 547 (1975), and references therein.
257. Kopinga, K., de Neef, T., de Jonge, W. J. M., and Gerstein, B. C., *Phys. Rev. B* **13**, 3953 (1976).
258. Dupas, C., and Renard, J. P., *J. Chem. Phys.* **61**, 3871 (1974).

259. Hurley, M., and Gerstein, B. C., *J. Chem. Phys.* **59**, 6667 (1973).
260. Gerstein, B. C., Gehring, F. D., and Wittet, R. D., *J. Appl. Phys.* **43**, 1932 (1972).
261. Dupas, C., and Renard, J. P., *J. Chem. Phys.* **61**, 3871 (1974).
262. Brener, R., Ehrenfreund, E., Shechter, H., and Makovsky, J., *J. Phys. Chem. Solids* **38**, 1023 (1977), and references therein.
263. McPherson, J., and Chang, J. R., *Inorg. Chem.* **12**, 1196 (1973).
264. Zandbergen, H. W., *Dissertation*, University of Leiden, 1981.
265. Morelock, M. M., Good, M. L., Trefonas, T. M., Karraker, D., Maleki, L., Eichelberger, H. R., Majeste, R., and Dodge, J., *J. Am. Chem. Soc.* **101**, 4858 (1979).
266. Nardelli, M., Gusparri, G. F., Battistini, G. G., and Domiano, P., *Acta Crystallogr.* **20**, 349 (1966).
267. Emori, S., Inoue, M., and Kubo, M., *Bull. Chem. Soc. Jap.* **44**, 3299 (1971), and references therein.
268. Nardelli, M., Gasparri, G. F., Musatti, A., and Manfredotti, A., *Acta Crystallogr.* **21**, 910 (1966).
269. Sharov, V. A., *Zh. Neorg. Khim.* **22**, 2762 (1977).
270. van Kralingen, C. G., van Ooijen, J. A. C., and Reedijk, J., *Transition Met. Chem.* **3**, 90 (1978).
271. Drew, M. G. B., Goodgame, D. M. L., Hitchman, M. A., and Rodgers, D., *Chem. Commun.*, 477 (1965).
272. Goodgame, D. M. L., Hitchman, M. A., and Marsham, D. E., *J. Chem. Soc. A*, 259 (1971).
273. Tedenac, J. C., Phung, N. D., Auinens, C., and Maurin, M., *J. Inorg. Nucl. Chem.* **38**, 85 (1976).
274. Nicolini, C., and Reiff, W. M., *Abstracts 181*st *American Chemical Society National Meeting, Atlanta*, March 30, 1981, paper 41.
275. Jotham, R. W., *J. Chem. Soc. Chem. Commun.*, 178–179 (1973); *Phys. Status Solidi B* **55**, K125–K129 (1973).
276. Duffy, Jr., W., and Barr, K. P., *Phys. Rev.* **165**, 647–654 (1968).
277. Bonner, J. C., Friedberg, S. A., Kobayashi, H., and Meyers, B. E., *Proceeding of the Twelfth International Conference on Low Temperature Physics, Kyoto, Japan*, E. Kanda, Ed., Keigaku, Tokyo (1971), p. 691.
278. Bonner, J. C., and Friedberg, S. A., *Proceedings of the International Conference on Phase Transitions and Their Applications in Material Science*, H. K. Hanish, R. Roy, and L. E. Cross, Ed., Pergamon Press, New York (1973), p. 429.
279. Diederix, K. M., Blote, H. W. J., Groen, J. P., Klaasen, T. O., and Poulis, N. J., *Phys. Rev. B* **19**, 420–431 (1979).
280. Bonner, J. C., Blote, H. W. J., Bray, J. W., and Jacobs, I. S., *J. Appl. Phys.* **50**, 1810 (1979).
281. Hall, J. W., Marsh, W. E., Weller, R. R., and Hatfield, W. E., *Inorg. Chem.* **20**, 1033–1037 (1981).
282. Hatfield, W. E., *J. Appl. Phys.* **52**, 1985 (1981).
283. Klein, D. J., private communication.
284. Ananthakrishna, G., Weiss, L. F., Foyt, D. C., and Klein, D. J. *Physica* (*Utr.*) **81B**, 275 (1976).
285. Weller, R. R., Ph. D. Dissertation, University of North Carolina, Chapel Hill, 1981.
286. Morosin, B., *Acta Crystallogr. B* **31**, 632 (1975).
287. Takeda, K., Matsukwa, S., and Haseda, T., *J. Phys. Soc. Jap.* **30**, 1330 (1971).
288. Jeter, D. Y., and Hatfield, W. E., *J. Inorg. Nucl. Chem.* **34**, 3055–3060 (1972).
289. Duffy, W., Venneman, J. E., Strandberg, D. L., and Richards, P. M., *Phys. Rev. B* **9**, 2220–2227 (1974).
290. Hatfield, W. E., *Comm. Inorg. Chem.* **1**, 105 (1981).
291. Swank, D. D., Landee, C. P., and Willett, R. D., *Phys. Rev. B* **20**, 2154–2162 (1979).
292. Marsh, W. E., Valente, E. J., and Hodgson, D. J., *Inorg. Chim. Acta* **51**, 49 (1981).
293. Morosin, B., *Acta Crystallogr. B* **26**, 1203–1208 (1970).

294. For early work, see Myers, B. E., Berger, L., and Friedberg, S. A., *J. Appl. Phys.* **40**, 1149–1151 (1969).
295. Diederix, K. M., Groen, J. P., Henkens, L. S. J. M., Klaassen, T. O., and Poulis, N. J., *Physica* (*Utr.*) **93B**, 99–113 (1978).
296. Taylor, M. R., Glusker, J. P., Gabe, E. J., and Minikin, J. A., *Bioinorg. Chem.* **3**, 189 (1974).
297. Hatfield, W. E., Weller, R. R., and Hall, J. W., *Inorg. Chem.* **19**, 3825–3838 (1980).
298. Brown, D. B., Donner, J. A., Hall, J. W., Wilson, S. R., Wilson, R. B., Hodgson, D. J., and Hatfield, W. E., *Inorg. Chem.* **18**, 2635–2641 (1979).
299. Santoro, A., Mighell, A. D., and Reimann, C. W., *Acta Crystallogr. B* **26**, 979–984 (1970).
300. Villa, J. F., and Hatfield, W. E., *J. Am. Chem. Soc.* **93**, 4081 (1971).
301. Losee, D. B., Richardson, H. W., and Hatfield, W. E., *J. Chem. Phys.* **59**, 3600–3603 (1973).
302. Koyama, M., Suzuki, H., and Watanabe, T., *J. Phys. Soc. Jap.* **40**, 1564–1569 (1976).
303. Boyd, D. W., and Mitra, S., *Inorg. Chem.* **19**, 3547–3549 (1980).
304. Richardson, H. W., and Hatfield, W. E., *J. Am. Chem. Soc.* **98**, 835–839 (1976).
305. Inoue, M., *J. Magn. Res.* **27**, 159–167 (1977).
306. Delker, G. L., Ph. D. Dissertation, University of Illinois, Urbana, 1976; unpublished, quoted by Ref. 308.
307. Bray, J. W., Interrante, L. V., Jacobs, I. S., and Bonner, J. C., in *Extended Linear Chain Compounds*, Vol. 3, J. Miller, Ed., Plenum Press, New York, 1982. Jacobs, I. S., Bray, J. W., Hart, Jr., H. R., Interrante, L. V., Kasper, J. S., Watkins, G. D., Prober, D. E., and Bonner, J. C., *Phys. Rev. B* **14**, 3036–3051 (1976).
308. Moncton, D. E., Birgeneau, R. J., Interrante, L. V., and Wudl, F., *Phys. Rev. Lett.* **39**, 507–510 (1977).
309. Bulaevskii, L. N., *Sov. Phys.-JETP* **17**, 684 (1963) [*Zh. Eksp. Teor. Fiz.* **44**, 1008 (1963)].
310. Okazaki, A., and Suemune, Y., *J. Phys. Soc. Jap.* **16**, 176 (1961).
311. Hutchings, M. T., Samuelsen, E. J., Shirane, G., and Hirakawa, K., *Phys Rev.* **188**, 919–923 (1969).
312. Hutchings, M. T., Ikeda, H., and Milne, J. M., *J. Phys. C: Solid State Phys.* **12**, L739–L744 (1979).
313. Kadota, S., Yamada, I., Yoneyama, S., and Kirakawa, K., *J. Chem. Soc. Jap.* **25**, 751 (1967).
314. Ikeda, H., and Hirakawa, K., *J. Phys. Soc. Jap.* **35**, 722–728 (1973).
315. Iio, K., Hyodo, H., Nagata, K., and Yamada, I., *J. Phys. Soc. Jap.* **44**, 1393–1394 (1978).
316. Ikebe, M., and Date, M., *J. Phys. Soc. Jap.* **30**, 93–100 (1971).
317. Willet, R. D., and Chang, K., *Inorg. Chim. Acta* **4**, 447 (1970).
318. Estes, W. E., Hatfield, W. E., van Ooijen, J. A. C., and Reedijk, J., *J. Chem. Soc. Dalton Trans.*, 2121–2124 (1980).
319. Bandoli, G., Biagini, M. C., Clemente, D. A., and Rizzardi, G., *Inorg. Chim. Acta* **20**, 71–78 (1976).
320. Bream, R. A., Estes, E. D., and Hodgson, D. J., *Inorg. Chem.* **14**, 1672–1675 (1975).
321. Losee, D. B., McElearney, J. N., Siegel, A., Carlin, R. L., Khan, A. A., Roux, J. P., and James, W. J., *Phys. Rev. B* **6**, 4342–4348 (1972).
322. Stirrat, C. R., Dudzinski, S., Owens, A. H., and Cowen, J. A., *Phys. Rev. B* **9**, 2183–2186 (1974).
323. Algra, H. A., de Jongh, L. J., Huiskamp, W. J., and Carlin, R. L., *Physica* **92B**, 187–200 (1977).
324. Gerstein, B. C., Gehring, F. D., and Willett, R. D., *J. Appl. Phys.* **43**, 1932–1941 (1972).
325. Giuseppetti, G., and Mazzi, F., *Rend. Soc. Min. Ital.* **11**, 202 (1965).
326. Reiff, W. M., Wong, H., Dockum, B., Brennan, T., and Cheng, C., *Inorg. Chim. Acta.* **30**, 69–76 (1978).
327. Laing, M., and Horsfield, E., *Chem. Commun.*, 735 (1968).

328. Crawford, V. H., and Hatfield, W. E., *Inorg. Chem.* **16**, 1336–1341 (1977).
329. Laing, M., and Garr, G. J., *J. Chem. Soc. A*, 1141 (1971).
330. Estes, W. E., Gavel, D. P., Hatfield, W. E., and Hodgson, D. J., *Inorg. Chem.* **17**, 1415–1421 (1978).
331. Kroese, C. J., Masskant, W. J. A., and Verschoor, G. C., *Acta Crystallogr. B* **30**, 1053 (1974).
332. Woenk, J. W., and Spek, A. L., *Cryst. Struct. Commun.* **5**, 805 (1976).
333. Landee, C. P., and Willett, R. D., *Phys. Rev. Lett.* **43**, 463–466 (1979).
334. Clay, R. M., Murray-Rust, P., and Murray-Rust, J., *J. Chem. Soc. Dalton Trans.*, 595 (1973).
335. van Ooijen, J. A. C., and Reedijk, J., *Inorg. Chim. Acta* **25**, 131 (1977).
336. Jansen, J. C., van Koningsveld, H., and van Ooijen, J. A. C., *Cryst. Struct. Commun.*, in press.
337. van Ooijen, J. A. C., and Reedijk, J., *J. Chem. Soc. Dalton Trans.*, 1170 (1978).
338. Smit, J. J., DeJongh, L. J., van Ooijen, J. A. C., Reedijk, J., and Bonner, J. C., *Physica* **97B**, 229 (1979).
339. van Ooijen, J. A. C., Reedijk, J., Sonneveld, E. J., and Visser, J. W., *Transition Met. Chem.* **4**, 305–307 (1979).
340. Koisumi, H., Osaki, K., and Watanabe, T., *J. Phys. Soc. Jap.* **18**, 117 (1963).
341. Date, M., Yamazaki, H., Motokawa, M., and Tazawa, S., *Prog. Theor. Phys. Suppl.*, No. 46, 194–209 (1970).
342. Haseda, T., and Miedema, A. R., *Physica (The Hague)* **27**, 1102 (1961).
343. Ueki, T., Ashida, T., Sarada, Y., and Kakudo, M., *Acta Crystallogr.* **22**, 870 (1967).
344. Estes, W. E., and Hatfield, W. E., *Inorg. Chem.* **17**, 3226–3231 (1978).
345. Fujimoki, H., Oonishi, I., Muto, F., Nakahara, A., and Komiyama, Y., *Bull. Chem. Soc. Jap.* **44**, 28 (1971).
346. Myers, B. E., Berger, L., and Friedberg, S. A., *J. Appl. Phys.* **40**, 1149 (1969).
347. Morosin, B., *Acta Crystallogr. B* **32**, 1237–1240 (1976).
348. Bhatia, S. N., O'Connor, C. J., Carlin, R. L., Algra, H. A., and de Jongh, L. J., *Chem. Phys. Lett.* **50**, 353–357 (1977).
349. Anderson, O. P., Packard. A. B., and Wicholas, M., *Inorg. Chem.* **15**, 1613 (1976).
350. Landee, C. P., Wicholas, M., Willett, R. D., and Wolford, T., *Inorg. Chem.* **18**, 2317–2318 (1979).
351. Brown, D. B., Hall, J. W., Helis, H. M., Walton, E. G., Hodgson, D. J., and Hatfield, W. E., *Inorg. Chem.* **16**, 2675–2680 (1977).
352. de Meester, P., and Skapski, A. C., *J. Chem. Soc. Dalton Trans.*, 424 (1973).
353. Bukovska, M., and Porai-Koshitz, M. A., *Sov. Phys. Cryst.* **5**, 127 (1960) [*Kristallografiya* **5**, 137 (1960)].
354. Huang, T. Z., and Soos, Z. G., *Phys. Rev. B* **9**, 4981–4984 (1974).
355. Plumlee, K. W., Hoffman, B. M., Ibers, J. A., and Soos, Z. G., *J. Chem. Phys.* **63**, 1926–1942 (1975).
356. Nakano, T., Miyoshi, T., Iwamoto, T., and Sasaki, Y., *Bull. Chem. Soc. Jap.* **40**, 1297 (1967).
357. Kitaguchi, H., Nagata, S., and Watanabe, T., *J. Phys. Soc. Jap.* **38**, 998–1002 (1975).
358. Lingafelter, E. C., Simmons, G. L., Morosin, B., Scheringer, C., and Freiburg, C., *Acta Crystallogr.* **14**, 1222–1225 (1961).
359. Knauer, R. C., and Bartkowski, R. R., *Phys. Rev. B* **7**, 450–453 (1973).
360. Azevedo, L. J., Clark, W. G., Hulin, D., and McLean, E. O., *Phys. Lett. A* **58**, 255–256 (1976).
361. Azevedo, L. J., Clark, W. G., and McLean, E. O., in *Proceeding of the 14th International Conference on Low Temperature Physics, LT14, Otaniemi, Finland*, M. Krusius and M. Vuorio, Eds., North–Holland, Amsterdam (1975), p. 369.
362. Algra, H. A., de Jongh, L. J., and Carlin, R. L., *Physica B + C* **93B**, 24–34 (1978).

3

Ferromagnetism in Linear Chains

R. D. Willett, R. M. Gaura, and C. P. Landee

1. Introduction

Recent years have seen an intense interest on the part of inorganic chemists in magneto-structural correlations in transition-metal salts, particularly cluster systems. This is evidenced by the large fraction of papers in journals like *Inorganic Chemistry* which deal in one way or another with magnetic interactions between metal ions. A similar interest has been displayed by the solid-state physics community in low-dimensional magnetic materials. Scanning the papers presented at the International Conference on Magnetism in Münich, September 1979 (*J. Magn. Magn. Mater.*, volumes 15–18) verifies this. Thus, the synthesis and analysis of low-dimensional magnetic systems is an area of strong, common interest to both physicists and chemists. The combination of spin dimensionality, spatial dimensionality, and spin quantum number leads to a large magnetic "zoo" for which theoretical models and physical realizations must be found.

The scope of this chapter will be limited to the static properties of salts with dominant ferromagnetic interactions in one dimension. We would like to point out the existence of an excellent review of low-dimensional properties by de Jongh and Miedema.[1] An excellent review of the dynamic properties of one-dimensional magnets up to 1976 has been given by Steiner *et al.*[2] Many new, exciting developments have occurred since that time in the area of dynamics, especially related to the observation of solitons.[3] We will not seek to review these rapid moving developments here. We caution the

R. D. Willett, R. M. Gaura, and C. P. Landee • Program in Chemical Physics, Department of Chemistry, Washington State University, Pullman, Washington 99164. Mr. Gaura's present address: IBM Instruments, Inc., Orchard Park, P.O. Box 332, Danbury, Connecticut 06810. Dr. Landee's present address: Department of Physics, Clark University, 950 Main Street, Worcester, Massachusetts 01610.

reader at the onset of a piece of terminology which is now commonplace although strictly incorrect. The term "one-dimensional ferromagnet" has come to simply mean a linear chain system with ferromagnetic exchange and is not taken in this context to imply ferromagnetic long-range order, which is in fact unobtainable for a strictly one-dimensional system.

The Section 2 gives a short statement of relevant magnetic phenomena, with concentration on those topics that can be used to define the spin dimensionality and the extent of "one-dimensionality" of the system. This will be followed in Section 3 by a general discussion of the synthetic strategies to be used in building new one-dimensional magnetic materials. Section 4 reviews recent theoretical and experimental work which has led to an understanding of the structural-magnetic correlations in simple binuclear species. This work has shown that the strength (and sign) of the exchange interaction is sensitive to structural parameters such as the metal–ligand–metal angle. These correlations will be used to present a statement of the necessary structural prerequisites for the formation of a ferromagnetic chain. Section 5 reviews the theoretical calculations available for the static properties of the one-dimensional magnetic systems, while Section 6 reviews the actual experimental one-dimensional ferromagnetic systems that have been investigated.

It is appropriate to ask why the study of one-dimensional systems with ferromagnetic (FM) interactions should be emphasized:

1. One dimensional materials present the simplest systems with extended interactions. Theoretically, the static magnetic properties have been well studied, but the dynamics of even the one-dimensional systems are still largely not understood.

2. One-dimensional FM systems are a class of compounds that have not been systematically investigated (although several specific compounds have been studied in depth). Yet evidence points to the existence of a wide variety of systems with varying spin anisotropy, spin quantum numbers, and degree of one dimensionality. Further insight into the physics of one-dimensional systems can be anticipated by a systematic investigation of the effects of these various parameters.

3. One-dimensional systems are candidates for nonlinear excitations, such as solitons or kinks. The existence of solitons may have been confirmed in an $S = 1$ XY FM ($CsNiF_3$). It is desirable to have a series of FM salts with different spin quantum numbers and anisotropy to investigate the generality (and limitations) of the existence of solitons.

4. With the proliferation of $S = 1/2$ FM linear chain systems, it will be possible to study the role of quantum effects in detail.

5. It appears that several isostructural systems exist with various metal ions. Thus, the static and dynamic properties of mixed one-dimensional FM

systems can be examined. This will provide a very good opportunity to study the effects of competing anisotropies on magnetic phase diagrams and relaxation processes.

6. Dimensionality crossover effects are especially pronounced in FM materials.

2. Background

Magnetic interactions reflect differences in energy of various electronic states within a system due to electrostatic interactions. By definition, we are interested in different spin states of the system whose energies lie very close together, e.g., that lie within an energy $E \approx k_B T$ of the ground state. These energy differences are very small compared with the total energy of the system. Thus, for a small transition-metal cluster, even if we can write down the Hamiltonian for the system exactly, computational methods are not sufficiently accurate to calculate the desired energy differences precisely enough. Rather than despair, one instead follows the lead of Van Vleck and parameterizes the problem through the use of a phenomenological spin Hamiltonian. Thus, for a simple two-center, two-electron system with only Coulomb potentials, the difference in energy of the singlet and triplet states can be expressed as

$$H = -2J\mathbf{S}_1 \cdot \mathbf{S}_2 \tag{1}$$

where $\mathbf{S}_1$ and $\mathbf{S}_2$ are spin operators for the two electrons. The energies of the singlet and triplet states are thus $(3/2)J$ and $(1/2)J$, respectively, with a level separation of $E = 2J$. The complex, true Hamiltonian, expressed in real-space coordinates, is thus replaced by a spin Hamiltonian involving spin operators. Since the eigenvalues of spin operators are (relatively) easy to obtain, the problem of obtaining the energy-level spectrum has been greatly simplified.

In actual practice, the spin Hamiltonian used is much more complex than the Hamiltonian in Eq. (1). This is because additional potential-energy terms are present in addition to the simple electron–electron repulsion term between the magnetic electrons. The main additional terms leading to complications are the spin–orbit coupling and crystal-field terms. Thus, it is necessary to express the Hamiltonian as

$$\begin{aligned} \mathrm{H} = & -2(J_x S_1^x S_2^x + J_y S_1^y S_2^y + J_z S_1^z S_2^z) + \mathbf{d}_{12} \cdot \mathbf{S}_1 \times \mathbf{S}_2 \\ & + \mathbf{S}_1 \cdot \mathbf{D} \cdot \mathbf{S}_1 + \mathbf{S}_2 \cdot \mathbf{D} \cdot \mathbf{S}_2 & \text{(2a)} \\ = & -2J\mathbf{S}_1 \cdot \mathbf{S}_2 + D(S_1^x S_2^x + S_1^y S_2^y) + E(S_1^x S_2^x - S_1^y S_2^y) \\ & + \mathbf{d} \cdot \mathbf{S}_1 \times \mathbf{S}_2 + \mathbf{S}_1 \cdot \mathbf{D} \cdot \mathbf{S}_1 + \mathbf{S}_2 \cdot \mathbf{D} \cdot \mathbf{S}_2 & \text{(2b)} \end{aligned}$$

where D and E are the anisotropic exchange contributions, $\mathbf{d}$ is the antisymmetric exchange interaction, and the $\mathbf{S}_1 \cdot \mathbf{D} \cdot \mathbf{S}_1$ terms are the single-ion anisotropy terms. The nonsymmetric components (D, E, and $\mathbf{d}$) express the preference of the spins to align in particular orientations in the crystal due to the spin–orbit and crystal-field interactions. In particular, D and E give rise to preferred collinear orientations, while $\mathbf{d}$ tends to align the spins at right angles to each other. The following limiting cases are commonly identified: (i) $J_x = J_y = J_z$, Heisenberg model; (ii) $J_x = J_y$, $J_z = 0$, XY model; (iii) $J_x = J_y = 0$, $J_z \neq 0$, Ising model. The single-ion terms are kept only if $|\mathbf{D}| \leq |J|$. If $|\mathbf{D}| > |J|$, then at low temperatures ($k_B T < \mathbf{D}$), the true spin value can be replaced by an effective spin of smaller magnitude.

For a crystal composed of arrays of chains, the Hamiltonian now becomes (omitting the single-ion anisotropy terms)

$$\begin{aligned} H = &-2J \sum_i \mathbf{S}_i \cdot \mathbf{S}_{i+1} + D \sum_i (S_i^x S_{i+1}^x + S_i^y S_{i+1}^y) \\ &+ E \sum_i (S_i^x S_{i+1}^x - S_i^y S_{i+1}^y) + \sum_i \mathbf{d}_{i,i+1} \cdot \mathbf{S}_i \times \mathbf{S}_{i+1} \\ &- \sum_{i,k} 2J'_{ik} \mathbf{S}_i \cdot \mathbf{S}_k - \sum_{i,m} 2J''_{im} \mathbf{S}_i \cdot \mathbf{S}_m \end{aligned} \qquad (3)$$

where the sum over i includes just nearest neighbors along the chains (due to the short-range nature of the superexchange interactions), the sum over i, k takes into account the interactions between nearest-neighbor chains, and the sum of i, m takes into account the interactions between next-nearest-neighbor chains (if necessary to describe the magnetic behavior). These latter interactions determine the degree of "one-dimensionality" and are important in the determination of the low-temperature properties of the system. In particular, they are responsible for triggering the onset of long-range order at some critical temperature, T_c.

In interpretation of the experimental data, it is frequently convenient to talk about effective fields, rather than the spin Hamiltonian parameters. The exchange fields are defined, in the mean-field approximation, as

$$\mathscr{H}_{\text{ex}} = 2z|J|S/g\mu_B \qquad \mathscr{H}'_{\text{ex}} = 2z'|J'|S/g\mu_B \qquad \mathscr{H}''_{\text{ex}} = 2z''|J''|S/g\mu_B \qquad (4)$$

where z, z', and z'' refer to the number of nearest neighbors for the respective exchange pathways. The anisotropy fields (neglecting contributions from antisymmetric exchange) are given as

$$H_{A_1} = DH_{\text{ex}} \qquad H_{A_2} = EH_{\text{ex}} \qquad (5)$$

These fields are obtained from the magnetization data in terms of the fields required to saturate the system (at $T = 0$ K), H_{sat}^i, as

$$\begin{aligned} H_{A_1} &= H_{\text{sat}}^1 - H_{\text{sat}}^3 \\ H_{A_2} &= H_{\text{sat}}^2 - H_{\text{sat}}^3 \end{aligned} \qquad (6)$$

where for H^i_{sat}, $i = 1$ refers to the hard axis, $i = 2$ to the intermediate axis, and $i = 3$ to the easy axis. Alternatively, these quantities may be determined from the magnetic phase diagram in the ordered state, as described below. It is desired to obtain reliable estimates of these quantities so that the degree of ideality of the systems can be ascertained.

The dominant intrachain exchange interactions will lead to large one-dimensional ferromagnetic correlations along the chain. The magnitude of these correlations will increase as the temperature is lowered, becoming long range only at $T = 0$ K for a truly one-dimensional system.[4] Because of the ever present interactions between chains (even if only dipolar in nature), long-range order will set in at some finite temperature. Neglecting for the moment the effects of antisymmetric exchange in our discussion, three types of magnetically ordered state can be identified: (i) ferromagnetic ordering between all chains (J', $J'' > 0$); (ii) ferromagnetic ordering between nearest-neighbor chains ($J' > 0$) but antiferromagnetic ordering between next-nearest-neighbor chains ($J'' < 0$); (iii) antiferromagnetic ordering between nearest-neighbor chains ($J' < 0$).

Case (i) has not been observed to date; surprisingly a substantial number of systems fall into class (ii).

The ordered magnetic states are commonly characterized by one of two types of behavior under the influence of an applied field: spin flop or metamagnetic. We denote the strongest antiferromagnetic exchange field by H_{AF}. Then if H_{AF} is large compared to H_{A_1}, spin-flop behavior will occur, while if the anisotropy fields are large compared to H_{AF}, metamagnetic behavior will be found. In a metamagnetic material, when a magnetic field is applied parallel to the direction of preferred spin alignment, the antiferromagnetic alignment is broken up at some critical external field, $H_{c-}(T)$. At this point, a first-order transition to a ferromagnetic state occurs. Because of demagnetization effects, the internal field is reduced to a value below the critical value as soon as some of the spins flip over. The transition is complete at some upper critical field, $H_{c+}(T)$. Both $H_{c-}(T)$ and $H_{c+}(T)$ are temperature dependent. The antiferromagnetic interchain exchange constant (J' or J'') can be determined for $H_{c-}(0)$ by the mean-field relationship

$$-2z_{AF}J_{AF}S^2 = g\mu_B SH_{c-}\,. \tag{7}$$

As T is raised, H_{c+} and H_{c-} gradually coalesce until the tricritical point is reached when $H_{c+}(T_t) = H_{c-}(T_t)$. For type (ii) systems ($J' > 0$, $J'' < 0$), mean-field theory gives

$$T_t/T_c - 1 = z''J''/3z'J' \tag{8}$$

The situation is a bit more complex for materials that undergo spin-flop transitions, since it is the anisotropy and not the antiferromagnetic exchange field which is crucial in determining the nature of the phase transition. Due to the anisotropy, the spin state with the spins aligned along the easy axis (parallel orientation) lies lower in energy than the spin state in which the spins are aligned perpendicular to the easy axis. The difference in energy between these two orientations is the anisotropic energy K. When a field is applied parallel to the easy axis, the energy of the parallel to the easy axis, the energy of the parallel spin state, is lowered by an amount $-\chi_{\parallel} H^2$ and the perpendicular orientation $-\chi_{\perp} H^2$. However, since $\chi_{\perp} > \chi_{\parallel}$, the energy of the perpendicular orientation decreases more rapidly, and eventually becomes the stable configuration. At this point, the spin system "flops" over to the perpendicular orientation. The anisotropy field is related to the spin-flop field, H_{SF}, by

$$H_{SF} = [2K/(\chi_{\perp} - \chi_{\parallel})]^{1/2} \tag{9a}$$

or

$$H_{SF}(0) = (2H'_{ex} H_{A_1} - H^2_{A_2})^{1/2} \tag{9b}$$

and to the field necessary to saturate the system, H_c, by

$$H_c(0) = 2H'_{ex} - H_{A_1} \tag{10}$$

The value of J' can be obtained from the value of ($T = 0$ K). The value of H_{A_2} can be obtained from the angular dependence of the spin-flop field.

Finally, we comment on the role of the antisymmetric exchange, which we have ignored in most of the previous discussion. This type of interaction gives rise to noncollinear (canted) spin arrangements. In a simple two-sublattice antiferromagnet this leads to the phenomenon of weak ferromagnetism. Since the moments on one sublattice are not directly opposed to those on the second sublattice, both sublattices will have a component of magnetization in a particular direction, leading to the existence of a bulk magnetization in the ordered state. For systems with more sublattices, this spin canting is not necessarily revealed in the bulk magnetic properties. For example, in a four-sublattice system, sublattices 1 and 2 may be canted with respect to each other, as well as 3 and 4. However, sublattices 1 and 3 may be strictly antiparallel, and the same for sublattices 2 and 4. This is referred to as "hidden" canting.

3. *Synthetic Strategies*

The secret of designing new one-dimensional materials is to find a "natural" system for which well-defined exchange pathways exist in only one direction in the solid. The exchange pathway will generally involve one or more ligands that coordinate to the adjacent metal ions. Occasionally the exchange pathways may involve hydrogen bonds or direct ligand–

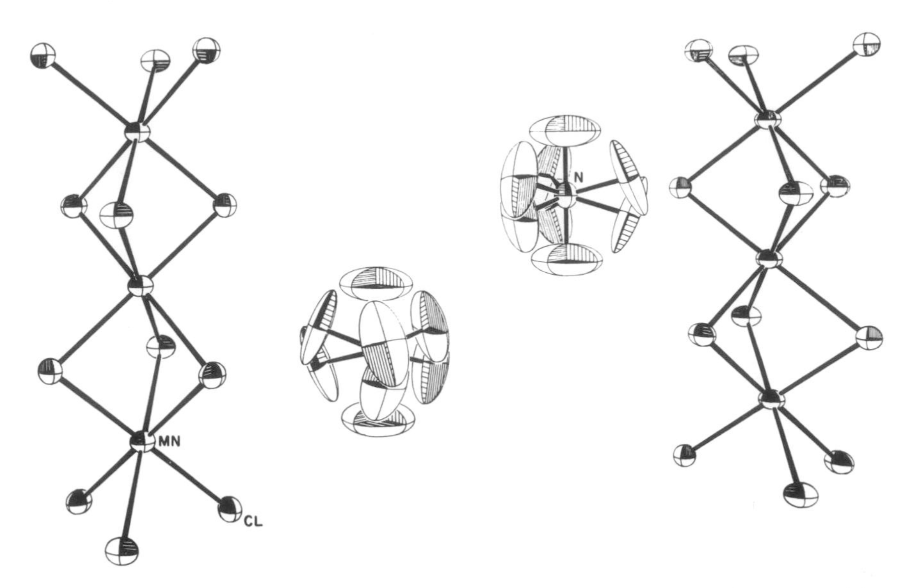

FIGURE 1. Crystal structure of $(CH_3)_4NMnCl_3$, TMMC. The $MnCl_3^-$ chains are parallel to the crystallographic *c* axis. Figure drawn from information contained in Reference 6.

ligand contacts. When the pathway involves a bridging ligand, the stoichiometric requirements of the metal ion must be such as to prohibit the formation of linkages to more than two neighboring metal ions. Systems highly one-dimensional in nature are attained by the introduction of large organic counterions or through the use of additional bulk ligands. The former is illustrated by the classic one-dimensional antiferromagnetic system $N(CH_3)_4MnCl_3$, TMMC,[5] where the large tetramethylammonium ion separates the $(MnCl_3)_n^{n-}$ chains from each other (Figure 1).[6] The ratio of inter to intrachain coupling, $|J'/J|$, is $\sim 10^{-4}$. Similarly, the use of pyridine, py, ligands in $CuCl_2 \cdot 2py$,[7] in place of water ligands in $CuCl_2 \cdot 2H_2O$,[8] converts a system with extensive three-dimensional interactions[9] into a good quasi-one-dimensional system.[10]

Initially, before specifically discussing ferromagnetic systems, let us briefly consider the structural strategies with respect to synthesis in the one-dimensional antiferromagnetic divalent metal halide systems. The M^{2+} ion in these systems generally assumes an octahedral coordination. Three types of stoichiometric relationships can lead to one-dimensional systems. One has already been mentioned in TMMC. Here, a salt with three ligands, such as the AMX_3 or $MX_2 \cdot L$ stoichiometries, leads to a tribridged linear chain system if A or L is large enough to prevent formation of a three-dimensional system. A bibridged system can be obtained if four ligands are present, either with a stoichiometry of $AMCl_3 \cdot L$, as seen in Figure 2 for

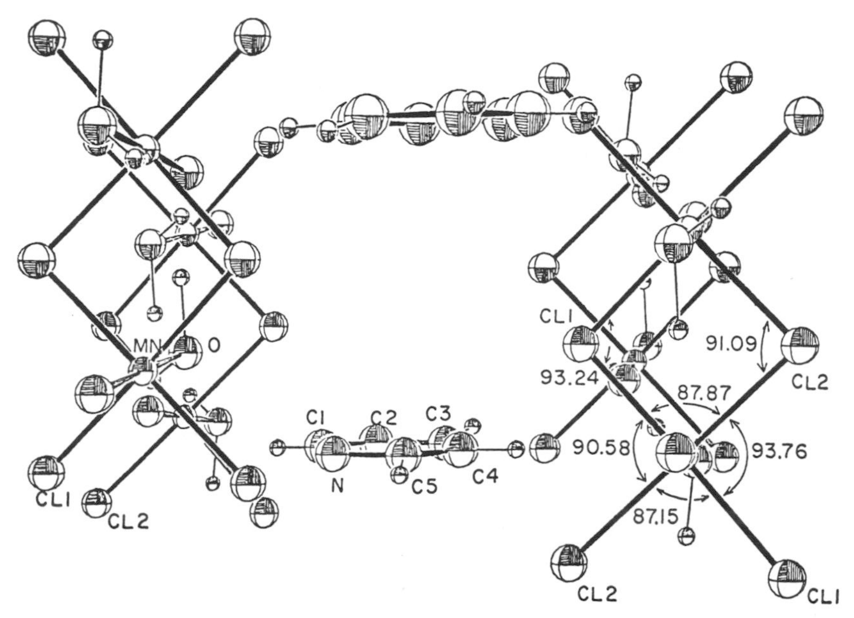

FIGURE 2. Crystal structure of $[C_5H_5NH]MnCl_3 \cdot 2H_2O$. The chains that run vertically in the *b* direction are hydrogen bonded to the chains behind them. The pyridinium, pyH^+, groups separate the planes of chains. (From Reference 11.)

$(pyH)MnCl_3 \cdot H_2O$,[11] or $MCl_2 \cdot 2L$, as in the previously mentioned $CuCl_2 \cdot$ 2py. In principle, bibridged chains could be obtained from A_2MX_4 stoichiometry also, but such systems do not appear to exist. Finally, with five ligands, a monobridged series can be attained in $AMX_3 \cdot 2L$ salts, such as $CsMnCl_3 \cdot 2H_2O$,[12] as seen in Figure 3. Again, $A_2MX_4 \cdot L$ or A_3MX_5 systems could lead to monobridged chains.

Caution should be exercised, of course, in assuming these stoichiometries will always give these structures. Thus, $(CH_3)_3NHMnCl_3 \cdot 2H_2O$ contains a bibridged chain, since one chloride ion does not coordinate to the metal ion.[13] In $(CH_3)_2CHNH_3MnCl_3 \cdot 2H_2O$, discrete $Mn_2Cl_6(H_2O)_4^{2-}$ dimers are formed, rather than a chain system.[14] Nevertheless, this constitutes a convenient starting point for searching for new systems. Similar arguments can be constructed if the metal is 4- or 5-coordinate.

This approach was used in seeking to modify the one-dimensional properties in a series of Mn(II) salts. Thus, the tetramethylammonium ion in TMMC was replaced by other organic cations. The most successful attempt resulted in the formation of $(CH_3)_2NH_2MnCl_3$, DMMC.[15] The tribridged linear chain system of TMMC was retained, ensuring the retention of quasi-one-dimensional behavior. Because of the smaller volume of the dimethylammonium cation (as compared to the tetramethylammonium

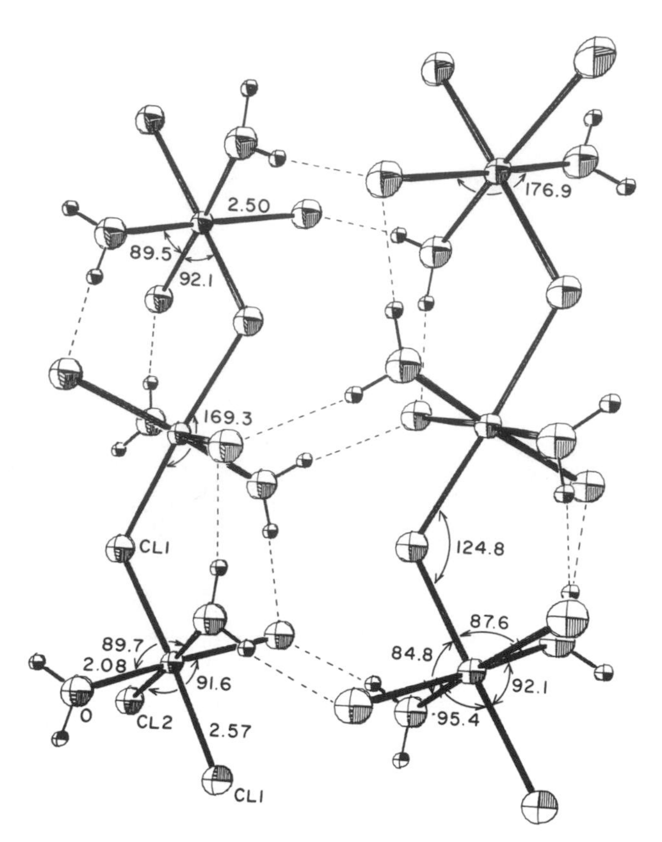

FIGURE 3. Crystal structure of $CsMnCl_3 \cdot 2H_2O$. The monobridged chains run parallel to the crystallographic *a* direction and are bonded together in the *b* direction. The cesium ions are not shown. Figure drawn from information contained in Reference 12.

cation), the $(MnCl_3)_n^{n-}$ chains are closer together. More importantly, however, the partially substituted ammonium ion is capable of hydrogen bonding to the chloride ions (Figure 4), and, in fact, forms a hydrogen bonding bridge (e.g., exchange pathway) between adjacent chains. Thus, the degree of one dimensionality has been altered, and $|J'/J|$ is now 10^{-2} compared to $|J'/J| < 10^{-4}$ for TMMC. This property has been exploited nicely by de Jonge in his series of studies on the magnetic behavior of quasi-one-dimensional antiferromagnets.[16]

Of course, in these attempts, nature is not always cooperative. With $A = CH_3NH_3^+$, the structurally very stable $(CH_3NH_3)_2MnCl_4$ two-dimensional salt is obtained. With $A = (CH_3)_3NH^+$, a real bonus was obtained. Instead of obtaining the expected $AMnCl_3$ salt, a salt with stoichiometry $[(CH_3)_3NH]_3Mn_2Cl_7$, TTMMC, resulted.[17] This hexagonal salt

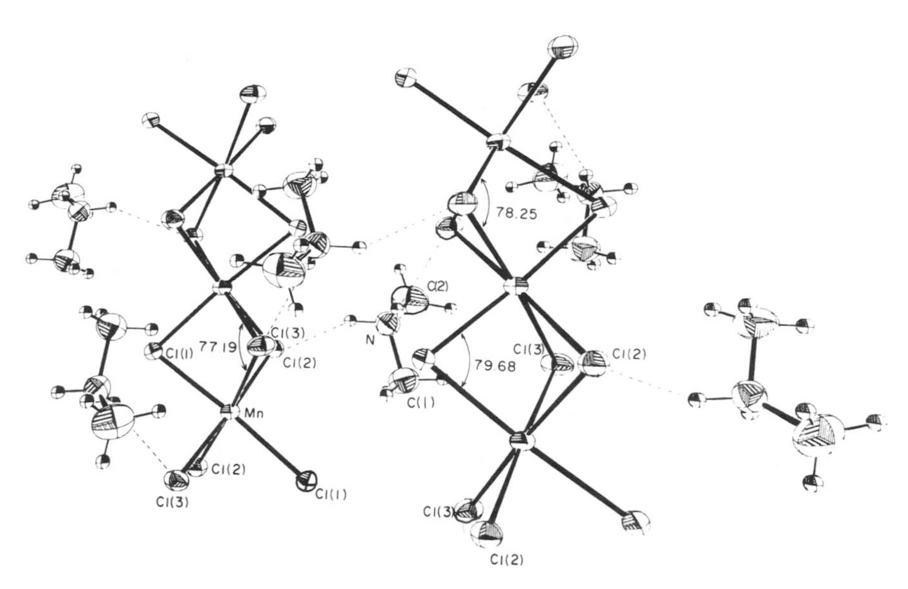

FIGURE 4. Crystal structure of $(CH_3)_2NH_2MnCl_3$, DMMC. The $MnCl_3^-$ chains that run vertically in the *c* direction are loosely connected into corrugated sheets along the *a* direction. (From Reference 15.)

contains, in addition to the TMMC-like tribridged chains, columns of tetrahedral $MnCl_4^{2-}$ anions. The trimethylammonium ions hydrogen bond to the columns, separating them from each other and from the $(MnCl_3)_n^{n-}$ chains. Thus, a system containing two types of linear chains was obtained. The challenge to the chemists of attaining a stoichiometric mixed salt of the type $[(CH_3)_3NH_3]_3MM'Cl_7$ has not been successfully taken up.

4. Magneto-Structural Correlations in Dimeric Systems

Magnetic exchange interactions between metal ions in binuclear transition-metal salts continue to be a subject of wide interest, with particular emphasis on determining magneto-structural correlations. These correlations are a manifestation of rather intricate details of the interactions between the unpaired electrons on the magnetic orbitals of the metal atoms. Thus, their elucidation and understanding present fundamental experimental and theoretical challenges to inorganic and theoretical chemists.

A number of theoretical treatments have been developed with which to predict and interpret strengths of magnetic interactions between metal ions in dimeric systems. We will wish to examine these first so that the relevant factors that cause the interactions to be FM can be identified. These factors can then be used as a guide for synthesis of new ferromagnetic linear chain systems, since it has been shown that these concepts can be transferred to extended systems.

Initial discussions of magnetic exchange were based on valence bond models,[18] and these have been very useful in the interpretation of magnetic interactions. In recent years, the use of valence bond theory has fallen into disfavor among chemists, who normally use molecular orbital (MO) theory for interpretation of spectra, bonding, etc. Thus, several authors have recast the superexchange arguments in terms of MO theory. It has been shown by Hay *et al.*[19] that the exchange energy $2J$ for the interaction of two spin-1/2 ions can be expressed by the combination of two terms,

$$2J = 2K_{ab} - 2(\varepsilon_1 - \varepsilon_2)^2/(J_{aa} - J_{bb}), \tag{11}$$

where the exchange and Coulomb integrals, $K_{ab} > 0$, $J_{aa} > J_{bb}$, are slowly varying functions of the molecular orbitals formed from the magnetic orbitals on the two metal ions. The first term gives a FM contribution; the second an antiferromagnetic (AFM) contribution. Semiempirical techniques[19, 20] have been used to calculate $\varepsilon_1 - \varepsilon_2$ as a function of various molecular parameters. Hence, it is possible to predict qualitatively the effect that a particular change in molecular geometry will have on the strength of the exchange coupling. A similar approach has been used by Kahn and Briat.[21]

If we consider copper(II) dimeric species, several experimentally observed trends can be rationalized in terms of Eq. (1). Consider the symmetrical bibridged system illustrated in Figure 5. Three molecular parameters can be defined: the Cu–L–Cu bridging angle ϕ, the dihedral angle δ between the CuX_2L_2 planes defined by the two halves of the dimer, and the twist angle τ between the Cu_2L_2 plane and the CuX_2 planes. It has been shown by Crawford *et al.*[22] for a series of compounds where $L = OH^-$ (with $\tau = 0$ and $\delta = 0$) that the exchange coupling obeys the relationship

$$2J = -74.53\phi + 7270 \text{ cm}^{-1}$$

with the coupling being FM for angles less than 97.5°. An equally good correlation can be made with the Cu–Cu distance, R:

$$2J = -4508R + 13\,018 \text{ cm}^{-1}$$

Calculations by both Hay *et al.*[19] and Bencini and Gatteschi[20] show that $\varepsilon_1 - \varepsilon_2$ varies strongly with angle, with $\varepsilon_1 - \varepsilon_2 = 0$ at an angle slightly

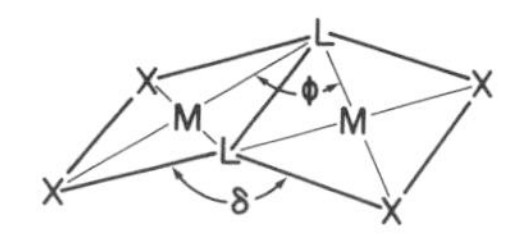

FIGURE 5. Diagram of a symmetrical bibridged dimeric system. Only the bridging angle ϕ and the dihedral angle δ are shown.

greater than 90°. The corresponding calculation for the $Cu_2Cl_6^{2-}$ dimer predicts the strongest FM interaction to be at a bridging angle near 88°. We will see later that many one-dimensional copper halide systems have bridging angles near this value.

Similarly, it has been shown experimentally that twisting the dimer ($\tau \neq 0$) causes the exchange interaction to become more FM in nature.[23, 24] Thus, in the $Cu_2Cl_6^{2-}$ dimer series, $2J/k_B$ changes from −55 K in $KCuCl_3$ ($\tau = 0$) to +56 K in $(C_6H_5)_4AsCuCl_3$ ($\tau = 48°$). Again the semi empirical molecular orbital calculations confirm this trend, predicting a maximum FM interaction near $\tau = 30°$, with AFM coupling at $\tau = 0°$ and 90°.[19] We have recently verified that AFM coupling does exist at $\tau = 86°$ in the bibridged tetrahedral $(CuBr_2)_n$ chain found in $[(C_2H_5)_2NH_2]_2Cu_4Br_{10} \cdot$ EtOH.[25]

Recently, an investigation of a "roof-shaped" dimer has been made by Charlot and co-workers.[26] They have shown that changing the dihedral angle δ between the CuX_2L_2 planes defined by the two halves of the dimers causes the interaction to become more FM. Again, we will see the importance of this correlation in some one-dimensional copper halide salts.

We have recently proposed an additional correlation in the dimeric systems $Cu_2Br_4L_2$.[27] In $KCuBr_3$[28] and NH_4CuBr_3,[29] which contain planar $Cu_2Br_4^{2-}$ anions, the value $2J/k_B \sim -190$ K is obtained. However, in the planar $Cu_2Br_4(py)_2$ dimer, where the pyridine ligands assume *trans* terminal positions, $2J/k_B \sim -20$ K. Structural characteristics suggest that the structures of the two dimers should be virtually identical except for the replacement of two nonbridging bromide ions by the more electronegative pyridine ligands. Nevertheless, the singlet–triplet splitting has changed by more than 150 K. We have suggested that this is evidence for a *trans* effect on the superexchange interaction; the pyridine removing electron density from the very polarizable bromide ion, leading to a sharp decrease in the AFM contribution to the exchange.

Finally, we conclude our discussion of interaction within simple dimers by referring to some recent work in Ni^{2+} dimers. A series of chloro-bridged salts have been studied[30]: with octahedral coordination, the exchange is invariably FM; with square pyramidal or trigonal bipyramidal coordination geometry, the compounds are AFM. No other structural trends are discernable. The theoretical arguments presented above can be extended to these many electron systems:

$$2J = \frac{1}{m^2} \sum_{i=1}^{m} \sum_{j=1}^{m} K_{ij} - \frac{1}{m^2} \sum_{i=1}^{m} \frac{\frac{1}{2}(\varepsilon_{2i} - \varepsilon_{2i-1})^2}{J_{ai,ai} - J_{ai,bi}} \tag{12}$$

where a pair of m is the number of unpaired electrons on each metal ion, the first sum runs over all m^2 combinations of electron i on metal a and electron j on metal b, and the second sum is over all distinct pairs of MOs formed from equivalent pairs of localized orbitals on the two metal sites. Again, the first term gives a FM contribution, the second an AFM interaction.

For Ni^{2+}, the localized orbitals are essentially metal d_{z^2} and $d_{x^2-y^2}$ orbitals. Because of the axially symmetric nature of the d_{z^2} orbital, the difference in energy between the MOs formed from that pair will be quite insensitive to small geometrical changes.[31] Thus, the AFM contributions in Eq. (12) will be dominated by the $d_{x^2-y^2}$ orbital contributions just as for the Cu^{2+} dimers, but reduced by a factor of 4. Hence, the Ni^{2+} case will be a much less-sensitive test of these superexchange arguments, in addition to being subject to two types of AFM contributions. Nevertheless, Bencini and Gatteschi have shown that a square pyramidal coordination around the metal ion leads to a more AFM interaction.[20] This is precisely the experimental correlation we have pointed out.[30]

The concept of orthogonality leading to FM interactions shows up more explicitly in the formulation of Kahn and Briat.[21] The exchange constant is written as

$$J = J_F + J_{AF} \tag{13}$$

where

$$J_F = \frac{2}{m^2} \sum_{\mu, \nu} J_{\mu\nu} \tag{13a}$$

and

$$J_{AF} = \frac{2}{m^2} \sum_{\mu} |\Delta_\mu S_{\mu\mu}| \tag{13b}$$

with $\Delta_\mu = \varepsilon_1 - \varepsilon_2$ and $S_{\mu\mu}$ is the overlay integral of the two equivalent magnetic orbitals on the two metal atoms. Thus, in this formalism, the AFM contribution clearly disappears when the orbitals are orthogonal. This has been very cleverly demonstrated by Kahn and co-workers in a pair of mixed-metal binuclear species.[32,33] In particular, in a mixed Cu–Co complex (both spin-1/2 ions) where the magnetic orbitals are restricted to be orthogonal by symmetry, the magnetic properties are characteristic simply of an isolated triplet state, implying that the singlet–triplet energy splitting is nearly an order-of-magnitude larger than $k_B T$.

5. *Theoretical Predictions*

For strictly one-dimensional systems with short-range interactions, it has long been recognized that there exists no transition to a state of long-range order at temperatures above $T = 0$ K.[4] Such systems are thus best described as paramagnets with greatly enhanced correlations at low relative temperatures $k_B T/J$. This section is devoted to a review of the available theoretical predictions for the susceptibilities and specific heats of ideal one-dimensional magnets of varying spin dimensionality and type of exchange interaction. These properties are emphasized since they are the most frequently used in determining the nature and magnitude of the exchange and the lattice dimensionality of new magnetic materials. Resonance properties and neutron scattering cross sections are held to be of less critical importance and will not be treated here. Information on one-dimensional dynamics is reviewed by Steiner *et al.*[2] We will concentrate on $S = 1/2$ and $S = 1$ linear chains. Most experimental and theoretical work has been done on systems with these low spin values; indeed, no examples are known of linear ferromagnets with a value of $S > 2$. While ferromagnetic compounds with $S > 1$ are known, the profusion of energy levels contributing to the effective moments makes interpretation of their behavior uncertain.

The theoretical evaluations of the static properties of the one-dimensional magnetic systems have been based upon the following Hamiltonian

$$\mathcal{H} = -2 \sum_{\langle i,j \rangle} [J_{xy}(S_i^x S_{i+1}^x + S_i^y S_{i+1}^y) + J_z S_i^z S_{i+1}^z] + \mathbf{D} \sum_i [(S_i^z)^2 - \tfrac{1}{3}S(S+1)] \tag{14}$$

This Hamiltonian is simpler than the general expression in Eq. (3) because interchain interactions are ignored and the antisymmetric term is neglected. The exchange term follows the usual convention that a positive J indicates parallel alignment of the spins (ferromagnetism). As indicated previously in Section 2, the limiting ratios of exchange parameters are characterized by historical names: $J_{xy} = J_z$, the Heisenberg model; $J_z/J_{xy} = 0$, the XY model; and $J_{xy}/J_z = 0$, the Ising model. The majority of calculations have been performed on one of these three limiting cases although more recent work has been done for variable ratios of J_{xy}/J_z.

The second term is due to single-ion anisotropy and vanishes exactly for $S = 1/2$. For an $S = 1$ state, the splitting between the $m_s = 0$ state and $m_s = \pm 1$ pair of states will equal $\mathbf{D}/R$. The sign of $\mathbf{D}$ has lead to an unfortunate amount of confusion since the Hamiltonian is frequently defined with the opposite sign to that given above. Citations of $\mathbf{D}$ for $S = 1$ compounds are thus ambiguous as to whether the doublet pair or singlet state is lower without knowledge of the author's convention. In this chapter we shall discuss all single-ion anisotropies in terms of the above definition for which positive $\mathbf{D}$ means the singlet ($m_s = 0$) state lies below the doublet ($m_s = \pm 1$) states. As a further means of reducing ambiguity the sign and magnitude of $\mathbf{D}$ will be followed by either (singlet low) or (doublet low) whenever an experimental value of **D** is cited. As an example, the zero-field splitting term in $NiCl_2 \cdot 2py$ was found by Klaaijson to be -27 K (doublet low).[34]

These static properties of the $S = 1/2$ magnetic linear chain will be discussed followed by a similar treatment for the $S = 1$ case. Within each subsection it has been found convenient to treat each of the limiting Hamiltonians (Heisenberg, XY, Ising) separately since the theoretical procedures involved are quite different.

5.1. Static Properties of the S = 1/2 Ferromagnetic Chain

5.1.1. S = 1/2 Heisenberg Exchange

The model Hamiltonian for the one-dimensional $S = 1/2$ Heisenberg magnet has proven to be particularly difficult to analyze. While the one-dimensional Ising and XY models can be solved exactly, results for the thermodynamics of the Heisenberg chain have been obtained only approximately. The susceptibility and specific heat of this model were first obtained in 1964 by Bonner and Fisher[35] who used machine calculations on finite rings and chains to extrapolate the behavior of infinite chains with both ferromagnetic and antiferromagnetic exchange.

An alternative approach was used by Baker *et al.*,[36] who obtained expressions for the energy and susceptibility of the ferromagnetic system using Padé approximation techniques and high-temperature series expansions. Their expression for the susceptibility χ is in good agreement[37] with the results of Bonner and Fisher and appears to be reliable to reduced temperatures ($k_B T/J$) as low as 0.05 where it predicts a divergence in χ as $T^{-2.7}$. This expression, and the results of Bonner and Fisher, have recently been used widely in the evaluation of experimental data (see Section 6.1).

A more general treatment of the susceptibility of the ferromagnetic chain is found in Obokata *et al.*[38] They have obtained a high-temperature series for the susceptibility as a function of both $k_B T/J$ and the ratio J_{xy}/J_z. The corresponding series for the special cases are found from the appropriate limiting ratio. Unfortunately, the series only contains seven terms and approximates the susceptibility poorly below $k_B T/J < 1.5$.

Predictions for the temperature dependence of the specific heat of the $S = 1/2$ Heisenberg chain are also found in the papers by Bonner and Fisher[35] and Baker *et al.*[36] Unlike the corresponding Ising and XY chains, the ferro and antiferromagnetic specific heats have different behavior. The results of Bonner and Fisher have recently been used[39] to analyze the specific heat, C_p, of cyclohexylammonium trichlorocopperate(II), $(C_6H_{11}NH_3)(CuCl_3)$, CHAC. Baker *et al.*[36] have derived a 21-term series expansion for the specific heat which agrees well with the Bonner–Fisher results when extended by Padé approximation techniques.

de Neef and co-workers[40] have published a number of articles on the specific heats of Heisenberg chains with $S \leq 5/2$. Their results were obtained using series-expansion techniques and finite-chain approximations. The most complete article[40d] presents the results in the useful form of polynomials of the variable $J(S)(S+1)/k_B T$ for ease in fitting experimental data. The actual numerical results are available only in an internal report.[40b]

A very thorough and useful pair of papers concerning the specific heat of linear chains has been published by Blöte.[41,42] The magnetic specific heats were calculated by the Bonner–Fisher method for both ferro and antiferromagnetic exchange for all $S \leq 5/2$ while in the presence of single-ion anisotropy. The effect of intermediate exchange interactions (variable ratios for J_{xy}/J_z) have been calculated for the $S = 1/2$ and $S = 1$ cases. The numerical results are tabulated. In addition, the problem of a linear chain with both single-ion anisotropy and intermediate exchange is treated. Since the number of variables (J_{xy}/J_z, **D**, S) becomes too large to cover completely, the author invites requests for the specific heat of a particular set of parameters to be directed to himself.

The specific heat C_p of linear chains in the low-temperature limit has been the subject of several studies.[43,44] Since that limit is usually experimentally unavailable due to the presence of three-dimensional interactions, we mention these works only because they provide a test for the results derived by other means. Takahashi[43] has calculated the specific heat of a $S = 1/2$ anisotropic Heisenberg ring at low temperatures and has conjectured the limiting analytical result:

$$\lim_{T \to 0} C = \frac{Nk}{3}(k_B T/J)$$

This limit is consistent with the results of Bonner and Fisher[35] and of Blöte[41] but disagrees with the results of de Neef[40b] at the lowest reduced temperatures ($k_B T/J < 0.3$ for the $S = 1/2$ ferromagnetic chain).[45]

5.1.2. $S = 1/2$ Ising Exchange

The extreme exchange anisotropy of the Ising interaction ($J_{xy} = 0$) renders it a simple model for theoretical investigation. Exact solutions are known for both the parallel and perpendicular susceptibilities and the specific heat of the $S = 1/2$ linear chain. The model was first studied by Ising[46] who evaluated $\chi_{\parallel}$. Fisher[47] and Katsura[48] independently arrived at the expression for $\chi_{\perp}$, and the magnetic specific-heat equation is found in Domb.[49]

For the system with an exchange Hamiltonian intermediate between Ising and Heisenberg, several results are available. Johnson[50] has presented analytical results for the entropy and magnetic specific heat as functions of temperature for the ratio $J_z/J_{xy} = 2$ which agree well with the results obtained from the Bonner–Fisher extrapolations.[35] Blöte[42] has calculated and tabulated the magnetic specific heats for varying ratios of J_{xy}/J_z for both $S = 1/2$ ferro and antiferromagnetic chains. More recently Johnson and Bonner[51] have presented new analytical results for the low-temperature thermodynamics of the Ising–Heisenberg linear ferromagnet in small fields. Expressions are derived for both specific heat and susceptibility as a function of J_{xy}/J_z and show crossovers with different magnetic and thermal dependences as the anisotropy ratio changes.

5.1.3. $S = 1/2$ *XY Exchange*

The limiting case of $J_z/J_{xy} = 0$ is known as the *XY* model (or planar Heisenberg model if the spins are constrained to lie within the plane). Like the Ising limit, it is amenable to exact solution: the spectral excitations were obtained by Lieb *et al.*[52] Katsura[48] has obtained exact integral equations for both the specific heat and axial susceptibility χ_z. Blöte[42] has tabulated the numerical specific-heat predictions of Katsura for the limiting *XY* model as well as for several other ratios of J_z/J_{xy}. The planar susceptibility χ_{xy} of the $S = 1/2$ *XY* antiferromagnetic chain has recently been calculated by Duxbury *et al.*[53]

5.2. *Static Properties of the* $S = 1$ *Ferromagnetic Chain*

5.2.1. $S = 1$ *Heisenberg Exchange*

The susceptibility of the $S = 1$ Heisenberg ferromagnet has been calculated by Weng[54] in his thesis using the Bonner–Fisher technique. His calculations were made in the absence of single-ion anisotropy. Using his results for $S = 1$, those of Bonner and Fisher for $S = 1/2$, and the exact expression of Fisher[55] for $S = \infty$, Weng devised an interpolation scheme to predict the thermodynamic properties of a Heisenberg chain of general S.

Incorporation of single-ion anisotropy into the Heisenberg $S = 1$ susceptibility has been accomplished by de Neef in his thesis.[40c] High-temperature series expansions were obtained as a function of both $J/k_B T$ and $\mathbf{D}/k_B T$ for both ferro and antiferromagnetic exchange. The coefficients of the series are tabulated in the thesis but have unfortunately not yet been published.

An early expression for the susceptibility of this model as a function of both J and $\mathbf{D}$ was derived in the molecular field approximation by Watanabe.[56] In spite of inappropriateness of this model to low-dimensional systems, this expression has been used fairly successfully in obtaining J and $\mathbf{D}$ parameters in cases for which $\mathbf{D} \gg J$.

The basic study of the specific heat of this model as a function of both J and **D** is that of Blöte[42] as mentioned above. Blöte also considers the case of variable exchange anisotropy but in the absence of **D**. De Neef and de Jonge[57] have carried out similar calculations that agree well with those of Blöte. There has also been a low-temperature calculation of the isotropic $S = 1$ antiferromagnetic specific heat by Moses *et al.*,[58] who find a linear contribution to C_p for $k_B T/J < 0.6$. This agrees well with Blöte's calculations.

5.2.2. $S = 1$ Ising Exchange

The one-dimensional Ising problem in the absence of single-ion anisotropy has been solved exactly by Suzuki *et al.*[59] for both the $S = 1$ and $S = 3/2$ cases. Expressions for the specific heats and susceptibilities were obtained. Under the addition of single-ion anisotropy, the specific heat and susceptibility have been calculated by Kowalski.[60] An independent calculation of the susceptibility as a function of J and **D** has recently been made by O'Brien *et al.*[61]

5.2.3. $S = 1$ XY Exchange

Physical realizations of the XY model are found in compounds of magnetically exchanged ions with ground-state Kramers doublets which lie well below other doublets in energy: these ground states can be treated as effective spin doublets at sufficiently low temperatures. Consequently little attention has been paid to the evaluation of the properties of XY model systems with $S > 1/2$. The only work we are aware of which treats the $S = 1$ XY linear chain is the aforementioned study by Blöte[42] of the magnetic specific heat of the XY chain itself and also on the $S = 1$ chain with interactions intermediate between XY and Heisenberg.

6. Recent Progress in Ferromagnetic Linear Chain Systems

We now turn to a survey of the known first-row transition-metal linear chain ferromagnets. The results of the magnetic and thermal studies are discussed in terms of the crystal structures (where known) to illustrate the principles underlying one dimensionality and ferromagnetic exchange as analyzed in previous sections. This section begins with a review of Cu^{2+} salts followed by shorter discussions on Ni^{2+}, Co^{2+}, and Fe^{2+} salts.

6.1. Copper(II) Salts

The first example of a FM spin-1/2 Heisenberg linear chain was $[(CH_3)_3NH]CuCl_3 \cdot 2H_2O$. Susceptibility and heat-capacity measurements have been interpreted in terms of a one-dimensional system with very weak coupling; $J/k_B = 0.85$ K.[62] Structurally, it contains planar $CuCl_2(H_2O)_2$ ions which stack so as to form an asymmetrical bibridged linear chain[63]:

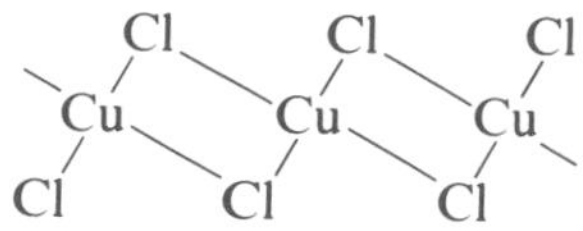

Cl Cl Cl
Cu Cu Cu
Cl Cl Cl

The water molecules lie above and below the plane defined by the copper and chlorine atoms. Adjacent chains are hydrogen bonded into sheets via the noncoordinated chloride ion, and these sheets are separated by the bulky trimethylammonium ions. The bridging angles, ϕ and ϕ', are 88.96° and 95.87°. The semiempirical MO calculations show that $\varepsilon_1 - \varepsilon_2$ is always small, but should be ferromagnetic at $\phi = 90°$ and have an increasingly AFM component as ϕ deviates from 90°. Thus, the FM intrachain interaction is not surprising (although in contrast to similar stacking arrangements of the antiferromagnetically coupled chains in $CuCl_2 \cdot 2H_2O$[9,64] and $CuCl_2 \cdot 2py$[65]).

The small value of $J_{FM}/k_B = 0.85$ K has made it an excellent candidate for very-low-temperature C_p and χ studies. However, as a consequence, the ratio of interchain coupling to intrachain coupling is large (0.05), as is the spin anisotropy (~10%). The Néel temperature is 0.165 K, so the value $k_B T_c/J \sim 0.2$ is quite large. This means the system is far from being representative of an ideal one-dimensional system. However, it is just these qualities that make the results of the studies of this system extremely interesting. The magnetic parameters of this and other ferromagnetic chains are summarized in Table 1.

The magnetic heat capacity (Figure 6) exhibits a sharp spike at T_c and then, at higher temperatures, a broad maximum characteristic of a low-dimensional system. The magnetic entropy change reported for a powdered sample of $\Delta S/R = 0.69$ is equal to the theoretical value of $\Delta S/R = \ln 2 = 0.693$. Only 28% of the entropy change occurs below T_c, indicating the presence of substantial short-range order at temperatures above T_c.

TABLE 1. Magnetic Parameters of Ferromagnetic Copper Chains

Compound	J/k_B (K)	T_N (K)	J'/J	Reference
$[(CH_3)_3NH]CuCl_3 \cdot 2H_2O$	0.85	0.165	5×10^{-2}	62
$[C_6H_{11}NH_3]CuCl_3$, CHAC	70	2.18	1.1×10^{-3}	67
	45			39
$CuCl_2 \cdot DMSO$	45	4.8	3.6×10^{-2}	68, 69
$CuCl_2 \cdot TMSO$	39	~3	2×10^{-2}	68
$N(CH_3)_4CuCl_3$, TMCuC	30	<2	$\sim 10^{-4}$	71
$[N(CH_3)_3H]_3Cu_2Cl_7$, TTMCuC	50	<2	$< 10^{-3}$	71
$(C_3H_6N_2H_6)Cu(ox)_2$	19	—	0.6	77

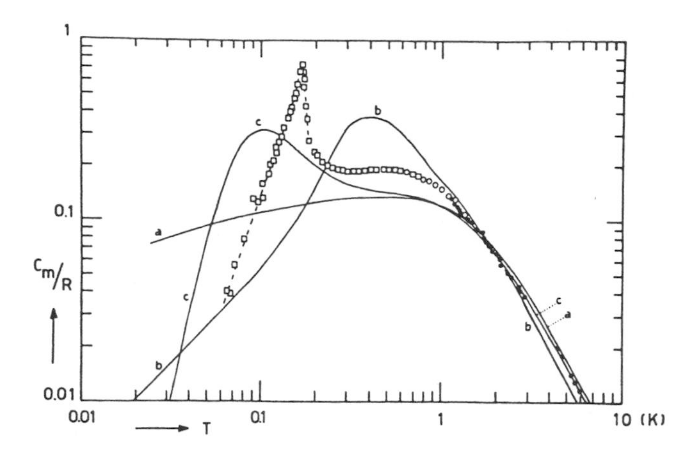

FIGURE 6. Magnetic specific heat of $[(CH_3)_3NH]CuCl_3 \cdot 2H_2O$ vs temperature on a double logarithm scale. See the text for details. (Reprinted from Reference 62, with permission.)

Detailed analysis revealed the data cannot be accounted for by a simple one-dimensional system. Curve a, which represents the C_p of a one-dimensional ferromagnetic Heisenberg chain, clearly does not fit the data below 1 K. Neither does the two-dimensional Heisenberg prediction (curve b). However, the observed values of C_p lie intermediate between the two curves, indicating substantial interchain coupling. An additional complication is the presence of spin anisotropy, which causes C_p to rise again at

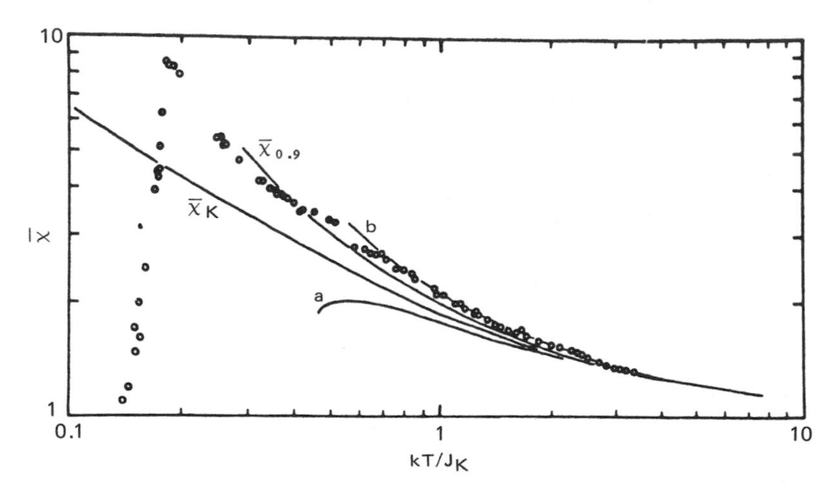

FIGURE 7. Single-crystal magnetic susceptibility of $[(CH_3)_3NH]CuCl_3 \cdot 2H_2O$ measured parallel to the chain axis. Data are plotted as $\chi T/C$ vs $k_B T/J$. See the text for details. (Reprinted from Reference 62, with permission.)

lower temperatures (curve c, for 10% Ising-like anisotropy). However, the data clearly show the effects of spin dimensionality crossover. For $T > 1$ K, one-dimensional behavior is observed. Between 0.2 and 1 K, two-dimensional behavior is seen; and three-dimensional behavior is seen below 0.2 K.

Quantitative estimates of the extent of interchain coupling and spin anisotropy were obtained from single-crystal susceptibility data. Analysis of the perpendicular susceptibility (χ_b and χ_c) indicated $J_{xy}/J_z \sim 0.9$, while an antiferromagnetic coupling between chains exists with $z_{\mathrm{AF}} J_{AF} = 0.024$ K. Figure 7 shows the reduced susceptibility data along the crystallographic a axis, $\bar{\chi} = \chi_a T/C_a$ versus reduced temperature, $k_B T/J$. In this figure, $\bar{\chi}_H$ is the Heisenberg susceptibility, while $\bar{\chi}_{0.9}$ is the susceptibility for $J_{xy}/J_z = 0.9$. Since $\bar{\chi}_{0.9}$ lies below the experimental data for $T > 0.4$ K, it is concluded that J' (the coupling between the nearest-neighbor chains) is FM. Curve b is the prediction assuming $z'J'/zJ = 0.05$, while curve c is for $z'J'/zJ = -0.05$ (z is the number of nearest neighbors in the chain and z' is the number of nearest-neighbor chains). Thus, FM layers exist. The effect of the AFM interchain coupling, J'', which couples the layers three dimensionally, is observable for $T < 0.4$ K. Again, dimensionality crossover behavior is observed: one-dimensional above 3 K, two-dimensional between 0.4 and 3 K, and three-dimensional below 0.4 K. It is noted that dimensionality crossover effects occur over a larger temperature range in the susceptibility data than in the heat-capacity data,[66] and also that the effects of spin anisotropy are more clearly seen.

Following the studies on $(CH_3)_3NHCuCl_3 \cdot 2H_2O$ by Carlin and coworkers, we have reported on a series of one-dimensional FMs with much stronger interchain coupling and which are much better realizations of a quantum one-dimensional ferromagnet. The investigation of these systems was all based on the recognition that symmetric Cu–Cl–Cu bridging angles in the range 85°–90° would lead to FM coupling. Fortunately, a large number of such systems were either available in the literature or were being synthesized in our laboratory.

The best characterized member of this series to date is cyclohexylammonium copper trichloride,[67] $C_6H_{11}NH_3CuCl_3$, CHAC. The structure of the orthorhombic salt is shown in Figure 8. It contains symmetrically bibridged linear chains parallel to the c axis, with $\phi = 86°$ and $\delta = 152°$. The copper ions assume a nearly perfect square pyramidal geometry. The $C_6H_{11}NH_3^+$ ions hydrogen bond the chains together into sheets in the bc plane through the apical chloride ions. The bulky $C_6H_{11}NH_3^+$ cations effectively separate these sheets from each other. Based on these structural characteristics, we anticipate the following magnetic exchange pathways, as shown schematically on Figure 9: strong FM coupling J_1 along the chain; weak coupling J_2 between chains in the bc plane, since the

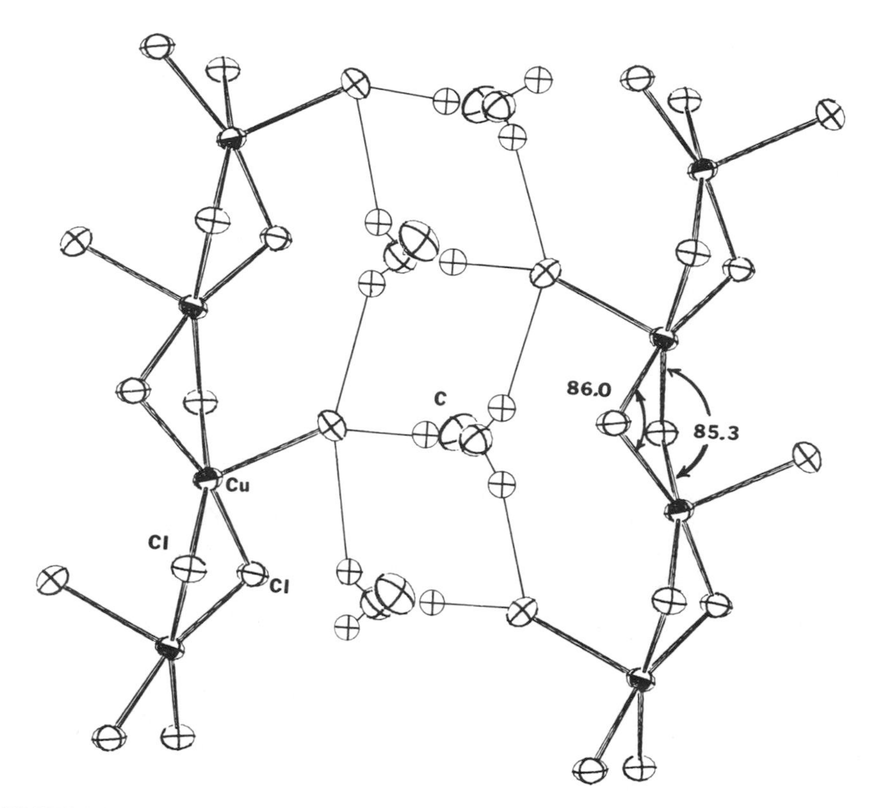

FIGURE 8. Projection of the crystal structure of CHAC onto the *bc* plane. The chains along the *c* axis are hydrogen bonded together (faint lines) in the *b* direction. (From Reference 67.)

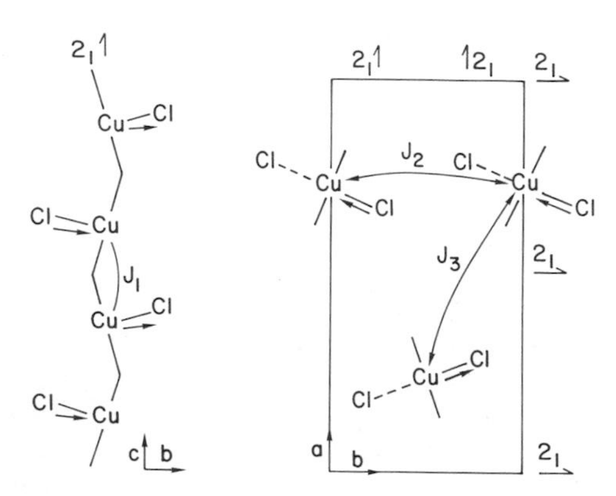

FIGURE 9. Exchange pathways and spin structure of $(C_6H_{11}NH_3)CuCl_3$, CHAC, below T_c. The dominant exchange coupling is J_1, the ferromagnetic interaction within the chains.

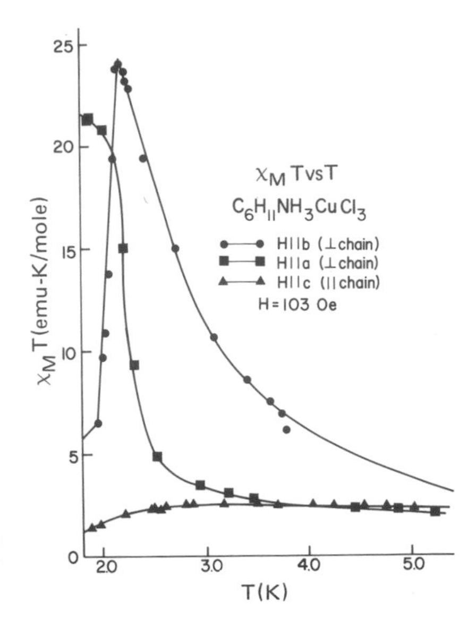

FIGURE 10. Single-crystal magnetic susceptibility data for CHAC. Data are plotted as $\chi_m T$ vs T. (From Reference 67.)

axial chloride ions involved in the hydrogen bonding network will have essentially no spin density on them; finally, much weaker coupling J_3 of a dipolar nature between the sheets.

The powder susceptibility of CHAC diverges much faster than Curie behavior, indicative of strong FM coupling. At $T = 2.2$ K, the ratio of χ_{exp}/χ_{Curie} is approximately 30. The susceptibility diverges as $1/T^{2.6}$, in good agreement with theoretical predictions for a one-dimensional ferromagnet of 2.7–2.8.[35, 36] An isotropic exchange constant of $J/k_B = 70$ K is extracted from the data. This exchange constant is nearly two orders of magnitude larger than that for $(CH_3)_3NHCuCl_3 \cdot 2H_2O$. Thus, while the latter was ideal for very low temperature investigations, CHAC will be an excellent candidate for the study of spin dynamics since the large exchange constant will lead to strong spin correlations over a very large temperature range.

Single-crystal magnetic susceptibility studies have been undertaken which show the onset of long-range order at $T_c = 2.18$ K. The data exhibit the onset of an Ising-like anisotropy at $T \sim 10T_c$. This is shown in Figure 10. For $T > 4$ K, the value of χ_a is nearly indistinguishable from χ_c (lower set of data, Figure 10).

The magnetization data (Figure 11) show two unusual features. The value of the magnetization when the field is along the b axis, M_b, increases smoothly with field for $T > T_c$. However, below T_c, M_b rises rapidly to nearly its saturation value when the applied field reaches a critical field H_c of approximately 100 Oe. This behavior is characteristic of a metamagnet.

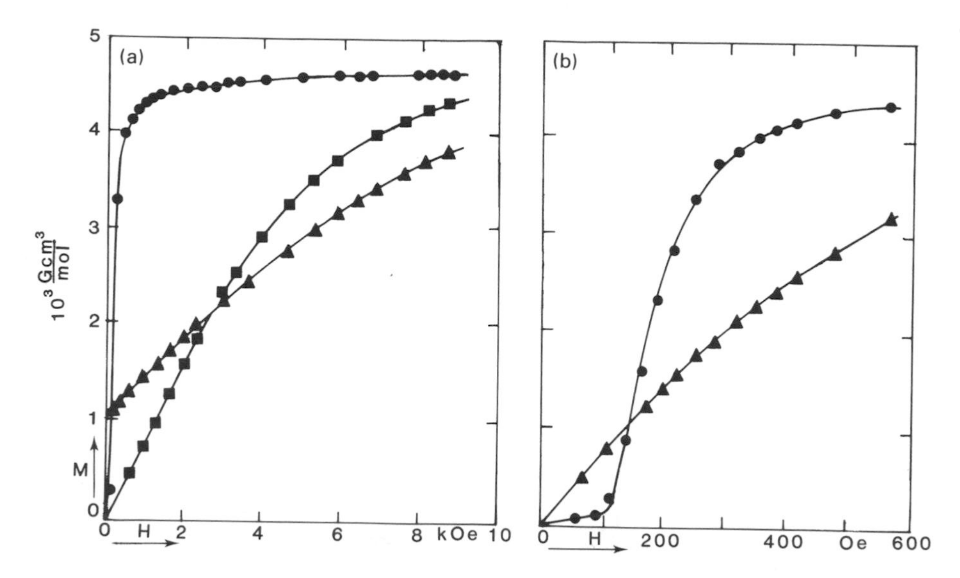

FIGURE 11. (a) Magnetization of CHAC vs external field at $T = 2.0$ K. Triangles indicate the field parallel to the *a* axis; circles, field parallel to *b* axis; and squares, field parallel to *c*. (b) Magnetization of CHAC with the field applied parallel to the *b* axis. Filled circles were data taken at 1.91 K; filled triangles were taken above the Curie temperature, at 2.47 K. Note the expanded *x*-axis scale.

For $H < H_c$, antiferromagnetically coupled spins lie (nearly) parallel and antiparallel to the *b* axis. For $H > H_c$, the spin arrangement goes directly to a FM state, e.g., all spins parallel.

In contrast, the value of M_a goes smoothly to zero as $H \to 0$ for $T > T_c$, but attains a nonzero value at $H = 0$ for $T < T_c$. The existence of a residual moment parallel to the *a* axis implies a noncollinear spin arrangement in the antiferromagnetic state. In particular, it means the exchange coupling J_2 must be FM also, since copper atoms coupled by J_2 are related by a simple unit cell translation. The AFM coupling thus must be J_3, yielding the spin arrangement shown in Figure 9. From the magnitude of the residual moment along the *a* axis, it is estimated that the spins make an angle of $\sim 17°$, and thus the moment is oriented nearly parallel to the projection of the axial Cu–Cl vector onto the *b* axis.

Magnitudes of J_2 and J_3 can be obtained from the metamagnetic phase diagram, which yields $J_{AF} = J_3 = -3.6 \times 10^{-3}$ K ($z_{AF} = 4$), so that $|J_3/J_1| < 10^{-4}$. Similarly, the tricritical temperature T_t yields $J_F = J_2 = 0.07$ K. Thus, $J_2/J_1 \sim 10^{-3}$. In addition, the spin anisotropy field can be estimated from the value of χ ($T = 0$ *K*) in the mean-field formalism to be $H_A = 9.0 \times 10^3$ Oe. This corresponds to a spin anisotropy of $J_{xy}/J_z = 0.99$.

The heat capacity of this system has been measured by de Jonge and co-workers.[39] This confirmed the presence of strong one-dimensional FM interactions, but yielded a lower value of J/k_B (45 K). The data agree well with the predicted curve for a spin-1/2 Heisenberg ferromagnet above 5 K. However, in the region from 3 to 5 K, the observed data are substantially larger than the theory. Calculations by Blöte[42] have predicted that an Ising-like anisotropy will lead to a "bump" in the heat capacity over the region $k_B T/J \sim 0.01$–0.1. Thus, the heat-capacity data would appear to confirm the estimate of 1% Ising-like anisotropy. (Detailed analysis, just transmitted, give 2% anisotropy from C_p data.)

The next series of one-dimensional spin-1/2 salts with FM intrachain coupling to be investigated were $CuCl_2 \cdot DMSO$ and $CuCl_2 \cdot TMSO$ [DMSO is dimethylsulfoxide, $(CH_3)_2SO$, and TMSO is tetramethylenesulfoxide, C_4H_8SO; both coordinate through the oxygen].[68] The complex structure of the TMSO salt is illustrated in Figure 12. Each pair of copper atoms is bridged by three ligands: (1) a symmetrical Cu–Cl–Cu bridge with $\phi = 86.7°$; (2) an asymmetrical Cu–Cl–Cu bridge with $\phi = 81.0°$; and (3) an asymmetrical Cu–O–Cu bridge with $\phi = 79.7°$. The values for the DMSO are very similar. (It may help to visualize this bridging arrangement as a very severely distorted tribridged system similar to TMMC.) The symmetric

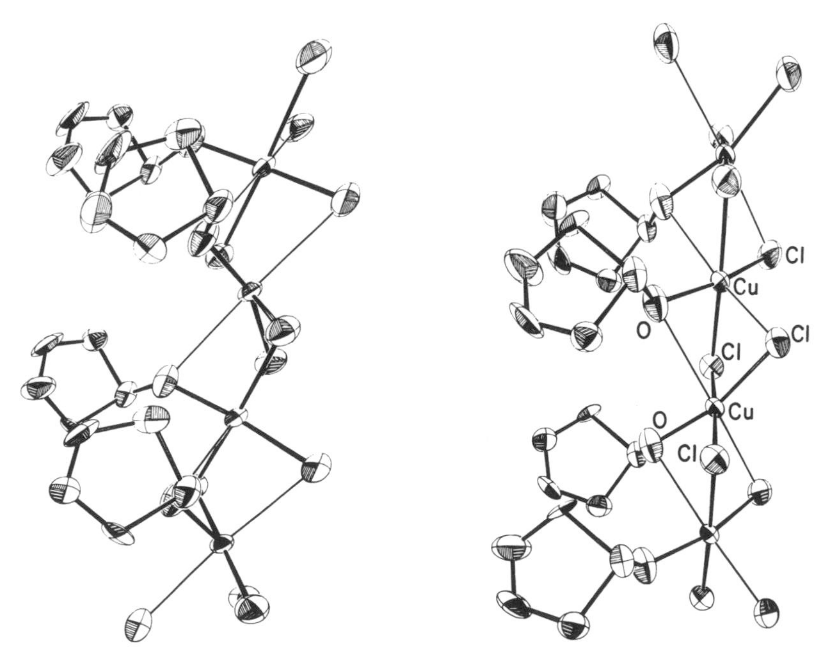

FIGURE 12. Crystal structure of $CuCl_2 \cdot TMSO$. Each copper ion is bridged by two chlorines and one sulfoxide ligand. (From Reference 68.)

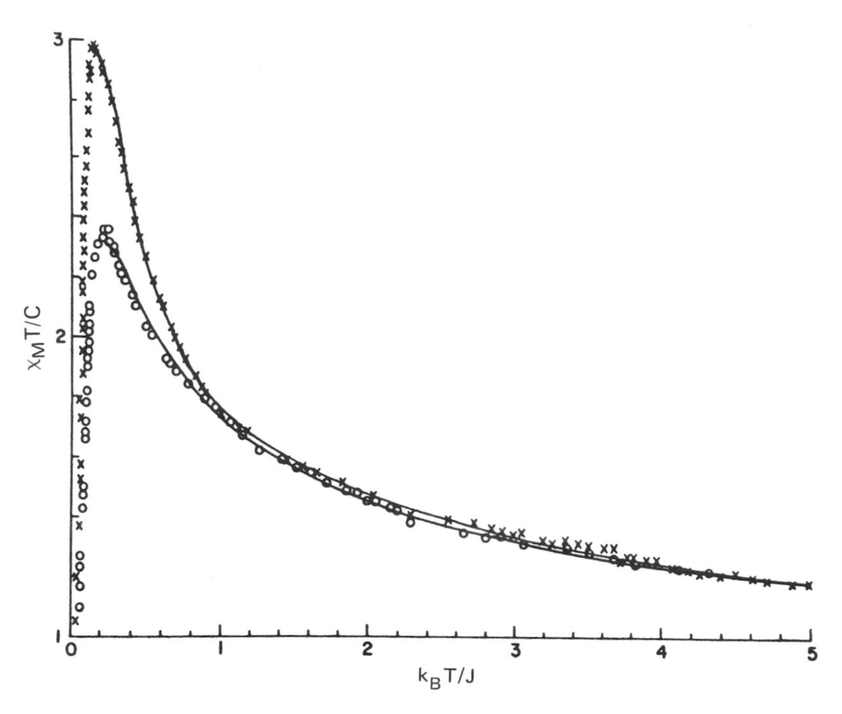

FIGURE 13. Powder magnetic susceptibility data plotted as $\chi_m T/C$ vs reduced temperature $k_B T/J$ for $CuCl_2 \cdot TMSO$ (×) and $CuCl_2 \cdot TMSO$ (○). (From Reference 68.)

bridge should be strongly FM, since ϕ is near the optimum value for FM exchange. The work by Algra *et al.*[62] on $(CH_3)_3NHCuCl_3 \cdot 2H_2O$ indicates that the exchange in asymmetric bridges is weak compared to what might be expected for symmetric bridges. Thus, the FM coupling would be expected to dominate. Powder susceptibility studies down to 2 K (Figure 13) confirmed this conjecture. The FM interactions within the chains are large ($J/k_B = 39$ and 45 K, respectively, for the TMSO and DMSO salts). The powder results also indicate the presence of rather substantial AFM interchain coupling ($|J'/J| \sim 10^{-2}$).

Single-crystal magnetic studies on $CuCl_2 \cdot DMSO$ confirm these results.[69] The salt orders AFM at 4.80 K ($k_B T/J = 0.12$) and undergoes a spin-flop transition with $H_{SF} = 3.9$ kOe at 0 K. A mean-field theory analysis of these data confirms that there is substantial interchain coupling ($H_{AF}/H_{FM} = 0.036$), so the system, while better than $(CH_3)_3NHCuCl_3 \cdot 2H_2O$, is hardly an ideal one-dimensional system. More interestingly, however, is the estimate of the spin anisotropy of only 0.05%. Similar studies underway on $CuCl_2 \cdot TMSO$ indicate an even smaller amount of anisotropy and smaller interchain coupling. Thus, these salts are nearly ideal spin-1/2

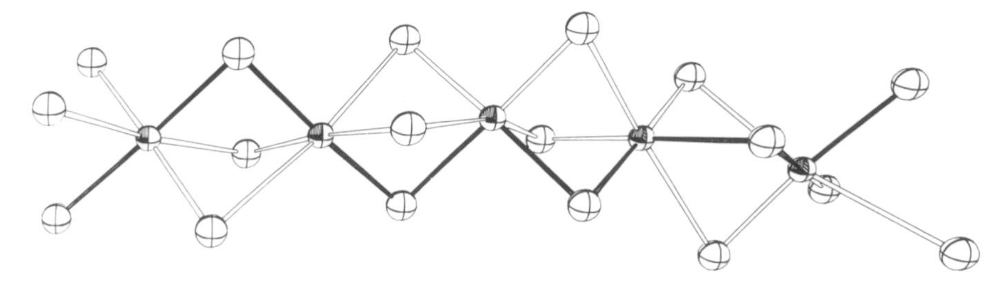

FIGURE 14. Chain structure of $(CH_3)_4NCuCl_3$, TMCuC. The five different symmetrical Cl–Cu–Cl bridges are drawn with bold lines. The segment shown is one half of one unit cell. Figure drawn according to crystallographic data in Reference 70.

Heisenberg systems. It is clear that a wide variety of studies would be desirable on these two systems.

At this stage, a search for other salts with similar structures was undertaken. The structure of the copper analog of TMMC, $[N(CH_3)_4]CuCl_3$, TMCuC, had been reported by Woenk and Spek[70] in 1976, but no one had attempted a magnetic study of the system, perhaps with some justification. The chain structure, illustrated in Figure 14, again is a severely distorted tribridged chain, quite similar to that in the DMSO and TMSO salts. However, there are five crystallographically different copper atoms in the chain, as well as five different symmetrical Cu–Cl–Cu bridges, and hence five different exchange constants. The results,[71] shown in Figure 15, clearly

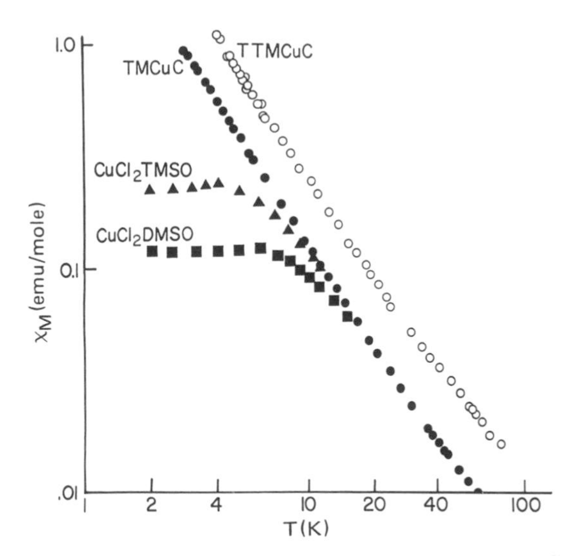

FIGURE 15. Powder susceptibility vs temperature for a series of quasi-one-dimensional ferromagnets.

show the much better extent of isolation of the chains. The value of J/k_B is 30 K and the ratio $|J'/J|$ is estimated to be about 10^{-4}. This is a very good one-dimensional system magnetically; however, the structural characteristics leave much to be desired.

Another related salt $[(CH_3)_3NH]_3Cu_2Cl_7)$, TTMCuC, also contains the requisite linear chain system[72] to lead to FM intrachain coupling. However, in addition, it contains stacks of tetrahedral $CuCl_4^{2-}$ anions. These data have also been analyzed in terms of a one-dimensional Heisenberg ferromagnet, yielding $J_F/k_B = 50$ K.[71] Unfortunately, the stacks of tetrahedra interact magnetically, so the analysis is complicated. The susceptibility data is presented in Figure 15.

This structural feature of symmetrical Cu–Cl–Cu linkages with bridging angles of 85°–90° is clearly a well-documented criteria for FM interactions. We have used it recently in two other instances. The salt $(CH_3)_2CHNH_3CuCl_3$ contains linear chains of AFM coupled $Cu_2Cl_6^{2-}$ dimers.[73] However, it undergoes a structural phase transition at 56°C to the distorted tribridged structure which characteristically lends to FM coupling. A sample was heated above the phase transition, quenched, and its susceptibility measured. The results, shown in Figure 16, clearly demonstrate the presence of FM interactions (albeit, modified by the presence of relatively strong AFM interactions at lower temperatures). The system represents a model for the magnetic behavior during a spin–Peirels type phase transition.

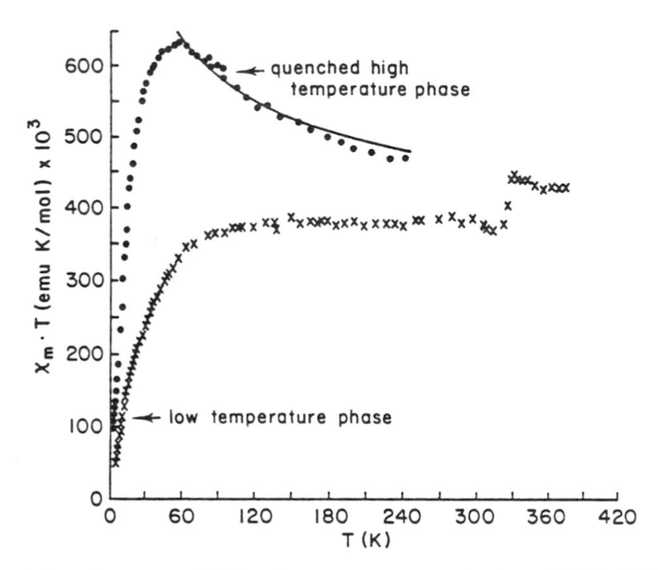

FIGURE 16. Powder susceptibility data vs temperature for $[(CH_3)_2CHNH_3]CuCl_3$, IPACuCl$_3$, plotted as $\chi_m T$ vs T. The quenched high-temperature phase (circles) shows evidence of ferromagnetic exchange while the normal low-temperature phase (×) behaves as a chain of antiferromagnetic dimers. (From Reference 73.)

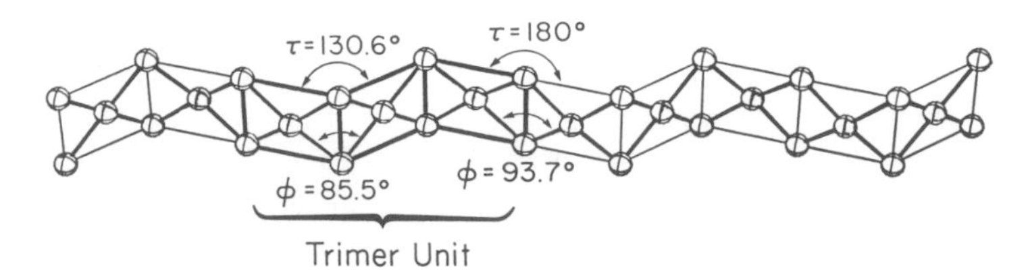

FIGURE 17. Structure of the bibridged chains in $3CuCl_2 \cdot 2dx$, dx = 1,4-dioxane. Within each trimer, the coppers are coupled together with strong ferromagnetic exchange. Figure drawn from crystallographic information in Reference 75.

The application of these magneto-structural correlations to extended systems is shown most dramatically in a recent study in our laboratory on $3CuCl_2 \cdot 2dx$, where dx = 1,4-dioxane.[74] The structure,[75] illustrated in Figure 17, consists of a symmetrically bibridged linear chain with an unusual repeat sequence of ϕ and δ values. Pairs of bridges with $\phi_2 = 85.5°$ and $\delta_2 = 130.6°$ are separated by bridges with $\phi_1 = 93.7°$ and $\delta_1 = 180°$. Clearly the former interaction should be strongly FM; the latter could be either weakly FM or AFM. Each pair of ϕ_2, δ_2 bridges defines a strongly FM-coupled group of three copper ions. Thus, a linear chain of weakly coupled FM trimers is anticipated. Figure 18 shows the plot of $\chi_M T$ vs T obtained.

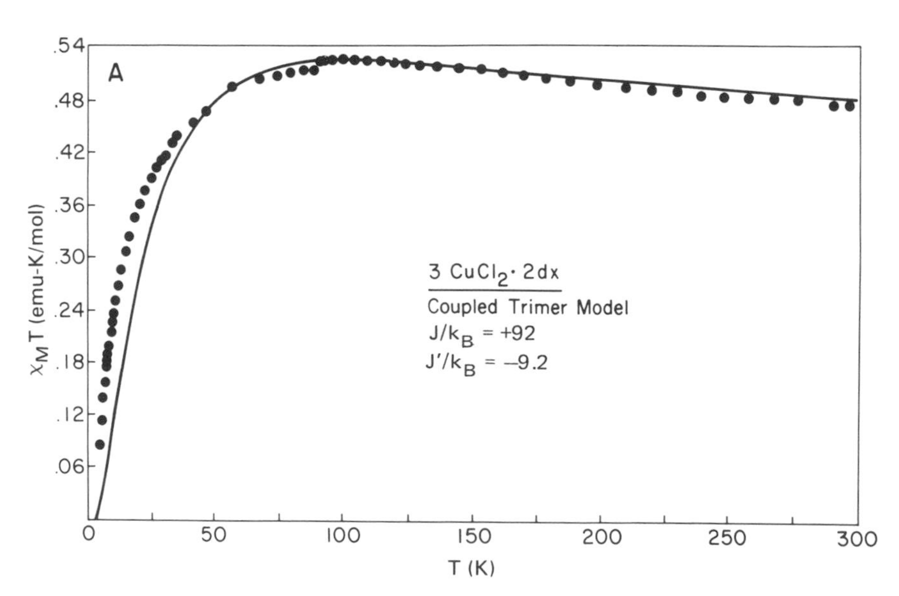

FIGURE 18. Powder susceptibility data plotted as $\chi_m T$ vs temperature for $3CuCl_2 \cdot 2dx$, dx = 1,4-dioxane. Solid curve is a theoretical fit to the high-temperature data ($T > 50$ K) based on a model of ferromagnetic Heisenberg trimers ($J/k_B = 92$ K) coupled together with antiferromagnetic Ising exchange ($J'/k_B = -9.2$ K). (From Reference 74.)

A dominant FM behavior is observed, which is influenced by weaker AFM coupling. Since the chains are well isolated, this latter is assumed to be due to the interactions between trimeric units. The solid curve gives the theoretical fit, using an approximate expression derived in this laboratory. The high-temperature fit is excellent, with $J_{FM}/k_B = 92$ K and $J_{AFM}/k_B = -9.2$ K. The low-temperature fit fails because of the Ising nature of the approximation. However, we can treat the low-temperature data independently. For $T \leq 20$ K, only the low-lying quartet of each trimer will be occupied. Thus the low-temperature data are well fit to the expression for the susceptibility of a $S = 3/2$ linear chain, yielding $J_{AFM}/k_B = -7.6$ K. We believe these results clearly indicate the usefulness of magneto-structural correlations in treating extended systems.

Recently, there has been much interest in the superexchange pathways provided by the organic ligand, the oxalate ion[76]:

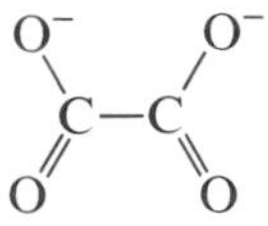

```
O⁻    O⁻
 \    /
  C—C
 //   \\
O      O
```

The compounds studied to date have all had dominant AFM interactions. We have recently studied the salt $(C_3H_6N_2H_6)Cu(ox)_2$ (Figure 19).[77] Note that the bridging angle is 93.1°. Based on the structural characteristics, the susceptibility data have been analyzed in terms of a linear chain system ($J/k_B = 19$ K) with strong AFM coupling between chains ($J/k_B = -11$ K). Because of the lack of suitable calculations for a $S = 1/2$ Heisenberg rectangular lattice, it was necessary to resort to calculations for an Ising system. The challenge to structural chemists is to separate the chains (reducing J') without disrupting the chain arrangement.

Mention will be made of one other compound found to behave as an $S = 1/2$ ferromagnetic linear chain although it does not contain copper ions. The series of donor–acceptor compounds $TTF^{\dot{+}} \cdot MS_4C_4(CF_3)_4^{\dot{-}}$ (TTF = tetrathiafulvalene) crystallize as chains of flat cations and anions stacked alternately.[78] The electrical properties are highly anisotropic. When M = Cu or Au, only the $TTF^{\dot{+}}$ ions possess moments that couple together forming antiferromagnetic chains which undergo spin–Peierls dimerizations at low temperatures. When M = Pt, both types of ions act as $S = 1/2$ entities with $g \approx 2$ and are coupled together within the chains ferromagnetically. Analysis of the magnetic specific heat showed $J/k_B = 66$ K with 10% Ising-like anisotropy.[79]

The two most useful systems in this array of quasi-one-dimensional ferromagnets would appear to be $CuCl_2 \cdot DMSO$ and CHAC. Large crystals, suitable for neutron scattering, nuclear quadrupole resonance, nuclear magnetic resonance studies, etc, can be grown readily. The DMSO salt is nearly

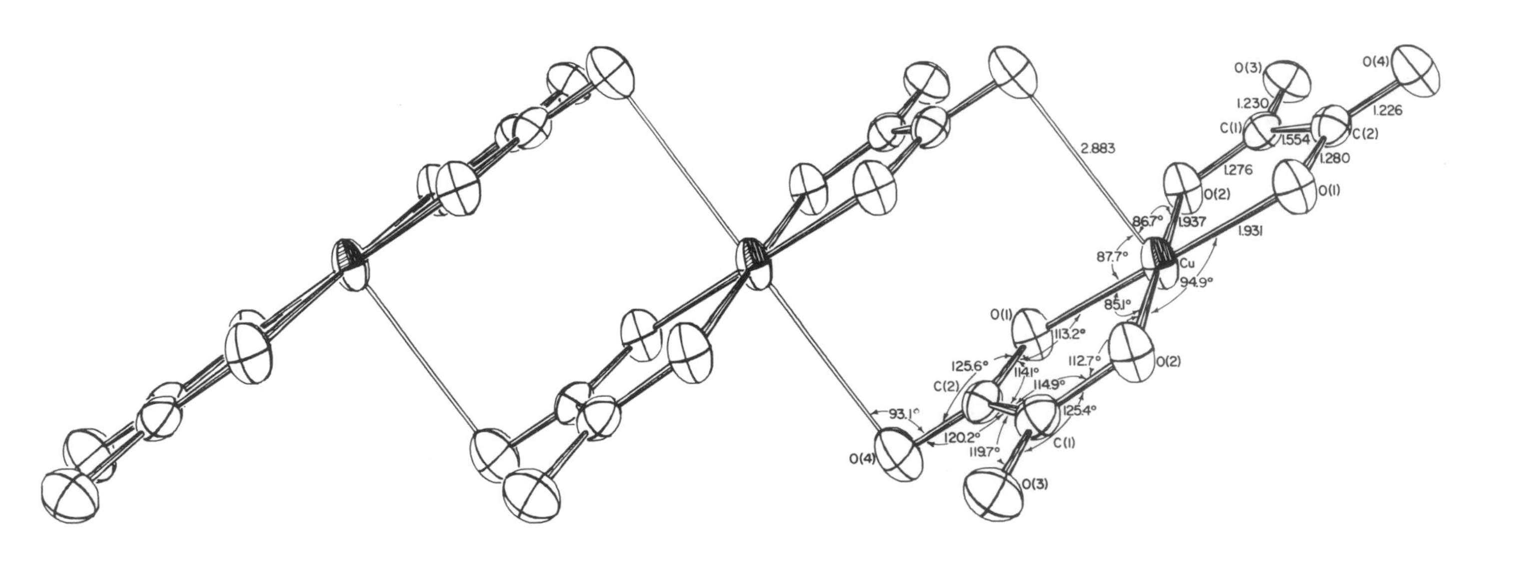

FIGURE 19. Structure of the chains of $Cu(ox)_2^{2-}$ ions in $(C_3H_6N_2H_6)Cu(ox)_2$, ox = $C_2O_4^{2-}$. Each copper ion has four short and two long oxygen bonds. (From Reference 77.)

a perfect Heisenberg system, but with significant two-dimensional interactions. It should be ideal for studying spatial dimensionality crossover. CHAC, in contrast, has well-isolated chains to study, but a significant spin anisotropy.

6.2. *Nickel(II) Salts*

The most widely studied of all the one-dimensional $S = 1$ ferromagnets is $CsNiF_3$.[2] The Ni^{2+} ions form nearly isolated tribridged chains along the c axis of the hexagonal crystal.[80] The crystal-field anisotropy favors spin orientation in the plane perpendicular to the c axis. The chains are well separated by the bulk of the Cs^+ cations which prevents the formation of long-range order above 2.61 K.[81] The three-dimensional order is built up of ferromagnetic planes which are coupled antiferromagnetically[82] in such a manner as to minimize the dipolar energy. Metamagnetic phase transitions are observed in applied fields of several kilogauss.[83] The properties of this compound are reviewed in Reference 2 where it is concluded $CsNiF_3$ is a good example of the one-dimensional, $S = 1$ (easy plane) ferromagnet.

A wide variety of values have been reported for J and $\mathbf{D}$ of this compound, due to both experimental difficulties (the magnetic susceptibility is highly anisotropic at low temperatures) and the utilization of inappropriate models. The most reliable values appear to be those of Dupas and Renard[84] based upon the analysis of their careful single-crystal susceptibility data in terms of de Neef's analysis of the $S = 1$ Heisenberg chain with single-ion anisotropy. The values obtained are $J/k_B = 10 \pm 0.5$ K and $\mathbf{D}/k_B = 8.5 \pm 0.5$ K (singlet low). This value of $\mathbf{D}$ is in good agreement with that obtained by Steiner and Kjems[85] ($\mathbf{D}/k_B = 8.9 \pm 0.2$ K, singlet low) from measurements of the spin-wave dispersion in $CsNiF_3$ in an external field, but the exchange parameter thus derived, $J/k_B = 11.5 \pm 0.05$ K, is somewhat higher. See Table 2 for a listing of the relevant parameters of the ferromagnetic nickel chains.

The great interest in $CsNiF_3$ has been stimulated by the statement of Mikeska that the Hamiltonian of a one-dimensional XY ferromagnet in an external field can be mapped onto the sine-Gordon equation[86] and thus be a good candidate to show nonlinear excitations. Kjems and Steiner[87] have detected excitations in the neutron spectrum of $CsNiF_3$ which were interpreted as thermally activated solitons. Recent theoretical work[88] has cast doubt upon this interpretation, however, since at the relatively high temperatures (compared to $\mathbf{D}$) where the experiment was conducted, one expects isotropic spin dynamics to be a better approximation.

Two other tribridged Ni^{2+} systems less extensively studied are $N(CH_3)_4NiCl_3$, TMNC, and $N(CH_3)_4NiBr_3$, TMNB. These salts both form the TMMC structure, Figure 1, and are thus good realizations of

TABLE 2. Magnetic Parameters[a] of Ferromagnetic Nickel Chains

Compound	J/k_B	$\mathbf{D}/k_B$	T_N	Technique	Reference
Tribridged chains					
$CsNiF_3$	10 ± 0.5	8.5 ± 0.5	2.61	χ	84
	11.5 ± 0.05	8.9 ± 0.2		spin-wave	85
$N(CH_3)_4NiCl_3$, TMNC	1.7	3.4	1.20	χ	84
	1.7 ± 0.3	3.3 ± 0.5		C_p	90
$N(CH_4)_4NiBr_3$, TMNB	4.8	3.4	2.68	χ	84
$pyHNiCl_3$	3.5	10.5	1.71	C_p	91
Bibridged chains					
$NiCl \cdot 2pz$	7.3	−29	6.05	C_p	34
$NiBr_2 \cdot 2pz$	2.7	−33	3.35	C_p	34
$NiCl_2 \cdot 2py$	5.35	−27	6.41, 6.75	C_p	34
$NiBr_2 \cdot 2py$	1.9	−30	2.85	C_p	34
$NiCl_2 \cdot 4ph\text{-}py$	8.5	−52	9.1	χ	98
$N(CH_3)_3NHNiCl_3 \cdot 2H_2O$, TMAN	13.4	11.3	3.6	χ	61
$(C_9H_7NH)NiCl_3$, QNC	14	14.7	—	χ	61

[a] All units are in degrees Kelvin. A positive **D** implies a singlet low, easy-plane anisotropy.

well-isolated one-dimensional magnets. Both chains have been found to contain ferromagnetic coupling and also to have the planar, XY anisotropy. Dupas and Renard[84] have found the magnetic parameters J/k_B and **D** to be 1.7 and 3.4 K (singlet low) for the chloride and 4.8 and 3.4 K (singlet low) for the bromide. Spontaneous sublattice magnetization has been found in the ordered state for both compounds but not for $CsNiF_3$. This magnetization is undoubtably due to the reduction in symmetry from hexagonal to monoclinic which occurs at the crystalline phase transition in TMMC-type salts.[89] Specific-heat measurements of TMNC by Kopinga *et al.*[90] have yielded the same magnetic parameters (1.7 ± 0.3 K and 3.3 ± 0.5 K, singlet low).

Recently, both our laboratory and that of de Jonge in Eindhoven, the Netherlands, have begun susceptibility (WSU) and specific-heat (Eindhoven) studies on another (presumably) tribridged system, $pyHNiCl_3$, $pyH = C_5H_5NH^+$. The specific-heat results[91] show this compound to behave as a ferromagnetic chain with an easy-plane anisotropy and good one-dimensional isolation. The material orders at 1.710 K and has J, **D** values of 3.5 and +10.5 K (singlet low). The $\mathbf{D}/J$ ratio is three times larger for this material as for $CsNiF_3$ and shows promise as a model system for the one-dimensional XY ferromagnet (Figure 20).

It is interesting to note that every tribridged Ni^{2+} chain that is ferromagnetic has been found to have an XY single-ion anisotropy. The well-known tribridged antiferromagnetic chains, such as $CsNiCl_3$,[92,93] $RbNiCl_3$,[92] and $CsNiBr_3$[94] have single-ion anisotropies of the opposite

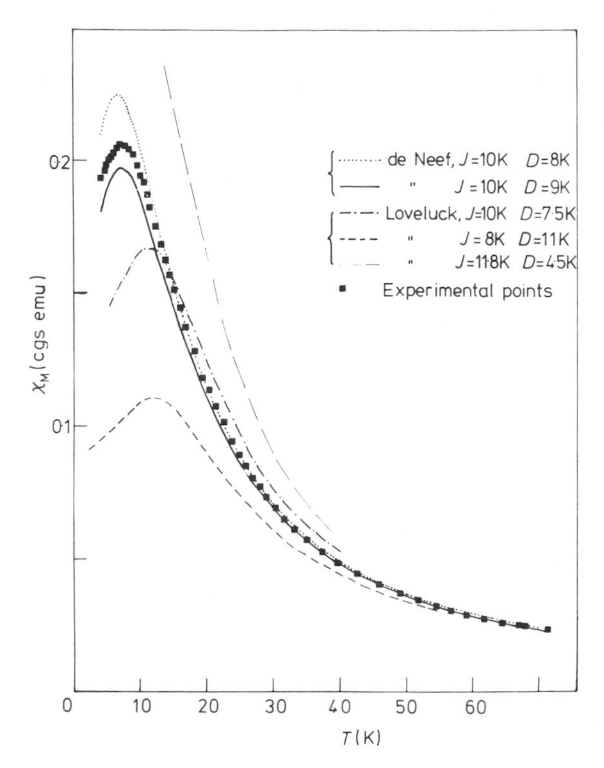

FIGURE 20. Magnetic susceptibility of $CsNiF_3$ parallel to the chain axis vs temperature. (Reprinted from Reference 84, with permission.)

sign. Full structural information does not yet exist for all of these compounds, thus hindering the discovery of magneto-structural correlations such as have been developed for copper. The problem is more severe for bibridged nickel halide chains for which structural data are almost completely lacking.

The bibridged nickel halide chains have offered another fertile source of ferromagnetic chains. These compounds of formula NiX_2L_2 are built upon a spine as shown in Figure 21 where X is a chloride or bromide and L is a monodentate ligand such as pyrazole or a substituted pyridine. Four members of this class of compounds are $NiX_2 \cdot py$ and $NiX_2 \cdot pz$: X = Cl, Br;

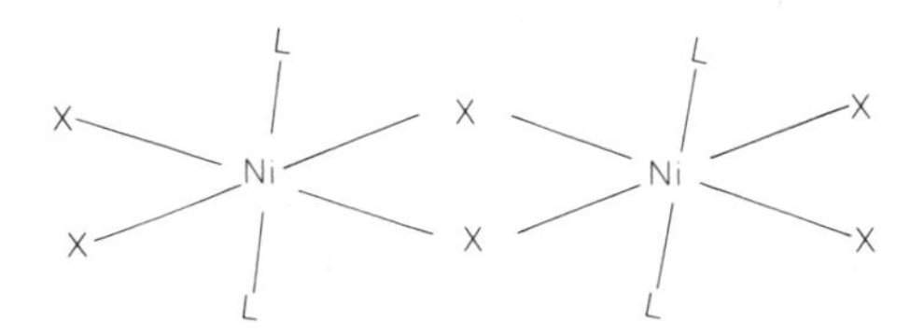

FIGURE 21. Model of the skeleton of the one-dimensional NiX_2L_2 chains.

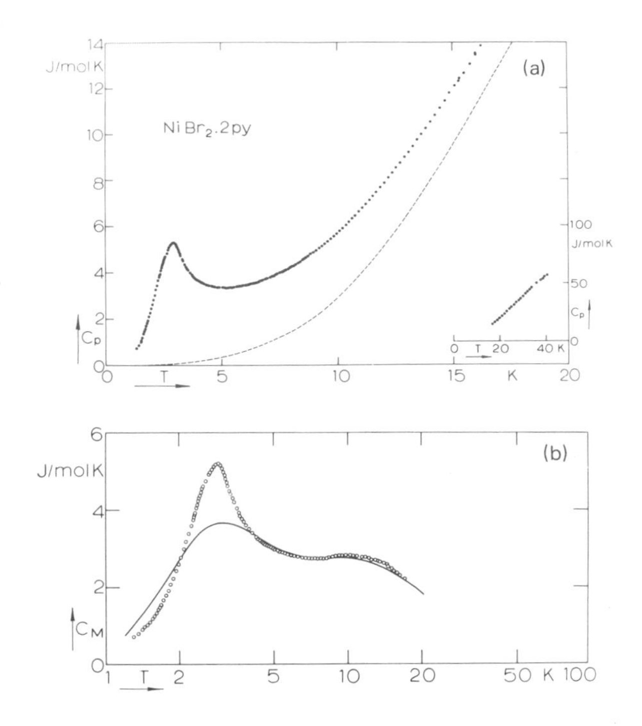

FIGURE 22. (a) Specific heat vs temperature of $NiBr_2 \cdot 2py$. Lower dashed line is the estimated lattice contribution. (Reprinted from Reference 34, with permission.) (b) Magnetic specific heat of $NiBr_2 \cdot 2py$. Solid curve is the theoretical specific heat of a one-dimensional $S = 1$ chain with $J/k_B = 1.9$ K and $\mathbf{D}/k_B = -29.8$ K. (Reprinted from Reference 34, with permission.)

pz = pyrazole ($N_2C_3H_4$), and py = pyridine (NC_5H_5). These compounds have been investigated by specific heat[34] and powder susceptibility measurements[95] as well as powder magnetization studies.[96] The compounds all consist of ferromagnetically coupled chains relatively well isolated $[J'/J = (1\text{–}5) \times 10^{-2}]$.[34] The specific-heat data have been analyzed in terms of Blöte's calculations[42] for the specific heat of $S = 1$ linear chains with isotropic exchange and single-ion anisotropy. The total specific heat and magnetic specific heat of $NiBr_2 \cdot 2py$ are shown in Figures 22a and 22b, respectively. The exchange parameters (Cl: 7.3 K, 5.35 K; Br: 2.7 K, 1.9 K) for the pz and py compounds, respectively, are comparable to those seen in the tribridged chains with the bromides having significantly weaker exchange than the chlorides. This trend is opposite to that seen in TMNC and TMNB. All four of these bibridged chains have large negative (doublet low) single-ion anisotropies, with values near -30 K. It is not understood why the magnitude of $\mathbf{D}$ is so large. The compounds undergo transitions to antiferromagnetic long-range order in the helium region and all show evidence of metamagnetic phase transitions in small applied fields.[95,96] Since these materials have only been obtained as powders, detailed structural information and single-crystal studies are entirely lacking.

Ferromagnetic exchange has also been reported in the similar compounds $Ni(4\text{-vpy})_2X_2$ ($X = Cl$, Br; 4-vpy = 4-vinylpyridine)[97] and $Ni(4\text{-ph-py})_2Cl_2$ (4-ph-py = 4-phenylpyridine).[98] Little is known about the 4-vpy compound but the 4-ph-py has been found to order near 9 K and possess a low-field metamagnetic transition similar to those of the other bibridged nickel chains. The powder susceptibility has been analyzed using Watanabes'[56] model to obtain values of $J/k_B = 8.5$ K and $\mathbf{D}/k_B = -52$ K (doublet low). This extremely large value of **D** cannot be simply attributed to the use of an inappropriate model since the same model was used to analyze the susceptibilities of the $Ni(py)_2X_2$ and $Ni(pz)_2X_2$ compounds[95] and reasonable agreement was found with the J and **D** parameters obtained from the specific-heat data.

Nickel halide chains of the type NiX_2L_2 thus act as ferromagnetic one-dimensional systems. In all known cases the single-ion anisotropy is of the easy-axis type (doublet low) and is much greater than the exchange parameter. The sign of **D** can be understood by comparing the two sets of compounds $Ni(pz)_2X_2$ and $Ni(pz)_4X_2$, $X = Cl$, Br for both sets.[99] For these latter, nonpolymeric compounds, Figure 23, the nickel ion has an axially elongated octahedral coordination with the four ligands bonding in the equitorial plane and the less tightly binding halides in the axial positions. This configuration corresponds to an elongation of the nickel coordination octahedron and causes the singlet to be low: $\mathbf{D}/k_B$ was found to be 7.2 K and 5.4 K for the chloride and bromide, respectively.[99] Comparison of Figures 21 and 23 shows the ligands and halides have reversed positions in the polymeric compounds so the reversal of the sign of **D** is not surprising. However, it is not apparent why **D** should be 6–10 times larger in the NiX_2L_2 structures.

Two recent bibridged chains have been discovered[61] in our laboratory which combine ferromagnetic exchange with positive single-ion anisotropy similar to that seen in $CsNiF_3$. These compounds are $[(CH_3)_3NH]NiCl_3 \cdot 2H_2O$, TMAN, and $(C_9H_7NH)NiCl_3 \cdot 1.5H_2O$, ONC. X-ray diffraction has

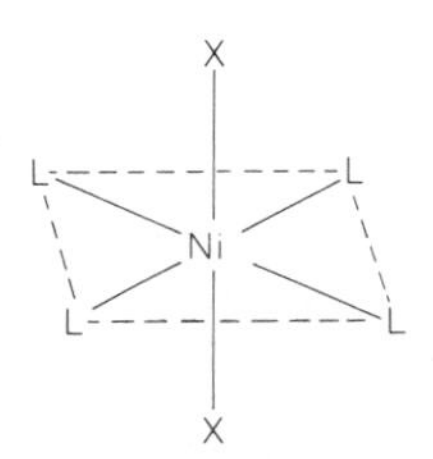

FIGURE 23. Structure of the NiX_2L_4 momomer.

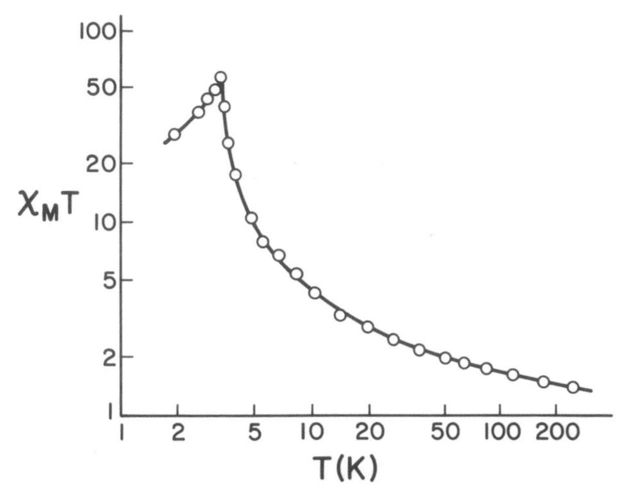

FIGURE 24. Powder susceptibility vs temperature for $[N(CH_3)_3H]NiCl_3 \cdot 2H_2O$. Solid line is a guide to the eye only.

confirmed TMAN to be isomorphous with the corresponding Co^{2+} and Mn^{2+} salts.[100, 101] The structure of QNC is not yet known but it is presumed to be isostructural with the corresponding Mn^{2+} salt,[11] Figure 2. Both structures thus contain well-separated bibridged nickel chloride chains. Powder susceptibility data for both compounds show the existence of strong ferromagnetic correlations, Figure 24. Analysis of the data in terms of de Neef's predictions leads to values of J/k_B and $\mathbf{D}/k_B$ of 14 and 14.7 K (singlet low) for QNC and 13.4 and 11.3 (singlet low) for TMAN. The exchange parameters are somewhat larger than those seen in other bibridged nickel halide chains and the anisotropies smaller with positive sign.

Both TMAN and QNC are thus XY-like $S = 1$ ferromagnetic chains, similar to $CsNiF_3$. However, contrary to the case for the hexagonal $CsNiF_3$, the z axes of these new spin systems should be normal to the chain axes. Because of their orthorhombic character, they will contain a preferred direction within the easy plane which leads to the conjecture of the existence of solitons in these salts in zero field. Further investigations of these compounds are in progress.

6.3. Cobalt(II) Salts

Only two ferromagnetic Co^{2+} salts are known: $(CH_3)_3NHCoCl_3 \cdot 2H_2O$ and $CoCl_2 \cdot 2py$ (besides the pseudo-one-dimensional $CoCl_2 \cdot 2H_2O$ system). In both salts, Co^{2+} has a distorted octahedral coordination geometry, so the lowest electronic state is an orbital singlet, spin quartet, with a large single-ion anisotropy. Thus the $M_S = \pm 3/2$ sublevels are lowest in energy, and the system may be treated with the $S' = 1/2$ Ising model.

The salt $CoCl_2 \cdot 2py$ has symmetric, bibridged -MCl_2-chains with the pyridine molecules coordinating in the fifth and sixth octahedral positions.[102] Two unfortunate structural features occur. First, the packing of the chains is such that there are three different interchain exchange pathways which must be considered. Second, a structural phase transition occurs at approximately 150 K.[103] The first leads to an extraordinarily complicated (rich) magnetic phase diagram in the ordered state, while the second leads to problems of crystal integrity (rather minor, fortunately). In addition, it is difficult to grow single crystals of appreciable size due to the lack of interactions holding the chains together in the solid state.

The magnetic properties of $CoCl_2 \cdot 2py$ were first reported by Takeda *et al.* in 1971,[104] who measured χ and C_p on powder samples down to 2 K. The C_p data show a broad maximum at 4.3 K and the onset of long-range order at 3.17 K. Both the χ and C_p data were treated on the basis of a spin-1/2 Ising chain. The C_p data yielded $J/k_B = 9.5$ K, while the exchange constant derived from the χ data was somewhat larger (11.7 K). This discrepancy is not unreasonable, considering that Co^{2+} will not be a perfect Ising system, and that interchain interactions were included in the χ analysis but not in the C_p analysis. Of interest is the fact that 85% of the total magnetic entropy, $R \ln 2$, remained above T_c, indicating that a very high degree of spin correlation exists at temperatures well above T_c.

A single-crystal field dependence study of $CoCl_2 \cdot 2py$ has been reported by Foner *et al.*[105] This revealed the existence of quite complex behavior in the ordered state. A single metamagnetic phase transition is observed with $H_0 \parallel a(H^a_{cr} = 700$ Oe), a two-step metamagnetic phase transition occurs with $H_0 \parallel b(H^b_{cr,1} = 800$ Oe, $H^b_{cr,2} = 1150$ Oe), and a single transition with $H_0 \parallel c(H^c_{cr} = 4200$ Oe). These results were interpreted in terms of a spin structure as proposed in Figure 25 with $J_1 < 0$ and J_0, J_2, J_3, and $J_4 > 0$. Based on values of the critical fields, it was deduced that $|J_1/J_2| = 1.25$. The behavior is very reminiscent of that found for $CoCl_2 \cdot 2H_2O$.[106]

In contrast to these results, de Jonge and co-workers[107, 108] claim the samples they studied remain in the high-temperature phase upon cooling to liquid-helium temperatures. On the basis of ^1H and ^{35}Cl nuclear magnetic resonance (nmr) data,[107] spin cluster resonance data,[108] and additional magnetization data, they conclude that $J_3/k_B = -15 \times 10^{-2}$ K, $J_2/k_B = 0$ K, $J_1/k_B = -3.9 \times 10^{-2}$ K, and $J_4/k_B = -3.2 \times 10^{-2}$ K. These values are in sharp contrast with those of Foner *et al.*[105] The complex magnetization behavior arises from the competing nature of the various interchain exchange couplings. Since $|J_3|$ is larger than the other constants, it is argued that the system can be treated as a two-dimensional rectangular lattice and, within this approximation, it is possible to quantitatively account for the observed ordering temperature, magnetization versus temperature, and

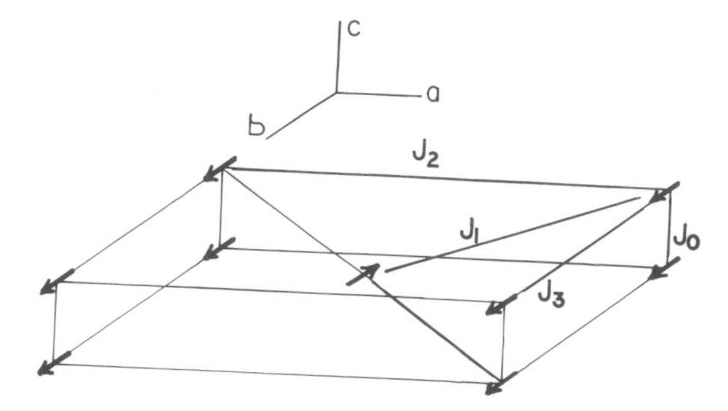

FIGURE 25. Proposed spin structure for $CoCl_2 \cdot 2py$. (Reprinted from Reference 105, with permission.)

magnetization versus field orientation in the *ab*-plane. In the nuclear magnetic resonance data an observed weak signal, which indicates a $T_c = 2.95$ K, is attributed to the presence of a small quantity of the low-temperature phase.

The compound $[(CH_3)_3NH]CoCl_3 \cdot 2H_2O$ consists of bibridged chains of edge-sharing *trans*-$[CoCl_4(OH_2)_2]$ octahedra, with chains running parallel to the *b*-axis of the orthohombic unit cell. An additional chloride ion in the lattice links the chains together via hydrogen bonds, forming sheets in the *bc*-plane. These sheets are separated by the trimethylammonium ions. The structure of the chain is very similar to that in $CoCl_2 \cdot 2H_2O$, except that the symmetry allows the presence of spin canting.

Carlin and co-workers measured C_p and single-crystal magnetic susceptibility on both the normal hydrated[100] and deuterated[109] species. The magnetic behavior was virtually unchanged by deuteration of the water and ammonium ion protons. In zero field, the heat capacity exhibits a sharp spike at $T_c = 4.135$ K (Figure 26). Upon subtraction of the lattice heat

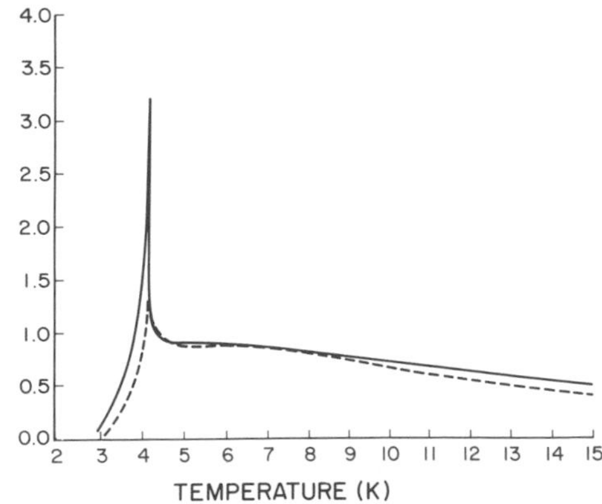

FIGURE 26. Magnetic specific heat of $TMACoCl_3 \cdot 2H_2O$, TMA = $[(CH_3)_3NH]^+$. Solid curve is the theoretical fit to the rectangular two-dimensional $S = 1/2$ Ising model, $|J/k_B| = 7.7$ K, $|J'/k_B| = 0.09$ K. (Reprinted from Reference 100, with permission.)

capacity, the magnetic contribution to C_p shows a long high-temperature tail, with only 8% of the $R \ln 2$ magnetic contribution being retained below T_c. This indicates the better one dimensionality of this salt as compared to $CoCl_2 \cdot 2py$, where 15% remained. Because of the structural characteristics, the C_p data were fit to Onsager's classical solution of the anisotropic two-dimensional Ising model[110] (dashed line, Figure 26), with $|J/k_B| = 7.7$ K and $|J'/k_B| = 0.09$ K.

The single-crystal ac susceptibility showed very anisotropic behavior, as illustrated in Figure 27. In the paramagnetic region, the easy axis of magnetization is clearly the crystallographic c-axis (which is approximately the direction in which the Co–O bonds lie). Also, χ_b shows typical $\chi_\perp$ behavior (the small peak in χ_b at T_c is due to a slight misorientation of the crystal). The behavior of χ_a is more complex, and it was necessary to include antisymmetric exchange effects to interpret the data. At very low temperatures, $\chi_a \sim \chi_b$ and $\chi_c \to 0$ as $T \to 0$ K, in accord with the interpretation of χ_c as $\chi_\parallel$.

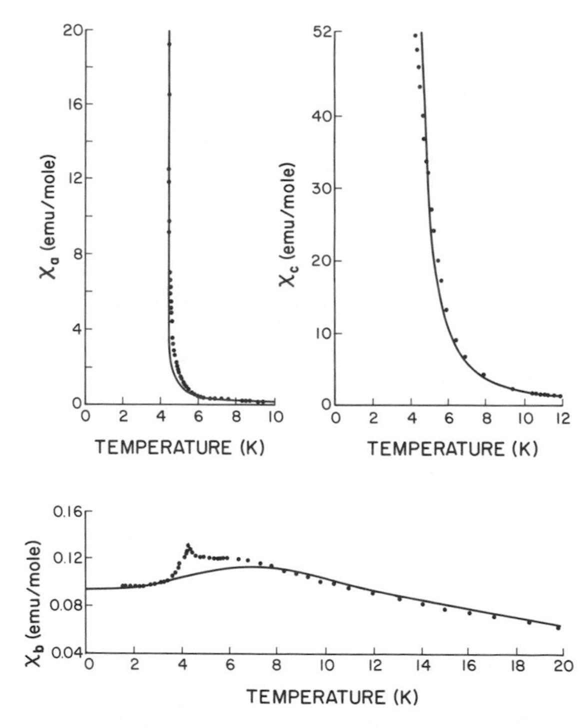

FIGURE 27. Single-crystal susceptibility data for $[N(CH_3)_3H]CoCl_3 \cdot 2H_2O$. Curve at upper left is for the a axis; curve at upper right is for the c axis; and the lowest set of data is for the b axis. (Reprinted from Reference 100, with permission.)

Quantitative interpretation of the χ data was made using the results for the one-dimensional $S = 1/2$ Ising model. The parallel susceptibility is given as

$$\chi_{\parallel} = (Ng^2\mu_B^2/4k_BT)\exp(2J/k_BT)$$

and the perpendicular susceptibility as

$$\chi_{\perp} = (Ng^2\mu_B^2/8J)[\tanh(|J|/k_BT)\,\mathrm{sech}^2(|J|/k_BT]$$

which gave good agreement with the χ_c and χ_b data, respectively (solid lines, Figure 27) with J/k_B fixed at 7.7 K, and $g_c = 6.54$ and $g_b = 3.90$. A correction for the interchain field was found to be a sensitive function of the demagnetization correction made and has been determined to be $zJ'/k_B = -0.09$ K.[109] As noted above, it was clear that the $\chi_{\perp}$ expression could not fit the χ_a data. However, the susceptibility for a weak ferromagnet can be expressed as[111]

$$\chi = Ng^2\mu_B^2(T - T_0)/[4k_B(T^2 - T_c^2)]$$

where $T_0 = -J/2k_B$ and $T_c = -(J/2k_B)[1 + (D/J)]^{1/2}$. Fixing $T_c = 4.30$ K, the following parameters reproduce that data very satisfactorily: $g_a = 2.95$ and $|\mathbf{D}/k_B| = 4.9$ K. The presence of spin canting has been confirmed by nmr studies.[112]

More recently, the magnetic phase diagram has been investigated by Groenendijk and van Duyneveldt,[113] using ac susceptibility techniques. From their analysis, they obtain $J/k_B = 13.3$ K, $J_1/k_B(=J'/k_B) = 0.16$ K, and $J_2/k_B = -0.0080$ K. The latter is the weak antiferromagnetic interlayer coupling. The metamagnetic phase diagram (Figure 28) has the following parameters: $T_c = 4.18$ K, $T_t = 4.12$ K, $H_t = 130$ Oe, and H_c $(T = 0\text{ K}) = 64$ Oe.

In summary, we see that $[(CH_3)_3NH]CoCl_3 \cdot 2H_2O$ is a very good realization of a one-dimensional $S = 1/2$ Ising system with ferromagnetic intrachain correlations. The ratio of $|J'/J| \sim 10^{-2}$ indicates quite good isolation of the chains. The strongest interchain interaction is also ferromagnetic, leading to the presence of ferromagnetic layers. In the metamagnetic transition, the spins of alternating layers, which are antiferromagnetically coupled in the ground state, become aligned parallel. The presence of a substantial antisymmetric contribution to the exchange coupling ($|\mathbf{d}/J| = 0.64$) can be related to the low symmetry of the chains, in which the Co–O bonds on adjacent octahedra are not colinear.

6.4. *Iron(II) Salts*

Investigations of linear chains containing Fe^{2+} ions have revealed ferromagnetic interactions in eight separate compounds but there is no detailed understanding of any of them. The difficulties inherent in treating

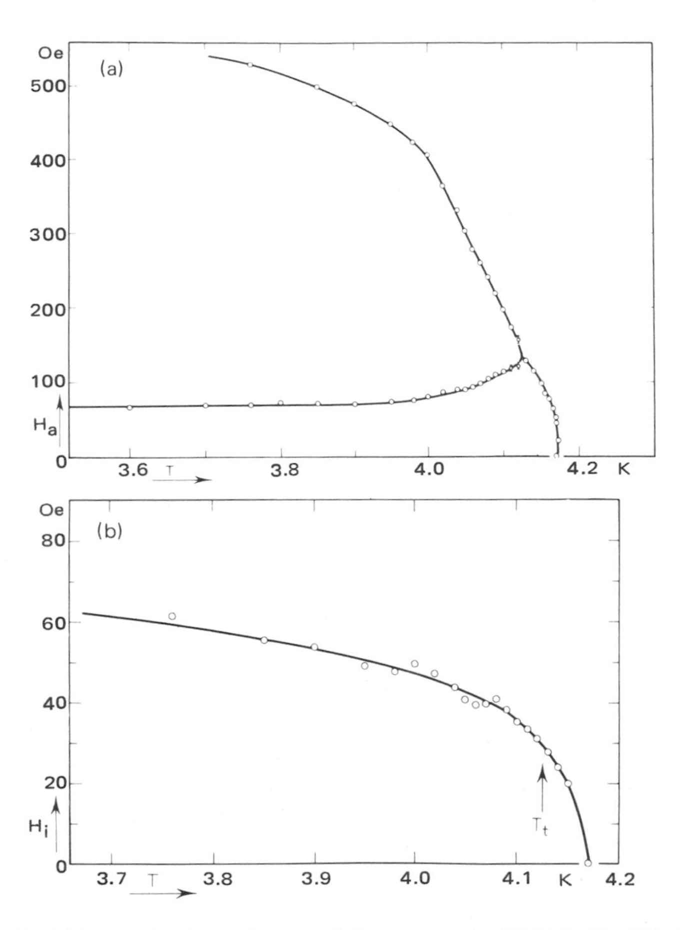

FIGURE 28. (a) Magnetic phase diagram of the metamagnet $TMACoCl_3 \cdot 2H_2O$. The ordinate is the applied field. (Reprinted from Reference 113, with permission.) (b) Magnetic phase diagram of $[N(CH_3)_3H]CoCl_3 \cdot 2H_2O$. The ordinate is now the internal magnetic field. (Reprinted from Reference 129, with permission.)

orbitally unquenched ions are so great that early reports on $RbFeCl_3$ failed to agree on even the *sign* of the exchange interaction.[92, 114] Even the most careful of treatments yet done on an Fe^{2+} chain[115] fails to offer a quantitative explanation of the magnetic behavior at low temperature. Further theoretical development is clearly needed to understand the interesting behavior observed in iron chains.

As with the nickel chains, there is a natural division of the iron compounds into the tribridged and bibridged chains, each group sharing common structural properties. The tribridged chains will be discussed first.

The three tribridged chlorides $AFeCl_3$, $A = Rb^+$, Cs^+, and K^+, all possess intrachain ferromagnetic exchange. The little-studied $KFeCl_3$ crystallizes in a double-chain structure[116] which permits relatively large interchain exchange: $KFeCl_3$ orders antiferromagnetically near 16 K.[116] More attention has been given to the isostructural hexagonal crystals $RbFeCl_3$ and $CsFeCl_3$. These have the basic TMMC structure with the $Rb^+(Cs^+)$ and Cl^- ions forming an hcp lattice and the Fe^{2+} ions occupying the (slightly distorted) octahedral interstices parallel to the hexagonal axis. Isolation of the chains appear to be considerably better as $RbFeCl_3$ remains paramagnetic above 2.5 K,[117] and $CsFeCl_3$ fails to order above 0.8 K.[118] See the discussion below.

The low-temperature magnetic behavior of both $RbFeCl_3$ and $CsFeCl_3$ can be qualitatively understood as arising from ferromagnetic chains of effective spins $S' = 1$ with easy-plane single-ion anisotropies which are larger than the anisotropic exchange couplings. In this sense there is an analogy between these two chains and $CsNiF_3$ which has not been exploited. Part of this neglect is surely due to the failure to achieve understanding of these materials due to the difficulties of treating exchange coupled ions of unquenched orbital angular momenta which reside in trigonally distorted sites. The magnetic properties depend strongly on the relative magnitudes of J, the spin–orbit coupling parameter λ, and the magnitude of the crystal-field distortion. Various simplified models have been proposed to explain the observed behavior: these various predictions for J are in strong disagreement. Using molecular field arguments, Achiwa[92] concluded that the exchange in $RbFeCl_3$ was anisotropic and antiferromagnetic: $J_{xy}/k_B = -3.8$ K, $J_z/k_B = -11.5$ K, $\mathbf{D}/k_B = 18.4$ K (singlet low). Montano *et al.*[114, 119] treated the chain as a collection of ferromagnetic dimers: $J_{xy}/k_B = 16$ K, $J_z/k_B = 7$ K. A careful analysis by Eibschütz *et al.*[115] based upon crystal-field theory and the correlated-effective-field statistical approximation led to a much smaller estimate of the isotropic exchange constant while also concluding interchain coupling and non-Heisenberg exchange are significant.

The most recent work has been inelastic neutron scattering by Yoshizawa *et al.*[120] at 5 K from which they obtained estimates of $J_{xy}/k_B = 3.1$ K, $\mathbf{D}/k_B = 28$ K (singlet low) with a surprisingly large value of J'/J of ~ 0.1. For $CsFeCl_3$, the corresponding values were found to be 2.6 K, 25 K (singlet low), and -0.05. No gap was found in the excitation spectrum of the Rb salt, preserving its XY character. A gap was observed for $CsFeCl_3$ which could explain its failure to order: the gap is due to the relative weakness of the exchange energy compared to the single-ion anisotropy. Such weak exchange is known as subcritical since it is incapable of inducing long-range order at any temperature.

The bibridged ferrous halides and pseudohalides also appear to be a potentially very fruitful area for the synthesis and study of one-dimensional systems with ferromagnetic intrachain correlations. A number of such salts have been reported by Reiff and co-workers[121–124, 127, 128] as well as others. These salts include

$FeCl_2 \cdot 2py$ ($T_c = 3.8$ K),[122, 126, 127]
$FeCl_2 \cdot bipy$ ($T_c = 4$ K),[121, 122, 124]
$FeCl_2 \cdot phen$ ($T_c = 5.0$ K)[122, 126, 127] (phen = *o*-phenanthroline),
$FeCl_2 \cdot 2pz$ ($T_c = 10$ K),[125]

and

$Fe(SCN)_2 \cdot 2py$ ($T_c > 4.2$ K).[123]

Unfortunately, these have not been studied in detail and little is known about their intra- and interchain exchange coupling. The principal method of investigation has been Mössbauer spectroscopy and powder susceptibility measurements. $FeCl_2 \cdot 2py$ and $Fe(SCN)_2 \cdot 2py$ exhibit metamagnetic behavior[128] while it is claimed that the ground states of $FeCl_2 \cdot bipy$ and $FeCl_2 \cdot phen$ are ferromagnetic.[122] These systems are certainly worthy of further investigation to elucidate more fully the exchange coupling and degree of one dimensionality of the compounds.

ACKNOWLEDGMENTS

It is a pleasure to acknowledge those people who have aided us in the preparation of this article. We thank Jill Bonner for helpful conversations and for her careful reading of a section of the manuscript. One of us (RDW) thanks the University of Zurich for its hospitality while part of this manuscript was being prepared. We also wish to thank those authors whose figures we have reproduced in the article. The support of the National Science Foundation is gratefully acknowledged (Grant No. CHE-77-08610).

Notation

Latin

AFM	Anti ferromagnetic
CHAC	Cyclohexylammonium trichlorocuperate (II), $[C_6H_{11}NH_3]CuCl_3$
d	Antisymmetric exchange vector
D	Anisotropic exchange parameter
D	Single ion anisotropy energy
DMMC	Dimethylammonium trichloromanganate(II), $[N(CH_3)_2H_2]MnCl_3$
E	Rhombic anisotropic exchange parameter
FM	Ferromagnetic

H_A	Effective anisotropy field
H_{AF}	Effective antiferromagnetic exchange field
H_{c-}, H_{c+}	Lower and upper metamagnetic critical fields
$H_c(O)$	Field required to saturate the system
H_{ex}	Effective exchange field
H^i_{sat}	Field required to saturate the system along the *i*th axis
H_{SF}	Spin-flop field
J	Intrachain exchange parameter
J'	Interchain exchange parameter
J_{aa}	Exchange integral
K	Magnetic anisotropy energy
K_{ab}	Coulomb integral
T_c	Critical temperature
T_N	Néel temperature
CHA	Cyclohexylammonium, $C_6H_{11}NH_3^+$
DMSO	Dimethylsulfoxide, $(CH_3)_2SO$
dx	1,4-dioxane, $C_4H_8O_2$
IPA	Isopropylammonium, $(CH_3)_2CHNH_3^+$
ox	Oxalate, $C_2O_4^{2-}$
py	Pyridine, C_5H_5N
pyH	Pyridinium, $C_5H_5NH^+$
pz	Pyrazole, $N_2C_3H_4$
QNC	Quinolinium trichloronickelate(II), $[C_9H_7NH]NiCl_3$
TMA	Trimethylammonium, $(CH_3)_3NH^+$
TMCuC	Tetramethylammonium trichlorocuperate(II) $[N(CH_3)_4]CuCl_3$
TMAN	Trimethylammonium trichloronickelate(II), $[N(CH_3)_3H]NiCl_3$
TMMC	Tetramethylammonium trichloromanganate(II), $[N(CH_3)_4]MnCl_3$
TMNB	Tetramethylammonium tribromonickelate(II), $[N(CH_3)_4]NiBr_3$
TMNC	Tetramethylammonium trichloronickelate(II), $[N(CH_3)_4]NiCl_3$
TMSO	Tetramethylene sulfoxide, C_4H_8SO
TTF	Tetrathiafulvalene
TTMCuC	Trimethylammonium heptachlorodicopperate(II), $[N(CH_3)_3H]_3Cu_2Cl_7$
TTMMC	Trimethylammonium heptachlorodimanganate(II), $[N(CH_3)_3H]_3Mn_2Cl_7$

Greek

ϕ	M–L–M bridging angle
δ	Dimeric dihedral angle
τ	Dimeric twist angle
χ	Magnetic susceptibility

References

1. De Jongh, L. J., and Miedema, A. R., *Adv. Phys.* **23**, 1 (1974).
2. Steiner, M., Villain, J., and Windsor, C. G., *Adv. Phys.* **25**, 87 (1976).
3. Steiner, M., *J. Appl. Phys.* **50**, 7395 (1979).
4. Mermin, N. D., and Wagner, H., *Phys. Rev. Lett.* **17**, 1133 (1966).
5. Hutchings, M. T., Shirane, G., Birgeneau, R. J., and Holt, S. L., *Phys. Rev. B* **5**, 1999 (1972).
6. Morosin, B., and Graeber, E. J., *Acta Crystallogr.* **23**, 766 (1969).

7. Dunitz, J. D., *Acta Crystallogr.* **10**, 307 (1957).
8. Harker, D., *Z. Kristallogr.* **93**, 136 (1936).
9. Poulis, N. J., and Hardeman, G. E. G., *Physica* **18**, 201 (1952).
10. Duffy, Jr., W., Venneman, J. E., Strandburg, D. L., and Richards, P. M., *Phys. Rev. B* **9**, 2220 (1974).
11. Caputo, R. E., Willett, R. D., and Morosin, B., *J. Chem. Phys.* **69**, 4976 (1978).
12. Jensen, S. J., Andersen, P., and Rasmussen, S. E. *Acta Chem. Scand.* **16**, 1890 (1962).
13. Caputo, R. E., Willett, R. D., and Muir, J. A., *Acta Crystallogr. B* **32**, 2639 (1976).
14. Willett, R. D., *Acta Crystallogr. B* **35**, 178 (1979).
15. Caputo, R. E., Willett, R. D., and Gerstein, B. C., *Phys. Rev. B* **13**, 3956 (1976).
16. de Jonge, W. J. M., Boersma, F., and Kopinga, K., *J. Magn. Magn. Mater.* **15-18**, 1007 (1980).
17. Caputo, R. E., Roberts, S. A., and Willett, R. D., *Inorg. Chem.* **15**, 820 (1976).
18. Anderson, P. W., *Phys. Rev.* **115**, 2 (1959); Goodenough, J. B., *Phys. Rev.* **100**, 564 (1959); Kanamori, J., *J. Phys. Chem. Solids* **10**, 87 (1959).
19. Hay, P. J., Thibeault, J. C., and Hoffman, R., *J. Am. Chem. Soc.* **97**, 4884 (1975).
20. Bencini, A., and Gatteschi, D., *Inorg. Chem. Acta* **31**, 11 (1978).
21. Kahn, O., and Briat, B., *J. Chem. Soc. Faraday II* **72**, 268 (1976).
22. Crawford, V. H., Richardson, H. W., Wasson, J. R., Hodgson, D., and Hatfield, W. E., *Inorg. Chem.* **15**, 2107 (1976).
23. Davis, J. A., and Sinn, E., *J. Chem. Soc.*, 165 (1976).
24. Willett, R. D., *Proc. N.Y. Acad. Sci.* **313**, 113 (1978).
25. Fletcher, R., Livermore, J., Hansen, J. J., and Willett, R. D., *Inorg. Chem.*, submitted.
26. Charlot, M. F., Kahn, O., Jeannin, S., and Jeannin, Y., *Inorg. Chem.* **19**, 1410 (1980).
27. Swank, D. D., and Willett, R. D., *Inorg. Chem.* **19**, 2321 (1980).
28. Inoue, M., Kishita, M., and Kubo, M., *Inorg. Chem.* **6**, 900 (1967).
29. O'Bannon, J., and Willett, R. D. *Inorg. Chem. Acta* **53**, L131 (1981).
30. Landee, C. P., and Willett, R. D., *Inorg. Chem.*, **20**, 2521 (1981).
31. Kahn, O., Tola, P., Galy, J., and Coudanne, H., *J. Am. Chem. Soc.* **100**, 3931 (1978).
32. Journaux, Y., and Kahn, O., *J. Chem. Soc.*, 1575 (1979).
33. Kahn, O., Claude, R., and Coudanne, H., *J. Chem. Soc. Chem. Commun.*, 1012 (1978).
34. Klaaijsen, F. W., Dokoupil, Z., and Huiskamp, W. J., *Physica* **79B**, 547 (1975).
35. Bonner, J. C., and Fisher, M. E., *Phys. Rev.* **135**, A640 (1964).
36. Baker, Jr., G. A., Rushbrooke, G. S., and Gilbert, H. E., *Phys. Rev.* **135**, A1272 (1964).
37. Swank, D. D., Landee, C. P., and Willett, R. D., *Phys. Rev. B* **20**, 2154 (1979).
38. Obokata, T., Ono, I., and Oguchi, T., *J. Phys. Soc. Jap.* **23**, 516 (1967).
39. Schouten, J. C., van der Geest, G. J., de Jonge, W. J. M., and Kopinga, K., *Phys. Lett. A* **78**, 398 (1980).
40. (a) De Neef, T., Kuipers, A. J. M., and Kopinga, K., *J. Phys. A* **7**, L171 (1974). (b) de Neef, T., Eindhoven University of Technology, Report THE/VVS/52075, Eindhoven, The Netherlands, May 1975. (c) de Neef, T., "Some Applications of Series Expansions in Magnetism," Doctoral thesis, Eindhoven University of Technology, 1975. (d) de Neef, T., *Phys. Rev. B* **13**, 4141 (1976).
41. Blöte, H. W. J., *Physica* **78**, 302 (1974).
42. Blöte, H. W. J. *Physica* **79B**, 427 (1975).
43. Takahashi, M., *Prog. Theor. Phys.* **50**, 1519 (1973).
44. Johnson, J. D., and McCoy, B., *Phys. Rev. A* **6**, 1613 (1972).
45. Blöte, H. W. J., *J. Appl. Phys.* **50**, 1825 (1979).
46. Ising, Z., *Z. Phys.* **31**, 253 (1925).
47. Fisher, M. E., *Physica* **26**, 618 (1960); *J. Math. Phys.* **4**, 124 (1963).
48. Katsura, S., *Phys. Rev.* **127**, 1508 (1962).

49. Domb, C., *Adv. Phys.* **4**, 149 (1960).
50. Johnson, J. D., *Phys. Rev. A* **9**, 1743 (1974).
51. Johnson, J. D., and Bonner, J. C., *Phys. Rev. Lett.* **44**, 616 (1980); *Phys. Rev. B* **22**, 251 (1980).
52. Lieb, E., Schultz, T., and Mattis, D., *Ann. Phys.* (*N.Y.*) **16**, 407 (1961).
53. Duxbury, P. M., Oitmaa, J., Barber, M. N., van der Bilt, A., Joung, K. O., and Carlin, R. L. *Phys. Rev. B*, **24**, 5149 (1981).
54. Weng, C., "Finite Exchange-Coupled Magnetic Systems," Doctoral thesis, Carnegie–Mellon University, 1968.
55. Fisher, M. E., *Am. J. Phys.* **32**, 343 (1964).
56. Watanabe, T., *J. Phys. Soc. Jap.* **17**, 1856 (1962).
57. de Neef, T., and de Jonge, W. J. M., *Phys. Rev. B* **11**, 4402 (1975).
58. Moses, D., Shechter, H., Ehrenfreund, E., and Makovsky, J., *J. Phys. C* **10**, 433 (1977).
59. Suzuki, M., Tsujiyana, B., and Katsura, S., *J. Math. Phys.* **8**, 124 (1967).
60. Kowalski, J. M., *Acta Phys. Pol. A* **41**, 763 (1972).
61. O'Brien, S., Gaura, R. M., Landee, C. P., and Willett, R. D., *Solid State Commun.*, **39**, 1333 (1981).
62. Algra, H. A., de Jongh, L. J., Huiskamp, W. J., and Carlin, R. L., *Physica* **92B**, 187 (1977).
63. Losee, D. B., McElearney, J. N., Siegel, A., Carlin, R. L., Kahn, A. A., Roux, J. P., and James, W. J., *Phys. Rev. B* **6**, 4342 (1972).
64. Gorter, C. J., *Rev. Mod. Phys.* **25**, 332 (1953).
65. Takeda, K., Matsukawa, S., and Haseda, T., *J. Phys. Soc. Jap.* **30**, 1330 (1971).
66. Liu, L. L., and Stanley, H. E., *Phys. Rev. Lett.* **29**, 927 (1972); *Phys. Rev. B* **8**, 2279 (1973).
67. Willett, R. D., Landee, C. P., Gaura, R. M., Swank, D. D., Groenendijk. H. A., and van Duyneveldt, A. J., *J. Magn. Magn. Mater.* **15-18**, 1055 (1980); Groenendijk, H. A., van Duyneveldt, A. J., Blöte, H. W. J., Gaura, R. M., Landee, C. P., and Willett, R. D., *Physica* **106B**, 47 (1981).
68. Swank, D. D., Landee, C. P., and Willett, R. D., *Phys. Rev. B* **20**, 2154 (1979).
69. Landee, C. P., and Willett, R. D., *J. Appl. Phys.* **52**, 2240 (1981); Willett, R. D., Barberis, D., and Waldner, F., *J. Appl. Phys.* **53**, 2077 (1982).
70. Woenk, J. W., and Spek, A. L., *Cryst. Struct. Commun.* **5**, 805 (1976).
71. Landee, C. P., and Willett, R. D., *Phys. Rev. Lett.* **43**, 463 (1979).
72. Clay, R. M., Murray-Rust, P., and Murray-Rust, J., *J. Chem. Soc. Dalton Trans.*, 595 (1973).
73. Roberts, S. A., Bloomquist, D. R., Willett, R. D., and Dodgen, H. W., *J. Am. Chem. Soc.*, **103**, 2603 (1981).
74. Livermore, J. C., Willett, R. D., Gaura, R. M., and Landee, C. P., *Inorg. Chem.*, in press.
75. Barnes, J. C., and Weakly, T. J. R., *Acta Crystallogr. B* **33**, 921 (1977).
76. McGregor, K. T., and Soos, Z. G., *Inorg. Chem.* **15**, 2159 (1976). Jotham, R. W., *J. Chem. Soc. Chem. Commun.*, 178 (1973); Girard, J. J., Kahn, O., and Verdaguer, M., *Inorg. Chem.* **19**, 274 (1980).
77. Bloomquist, D. R., Hansen, J. J., Landee, C. P., Willett, R. D., and Buder, R., *Inorg. Chem.*, **20**, 3308 (1981).
78. Interrante, L. V., Browall, K. W., Hart, Jr., H. R., Jacobs, I. S., Watkins, G. D., and Wee, S. H., *J. Am. Chem. Soc.* **97**, 889 (1975).
79. Bonner, J. C., Wei, T. S., Hart, Jr., H. R., Interrante, L. V., Jacobs, I. S., Kasper, J. S., Watkins, G. D., and Blöte, H. W. J., *J. Appl. Phys.* **49**, 1321 (1978).
80. Babel, D., *Z. Anorg. Allg. Chem.* **369**, 117 (1969).
81. Lebesque, J. V., Snel, J., and Smit, J. J., *Solid State Commun.* **13**, 371 (1973).
82. Steiner, M., *Solid State Commun.* **11**, 73 (1972).
83. Steiner, M., and Dachs, H., *Solid State Commun.* **14**, 841 (1974).

84. Dupas, C., and Renard, J.-P., *J. Phys. C* **10**, 5057 (1977).
85. Steiner, M., and Kjems, J. K., *J. Phys. C* **10**, 2665 (1977).
86. Mikeska, H. J., *J. Phys. C* **11**, L29 (1978).
87. Kjems, J. K., and Steiner, M., *Phys. Rev. Lett.* **41**, 1137 (1978).
88. Reiter, G., *Phys. Rev. Lett.* **46**, 202 (1980). Loveluck, J. M., Schneider, T., Stoll, E., and Jauslin, H. R., *Phys. Rev. Lett.* **45**, 1505 (1980).
89. Peercy, P. S., Morosin, B., and Samara, G. A., *Phys. Rev. B* **8**, 3378 (1973).
90. Kopinga, K., de Neef, T., de Jonge, W. J. M., and Gerstein, B. C., *Phys. Rev. B* **13**, 3953 (1976).
91. Schouten, J. C., Hadders, H., Kopinga, K., and de Jonge, W. J. M., *Solid State Commun.*, to be published.
92. Achiwa, N., *J. Phys. Soc. Jap.* **27**, 561 (1969).
93. Smith, J., Gerstein, B. C., Liu, S. H., and Stucky, G., *J. Chem. Phys.* **53**, 418 (1970).
94. Brener, R., Ehrenfreund, E., Shechter, H., and Makovsky. J., *J. Phys. Chem. Solids* **38**, 1023 (1977).
95. Witteveen, H. T., Rutten, W. L. C., and Reedijk, J., *J. Inorg. Nucl. Chem.* **37**, 913 (1975).
96. Foner, S., Frankel, R. B., Reiff, W. M., Little, B. F., and Long, G. J., *Solid State Commun.* **16**, 159 (1975).
97. Agnew, N. H., Collin, R. J., and Larkworthy, L. F., *J. Chem. Soc. Dalton Trans.*, 272 (1974).
98. Estes, W. E., Weller, R. R., and Hatfield, W. E., *Inorg. Chem.* **19**, 26 (1980).
99. Klaaijsen, F. W., Reedijk, J., and Witteveen, H. T., *Z. Naturforsch. A* **27**, 1532 (1972).
100. Losee, D. B., McElearney, J. N., Shankle, G. E., Carlin, R. L., Cresswell, P. J., and Robinson, W. T., *Phys. Rev. B* **8**, 2185 (1973).
101. Caputo, R. E., Willett, R. D., and Muir, J. H., *Acta Crystallogr. B* **32**, 2639 (1976).
102. Dunitz, J., *Acta Crystallogr.* **10**, 207 (1957).
103. Clark, P. J., and Milledge, H. J., *Acta Crystallogr. B* **31**, 1543 (1975); **31**, 1554 (1975).
104. Takeda, K., Matsukawa, S., and Haseda, T., *J. Phys. Soc. Jap.* **30**, 1330 (1971).
105. Foner, S., Frankel, R. B., Reiff, W. M., Wong, H., and Long, G. L., *J. Chem. Phys.* **68**, 4781 (1978).
106. Narath, A., *J. Phys. Soc. Jap.* **19**, 2244 (1964); *Phys. Rev.* **136**, A766 (1964).
107. de Jonge, W. J. M., van Vlimmeren, Q. A. G., Hijmans, J. P. A. M., Swüste, C. H. W., Buys, J. A. H. M., and van Workum, G. J. M., *J. Chem. Phys.* **67**, 751 (1977).
108. Swüste, C. H. W., Phaff, A. C., and de Jonge, W. J. M., *Physica*, to be published.
109. Bhatia, S. M., O'Connor, C. J., and Carlin, R. L., *Inorg. Chem.* **15**, 2900 (1974).
110. Onsager, L., *Phys. Rev.* **65**, 117 (1944).
111. Moriya, T., *Phys. Rev.* **120**, 91 (1960).
112. Spence, R. D., and Botterman, A. C., *Phys. Rev. B* **9**, 2993 (1974).
113. Groenendijk, H. A., and van Duyneveldt, A. J., *J. Magn. Magn. Mater.* **15–18**, 1032 (1980).
114. Montano, P. A., Cohen, E., Shechter, H., and Makovsky, J., *Phys. Rev. B* **7**, 1180 (1973).
115. Eibschütz, M., Lines, M. E., and Sherwood, R. C., *Phys. Rev. B* **11**, 4595 (1973).
116. Gurewitz, E., Makovsky, J., and Shaked, H., *Phys. Rev. B* **9**, 1071 (1974).
117. Davidson, G. R., Eibshütz, M., Cox, D. E., and Minkiewicz, V. J., *AIP Conf. Proc.* **5**, 436 (1971).
118. Takeda, K., unpublished results. Cited in Reference 121.
119. Montano, P. A., Shechter, H., Cohen, E., and Makovsky, J., *Phys. Rev. B* **9**, 1066 (1974).
120. Yoshizawa, H., Kozukue, W., and Hirakawa, K., *J. Phys. Soc. Jap.* **49**, 144 (1980).
121. Reiff, W. M., Dockum, B., Torardi, C., Foner, S., Frankel, R. B., and Weber, M. A., in *Extended Interactions between Metal Ions in Transition Metal Complexes*, ACS Symposium Series, 1974.

122. Reiff, W. M., Dockum, B., Weber, M. A., and Frankel, R. B., *Inorg. Chem.* **14**, 800 (1975).
123. Reiff, W. M., Frankel, R. B., Little, B. F., and Long, G. L., *Inorg. Chem.* **13**, 2153 (1974).
124. Reiff, W. M., and Foner, S., *J. Am. Chem. Soc.* **95**, 260 (1973).
125. Gubbens, P. C. M., van der Kraan, A. M., Ooijen, J. A. C., and Reedijk, J., *J. Magn. Magn. Mater.* **15–18**, 635 (1980).
126. Little, B. F., and Long, G. L., *Inorg. Chem.* **17**, 3401 (1978).
127. Foner, S., Frankel, R. B., McNiff, Jr., E. J., and Reiff, W. M., *AIP Conf. Proc.* **24**, 363 (1975).
128. Reiff, W. M., Long, G. J., Little, B. F., Foner, S., and Frankel, R. B., *Solid State Commun.* **16**, 159 (1975).

4

Magnetic Resonance in Ion-Radical Organic Solids

Zoltán G. Soos and Stephen R. Bondeson

1. Introduction

The remarkable physical properties of ion-radical molecular solids have opened up a new area of solid-state chemistry and physics. The synthetic problem is to stabilize open-shell and mixed-valent molecular arrays while suppressing the recombination of adjacent radicals. Recent chemical studies emphasize mixed-valent systems based on π-molecular cation and anion radicals, on transition-metal complexes, on macrocyclic ligands, and on doped polymers. The subsequent characterization of conducting, semiconducting, or paramagnetic molecular solids draws on a bewildering array of conventional and novel techniques, including electron paramagnetic resonance (epr) and nuclear magnetic resonance (nmr) results discussed in this review. The problem is to determine, in the solid state, the delocalization, interactions, or relaxation of unpaired electronic moments. Solid-state models for ion-radical solids are still fragmentary but have been successfully applied in specific cases. Magnetic resonance methods have been notably useful in delineating the proper starting point for physical models.

We will illustrate the flexibility and variety of magnetic studies. The nmr detection, for example, of low-dimensional electronic motion probes weak interactions and can yield detailed information about largely uncharacterized paramagnetic solids. The theoretical analysis of magnetic insulators by Heisenberg exchange Hamiltonians, on the other hand, demonstrates continuing progress in unraveling many-body problems encountered

Zoltán G. Soos and Stephen R. Bondeson • Department of Chemistry, Princeton University, Princeton, New Jersey 08544. Dr. Bondeson's present address: Department of Chemistry, University of Wisconsin, Stevens Point, Wisconsin 54481.

in solids. The correlated motion of triplet spin excitons in dimerized stacks illustrates still another aspect of magnetic studies—that of providing a starting point for understanding complicated systems before they are extensively characterized. These different applications of magnetic studies must naturally meet quite different criteria. Modeling the magnetic, optical, and electric properties, which are inextricably linked in ion-radical molecular solids, involves intermediate problems that have been of particular interest to us.

Several reviews[1–5] and conference proceedings[6–10] describe the wealth of static and dynamic (nmr, epr) magnetic phenomena in low-dimensional inorganic insulators and conductors, in π-molecular charge-transfer (CT) and ion-radical salts, and in organic or polymeric conductors. We focus here on the relationship between magnetic information and the microscopic parameters of theoretical models, rather than on broadly applicable and standard susceptibility, epr, or nmr methods. Our approach is selective and emphasizes basic considerations rather than detailed comparisons. No futile attempts are made to be comprehensive, as there are numerous reviews[1–10] and many current magnetic investigations.

The wide range of external variables in magnetic studies is particularly useful in developing physical models. These include the applied field, temperature, pressure, or the orientation of a crystal. Additional variables are the natural or controlled abundance of isotopes and the intrinsic or induced levels of chemical impurities or mechanical defects. Induced deviations of the spins from thermal equilibrium and time-resolved studies using pulse sequences afford still another class of variables. A complete analysis of such diverse approaches is fortunately not required in developing physical models, which are initially guided by qualitative considerations. The variety of magnetic methods is then quite advantageous.

2. *Triplet Spin Excitons: Dimerized Ion-Radical Stacks*

2.1. *Wurster's Blue Perchlorate, (TMPD)(ClO_4), Revisited*

Triplet spin excitons are characteristic of solids containing ion-radical dimers[11] with a singlet ground state and a thermally accessible triplet excited state. A wide variety of organic ion-radicals[1] and of transition-metal complexes[4,12] satisfy the structural requirements of dimers. Truly isolated dimers[13] are signaled by hyperfine splittings in the epr spectrum, when the immobile triplet interacts with nuclear spins. This situation is commonly, but not exclusively,[14] found in transition-metal dimers. On the other hand, narrow epr lines without hyperfine splittings signify mobile excitations, or spin excitons, that are commonly observed in organic ion-radical solids.

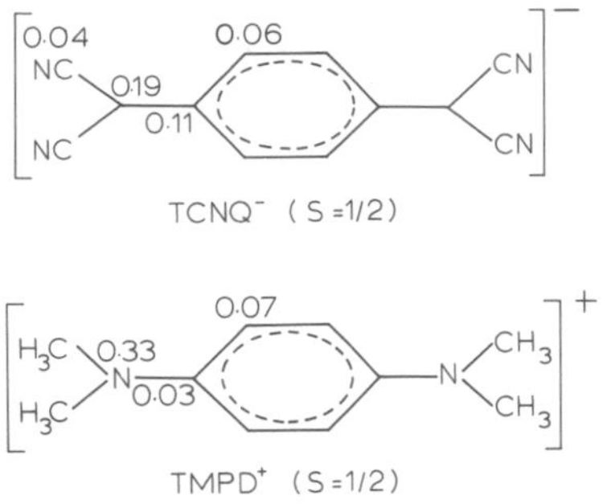

FIGURE 1. Structure and spin densities of $(TCNQ)^-$ and $(TMPD)^+$ ion radicals, both with D_{2h} symmetry.

Triplet spin excitons have been reviewed previously.[1, 11] As an illustrative example, we update the analysis of an early and famous triplet-exciton system, Wurster's blue perchlorate. The strong π-donor (D), *NNN′N′*-tetramethyl-*p*-phenylenediamine (TMPD), is shown in Figure 1, together with the unpaired spin densities of the D^+ cation radical. The complex $(TMPD)(ClO_4)$ forms[15] a regular $\cdots D^+D^+D^+ \cdots$ stack for temperature T above 186 K and a dimerized $\cdots (D^+D^+)(D^+D^+) \cdots$ stack for $T < 186$ K. Hausser and Murrell[16] interpreted the strong charge-transfer (CT) absorption in the dimerized phase in terms of self-complexes, while Thomas *et al.*[17] analyzed the epr in terms of thermally activated triplets. The Mulliken CT integral, alias the transfer integral $|t|$ of solid-state physics or Huckel β of theoretical chemistry, connects the magnetic and optical excitations. This key connection was recognized at the outset[11] and remains central to organic ion radicals.

Triplet excitons highlight the power of magnetic methods in probing both electronic structure and spin dynamics. Their dimeric structure is unusually favorable and bypasses many limitations encountered in regular stacks. Although the rich hyperfine spectrum of isolated free radicals is lost, the two unpaired electrons of a triplet produce a resolved epr fine structure whose analysis yields decisive information about the triplet wave function. No such splittings are found or expected in crystals based on regular chains. Instead, their single, exchange-narrowed epr line contains information about the dynamics of spin fluctuations. Motional narrowing of hyperfine structure at low temperature and exchange narrowing at high temperature afford such dynamical information in triplet-exciton systems.

The energy-level diagram in Figure 2 for a D^+D^+ dimer was originally discussed by Coulson and Fisher[18] for two H atoms at large separation R. It is equally appropriate for A^-A^- ion radicals of, for example, the acceptor, (A), 7,7,8,8-tetracyano-*p*-quinodimethane (TCNQ), whose spin densities are also shown in Figure 1. Minor modifications yield the Mulliken DA dimer with either largely neutral or largely ionic ground state.[1] The basic

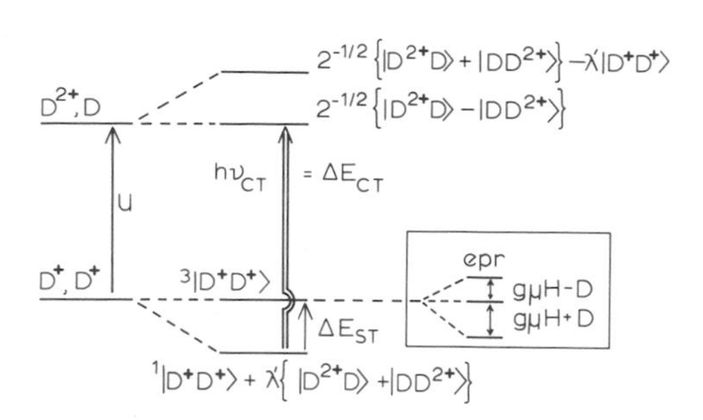

FIGURE 2. Schematic energy-level diagram for an ion-radical pair, whether $(D^+)_2$ or $(A^-)_2$ or even H_2, showing the allowed CT transition and the singlet–triplet gap arising from configuration interactions. The indicated epr fine-structure transitions in $^3|D^+D^+\rangle$ are not drawn to scale.

idea is to consider the exact molecular state $|D^+\alpha\rangle$ or $|D^+\beta\rangle$ of a radical. The highest-occupied molecular orbital (HOMO) is half-filled. Two weakly interacting radicals are then described by the Heitler–London or valence bond functions $^1|D^+D^+\rangle$ and $^3|D^+D^+\rangle$. The two electrons are perfectly correlated, one in each HOMO. The CT integral

$$t = \langle D^{2+}D|\mathscr{H}|D^+D^+\rangle \tag{1}$$

describes configuration interactions (CI) with singlets $|D^{2+}D\rangle$ or $|DD^{2+}\rangle$ in which both electrons are in the same HOMO, as in the ionic structures for H_2. The excitation U to $|D^{2+}D\rangle$ in Figure 2 clearly involves the ionization potential (I_P) and electron affinity (E_A) of D^+, together with various electrostatic interactions. The point is that the $^1|D^+D\rangle$ energy is lowered relative to $^3|D^+D^+\rangle$ by CI with $|D^{2+}D\rangle + |DD^{2+}\rangle$. The triplet may be stablilized by higher-energy triplets, $^3D^*$, involving molecular (e.g., $\pi \rightarrow \pi^*$) excitations whose possible contributions in extended systems are far from understood.

Neglecting higher excited states amounts to restricting the basis to one valence state (HOMO) per site. Such a restriction has generally been used in solid-state models, whose second quantized representations are based on fermion operators ($a^\dagger_{n\sigma}$, $a_{n\sigma}$) that create, annihilate an electron with spin σ in the nth HOMO, which can be empty, singly occupied, or doubly occupied. The CI problem for the D^+D^+ dimer in Figure 2 thus leads to a singlet-triplet splitting, ΔE_{ST}, given by

$$2\Delta E_{ST} = (U^2 + 16t^2)^{1/2} - U \tag{2}$$

where U is the Coulomb repulsion. The static susceptibility, or epr intensity, affords a direct experimental determination of ΔE_{ST}. The thermal equilibrium density $\rho(T)$ of triplets is simply

$$\rho(T) = [1 + \tfrac{1}{3}\exp(\Delta E_{ST}/k_B T)]^{-1} \tag{3}$$

Noninteracting triplets yield a susceptibility proportional to $\rho(T)/T$ and many such fits have been demonstrated.[1, 12] The allowed CT excitation in Figure 2 to $|D^{2+}D\rangle - |DD^{2+}\rangle$ has an energy ΔE_{CT} given by

$$2\Delta E_{CT} = U + (U + 16t^2)^{1/2} \tag{4}$$

Such CT absorptions are in the region of 1–1.5 eV for π radicals and were the original meaning of "charge-transfer" complexes. This designation has recently been broadened to include charge transfer in the ground state and charge transport in organic conductors.

Table 1 gives the (TMPD)(ClO_4) values of U and $|t|$ obtained via Eqs. (2) and (4) from the low-temperature values[17] of $\Delta E_{ST} \simeq 0.030$ eV and[19] $\Delta E_{CT} = 1.43$ eV. Other dimers with known ΔE_{ST} and ΔE_{CT} can be treated similarly. The dimer energies define phenomenological parameters U and $|t|$ that may, and probably do, contain important electron–phonon contributions as well as intrasite electronic correlations. These topics have their own considerable literature.

The intrinsic antiferromagnetism of CT stabilization, or of kinetic exchange, is also well documented.[1, 20] For $U \gg |t|$ in Eq. (2), the singlet-trip' ' splitting can be described by a Heisenberg spin Hamiltonian, $\mathscr{H}_{12}$,

$$\mathscr{H}_{12} = 2J(\mathbf{s}_1 \cdot \mathbf{s}_2 - \tfrac{1}{4}) \tag{5}$$

where $\Delta E_{ST} = 2J = 4t^2/U$ is necessarily positive and $\mathbf{s}_1$, $\mathbf{s}_2$ describe the spins of the D^+ radical ions. Such Heisenberg Hamiltonians are routinely adopted for magnetic insulators,[4] even when the origin of the CT state at U is not directly known. The exchange constants J then become adjustable model parameters. These dimer results are consequently readily generalized to a host of extended lattices.

The thermally populated triplet in Figure 2 is further split by an applied magnetic field H_0 and fine-structure interactions D and E between the unpaired electrons. The standard result[21] for any $S = 1$ state is

$$\mathscr{H}_T = \mu_B \mathbf{H}_0 \cdot g \cdot \mathbf{S} + D(S_Z^2 - \tfrac{2}{3}) + E(S_X^2 - S_Y^2) \tag{6}$$

TABLE 1. CT Matrix Element $|t|$, Effective On-Site Correlation U, and Interplanar Separation d for (TMPD)(ClO_4) in the Dimerized Phase for $T < 186$ K and regular phase for $T > 186$ K

Parameter	Dimerized	Regular	Comment
$\|t\|$ (eV)	0.11	0.059	Based on $\chi(T)$, ΔE_{CT} plus intensity of CT band
U (eV)	1.56	1.43	
d (Å)	3.10	3.55	Reference 15

where μ_B is the Bohr magneton, g is the g tensor, and (X, Y, Z) are the principal axes of the dipolar interaction. The small spin–orbit coupling in π radicals yields g tensors close to the isotropic value 2.002 32 for free spins and negligible contributions to D and E. Observed fine-structure splittings are in good agreement[1, 22, 23] with calculated values based on dipole–dipole interactions of the spin densities in Figure 1 and the observed dimer geometry in the crystal. Fine-structure splittings prove that the excitation is a triplet and strongly support the occurrence of essentially unperturbed ion radicals $D^{\dot{+}}$ or $A^{\dot{-}}$ in the solid state. This approximation, which is widely assumed for organic molecular solids, is hardly obvious in view of the π overlaps in organic conductors and semiconductors. The straightforward demonstration of essentially unperturbed ion radicals in triplet-exciton systems thus has important implications for other organic solids with regular stacks.

The absence of hyperfine structure and the narrow ($\lesssim 1$ G) fine-structure lines demonstrate rapid triplet motion even at the lowest temperatures. The motion must be rapid compared to typical π-electron hyperfine constants of 10^7–10^8 Hz. As proposed on the basis of exciton–phonon interactions[24] and subsequently verified by mixed-crystal studies,[25] the motion is diffusive. As the density $\rho(T)$ of triplets increases, exciton collisions broaden and ultimately destroy the fine-structure splittings. At still higher temperature, a single exchange-narrowed Lorentzian line at $g\mu_B H_0$ is found. The linewidth decreases as $\rho^{-1}(T)$ with increasing T. While these features are readily understood in terms of one-dimensional diffusion and exciton–exciton collisions, the wealth of angular and temperature dependences observed in single-crystal epr studies[26, 27] remains to be analyzed quantitatively. Recent advances in understanding and probing π-radical solids suggest numerous extensions of the early work.

The CT absorption in Figure 2 originates in the singlet ground state. Thermal population of triplets at ΔE_{ST} can be followed via epr. The concomitant depopulation of the singlet, on the other hand, should decrease the intensity of the CT absorption. Sakata and Nagakura[19] fit CT intensities in (TMPD)(ClO_4) with a singlet-triplet gap of 300 cm^{-1}, in reasonable agreement with the epr value of 250 cm^{-1}. The latter reflect $T < 100$ K data, where there are few excitons and little collisional broadening, while the former is based on $T > 100$ K, where there is appreciable thermal depopulation of the singlet. Since either estimate of ΔE_{ST} has an experimental uncertainty of $\pm 10\%$, they both support the simple picture of an ion-radical dimer.

The 110-K structure[15] of (TMPD)(ClO_4) has intradimer separations of 3.10 Å and interdimer separations of 3.62 Å. While the face-to-face overlap of the $(TMPD)^{\dot{+}}$ ion radicals must be examined in detail to assess the stronger interaction, it is natural here to choose the shorter separation for the dimer. The model parameters $|t|$ and U in Table 1 should ultimately be computed for this geometry. The larger separation is easily strong enough to average away hyperfine splittings, as an interdimer exchange of $J' \sim 10$ $\text{cm}^{-1} \ll \Delta E_{ST} = 250\ \text{cm}^{-1}$ is quite sufficient.[24] An examination of exciton–phonon interactions rationalizes the large differences between J' and ΔE_{ST} commonly found in triplet-exciton systems. However, the precise value of J' is rarely known.

Above its $T = 186$-K first-order phase transition, (TMPD)(ClO_4) crystallizes in a regular stack. The 300-K spacings[15] are 3.55 Å. The extended system now has uniform exchange interactions $J = 2|t|^2/U$ suggested by Eq. (5) when U exceeds $4|t|$. The static susceptibility[28] was measured before the numerical solutions of Bonner and Fisher[29] were widely applied to regular Heisenberg antiferromagnets. The absolute susceptibility between $200 < T < 350$ K is adequately fit[30] by $J = 53\ \text{cm}^{-1}$ (0.0066 eV).

The CT absorption is far weaker in the regular phase, as expected on reducing the intradimer overlap on forming a regular chain. Its position[19] at 190 K is $\Delta E_{CT} = 1.43$ eV, which remains a shoulder at higher T. The direct analysis of CT transitions in extended systems requires[31] far larger CI matrices than for the dimer. The absorption remains around U in a half-filled regular stack like $\cdots D^+D^+D^+ \cdots$ close to the atomic limit ($U \gg 4|t|$). The CT absorption and static susceptibility thus again yield $|t|$ and U, but now for the infinite regular stack. (TMPD)(ClO_4) results are listed in Table 1. The decrease in $|t|$ for $T > 186$ K is consistent with structural data. However, the effects of thermal expansion must be included in a quantitative analysis. Such structural complications[32] have been discussed in some detail in connection with the energetics of the (TMPD)(ClO_4) phase transition.

An independent, but qualitative, estimate for the smaller $|t|$ can be obtained from the calculated CT intensities. At 180 K, thermal depopulation reduces the dimer intensity by $1 - \rho(180) = 0.71$ from its $T < 100$ K value. Above the transition, the total spin S of the infinite stack is conserved in CT transitions, which now superimpose in $S = 0, 1, 2, \ldots$ manifolds. This *increases* the CT intensity of $|t|$ on going to a regular stack.[31] The (TMPD)(ClO_4) values in Table 1 thus rationalize the rather small intensity change in the immediate vicinity of the 186-K transition and illustrate possible future tests of phenomenological models. The simultaneous treatment of CT absorptions, intensities, and of paramagnetic susceptibilities offers a direct approach to several microscopic parameters and is the basis of current studies[33,34] on dimers. Meanwhile, there are few reliable results on CT intensities in solids and their temperature dependences.

2.2. *Supermolecular Radicals: Complex TCNQ Salts*

Chesnut and Phillips[35] first identified triplet spin excitons in complex TCNQ salts like $Cs_2(TCNQ)_3$, $[Ph_3AsCH_3](TCNQ)_2$, and

$[Ph_3PCH_3](TCNQ)_2$, which formally contain both neutral TCNQ molecules and $TCNQ^{1\dot{-}}$ anion radicals along segregated acceptor stacks. The systematically higher conductivities[36, 37] of complex salts is readily understood[1, 38] in terms of electronic correlations, since charge motion is then possible without high-energy A^{-2} sites. The actual charge distribution of complex segregated stacks depends on the crystalline environment. Crystallographically equivalent sites imply fractional ionicities (γ), or $(TCNQ)^{\gamma-}$, where γ is often related to the stoichiometry. Highly inequivalent environments may stabilize $(TCNQ)^{1\dot{-}}$ and TCNQ sites. Even the inequivalent[39] $Cs_2(TCNQ)_3$ sites, however, are not fully neutral and ionic, since the fine-structure splittings are considerably smaller[1] than expected from Eq. (6) for an $A^{\dot{-}}A^{\dot{-}}$ dimer. There has been an unfortunate tendency to assign neutral and ionic species for crystallographically inequivalent sites. The available fine-structure data[22] demonstrate fractional charges in most complex salts, even when the sites are not equivalent by symmetry. Fine-structure splittings thus clarify and complement the difficult assignment of ionicities based on small changes in bond lengths and angles.[40] The solution or theoretical spin densities of Figure 1 are adequate for the solid state, in support of a molecular-solid picture.

Several well-studied 1 : 2 TCNQ salts form similar segregated stacks shown schematically in Figure 3. They include the closely related (Ph_3AsCH_3) and (Ph_3PCH_3) complexes and the triethylammonium (Et_3NH) complex, which shows[41] triplet excitons at low temperature. The intermediate conductivity[42] and large single crystals of $(Et_3NH)(TCNQ)_2$ make it an attractive candidate for detailed characterization. The (Ph_3AsCH_3) and (Ph_3PCH_3) complexes, on the other hand, allow mixed-crystal studies[25] and are convenient for both epr and nmr investigations.

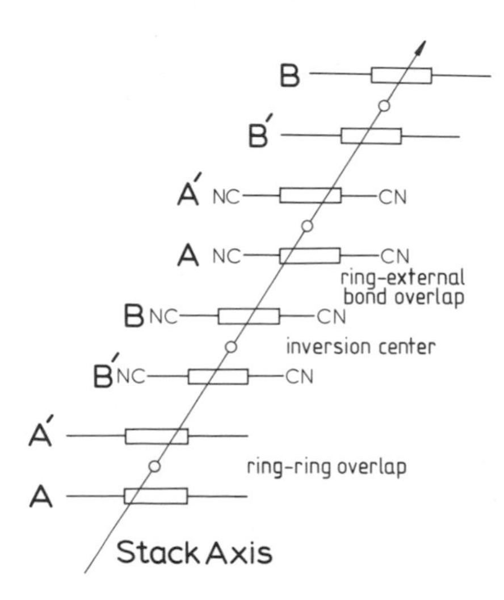

FIGURE 3. Schematic representation of $(TCNQ)^{0.5-}$ ion radicals seen edge on, forming a tetrameric stack $\cdots(AB)(B'A')(AB)(A'B')\cdots$ in $D^+(TCNQ)_2^-$ complexes. The molecular planes are tilted relative to the stack axis. The indicated centers of inversion (○) lead to equivalent overlaps for AB and B′A′ pairs, but do not impose any relation between AA′ and BB′ overlap; AA′ illustrates ring–ring overlap, AB and B′A′ have ring-to-external-bond overlap, and BB′ is in between.

We focus here on the characteristic stacking of Figure 3, whose inversion centers produce a *tetrameric* unit cell with *two* inequivalent sites and *three* potentially inequivalent overlaps. The closest interplanar separation and the favorable[43] ring-to-external-bond overlap of either AB or B′A′ probably define the largest CT matrix element $|t|$. There are no symmetry restrictions on the relative site energies within AB or B′A′ dimers, nor on the smaller CT interactions $|t'|$ and $|t''|$ for forming a tetrameric stack. The reduced epr fine-structure splittings[22] point to A_2^- supermolecular radicals associated with one unpaired electron per AB or B′A′ dimer. The dimerized stack required for triplet spin excitons then reflects different CT interactions $|t'| > |t''|$ along the chain of supermolecular radicals in Figure 3. The stronger of the weak CT interactions can in principle be identified[22] by relating the angular dependence of the fine-structure splittings in Eq. (6) to the crystal structure. A complete single-crystal epr study is needed, however, when the spin distribution within an AB or B′A′ dimer must also be determined.

Supermolecular radicals probably occur in 1 : 2 complex salts that do *not* show fine structure. The simple condition $|t'| = |t''|$ in Figure 3, for example, produces a regular chain and precludes triplet spin excitons. $(TCNQ)^{0.5-}$ sites may then be established by their characteristic vibrational frequencies or by proton nmr measurements[44] of the local susceptibility. The following elementary argument rationalizes supermolecular radicals whenever $|t|$ is sufficiently large. We consider in Figure 4a the correlation diagram for H_2^+, the simplest one-electron dimer. The restriction to one valence state per site, the HOMO of A^- or of D^+, leads to a parallel development for constructing bonding and antibonding molecular obitals (MOs). The resulting splitting of $2|t|$ in Figure 4a gives a stabilization of $|t|$

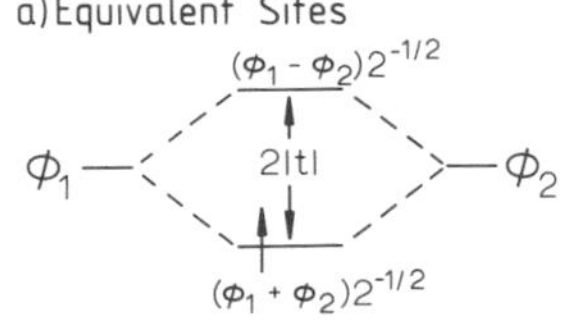

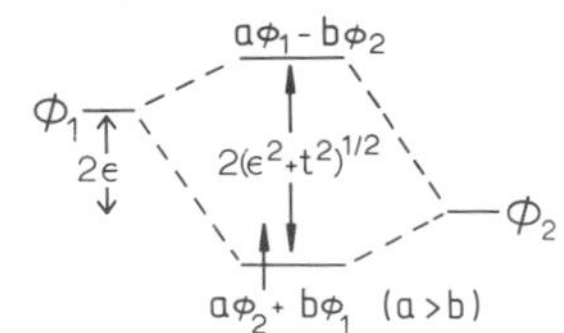

FIGURE 4. Correlation diagram for A_2^-, D_2^+, or H_2^+ supermolecular radicals, with one electron on two sites, for (a) equivalent sites and (b) inequivalent sites. The excitation from the bonding to the antibonding MO is dipole allowed.

per electron. The case of inequivalent sites in Figure 4b leads to a stabilization of

$$\Delta E = (t^2 + \varepsilon^2)^{1/2} - \varepsilon \tag{7}$$

where 2ε is the difference in the site (HOMO) energies due to the different crystallographic environment. For $\varepsilon \neq 0$, the lower-energy site has the larger coefficient in the bonding MO and the smaller coefficient in the

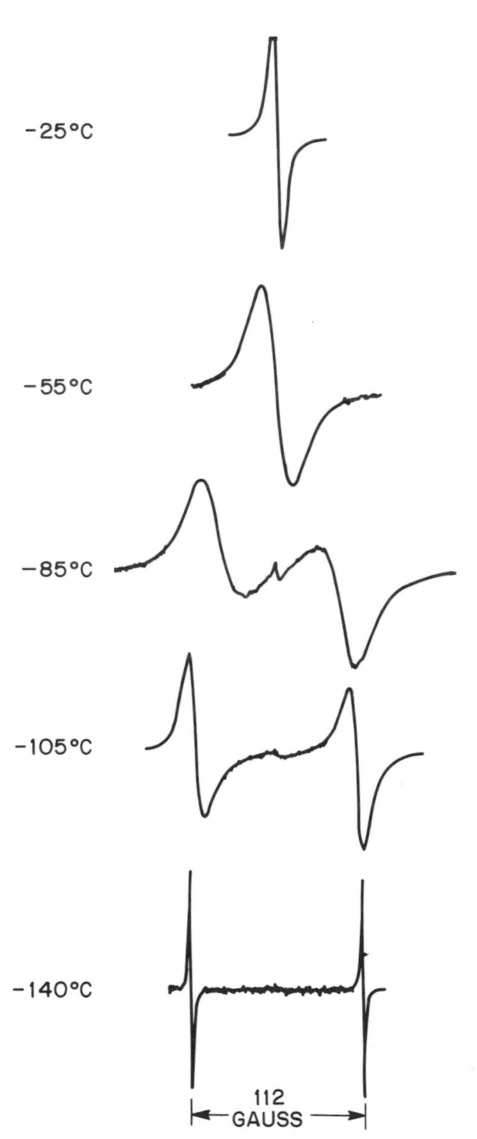

FIGURE 5. Temperature dependence of triplet spin-exciton epr in single crystal $(Ph_3PCH_3)(TCNQ)_2$ or $(Ph_3AsCH_3)(TCNQ)_2$. Increasing T increases the triplet concentration and collapses the fine structure. The orientation is fixed, but the spectrometer gain settings are different at each T. (Reproduced from Reference 35, with permission.)

antibonding MO. The CT excitation around $2(t^2 + \varepsilon^2)^{1/2}$ of a supermolecular radical is dipole allowed, a possibility that may better establish supermolecular radicals. The analysis[45] of low-energy CT transitions in partly filled (complex) organic systems will undoubtedly receive further attention.

We have emphasized the structural information derived from single-crystal epr fine-structure studies. The spin distributions of noninteracting $A^{\dot{-}}$ or $D^{\dot{+}}$ ion radicals and the observed structure usually suffice for solid-state data. Complex salts merely require fractional ionicities $\gamma < 1$. The corresponding treatment of exciton dynamics has long been understood qualitatively,[11, 24] although a number of outstanding questions remain. The classic[35] fine-structure spectra of $[Ph_3AsCH_3](TCNQ)_2$ or $[Ph_3PCH_3](TCNQ)_2$ are shown in Figure 5. The crystal orientation is fixed, but the spectrometer gain setting is varied. This suppresses the intensity variation of $\rho(T)/T$, where the triplet-exciton density is given by Eq. (3) with $\Delta E_{ST} = 0.065$ eV. The first problem is that ΔE_{ST} is itself a weakly decreasing function of temperature. The singlet-triplet model is, of course, a limiting case of an alternating chain and thermal expansion may also reduce the activation energy. The precise contributions of these two effects, or of exciton–phonon contributions to ΔE_{ST}, are not known.

The qualitative explanation for Figure 5 is that heating increases $\rho(T)$ and, thus, the rate of exciton–exciton collisions, thereby broadening and eventually collapsing the fine-structure splitting. When the exchange $J'\rho(T)$ between excitons exceeds the splitting, the exchange-narrowed epr line narrows as $[J'\rho(T)]^{-1}$. The usual analysis[26] can be improved via density-matrix methods,[46] although the one-dimensional nature of the exciton diffusion is severely approximated. The point[26] is that the rates of line broadening, of line merging, and of line narrowing in Figure 5 provide three

TABLE 2. Activation Energies in Triplet Exciton Salts.

Salt	ΔE_{ST}[b] (eV)	$\Delta E(1)$[c] (eV)	$\Delta E(2)$[d] (eV)	$\Delta E(3)$[e] (eV)	Reference
$(Ph_3AsCH_3)(TCNQ)_2$	0.065	0.050	0.11	0.16	26
$(Ph_3PCH_3)(TCNQ)_2$	0.065	0.042	0.13	0.19	26
$Cs_2(TCNQ)_3$	0.16	0.054	0.22	0.22	26
(M)(TCNQ)[a]	0.36	0.11	0.65	—	47
$(M)_2(TCNQ)_3$	0.31	0.069	0.35	0.38	47
$(M)_3(TCNQ)_4$	0.33	—	0.38	—	47
$(TMPD)_2Ni(S_2C_2(CN)_2)_2$	0.24	0.24	0.53	—	48

[a] M = morpholinium.
[b] ΔE_{ST} is based on the magnetic susceptibility.
[c] $\Delta E(1)$ involves the slow-exchange line separation.
[d] $\Delta E(2)$ involves the slow-exchange linewidth.
[e] $\Delta E(3)$ involves the fast-exchange linewidth.

additional activation energies $\Delta E(2)$, $\Delta E(1)$, and $\Delta E(3)$, respectively, that are seen in Table 2 to differ from the ΔE_{ST} obtained from the magnetic susceptibility data. Various (morpholinium)(TCNQ) complexes[47] show even larger variations among these activation parameters. Similar effects are also observed[48] for dimerized $(TMPD)^+$ stacks. Once again, there are several obvious candidates for improved treatments, but no convincing explanation for the different activation energies in Table 2. We expect similar puzzles for many other, more recent triplet spin exciton systems. These data are reminders of outstanding questions even in the relatively well-understood area of dimerized ion-radical stacks.

2.3. Charge-Transfer and σ-Bonded Dimers

Triplet spin excitons reflect pairwise correlations of two unpaired electrons. We have so far considered thermally activated triplets arising from π overlaps in dimerized stacks of cation radicals $(D^+)_2$ in $(TMPD)(ClO_4)$, of anion radicals $(A^-)_2$ in some 1 : 1 TCNQ salts, and of supermolecular radicals $(A_2^-)_2$ in several 1 : 2 TCNQ salts. These possibilities hardly exhaust the available triplet-exciton systems. The motion of optically produced triplets $^3M^*$ in neutral molecular crystals like anthracene has been extensively studied.[49, 50] Triplet motion over magnetically inequivalent molecules dominates the spin relaxation[51] and is a useful probe[52] for interchain motion. The high energies ($>10{,}000$ cm^{-1}) needed to produce $^3M^*$ sites necessitate optical, rather than thermal, generation. Another large class of organic solids,[53, 54] the largely neutral CT complexes of weak donors and acceptors, frequently crystallize in mixed regular $\cdots$DADA$\cdots$ stacks. The triplet excited state $^3|D^+A^-\rangle$ again involves too large an excitation energy for thermal detection, but can be populated by intersystem crossing from $^1|D^+A^-\rangle$. Fine-structure studies[55] of weak complexes show the unpaired electrons to be on adjacent molecules, although CI contributions from nearby $^3D^*$ or $^3A^*$ states are sometimes important. Comparison with fine-structure computations based on the observed crystal structure and theoretical molecular spin densities yields[56] the degree of ionicity in the excited triplet state. Such optically produced triplets in organic molecular solids or in weak CT complexes are beyond the scope of this chapter. We turn instead to some unusual, thermally activated triplet spin exciton systems.

Substituted phenazines, P,[57] form both cation-radical salts and CT complexes with TCNQ. The 1 : 1 complex of dimethylphenazine (Me_2P), shown in Figure 6, and TMPD crystallizes[58] in mixed dimerized stacks and shows[59] triplet spin excitons with $\Delta E_{ST} \simeq 0.60 \pm 0.06$ eV in Eq. (3). The magnitudes of D and E and the orientation of the principal axes X, Y, Z in Eq. (6) are consistent[59] with a triplet with unpaired electrons on adjacent D^+ and A^- radicals. The absence of hyperfine structure again points to

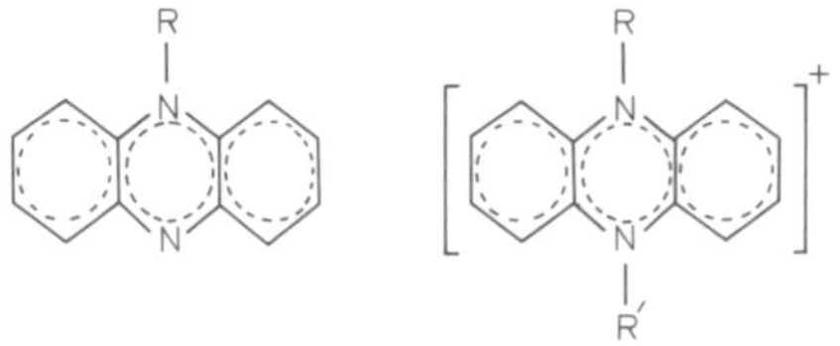

NMP ($R=CH_3$)
NEP ($R=CH_2CH_3$)
DMP ($R=CD_3$)

Me_2P^+ ($R=R'=CH_3$)
$NMPH^+$ ($R=H, R'=CH_3$)

FIGURE 6. Representative alkyl-substituted neutral and ionic phenazine radicals based on 15π electrons.

exciton motion. Triplet spin excitons have subsequently been observed[60] in other phenazine : TCNQ complexes. Such triplet-exciton CT crystals are probably fairly common. They were not previously found primarily because attention centered on such highly symmetric donors as TMPD, *p*-phenylenediamine (PD), or tetrathiofulvalene (TTF) and acceptors as TCNQ or chloranil, which usually form mixed regular and segregated stacks.

The restriction to one valence state per site requires a purely ionic triplet $^3|D^+A^-\rangle$. The singlet state, by contrast, can be partly ionic due to CI between $^1|D^+A^-\rangle$ and $|DA\rangle$, as can be seen by slightly modifying Figure 2 to a D^+A^- dimer. Weak complexes have a largely neutral ground state and largely ionic excited states. The CT excitation ΔE_{CT} to the singlet and ΔE_{ST} should then be comparable. Strong complexes, with largely ionic ground states, have $\Delta E_{ST} \ll \Delta E_{CT}$ since $^3|D^+A^-\rangle$ and $^1|D^+A^-\rangle$ are degenerate in the limit of zero overlap. Comparison of $\Delta E_{CT} \simeq 0.89$ eV[61] and $\Delta E_{ST} \simeq 0.60$ eV for (Me_2P)(TCNQ) suggests an intermediate ionicity. The situation is far from clear, as the crystal-structure analysis[58] points to a neutral complex while optical data have been interpreted[62] in terms of an ionic ground state. The large value of ΔE_{ST} seems inconsistent with a completely ionic dimer, since $|t| \sim 0.1$–0.3 eV produce splittings of at most $|t|^2/\Delta E_{CT}$ according to Eq. (2). Partial ionicity[63] is discussed in greater detail in Section 3. The point here is that 1 : 1 phenazine: TCNQ complexes demonstrate triplet spin excitons in CT solids with mixed dimerized stacks.

Several early studies[16, 19] focused on the physical properties of the solid free radical *N*-ethylphenazine (NEP) in Figure 6. The 1 : 1 complex of NEP and TCNQ has typical mixed stacks,[64] except for a long (1.63 Å) σ bond between $(TCNQ)^-$ on adjacent stacks. The σ-bonded $(TCNQ)_2^-$ dimer is shown in Figure 7. It closely resembles the first such dimer found[65] in the $(dipyridyl)_2Pt(II)$ $(TCNQ)_2$ complex. Now $\Delta E_{ST} = 0.27$ eV[66] corresponds to the thermal activation of the σ bond in Figure 7. This assignment is confirmed by the orientation of the fine-structure principal axes shown in Figure 7. The fine-structure splittings are fit by Eq. (6) for $\hat{Z}$ close to the C–C bond and $\hat{X}$ precisely along the projection of the C–C bond.

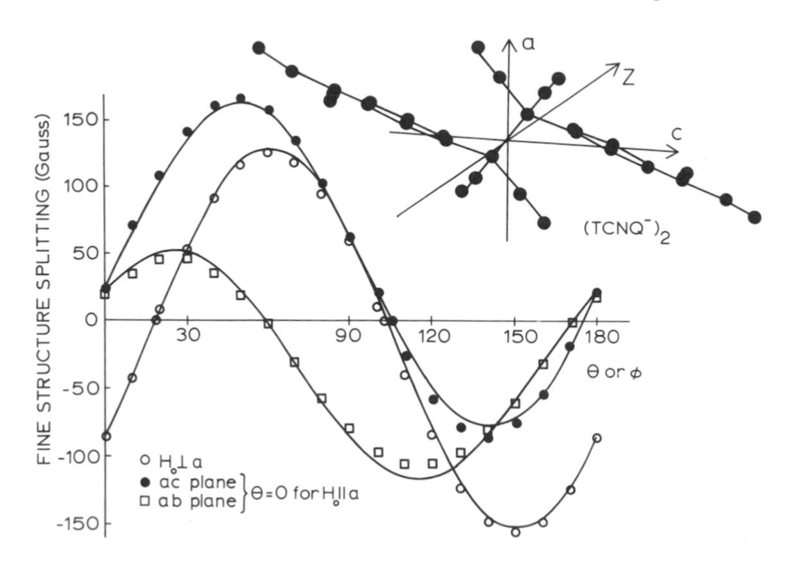

FIGURE 7. The σ-bonded $(TCNQ)_2^-$ dimer and epr fine-structure splittings in crystalline (NEP)(TCNQ). The XYZ principal axes in Eq. (6) indicate a triplet state on activating the σ bond. The detailed relation between the fine-structure and crystal axes is given in Reference 66.

The (NEP)(TCNQ) triplet thus consists of $TCNQ^-$ ion radicals on adjacent stacks. Breaking the σ bond should increase the bond length and yield more planar radicals. The D and E values are in fact smaller[66] than obtained from the observed geometry in Figure 7, which overwhelmingly reflects singlets in view of the large ΔE_{ST}. We note that these rare $(TCNQ)_2^-$ dimers illustrate deviations between solution and solid-state spin densities. The picture[1] of essentially unperturbed molecular ions in the solid state must clearly be generalized when specific bonds are formed. The theoretical treatment of π complexes is thus generally simpler than that of σ-bonded polymers like poly(sulfidenitride), $(SN)_x$, or polyacetylene, $(CH)_x$.

Large geometrical changes expected on creating a triplet in (NEP)(TCNQ) hinder exciton motion along the mixed stack. While hyperfine structure is not observed, the broad epr lines and their temperature dependence are quite different[66] from usual triplet-exciton spectra in Figure 5. The triplets are probably immobile and their linewidths are not understood. Strictly speaking, they are not excitons at all. Nevertheless, they reflect the large variety of ion-radical dimers that have already been identified in organic solids. The activation energy $\Delta E_{ST} \simeq 0.27$ eV for the σ bond also suggests that comparable or larger ΔE_{ST} in π complexes, at larger separations of 3.1–3.5 Å, cannot be entirely due to a phase reversal.

3. Charge-Transfer Complexes with Mixed Regular Stacks

3.1. Paramagnetism and Ionicity of (TMPD)(TCNQ)

Strong π donors like TMPD and strong π acceptors like TCNQ are paramagnetic semiconductors[1] that crystallize in mixed regular $\cdots D^{\gamma+}A^{\gamma-}D^{\gamma+}A^{\gamma-}\cdots$ stacks. The degree of ionicity γ is restricted[67] to $\gamma = 0$ or $\gamma = 1$ in the limit $|t| \to 0$, when CI is suppressed. The ionic lattice is favored if the Madelung energy, M, exceeds the energy I_P-E_A for creating a D^+A^- pair and vice versa. The paramagnetism[68] of (TMPD)(TCNQ) as well as infrared and optical spectra[69] of strong complexes suggests a largely ionic, or $\gamma \sim 1$, ground state. Conversely, the diamagnetism of many CT complexes of weak donors and acceptors[53] points to largely neutral, or $\gamma \sim 0$, ground states. The sharp early separation[11,67] of CT complexes into neutral and ionic solids fails for borderline cases in which $|t|$ is comparable to I_P-E_A-M and even strong π complexes like (TMPD)(TCNQ) turn out to be borderline. The energetics of partly ionized molecular sites must also be revised.[63] Partly ionized, or mixed-valent, or partly charge-transferred, segregated stacks are in fact common to all presently known molecular conductors.

In addition to these general considerations, the structure[70] and paramagnetism[68] of (TMPD)(TCNQ) preclude an ionic ($\gamma = 1$) lattice and require[71] $\gamma \sim 0.70$. The centers of inversion at each D and A site lead by symmetry to a single $|t| = \langle DA|\mathscr{H}|D^+A^-\rangle$ along the mixed stack. The $\gamma = 1$ regular ionic lattice then reduces[11] to a regular Heisenberg antiferromagnetic chain

$$\mathscr{H}_{\text{Heis}} = 2J \sum_n \mathbf{s}_n \cdot \mathbf{s}_{n+1} \tag{8}$$

with $J = |t|^2/\Delta E_{CT}$ for $\Delta E_{CT} \gg |t|$. The magnetic susceptibility[29] $\chi(T)$, of Eq. (8) is finite at 0 K and has a weak maximum around $T \sim J/k_B$. The susceptibility of (TMPD)(TCNQ) is quite different[68] in the region $200 < T < 350$ K

$$\chi(T) \propto T^{-1} \exp(-\Delta E_m/k_B T) \tag{9}$$

with a magnetic gap of $\Delta E_m = 0.07$ eV. The related complexes (TMPD)(Chloranil)[72] and (PD)(Chloranil)[73] both have $\Delta E_m = 0.13$ eV. The structure[74] of the former also requires a single $|t|$ along the mixed stack. These $\chi(T)$ data and the crystal structure thus rule out a $\gamma = 1$ ionic lattice.

Soos and Mazumdar[71] developed a diagrammatic valence bond (VB) method for exact CI computations in $\cdots$DADA$\cdots$ rings of up to ten sites. Admixtures of $|D\rangle$ and $|A\rangle$ states reduce γ in the ground state and produce polar covalent π bonding. Neutral states must in fact dominate for parameters describing weak donors and acceptors. As in the dimer described in

Figure 2, CI depends on the relative value of $|t|$ and 2δ. The latter involves the energy difference between an adjacent DA and D^+A^- pair in the solid. It can be related to I_P-E_A and various electrostatic contributions.[75] Alternately, it can be taken from the experimental CT absorption[76] at $\Delta E_{CT} =$ 0.95 eV for (TMPD)(TCNQ). VB computations[71] demonstrate a symmetry change in the ground-state wave function at $z_c = \delta/\sqrt{2}|t| = 0.53 \pm 0.01$ for the infinite chain, with a magnetic gap for $z < z_c$ and $\Delta E_m = 0$ for $z > z_c$. The crossover from a diamagnetic neutral lattice to a paramagnetic ionic lattice occurs at $\gamma_c = 0.68 \pm 0.01$ in the simplest model of retaining only $|t|$ and 2δ. As shown below, the crossover is slightly changed in more complicated models. The point is that a small $\Delta E_m = 0.07$ eV in (TMPD)(TCNQ) is consistent with a ground-state ionicity γ close to, but slightly smaller than, $\gamma_c \sim 0.7$.

Resonance Raman spectroscopy[77] offers an independent value of the ionicity. The idea is to identify a vibration, such as the exocyclic TCNQ stretch in Figure 8, that is sensitive to the charge. A variation of 64 cm^{-1} is found[77,78] in Figure 8 on going from $\gamma = 0$ in solid TCNQ to $\gamma = 1$ in $(TCNQ)^{1-}$ ions in solution, with $\gamma = 0.5$ for $(Et_3NH)(TCNQ)_2$ falling precisely in the middle. The estimated uncertainties for (TTF)(TCNQ) reflect[77] linear interpolations for several vibrations of each molecule. The (TMPD)(TCNQ) result of $\gamma = 0.7$ thus strongly supports the partial ionicity indicated by a small finite ΔE_m. Slightly higher $\gamma \sim 0.8$ are obtained[79] when the $\gamma = 1$ model compound is chosen to be an (alkali)(TCNQ) or when infrared data for the CN vibrations are used. Systematic investigations of molecular vibrations in solids should yield more direct and reliable ionicities over the next several years.

Table 3 relates the (TMPD)(TCNQ) data for γ, ΔE_m, and ΔE_{CT} to the microscopic parameters $|t|$ and 2δ, the energy for $DA \rightarrow D^+A^-$. CT transitions are modeled[31] by exact results for finite chains. As discussed in connection with $(TMPD)(ClO_4)$, the simultaneous analysis of magnetic, optical,

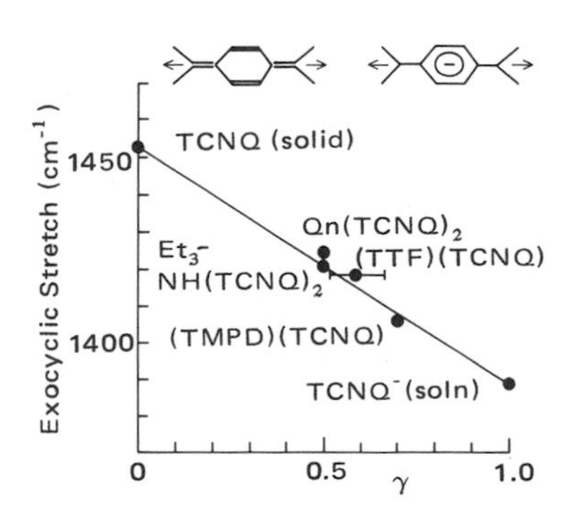

FIGURE 8. Variation of the exocyclic TCNQ stretch with ionity, as observed by resonance Raman. (Data from References 77 and 78.)

TABLE 3. (TMPD)(TCNQ) Parameters: CT Matrix Element $|t|$, Ionicity γ, and Effective Splitting $2\delta = E(D^+A^-) - E(DA)$

Parameter	Theory, references 31 and 71	Comment
$\|t\|$ (eV)	0.25	Based on ΔE_m, ΔE_{CT}
γ	0.67	0.7–0.8 from vibrational data
-2δ (eV)	-0.4	$I_D - E_A - M = 6.2 - 2.8 - 4 = -0.6$

and structural data provides the most convincing applications of theoretical models. Table 3 suggests $|t|$ values of the order of 0.2–0.3 eV. (TMPD)(TCNQ) is evidently a borderline complex whose ionicity of ~0.7 is still on the diamagnetic side of the diamagnetic–paramagnetic interface. Direct estimates[75] of the ionization potential I_P, electron affinity E_A, Madelung sum M, and other small contributions also indicate borderline stability for mixed ionic stacks.

Perfluorinated TCNQ, $TCNQF_4$, has a fluorinated phenyl ring in Figure 1. It is a stronger acceptor[80] than TCNQ and should produce more ionic complexes. The crystal structures of (D)(TCNQ) and (D)($TCNQF_4$) may not be isomorphous, however. (Me_2P)($TCNQF_4$) has recently been found[81] to crystallize in a mixed regular stack that is similar to (TMPD)(TCNQ) rather than to the alternating (Me_2P)(TCNQ) stacks. The susceptibility $\chi(T)$ of (Me_2P)($TCNQF_4$) accurately follows Eq. (8) for $J = 175$ cm^{-1} and $T > 121$ K. A structural change, probably to a dimerized phase, reduces $\chi(T)$ for $T < 121$ K. Thus (Me_2P)($TCNQF_4$) is the first realization of a truly "ionic" complex and confirms the early prediction[67] that Eq. (8) describes the magnetic excitations. Its ionicity clearly exceeds $\gamma_c \sim 0.7$ and is probably still higher[81] ($\gamma \gtrsim 0.9$) in view of the good $\chi(T)$ fit. A typical estimate of $\Delta E_{CT} \sim 1$ eV in $J = |t|^2/\Delta E_{CT}$ yields $|t| \sim 0.15$ eV for (Me_2P)($TCNQF_4$), which is slightly smaller than $|t|$ for (TMPD)(TCNQ). While such comparisons of microscopic parameters are still quite approximate, they will undoubtedly be improved and extended in future work.

The (TMPD)($TCNQF_4$) complex is also strongly paramagnetic,[81] but single crystals suitable for x-ray analysis could not be prepared. The observed $\chi(T)$ goes approximately as $T^{-\alpha}$, with $\alpha \sim 0.75$, in the interval $4 < T < 300$ K. Such high $\chi(T)$ at low temperature rules out any magnetic gap. The value of $\chi(300 \text{ K})$ requires an ionic lattice, with $\gamma > 0.7$ even assuming noninteracting spins. Since a power-law susceptibility indicates antiferromagnetic exchange, the ionicity must be still higher and probably satisfies[81] $\gamma > 0.9$. On the other hand, $\chi(T)$ is not consistent with Eq. (8), but is symptomatic of disordered exchange distributions. We consequently do not expect (TMPD)($TCNQF_4$) to form mixed regular or alternating stacks, but suggest instead the occurrence of some novel ionic lattice.

The structural variety of these phenazine:TCNQ complexes is particularly striking.[57b] Thus (Me_2P)(TCNQ) and (Me_2P)($TCNQF_4$) form mixed alternating and regular stacks, respectively, while (NEP)(TCNQ) has long σ bonds between $TCNQ^-$ anions in adjacent mixed stacks. (NMP)(TCNQ), (NMP = *N*-methylphenazinium), the first good organic conductor[82] crystallizes in partly filled segregated stacks. Segregated stacking and high conductivity are again found[83] in the alloy $[(Me_2P)_xP_{1-x}]$(TCNQ), with $x = 0.5 \pm 0.1$ (P = phenazine), which has essentially the same composition and structure as (NMP)(TCNQ). The main difference is that the single methyls of NMPs in Figure 6 occur in pairs on Me_2P and are missing on P. (NMP)(TCNQ) can also be doped with phenazine[84] to vary the ionicity. These organic alloys are naturally more difficult to characterize, since the actual composition of the lattice, the random distribution of the cations, and their ionicity must all be established. As there are many additional substituted phenazines,[57] still other structural, physical, and transport properties can be anticipated for these versatile CT complexes with TCNQ or $TCNQF_4$.

3.2. Structure of Spin Excitations

Finite CT integrals $|t|$ thus permit a continuous range of ionicity, or of polar covalent π bonding, in the ground state of mixed DA stacks. Long-range Coulomb interactions facilitate[11] ionization of partly charged lattices and can drive[71] a discontinuous change in γ. Such a neutral-ionic transition has been proposed[85] in the 1 : 1 CT complex (TTF)(Chloranil) under applied pressure. Long-range interactions produce a narrower crossover region in the solid then in molecules on varying I_P, E_A, or the nearest-neighbor Coulomb interaction $M_1 \sim e^2/R \sim$ 3–4 eV. As shown above, there is a sharp diamagnetic–paramagnetic interface at a particular ionicity γ_c. The finite ΔE_m in (TMPD)(TCNQ), (TMPD)(Chloranil), or (PD)(Chloranil) indicates ground-state ionicities $\gamma < \gamma_c$. Fitting γ in addition to ΔE_m and ΔE_{CT} is a stringent test for model parameters.

We now consider the internal structure of the triplets at ΔE_m. In $\cdots$DADA$\cdots$ ground states of neutral complexes, a D^+A^- triplet is clearly stabilized by the Coulomb interaction M_1. The unpaired electrons are consequently adjacent and, as mentioned in Section 2.3, epr fine-structure splittings are observed. Preferential Coulomb stabilization of the triplet is lost when all the sites are ionic in the $\gamma = 1$ lattice. The thermally activated unpaired electrons then move independently. A single exchange-narrowed epr line is expected[86] at $\mu_B g \cdot \mathbf{H}_0$, with different temperature and pressure dependences than in triplet-exciton crystals. The angular dependence of the (TMPD)(Chloranil) linewidth demonstrates[87] dipolar broadening. The temperature, frequency, pressure, and angular dependence of single-crystal epr in[88] (TMPD)(TCNQ) and[73] (PD)(Chloranil) also support a picture of independent S = 1/2 excitations. The origin of a finite ΔE_m in mixed regular stacks, which is now understood in terms of partial ionicity, is not required for phenomenological spin Hamiltonians contrasting triplets with adjacent and independent spins.

The internal structure of triplet excitations for intermediate ionicity thus depends on the Coulomb interaction M_1 of adjacent D^+A^- sites. These must be explicitly included in any treatment of possible fine-structure splittings. A nearest-neighbor M_1 yields a CT excitation $DA \rightarrow D^+A^-$ around

$$\Delta E_{CT} = I_P - E_A - M_1 + \cdots \tag{10}$$

where we have neglected additional polarization, induction, relaxation, and long-range Coulomb interactions. Theories of intermolecular interactions,[75,89] which are far from quantitative, may eventually yield these terms. Such contributions can be combined approximately with the $I_P - E_A$ term, which defines a parameter $2\delta_0$. Nearest-neighbor interactions M_1 are included as[31]

$$\mathscr{H}_1 = -M_1 \sum_p n_p n_{p+1} \tag{11}$$

The number operators $n_p = a_{n\alpha}^\dagger a_{n\alpha} + a_{n\beta}^\dagger a_{n\beta}$ describe the charges at D or A, and are restricted to $n_p = 0, 1$. Such δ_0, M_1 models involve the same VB diagrams and solutions as the simple U models.[31,71] The paramagnetic–diamagnetic interface is accurately given by[90]

$$\delta_0^c = -1.24\ M_1 + 0.53\sqrt{2}|t| \tag{12}$$

which reduces to the previous result for $M_1 = 0$. As M_1 increases, $\gamma_c(M_1)$ increases slightly from 0.68 at $M_1 = 0$ to ~ 0.75 and the variation of γ vs $\delta_0/\sqrt{2}|t|$ becomes more rapid. The straightforward interpretation is that an attractive M_1 makes it easier to ionize next to an ionic site. The crossover thus becomes more rapid and even small long-range interactions may produce a discontinuity in γ.

Only the lowest triplets are thermally populated at 300 K for $\Delta E_m \sim 10^3$ cm^{-1} in Eq. (9). Their internal structure can be found exactly in finite rings and chains for any value of U_0, M_1. We are particularly interested in the stabilization of adjacent unpaired spins by M_1 for various ionicities, since the r^{-3} decrease of D and E essentially precludes fine-structure splittings for more distant spins. Exact VB results[90] are shown in Figure 9 for the probability of adjacent spins in the lowest triplet of a mixed regular ring of eight sites. Since a triplet necessarily contains a D^+ and an A^- site, the most neutral triplet for N sites has $\gamma = 2/N$, or 0.25 in Figure 9. Furthermore, each D^+ has two adjacent A^- neighbors out of $N/2$. The purely

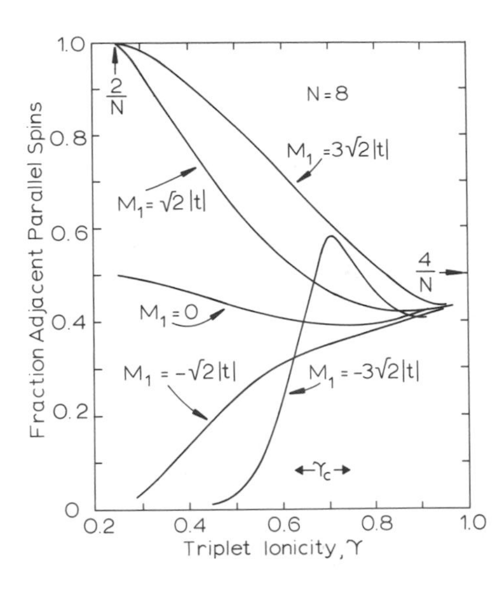

FIGURE 9. Internal spin structure of the lowest triplet in a mixed regular ring of eight sites. Attractive interactions $M_1/\sqrt{2}\ |t| > 0$ in Eq. (11) stabilize parallel spins on adjacent sites for neutral lattices, but not for largely ionic lattices with $\gamma \gtrsim \gamma_c$.

statistical chance for adjacent spins in the triplet is consequently $4/N$, as indicated in Figure 9. The exact admixture of VB diagrams in the lowest triplet state for $N = 8$ is close to, but not quite, statistical in the absence of nearest-neighbor interactions ($M_1 = 0$).

Attractive interactions clearly stabilize $^3|D^+A^-\rangle$ in largely neutral lattices, as shown by the $M_1 = \sqrt{2}|t|$ and $3\sqrt{2}|t|$ curves in Figure 9. Conversely, repulsive interactions $M_1 = -\sqrt{2}|t|$ or $-3\sqrt{2}|t|$ destabilize adjacent spins in the triplet. The peak around $\gamma \sim 0.6$ for $M_1 = -3\sqrt{2}|t|$ is a finite ring effect: unphysical *repulsive* interactions favor VB structures with distant neutral D and A sites, thereby favoring adjacent spins in the triplet for $\gamma \sim 1\text{–}2/N$. The difference between statistical and repulsive correlations disappears in the limit $N \to \infty$, when two spins have vanishing chance of being neighbors without an attractive M_1. Thus Figure 9 demonstrates that adjacent triplets are favored for $\gamma \lesssim 0.5$ and attractive D^+A^- interactions, but not for $\gamma > 0.5$ when the lattice is already substantially ionic. As expected, triplet excitons occur in mixed regular stacks with small γ, but not with large γ. The explicit computation is consistent with $\gamma \sim 0.7$ and no fine structure in (TMPD)(TCNQ). But the appropriate value of M_1 and the finite N restrictions preclude an accurate analysis of the decrease of D and E with increasing γ.

These results are not directly applicable to (Me_2P)(TCNQ), a triplet-exciton system with a dimerized mixed stack discussed in Section 2.3. The relative values of $|t_1|$ and $|t_2|$ along the stack are not known and introduce another parameter in the VB analysis. The crystal structure[58] suggests modest alternation, leading[59] to similar D and E values. As illustrated by triplet excitons in ion-radical salts with segregated stacks, dimerization leads to adjacent unpaired electrons even in a fully ionic lattice. It is consequently the large value $\Delta E_m \sim 0.60$ eV rather than fine-structure splittings that suggest $\gamma \sim 0.4 \pm 0.1$. Attractive Coulomb interactions would then stabilize adjacent $^3|D^+A^-\rangle$ states with observable fine structure even if the $|t|$ alternation is removed. The characterization of other CT complexes with intermediate ionicity is needed to distinguish among these possibilities.

Our discussion of partial ionicities in solid CT complexes has focused on magnetic evidence such as $\chi(T)$ or fine-structure splittings. The position and intensity of the CT absorption around 1 eV is also sensitive to the CT integral $|t|$. The resulting physical picture contains molecular parameters like I_P or E_A and crystal parameters like M, M_1, or polarization energies. Some of these are considerably larger than $|t|$, in the range 1–10 eV, and are potentially amenable to measurement by nonmagnetic techniques and to direct computation that are outside the scope of this chapter. All such information must eventually be included in assessing organic CT solids. We note that magnetic methods often depend on effective parameters that may include complicated many-electron and relaxation effects. Furthermore, the large molecular and crystal parameters must inevitably contain important cancellations for CT integrals $|t| \sim 0.1$–0.3 eV to produce intermediate ionicities. Thus accurate values for many parameters will be needed.

4. NMR Studies of Ion-Radical Solids

4.1. Nuclear Spins as Solid-State Probes

Magnetic nuclei provide a rich variety of low-energy probes in virtually all solids. Nuclear magnetic resonance (nmr) transitions at $\omega = \gamma H$, where γ is the gyromagnetic ratio and H is the applied field, quite generally give information about the electronic environment. The ^{1}H spectra of organic ion-radical solids rarely show resolved structure. These dense magnetic systems yield instead information through the position, widths, and relaxation characteristics of a single transition.

The most abundant nuclear spins in organic systems have $I = 1/2$. These include ^{1}H, ^{13}C, ^{19}F, and ^{31}P among others. Except for ^{13}C, whose natural abundance is 1.1%, the others are at, or close to, 100% natural abundance. The low natural abundance of ^{13}C is actually advantageous because it greatly reduces dipole–dipole interactions. In diamagnetic systems, resolved spectra due to chemically inequivalent carbons are then possible, especially when combined with techniques like magic-angle spinning and/or decoupling pulses. These direct approaches to high-resolution solid-state nmr are rapidly being implemented[91,92] for diamagnetic solids.

We restrict discussion here to paramagnetic solids and to $I = 1/2$ nuclei, thereby excluding quadrupolar studies such as ^{14}N resonances[93] in $TCNQ^{\gamma -}$.

The spin Hamiltonian for a nucleus **I** can be written, for $I = 1/2$, as

$$\mathscr{H}_I = -\gamma\hbar \mathbf{H}_0 \cdot (1 - \sigma) \cdot \mathbf{I} + \mathbf{I} \cdot D \cdot \mathbf{I}' + \mathbf{I} \cdot A \cdot \mathbf{S} \tag{13}$$

The chemical shift tensor σ will be taken as a scalar, so that the nuclear Zeeman transition is at $\omega = \gamma(1 - \sigma)H_0$. The nuclear dipole–dipole term D is a tensor that must be found for every nearby nucleus $\mathbf{I}'$. The tensor A describes interactions with an electronic spin **S** and may involve both hyperfine and dipolar contributions. Spin–orbit contributions to A are negligible for such light atoms as H, C, or F. The effects of D and A on the nmr spectrum, although formally similar, are quite different. Electronic couplings are much larger and electronic relaxation rates are generally much faster. The resulting modulation dominates nuclear relaxation, except in solids with rapid molecular reorientations. As the notation implies, the internuclear coupling is dipolar and thus depends on the relative orientations of the various magnetic nuclei. D makes a major contribution to the linewidth. Molecular motions that modulate D also contribute to nuclear relaxation.

We consider the second and third terms of Eq. (13) and discuss both static and dynamic properties. In the first category, we include broadband spectral features usually obtained by continuous wave (cw) measurements that were emphasized in early nmr studies.[94] The nature of local fields and the extent of electron delocalization are obtained. Various pulsed nmr experiments allow the measurement of the time dependence of the local fields produced by $\mathbf{S}(t)$. Of particular interest here is the modulation of the hyperfine coupling by the motion, relaxation, or exchange of the electronic spin.

Fortunately, the dynamics of $\mathbf{S}(t)$ can be probed without having detailed knowledge of the magnitudes and orientations of the tensors in Eq. (13). Powdered samples of systems with many inequivalent protons can be used, especially if augmented by selective deuteration. The sensitivity requirements of nmr often imply larger samples than the available single crystals of ion-radical conductors. The temperature, pressure, and field dependences of powders may, nevertheless, provide detailed information about dimensionality, local susceptibilities, transport mechanisms, or small transverse interactions in these highly anisotropic solids. Since fully quantitative analyses must eventually depend on direct information about D and A in Eq. (13), the powder results reviewed later typically rely on additional nonmagnetic information.

We first discuss the single-crystal nmr of an unusually favorable one-dimensional magnetic insulator, α-bis(*N*-methylsalicylaldiminato)Cu(II), α-CuNSal. The magnetic properties of α-CuNSal are given by the regular

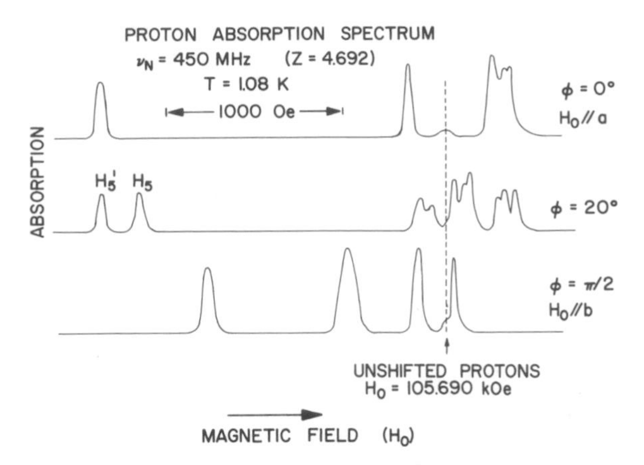

FIGURE 10. Single-crystal proton nmr of α-CuNSal. All magnetically inequivalent protons are resolved for H_0 along the crystallographic *a* and *b* axes. (Reproduced from Reference 96a, with permission.)

Heisenberg antiferromagnet, Eq. (8), with[95, 96] $J/k_B = 3.04$ K. Interchain exchange is small, since three-dimensional ordering occurs[97] at 0.044 K. The three principal advantages of α-CuNSal are: (1) large single crystals can be prepared; (2) the small value of J allows the imposition of a ferromagnetic ground state near 0 K by a field $H_0 > 4J/g\mu_B = 88$ kOe; (3) the spin-1/2 regular Heisenberg antiferromagnet is one of the best understood solid-state models.

The single-crystal α-CuNSal spectrum[96] shown in Figure 10 resolves all five inequivalent protons per unit cell for $\mathbf{H}_0$ along the crystallographic *a* and *b* axes, and partly resolves the ten protons for a general orientation H_0. The indicated spectra have $g\mu_B H_0/J \equiv Z = 4.692$, when at 1.08 K the Cu(II) spins are essentially completely aligned with $H_0\hat{\mathbf{z}}$. The enormous Knight shift[98] of the H_5 protons in Figure 10 reflects the magnetization $g\mu_B\langle S_z\rangle H_0 \approx -g\mu_B H_0/2$ of the two magnetically inequivalent $S = 1/2$ CuNSal complexes. According to Eq. (13), the proton resonance is shifted by $A_{zz}\langle S_z\rangle$ in a large field $H_0\hat{\mathbf{z}}$. The field and temperature dependence of the shifts yields[96] a magnetization curve in excellent agreement with theory[29] for $J/k_B = 3.04$ K. The angular dependence of the H_5 shifts in Figure 10 define the coupling constant A in Eq. (13). The hyperfine interaction is dominated[96] by an isotropic $a\mathbf{I}\cdot\mathbf{S}$ contact interaction with $a = 5.4 \times 10^{-4}$ cm^{-1}, and a smaller dipole interaction of 1.3×10^{-4} cm^{-1} between the electronic moment and the H_5 proton. The electron–nuclear coupling constants are thus completely determined by the positions of the resolved single-crystal spectra.

Now spin–lattice relaxation (T_1^{-1}) measurements[96] of the H_5 protons require no additional parameters since $J/k_B = 3.04$ K completely specifies the Cu(II) linear chain. The strong field dependence of T_1^{-1} around the critical field $H_c = 4J/g\mu_B = 88$ kOe is described quantitatively.[96] Thus Heisenberg exchange models, which are usually parametrized in terms of static properties, also account for the spin dynamics of α-CuNSal. The ideal $s = 5/2$ Mn(II) antiferromagnetic chain in $N(CH_3)_4MnCl_3$ (TMMC) leads to similar dynamical conclusions for the epr absorption[99] and neutron diffraction.[100] The consistency of static and dynamic properties remains to be demonstrated in many other magnetic insulators. The situation is far more complicated in open-shell ion-radical solids whose electronic structure have not been modeled quantitatively.

4.2. Internuclear Dipolar Local Fields

The nmr line shape in solids is usually determined by the nuclear-dipolar interactions,[94,101] the second term in Eq. (13). A nucleus $\mathbf{I}_j$ at $\mathbf{r}$ produces a dipolar field $\mathbf{H}_j = D \cdot \mathbf{I}_j$ decreasing as r^{-3} at some reference spin. For protons separated by 2 Å, $|\mathbf{H}_j|$ is roughly 1 gauss and typically $\sum_j |\mathbf{H}_j|^2 \simeq 5$–$10$ G^2. Hence, nuclear-dipolar fields broaden spectral lines by several gauss and lead to a Gaussian distribution for many neighbors. Because $\mathbf{H}_j$ depends on the relative orientations of the external (Zeeman) field and the internuclear coupling vector, it is anisotropic and the line shape is asymmetric.[101] The anisotropy of $\mathbf{H}_j$ usually exceeds the separation in resonance fields for inequivalent nuclei and a single broad line is obtained.

The nmr spectrum in solids can rarely be calculated in detail. Rather, it is analyzed in terms of the characteristic moments[101] which can be easily measured for any reasonably intense spectrum. The nth moment about the mean for an arbitrary line shape $g(\omega)$ is simply

$$M_n = \langle \Delta\omega^n \rangle = \int (\omega - \bar{\omega})^n \, g(\omega) \, d\omega \Big/ \int g(\omega) \, d\omega \tag{14}$$

where $\bar{\omega}$ is the first moment about zero. The second moment M_2 is by far the most important in the nmr analysis and is given by the Van Vleck formula[102] for nuclei with the same gyromagnetic ratio γ_I

$$\langle \Delta\omega^2 \rangle = \tfrac{3}{4}\gamma_I^4 \hbar^2 I(I+1) \sum_{j,k}{}' (1 - 3\cos^2\theta_{jk})^2 r_{jk}^{-6} \tag{15}$$

where $j = k$ is excluded and the internuclear vector $\mathbf{r}_{ij}$ makes an angle θ_{ij} with the magnetic field (taken as the z axis). The average of Eq. (15) over θ_{jk} yields the standard result for powders.[101] Nuclei with gyromagnetic ratios different from the resonant nucleus require a numerical factor of $(1/3)\gamma_I^2\gamma_{I'}^2\hbar^2 I'(I'+1)$ in Eq. (15).

Averages of the geometrical factor must be taken over all the orientations for systems with molecular motion on the nmr time scale. The classic study[103] of solid benzene and its deuterated analogs demonstrates the power of Eq. (15). The distance between the ring protons can be computed from the measured second moment and Eq. (15). Such calculations showed a remarkable accuracy and precision; 2.50 ± 0.02 Å was found for adjacent ring protons, while x-ray measurements gave 2.47 ± 0.02 Å. Perhaps the most interesting result was that the intramolecular contribution to $\langle \Delta\omega^2 \rangle$ decreased from 3.10 to 0.77 G^2 as the temperature was raised from 90 to 120 K. This is just the factor of 1/4 which arises from averaging the cosine term in Eq. (15) over the rotations of the benzene molecule about its sixfold axis. The molecule evidently rotates faster than the nmr frequency (>50 MHz) at temperatures above 90 K. Similar results are obtained from ^{13}C nmr studies.[104] In naphthalene, $\langle \Delta\omega^2 \rangle$ computations can distinguish[105] between trial structures whose CH distances differ by as little as ± 0.05 Å.

The nuclear contributions to the nmr second moment can readily be estimated via Eq. (15), as illustrated[106] for the strongly dimerized linear chain in $Cu(NO_3)_2$(pyrazine). This inorganic insulator is suitable for single-crystal nmr and the few protons per unit cell are resolved. The single nmr absorption[107] of $[Ph_3(PCH_3)](TCNQ)_2$, the complex salt whose triplet-exciton behavior was sketched in Section 2.2, shows a discontinuity in M_2 at the 315 K structural transition. Local nuclear fields were probed[108] by cw nmr of the organic conductor (NMP)(TCNQ) and its deuterated analog, (DMP)(TCNQ), where the CH_3 group in Figure 6 was replaced by CD_3. The variation of M_2 with temperature is shown in Figure 11 for both systems. Equation (15) immediately gives the (DMP)(TCNQ) second moment at high temperature. The indicated range of M_2 for various fixed CH_3 geometries demonstrates rapid rotation (compared to 28 MHz) for all $T \geq 1.5$ K, as expected for small sixfold barriers. The (NMP)(TCNQ) second moment at high temperature is again consistent with the (1/4 contribution of a rotating methyl. The parallel increase of $M_2(T)$ for both com-

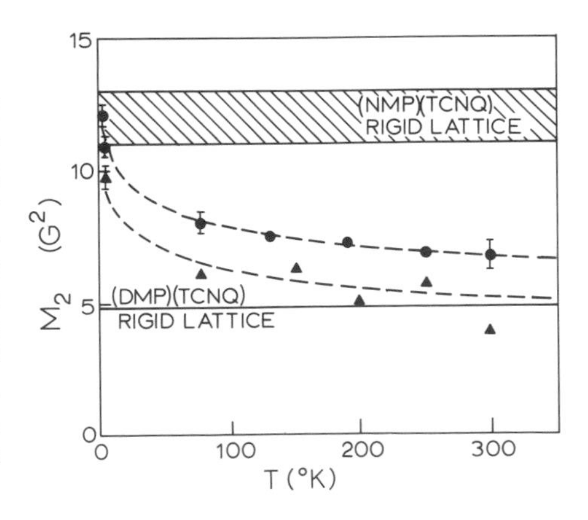

FIGURE 11. Temperature dependence of the proton second moment of powder (NMP)(TCNQ) (●) and (DMP)(TCNQ) (▲). The calculated value for (DMP)(TCNQ) is based on the $P\bar{1}$ structure. Several hypothetical $P\bar{1}$ structures with nonrotating methyls are used for (NMP)(TCNQ). The dashed lines are smoothed curves separated by 1.5 G^2 through both sets of data points and indicate a common electronic contribution. (Reproduced from Reference 108, with permission.)

plexes below 150 K in Figure 11 supports[108] a common electronic contribution that increases as cooling decreases the modulation rate of $\mathbf{S}(t)$. Other evidence[109] based on T_1^{-1} data leads to the same result. In addition to molecular reorientation, thermal motions of atoms about their equilibrium positions can modulate[110] M_2 significantly and Sjöblom[111] has developed a convenient formulation for calculating such relaxation rates.

Molecular reorientations in solids have been extensively studied by nmr.[112] Methyl group rotations have been particularly interesting in connection with cross relaxation[113] and low-temperature tunnelling.[114, 115] Since the motion of the three CH_3 protons is not independent, the usually small cross-relaxation terms can become important and may produce deviations from exponential relaxation even in a one-component spin system, as observed.[116] However, nonexponential decay is not a general consequence of methyl rotation.[117, 118] At high temperature, classical diffusion leads to an activation barrier for rotation and the characteristic Bloembergen–Purcell–Pound (BPP) relation[101] is found between T_1^{-1} and temperature. In some systems, tunnelling[114, 115] persists to a very low temperature and produces a variety of quantum-mechanical effects. The Pauli principle for three $I = 1/2$ particles correlates spin and rotational wave functions. In the absence of other (e.g., electronic) moments, processes that change $I = I_1 + I_2 + I_3$ become very slow indeed.[119]

The point is that molecular motions which modulate dipolar fields can usually be distinguished in T_1^{-1} experiments from electronic processes such as exchange or carrier mobility. The $(NH_4)^+(TCNQ)^-$ and $K^+(TCNQ)^-$ relaxation data[120] in Figure 12 illustrate such differences. The broad T_1 vs

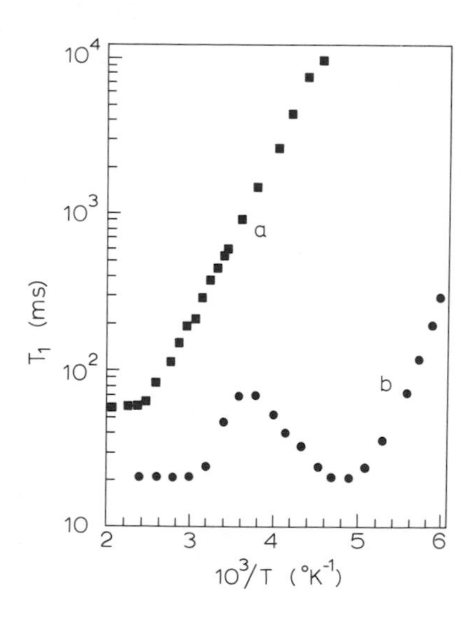

FIGURE 12. Temperature dependence of the 54-MHz proton spin-lattice relaxation time T_1 in powdered samples of (a) $(K)^+(TCNQ)^-$ and (b) $(NH_4)^+(TCNQ)^-$. The triplet exciton density $\rho(T)$ contributes in both cases, as does $(NH_4)^+$ tumbling in $(NH_4)^+(TCNQ)^-$. (Reproduced from Reference 120, with permission.)

T^{-1} minimum of the NH_4^+ salt is a typical BPP temperature dependence, with a minimum[98, 101] when the NH_4^+ reorientation rate plus possible electronic contribution match the nmr frequency $\omega = \gamma(1 - \sigma)H_0$ in Eq. (13). On the other hand, T_1 for $K^+(TCNQ)^-$ has a strikingly different, monotonic temperature dependence in Figure 12. There are no molecular reorientations and T_1 is from 10 to 10^3 times longer than $(NH_4)^+(TCNQ)^-$. Two reorientations are important in the relaxation mechanisms in $[N(CH_3)_3H]^+(I_3)^-_{0.33}(TCNQ)^-_{0.67}$.[121] The cation both rotates about the C_3 symmetry axis and tumbles. Two minima in the T_1 vs T plots were observed[121] and activation energies determined from the nmr data were consistent with those of $N(CH_3)_4^+X^-$ salts, where X^- is a halide. Again, contributions from electron–nuclear coupling dominate the relaxation at high temperatures.

It is perhaps surprising that M_2 measurements, which were so important in early nmr work,[94] are not routinely available for ion-radical solids. In particular, for compounds in which local nuclear fields contribute to the relaxation, it would be of interest to investigate the dependence of M_2 on temperature and deuterium substitution. A great deal of information can readily be obtained from a straightforward powder measurement.

4.3. Electronic Local Fields

Although formally analogous to nuclear fields, electronic local fields produce quite different nmr effects. The electron spin S has a magnetic moment $g\mu_B$ that is some 660 times larger than the proton moment. The time variations of $\mathbf{S}(t)$ due to exchange, conduction, exciton hopping, etc., are unique to the electronic moments and often dominate the nuclear relaxation. We note that a combination of spin delocalization, which reduces the effective local field, and rapid modulation, which again decreases the average field, has made possible high-resolution nmr studies of paramagnetic molecules.[122] These effects are less central in ion-radical organic solids, where instead the temperature and field dependence of the nuclear relaxation are probes for the localization and motion of electronic moments.

In high fields, the $A_{zz}I_zS_z$ term in Eq. (13) gives a local field $A_{zz}\langle S_z \rangle/\gamma_I$ that shifts and broadens the nuclear resonance, as shown in Figure 10 and discussed under Knight shifts in standard texts.[98, 101] Varying the orientation of $\mathbf{H}_0$ in single-crystal studies thus yields, in principle, the anisotropy of the hyperfine interaction, as shown for[96] α-CuNSal and for[123] $(Ph_3AsCH_3)(TCNQ)_2$. The nmr shifts[124] in WBP provided an early confirmation of the susceptibility change at the dimerization transition. Although Knight shifts were originally associated with conduction electrons in metals, the preceding examples based on magnetic insulators and triplet-exciton

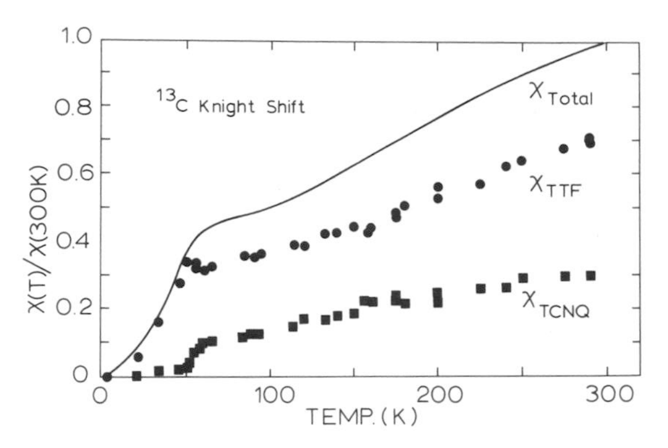

FIGURE 13. Local susceptibilities of TTF and TCNQ chains in (TTF)(TCNQ) obtained from ^{13}C nmr Knight shifts and the total susceptibility. (Reproduced from Reference 125, with permission.)

salts merely require a more general approach to the coupling constant A in Eq. (13) and to the calculation of $\langle S_z \rangle$. All such paramagnetic shifts yield, via $\chi \propto g\mu_B \langle S_z \rangle$, the electronic susceptibility at the nuclear site.

The possibility of measuring *local* susceptibilities is quite advantageous when there are several paramagnetic subsystems, as in (TTF)(TCNQ). The $TCNQ^{\gamma-}$ and $TTF^{\gamma+}$ susceptibilities inferred[125] from ^{13}C Knight shifts are shown in Figure 13. The cyano groups of TCNQ were selectively enriched. It is immediately seen from Figure 13 that, for any $T \gtrsim 70$ K, the paramagnetism is two times larger on the TTF stack. A nearly equal decomposition is deduced from epr studies[126] of the position and relaxation of the single narrow resonance. Both methods suggest similar temperature dependences for $T > 70$ K, and thus imply similar pictures for the two stacks. This immediately undermines any model based on substantially different $\chi(T)$ for each stack. The exact decomposition is still open, as are the relative magnitudes of the microscopic CT integral $|t|$ and correlation energies U for the two stacks.[10,45] Different magnetic, optical, and transport measurements suggest comparable magnitudes, but disagree about which stack has the larger parameters.

Similar studies for the proton nmr spectra of several TCNQ and TTF salts allowed measurement of the Fermi contact and dipolar coupling.[127] Dipolar hyperfine coupling were found to be significant for both the TTF and TCNQ protons. The field dependence[128] of M_2 for (NMP)(TCNQ) and (DMP)(TCNQ) reflects unresolved Knight shifts. The far stronger M_2 vs H behavior at $T \leq 4.2$ K for (NMP)(TCNQ) relative to (DMP)(TCNQ) indicates that the unpaired spins are close to the methyl group and are thus on the phenazine moiety. Since $(NMP)^+$ is diamagnetic, a formulation like $(NMP)^{\gamma+}(TCNQ)^{\gamma-}$ is indicated,[128] with a partial charge transfer of

$\gamma = 0.94$. Other experiments[129] support this result, most notably here a quantitative study[109] of ^1H relaxation. Unfortunately, the unavoidable[130] contamination by $(NMPH)^+$, in which the second NMP nitrogen in Figure 6 is protonated, somewhat obscures these results, and rather lower γ is inferred[131] from diffraction studies. The long and still elusive preparation of pure (NMP)(TCNQ) does not concern us here.

Most nmr studies of organic ion-radical solids have focused on the nuclear spin-lattice relaxation rate, T_1^{-1}, rather than on Knight shifts. The modulation of $\mathbf{S}(t)$ dominates T_1^{-1} in solids without mobile molecular fragments. The relaxation rate is quite generally given by[132]

$$T_1^{-1} = \Omega_z f(\omega_N) + \Omega_+ f(\omega_e) \tag{16}$$

where Ω_z and Ω_+ are geometrical factors related to the I_z and $I^{\pm}$ components, respectively, of the electron–nuclear interaction $\mathbf{S} \cdot A \cdot \mathbf{I}$ in Eq. (13). They are usually anisotropic, but are independent of field and temperature. The time dependence of $\mathbf{S}(t)$ is contained in the spectral density function

$$f(\omega) \propto \int_0^\infty dt \langle S_z(t) S_z(0) \rangle \cos \omega t \tag{17}$$

Thus, theoretical analyses[133,134] of the two-spin correlation function $\langle S_z(t)S_z(0)\rangle$ are directly probed by nmr at the nuclear (ω_N) and electronic ($\omega_e \gg \omega_N$) frequencies. In contrast to the four-spin correlations encountered[135,136] for the epr linewidth, only the simpler two-spin correlations occur in Eq. (16) due to the generally slow relaxation of nuclear spins. Varying ω_N and ω_e by changing H_0 then affords a direct nmr test for $f(\omega)$, as illustrated in greatest detail[134] for $N(CH_3)_4MnCl_3$(TMMC). We emphasize that Eq. (16) is quite general. As has already been suggested, it can be applied cautiously without knowing the precise values of Ω_z and Ω_+.

The anisotropies of Ω_z and Ω_+ in TCNQ salts are usually only known approximately. Motional narrowing averages out all hyperfine structure, as shown in Figure 5, and has so far precluded experimental determinations as in α-CuNSal. Computations of Ω_z and Ω_+ are based on the observed crystal structure and theoretical or solution spin densities, which at present still contain some inconsistencies.[137] The resulting Ω_z and Ω_+ are only moderately successful[138,139] in fitting the observed anisotropy of T_1^{-1} in $(Ph_3AsCH_3)(TCNQ)_2$. Other experimental checks of the solid-state spin densities include fitting[1,22,23] epr fine-structure constants, as summarized in Section 2, and fitting $TCNQ^-$ hyperfine splitting in liquid crystals.[140] Fitting the T_1^{-1} anisotropy in a complicated solidlike $(Ph_3AsCH_3)(TCNQ)_2$ is a far more stringent test of spin densities and coupling constants.

The triplet-exciton system $(Ph_3AsCH_3)(TCNQ)_2$ has been the subject of several nmr relaxation studies, partly because large single crystals can be prepared but also because of the small $\Delta E_{ST} \simeq 0.065$ eV. Initial measurements[107] of T_1 vs T on the isomorphous phosphonium salt suggested that molecular motions, presumed to be methyl rotation, provide the dominant relaxation pathway. However, more precise studies on both the phosphorous and arsenic compounds subsequently confirmed that excitons were directly involved in the proton relaxation.[141, 142] A lower bound of $\sim 10^9$ hops/s for triplet excitons results from the narrow epr line. These fluctuating magnetic fields control T_1^{-1}. The temperature dependence of $\rho(T)$ according to Eq. (3) then allows experimental variations of the internal field. The strong temperature dependences due to magnetic dilution yield many interesting effects. Magnetic dilution complicates the analysis, however, since rigorous results usually hold in the high-temperature limit.

4.4. Dimensionality of Spin Dynamics

The autocorrelation function $\langle S_z(t) S_z(0) \rangle$ in Eq. (17) is quite different when fluctuations are restricted to one, two, or three dimensions. Spin diffusion at high temperature leads asymptotically to $t^{-d/2}$ decay, where d is the dimensionality. The $\omega = 0$ component of Eq. (17) then diverges for $d = 1, 2$ and results in enhanced relaxation.[99, 133–136] The normalized $d = 1$ autocorrelation function $\langle S_z(t) S_z(0) \rangle / \langle S_z^2(0) \rangle$ is sketched in Figure 14, together with a long-time "cutoff" ν_c defined as

$$C(t) = C_{1d}(t)\, e^{-\nu_c t} \tag{18}$$

Here $C_{1d}(t) \propto t^{-1/2}$ is the ideal behavior that is eventually lost due to weak $d = 3$ interactions, collisions, or other processes. The short-time behavior

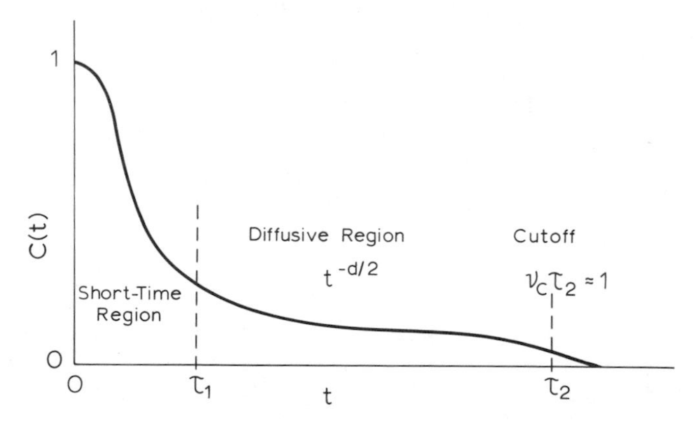

FIGURE 14. Schematic time dependence of the spin-autocorrelation function. The short-time decay does not depend on dimensionality. The $d = 1$ or 2 diffusive behavior is eventually cut off at long times, as indicated in Eq. (18).

indicated in Figure 14 does not depend on d. Soos[143] introduced such cutoffs for the one-dimensional (1d) random walk of triplet excitons and showed that even modest ν_c can restore a Lorentzian line shape and suppress the $\omega = 0$ divergence. The subsequent observation[144] of a non-Lorentzian epr absorption in TMMC led to many investigations[99, 132–136, 145] of low-dimensional magnetic insulators, but it remains difficult[146] to determine and interpret cutoffs accurately.

The $\nu_c = 0$ limit of Eq. (18) leads to an $\omega^{-d/2}$ divergence of $f(\omega)$ as $\omega \to 0$. Both $d = 1$ and $d = 2$ diverge, while $d = 3$ converges. The strong $d = 1$ divergence clearly increases T_1^{-1} in Eq. (16), although the magnitude of the effect cannot be properly assigned unless Ω_z and Ω_+ are known. It is far simpler to demonstrate the appropriate H_0 dependences, with T_1^{-1} proportional to $\omega^{-1/2}$ for $d = 1$ and to ln ω for $d = 2$. Additional information about spin fluctuations may indicate, still without evaluating Ω_z or Ω_+, whether the ω_N or ω_e term is appropriate. The relevant frequency can also be checked[147] via an Overhauser experiment, preferably on the same system. The highly suggestive T_1^{-1} vs $\omega^{-1/2}$ or ln ω data are central, rather than a fully quantitative analysis.

As summarized in Section 2.1, the diffusive nature[24] of triplet excitons in $(Ph_3XCH_3)(TCNQ)_2$, with X = As or P, is supported by mixed-crystal work. Magnetic resonance thus yields information about coherence, thereby potentially distinguishing between wave and diffusive motion.[148] The one-dimensional random walk of a small, temperature-dependent concentration of triplets leads to $\omega^{-1/2}$ behavior. An exponential cutoff occurs naturally if electronic relaxation and exchange are explicitly included.[149] Transverse exciton motion could also be approximated[150] by an exponential cutoff in Eq. (18). Devreux and Nechtschein[151] plotted T_1^{-1} data for $(Ph_3AsCH_3)(TCNQ)_2$ vs $\omega^{-1/2}$ and concluded that the motion was one-dimensional between 7 and 100 MHz and between 100 and 295 K. A cutoff $\nu_c = 10^7$ Hz was determined at 145 K and interpreted as separate regimes in which different terms of Eq. (16) dominate T_1^{-1}. The same data, in Figure 15, were plotted against ln ω by Butler *et al.*[152] Their analysis of highly anisotropic three-dimensional random walks indicates that modest transverse motion results, on the slow nmr time scale, in higher-dimensional $C(t)$. The entire 145-K range of 7–100 MHz in Figure 15 is now seen to involve two-dimensional triplet motion. The $(Ph_3AsCH_3)(TCNQ)_2$ crystal structure[153] shows $(TCNQ_2)^-$ stacks forming sheets separated by diamagnetic $Ph_3AsCH_3^+$ cations. Spin diffusion across sheets is slower than 7 MHz, while spin motion along the stack is approximately known from epr to be around 10^4 MHz at 145 K. The nmr data imply a stack-to-stack diffusion of at least 100 MHz within a given sheet. To prove crossovers to three- or one-dimensional motion would require at least $\nu < 7$ MHz and $\nu > 100$ MHz, respectively. Deviations from $\omega^{-1/2}$ or ln ω behavior signal dimen-

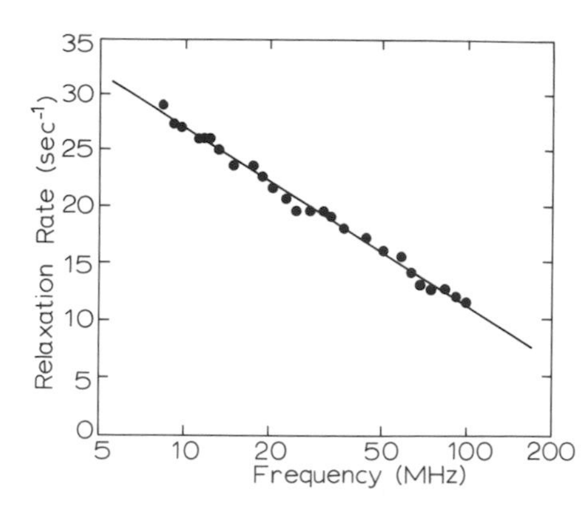

FIGURE 15. Proton spin-lattice relaxation rate in $(Ph_3AsCH_3)(TCNQ)_2$ at 145 K, as reported by Devreux and Nechtschein, Reference 151. A straight line in a semilogarithmic plot is expected for two-dimensional diffusion. (Reproduced from Reference 152, with permission.)

sionality crossovers whose analysis should be particularly interesting in ideal cases with accurately known coupling constants Ω_z and Ω_+.

In all TCNQ salts studied so far, a single relaxation time can unambiguously be defined for all protons in the sample. For compounds with protic cations, it is then necessary to assume that the protons are tightly coupled and have a single spin temperature. When hyperfine modulation is the relaxation mechanism even protons on diamagnetic cations are relaxed by the electronic local fields. These aspects were studied in the $(NH_4)^+(TCNQ)^-$ salt.[154] Both spin–spin coupling and electron–cation proton interactions were found to be important. At temperatures where the exciton concentration, $\rho(T)$ in Eq. (3), was low, there are additional contributions from NH_4^+ reorientation. The investigations of the ammonium salt are particularly complete and amply demonstrate the complications arising from multiple proton spin baths and relaxation mechanisms.

The nmr of low-dimensional conductors parallels that of triplet-exciton systems, but with an entirely different origin for $\langle S_z(t)S_z(0)\rangle$. Theoretical expressions[155] based on the strongly correlated (large U) Hubbard Hamiltonian describe the scattering of conduction electrons. The inevitable $\omega^{-1/2}$ dependence of the spectral density function $f(\omega)$ is again found for a one-dimensional chain, as well as for arbitrary filling[156] of one-dimensional Hubbard models. Neither Hubbard models nor two-dimensional diffusion yield the observed[157] Korringa temperature dependence for T_1^{-1} in $Qn(TCNQ)_2$, where Qn = quinolinium. The $\omega^{-1/2}$ dependence of T_1^{-1} is nonetheless valid at high temperature and frequencies[158] and recent nmr work[159] relates conduction and spin diffusion in $Qn(TCNQ)_2$.

Other relaxation mechanisms in conductors must be considered, but the lack of accurate electron–nuclear coupling constants hampers quantitative analysis.[160] An interesting application of the "cutoff" theory, Eq. (18), for (TTF)(TCNQ) gave 5.8 meV for the transverse transfer integral $t_\perp$ of electrons,[161] which is reasonable in light of conduction anisotropies of 10^3 and $t_{\parallel} \sim 0.1$–0.2 eV. Transverse motion in organic conductors is probably diffusive, while charge motion along the partly filled stacks may involve significant mean-free-paths, especially at low temperature. Magnetic resonance methods offer many approaches to such important questions about coherence. Recent nmr results[162] for doped and neat polyacetylene $(CH)_x$ establish remarkably high one-dimensional behavior in this noncrystalline organic polymer.

5. *Paramagnetism of Ion-Radical Solids*

5.1. *Organic Conductors: Susceptibility of (TTF)(TCNQ)*

The unusual physical properties of organic conductors and of paramagnetic semiconductors are associated with the narrow, partly filled electronic bands resulting from π-electron overlap. Thus paramagnetism is intrinsic, and typical magnetic measurements directly probe the electronic states near the Fermi level that also dominate transport properties. On the other hand, no quantitative analysis of the paramagnetism can be expected until the appropriate solid-state model is established for these partly filled bands. First, in contrast to the nuclear probes in Section 4, the electronic moments function both as probes and as the primary system. The former suggests magnetic experiments in conjunction with proposed models of optical or transport properties, while the latter involve intrinsic characterizations of the interacting spin system. Second, various Heisenberg exchange networks can be invoked for insulators, where the magnetism and conductivity are modeled independently. Magnetic insulators remain active research areas; they are far better characterized than organic conductors, whose magnetic, optical, and transport properties are interconnected.

The statistical mechanics of infinite networks of exchange-coupled spins have been extensively discussed[163] in connection with high-temperature expansions, spin correlation functions, and various phase transitions. Applications to the static magnetic properties of transition-metal ions are reviewed by de Jongh and Miedema[4] and by Carlin and van Duyneveldt.[164] The $S = 5/2$, ${}^6A$ ion Mn(II) and $S = 1/2$, ${}^2D$ ion Cu(II) have particularly simple spin Hamiltonians, with minimal interference from spin–orbit interactions. Thus ideal extended systems usually involve these ions, as already mentioned for the one-dimensional TMMC[99, 134, 145] and α-CuNSal systems.[96] Two-dimensional antiferromagnetic lattices based on R_2MnX_4 can be systematically studied[165–170] by varying X = F, Cl, Br, and R = alkali metal or various amines. Spin dynamics based on the epr linewidth are less well characterized, although[171] K_2MnF_4 and[172]

(1,3-propanediammonium)$MnCl_4$ are consistent with two-dimensional Heisenberg exchange and a cutoff in Eq. (18).

The corresponding R_2CuX_4 systems indicate two-dimensional ferromagnetic exchange.[172–175] Here again, varying the length of the amines separating the MX_4^{-2} layers provides a systematic approach to weak interplanar interactions. The approximation of isotropic exchange $JS_1 \cdot S_2$ is no longer adequate for Cu(II). Anisotropic and antisymmetric corrections[176,177] must be included. These lead to additional epr line broadening, as shown[178] for anisotropic contributions in antiferromagnetic Cu(II) linear chains and for both types of corrections[179,180] in ferromagnetic Cu(II) layers. Such spin–orbit (so) corrections are inevitable for heavier ions, but are poorly understood quantitatively even for high-symmetry sites. Thus even 5–10% spin delocalization to bromide ligands may greatly broaden and effectively spoil the epr spectrum. The versatile organic donor[181] tetrathiofulvalene (TTF) in Figure 16 provides a large family (6–10) of organic metals, either as two-stack 1 : 1 systems with TCNQ or as one-stack, nonstoichiometric 1 : γ halides. The broader epr lines of the selenium derivatives from Figure 16 are discussed in Section 5.2. Such molecular spin–orbit contributions are even less understood. Only qualitative discussions are consequently possible for the convenient linewidth comparisons of conductors based on substituted TTF.

The distance dependences of the exchange constant $J(R)$ in Eq. (5) or of the CT integral $|t(r)|$ yield other, biquadratic corrections to isotropic exchange. Exchange striction[182] is a general phenomenon that has been identified in substitutional Mn(II) dimers,[183] in Cr(III) trimers,[184] and in diffusional triplet excitons.[24] The relatively high thermal expansion[185] and pressure sensitivity[10,186] of organic conductors clearly point to such dis-

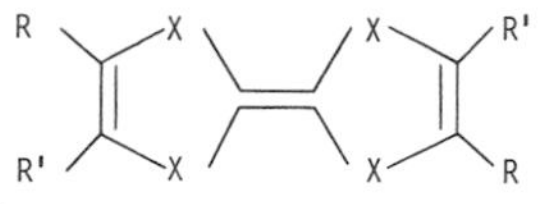

R, X, X, R'
R', X, X, R

X=S		X=Se	
TTF	H=R=R'	TSF	H=R=R'
DMTTF*	H=R; CH_3=R'	DMTSF*	H=R; CH_3=F'
TMTTF	CH_3=R=R'	TMTSF	CH_3=R=R'
HMTTF	$(CH_2)_3$=R+R'	HMTSF	$(CH_2)_3$=R+R'
DEDMTTF*	CH_3=R; C_2H_5=R'	DEDMTSF*	CH_3=R; C_2H_5=R'

*cis, trans isomers possible

FIGURE 16. Representative TTF related donors. Mixed Se, S, and unsymmetrically substituted donors lead to chemically disordered stacks.

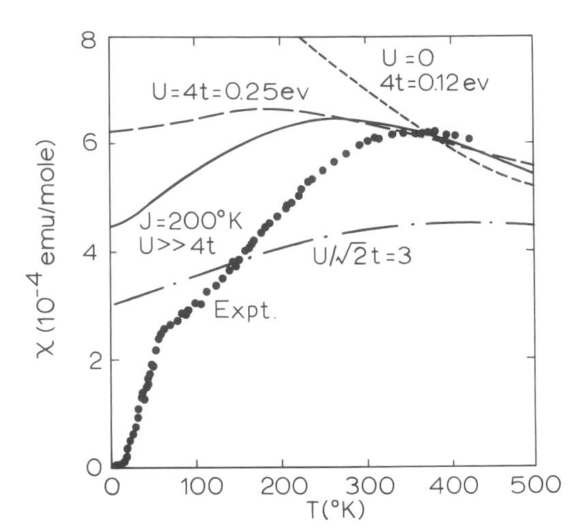

FIGURE 17. Four computations for the (TTF)(TCNQ) paramagnetic susceptibility for $T > 60$ K, ranging from band theory ($U = 0$) to intermediate correlations ($U/\sqrt{2}|t| = 3$), to the atomic limit $U \gg 4|t|$. (Adapted from References 189 and 199.)

tance dependences, as proposed[187] for improved susceptibility fits. As with spin–orbit contributions to g values and exchange corrections, striction effects are far from quantitatively understood.

We have deliberately omitted the largest class of paramagnetic salts, those based on three-dimensional networks of transition-metal and rare-earth ions. Such systems illustrate both ferromagnetic and antiferromagnetic coupling. They are not directly relevant to ion-radical organic solids, which are generally anisotropic due to the stacking of planar subunits. We are not aware of any reported ferromagnetism in extended organic systems. The kinetic exchange[20] in Eq. (5) describes CI stabilization of spin-paired states and is intrinsically antiferromagnetic. Other, potentially ferromagnetic exchange contributions may be small in organic systems, but are no means impossible. In any case, low-dimensionality is no longer a problem, now that ferromagnetic exchange in Cu(II) linear chains has been identified.[188]

The parent compound in the (TTF)(TCNQ) family has now been extensively characterized. The magnetic susceptibility[189] of (TTF)(TCNQ) in Figure 17 is but one measurement among many, as reviewed by Toombs.[190] The same data can be obtained from the epr intensity[191] and, as already shown in Figure 13, from ^{13}C nmr shifts. The less precise epr intensity avoids the diamagnetic contributions that have been subtracted from Figure 17. The metal–insulator transition around 54 K produces a knee, as in other salts,[192] and the system becomes a diamagnetic insulator at low temperature. Several epr signals are seen[193] for $T < 15$ K, presumably due to isolated paramagnetic misfits. They correspond to molecular $(TTF)^+$ and $(TCNQ)^-$ g tensors, as found in salts with a single paramagnetic center or from the isotropic solution g value. A single narrowed epr resonance occurs at high T. Its position at $\omega = g\mu_B H_0$, width ΔH, and intensity constitute the principal epr data for (TTF)(TCNQ) and other organic conductors.

(TTF)(TCNQ) is a two-stack conductor. The different g values of $(TTF)^+$ and $(TCNQ)^-$ are averaged in the observed resonance. The simplest decomposition[126, 191] is to weight their contributions by the fractional susceptibilities χ_A/χ and χ_D/χ on the acceptor and donor stacks,

$$g = g(A^-)\chi_A/\chi + g(D^+)\chi_D/\chi \tag{19}$$

Here $\chi(T)$ is the observed total susceptibility in Figure 17 and $g(A^-)$, $g(D^+)$ are characteristic molecular g values associated with $(TCNQ)^-$ and $(TTF)^+$. The observed g value in Eq. (19) then yields χ_A and $\chi_D = \chi - \chi_A$ as a function of temperature.[126, 191] The fact that g is close to $g(TTF)^+$ for $20 < T < 40$ K demonstrates that a magnetic gap opens at 54 K on the $(TCNQ)^-$ stack, thereby reducing χ_A. The gap on TTF stack opens at lower $T \sim 38$ K, and the magnitude of both gaps can be estimated from the intensity. Above $T \sim 54$ K, the observed g value[126, 191] in the conducting regime is temperature independent and close to the average of $g(TTF)^+$ and $g(TCNQ)^-$. The contributions of the two stacks are comparable. The similar behavior of $\chi_A(T)$ and $\chi_D(T)$ strongly suggests that similar microscopic models are required, a conclusion that is quite independent of any model.

These epr measurements complement the more extensive and precise structural and transport characterization of Peierls transitions. Their value lies in readily identifying the individual stacks through Eq. (19), as illustrated[194] in neutron-irradiated (TTF)(TCNQ) samples. Irradiation produces paramagnetic defects and lowers the metal–insulator transition temperature, T_c. The 300-K lattice constants in a heavily damaged sample suggest[194] that lowering T_c involves more than structural changes. In a moderately irradiated sample, χ_D definitely decreases for $T < T_c$ but remains finite in the limit $T \to 0$, as expected for localized paramagnetic defects. On the other hand, χ_A vanishes smoothly as $T \to 0$. The paramagnetic defects on the two stacks are evidently different. A decomposition similar to Eq. (19) can be attempted for any property that can be estimated separately for each stack.

Such g-value decompositions have been questioned[195] and some caution is clearly indicated, since the g shifts are usually small in organic ion radicals and crystal contributions[196] could also produce small changes. The basic hypothesis[1] of weakly perturbed molecular units can in effect be tested by such decompositions, which have indeed been reliable for many ion-radical solids. We estimate $\pm 10\%$ decompositions in most cases. The simplicity of the analysis contrasts favorably with the extensive modeling that is often required to interpret transport measurements.

Since $\chi(T)$ is a static property, its modeling depends only on the energy spectrum of the narrow bands. The analysis of transport properties is inherently more difficult, as it also requires knowledge of relaxation processes. The modeling[189, 197] of $\chi(T)$ for $T > 54$ K is consequently a simple but stringent test for any overall picture of (TTF)(TCNQ). The analysis is complicated by intermediate correlations in narrow organic bands,[38, 197] whose width $4|t|$ is comparable to the effective on-site repulsion $U_{\text{eff}} \sim 1$–1.5 eV.

Torrance[189] has emphasized that the magnitude of $\chi(300\text{ K})$ rules out a simple metal, as indicated by the $U = 0$, $4|t| = 0.12$ eV curve in Figure 17. Such low U is incompatible with optical data and with the basic structural fact of π overlaps in face-to-face stacks at 3.17 Å. The $4|t| = U = 0.25$ eV curve in Figure 17 is a clear improvement, as is the large U (Heisenberg) case of Eq. (8), but neither is quantitative. The effects of partial filling[198, 199] on the exchange constant J have been included for the $(\text{TTF})^{\gamma+}(\text{TCNQ})^{\gamma-}$ value of $\gamma = 0.59$. The explicit VB computation[197] for intermediate correlations $U = 3\sqrt{2}|t|$, $4|t| = 0.56$ eV in Figure 17 still yields only qualitative agreement, although now for typical microscopic parameters. A gentle decrease of $|t(R)|$ with increasing R due to thermal expansion[187] could produce a quantitative fit, and approximate data are independently available from high-pressure work.[186] A single precise value of $|t|$ and of U_{eff} is not yet available for (TTF)(TCNQ) or other organic conductors. But optical, transport, and magnetic data are converging to $4|t| \sim 0.7 \pm 0.2$ eV, $U_{\text{eff}} \simeq \Delta E_{CT} \sim 1.2 \pm 0.3$ eV, and $U_1 \sim 0.4 \pm 0.2$ eV for the low-energy optical transition[45] of partly filled regular stacks. Such values represent intermediate correlations, which should ultimately modify current free-electron discussions of various transport properties.

5.2. *EPR Linewidth of Organic Conductors*

Line shapes for epr absorption in metals are sensitive to various characteristic times of conduction electrons. In large samples where a relatively long time is required for the conduction electrons to traverse the sample, the asymmetric Dyson[200] line shape is observed.[201] Thin samples, on the contrary, show Lorentzian line shapes, other things being equal. Standard texts summarize such epr measurements for metals.[202] In organic conductors the epr line shape and, in particular, its width and position yield valuable information about the electron localization, dimensionality, and interaction with the lattice or other spins. In addition to temperature and pressure effects, the spectrum is sensitive to the composition of organic alloys and to substitutions on the donors or acceptors. A large number of experimental studies have been performed on systems with the TTF related donors in Figure 16. A relevant review[190] provides background information on (TTF)(TCNQ). Complications arising from spin density on the two dissimilar chains (i.e., different g values) can be resolved by decomposition techniques as discussed above. The single epr peak has a Dyson line shape for high-conductivity crystals and appropriate field orientations.

The most important consideration in the TTF family of organic conductors is the spin–orbit interaction of the sulfur, selenium, or perhaps halogen atoms. In metals, spin–orbit coupling admixes Zeeman states in the Bloch functions. As shown by Elliott[203] and later developed by Yafet,[204] scattering between k and k' Bloch states results in electron spin flips. The transverse spin-relaxation rate T_2^{-1} becomes large, broadening the epr line. An "orbit–phonon" broadening mechanism exists for phonons with wave vectors k-k', and T_2^{-1} is proportional to the square of the deviation of g from the free-electron value, since Δg is a function of spin–orbit coupling. A recent experimental test of the Elliott mechanism in a pure three-dimensional metal demonstrated remarkable agreement with theory over almost four orders of magnitude for the linewidth.[205]

The Elliott mechanism has also been invoked[206] in two-dimensional systems based on intercalated graphite.[207] Doping with donors like alkali metals or acceptors like nitric acid shifts the Fermi energy. The epr linewidth in several cases can be related[206] to the scattering rate τ^{-1}. Since ΔH is proportional to τ^{-1}, while the conductivity σ goes as τ, the product $\sigma\Delta H$ should suppress any temperature dependence of τ. A variety of tests involving magnetic and transport properties are then suggested by simple band theory. While no doubt oversimplified, such considerations are useful guides for additional studies. The different $\sigma(T)$ and $\Delta H(T)$ dependences of the conducting organic metals cannot all be assigned to the Elliott or any other mechanism. As shown in Section 5.4, extremely narrow bands point to diffusive motion for localized charges and spins. Now increasing the rate of hopping increases both ΔH and σ, in contrast to the inverse relation for the Elliott mechanism. Sorting out and parametrizing such mechanisms in molecular conductors is difficult but necessary for any microscopic picture that does justice to both magnetic and transport properties.

The narrow epr ΔH for TTF-related conductors is surprising in view of the Elliott mechanism and is critically dependent on dimensionality. The isostructural $(TSF)_x(TTF)_{1-x}(TCNQ)$ family of organic alloys is particularly convenient for varying ΔH through the spin–orbit of Se and S, as shown in Figure 18. The strong increase of ΔH at 300 K on increasing x immediately implicates the donor stack and spin–orbit as controlling the relaxation.[208] The same conclusion follows from a T_2^{-1} decomposition similar to Eq. (19) and from Δg measurements. The Elliott mechanism for $\Delta H \sim T_2^{-1}$ gives

$$T_2^{-1} = \alpha_D(\Delta g_D)^2 \tag{20}$$

where α_D is a proportionality constant that increases rapidly with x. The magnitude of the spin–orbit coupling is contained in the Δg factor of Eq. (20), while α_D is a function of the matrix element connecting the Bloch states

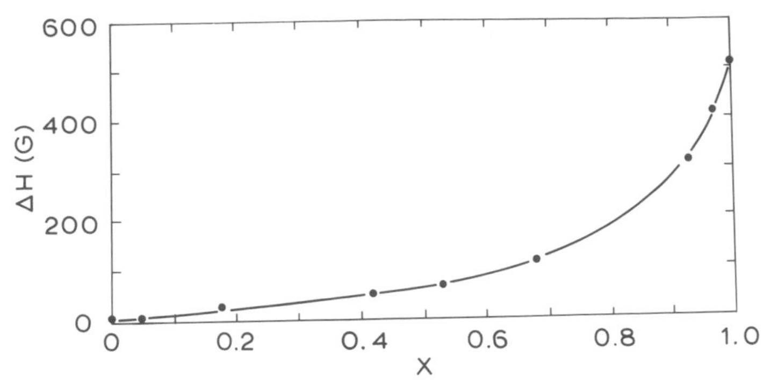

FIGURE 18. The epr linewidth ΔH of $(TSF)_x(TTF)_{1-x}TCNQ$ at 300 K. The larger spin–orbit of Se in TSF also increases the g shift, Δg_D in Eq. (20). The order-of-magnitude increase of $\Delta H/(\Delta g)^2$ between $x = 0$ and $x = 1$ is interpreted as increasing interchain coupling with TSF content. (Reproduced from Reference 208a, with permission.)

involved in the spin flip. Since α_D changes with x, the spin-flip matrix element also depends on x.

Tomkiewicz *et al.*[209] sketched the following dimensionality argument for the Elliott mechanism. In one-dimension the phonon-induced scattering is from k to $k' \simeq \pm k$. Coupling to $q \sim 0$ acoustic modes is weak and time reversal symmetry leads to vanishing $k' = -k$ matrix elements. Hence, in strictly one-dimensional conductors the spin-flip matrix element is nearly zero and the epr line narrows. A more complete analysis would presumably quantify the degree of narrowing. Increasing ΔH is interpreted as increasing multidimensional behavior with increasing TSF content. We note that, on the contrary, a broad epr line in insulators is indicative of a high degree of unidimensionality, since the modulation of the dipolar or hyperfine interactions is then less effective. Quite generally, dimensionality effects require prior elucidation of the relaxation mechanism. The variety of possible mechanisms is then a hindrance.

Bloch *et al.*[210] have also associated ΔH in conductors with other physical properties that are sensitive to the dimensionality. Table 4 contains an expanded list of ΔH values, metal–insulator transition temperature T_c, and shortest interchain contacts d for TTF related conductors. Compounds have been grouped according to cation similarity (viz. approximately constant spin–orbit coupling) and increasing linewidth within each group. An increasing ΔH correlates with decreasing T_c, or greater stability toward a Peierls distortion and hence two or three-dimensional interactions. As indicated in Table 4, short interchain contacts also parallel decreasing unidimensionality. Caution must be taken when comparing systems with different spin–orbit couplings or crystal parameters (other than dimensionality considerations) which might affect the spin-flip matrix element.

TABLE 4. EPR Linewidth ΔH, Metal–Insulator Transition T_c, and Smallest Interchain Contact d for Organic Conductors Based on TTF Related Donors Defined in Figure 16 and in the Notation

Compound	ΔH (G)	T_c (K)	d (Å)	Reference
(DEDMTTF)(TCNQ)	2–3	—	3.60	211
(TMTTF)(TCNQ)	3–4	34	3.45	210, 212
(DMTTF)(TCNQ)	5	35 (broad)	—	210
(TTF)(TCNQ)	6–7	53	3.20	210, 214
(HMTTF)(TCNQ)	11	50	3.25	213
$(TTF)Cl_x$	9	—	—	215
$(TTF)_{11}(SCN)_6$	11	175	—	192
$(TTF)_{11}(SeCN)_6$	15	187	—	192
$(TTF)Br_{0.76}$	52	—	—	215
$(TTF)_7I_5$	200	230	—	192
(DEDMTSF)(TCNQ)	70	28	—	216, 217
(TMTSF)(TCNQ)	120	57	3.36	210, 217, 218
(TSF)(TCNQ)	550	28	3.16	208, 219, 220
(HMTSF)(TCNQ)	>4000	0	3.10	210, 221
(TMTSF)(DMeTCNQ)[a]	70	42	3.48	222, 217, 223
(TSF)(DEtTCNQ)[b]	175	100	3.39	224
$(TSF)_x(TTF)_{1-x}TCNQ$				225, 219
$x = 0.18$	29	47	—	
0.42	50	42	—	
0.68	117	41	—	
0.95	375	39	—	

[a] DMeTCNQ = 2,5-dimethyl-7,7,8,8-tetracyano-*p*-quinodimethane.
[b] DEtTCNQ = 2,5-diethyl-7,7,8,8-tetracyano-*p*-quinodimethane.

The TTF halides and pseudohalides are also interesting special cases. The ΔH dependence on the number of TTF stacks per unit volume suggests[215] a spin–spin (i.e., dipolar) origin, which is also consistent with the temperature dependence of ΔH. This surprisingly suggests that the spin–phonon mechanism is insignificant in the TTF halides, even though the spin–orbit coupling should be as large as in (TTF)(TCNQ) where it dominates ΔH. A proposed explanation[215] is based on estimating an upper limit for the interstack hopping rate. Since the TTF halides show a single epr line and have two inequivalent TTF chains per unit cell, an estimate of $\nu_\perp$ is inferred by a comparison of the difference in Larmor frequencies for the two stacks. Using this procedure $\nu_\perp$ is found to be an order of magnitude smaller in halides than in the (TTF)(TCNQ) family of conductors. The spin-flip matrix element may thus be significantly reduced in the halides due to a high degree of unidimensionality. In one-dimensional electronic systems, exchange and diffusion are less effective in narrowing a dipolar broadened epr line. It is not clear, however, that dipolar interactions account for the magnitude of the linewidth.

The temperature dependence of epr linewidths in organic conductors remains poorly understood. At least two competing processes must be considered. Peierls fluctuations increase the scattering rate leading to an increase in the spin–orbit mediated spin-flip rate. Conductors that show an increasing linewidth with decreasing temperature [e.g., (TTF)(TCNQ), (TSF)(TCNQ), (HMTTF)(TCNQ)] have a maximum linewidth at the Peierls transition temperature, consistent with this claim. Other compounds have a decreasing linewidth with decreasing temperature [(TMTTF)(TCNQ) and (TMTSF)(TCNQ)]. A decrease in the density of available scattering states for the electron as the temperature is lowered would result in such variations. Variations in the temperature dependence of different conductors and the observation that it does not parallel the resistivity (as required by the Elliott mechanism) place severe restrictions on any interpretations. Even the far better characterized magnetic insulators show a variety of temperature-dependent linewidths whose quantitative interpretation remains open.

5.3. *Random Exchange Heisenberg Antiferromagnetic Chains*

Random exchange Heisenberg antiferromagnetic chains, REHACs, were introduced by Bulaevskii *et al.*[226] for the low-temperature thermodynamics of TCNQ salts with asymmetric donors. The 1 : 2 TCNQ salts of quinolinium (Qn) and acridinium (Ad) have disordered cations,[227, 228] a single TCNQ per unit cell along the chain, and one unpaired electron per TCNQ dimer. The tetrameric stacks in Figure 3 reduce to such an arrangement when all overlaps and crystalline potentials are equal. Except for random potentials arising from the structural disorder, the crystal symmetry dictates equal ionicities $TCNQ^{\gamma -}$ close to the stoichiometric value $\gamma = 0.50$. The 1 : 1 (NMP)(TCNQ) salt also crystallizes[229] in partly filled regular segregated $TCNQ^{\gamma -}$ stacks, but as already noted the crystal preparation[130] and degree of CT have been problematical.

This REHAC behavior is signaled by power laws at low temperature for such static thermodynamic properties as the susceptibility $\chi(T)$, the magnetization $M(T, H)$, and the magnetic specific heat $C(T, H)$. These quantities are experimentally found to go as

$$\begin{aligned} \chi(T) &= AT^{-\alpha} \\ C(T, 0) &= BT^{1-\alpha} \\ M(T, H) &= CH^{1-\alpha}, \\ C(T, H) &= DTH^{-\alpha}, \end{aligned} \qquad k_B T \ll g\mu_B H \tag{21}$$

with a single exponent $\alpha \sim 0.8$. Clark and co-workers[230–234] have extensively documented Eq. (21) for $Qn(TCNQ)_2$ at $T < 20$ K. The behavior of

$Ad(TCNQ)_2$ is closely related.[235] The observed exponent α varies slightly with sample preparation and with deliberate exposure to neutron irradiation,[236] but different measurements on the same sample yield identical exponents within experimental error.

The REHAC Hamiltonian in an applied field H_0 is

$$\mathscr{H} = \sum_n J_n \mathbf{s}_n \cdot \mathbf{s}_{n+1} - g\mu H_0 \sum_n s_n^z \tag{22}$$

Here $J_n \geq 0$ are independent random antiferromagnetic exchanges between $s = 1/2$ sites along the $(TCNQ)_2^-$ chain. The normalized distribution function $f(J)$ is related, in principle, to the cationic disorder but is in practice simply chosen by fiat. The choice of $f(J)$ completely fixes the static thermodynamics and thus yields the exponent α and the observables in Eq. (21), as demonstrated in detail by Bondeson and Soos[237] for $Qn(TCNQ)_2$ and $Ad(TCNQ)_2$ for arbitrary T and H. While REHAC spin dynamics should also follow on including dipolar and hyperfine terms, such applications are still quite preliminary.

The connection between the microscopic exchanges $f(J)$ and the power laws in Eq. (21) is still open, due to an interesting recent development. Bulaevskii *et al.*[226] assumed a singular density of spin states to mimic a singular $f(J)$ with vanishing exchanges along the chain. Fermi-liquid theory then gave Eq. (21) and consistently good agreement for $T < 20$ K. Theodorou and Cohen[238, 239] deduced a singular $f(J)$ for a strongly disordered Hubbard model and argued[240] that a good $\chi(T)$ fit for (NMP)(TCNQ) followed for reasonable parameters and a half-filled ($\gamma = 1$) stack. The singular behavior of $f(J) = J^{-\alpha}$ was central to all early theories, with at most some concern[234] that the exponent in $f(J)$ might be different from those in Eq. (21). The singular nature of $f(J)$ is nevertheless open in view of renormalization calculations[237, 241, 242] showing that virtually any distribution $f(J)$ produces power laws at low temperature. Such behavior is consequently quasi-universal for REHACs.[231, 243]

The point is that a large J_n in Eq. (22) forms a singlet–triplet pair at sites n and $n + 1$. The singlet state will freeze out for $k_B T \ll J_n$, thereby attenuating the exchange between spins at sites $n - 1$ and $n + 1$ to $J_{n-1}J_{n+1}/J_n$. Such a decimation procedure, introduced by Ma *et al.*,[241] systematically pairs up the most strongly interacting spins and produces small effective exchanges among the surviving paramagnetic sites. A different approach was taken by Hirsch and José,[242] who renormalize by retaining at each step the two lowest states of trimers produced by neglecting every third J_n in Eq. (22). Successive renormalizations are clearly shown to generate small effective exchanges with a singular distribution. A single renormalization was sufficient for the ten-spin segments treated exactly by Bondeson and Soos.[237] These different methods all generate power laws for both singular and nonsingular $f(J)$, but may differ in whether the exponent α is slightly temperature dependent or is the same for $\chi(T)$ and $C(T, 0)$.

Any direct connection between the distribution $f(J)$ and the structural disorder is consequently speculative. We further note that the degree of disorder is also open. The asymmetric $(NMP)^+$ cation in Figure 6 is similar to $(Qn)^+$ and $(Ad)^+$ in forming stacks with a substituent (the methyl group) in two equally probable positions, as found by x rays. Quite aside from steric effects, the substituents strongly affect the cationic dipole moments[244] and thus modulate the $(TCNQ)^{\gamma-}$ site energies. Unfortunately x-ray evidence about equal populations in both orientations does not fix the correlation length. Thus estimates[245] of disorder energies based on truly random orientations, as apparently assumed in most theoretical discussions, are at best upper bounds that may be orders of magnitude too large for a sensible structure. In a truly random (NMP)(TCNQ) structure, for instance, 25% of the methyl groups from adjacent stacks would be unphysically close to each other. There are probably only a few percent of such high-energy contacts, where new correlated sequences are started.

It is nevertheless instructive to adopt a physical picture for REHACs, bearing in mind that experimental fits for static properties may not corroborate the choice for $f(J)$. We consider a few misoriented cations that polarize the spin distribution on nearby supermolecular $(TCNQ)_2^-$ ion radicals. The reduced spin density at one TCNQ introduces a weaker exchange to the next $(TCNQ)_2^-$ along the chain. We postulate[237] a random concentration c of weak exchanges εJ among otherwise regular Heisenberg chains with exchange J. The result is a bimodal distribution

$$f_2(x) = c\delta(\varepsilon - x) + (1 - c)\delta(1 - x) \tag{23}$$

where x is in units of J. The weak exchanges $x = \varepsilon$ define segments with equal exchanges $x = 1$. The distribution is not singular, since there are no exchanges with $x < \varepsilon$.

In any antiferromagnetic system, it is convenient to define the effective number of spins, which is simply $\chi(T)/\chi_c \leq 1$. For $S = 1/2$, the Curie contribution $\chi_c = g^2\mu_B^2/4k_BT$ is an upper bound for high temperatures that exceed all exchange interactions. In Figure 19, $\chi(T)/\chi_c$ data for $Qn(TCNQ)_2$ is fit[237] over some four decades with $J/k_B = 230$ K, $\varepsilon = 0.30$, and $c = 0.10$ in $f_2(x)$. The inset shows the flat $\chi(T)$ behavior of both $Qn(TCNQ)_2$ and $Ad(TCNQ)_2$ for $T > 50$ K. The high value of $\chi(300\ \mathrm{K})/\chi_c$ and this flat behavior reflect a single characteristic exchange J, around $J/k_B = 230$ K. Some 90% ± 10% of the exchanges in realistic distributions will consequently have to be comparable. Thus the singular distribution $f_1(x) = J^{-\alpha}$ in Figure 19, with $\alpha = 0.75$ and an enormous $J/k_B = 2 \times 10^7$ K, fails at high temperature. But the crucial low-energy end of $f(J)$, the $c\delta(\varepsilon - x)$ term of Eq. (23), remains open and a singular distribution for the 10% weak ex-

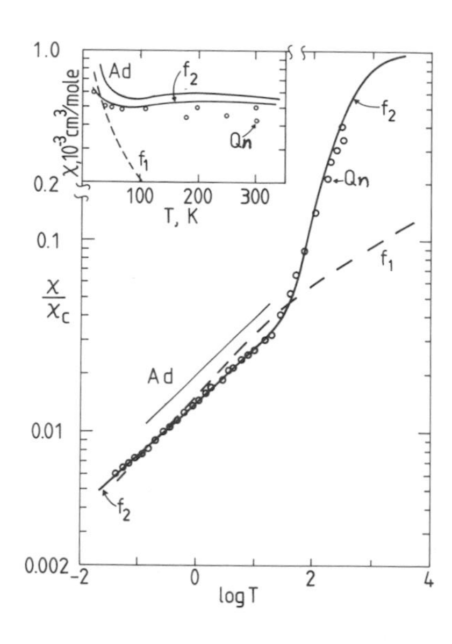

FIGURE 19. Absolute $\chi(T)/\chi_c$ and (inset) $\chi(T)$ data from Reference 233 for $Qn(TCNQ)_2$ and from Reference 226 for $Ad(TCNQ)_2$. The $f_2(x)$ fit for weakly interacting segments is based on $\varepsilon = 0.30$, $c = 0.10$, and $J = 230$ K in Eq. (23). The singular distribution $f_1(x)$ has exponent $\alpha = 0.75$ and $J = 2 \times 10^7$ K. (Reproduced from Reference 237a, with permission.)

changes would produce comparable fits for static properties. Available $\chi(T)$, $M(T, H)$, $C(T, 0)$, and $C(T, H)$ data for $Qn(TCNQ)_2$ and Ad(TCNQ) can be described[237] down to 20 mK by the same parameters in $f_2(x)$. At $T \lesssim 7$ mK,[246] $\chi(T)$ for $Qn(TNCQ)_2$ begins to deviate from a pure power law and an internal field builds up. Both observations require going beyond the simple REHAC Hamiltonian, Eq. (22).

The resulting physical picture has regular segments of average length $c^{-1} \sim 10$ whose ends are weakly coupled by exchanges $\varepsilon J/k_B = 70$ K. Since internal excitations decrease with increasing length, finite Heisenberg segments have energy gaps of the order of $cJ/k_B \sim 23$ K for the parameters above. Finite segments consequently freeze out at low temperature ($k_B T < cJ$) to diamagnetic singlet states for segments with an even number of spins and $S = 1/2$ doublets for odd-length segments. Randomly placed weak exchanges produce a density $c(2-c)^{-1}$ of odd segments, each behaving as a free spin in the $\varepsilon \to 0$ limit of noninteracting segments. In this case, $\chi(T)/\chi_c$ would become temperature independent and equal to $c(2-c)^{-1}$ for $k_B T < cJ$. Interacting ($\varepsilon > 0$) odd segments again produce a REHAC, with Eq. (22) now describing exchanges $Jx_k > 0$ between odd segments k and $k+1$ in the infinite chain. The value of x_k depends[237] on the known terminal spin densities $\rho(O_1)$ for an odd segment with O_1 sites and on the attenuation factor $g(e_1)$ for each intervening even segment. The general result for $m = 0, 1, \ldots$ even segments is

$$2x(O_1 e_1 \ldots e_m O_2) = (2\varepsilon)^{m+1} \rho(O_1)\rho(O_2) g(e_1) g(e_2) \cdots g(e_m) \tag{24}$$

The distribution of renormalized exchanges between odd segments are shown in Figure 20. They represent Eq. (24) for 50 000 segments satisfying the law $c(1-c)^p$ for a segment of length $p+1$. The actual values are for $\varepsilon = 0.30$ and $c = 0.10$, as required in the $Qn(TCNQ)_2$ and $Ad(TCNQ)_2$ fits for $f_2(x)$. Other types of weak interactions could readily be incorporated into Eq. (24) by computing the resulting splitting regular segments.

Renormalization thus generates weak effective exchanges among odd segments in their ground states. The area under the curve in Figure 20 is $Nc/(2-c)$, the total number of odd segments. The distribution $g(x)$ evidently satisfies a power law, x^δ, with $\delta = 0.29$ for the parameters chosen. Another renormalization would certainly shift the remaining interactions to still lower energies, as shown explicitly for trimers.[242] For example, adjacent O_1O_2 pairs have the largest exchanges in Eq. (23) and all such direct contacts between odd segments can be renormalized. The enormous variety of segment lengths and topologies leads to more complicated statistics than the initial renormalization of finite Heisenberg segments. The dashed line in Figure 20 represents the exchanges retained[237] in ten-spin replicas, neglecting all $m \geqslant 3$ interactions in Eq. (24). The area between the two curves is the 10% of weak interactions discarded in going from an infinite sequence of odd segments to replicas with ten odd segments. The similarity of the two distributions for $x \gtrsim 10^{-4}$, or $Jx/k_B \sim 0.023$ K, suggests that a single ten-spin renormalization is adequate down to 20–30 mK. We are effectively treating exactly more than $20c^{-1} \sim 200$ sites of the original chain with the distribution $f_2(x)$ in Eq. (23). Longer twelve-spin relicas are feasible for still lower temperatures. Finally, we emphasize that the effective exchanges in

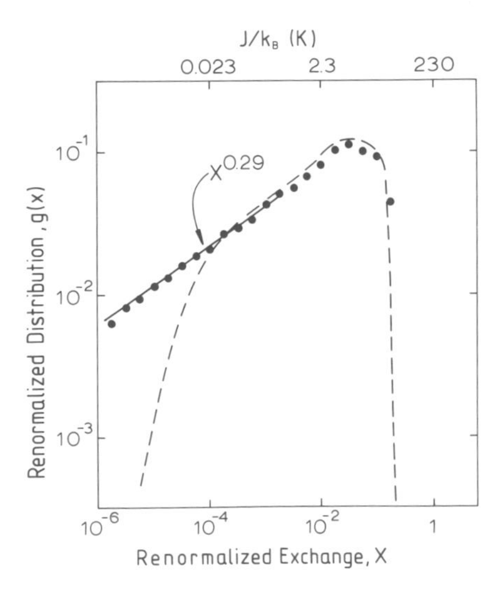

FIGURE 20. Distribution of effective exchanges among odd-length, $S = 1/2$ segments in their ground state according to the renormalization, Eq. (24), with $\varepsilon = 0.30$ and $c = 0.10$. Also shown are the power-law fit with exponent $\delta = 0.29$ and (dashed line) the exchanges retained for the ten-segment replicas used in Reference 237.

Eq. (24) do not apply for $k_BT > cJ$ when the segments are no longer in their ground states. The exchanges then revert to εJ and J, as indicated in $f_2(x)$.

The renormalization procedures for successive decimation[241] or for trimers[242] yield power laws for a wide variety of $f(J)$. These computations are probably best for $T \to 0$ extrapolations, which may account for the absence of experimental fits. Direct numerical analysis is better adapted for explicit comparisons at higher temperature and arbitrary magnetic fields, but fails at sufficiently low temperature. There are additional contrasts among these approaches, beyond the observation that neither decimation nor trimers is suitable for distributions involving segments with relatively constant exchanges. The former emphasizes the idea of treating the strongest interactions first, while the latter does not. Both methods contain sufficient approximations for many successive renormalizations that eventually almost wash out the original $f(J)$. Our ten-spin computations, on the other hand, suggests that the experimentally relevant regime is covered in the first renormalization; $f(J)$ then still counts and so do many of the excited states that are neglected in the other methods. Each approach consequently has advantages and disadvantages for analyzing the REHAC Hamiltonian, Eq. (22).

The REHACs also have interesting spin dynamics. The early T_1^{-1} proton relaxation in $Qn(TCNQ)_2$ powder by Azevedo and Clark[232] revealed strong temperature and field effects below 1 K. Their interpretation is still open. More recent epr studies on $Qn(TCNQ)_2$ by Tippie and Clark[233, 247] show that the exchange-narrowed line has an unusual ln T_0/T width below $T \gtrsim 20$ K, with $T_0 \sim 50$–100 K, as shown in Figure 21. A three-reservoir model[247] based on exchange, Zeeman, and lattice reservoirs offers a qualitative description, but without a microscopic picture relating the random exchanges and dipolar or hyperfine interactions to observed linewidths or relaxation rates.

Antiferromagnetic interactions reduce the effective number of spins, $n_e(T) = \chi(T)/\chi_c$. Strong dilution $n_e \lesssim 10^{-2}$ is indicated for $Qn(TCNQ)_2$ in Figure 19 for $T < 1$ K. The dipolar second moment for random magnetic

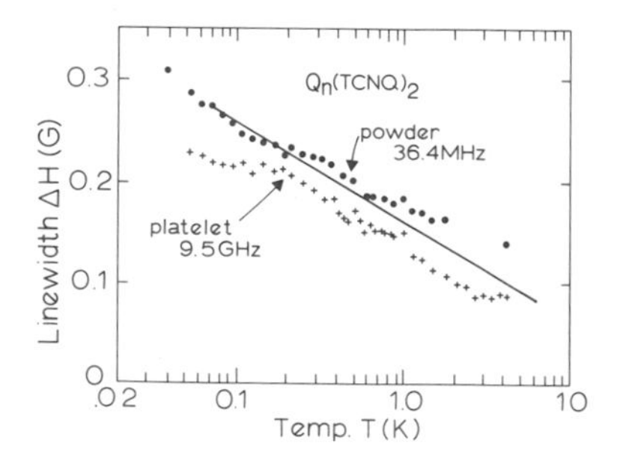

FIGURE 21. The epr linewidth of a powder and single platelet sample of $Qn(TCNQ)_2$, from Reference 233b. The theoretical curve involves the weighted superposition of Lorentzian absorptions for the ten-segment replicas used in Reference 237 for static properties. Uniform spin dilution and weak interchain exchange are assumed, as discussed in the text, to fit the magnitude of ΔH at 1 K.

dilution goes[86] as $n_e M_2$, where M_2 is the usual result in Eq. (15) for a spin at each lattice site. Delocalizing an unpaired electron in a D^+ or A^- ion-radical HOMO, as indicated in Figure 1 for $(TMPD)^+$ and $(TCNQ)^-$, reduces dipolar interactions by another order of magnitude. The (TMPD)(Chloranil) estimate[87] of $M_2 \sim 10^5$ G^2 is typical for such solids. The lower spin density of $(TCNQ)^{0.5-}$ sites in $Qn(TCNQ)_2$ reduces M_2 by a factor of 4, quite aside from smaller changes due to stacking differences. The effective second moment for $n_e \sim 10^{-2}$ is consequently of the order of 2×10^2 G^2. The resulting internal dipolar fields, $M_2^{1/2}$, in $Qn(TCNQ)_2$ are about 10–20 G for $T < 1$ K. Hyperfine fields are probably even smaller, although their magnitude is more sensitive to the postulated delocalization of the electronic spin density.

The usual estimate for the exchange-narrowed Lorentzian linewidth is, neglecting factors of order unity,

$$\Delta H \sim M_2/\omega_e \tag{25}$$

This high-temperature, three-dimensional result has been extended[132, 134, 146] to low-dimensional, low-temperature systems with uniform exchanges. The wide range of exchanges in REHACs necessitates further extensions. At 1 K an epr linewidth of about 100 mG suggests an effective exchange of 2×10^3 G for $M_2 \sim 2 \times 10^2$ G^2. The effective exchange $\sim$0.3 K is then somewhat smaller than T. Such effective exchanges are around $x \sim 10^{-3}$ in Figure 20.

It is now straightforward to demonstrate substantial interchain exchange in $Qn(TCNQ)_2$. There are two magnetically inequivalent $(TCNQ)_2^-$ stacks in the structure,[227] with a dihedral angle of 65° between the TCNQ planes. The nearest neighbors of each stack have opposite tilt. The $(TCNQ)^-$ g tensors in a variety of solids are approximately $g_{\parallel} \simeq 2.0023$ and $g_{\perp} \simeq 2.0030$, respectively, for a field normal to and in the molecular plane. The resulting splitting $|g_{\parallel} - g_{\perp}|\mu_B H$ of several gauss for $H \sim 3000$ G should readily be resolved in single-crystal spectra. A single narrow ($\Delta H \lesssim 0.2$ G) epr line is found[233] in single platelets down to the lowest reported temperatures of 50 mK. The interchain exchange fields must consequently exceed $|g_{\parallel} - g_{\perp}|\mu_B H$ by at least an order of magnitude to produce such a narrow line, as found from Eq. (25) with $M_2 = |g_{\parallel} - g_{\perp}|^2 \mu_B^2 H^2$. The interchain exchange is at least 10^2 G, almost comparable to the effective intrachain exchange at 1 K. Although interchain dipolar interactions may suppress g-tensor splittings,[196] the small internal fields here would not suffice.

We emphasize that the estimate for interchain exchange depends on the observed structure and the single-crystal epr rather than on any postulated model. The latter is needed for a precise value instead of a lower bound. In the segment model, spin density on odd segments is generated by an occasional misoriented Q_n^+ dipole. Such a defect produces weak exchanges in adjacent $(TCNQ)^{0.5-}$ stacks, some of which are nearest neighbors and magnetically inequivalent. Weak exchange between odd segments on adjacent stacks seems quite natural when the intrachain neighbors are diamagnetic even segments. It is precisely such isolated odd segments that dominate $\chi(T)$ at low temperature where magnetic inequivalences should be resolved in the absence of interchain exchange. The one dimensionality need not be entirely suppressed, however, since the blocking effect of even spacers may preserve the topology of the random exchanges in Eq. (24) even if an x_k sometimes connects odd segments in adjacent chains. In any case, interchain exchange fields exceeding 10^2 G must clearly be included at low temperature in future models.

Thermal decoupling at weak exchanges $x_k J \lesssim k_B T$ reduces REHACs to subsystems that differ in their segment lengths and topologies. For example, the distribution $f_2(x)$ in Eq. (23) produces sequences of odd and even segments whose lengths are controlled by c and whose effective exchanges are given in Eq. (24). The $Qn(TCNQ)_2$ parameters suggest that sequences of ten odd segments, or some $20_c^{-1} \sim 200$ sites, are decoupled around 20 mK. Physical properties then depend on appropriately weighing independent contributions from such sequences, just as in the case of inhomogeneously broadened lines. We are applying this idea to the epr linewidth[248] and to nmr T_1^{-1} data.[249] The two cases involve somewhat different considerations.

For the epr ΔH in Figure 21, we adopt a one-dimensional version[143] of Eq. (25) in which ω_e is the geometric mean of the intrachain and interchain exchange. The weight of a ten-segment replica is simply $n_e = \chi(T)/\chi_c$ and is accurately known from fits to $\chi(T)$ shown in Figure 19. The temperature dependence of interchain dipolar fields in M_2 is also given by $n_e(T)$, which is taken to be the crystal average because every replica samples all environments in the solid. The most serious approximations[248] are the Boltzmann average over the exchanges x_k and the choice of an interchain exchange $\omega_\perp$. The resulting $\Delta H(T)$ for the 60 replicas used in the static thermodynamics is shown in Figure 21 for $\hbar\omega_\perp/k_B = 30$ mK and $M_2 = 10^4$ G. The unusual $\ln(T_0/T)$ temperature dependence is clearly observed. It is a consequence of the inhomogeneous broadening of independent contributions. The interchain exchange $\omega_\perp$ is of the correct order of magnitude and apparently controls T_0 rather than the logarithmic behavior. Its interpretation is still incomplete.

Our approach[249] to the nmr T_1^{-1}, on the other hand, is an adaptation of the α-CuNSal treatment[96] mentioned in Section 4.1. There are few phonons below 1 K and the electronic moments must provide energy conservation for a nuclear transition $\hbar\omega_N$ detected as T_1^{-1} relaxation. In the uniform α-CuNSal case, the relevant quantity was the field-dependent density of spin-flip states. This corresponds in decoupled replicas to $\Delta S = 1$

crossovers around $\hbar\omega_N \sim 0$ as a function of applied field. For example, the $S = 0$ ground state for $H = 0$ becomes an excited state as H increases and various S_Z components of higher S are lowered. There is an important and uncontrolled assumption, namely, that each crossover contributes equally. On the other hand, this completely reduces the computation to the replica energies used for the static thermodynamics. In terms of the Golden Rule, the matrix element for all allowed transitions is taken to be the same and only the density of states is considered.

Spin dynamics in REHACs will undoubtedly receive further attention. The wealth of recent low-temperature observations coincides with the successful analysis of static magnetic properties. The preliminary applications summarized above are subject to many refinements, including the explicit computation of the time, temperature, field, and dimensionality dependences of REHAC spin-correlation functions. The need for independent random subsystems, or for inhomogeneous broadening, will in our opinion emerge as a general low-temperature feature of REHACs.

5.4. *Atomic Limit and Spin Dynamics*

We have emphasized that the same partly filled bands are probed in magnetic, transport, and optical studies of organic conductors. Hubbard models rationalize the special semiconductive case of half-filled regular stacks and account qualitatively for most physical properties. Consistent values for the on-site correlation U and the bandwidth $4|t|$ are in fact emerging for paramagnetic semiconductors, as illustrated for (TMPD)(ClO_4) in Section 2.1 and for (TMPD)(TCNQ) in Section 3.1. Many other systems should yield comparable fits. Complex TTF or TCNQ salts and partly filled conductors like (TTF)(TCNQ) have additional spatial degrees of freedom. The description of organic conductors has, curiously enough, tended to diverge as more complete transport and magnetic studies have become available. Weger[250] summarizes transport in organic conductors in terms of small or negligible U and notes the absence of even "handwaving" explanations based on large U. The situation is reversed for the static susceptibility. Now serious analysis begins with $U \gtrsim 4|t|$ and intermediate correlations or various Heisenberg exchange models are apparently needed for the magnetic properties. Even partial descriptions of either transport or magnetism remain challenging, given the variety of modern experimental methods. Such quantitative models must nevertheless ultimately converge to similar microscopic parameters, since the same electronic states are involved.

The difficulties of intermediate correlations $U \sim 4|t|$ are avoided in systems with unusually small $U \ll 4|t|$ or $4|t| \ll U$. The former yields simple band theory. The latter is the atomic limit[251–253] and is exactly soluble for any degree of filling. Charge and spin excitations are decoupled. The spatial degrees of freedom involves a simple $-2|t|\cos k$ band with but one electron per k state. The spin excitations involve noninteracting moments with at most small exchange corrections $J \sim 2|t|^2/U$. Except for half-filled bands, the atomic limit combines metallic conductivities with noninteracting spins. Its mathematical simplicity[252, 253] makes it a useful starting point even when its premises are not met. Thus TCNQ salts, for example, have strongly coupled spins whose susceptibility hardly resembles a Curie law.

Hoffman and co-workers[254–258] have prepared a variety of molecular conductors based on macrocyclic porphyrin and phthalocyanine donors, various transition metals, and partial oxidation with iodine or bromine. Representative donors are shown in Figure 22. Mixed-valent systems of the type $D^{\gamma+}(I_3^-)_\gamma$, with $\gamma \sim 0.33$, contain one-dimensional ($d = 1$) stacks of donors separated to iodine columns. Such columnar structures are quite common with planar d^8 systems[5, 259, 260] while the stabilization of mixed-valent complexes is distinctly rare and may well involve the flexibility of I_n^- ions for $n = 3, 5, \ldots$, as found by resonance Raman spectroscopy.[261]

The sterically crowded OMTBP (octamethyltetrabenzporphinato) ligand in Figure 22 is of particular interest in connection with the atomic limit, since sharply reduced bandwidths $4|t|$ are expected. A complete x-ray investigation has been carried out[255] for the $[\mathrm{Ni(OMTBP)}]^{\gamma+}(I_3^-)_\gamma$ complex with $\gamma = 0.37$. The closely related $\gamma = 0.97$ system did not yield crystals suitable for a complete structural investigation, a situation that unfortunately holds for many molecular conductors. The OMTBP ligands are slightly puckered and have relatively large interplanar separations of 3.78 Å between adjacent π systems. Single-crystal epr demonstrates[254, 256] that the unpaired electrons are ligand, rather than metal, centred in such macrocylic stacks. The partly filled $D^{\gamma+}$ bands thus involve holes in the porphyrin or phthalocyanine π systems.

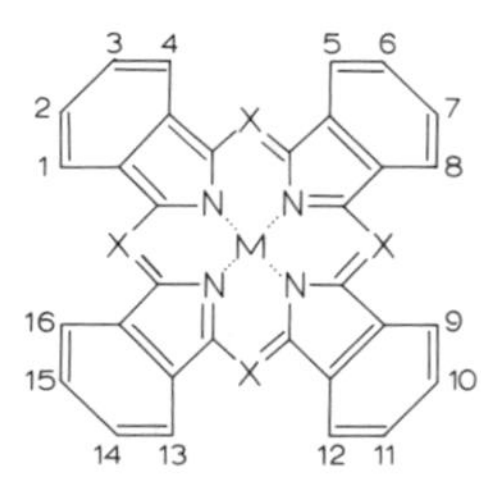

M (phthalocyanine) N = X, H = 1-16
M (tetrabenzporphyrin) C = X, H = 1-16
M (OMTBP) C = X, CH_3 = 1-16
M = Transition Metal

FIGURE 22. Representative macrocyclic donors based on substituted porphorins and phthalocyanines.

Both $Ni(OMTBP)^{\gamma+}(I_3^-)_\gamma$ complexes are essentially at the atomic limit.[262] Their $\chi(T)$ follows a Curie law for a concentration γ of spins that is independently known from the stoichiometry. Lower-temperature $1.6 < T < 20$ K epr[263] on the larger crystals of the $\gamma = 0.37$ complex place an upper bound of $J \lesssim 1$ cm^{-1} on the exchange. The coupling in the $\gamma = 0.97$ complex is perhaps an order of magnitude larger, as estimated from deviations from Curie behavior below 78 K. The on-site correlation U is still expected to be about 0.5–1 eV for the somewhat larger π systems in Figure 22. The CT integral $|t|$ is then obtained from the lowest-order corrections in t/U from the atomic limit, through the exchange[198–199]

$$J(\gamma) = \frac{2|t|^2}{U}\gamma\left(1 - \frac{\sin 2\pi\gamma}{2\pi\gamma}\right) \tag{26}$$

for band-filling γ. Numerical results[197] confirm that for $U \gg |t|$ each orbital manifold reduces to the regular Heisenberg chain, Eq. (8), but with different exchanges. The coupling in the ground-state manifold is accurately given by Eq. (26). Now $\gamma = 0.37$, $\gamma \sim 1$ cm^{-1}, and $U \sim 1$ eV yield $|t| < 100$ cm^{-1} at the large separation of 3.78 Å. The smaller spacings ($\sim$3.2–3.4 Å) in $(TCNQ)^{\gamma-}$ or $(TTF)^{\gamma+}$ systems yields $|t| \sim 0.1$–0.2 eV ($\gtrsim 10^3$ cm^{-1}), which is larger by over an order of magnitude. The accurate measurement of very small bandwidths is generally possible through Eq. (26) in systems close to the atomic limit. Other complexes with different π-electron overlaps at different separations should provide experimental information about the distance and overlap dependence of $|t|$.

Both $[Ni(OMTBP)]^{\gamma+}(I_3^+)_\gamma$ complexes combine essentially noninteracting spins with high conductivity along the macrocyclic stack of the order of 1–10 ohm^{-1} cm^{-1} at 300 K. The conductivity $\sigma(T)$ initially increases with decreasing temperature, thus demonstrating that no activation energy is involved, and then decreases on further cooling. The single-crystal conductivity along the stack goes as[255]

$$\sigma(\gamma,T) = A(\gamma)T^{-\alpha}\exp(-\Delta_0/k_BT) \tag{27}$$

The γ dependence of $A(\gamma)$ is close to the expected[262] value of $\gamma(1-\gamma)$ for a concentration γ of carriers whose motion is constrained with increasing filling due to the exclusion of doubly occupied sites. The resulting many-body dynamics have been treated numerically.[264] While individual motion is nondiffusive, the overall mobility $\mu(T)$ and the mean-field diffusion constant $D(T) = (1-\gamma)c^2\omega_h(T)$ still satisfy[264] the Einstein relation $\mu = DT^{-1}$. Since the lattice spacing c is known, $\sigma(\gamma,T)$ is directly linked to the hopping rate $\omega_h(T)$. The latter controls any motional modulation of $S(t)$ and consequently contributes to the epr linewidth ΔH. The connection between transport and magnetic properties is straightforward in the atomic limit, where there are rigorous results about each type of excitation. Such connections persist for intermediate correlation $U \sim 4|t|$, when the stronger exchange $J \sim 2|t|^2/U \sim 0.015$–$0.05$ eV must be included and neither space nor spin excitations follow simple models.

Although the spin-flip mechanism due to one-dimensional hopping $\omega_h(T)$ in Ni(OMTBP $(I_3^-)_\gamma$ is not accurately known, $\Delta H(T)$ is proportional to $T\omega_h(T)$. Since $\sigma(T)$ goes as ω_h/T, it follows[262] that the ratio $\Delta H/T^2\sigma$ is independent of T, as found for $T \gtrsim 150$ K in both the $\gamma = 0.37$ and $\gamma = 0.97$ complexes. The narrow epr linewidth at lower T has additional small ($\sim$0.1 G) contributions. The conductivity is also sharply reduced, as seen from Eq. (27) for typical values of $\alpha \sim 3$ and $\Delta_0/k_B \sim 10^3$ K. This suppresses the proposed spin-lattice relaxation via spin–orbit coupling due to delocalization to I_3^-, as measured independently[255] from the g tensor. Several additional complexes support[265] such a model for ΔH. The g value of the metal-free system (OMTBP) $(I_3)_\gamma$, with $\gamma = 0.43$, shows very little delocalization to I_3^-; as expected, ΔH remains less than 0.3 G even in the high-conductivity regime above 200 K. The smaller spacings expected in Ni(TBP)$(I_3)_\gamma$ on removing the eight methyls, on the other hand, should increase both $\sigma(T)$ and ΔH, as observed. The larger g-tensor deviations in fact suggest a larger linewidth than the broad $\sim$100 G line at 300 K.

We believe that there will be many additional comparisons of magnetic and transport properties in related systems. Such a strategy initially avoids computing various small microscopic parameters or indeed identifying the precise relaxation mechanism. Thus the angular dependence of ΔH is not understood in the Ni(OMTBP)$(I_3)_\gamma$ systems, nor is there a direct evaluation of $\omega_h(T)$. The emphasis is instead on related temperature or field dependences in structurally and chemically similar systems. The chief advantages of such a chemical approach are that common trends are readily identified, thereby possibly minimizing suggestive but erroneous analyses based on "reasonable" parameters. The complexity of organic ion-radical solids has so far defeated most detailed parametrizations of transport properties, although various mechanisms and estimated parameters are probably correct. The chief disadvantages of partial information about related systems are equally clear. Quite aside from eventually accounting for parameters in detail, it is tedious to prepare many systems that may actually not represent small systematic variations.

The atomic limit $U > 4|t|$ for partly filled bands thus associates[262] the temperature dependence of $\sigma(T)$ in Eq. (27) to the hopping $\omega_h(T)$ of localized charges along the macrocyclic stack. Polaron theories[266–268] describe various possible motions of electronic excitations in solids. There are many special cases due to variations in the electronic bandwidth, the possible distortions of molecular sites, various short- and long-range coupling to the lattice, and different temperature regimes. The observed $T^{-\alpha}\exp(-\Delta_0/k_BT)$ dependence in Eq. (27) arises frequently in model calculations, but of course the solutions are approximate and the relevant microscopic parameters are estimated.

The functional form of $\sigma(T)$ in Eq. (27) also describes[84, 269] TCNQ salts like $Qn(TCNQ)_2$, (NMP)(TCNQ), or phenazine-doped (NMP)(TCNQ). Even the parameters $\alpha \sim 3$ and $\Delta_0/k_B \sim 10^3$ K are similar. These systems do not satisfy the atomic limit and probably illustrate intermediate correlations $U \sim 4|t|$. The good fits based on Eq. (27) for $T > 100$ K were originally rationalized[84] in terms of simple semiconductors with a filled and empty band separated by $2\Delta_0$. The Curie susceptibilities of the $Ni(OMTBP)(I_3)_y$ complexes immediately rule out any such gap. The $\chi(T)$ behavior of these TCNQ salts is also inconsistent with a filled band and an energy gap $2\Delta_0$ for spin excitations, but the model was primarily aimed at transport properties. The large $|t|$ expected for closer contacts in TCNQ salts may well change the strong charge localization described above. The similarity of the parameters α and Δ_0 is nevertheless suggestive of a related behavior. Recent nmr work[159] on $Qn(TCNQ)_2$ offers evidence of similar temperature dependences for $\sigma(T)$ and the diffusion constant $D(T)$ along the chain. The interesting point for the present discussion is that neither the conduction nor the spin diffusion need be modeled accurately in such studies.

Polymeric organic solids also illustrate the power of magnetic resonance methods for preliminary characterizations. Polyacetylene $(CH)_x$ yields a versatile new class[270] of organic conductors on doping with 1–2% of either donors or acceptors. The basic unit is a CH radical whose $2p_z$ atomic orbital is available for π bonding. The resulting infinite polyene is unstable[271] for a regular carbon backbone and distorts to alternating double and single bonds. The two Kekulé structures contain partial rather than pure single and double bonds and the precise degree of alternation remains open.[272, 273] The optical transitions of π molecules or finite polyenes have long been of central interest[274, 275] to theoretical chemists. The C–C separations along the backbone are now around 1.35–1.45 Å, rather than the 3.2–3.5 Å interplanar separation along TCNQ or TTF ion-radical solids. The relevant parameters[275, 276] $|t|$ and U are consequently larger, with $|t| \sim 2.5$ eV and $U \sim 11$ eV for two electrons in a $2p_z$ atomic orbital. Interestingly, such values again imply $4|t| \sim U$, or intermediate correlations. Recent experimental[277] and theoretical[278] work in finite polyenes like octatetraene has centered on the lowest ^1B_u and ^1A_g excited states, whose order is reversed from the uncorrelated ($U = 0$) Hückel result. The complete CI for the 1764 VB structures of octatetraene yields[279] the correct order for the usual[276] parameters. Correlations will consequently be needed in future models for $(CH)_x$.

We discuss briefly magnetic resonance studies on $(CH)_x$. Direct structural and physical investigations are hampered by the polycrystallinity of all current samples.[270] Proton relaxation[162] in such samples as a function of applied field is consistent with highly one-dimensional spin transport, while the epr linewidth[280] of both *cis* and *trans* $(CH)_x$ suggests motional narrowing of hyperfine interactions. The paramagnetism of about one spin per 3000 carbons is associated with a delocalized π-radical occurring for example on an odd-length $(CH)_x$ fragment, as anticipated by Pople and Walmsley.[281] These radicals have recently been described as solitons[282, 283] and more accurate determinations of their degree of delocalization, binding, and dynamics have been attempted. Since small odd-length polyenes have regular C–C bonds[274, 275] and a nonbonding half-filled atomic orbital, an unpaired spin should produce a more regular region, similar to a polaron, in an infinite polyene with alternating double and single bonds. Furthermore, doping with donors and acceptors should doubly occupy or empty this HOMO, respectively, thereby decreasing the paramagnetism as observed.[270] Different interpretations[284] are possible for the epr data, based on the idea that doping by several percent is hardly likely to be uniform.

The role of crosslinks between $(CH)_x$ strands, especially after exposure to an acceptor like iodine, has been frequently raised by Wudl.[285] The soliton picture of a perfect infinite polyene is clearly idealized. Either crosslinks or CH_2 units yield sp^3 hybridization whose chemical shifts are different from the sp^2 hybridization in the polyene. Solid-state nmr on ^{13}C-enriched samples indicates up to 5% sp^3 carbons.[286] If these are crosslinks, then there are only about 20 carbons in between.

These illustrations demonstrate different types of magnetic resonance information for $(CH)_x$. Many additional studies can be expected in the important emerging field of polymeric organic conductors. The resulting description of magnetism and transport will undoubtedly go far beyond current models, although there have been few changes in the basic questions about the structure and motion of electronic excitations. The present juxtaposition of macrocylic donors and $(CH)_x$ polymers, which both differ from the organic ion-radical solids treated in this chapter, also emphasizes the different possible magnitudes of microscopic parameters in otherwise similar models.

6. *Discussion*

Throughout we have emphasized the complementary nature of magnetic studies in modeling organic ion-radical solids whose CT excitations and transport properties also involve the D^+ and/or A^- HOMOs. In half-filled magnetic insulators like (TMPD)(ClO_4) in Section 2.1, both the exchange and the CT excitation are related the CT integral $|t|$ and an effective on-site correlation U. The degree of ionicity, γ, adds another variable for the half-filled mixed regular stack illustrated by (TMPD)(TCNQ) and $(Me_2P)(TCNQF_4)$ in Section 3. The optical and magnetic properties of partly filled segregated stacks yield additional parameters. The central experimental feature[1, 38] is a new, low-energy CT absorption whose modeling[45] requires explicit consideration of nearest-neighbor correlations. Partly filled segregated regular stacks are found in molecular conductors. The resulting connections between transport and magnetism are sketched in Section 5 primarily in terms of relaxation mechanisms. There are now many different types of related experimental observations for any satisfactory theoretical picture.

More rigorous analysis has tended to increase the complexity of theoretical models as well, even on retaining severe physical simplifications. For example, all D^{2+} or A^{2-} sites are excluded in the atomic limit. The γN electrons on a chain of N sites undergo highly correlated motions in which electrons cannot pass each other. This situation may be better realized in superionic conductors.[287, 288] Ion migration in some of these inorganic solids is restricted to channels, thus leading to one-dimensional models. The motion of individual ions is not diffusive,[264, 289] but is even slower on account of the confinement produced by the neighbors. Various theoretical[289, 290] and numerical[264] methods are beginning to unravel such correlated dynamics. The interesting point for the present discussion is that transport and nmr probe different rates.[287] The former depends on the net distance traveled by all charges, while the latter depends on the residence time of a given nuclear or electronic spin at a site. A distinction must consequently be made between total transport for indistinguishable particles and relaxation due to distinguishable particles. The intermediate correlations $U \gtrsim 4|t|$ in most organic conductors effectively exclude A^{2-} or D^{2+} sites and the resulting correlated motion may be even more complicated. The emergence of a satisfactory physical model, however, will inevitably require a more thorough analysis.

We have also neglected distinctions for motion that is slow or fast compared to the Larmor frequency, $\omega_0 = g\mu H/\hbar$. These "10/3" effects[101] due to nonsecular terms have been studied in inorganic magnetic insulators,[99, 132, 135] described by Heisenberg antiferromagnetic chains. They are in fact useful probes whenever the Zeeman energy $\hbar\omega_0$ can be made comparable to the exchange. When neither the spin Hamiltonian nor the broadening mechanisms is known, however, these additional possibilities are a hindrance. The cutoffs discussed in Section 4.4 are another source of complications. Their magnitudes are rarely known with confidence and even their interpretations as motion in a higher dimension is tentative when, as in triplet-exciton solids or in partly filled conductors, intrachain collisions also limit one-dimensional diffusion. The optimistic view is that such special effects will eventually allow extremely detailed comparisons between theory and magnetic resonance experiments.

Disorder is particularly important in one-dimension, and ion-radical solids are noted for their chemical purity or structural perfection. While the structurally disordered TCNQ salts in Section 5.3 are leading candidates, disorder is probably important at low temperature in many other organic ion-radical solids. Disorder raises many interesting questions.[210, 231, 291] It remains to be seen which of these are actually realized in organic solids. In connection with charge and spin dynamics, we note that Monte Carlo studies[292] and exact one-dimensional models[293] show that disorder may modify $t^{-d/2}$ diffusive behavior. The relevance of interrupted strands or of the model parameters for actual solids is neither known nor important. Rather the point is that diffusive behavior may be suppressed by disorder, thereby potentially spoiling the $\omega^{-1/2}$ divergences of the spin-correlation functions discussed in Section 4.4. Yet these field dependences remain extremely useful even without detailed knowledge of the coupling constants.

Even the simplest models for spin dynamics and relaxation thus pose difficult theoretical questions. Occasional exact results have usually contained surprises. Numerical methods for one-dimensional models should extend the basic ideas of exact results to more realistic models, thereby guiding approximations in general treatments. Since several models have been intrinsically interesting, their mathematical evolution is likely to proceed independently. The obvious challenge is to recognize when and to what extent such models mimic real systems. The description of spin and charge dynamics in organic conductors is rather rudimentary, in spite of occasional success, while spin dynamics in magnetic insulators may be known with some confidence.

Magnetic resonance methods often yield precise and extensive information that is fairly directly related to such underlying microscopic parameters as $|t|$ or U. Still, the parameters of spin Hamiltonians summarizing many observations may not easily be derived. Single-crystal work, for example, may fix various hyperfine and fine-structure coupling constants that are neither amenable to computation nor particularly important in terms of theoretical models. The magnitudes of microscopic parameters or the measurement of solid-state properties like partial ionicity tend to be uncertain on account of the proposed model, rather than any lack of accurate data. There is an additional complication that such measured parameters involve effective parameters $|t|$ or U that already include averaged vibronic or electrostatic or molecular properties.

It is consequently an enormous advantage to observe a resolved fine-structure, or a clear $\omega^{-1/2}$ process, or a Heisenberg susceptibility, etc. that point unambiguously to spin correlation, one-dimensional motion, or a Heisenberg chain, respectively. Such magnetic information restricts possible models to a more manageable few, thereby suggesting the starting point for more complete analyses of optical and transport properties. The fact that parameter values often remain elusive, or even get blurred, in more ambitious subsequent studies is secondary. After all, the electronic structure of organic ion-radical solids is complicated. Magnetic resonance methods have already been vital in delineating the basic features of triplet spin excitons, lower-dimensional motion, supermolecular radicals in complex salts, solid CT complexes, random-exchange chains, and even organic conductors. There is every indication that they will be equally important in sorting out the next generation of semiquantitative models aimed at transport and optical, as well as magnetic, properties.

ACKNOWLEDGMENT

We gratefully acknowledge support for this work by the National Science Foundation through NSF-DMR-7727418 A01.

Notation

Latin

A	Acceptor molecule
A_{zz}	*zz* component of the hyperfine coupling tensor
A^-	Anion radical derived by adding an electron to A
$\|A^-\sigma\rangle$	HOMO of anion radical with spin $\sigma(=\alpha, \beta)$
Ad	Acridinium
$a_{n\sigma}$	Annihilation operator for electron of spin σ $(=\alpha, \beta)$ on site n
$a_{n\sigma}^{\dagger}$	Creation operator for electron of spin σ $(=\alpha, \beta)$ on site n
a	Contact hyperfine coupling
c	Fraction of weak exchanges; lattice spacing
$C(t)$	Time correlation function
$C_{1d}(t)$	Time correlation function for one-dimensional (1*d*) motion
$C(T, H)$	Temperature and magnetic field dependent specific heat
$(CH)_x$	Polyacetylene
CI	Configuration interaction
cm^{-1}	Wave number (energy/*hc*)
CT	Charge transfer
cw	Continuous wave
D	Donor molecule
D	Dipolar tensor; fine structure constant
d	Dimensionality
D^+	Cation radical derived from ionization of D
$D(T)$	Temperature-dependent diffusion constant
DA	Donor–acceptor complex
$\|D^+\sigma\rangle$	HOMO of a donor cation radical with spin $\sigma(=\alpha, \beta)$
DEDMTSF	Diethyldimethyltetraselenafulvalene

DEDMTTF	Diethyldimethyltetrathiafulvalene
DEtTCNQ	2,5-Diethyl-7,7,8,8-tetracyano-*p*-quinodimethane
DMP	*N*-methylphenazinium with deuterated methyl
DMeTCNQ	2,5-Dimethyl-7,7,8,8-tetracyano-*p*-quinodimethane
DMTTF	Dimethyltetetrathiafulvalene
E	Fine structure constant
E_A	Electron affinity of A
epr	Electron paramagnetic resonance
e_n	Segment of n spins where n is even
$f(J)$	Distribution function of coupling constants
$f(\omega)$	Frequency correlation function (spectral density)
g	g Factor
$g(e_n)$	Spin attenuation factor due to segement e_n
$g(\omega)$	Line-shape function
$\hbar$	Planck's constant divided by 2π
$\mathscr{H}$	Hamiltonian
$\mathscr{H}_{12}$	Spin Hamiltonian for spins 1 and 2
$\mathscr{H}_I$	Hamiltonian for nucleus with spin I
$\mathscr{H}_T$	Triplet-state Hamiltonian
H_c	Critical magnetic field in α-CuNSal
H_j	Local magnetic field due to spin j
H_0	Applied, static magnetic field
HMTSF	Hexamethyltetraselenafulvalene
HMTTF	Hexamethyltetrathiafulvalene
HOMO	Highest occupied molecular orbital
I_P	Ionization potential
$I_{x,y,z}$	Nuclear spin components
J, J'	Dimer and interdimer spin coupling
k	Lattice wave vector
k_B	Boltzmann constant
LUMO	Lowest unoccupied molecular orbital
M	Madelung sum; morpholinium
$^3M^*$	Excited molecular triplet state
M_1	Nearest-neighbor Coulomb interaction
Me_2P	Dimethylphenazine
$(Me_2P)_xP_{1-x}$	Alloy of Me_2P and P
M_n	*n*th Moment about the mean
mnt	Maleonitriledithiolate
$M(T, H)$	Magnetization as a function of temperature and magnetic field
N	Number of sites
$n_e(T)$	Effective number of spins at T
n_p	Number operator for site p
NEP	*N*-ethylphenazine
NMP	*N*-methylphenazine
$(NMPH)^+$	Cation radical of 5,10-Dihydro-5-Methylphenazine
nmr	Nuclear magnetic resonance
o_n	Segment of n spins where n is odd
OMTBP	Octamethyltetrabenzporphyrin
P	Phenazine
$(Ph_3AsCH_3)^+$	Triphenylmethylarsonium ion
$(Ph_3PCH_3)^+$	Triphenylmethylphosphonium ion

PD	Phenylenediammine
$(Qn)^+$	Quinolinium ion
r_{jk}	Separation of spins j and k
REHAC	Random exchange Heisenberg antiferromagnetic chain
S	Total electronic spin vector
s_1, s_2	Individual electronic spins
$S_{x,y,z}$	Electronic spin components
SCN	Thiacyanide
SeCN	Selenacyanide
$(SN)_x$	Polysulfurnitride
so	Spin–orbit
t, $t(r)$	Longitudinal electronic transfer integral, large in one-dimension
t', t''	Transverse transfer integrals, small in one-dimension
T	Temperature, K
T_1^{-1}	Spin-lattice relaxation rate
T_2^{-1}	Spin–spin relaxation rate
T_c	Metal–insulator transition temperature
TCNQ	Tetracyanoquinodimethane
$TCNQF_4$	Tetraflouro(TCNQ)
TMMC	Tetramethylammonium manganese chloride
TMPD	$NNN'N'$-tetramethyl-p-phenylenediammine
TMTSF	Tetramethyltetraselenafulvalene
TMTTF	Tetramethyltetrathiafulvalene
TSF	Tetraselenafulvalene
TTF	Tetrathiafulvalene
U, U_0	Correlation energy for doubly occupied site
U_1	Nearest-neighbor correlation energy
U_{eff}	Effective on-site correlation interaction
VB	Valence bond
WBP	Wurster's blue perchlorate
x	Reduced spin exchange constant, units of J
X, Y, Z	Principal axes of a tensor
z	$U/\sqrt{2}t$ or $\delta/\sqrt{2}t$ for segregated or mixed stacks, respectively
z_c	z at neutral-ionic interface

Greek

α	Electron spin state; power-law exponent for REHACs
α_D	Proportionality constant for linewidth dependence on spin-orbit coupling
α-CuNSal	α-bis(N-methylsalicyladiminato)Cu(II)
β	Electron spin state
γ	Ionicity; gyromagnetic ratio
γ_c	Ionicity at neutral-ionic interface
2δ	Energy difference between DA and D^+A^- pairs
δ_0	First-order contribution to δ; $\delta_0 = I_D - E_A$
Δ_0	Activation energy for conduction
ΔE_{CT}	Charge transfer excitation energy
ΔE_m	Magnetic gap
ΔH	epr linewidth

ε	Site energy; magnitude of weak exchanges in REHACs
θ_{jk}	Angle between the internuclear vector r_{jk} and the applied magnetic field
μ_B	Bohr magneton
$\mu(T)$	Temperature-dependent electron mobility
$\nu_\perp$	Transverse hopping rate
ν_c	Cutoff frequency limiting one-dimensional motion
$\rho(O_n)$	Spin density on terminal sites of O_n
$\rho(T)$	Temperature-dependent spin density
σ	Electronic spin (α or β); conductivity
τ^{-1}	Electron scattering rate
χ	Magnetic susceptibility
χ_A, χ_D	Magnetic susceptibility due to acceptor or donor chain, respectively
ω	Angular frequency of a radiofrequency field
$\bar{\omega}$	First moment about zero of an absorption line
$\omega_h(T)$	Temperature dependent hopping rate
ω_N, ω_e	Nuclear and electronic Larmor frequency for a given H_0
Ω_+, Ω_z	Geometrical factors deriving from $I^\pm$ or I^z terms in the relaxation equation (16)

References

1. Soos, Z. G., *Ann. Rev. Phys. Chem.* **25**, 121 (1974); Soos, Z. G., and Klein, D. J., in *Molecular Association*, Vol. 1, R. Foster, Ed. Academic Press, New York (1975), pp. 1–119. Electronic structure, classification, correlations, magnetic, and optical properties of mixed and segregated ion-radical solids.
2. Weill, G., Andre, J. J., and Bieber, A., *Mol. Cryst. Liq. Crist.* **44**, 237 (1978). Epr of organic ion-radical salts and conductors.
3. Zeller, H. R., *Adv. Solid State Phys.* **13**, 31 (1973); Shchegolev, I. F., *Phys. Status Solidi a* **12**, 4 (1972). Physical properties and interpretation of $K_2Pt(CN)_4Br_{0.3} \cdot 3H_2O$ and of conducting TCNQ salts, respectively.
4. DeJongh, L. S., and Miedema, A. R., *Adv. Phys.* **23**, 1 (1974) Comprehensive listing of static magnetic properties of inorganic systems, including low-dimensional crystals.
5. Mille· J. S., and Epstein, A. J., *Prog. Inorg. Chem.* **20**, 1 (1976) Optical and electrical properties of inorganic linear chains, with emphasis on $K_2Pt(CN)_4Br_{0.3} \cdot 3H_2O$.
6. Keller, H. J., Ed., *Chemistry and Physics of One-Dimensional Metals*, NATO-ASI Series B25, Plenum Press, New York (1976); also, Series B7 (1974).
7. Hatfield, W. E., Ed., *Molecular Metals*, Plenum Press, New York (1979).
8. Miller, J. S., and Epstein, A. J., Eds., *Synthesis and Properties of Low-Dimensional Materials, Ann. N.Y. Acad. Sci.* **313**, 442 (1978).
9. Barisic, S., Bjelis, A., Cooper, J. R., and Leontic, B. *Lecture Notes in Physics 96*, *Quasi-One-Dimensional Conductors II*, Springer, New York (1979); also, Ehlers, J., Hepp, K., Kippenhahn, R., Weidenmüller, H. A., and Zittartz, J., *Lecture Notes in Physics 65*, Springer, New York (1977).
10. Alcacer, L., Ed., *The Physics and Chemistry of Low-Dimensional Solids*, NATO-ASI at Tomar, Portugal, 1979, Reidel Publishing, Dordrecht, The Netherlands (1980).
11. Nordio, P. L., Soos, Z. G., and McConnell, H. M., *Ann. Rev. Phys. Chem.* **17**, 237 (1966).
12. Kokoszka G. F., and Gordon, G., in *Transition Metal Chemistry*, Vol. 5, R. L. Carlin, Ed., Marcel Dekker, New York (1969), pp. 181–277; Hatfield, W. E., and Whyman, R., *ibid.*, pp. 47–179.

13. Bleaney, B., and Bowers, K. D., *Proc. R. Soc. Lond. A* **214**, 451 (1952); Valentine, J. S., Silverstein, A. J., and Soos, Z. G., *J. Am. Chem. Soc.* **96**, 97 (1974).
14. McGregor, K. T., and Hatfield, W. E., *J. Chem. Phys.* **62**, 2911 (1975).
15. DeBoer, J. L., and Vos, A., *Acta Crystallogr. B* **28**, 835, 839 (1972).
16. Hausser, K., and Murrell, J. N., *J. Chem. Phys.* **27**, 500 (1957).
17. Thomas, D. D., Keller, H., and McConnell, H. M., *J. Chem. Phys.* **39**, 2321 (1963); Thomas, D. D., Merkl, A. W., Hildebrandt, A. F., and McConnell, H. M., *J. Chem. Phys.* **40**, 2588 (1964).
18. Coulson, C. A., and Fisher, I., *Philos. Mag.* **40**, 386 (1949).
19. Sakata, T., and Nagakura, S., *Bull. Chem. Soc. Jap.* **42**, 1497 (1969).
20. Anderson, P. W., *Solid State Physics*, Vol. 14, F. Seitz and D. Turnbull, Eds., Academic, New York (1963), pp. 99–214.
21. Carrington, A., and McLachlan, A. D., *Introduction to Magnetic Resonance*, Harper and Row, New York (1967), Chapter 8.
22. Silverstein, A. J., and Soos, Z. G., *Chem. Phys. Lett.* **39**, 525 (1976).
23. Flandrois, S., and Boissonade, J., *Chem. Phys. Lett.* **58**, 596 (1978).
24. Soos, Z. G., and McConnell, H. M., *J. Chem. Phys.* **43**, 3780 (1965).
25. Iida, Y., *J. Chem. Phys.* **59**, 1607 (1972); Suzuki, Y., and Iida, Y., *Bull. Chem. Soc. Jap.* **46**, 2056 (1973).
26. Jones, M. T., and Chesnut, D. B., *J. Chem. Phys.* **38**, 1311 (1963).
27. Brown, I. M., and Jones, M. T., *J. Chem. Phys.* **51**, 4687 (1969).
28. Okumara, K., *J. Phys. Soc. Jap.* **18**, 69 (1963).
29. Bonner, J. C., and Fisher, M. E., *Phys. Rev.* **135**, 640 (1964).
30. Bondeson, S. R., and Soos, Z. G., unpublished results.
31. Bondeson, S. R., and Soos, Z. G., *Chem. Phys.* **44**, 403 (1979).
32. Metzger, R. M., *J. Chem. Phys.* **64**, 2069 (1976).
33. Rice, M. J., Yartsev, V. M., and Jacobson, C. S., *Phys. Rev. B* **21**, 3437 (1980); Rice, M. J., Lipari, N. O., and Strässler, S., *Phys. Rev. Lett.* **39**, 1359 (1977).
34. Tanaka, J., Tanaka, M., Kawai, T., and Maki, O., *Bull. Chem. Soc. Jap.* **49**, 2358 (1976); Torrance, J. B., Scott, B. A., Welber, B., Kaufman, F. B., and Seiden, P. E., *Phys. Rev. B* **14**, 730 (1979); Cummings, K. D., Tanner, D. B., and Miller, J. S., *Bull. Am. Phys. Soc.* **25**, 218 (1980).
35. Chesnut, D. B., and Phillips, W. D., *J. Chem. Phys.* **35**, 1002 (1961).
36. Siemons, W. J., Bierstedt, P. E., and Kepler, R. G., *J. Chem. Phys.* **39**, 3523 (1963).
37. Sakai, N., Shirotani, I., and Minomura, S., *Bull. Chem. Soc. Jap.* **45**, 3314 (1972).
38. Torrance, J. B., *Acct. Chem. Res.* **12**, 79 (1979).
39. Fritchie, Jr., C. J., and Arthur, Jr., P., *Acta Crystallogr.* **21**, 139 (1966).
40. Flandrois, S., and Chasseau, D., *Acta Crystallogr. B* **33**, 2744 (1977).
41. Flandrois, S., Amiell, J., Carmona, F., and Delhaes, P., *Solid State Commun.* **17**, 287 (1975).
42. Farges, J. P., in *The Physics and Chemistry of Low-Dimensional Solids*, NATO-ASI Series C56, L. Acacer, Ed., Reidel, Dordrecht, The Netherlands (1980), p. 233.
43. Chesnut, D. B., and Moseley, R. W., *Theor. Chim. Acta* (*Berlin*) **13**, 230 (1969).
44. Kuindersma, P. I., Sawatzky, G. A., and Kommandeur, J., *J. Phys. C* **8**, 3008, 3016 (1975).
45. Mazumdar, S., and Soos, Z. G., *Phys. Rev. B* **23**, 2810 (1981).
46. Lynden-Bell, R., *Mol. Phys.* **8**, 71 (1964).
47. Bailey, J. C., and Chesnut, D. B., *J. Chem. Phys.* **51**, 5118 (1969).
48. Hove, M. J., Hoffman, B. M., and Ibers, J. A., *J. Chem. Phys.* **56**, 3490 (1972).
49. Avakian, P., and Merrifield, R. E., *Mol. Crystallogr.* **5**, 37 (1968); Ermolaev, V., *Sov. Phys. Usp.* **80**, 333 (1963).
50. Ern, V., Suna, A., Tomkiewicz, Y., Avakian, P., and Groff, R. P., *Phys. Rev. B* **5**, 3222 (1972).

51. Soos, Z. G., *J. Chem. Phys.* **51**, 2107 (1969); Sternlicht, H., and McConnell, H. M., *J. Chem. Phys.* **35**, 1793 (1961).
52. Zewail, A. H., and Harris, C. B., *Phys. Rev. B* **11**, 935 (1975); Harris, C. B., and Zwemer, D. A., *Ann. Rev. Phys. Chem.* **29**, 473 (1978).
53. Briegleb, G., *Electronen-Donator-Acceptor-Komplexe*, Springer-Verlag, Berlin (1961).
54. Foster, R., *Organic Charge Transfer Complexes*, Academic, New York (1969); Mulliken, R. S., and Person, W. B., *Molecular Complexes: A Lecture and Reprint Volume*, John Wiley, New York (1969).
55. Kuroda, H., Kunii, T., Hiroma, S., and Akamatu, H., *J. Mol. Spectrasc.* **22**, 60 (1967).
56. Haarer, D., Festkorperprobleme **XX**, 344 (1980).
57. (a) Soos, Z. G., Keller, H. J., Moroni, W., and Nöthe, D., *J. Am. Chem. Soc.* **99**, 504 (1977); (b) *Ann. N.Y. Acad. Sci.* **313**, 442 (1978).
58. Goldberg, I., and Shmueli, U., *Acta. Crystallogr. B* **29**, 421 (1973).
59. Nöthe, D., Moroni, W., Keller, H. J., Soos, Z. G., and Mazumdar, S., *Solid State Commun.* **26**, 713 (1978). See Soos, Z. G., in Reference 10 for the orientation of the principal axes.
60. Keller, H. J., and Nöthe, D., private communication.
61. Torrance, J. B., private communication and Reference 85.
62. Fujita, I., and Matsunaga, Y., *Bull. Chem. Soc. Jap.* **53**, 267 (1980).
63. Soos, Z. G., *Chem. Phys. Lett.* **63**, 179 (1979).
64. Morosin, B., Plastas, H. J., Coleman, L. B., and Stewart, J. M., *Acta Crystallogr. B* **34**, 540 (1978).
65. Dong, V., Endres, H., Keller, H. J., Moroni, W., and Nöthe, D., *Acta Crystallogr. B* **33**, 2428 (1977).
66. Harms, R. H., Keller, H. J., Nöthe, D., Werner, M., Gundel, D., Sixl, H., Soos, Z. G., and Metzger, R. M., *Mol. Cryst. Liq. Cryst.*, **65**, 179 (1981).
67. McConnell, H. M., Hoffman, B. M., and Metzger, R. M., *Proc. Natl. Acad. Sci. USA* **53**, 46 (1965).
68. Kinoshita, M., and Akamatu, H., *Nature* **207**, 291 (1965); Ohmasa, M., Kinoshita, M., Sano, M., and Akamatu, H., *Bull. Chem. Soc. Jap.* **41**, 1998 (1968).
69. Matsunaga, Y., *J. Chem. Phys.* **41**, 1609 (1964); *ibid.* **42**, 1982 (1965); Foster, R., and Thompson, T. J., *Trans. Faraday Soc.* **59**, 296 (1963); Kronick, P. L., Scott, H., and Labes, M. M., *J. Chem. Phys.* **40**, 890 (1964).
70. Hanson, A. W., *Acta Crystallogr.* **19**, 610 (1965).
71. Soos, Z. G., and Mazumdar, S., *Phys. Rev. B* **18**, 1991 (1978); *Chem. Phys. Lett.* **56**, 515 (1978).
72. Pott, G. T., and Kommandeur, J., *Mol. Phys.* **13**, 373 (1967).
73. Hughes, R. C., and Soos, Z. G., *J. Chem. Phys.* **48**, 1066 (1968).
74. DeBoer, J. L., and Vos, A., *Acta Crystallogr. B* **24**, 720 (1967).
75. Metzger, R. M., *Ann. N.Y. Acad. Sci.* **313**, 145 (1978); Metzger, R. M., and Bloch, A. N., *J. Chem. Phys.* **63**, 5098 (1975); Epstein, A. J., Lipari, N. O., Sandman, D. J., and Nielsen, P., *Phys. Rev. B* **13**, 1569 (1975); Torrance, J. B., and Silverman, B. D., *Phys. Rev. B* **15**, 788 (1977); Metzger, R. M., in *Crystal Cohesion and Conformational Energies*, Topics in Current Physics, Vol. 26, (Metzger, R. M., ed.), Springer Verlag, Berlin, 1981), pp. 80–107.
76. Kuroda, H., Hiroma, S., and Akamatu, H., *Bull. Chem. Soc. Jap.* **41**, 2855 (1968).
77. Suchanski, M. R., Ph.D. thesis, Northwestern University, 1977 (unpublished).
78. Soos, Z. G., Mazumdar, S., and Cheung, T. T. P., *Mol. Cryst. Liq. Cryst.* **52**, 93 (1979).
79. Bozio, R., and Pecile, C., in *The Physics and Chemistry of Low-Dimensional Solids*, NATO-ASI Series C56, L. Alcacer, Ed., Reidel, Dordrecht, The Netherlands (1980), p. 165. Pecile, C., private communication; Bloch, A., private communication.
80. Wheland, R. C., and Gillson, J. L., *J. Am. Chem. Soc.* **98**, 3916 (1976).
81. Soos, Z. G., Keller, H. J., Ludolf, K., Queckbronner, Q., Wehe, D., Flandrois, S., *J. Chem. Phys.* **74**, 5287 (1981).

82. Melby, L. R., *Can. J. Chem.* **43**, 1448 (1965).
83. Endres, H., Keller, H. J., Moroni, W., and Nöthe, D., *Acta Cryst. B* **36**, 1435 (1980).
84. Epstein, A. J., Conwell, E. M., and Miller, J. S., *Ann. N.Y. Acad. Sci.* **313**, 183 (1978);
85. Torrance, J. B., Girlando, A., Mayerle, J. J., Crowley, J. I., Lee, V. Y., Batail, P., and La Place, S. J., *Phys. Rev. Lett.* **47**, 1747 (1981); Hubbard, J., and Torrance, J. B., *Phys. Rev. Lett.* **47**, 1750 (1981); Torrance, J. B., Vasquez, J. E., Mayerle, J. J., and Lee, V. Y., *Phys. Rev. Lett.*, **46**, 256 (1981).
86. Soos, Z. G., *J. Chem. Phys.* **46**, 4284 (1967).
87. Huang, T. Z., Taylor, R. P., and Soos, Z. G., *Phys. Rev. Lett.* **28**, 1054 (1972).
88. Hoffman, B. M., and Hughes, R. C., *J. Chem. Phys.* **52**, 4011 (1970); *Solid State Commun.* **7**, 895 (1969).
89. Murrell, J. N., Randic, M., and Williams, D. R., *Proc. R. Soc. Lond. A* **284**, 566 (1965); Lippert, J. L., Hanna, M. W., and Trotter, P. J., *J. Am. Chem. Soc.* **91**, 4035 (1968).
90. Bondeson, S. R., and Soos, Z. G., unpublished work.
91. Vaughan, R. W., *Ann. Rev. Phys. Chem.* **29**, 397 (1978); Mehring, M., in *NMR, Basic Principles and Progress*, P. Diehl, E. Fluch, and R. Kosfeld, Eds., Springer, New York (1978), Vol. 11.
92. Cheung, T. T. P., Gerstein, B. C., Ryan, L. M., and Taylor, R. E., *J. Chem. Phys.*, **73**, 6059 (1980), **74**, 6537 (1981); Cheung, T. T. P., Worthington, L. E., Murphy, P. DuBois, and Gernstein, B. C., *J. Magn. Res.* **41**, 158 (1980).
93. Murgich, J., *J. Chem. Phys.* **70**, 1198 (1979).
94. Gutowsky, H. S., and Pake, G. E., *J. Chem. Phys.* **18**, 162 (1950); Andrew, E. R., *ibid.* **18**, 607 (1950); Andrew, E. R., and Bersohn, R., *ibid.* **18**, 159 (1950).
95. Knauer, R. C., and Bartkowski, R. R., *Phys. Rev. B* **7**, 450 (1973); Azevedo, L. J., Clark, W. G., Hulin, D., and McLean, E. O., *Phys. Lett. A* **58**, 255 (1976).
96. (a) Azevedo, L. J., Narath, A., Richards, P. M., and Soos, Z. G., *Phys. Rev. B* **21**, 2871 (1980); (b) *Phys. Rev. Lett.* **43**, 875 (1979).
97. Azevedo, L. J., Clark, W. G., and McLean E. O., *Proceedings of the XIVth International Conference on Low Temperature Physics* LT14, Helsinki, Finland, p. 369 (1975).
98. Slichter, C. P., *Principles of Magnetic Resonance*, Harper and Row, New York (1963).
99. Cheung, T. T. P., Soos, Z. G., Dietz, R. E., and Merritt, F. R., *Phys. Rev. B* **17**, 1266 (1978).
100. Hutchings, M. T., Shirane, G., Birgeneau, R. J., and Holt, S. L., *Phys. Rev. B* **5**, 1999 (1972).
101. Abragam, A., *The Principles of Nuclear Magnetism*, Oxford University Press, London (1961).
102. Van Vleck, J. H., *Phys. Rev.* **74**, 1168 (1948).
103. Andrew, E. R., and Eades R. G., *Proc. R. Soc. Lond. A* **218**, 537 (1953).
104. Gibby, M. G., Pines, A., and Waugh, J. S., *Chem. Phys. Lett.* **16**, 296 (1972).
105. Gorskaya, N. V., and Fedin, E. I., *J. Struct. Chem.* **9**, 488 (1968).
106. Inoue, M., *J. Magn. Reson.* **27**, 159 (1977).
107. Kawamori, A., *J. Chem. Phys.* **47**, 3091 (1967).
108. Butler, M. A., Wudl, F., and Soos, Z. G., *J. Phys. Chem. Sol.* **37**, 811 (1976).
109. Devreux, F., and Nechtschein, M., in *Lecture Notes in Physics* **96**, S. Barisić, A. Bjelis, J. R. Cooper, and B. Leontić, Eds., Springer, New York (1979), p. 153.
110. Shmueli, U., Sheinblatt, M., and Polak, M., *Acta Crystallogr. A* **32**, 192 (1976).
111. Sjöblom, R., *J. Magn. Reson.* **22**, 411 (1976).
112. Ailion, D. C., *Adv. Magn. Reson.* **5**, 177 (1971); Spiess, H. W., in *NMR: Basic Principles and Progress*, Vol. 15, P. Diehl, E. Fluck, and R. Kosfeld, Eds., Springer, New York (1978), p. 55.
113. Hilt, R. L., and Hubbard, P. S., *Phys. Rev. A* **134**, 392 (1964).
114. Press, W., Prager, M., and Heidemann, A., *J. Chem. Phys.* **72**, 5924 (1980); Clough, S., in *NMR: Basic Principles and Progress*, Vol. 13, P. Diehl, E. Fluck, and R. Kosfeld, Eds., Springer, New York (1976), p. 113.

115. Clough, S., Hill, J. R., and Punkkinen, M., *J. Phys. C: Solid State Phys.* **7**, 3413 (1974); Punkkinen, M., and Clough, S., *ibid.* **7**, 3403 (1974).
116. Punkkinen, M., and Ingman, L. P., *Chem. Phys. Lett.* **38**, 138 (1976).
117. Wind, R. A., Emid, S., Pourquié, J. F. J. M., and Smidt, J., *J. Chem. Phys.* **67**, 2436 (1977).
118. Rager, H., and Weiss, A., *Z. Phys. Chem. N. F.* **93**, 299 (1974).
119. Clough, S., Hill, J. R., and Hobson, T., *Phys. Rev. Lett.* **33**, 1257 (1974); Glätti, H., Sentz, A., and Sienkremer, M., *Phys. Rev. Lett.* **28**, 871 (1972).
120. Alizon, J., Berthet, G., Blanc, J. P., Gallice, J., and Robert, H., *Phys. Status Solidi B* **58**, 349 (1973).
121. Froix, M. F., Epstein, A. J., and Miller, J. S., *Phys. Rev. B* **18**, 2046 (1978).
122. LaMar, G. N., Horrocks, Jr., W. DeW., Holm, R. H., Eds. *NMR of Paramagnetic Molecules*, Academic, New York (1973); Orrell, K. G., in *Annual Reports on NMR Spectroscopy*, Vol. 9, G. A. Webb, Ed., Academic Press, New York (1979), p. 1.
123. Chesnut, D. B., unpublished results.
124. Kawamori, A., and Suzuki, S., *Mol. Phys.* **8**, 95 (1964).
125. Rybaczewski, E. F., Smith, L. S., Garito, A. F., Heeger, A. J., and Silbernagel, B. G., *Phys. Rev. B* **14**, 2746 (1976).
126. Tomkiewicz, Y., Taranko, A. R., and Torrance, J. B., *Phys. Rev. Lett.* **36**, 751 (1976); Tomkiewicz, Y., Taranko, A. R., and Engler, E. M., *ibid.* **37**, 1705 (1976).
127. Devreux, F., Jeandey, C., Nechtschein, M., Fabre, J. M., and Giral, L., *J. Phys. (Paris)* **40**, 671 (1979).
128. Butler, M. A., Wudl, F., and Soos, Z. G., *Phys. Rev. B* **12**, 4708 (1975).
129. Torrance, J. B., Scott, B. A., and Kaufman, F. B., *Solid State Commun.* **17**, 1369 (1975).
130. Keller, H. J., Nöthe, D., Moroni, W., and Soos, Z. G., *Chem. Commun.*, 331 (1978); Sandman, D. J., *Mol. Cryst. Liq. Cryst.* **50**, 235 (1979).
131. Pouget, J. P., Metgert, S., Comes, R., and Epstein, A. J., *Phys. Rev. B* **21**, 486 (1980).
132. See, for instance, Richards P. M., in *Proceedings of the International School of Physics, Enrico Fermi*, Course LIX, K. A. Muller and A. Rigamonti, Eds., North-Holland, Amsterdam (1976), p. 539.
133. Blume, M., and Hubbard, J., *Phys. Rev. B* **1**, 3815 (1970); McLean, F. B., and Blume, M., *Phys. Rev. B* **7**, 1149 (1973).
134. Boucher, J. P., Bakheit, M. A., Nechtschein, M., Villa, M., Bonera, G., and Borsa, F., *Phys. Rev. B* **13**, 4098 (1976).
135. Soos, Z. G., Huang, T. Z., Valentine, J. S., and Hughes, R. C., *Phys. Rev. B* **8**, 993 (1973); **9**, 4981 (1974).
136. Hennessy, M. J., McElwee, C. D., and Richards, P. M., *Phys. Rev. B* **7**, 930 (1973).
137. Lowitz, D. A., *J. Chem. Phys.* **46**, 4698 (1967); Hockman, H. T., Van der Velde, G. A., and Nieuwpoort, W. C., *Chem. Phys. Lett.* **25**, 62 (1974); Johansen, H., *Int. J. Quantum Chem.* **9**, 459 (1975).
138. Avalos, J., Devreux, F., Guglielmi, M., and Nechtschein, M., *Mol. Phys.* **36**, 669 (1978).
139. Bondeson, S. R., Ph.D. Dissertation, Duke University, 1978.
140. Corvaja, C., Farnia, G., and Lunelli, B., *J. Chem. Soc. Faraday Trans.* **71**, 1293 (1975).
141. Nyberg, G., Chesnut, D. B., and Crist, B., *J. Chem. Phys.* **50**, 341 (1969).
142. Semeniuk, G. M., and Chesnut, D. B., *Chem. Phys. Lett.* **25**, 251 (1974).
143. Soos, Z. G., *J. Chem. Phys.* **44**, 1729 (1966).
144. Dietz, R. E., Merritt, F. R., Dingle, R., Hone, D., Silbernagel, B. G., and Richards, P. M., *Phys. Rev. Lett.* **26**, 1186 (1971).
145. Hone, D. W., and Richards, P. M., *Ann. Rev. Mater. Sci.* **4**, 337 (1974).
146. Cheung, T. T. P., and Soos, Z. G., *J. Chem. Phys.* **69**, 3845 (1978).
147. Marticorena, B., and Nechtschein, M., in *Proceedings of the 17th Colloquium Ampere*, V. Hovi, Ed., North-Holland, Amsterdam, (1973), p. 505; Alizon, J., Berthet, G., Blanc, J. P., Gallice, J., Robert, H., Fabre, J. M., and Giral, L., *Phys. Statis Solidi B* **85**, 603 (1978).

148. Haken, H., and Schwarzer, E., *Chem. Phys. Lett.* **27**, 41 (1974).
149. Chesnut, D. B., and Bondeson, S. R., *J. Chem. Phys.* **68**, 5383 (1978).
150. Boucher, J. P., Ferrieu, F., and Nechtschein, M., *Phys. Rev. B* **9**, 3871 (1974).
151. Devreux, F., and Nechtschein, M., *Solid State Commun.* **16**, 275 (1975); *Mol. Cryst. Liq. Cryst.* **32**, 251 (1976).
152. Butler, M. A., Walker, L. R., and Soos, Z. G., *J. Chem. Phys.* **64**, 3592 (1976).
153. McPhail, A. T., Semeniuk, G. M., and Chesnut, D. B., *J. Chem. Soc. A*, 2174 (1971).
154. Alizon, J., Berthet, G., Blanc, J. P., Gallice, J., and Robert, H., *Phys. Status Solidi B* **65**, 577 (1974).
155. Devreux, F., *Phys. Rev. B* **13**, 4651 (1976).
156. Villain, J., *J. Phys. Lett.* (*Paris*) **36**, L173 (1975).
157. Ehrenfreund, E., and Heeger, A. J., *Solid State Commun.* **24**, 29 (1977).
158. Devreux, F., *Mol. Cryst. Liq. Cryst.* **32**, 241 (1976).
159. Devreux, F., Nechtschein, M., and Grüner, G., *Phys. Rev. Lett.* **45**, 53 (1980).
160. Devreux, F., and Nechtschein, M., in *Lecture Notes in Physics*, **96**, S. Barisić, A. Bjelis, J. R. Cooper, and B. Leontić, Eds., Springer, New York (1979), p. 145.
161. Soda, G., Jérome, D., Weger, M., Fabre, J. M., and Giral, L., *Solid State Commun.* **18**, 1417 (1976).
162. Nechtschein, M., Devreux, F., Green, R. L., Clark, T. C., and Street, G. B., *Phys. Rev. Lett.* **44**, 356 (1980).
163. See, for example, *Phase Transitions and Critical Phenomena*, Vol. 3, C. Domb and M. S. Green, Eds., Academic Press, New York (1974).
164. Carlin, R. L., and Van Duyneveldt, A. J., *Magnetic Properties of Transition-Metal Compounds*, Springer-Verlag, New York (1977).
165. Larsen, K. P., *Acta Chem. Scand. A* **28**, 194 (1974); Anderson, D. N., and Willett, R. D., *Inorg. Chim. Acta.* **8**, 167 (1974); Furgeson, G. L., and Zaslow, B., *Acta Crystallogr. B* **27**, 849 (1971).
166. Phelps, D. W., Losee, D. B., Hatfield, W. E., and Hodgson, *Inorg. Chem.* **15**, 3147 (1976); Morosin, B., Fallon, P., and Valentine, J., *Acta. Crystallogr. B* **31**, 3147 (1976).
167. Losee, D. B., McGregor, K. T., Estes, W. E., and Hatfield, W. E., *Phys. Rev. B* **14**, 4100 (1976); Losee, D. B., and Hatfield, W. E., *Phys. Rev. B* **10**, 1122 (1974); Losee, D. B., Hatfield, W. E., and Bernard, I., *Phys. Rev. Lett.* **35**, 1665 (1975).
168. Epstein, A., Gurewitz, E., Makovsky, J., and Shaked, H., *Phys. Rev. B* **2**, 3703 (1970).
169. Baberschke, K., Rys, F., and Avend, H., *Physica B* **86–88**, 685 (1977); Gerstein, B. C., Chang, K., and Willett, R. D., *J. Chem. Phys.* **60**, 3454 (1974).
170. Shimizu, M., and Ajiro, Y., *J. Phys. Soc. Jap.* **48**, 414 (1980); Lynch, M. W., Szydlik, P. P., and Kokoszka, G. F., *J. Phys. Chem. Solids*, **40**, 79 (1979).
171. Richards, P. M., and Salamon, M. B., *Phys. Rev. B* **9**, 32 (1974).
172. Cheung, T. T. P., McGregor, K. T., and Soos, Z. G., *Chem. Phys. Lett.* **61**, 457 (1979).
173. Losee, D. B., and Hatfield, W. E., *Phys. Rev. B* **10**, 212 (1974); Estes, W. E., Losee, D. B., and Hatfield, W. E., *J. Chem. Phys.* **72**, 630 (1980).
174. Dupas, A., LeDang, K., Renard, J. P., Veillet, P., Daoud, A., and Perret, R., *J. Chem. Phys.* **65**, 4099 (1976); Iwashita, T., and Uryu, N., *J. Phys. Soc. Jap.* **39**, 905 (1975).
175. Bellito, C., and Day, P., *J. Chem. Soc. Dalton Trans.*, 1207 (1978). (This is a $CrCl_4(RNH_3)_2$ system.)
176. Moriya, T., *Phys. Rev.* **120**, 91 (1960).
177. Abragam, A., and Bleaney, B., *Electron Paramagnetic Resonance of Transition Ions*, Oxford University Press, New York (1970), Chapter 9.
178. McGregor, K. T., and Soos, Z. G., *Inorg. Chem.* **15**, 215 (1976).
179. Soos, Z. G., McGregor, K. T., Cheung, T. T. P., and Silverstein, A. J., *Phys. Rev. B* **16**, 3036 (1977).

180. deJong, W. M., Rutten, W. L. C., and Verstelle, J. C., *Physica B* **82**, 303 (1976); **82**, 288 (1976).
181. Wudl, F., in *Chemistry and Physics of One-Dimensional Metals*, NATO-ASI Series B25, H. J. Keller, Ed., Plenum Press, New York (1977), p. 233; Wudl, F., Wobschall, D., and Hufnagel, E. F., *J. Am. Chem. Soc.* **94**, 670 (1972).
182. Kittel, C., *Phys. Rev.* **120**, 335 (1960).
183. Harris, E. A., *Proc. Phys. Soc. Lond.* **5**, 338 (1972).
184. Lines, M. E., *Solid State Commun.* **11**, 1615 (1975).
185. Schaefer, D. E., Thomas, G. A., and Wudl, F., *Phys. Rev. B* **12**, 5532 (1975).
186. Jerome, D., in *Molecular Metals*, W. Hatfield, Ed., Plenum Press, New York (1979), p. 105–121.
187. Conwell, E. M., in *Proceedings of the Conference Quasi-One-Dimensional Conductors*, S. Barisic, Ed., Springer, New York (1979).
188. Landee, C. P., and Willett, R. D., *Phys. Rev. Lett.* **43**, 463 (1979).
189. Torrance, J. B., in *Chemistry and Physics of One-Dimensional Metals*, NATO ASI B25, H. J. Keller, Ed., Plenum Press, New York (1977), p. 137. Torrance, J. B., Tomkiewicz, Y., and Silverman, B. D., *Phys. Rev. B* **15**, 4738 (1977).
190. Toombs, G. A., *Phys. Rep.* **40c**, 181 (1978).
191. Tomkiewicz, Y., Taranko, A. R., and Torrance, J. B., *Phys. Rev. B* **15**, 1017 (1977); Tomkiewicz, Y., Craven, R. A., Schultz, T. D., Engler, E. M., and Taranko, A. R., *Phys. Rev. B* **15**, 3643 (1977).
192. Wudl, F., Schafer, D. E., Walsh, Jr., W. M., Rupp, L. W., DiSalvo, F. J., Waszczak, J. V., Kaplan, M. L., and Thomas, G. A., *J. Chem. Phys.* **66**, 377 (1977).
193. Walsh, Jr., W. M., Rupp, L. W., Wudl, F., Kaplan, M. L., Schafer, D. E., Thomas, G. A., and Gemmer, R., *Solid State Commun.* **33**, 413 (1980).
194. Gunning, W. J., Chiang, C. K., Heeger, A. J., and Epstein, A. J., *Phys. Status Solidi B* **96**, 145 (1979).
195. Conwell, E. M., *Phys. Rev. B* **22**, 1761 (1980).
196. Soos, Z. G., *J. Chem. Phys.* **49**, 2493 (1968).
197. Mazumdar, S., and Soos, Z. G., *Synth. Met.* **1**, 77 (1979).
198. Klein, D. J., and Seitz, W. A., *Phys. Rev. B* **10**, 3217 (1974).
199. Ovchinnikov, A. A., Ukranskii, I. I., and Kventsel, G. V., *Sov. Phys. Usp.* **15**, 575 (1973).
200. Dyson, F. J., *Phys. Rev.* **98**, 349 (1955).
201. Feher, G., and Kip, A. F., *Phys. Rev.* **98**, 337 (1955).
202. Poole, Jr., C. P., *Electron Spin Resonance*, Wiley-Interscience, New York (1967); Poole, Jr., C. P., and Farach, H. A., *Relaxation in Magnetic Resonance*, Academic Press, New York (1971).
203. Elliott, R. J., *Phys. Rev.* **96**, 266 (1954).
204. Yafet, Y., in *Solid State Physics*, Vol. 14, F. Seitz and D. Turnbull, Eds., Academic Press, New York (1963).
205. Beuneu, F., and Monod, P., *Phys. Rev. B* **18**, 2422 (1978).
206. Lauginie, P., Estrade, H., Conard, J., Guérard, D., Lagrange, P., and El Makrini, M., *Physica* **99B**, 514 (1980).
207. Vogel, F. L., in *Molecular Metals*, W. E. Hatfield, Ed., Plenum Press, New York (1979), p. 261; Fisher, J. E., *ibid.*, p. 281.
208. (a) Tomkiewicz, Y., Engler, E. M., and Schultz, T. D., *Phys. Rev. Lett.* **35**, 456 (1975); (b) Schultz, T. D., and Craven, R. A., in *Highly Conducting One-Dimensional Solids*, J. T. Devreese, R. P. Evrard, and V. E. van Doren, Eds., Plenum Press, New York (1979), p. 147.
209. Tomkiewicz, Y., Schultz, T. D., Schultz, T. D., Engler, E. M., Taranko, A. R., and Bloch, A. N., *Bull. Am. Phys. Soc.* **21**, 287 (1976).

210. Bloch, A. N., Carruthers, T. F., Poehler, T. O., and Cowan, D. O., in *Chemistry and Physics of One-Dimensional Metals*, NATO ASI B25, H. J. Keller, Ed., Plenum Press, New York (1977), p. 47.
211. Keryer, G., Delhaes, P., Amiell, J., Flandrois, S., and Tissier, B., *Phys. Status solidi B* **100**, 251 (1980).
212. Phillips, T. E., Kistenmacher, T. J., Bloch, A. N., Ferraris, J. P., and Cowan, D. O., *Acta Crystallogr. B* **33**, 422 (1977).
213. Tomkiewicz, Y., Taranko, A. R., Schumaker, R., *Phys. Rev. B* **16**, 1380 (1977).
214. Kistenmacher, T. J., Phillips, T. E., and Cowan, D. O., *Acta Crystallogr. B* **30**, 763 (1974).
215. Tomkiewicz, Y., and Taranko, A. R., *Phys. Rev. B* **18**, 733 (1978).
216. Tomkiewicz, Y., and Andersen, J. R., unpublished results.
217. Jacobsen, C. S., Mortensen, M., Andersen, J. R., and Bechgaard, K., *Phys. Rev. B* **18**, 905 (1978).
218. Bechgaard, K., Kistenmacher, T. J., Bloch, A. N., and Cowan, D. O., *Acta Crystallogr. B* **33**, 417 (1977).
219. Etemad, S., Engler, E. M., Schultz, T. D., Penney, T., and Scott, B. A., *Phys. Rev. B* **17**, 513 (1978).
220. Laplaca, S. J., and Corfield, P. W., unpublished results.
221. Phillips, T. E., Kistenmacher, T. J., Bloch, A. N., and Cowan, D. O., *J. Chem. Soc. Chem. Commun* **1976**, 334 (1976).
222. Tomkiewicz, Y., Andersen, J. R., and Taranko, A. R., *Phys. Rev. B* **17**, 1579 (1978).
223. Andersen, J. R., Bechgaard, K., Jacobsen, C. S., Rindorf, G., Soling, H., and Thorup, N., *Acta Crystallogr. B* **37**, 1901 (1978).
224. Andersen, J. R., Craven, R. A., Weidenborner, J. E., and Engler, E. M., *J. Chem. Soc. Chem. Commun.* **1976**, 526 (1976).
225. Tomkiewicz, Y., *Phys. Rev. B* **19**, 4038 (1979).
226. Bulaevskii, L. N., Zvarykina, A. V., Karimov, Y. S., Lyubovskii, R. B., and Shchegolev, I. F., *Sov. Phys. JETP* **35**, 384 (1972).
227. Kobayashi, H., Marumo, F., and Saito, Y., *Acta. Crystallogr. B* **27**, 373 (1971).
228. Kobayashi, H., *Bull. Chem. Soc. Jap.* **47**, 1346 (1974).
229. Fritchie, Jr., C. J., *Acta Crystallogr.* **20**, 892 (1965); Morosin, B., *Phys. Lett. A* **53**, 455 (1975).
230. Tippie, L. C., and Clark, W. G., *Phys. Rev.* **B23**, 5846 (1981).
231. Clark, W. G., in *Proceedings of the International Conference on Physics in One-Dimensional, Fribourg*, Springer, New York (1980).
232. Azevedo, L. J., Clark, W. G., McLean, E. O., and Seligman, P. F., *Solid State Commun.* **16**, 1267 (1975).
233. (a) Clark, W. G., Hamman, J., Sanny, J., and Tippie, L. C., in *International Conference on Quasi-One-Dimensional Conductors, Dubrovnik*, Barisic, S., Bjelis, A., Cooper, J. R., and Leontic, B., Eds., Springer, New York (1979), p. 255; (b) Sanny, J., and Clark, W. G., *Solid State Commun.* **35**, 473 (1980).
234. Clark, W. G., and Tippie, L. C., *Phys. Rev. B* **20**, 2914 (1979).
235. Duffy, Jr., W., Weinhaus, F. M., Strandburg, D. L., and Deck, J. F., *Phys. Rev. B* **20**, 1164 (1979).
236. Sanny, J., Grüner, G., and Clark, W. G., *Solid State Commun.* **35**, 657 (1980).
237. (a) Bondeson, S. R., and Soos, Z. G., *Phys. Rev. B* **22**, 1793 (1980); (b) *Solid State Commun.* **35**, 11 (1980).
238. Theodorou, G., and Cohen, M. H., *Phys. Rev. Lett.* **37**, 1014 (1972).
239. Theodorou, G., *Phys. Rev. B* **16**, 2254 (1977).
240. Theodorou, G., *Phys. Rev. B* **16**, 2273 (1977).

241. Ma, S., Dasgupta, C., and Hu, C. K., *Phys. Rev. Lett.* **43**, 1434 (1979); Dasgupta, C., and Ma, S., *Phys. Rev. B* **22**, 1305 (1980).
242. Hirsch, J. E., and José, J. V., *J. Phys. C Lett.* **13**, L53 (1980); *Phys. Rev. B* **22**, 5339 (1980).
243. Hirsch, J. E., *Phys. Rev. B* **22**, 5353 (1980).
244. Mihaly, G., Holczer, K., Janossy, A., Gruner, G., and Miljak, M. in *Organic Conductors and Semiconductors, Lecture Notes in Physics 65*, Ehlers, J., Hepp, K., Kippenhahn, R., Weidenmüller, H. A., and Zittartz, J., Eds., Springer-Verlag Berlin, (1977), p. 553.
245. Papatriantafillou, C., *Phys. Status Solidi B* **88**, 663 (1978).
246. Bozler, H. M., Gould, C. M., and Clark, W. G., *Phys. Rev. Lett.* **45**, 1303 (1980).
247. Tippie, L. C., and Clark, W. G., *Phys. Rev. B, B* **23**, 5854 (1981).
248. Soos, Z. G., and Bondeson, S. R., *Solid State Commun.* **39**, 289 (1981).
249. Soos, Z. G., and Bondeson, S. R., unpublished work.
250. Weger, M., in *The Physics and Chemistry of Low Dimensional Solids*, NATO ASI Serie C57, L. Alcacer, Ed., Reidel, Dordrecht, The Netherlands (1980), p. 77.
251. Sokoloff, J. B., *Phys. Rev. B* **2**, 779 (1972).
252. Beni, G., Holstein, T., and Pincus, P., *Phys. Rev. B* **8**, 312 (1973).
253. Klein, D. J., *Phys. Rev. B* **8**, 3452 (1973).
254. Hoffman, B. H., Phillips, T. E., Schramm, C. J., and Wright, S. K., in *Molecular Metals*, W. E. Hatfield, Ed., Plenum Press, New York (1979), p. 393.
255. Phillips, T. E., Scaringe, R. P., Hoffman, B. M., and Ibers, J. A., *J. Am. Chem. Soc.* **102**, 3435 (1980).
256. Schramm, C. J., Scaringe, R. P., Stojakovic, D. R., Hoffman, B. M., Ibers, J. A., and Marks, T. J., *J. Am. Chem. Soc.*, **102**, 6702 (1980).
257. Schramm, C. T., and Hoffman, B. M., *Inorg. Chem.* **19**, 383 (1980).
258. Wright, S. K., Schramm, C. J., Phillips, T. E., Scholler, D. M., and Hoffman, B. M., *Synth. Metals* **1**, 43 (1979).
259. Krogmann, K., *Angew. Chem. Int. Ed. Eng.* **8**, 35 (1969).
260. Endres, H., Keller, H. J., Lehmann, R., van de Sand, H., Vu, D., and Poveda, A., *Ann. N.Y. Acad. Sci.* **313**, 617 (1978); see also Reference 5.
261. Teitelbaum, R. C., Ruby, S. L., and Marks, T. J., *J. Am. Chem. Soc.* **100**, 3215 (1978); **101**, 7568 (1979).
262. Hoffman, B. M., Phillips, T. E., and Soos, Z. G., *Solid State Commun.* **33**, 51 (1980).
263. Hoffman, B. M., and Martinsen, J., private communication.
264. Richards, P. M., *Phys. Rev. B* **16**, 1393 (1977).
265. Soos, Z. G., Hoffman, B. M., and Martinsen, J., unpublished results.
266. Trlifaj, M., *Czech. J. Phys.* **6**, 533 (1956); Holstein, T., *Ann. Phys.* (*N.Y.*) **8**, 343 (1959).
267. Austin, I. G., and Molt, N. F., *Adv. Phys.* **18**, 41 (1969).
268. Emin, D., in *Electronic and Structural Properties of Amorphous Semiconductors*, R. G. Le Combre and J. Mort, Eds., Academic Press, New York (1973), p. 261.
269. Epstein, A. J., Conwell, E. M., Sandman, D. J., and Miller, J. S., *Solid State Commun.* **23**, 355 (1977); Epstein, A. J., and Conwell, E. M., *ibid.* **24**, 627 (1977).
270. Heeger, A. J., and McDiarmid, A. G., in *The Physics and Chemistry of Low-Dimensional Solids*, NATO ASI C56, L. Alcacer, Ed., Reidel, Dordrecht, The Netherlands (1980), p. 353, and references therein.
271. Longuet-Higgins, H. C., and Salem, L., *Proc. R. Soc. Lond. A* **251**, 172 (1959). This is a special case of a Peierls instability.
272. Ooshika, Y., *J. Chem. Soc. Jap.* **12**, 1238, 1246 (1957); Labhart, H., *J. Chem. Phys.* **27**, 957 (1957); Grant, P. M., and Batra, I. P., *Solid State Commun.* **29**, 255 (1979).
273. Fincher, Jr., C. R., Ozaki, M., Tanka, M., Peebles, D. L., Lauchlan, L., Heeger, A. J., and MacDiarmid, A. G., *Phys. Rev. B* **20**, 1599 (1979).
274. Coulson, C. A., *Proc. R. Soc. Lond. A* **164**, 383 (1938); **169**, 413 (1939).

275. Salem, L., *The Molecular Orbital Theory of Conjugated Systems*, W. A. Benjamin, New York (1966).
276. Ohno, K., *Theor. Chem. Acta* **2**, 219 (1964); Mataga, M., and Nishimoto, K., *Z. Phys. Chem. (Frankfort)* **13**, 140 (1957).
277. Hudson, B., and Kohler, B. E., *Ann. Rev. Phys. Chem.* **25**, 437 (1974); Granville, M. F., Holton, G. R., and Kohler, B. E., *J. Chem. Phys.* **72**, 4671 (1980); Gavin, R. M., Weisman, C., McVey, J. K., and Rice, S. R., *J. Chem. Phys.* **68**, 522 (1978).
278. Schulten, K., Ohmine, J., and Karplus, M., *J. Chem. Phys.* **64**, 4422 (1976).
279. Miller, T. E., and Soos, Z. G., in *Proceedings of the 14th Jerusalem Symposium on Intermolecular Forces*, B. Pullman, Ed., Plenum Press, New York (1981), pp. 117–132; Ducasse, L. R., Miller, T. E., Soos, Z. G., *J. Chem. Phys.* **76**, 4094 (1982).
280. Weinberger, B. R., Ehrenfreund, E., Pron, A., Heeger, A. J., and McDiarmid, A. G., *J. Chem. Phys.* **72**, 4749 (1980).
281. Pople, J. A., and Walmsley, S. H., *Mol. Phys.* **5**, 15 (1962).
282. Rice, M. J., *Phys. Lett. B* **71**, 152 (1979).
283. Su, W. P., Schrieffer, J. R., and Heeger, A. J., *Phys. Rev. Lett.* **42**, 1698 (1979).
284. Tomkiewicz, Y., Schultz, T. D., Broom, H. B., Clarke, T. C., and Street, G. B., *Phys. Rev. Lett.* **43**, 1532 (1979).
285. Wudl, F., private communication.
286. Maricq, M. M., Waugh, J. S., MacDiarmid, A. G., Shirakawa, H., and Heeger, A. J., *J. Am. Chem. Soc.* **100**, 7729 (1978). Much less Sp^3 carbon is seen in more recent samples.
287. Richards, P. M., *Phys. Rev. B* **18**, 6358 (1978); in *Current Topics in Physics, Superionic Conductors*, Springer-Verlag, Berlin (1979).
288. Whittingham, M. S., and Silbernagel, B. G., in *Solid Electrolytes: General Principles, Characterization Materials, Applications*, P. Hagenmuller and W. van Gool, Eds., Academic Press, New York (1977).
289. Alexander, S., and Pincus, P., *Phys. Rev. B* **18**, 2011 (1978).
290. Sankey, D. F., and Fedders, P. A., *Phys. Rev. B* **15**, 3586 (1977); Fedders, P. A., *Phys. Rev. B* **17**, 40 (1978).
291. Bloch, A. N., Weisman, R. B., and Varma, C. M., *Phys. Rev. Lett.* **28**, 753 (1972); Papatriantafillou, C., Economou, E. N., and Eggarter, T. P., *Phys. Rev. B* **13**, 910, 920 (1976).
292. Rich, R. M., Alexander, S., Bernasconi, J., Holstein, T., Lyo, S. K., and Orbach, R., *Phys. Rev. B* **18**, 3048 (1978).
293. Alexander, S., Bernasconi, J., and Orbach, R., *Phys. Rev. B* **17**, 4311 (1978); Alexander, S., Bernasconi, J., Schneider, W. R., and Orbach, R., *Rev. Mod. Phys.* **53**, 175 (1981).

5

Salts of 7,7,8,8-Tetracyano-p-quinodimethane with Simple and Complex Metal Cations

Helmut Endres

1. Introduction

This chapter deals with salts (or adducts) of 7,7,8,8-tetracyano-*p*-quinodimethane (TCNQ) with metal ions or metal-containing compounds or ions. Reflecting the state of knowledge of such compounds, the greater portion of this review will focus on structural and physical properties of alkali salts of TCNQ. For reasons of analogy, ammonium (NH_4^+) salts will be considered, too, but not salts of substituted ammonium cations, i.e., $(R_nNH_{4-n})^+$.

The Section 2 will review the work on TCNQ salts of simple main-group and transition-metal ions. This section will be followed by one discussing TCNQ compounds of metal complexes, including organometallic compounds like $[(CH_3)_3Sn](TCNQ)$, where the stress is on the crystal structures.

Section 5 contains mention of possible applications of the compounds considered here. We will not include compounds that contain TCNQ salts attached to a polymeric backbone of any kind.

In the summary there is a survey in tabular form of the possible TCNQ arrangements in its salts with metal-containing compounds: stacks, no stacks; mixed stacks, segregated stacks; regular stacks, stacks of diads, triads, etc.; overlap patterns; and a list of lattice parameters of TCNQ compounds of metal complexes.

Helmut Endres • Anorganisch-Chemisches Institut der Universität, Im Neuenheimer Feld 270, 6900 Heidelberg 1, West Germany.

2. Alkali and Ammonium Salts of 7,7,8,8-Tetracyano-p-quinodimethane

2.1. Preparation

The 7,7,8,8-tetracyano-*p*-quinodimethane (TCNQ) acceptor can be prepared[1] from 1,4-cyclohexanedione and malodinitrile. It readily undergoes an electron-transfer reaction with metal iodides to form $(TCNQ)^-$ metal salts,[2] e.g.,

$$Li^+I^- + (TCNQ)^0 \longrightarrow Li^+(TCNQ)^- + \tfrac{1}{2}I_2$$

which is a general route to such compounds. This reaction is usually carried out with an excess of iodide to shift the equilibrium by the formation of I_3^-:

$$3\,Li^+I^- + 2(TCNQ)^0 \longrightarrow 2\,Li^+(TCNQ)^- + Li^+I_3^-$$

Li(TCNQ) is frequently the starting material for other TCNQ compounds, because it is readily soluble in common solvents, including water, in contrary to most of the other TCNQ salts. These salts may be prepared by a metathetical reaction of the kind

$$MX + Li(TCNQ) \longrightarrow M(TCNQ) + Li^+X^-$$

In this preparative method, M may signify a transition-metal complex as well.

Another synthetic approach is the direct oxidation of an appropriate zero-valent metal compound, e.g., (arene)$Cr(CO)_3$ with TCNQ.[2a, 114] Electrochemical synthesis has been applied for the preparation of several transition-metal TCNQ salts.[2b] The obtained products and their conductivities are listed in Table 1.

The reaction of metal iodides with TCNQ may be used to grow crystals by a diffusion process[3]: A test tube containing the metal iodide is placed into an Erlenmeyer flask containing TCNQ. The test tube and the flask are then filled with acetonitrile such that the solvent level in the Erlenmeyer flask exceeds the height of the test tube. The flask is covered and left standing at room temperature for two or three weeks. An analogous procedure can be used for growing crystals by a metathetical reaction starting from Li(TCNQ). An alternative apparatus was described by Pott and Kommandeur.[4]

2.2. Structural Investigations

Only unit-cell data are available for Li(TCNQ),[5] whereas full x-ray structure determinations have been performed with Na(TCNQ) (low-[6] and high-temperature[7] forms), K(TCNQ) (low-[8, 9] and high-temperature[9] forms), three polymorphs of Rb(TCNQ),[5, 10, 11] $Rb_2(TCNQ)_3$,[12, 13]

TABLE 1. Preparation, Composition, and Conductivities of TCNQ Salts Prepared by Electrochemical Synthesis (Reproduced with permission from Reference 26)

Reactant, ratio[b] (% yield)	Potential(V) vs. S.C.E.[c]	$t(h)$	Product	$\sigma(\Omega^{-1}\ cm^{-1})$	Reference 2 Product	Reference 2 $\sigma(\Omega^{-1}\ cm^{-1})$
KBr, 1(60)	−0.513	194	K(TCNQ)	2×10^{-4} pd[a]	K(TCNQ)	2×10^{-4}
$Mn(ClO_4)_2$, 2(42)	−0.504	49	$Mn(TCNQ) \cdot \frac{2}{3}MeCN$	4.4×10^{-3} pd	$Mn(TCNQ)_2 \cdot 3H_2O$	1.1×10^{-5}
$FeCl_2 \cdot 4H_2O$, 2(30)	−0.430	72	$Fe_2(TCNQ)_3 \cdot 5H_2O$	4.8×10^{-4} pd	$Fe(TCNQ)_2 \cdot 3H_2O$	2×10^{-5}
$Co(ClO_4)_2 \cdot 6H_2O$, 2(50)	−0.540	72	$Co(TCNQ)_2 \cdot 3H_2O$	2.6×10^{-4} pd	$Co(TCNQ)_2 \cdot 3H_2O$	1×10^{-5}
$NiCl_2 \cdot 6H_2O$, 2(48)	−0.790	102	$Ni_2(TCNQ)_3 \cdot 6H_2O$	2.6×10^{-4} pd	$Ni(TCNQ)_2 \cdot 3H_2O$	1×10^{-5}
$CuCl_2$, 2(21)	−0.513	10	Cu(TCNQ)	2.9×10^{-2} pd	Cu(TCNQ)	5×10^{-3}
			Cu(TCNQ)	31 ps		
$CuCl_2$, 2(72)	(g 221[d])	247	Cu(TCNQ)	784 gs		
$Zn(ClO_4)_2 \cdot 6H_2O$, 2(15)	−0.540	38	$Zn_2(TCNQ)_3(ClO_4)_2 \cdot 3H_2O$	2.1×10^{-2} pd		
$AgNO_3$, 1(61)	−0.001	35	$Ag_4(TCNQ)_3$	6×10^{-4} pd	Ag(TCNQ)	1.25×10^{-6}
			$Ag_4(TCNQ)_3$	8 ps		
$AgNO_3$, 1(83)	(g 990[d])	16	$Ag_4(TCNQ)_3$	44 gs		
$CrAc_3$, 3(41)	(g 31[d])	408	$Cr(TCNQ)_3$	5.1×10^{-3} gd	$Cr_2Ac_4OH(TCNQ) \cdot 6H_2O$	1×10^{-9}

[a] Abbreviations: p: potentiostat, g: galvanostat, s: single crystal, d: compacted disc.
[b] Amount of TCNQ per 1 reactant.
[c] Standard calomel electrode.
[d] Current density in $\mu A\ cm^{-2}$.

TABLE 2. *Structural Features of Alkali (TCNQ) Salts*

Compound, Temperature	Color; space group	Lattice parameters (Å)	Stack arrangement[a]	Type of stack	Overlap type[b]	Interplanar distance (Å)[c]	R σ_{bl}[d]
Li(TCNQ), R.T. (Ref. 5)	purple tetragonal (Ref. 2)	a = 12.3 c = 7.9		diads			
Na(TCNQ), R.T. (Ref. 6)	purple "$C\bar{1}$" (Ref. 2)	a = 6.993(1) b = 23.707(2) c = 12.469(2) α = 90.14(2) β = 98.58(1) γ = 90.76(1)	⊥	diads	R–R	3.223 3.200 3.505 3.480	0.046 0.008
Na(TCNQ), 353 K (Ref. 7)	$P2_1/n$	a = 3.512(1) b = 11.866(2) c = 12.465(3) β = 98.21(2)	⊥	regular	R–R	3.385	0.033 0.003
K(TCNQ), R.T. (Ref. 8)	red (Ref. 2) $P2_1/c$	a = 3.543(1) b = 17.784(5) c = 17.868(3) β = 94.96	⊥	regular		3.43 3.44	0.079 0.04
K(TCNQ), R.T. (Ref. 9)	$P2_1/n$	a = 7.0835(7) b = 17.773(3) c = 17.859(4) β = 94.95(1)	⊥	diads	R–R	3.240 3.234 3.554 3.580	0.044 0.006
K(TCNQ), 413 K (Ref. 9)	$P2_1/c$	a = 3.587 b = 12.676 c = 12.614 β = 96.44(3)	⊥	regular	R–R	3.479	0.052 0.013
Rb(TCNQ) (phase I), 213 K (Ref. 5)	reddish-purple $P2_1/c$ (Ref. 3)	a = 7.187(1) b = 12.347(2) c = 13.081(3) β = 98.88(2)	⊥	diads	R–R R–B (slipped)	3.159 3.484	0.066 0.004

Rb(TCNQ) (phase II), R.T. (Ref. 10)	dark purple $P\bar{1}$ (Refs. 3, 10)	$a = 9.914$ $b = 7.196$ $c = 3.390$ $\alpha = 92.70(10)$ $\beta = 86.22(11)$ $\gamma = 97.73(7)$	∥	regular	R–B	3.43	0.130 0.03
Rb(TCNQ) (phase III), R.T. (Ref. 11)	$P4/n$	$a = 17.645(11)$ $c = 3.864$	⊥	regular	R–B	3.33	0.065 0.03
$Rb_2(TCNQ)_3$, R.T. (Ref. 12)	$P2_1/c$	$a = 7.297(1)$ $b = 10.392(1)$ $c = 21.444(2)$ $\beta = 97.45(1)$	∥	triads	R–B R–R	3.26 3.12	0.036 0.008
$Rb_2(TCNQ)_3$, 113 K (Ref. 13)	dark purple $P2_1/c$ (Ref. 13)	$a = 7.296(2)$ $b = 10.264(3)$ $c = 21.328(6)$ $\beta = 97.11(2)$	∥	triads	R–B R–R	3.23 3.07	0.038 0.004
$Cs_2(TCNQ)_3$, R.T. (Ref. 14)	$P2_1/c$	$a = 7.34(1)$ $b = 10.40(2)$ $c = 21.98(4)$ $\beta = 97.18(5)$	∥	triads	R–B R–R	3.22 3.26	0.039 0.005
NH_4(TCNQ) (phase I), 293 K, 311 K (Ref. 18)	purple (Ref. 18)	stack axis = 7.19 3.61					
NH_4(TCNQ) (phase II) (Ref. 18)	bluish-purple $P4/mbm$ (Ref. 18)	$a = 12.50(2)$ $c = 3.82(4)$	⊥	regular	R–B	3.31	0.128 0.02

[a] ⊥ Projections of the long axes of TCNQ of adjacent stacks perpendicular and ∥ parallel.
[b] R–R: ring–ring overlap (cf. Figure 3). R–B: Ring–external bond overlap (cf. Figure 4).
[c] Two values in one line are interplanar distances in two independent stacks. The top value in a row is the distance between such units.
[d] σ_{bl} is the typical standard deviation of the bond distances.
R.T. stands for room temperature.

$Cs_2(TCNQ_3)$,[14] and low- and high-temperature forms of $NH_4(TCNQ)$ (phase II).[15] Phase I of $NH_4(TCNQ)$ is characterized by unit-cell data.[15]

Relevant structural information is summarized in Table 2. In all the simple (1 : 1) salts, the TCNQ radical ions form either regular stacks, or stacks composed of diadic units, whereas stacks of triads are formed in the complex (2 : 3) salts. The stacks may be arranged relative to each other in two ways: The long axes of the TCNQ anions, projected along the stack direction, may be parallel ($\parallel$) or perpendicular ($\perp$) to each other. This is illustrated by Figure 1 [$\parallel$ arrangement in Rb(TCNQ), phase II[10]] and Figure 2 [$\perp$ arrangement in K(TCNQ)[9]]. Adjacent TCNQ species in a stack usually exhibit either of two overlap patterns shown in Figures 3 and 4: the so-called ring–ring (R–R) overlap, where one quinoid bond lies over the center of the ring of the adjacent molecule, or the ring–external bond (R–B) overlap, where the exocyclic double bond lies over the ring of the adjacent molecule.

To characterize the accuracy of the structure determinations as summarized in Table 2, the typical standard deviation in bond lengths (σ_{bl}) is given together with the R factor. Caution should be noted since these are not the best criteria because they depend on the number of reflections, the threshold value for "unobserved reflections," weighting scheme, and so on. Care must also be taken in the comparison of interplanar distances. The first problem arises with the shape of the TCNQ ion, since the planes through the terminal $C(CN)_2$ groups often deviate a few degrees from the plane through the quinoid part of the molecule. So the molecule adopts

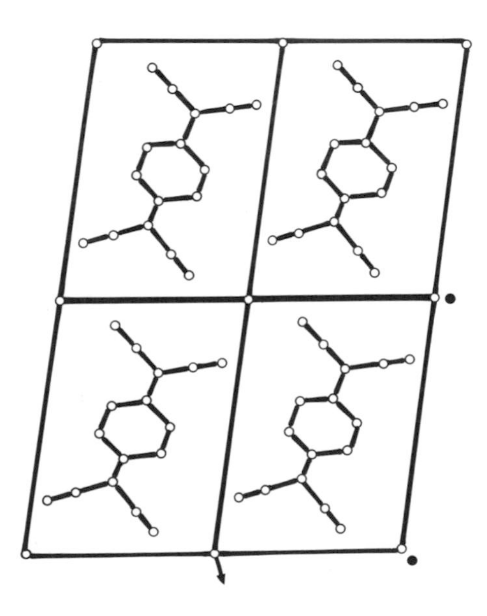

FIGURE 1. Projection of the structure of Rb(TCNQ) (phase II) along the stacking axis with parallel arrangement of the TCNQ axes. (Reproduced from Reference 10, with permission.)

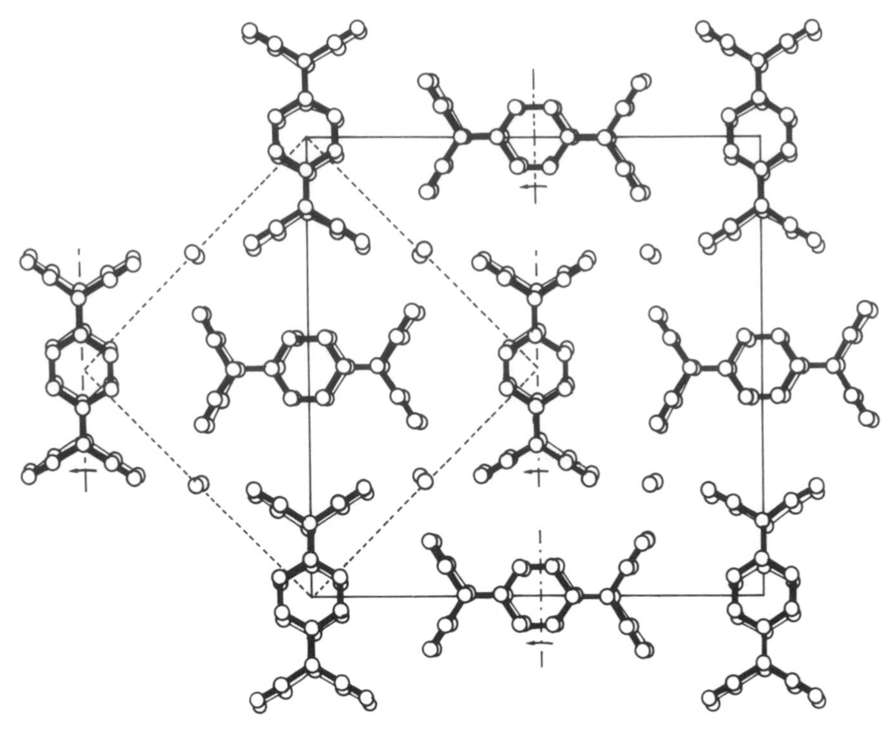

FIGURE 2. Projection of the structure of K(TCNQ) along the stacking axis with perpendicular arrangement of the TCNQ axes. The solid lines indicate the unit cell of the low-temperature form, the broken lines that of the high-temperature form. (Reproduced from Reference 9, with permission.)

either a shallow boat or chair conformation. Thus the molecular symmetry is not exactly D_{2h}, but rather C_{2v} or C_{2h}. Most authors honor this fact by defining the *interplanar distance* as the distance between the quinoid parts of the molecules. A second problem arises when the planes through (the quinoid parts of) adjacent molecules are not exactly parallel. This used to be the case in TCNQ triads, so before comparing distances in $M_2(TCNQ)_3$ compounds with those in M(TCNQ) salts (where the planes are parallel by symmetry), the definition of the interplanar distance should be considered. An example is given in Reference 13.

Table 2 contains most of the structural information, however, some comments remain to be made. The structure of low- and high-temperature forms have been determined for Na^+, K^+, and NH_4^+ salts of TCNQ. It is evident that the unit-cell length along the stack direction in the low-temperature form is approximately twice that of the high-temperature form, caused by a dimerization of TCNQ ions. (The phase transition phenomena

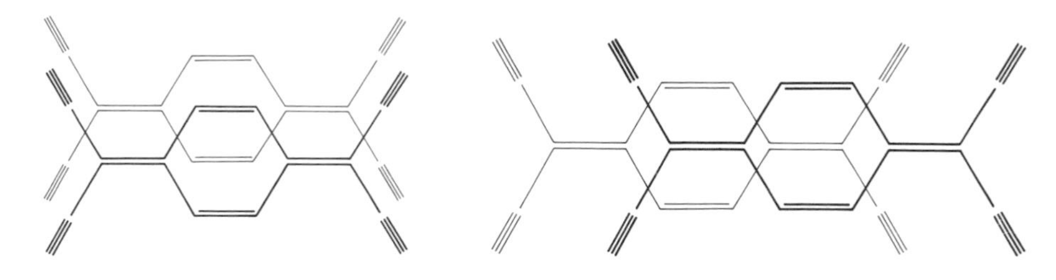

FIGURE 3. Typical ring–ring (R–R) overlap of two TCNQ species.

FIGURE 4. Typical ring–external bond (R–B) overlap of two TCNQ species.

will be discussed in the context Section 2.3.) Thus, it may be concluded that the structure of Li(TCNQ) at room temperature is a low-temperature form, and that Rb(TCNQ) (phase II) is in a high-temperature form at ambient temperature. The three forms of Rb(TCNQ) are polymorphs, and there is no phase transition between them. Phase III seems to be rare, as "just one crystal has been detected in a preparation of phase I."[11] The two structure determinations of $Rb_2(TCNQ)_3$ at different temperatures describe the same phase, isostructural with $Cs_2(TCNQ)_3$.

Two room-temperature studies have been carried out for K(TCNQ), one by Richard *et al.*[8] and the other one by Konno *et al.*[9] They agree in the values for b and c axes and the angle β, but for the a-axis value Richard *et al.* find exactly half the value of Konno *et al.* This might mean that there are two polymorphs for K(TCNQ) at room temperature, one with regular stacks and one with stacks of diads. Yet it is the impression of this author that Richard *et al.* have overlooked weak layers of reflections that indicate a doubling of the a axis. The rather high R factor and σ_{bl} support this feeling. It is also favored by the observation of a phase transition at 395 K by Richard *et al.*—the same value as reported by Konno *et al.*

The stoichiometry implies that the TCNQ species carry a full negative charge in the simple salts, whereas in the complex salts two negative charges are distributed over three TCNQ moieties. Comparison of bond lengths suggests that the central TCNQ of a triad is neutral, and the two outer ones are $(TCNQ)^-$. (Flandrois and Chasseau[16] point out that the charge on a TCNQ molecule may be estimated from the bond lengths, especially from the difference of the lengths of the "formal simple" and "formal double" bonds.) The occurence of both neutral and uninegative TCNQ in the 2 : 3 salts is also evident from Raman data,[17] and the interpretation of electronic spectra[18] is facilitated by the assumption of a charge transfer from $(TCNQ)^-$ to $(TCNQ)^0$.

2.3. Phase Transitions

Phase transitions in all the simple TCNQ salts listed in Table 2 [except Rb(TCNQ), phase III] and in $Cs_2(TCNQ)_3$ have been revealed by different techniques: measurements of conductivities[3, 20, 21] and thermopower,[21] magnetic properties and electron-spin-resonance spectra,[22, 23] ^{14}N-nuclear quadrupole resonance,[24] vibrational spectra,[25, 26] and x-ray powder[22] and single-crystal studies.[15, 27, 28] The phase-transition temperatures determined by different methods are listed in Table 3, including enthalpies, ΔH, for the transitions if known. It should be remarked that Afifi *et al.*[20] are the only

TABLE 3. Phase-Transition Temperatures and Enthalpies in Alkali TCNQ Salts

Compound	T(K)	ΔH (J/mole)	Method	Reference
Li(TCNQ)	420		conductivity	20
Na(TCNQ)	347–373		conductivity	20
	360		ir spectra	25
	338		conductivity	3
	348	unobservable	magnetism	29
	348		x ray	28
	345		magnetism	22
K(TCNQ)	347–373		conductivity	20
	395		x ray	8, 9, 28
	403		esr	23
	379		ir spectra	26
	391		conductivity	3
	395	251	magnetism	29
	396	251	magnetism	22
Rb(TCNQ)	374		conductivity	3
(phase I)	381		x ray	28
	381	4222	magnetism	29
	376	4222	magnetism	22
Rb(TCNQ)	220		ir spectra	25
(phase II)	220	unobservable	magnetism	22
	231		conductivity	3
	223		magnetism	29
Cs(TCNQ)	210		ir spectra	25
	254		conductivity	3
	217	unobservable	magnetism	22, 29
$Cs(TCNQ)_2$	347–373		conductivity	20
$Cs_2(TCNQ)_3$	397		conductivity	21
$NH_4(TCNQ)$	301		x ray	15
(phase I)	305		ir spectra	25
	299	84	magnetism	22
$NH_4(TCNQ)$ (phase II)	215		conductivity	15

authors to report a phase transition in Li(TCNQ). They are also the only ones who describe a salt $Cs(TCNQ)_2$.

The phase transitions and the physical properties in which they manifest themselves have been extensively discussed by Kommandeur,[30] showing that one-electron theory can be applied in most cases. A more recent summary of physicochemical properties of alkali TCNQ salts, including phase transitions and structural, electrical, and optical properties was given by Shirotani and Sakai.[31]

Vegter and Kommandeur[22] conclude from the behavior of the magnetic susceptibility and x-ray powder patterns around the transition temperature that the phase transitions in K(TCNQ), NH_4(TCNQ), Rb(TCNQ)

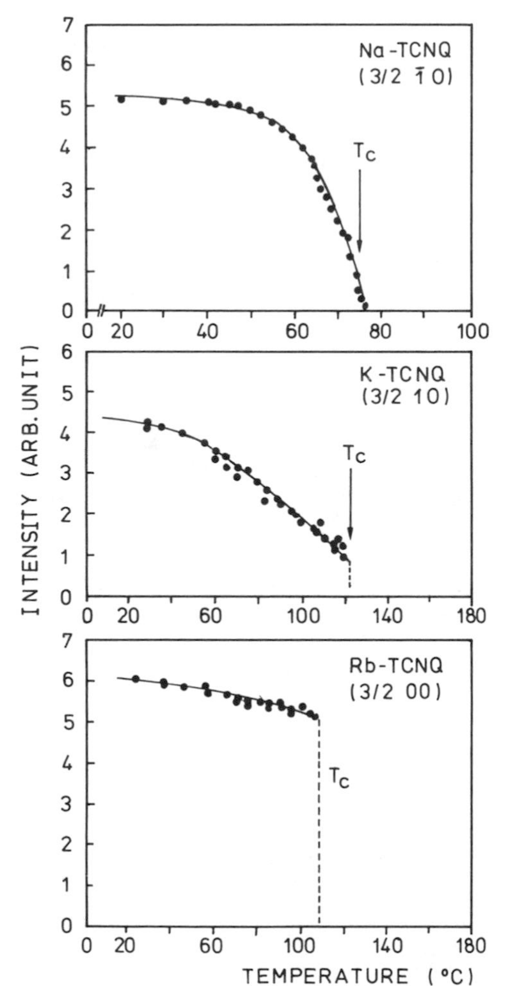

FIGURE 5. Temperature dependence of the intensity of a reflection of the low-temperature phase around the transition temperature. (Reprinted from Reference 28, with permission.)

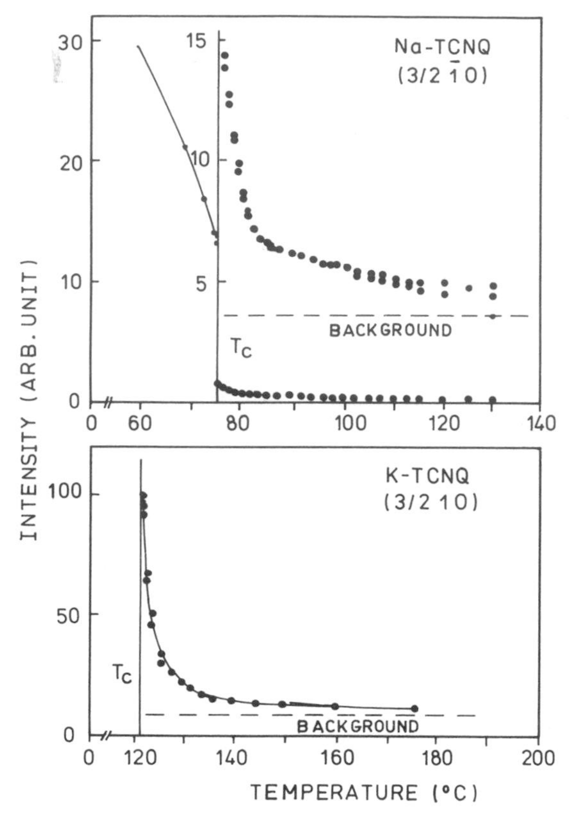

FIGURE 6. Temperature dependence of diffuse scattering above T_c, measured at the scattering vector of a low-temperature phase reflection. (Reprinted from Reference 28, with permission.)

(phase I), and Rb(TCNQ) (phase II) are of first order, whereas it is a continuous one in Na(TCNQ). This is in agreement with the single-crystal x-ray investigations of Terauchi,[28] who followed the intensities of reflections of the low-temperature phase when passing the transition temperature (Figure 5). The intensities diminish on approaching the transition temperature, and then disappear suddenly. For Na(TCNQ) the reduction in intensity is very pronounced and the abrupt drop very weak, so that the transition starts as a continuous one as reported by Vegter and Kommandeur,[22] but the final step has first-order character. This first-order character of the transition increases from Na(TCNQ) over K(TCNQ) to Rb(TCNQ), as evident from Figure 5. Terauchi[28] gives some theoretical background for the understanding of the transition. Observed hystereses were 0.2°, 0.8°, and 1.4° for Na(TCNQ), K(TCNQ), and Rb(TCNQ), respectively. Approaching the critical temperature from the high-temperature side, Na(TCNQ) and

K(TCNQ) develop a diffuse scattering, Figure 6, but not Rb(TCNQ), because the first-order phase transition is too strong.

From a structural point of view and speaking of the simple salts, the phase transitions manifest themselves in a distortion of the regular TCNQ stacks in the high-temperature phase to alternating stacks (composed of diads) in the low-temperature phase. This change is accompanied by a doupling of the unit-cell length in stack direction, observable in x-ray patterns. Tables 2 and 3 show that some of the salts are in the low-temperature phase at room temperature [Li^+, Na^+, K^+, Rb(TCNQ) (phase I), and

TABLE 4. Room-Temperature Conductivities σ and Activation Energies E_a of Alkali TCNQ Salts[a]

Compound	σ (ohm^{-1} cm^{-1})	E_a (eV)	Reference
Li(TCNQ)	5×10^{-6}		2
	5×10^{-5}	0.32	33
	1.8×10^{-5}	0.16	20
Na(TCNQ)	3.3×10^{-5}		2
	1×10^{-5}	0.33	33
	1×10^{-3}, 3.3×10^{-5}[b]	0.23–0.32	21
	1.1×10^{-6}	0.34	20
K(TCNQ)	2×10^{-4}		2
	1×10^{-4}	0.35	33
	2×10^{-4}, 5×10^{-6}[b]	0.15–0.45	21
	3×10^{-4}[d]		34
	1×10^{-4}	0.35	35
	5×10^{-7}	0.43	20
Rb(TCNQ) (phase I)	3.3×10^{-6}, 3.3×10^{-7}[b]	0.44–0.53	21
Rb(TCNQ) (phase II)	1×10^{-2}[b]	0.16–0.22	21
Cs(TCNQ)	3.3×10^{-5}	0.15	2
	3×10^{-4}	0.18	33
	1.7×10^{-3}[b]	0.16	21
NH_4(TCNQ)[c]	1.7×10^{-5}		2
Cs$(TCNQ)_2$	3.4×10^{-6}	0.25	20
$Cs_2(TCNQ)_3$	1.1×10^{-5}	0.36	2
	2×10^{-3}, 1.7×10^{-5}, 1.7×10^{-5}[b]		2
	1×10^{-5}		33
	1×10^{-3}, 4×10^{-5}, 4×10^{-5}[b]	0.30	33
	2×10^{-3}[b]	0.32–0.35	21
	1×10^{-3}	0.35	35
	1×10^{-3}		36

[a] Powder values are reported unless otherwise noted.
[b] Single crystal values along principal axes.
[c] It is usually not evident from the literature which polymorph has been investigated.
[d] Microwave measurement.

NH_4(TCNQ) (phase I)], some in the high-temperature phase [Rb(TCNQ) (phase II), Cs(TCNQ), and NH_4(TCNQ) (phase II)].

The structural phase transitions, driven by the electronic system, can be predicted theoretically using an elementary one-dimensional split-band model.[22] They are regarded as spin–Peierls transitions,[15, 32] and the low-temperature phase has been described as a nonuniform Heisenberg chain with a temperature-dependent order parameter.[32]

2.4. Electrical Conductivity and Other Transport Properties

All the alkali TCNQ salts behave as semiconductors above and below the phase transition, with different activation energies for each phase. Conductivities have been measured by many groups, so that the room-temperature data given in Table 4 are representative, but probably not complete. The effect of pressure on the conductivities will be discussed in Section 2.8. As an example of how the temperature dependence of the resistivity ρ looks like the curves measured by Afifi *et al.*[20] reproduced in Figure 7. To illustrate the change of the activation energies during the

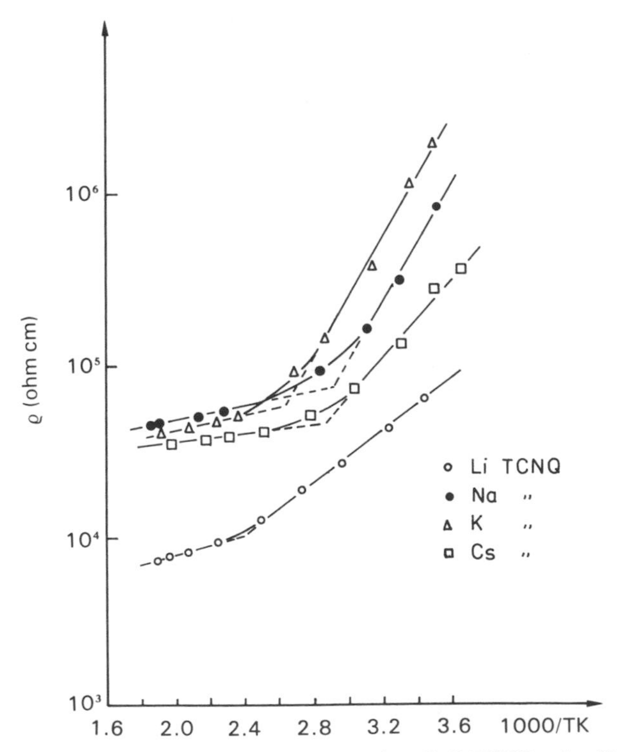

FIGURE 7. Electrical resistivity ρ against $1000/T$(K) for alkali TCNQ salts. (Reprinted from Reference 20, with permission.)

phase transition, the high-temperature values of Afifi *et al.*[20] are listed here: Li(TCNQ), 0.057 eV; Na(TCNQ), 0.041 eV; and K(TCNQ), 0.047 eV. It is interesting to note that Sakai *et al.*[21] find the inverse order, with higher activation energies above the phase transitions: Na(TCNQ), 0.6–1.1 eV; K(TCNQ), 0.4–0.6 eV; Rb(TCNQ) (phase II), 0.22–0.4 eV; $Cs_2(TCNQ)_3$, 0.78 eV. But in a subsequent paper[3] the same authors quote different values for the activation energies in the high-temperature phases: Na(TCNQ), 0.16–0.22 eV; K(TCNQ), 0.15–0.31 eV; Rb(TCNQ) (phase I), 0.28–0.37 eV. Sakai *et al.*[21] report an almost monotonous change of the resistivity of Cs(TCNQ) with the temperature. They also report that they found crystals with different habit in a batch of dark purple Rb(TCNQ) (phase II), needles and prisms, with different conductivities and activation energies: needles, 10^{-2} ohm^{-1} cm^{-1}, 0.16 eV; prisms, 10^{-4} ohm^{-1} cm^{-1}, 0.34 eV at room temperature.

A look at Table 4 shows some expected facts: the conductivity of the complex salt $Cs_2(TCNQ)_3$ is somewhat higher than that of the 1 : 1 salts, which are in the low-temperature phase at room temperature. This is probably due to a reduced Coulomb repulsion caused by the presence of neutral TCNQ molecules in the stacks. Yet the conductivity of $Cs_2(TCNQ)_3$ is lower than or approximately equal to that of Rb (TCNQ) (phase II) and Cs(TCNQ), which are in the high-temperature phase at room temperature. From this observation the impression may be derived that a single (large) exchange integral is more important for the conductivity than a reduced Coulomb repulsion. A comparison of activation energies leads to the same conclusions. The effect of the Coulomb repulsion on the conductivities has been discussed by LeBlanc.[37]

Blakemore *et al.*[38] described a study of electrical conductivity in connection with measurements of the thermoelectric power of $Cs_2(TCNQ)_3$ as a function of temperature, following an introduction in which they summarize previous results. Their observations about the sign of the thermoelectric power partially contradict the earlier results of Sakai *et al.*,[21] probably due to extrinsic effects. They note especially the sample dependence of the conductivity at lower temperatures. The transport properties of $Cs_2(TCNQ)_3$ are discussed on the basis of a deep level acceptor model. More data about the thermoelectric power of alkali TCNQ salts have been reported by Sakai *et al.*[21] and Siemons *et al.*[33] The Hall mobility of $Cs_2(TCNQ)_3$ is about 0.1 cm^2/V s in the highly conducting direction.[33] Measurements of Seebeck and Hall effects in compactions[39] indicate nearly intrinsic semiconductivity in narrow bands, with an average drift mobility of $\sim 10^{-2}$ cm^2/V s. Thermal conductivities have been determined as a function of temperature by Nurullaev *et al.*[40] and by Vlasova *et al.*[41] The activated behavior of the thermal conductivities has been related to the observation of triplet excitons (*vide infra*). Low-temperature specific heats[42,43] C_p (1.5–4.2 K) obey the Debye formula $C_p = \beta T^3$ with $\beta = 16.4$, 20.3, 35.7 kerg/mole K^3 for the simple Li^+, Rb^+, and Cs^+ salts, respectively. Debye temperatures for these three salts are 107, 100, 83 K.[43]

2.5. Magnetic Measurements

A summarizing discussion of magnetic and electron-spin-resonance properties has been given by Kommandeur,[30] including model calculations of the temperature dependence of the susceptibility. More recent susceptibility data [Li(TCNQ) between 2 and 300 K] have been measured by Siratori *et al.*[44] and discussed in connection with low-temperature specific heats. Vegter and Kommandeur[22] have published another detailed study about spin susceptibilities and phase transitions in simple alkali TCNQ salts. Surface effects in the magnetism of $Cs_2(TCNQ)_3$ (by pressing or grinding the specimens) have been analyzed by Holczer *et al.*[36]

Many authors report electron-spin-resonance (esr) spectra of alkali TCNQ salts[23,45–54] or give some theoretical remarks about them,[32,55–59] including the relaxation mechanism.[56,59] The influence of the solvent on esr spectra has been studied by Stösser and Siegmund[60]: The hyperfine structure (not resolved in the solid state) becomes more and more pronounced in going from acetone to hexamethylphosphorustriamide, dimethylsulfoxide, formamide, and water.

Most of the above cited papers deal with single compounds; e.g., Li(TCNQ),[49,53] K(TCNQ),[23,32,35,47,50,51,58] Rb(TCNQ),[51,52,54,55,57,59] $Cs_2(TCNQ)_3$,[35,45,46] and $NH_4(TCNQ)$.[56] The esr signal in $Cs_2(TCNQ)_3$ has been attributed by Chesnut *et al.*[45,46] to a thermally accessible triplet state lying 0.16 eV above the ground state. The exchange-narrowed line (values for the exchange frequency as a function of temperature are presented[46]) is interpreted as being caused by a triplet excitation that is delocalized over several TCNQ molecules and may be free to propagate as a triplet exciton. Kepler[35] reported singlet–triplet separation energies of 0.14 eV for $Cs_2(TCNQ)_3$ and 0.2 eV for K(TCNQ). Vlasova *et al.*[41] have measured simultaneously the temperature dependence of the esr intensity and thermal conductivity in Li, Na, and K(TCNQ). From the observation that the activation energies for both processes are identical, the authors conclude that they are related to the same excited states, and that the additional thermal conductivity is brought about by excited paramagnetic states capable of moving over the crystal. Observed activation energies are 0.05, 0.16, and 0.26 eV for Li, Na, K(TCNQ), respectively.

The observed dipolar splittings in K(TCNQ) and Rb(TCNQ) have been attributed to triplet excitons. In the case of Rb(TCNQ) an exciton model intermediate between the localized and the free-electron limit has been proposed.[52] Its application shows that a Heisenberg antiferromagnetic treatment is not valid in this case. Furthermore, it was concluded that the on-site electron interaction and the one-electron bandwidth are of the same order of magnitude for this salt.[55] The excitons are diffusive with an activation energy of 0.08 eV and an anisotropy of the exciton motion of the order of 10^5 for Rb(TCNQ) (phase I).[51] Possible exchange pathways for this triplet exciton have been discussed by Silverstein and Soos.[57] Spin-lattice relaxation times for Rb(TCNQ)] are the subject of a communication by Berthet *et al.*[59]

For K(TCNQ), too, it is suggested that the effective electron–electron repulsion and the one-electron bandwidth are comparable.[58] Two types of exciton have been discovered,[47, 50] corresponding to the two independent rows of $(TCNQ)^-$ ions in the crystal. From the small values of the zero-field splitting parameters D and E ($D = 164$ and 134 MHz, $E = 18.8$ and 14.0 MHz for the two stacks at 227 K), Hibma and Kommandeur[50] concluded that the exciton is large compared to a lattice parameter. In the same communication, the dipolar splitting is given as a function of the orientation of the crystal.

The relaxation mechanisms in NH_4(TCNQ) have been studied as well.[56] The experimental results (above the transition temperature) have been interpreted in terms of a model of Frenkel spin excitons localized on the TCNQ chains and with the dynamics of a one-dimensional triplet system.

Besides the dipolar split lines central, so-called "impurity lines" are usually observed in esr experiments. The impurity lines in Li(TCNQ) [and Cu(TCNQ) and $Cu(TCNQ)_2$] have been studied at liquid He temperature.[53] It is partially resolved at 1.5 K with g values of $g_{\parallel} = 2.008 \pm 0.001$ and $g_{\perp} = 2.0025 \pm 0.005$. It is concluded from the different behavior of the copper salts that the so-called impurity resonance is associated with the type of the compound at hand, and possibly a consequence of built-in $(TCNQ)^0$.

2.6. *Nuclear Magnetic Resonance*

It has already been mentioned that ^{14}N-nuclear-quadrapole resonance data are in agreement with the Peierls transition in K(TCNQ).[24] The data reveal eight nonequivalent N sites,[61] which fits to the structure determination by Konno *et al.*[9] Gillis *et al.*[62] have studied the nuclear magnetic relaxation in K(TCNQ) and have compared their results with qualitative agreement to the temperature dependence of the diffusion coefficient calculated theoretically in a one-dimensional Heisenberg alternating model. The dynamic nuclear polarization of the protons in alkali salts of TCNQ has been investigated by Alizon *et al.*[63] Similar measurements for NH_4(TCNQ) (deuterated samples) give hints at the dynamics of the excitons[64]: The excitons are localized on the TCNQ chains, and their dynamics reflect the one-dimensional character of the system.

2.7. Vibrational, Electronic, and Photoelectron Spectra

We will summarize investigations at ambient pressure in this section, delaying high-pressure studies to the following one.

Infrared (ir), Raman, and resonance Raman spectra of alkali TCNQ salts have been recorded by several groups,[17, 25, 26, 65–72] including studies of the monomer–dimer equilibrium of $(TCNQ)^-$ in solution.[66, 68] In an aqueous solution of Li(TCNQ) practically only dimers are found at concentrations as low as 0.1 *M*, whereas in dimethylsulfoxide solution the presence of monomers (together with dimers) is observed even at high concentrations.[68]

Some results of solid-state ir and Raman spectra have already been mentioned: $Cs_2(TCNQ)_3$ contains[17] both $(TCNQ)^-$ and $(TCNQ)^0$, and the phase transitions manifest themselves in the vibrational spectra.[25, 26] The latter effect is partly due to the fact that $(TCNQ)^-$ which are crystallographically independent in the low-temperature phase become related by symmetry in the high-temperature phase. This can be followed by the disappearance of a set of $C\equiv N$ stretching absorptions.[26] Another observation of phase transitions by means of ir spectra[25] is related to the vibronic activation of totally symmetric modes of TCNQ. Starting with the high-temperature phase and lowering the temperature, a striking intensity enhancement of such modes is observed, followed by an abrupt variation at nearly the same transition temperatures determined by other techniques. Thus Bozio and Pecile[25] suggest that the vibronic intensity enhancement follows the distortion of the regular TCNQ anion stack of the high-temperature phase to the alternating stack of the low-temperature one. These observations may signify that the underlaying electron–molecular vibration interactions play a relevant role in the stabilization of the low-temperature distorted phases.[25] The temperature dependence of the ir spectra of $Cs_2(TCNQ)_3$ has been studied by Kondow and Sakata.[65] The importance of the vibronic effect (the enhancement of vibrational modes by electron–molecular vibration interaction) for the interpretation of the vibrational spectra has been stressed by several authors.[67, 68, 71] Two papers may be quoted[69, 70] which deal with pre-resonance Raman and resonance Raman studies. Raman scattering investigations allowed the determination of lattice dynamics.[72]

Tanner *et al.*[71] discussed polarized reflectance studies from 300 to 23 000 cm^{-1} in the light of earlier papers. They found evidence for two superimposed electronic transitions along the chain axis for photon energies in the range 0.7–1.5 eV and conclude that 90% of the direct electronic transitions occur for photon energies above 0.8 eV. This implies that the optical gap is much larger than the triplet excitation energy and that Coulomb interactions are strong. These statements disagree with the view of Kommandeur and co-workers[50, 73] and of Torrance *et al.*[74]

Electronic spectra of the simple and complex alkali TCNQ salts have been a major field of interest in the last fifteen years.[18, 71, 73–85] The paper by Boyd and Phillips[75] studies the dimerization of $(TCNQ)^-$ in solutions of Li(TCNQ) by following the charge transfer (CT) band at 15 550 cm^{-1} (1.93 eV; 643 nm).

The solid-state spectra show three prominent peaks, as listed in Table 5. The low-frequency one is regarded as an intermolecular CT absorption, the high-frequency α and β peaks as local intramolecular excitations. An additional band (11 900 cm^{-1}, 1.48 eV, 840 nm) in crystalline Li(TCNQ) has been attributed to $(TCNQ)^-$ monomers at defect sites.[76] Another listing of transition energies and oscillator strengths can be found in a paper by Oohashi and Sakata.[79] Inspection of Table 5 shows that the energy of the CT band depends on the cation (via polarization of the TCNQ anions), whereas the local α and β bands are practically independent of the cation.[76] Nevertheless, there is a difference of the local excitation bands in solution and in the solid state.[78] The temperature dependence of the electronic absorption spectra of K(TCNQ) [together with $Ba(TCNQ)_2$ and $Ca(TCNQ)_2$] has been investigated by Michaud *et al.*[85] They found that the locally excited bands sharpen considerably at low temperature, revealing vibronic structure. The CT band remains broad and temperature independent. The effect of the particle size on the spectra has been studied by Papavassiliou and Spanov.[82, 83] They report that the low-frequency band is shifted to lower frequencies, but the high-frequency bands remain unshifted as the size (or the dielectric constant of the surrounding medium) increases.

TABLE 5. Electronic Absorption Peaks[a] of Alkali TCNQ Salts

Compound	CT Band	α Band	β Band	Reference
Li(TCNQ)	8.2	16.2	27.5	76
Na(TCNQ)	9.3	16.3	27.5	76
K(TCNQ)	8.5	16.4	27.8	18, 76
	7.8			78
	8.5, 11.9		25.3	84
$NH_4(TCNQ)$	8.9	16.3	27.8	76
	9.5	16.0	26.9	80
$Cs_2(TCNQ)_3$	5.5, 11.1	16.8	25.5	80
	4–6, 10.1	16.2	27.5	18

[a] In kK = 10^3 cm^{-1}.

An analysis of the spectra[73,81] has usually been tried on the basis of a Hubbard model[84,86] [or a modified one for $Cs_2(TCNQ)_3$.[87]] The intermediate-to-strong coupling limit[34] or the strong-coupling approximation have been applied.[88,89] Gasser and Hoefling[89] have calculated the dielectric function and give expressions for the plasma frequencies, the plasmon stiffness, the critical wave number, and for the real and the imaginary parts of the dielectric function. They compare their results with experimental findings on K(TCNQ).

Some spectroscopic and photoemission studies have been done on thin films. Films may be prepared by vapor reaction techniques[90,91] or by sublimation.[92–94] Optical absorption spectra of films of $NH_4(TCNQ)$ at 173 K[93] have been interpreted by the presence of isolated TCNQ species at high vacuum, replaced by dimers if the pressure increases. This happens also when the sample is warmed up to room temperature. A temperature-dependent monomer–dimer equilibrium has been observed in the electronic spectra of Li(TCNQ) films too.[95] Ultraviolet and x-ray photoelectron spectroscopy has been applied to film and bulk samples,[94,96–98] mostly to compare the spectra of the alkali salts with those of compounds with organic cations. The binding energy of the electrons in a series of metal TCNQ salts are listed in Reference 98. The peaks in the electron energy distribution curves are broad,[94] indicating strong coupling of the electrons to molecular modes. The energy distribution is dominated by the electronic structure of $(TCNQ)^-$ and not by delocalized energy bands.

The communication of Khatkale and Devlin[91] deserves attention, for the authors claim to have prepared and investigated by vibrational spectroscopy films not only of Na(TCNQ), but also of $Na_2(TCNQ)$ and $Na_3(TCNQ)$. These films should contain $(TCNQ)^-$, $(TCNQ)^{2-}$, and $(TCNQ)^{3-}$, respectively.

2.8. Pressure Effects

Young's modulus of[99] $Cs_2(TCNQ)_3$ is 1.4×10^{13} dynes/cm^2. The effect of pressure on the conductivities and the phase transitions has been studied by a number of authors.[3,100–103] There is a general agreement that the resistivity decreases with increasing pressure. A minimum is reached at pressures of about 150–200 kbars, where the resistivity is four to five orders of magnitude lower than at ambient pressure. But at higher pressures the resistivity rises again,[101] until constant values are reached at pressures around 400 kbars. As this increase in resistivity is not reversible, it is supposed to be due to a chemical reaction occuring under pressure. These observations are illustrated in Figure 8. The minimum resistivities and the pressures at which they are observed are listed in reference 101. The phase transitions are effected by pressure as well[3]: The transition temperature of

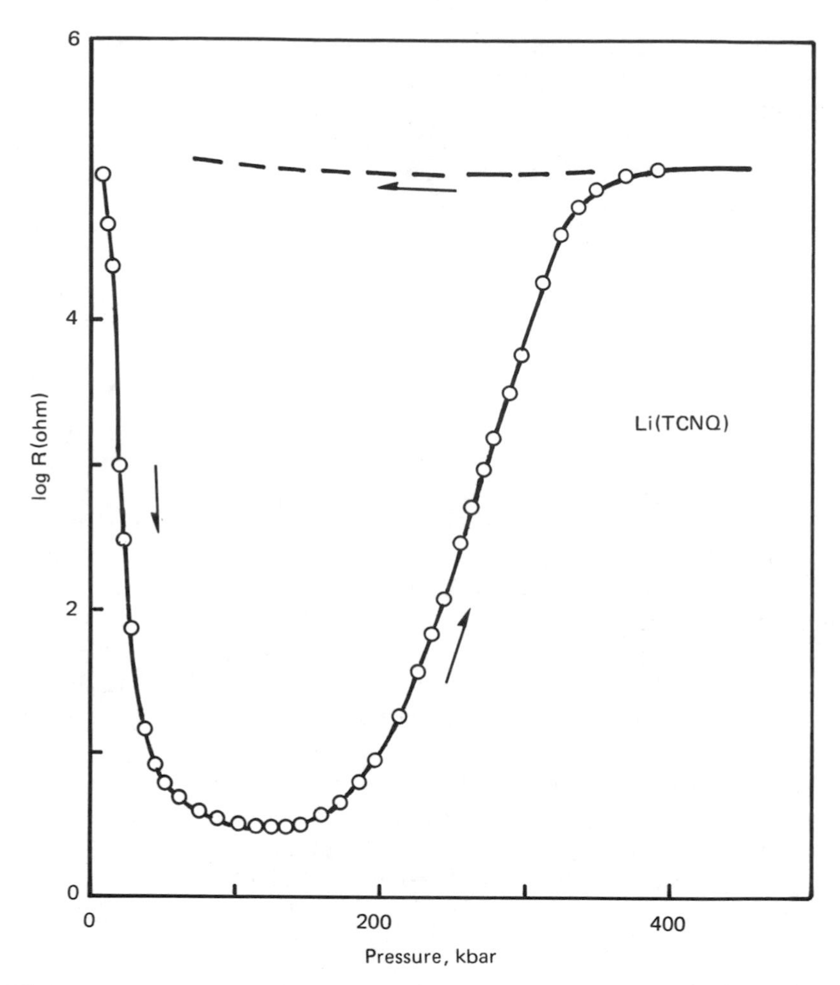

FIGURE 8. (a)–(c) Pressure dependence of the resistivity of several alkali TCNQ salts. (Reprinted from Reference 101, with permission.)

Rb(TCNQ) (phase I) increases with increasing pressure, and new pressure-induced phase transitions, even for Li(TCNQ), are observed.

Some papers deal with the influence of pressure on the electronic spectra.[102, 104, 105] There is a blue shift in the local excitation bands, but no evidence is reported for the behavior of the CT band. For Li(TCNQ) [K(TCNQ)] the band at 26 500 cm^{-1} (3.29 eV, 377 nm) [27 500 cm^{-1} (3.41 eV, 364 nm)] shifts by +86 cm^{-1}/kbar [+58 cm^{-1}/kbar], the band at 16 000 cm^{-1} (1.97 eV, 625 nm) [16 100 cm^{-1} (1.98 eV, 625 nm)] shifts by +51 cm^{-1}/kbar [+45 cm^{-1}/kbar].[104] It was suggested[104] that this effect is mainly caused by the crystal field of the positive ions.

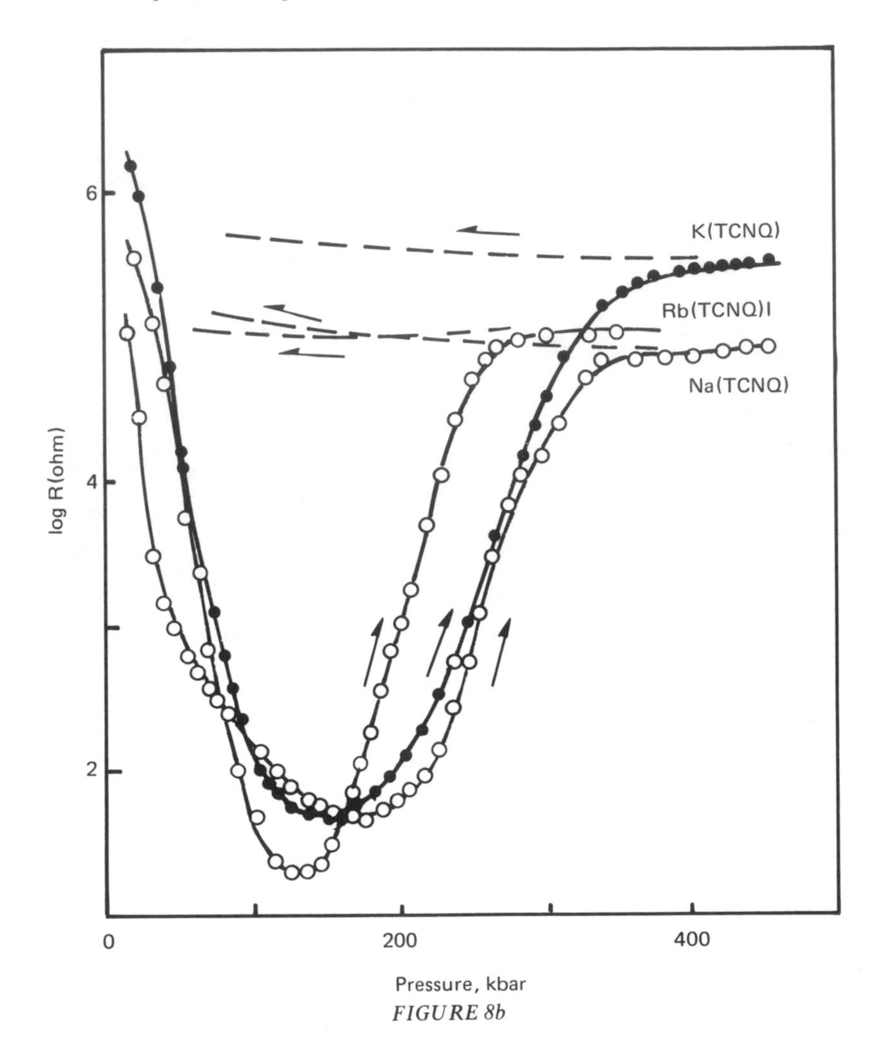

FIGURE 8b

2.9. *Miscellaneous*

Metzger has calculated Madelung energies of some alkali TCNQ salts.[19] Thermal and oxidative thermal stability of alkali TCNQ salts has been studied using differential thermal and thermogravimetric analysis.[106] The results suggest that increasing the ionization potential of the donor increases the thermal stability of the salt.

Sharp[107] measured the electrode potential of an ion-selective K(TCNQ) electrode. Nogami *et al.*[108] reported synthesis and electrical properties of some cation TCNQ crown ether complexes. The conductivities of simple salts (crown ether)$_m$[M(TCNQ)] are low ($\sigma \approx 10^{-9}$–10^{-11}

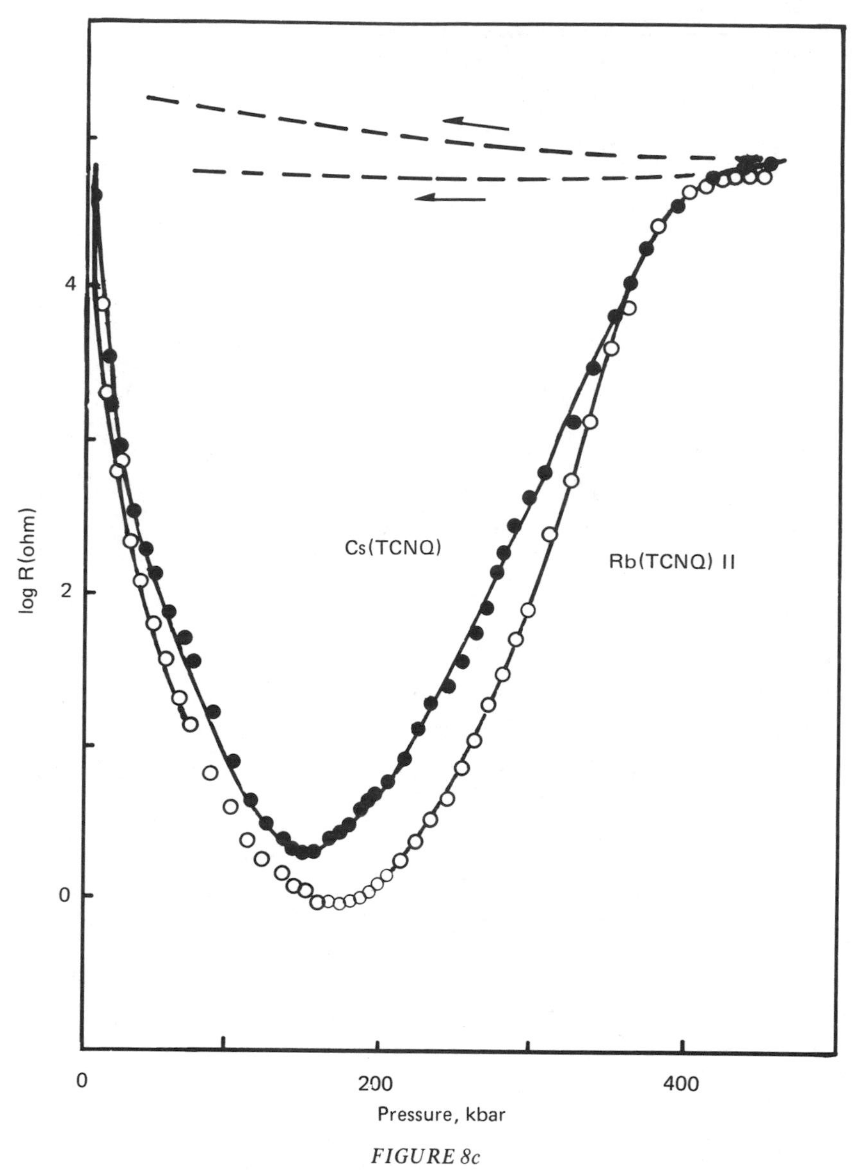

FIGURE 8c

ohm^{-1} cm^{-1}). Complex salts (crown ether)$_m$[M(TCNQ)]TCNQ exhibit higher conductivities of the order 10^{-4} ohm^{-1} cm^{-1}.

3. 7,7,8,8-Tetracyano-p-quinodimethane Salts of Other Simple Metal Ions

In addition to the alkali and ammonium salts, TCNQ forms salts with other simple metal ions. Melby *et al.*[2] report preparation and conductivities of $Ba(TCNQ)_2$, Cu(TCNQ), $Cu(TCNQ)_2$, Ag(TCNQ), $Pb(TCNQ)_2 \cdot 1.5H_2O$, $M(TCNQ)_3 \cdot 6H_2O$ (where M=Ce, Sm), and $M(TCNQ)_2 \cdot 3H_2O$ (where M=Mn, Fe, Co, Ni). With the exception of the copper salts, which have conductivities of order of 5×10^{-3} ohm^{-1} cm^{-1} and activation energies of 0.13 and 0.12 eV, the conductivities are of the order 10^{-4}–10^{-5} ohm^{-1} cm^{-1}. Melby *et al.*[2] conclude from the magnetic behavior of $Cu(TCNQ)_2$ that this salt must *not* be formulated as $Cu^{2+} \cdot 2(TCNQ)^-$, but that it is a complex salt of Cu(I). This view is supported by magnetic measurements of Sano *et al.*[109] who have also measured $Ba(TCNQ)_2$, $Ni(TCNQ)_2 \cdot 3H_2O$, $Fe(TCNQ)_2 \cdot 3H_2O$, and salts of *tris*-bipyridyl complexes of Fe(II), Ni(II), and Cu(II). It has been followed from x-ray photoelectron spectra[110] that the Cu species in Cu(TCNQ) and $Cu(TCNQ)_2$ are identical and that the latter compound contains two different TCNQ moieties.

Afifi *et al.*[20] have measured the conductivities of $M(TCNQ)_2$ salts (M=Mn, Fe, Cu, Ni, Co, Zn), which decrease in the given order from 3×10^{-6} ohm^{-1} cm^{-1} for the Mn salt to 1.4×10^{-8} ohm^{-1} cm^{-1} for the Zn^{2+} salt. Anomalies in the slopes of the resistivity versus temperature curves (Figure 9) indicate phase transitions in all of them, which occur in

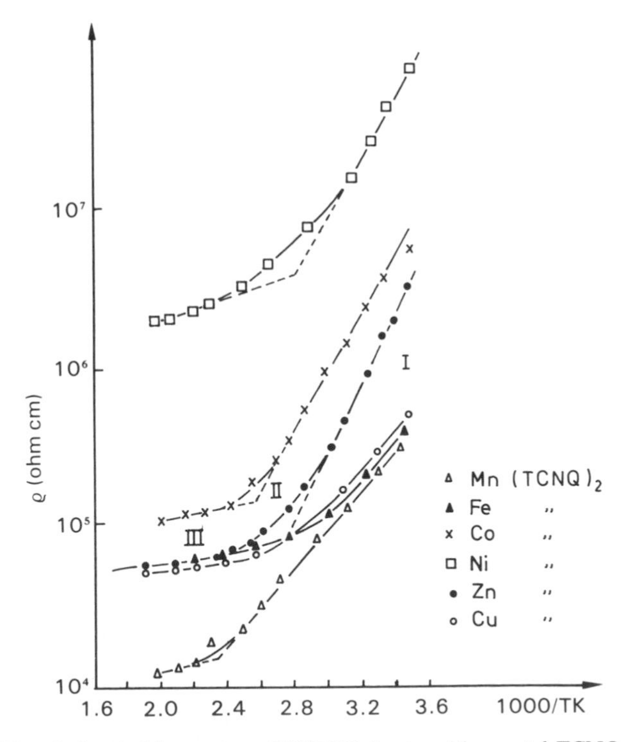

FIGURE 9. Electrical resistivity against 1000/T(K) for transition-metal TCNQ salts. (Reproduced from Reference 20, with permission.)

the range 340–370 K. Another anomaly in the magnetic and conductivity properties of $Mn(TCNQ)_2$ and its hydrates has been reported[111] to occur around 200 K. An abrupt change in the esr and magnetic behavior at 20 K suggests a change in the Mn–Mn interaction from antiferromagnetic to ferromagnetic. Other esr studies[112] on protonated and deuterated $Mn(TCNQ)_2$ from 1.5 to 375 K show that the esr resonance consists of a broad signal attributed to exchange-coupled Mn^{2+} ions, superimposed by a narrow impurity line. The effective magnetic moment of $4.66\mu_B$ (μ_B is the Bohr magneton) implies an antiferromagnetic coupling of the manganese ions.[112] A study of the low-temperature behavior of the so-called impurity resonance in the esr spectra of Cu(TCNQ) and $Cu(TCNQ)_2$ is included in the paper by Schwerdtfeger.[53] The resonance of Cu(TCNQ) is asymmetric, but not resolved [in contrary to Li(TCNQ)]. The line in $Cu(TCNQ)_2$ is not resolved as well, but seems to consist of five resonances.

The complex $Fe(TCNQ)_2 \cdot 3H_2O$ has been subjected to Mössbauer experiments.[113] They indicate that iron is in a 2+ state, and that a phonon shift (or a phase transition) occurs below 50 K.

Siedle *et al.*[114] have investigated magnetic susceptibilities and performed ir and x-ray photoelectron measurements on a series of transition-

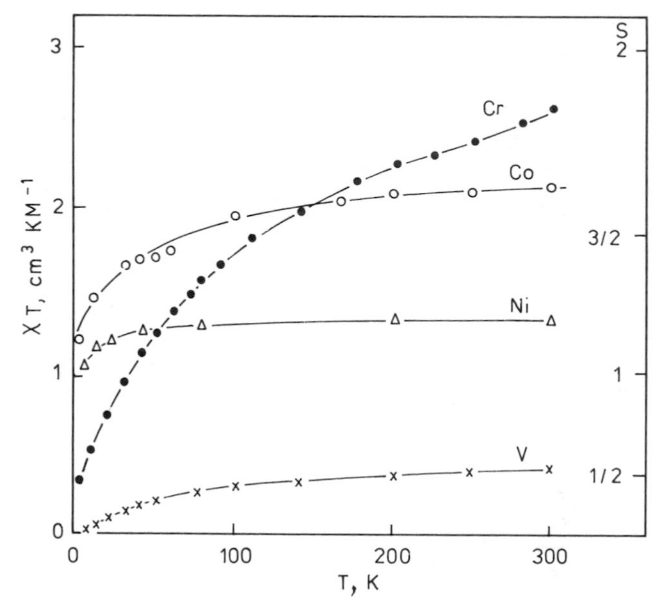

FIGURE 10. Magnetic susceptibility of $V(TCNQ)_2$, $Cr(TCNQ)_2(CH_3CN)_2$, $Co(TCNQ)(CH_3CN)_2$, and $Ni(TCNQ)_2$ as a function of temperature. The curves are denoted by the symbol of the metal. (Reproduced from Reference 114, with permission.)

TABLE 6. Magnetic, Conductivity, and Electronic Absorption Data of Transition-Metal TCNQ Salts[a]

Compound	μ_{eff} at 298 K, B.M.	$\sigma(\text{ohm}^{-1}\ \text{cm}^{-1}) \times 10^{-4}$	λ_{max} (cm^{-1})
2[b]	1.73	2.6	28 990; 23 260; 12 500(sh); 9175
3[c]	4.08	2.2	34 480; 31 750; 24 100; 11 900
4[d]	0.03	2.3	30 800; 24 400
5[e]	0.00	2.6	
6[f]	3.56	13.0	32 780; 26 320; 12 100
7[g]	3.23	1.7	30 300; 22 220; 13 300; 11 900

[a] Reproduced with permission from Reference 114.
[b] $V(TCNQ)_2$.
[c] $Cr(TCNQ)_2(CH_3CN_2)$.
[d] $Mo(CO)_2(TCNQ)(CH_3CN)_2$.
[e] $W(TCNQ)_2(CH_3CN)_2$.
[f] $Co(TCNQ)(CH_3CN)_2$.
[g] $Ni(TCNQ)_2$.

metal salts of $(TCNQ)^{2-}$. The mostly air-sensitive compounds include $V(TCNQ)_2$, $Cr(TCNQ)_2(H_3CCN)_2$, $Mo(TCNQ)(CO)_2(H_3CCN)_2$, $W(TCNQ)_2H_3CCN$, $Co(TCNQ)(H_3CCN)_2$, and $Ni(TCNQ)_2$. Some magnetic susceptibility data are reproduced in Figure 10. Magnetic, conductivity, and electronic absorption data are given in Table 6. The $(TCNQ)^{2-}$ species is identified by strong infrared absorption bands, one of them always around 2200 cm^{-1}, and one or two more around 2100 cm^{-1} or at about 2060 cm^{-1} and 2135 cm^{-1}. More detailed information is given in Table 7.

The electronic binding energies of $M(TCNQ)_2$, with M = Fe, Co, Ni, or Cu, have been determined by Ikemoto *et al.*[98] applying x-ray photoelectron spectroscopy. Their results are listed in Table 8. The temperature dependence of the electronic absorption spectra of $Ca(TCNQ)_2$ and $Ba(TCNQ)_2$ corresponds to that observed for K(TCNQ).[85] The $Ba(TCNQ)_2$ compound yields a monomer spectrum. A listing of transition energies and oscillator strengths of $M(TCNQ)_2$, with M = Ba, Mn, Fe, Co, or Ni, is included in the paper by Oohashi *et al.*[79] and are reproduced in Table 9. Experimental band parameters for TCNQ salts of copper and calcium have been determined by Vegter and Kommandeur.[22] The determination of the lattice dynamics of $Cu(TCNQ)_2$ by Raman scattering is reported in a paper quoted earlier.[72]

The temperature dependence of the thermopower of Cu(TCNQ), $Ba(TCNQ)_2$, and $Fe(TCNQ)_2 \cdot 3H_2O$ has been measured by Siemons *et al.*[33] They find room-temperature conductivities (activation energies) of 2×10^{-5} (0.24), 10^{-5} (0.16), 2×10^{-8} (0.45), 10^{-2} (0.16), and 2×10^{-5}

TABLE 7. Infrared Absorption Data of Transition-Metal TCNQ Compounds[a]

Compound	Absorption maxima (cm^{-1})
2[b]	3120(w), 2190(s), 2140(s), 2100(s), 1600(w), 1500(s), 1320(m), 1285(w), 1240(m), 1130(w), 1075(w), 1030(m), 845(s), 835(s), 560(m), 505(m)
3	3115(w), 2205(s), 2100(s), 1600(w), 1505(m), 1320(m), 1260(w), 1180(m), 1050(w), 1005(w), 835(s), 820(s), 745(m), 760(m), 565(m), 485(m), 400(m)
4	3120(w), 2190(s), 2090(s), 1590(m), 1510(sh), 1505(m), 1310(m), 1250(w), 1135(m), 1040(w), 1000(w), 825(m), 720(w),
5	3115(w), 2190(s), 2080(s), 1590(m), 1510(sh), 1500(m), 1335(w), 1185(w), 1005(w), 970(w), 920(w), 835(m), 815(m), 720(w), 560(w), 470(w)
6	3110(w), 2200(s), 2135(s), 2065(s), 1600(w), 1575(m), 1500(m), 1380(w), 1245(m), 1285(m), 1160(m), 985(m), 820(m), 810(sh), 480(w)
7	3120(w), 2205(s), 2135(s), 2060(s), 1600(m), 1505(m), 1305(w), 1255(w), 1180(w), 825(m), 815(m), 485(w)

[a] Reproduced with permission from Reference 114.
[b] See Table 6 for names of numbered compounds.

ohm^{-1} cm^{-1} (0.37 eV) for $Fe(TCNQ)_2 \cdot 3H_2O$, $Mn(TCNQ)_2 \cdot 3H_2O$, $Ba(TCNQ)_2$, Cu(TCNQ), and Ag(TCNQ), respectively.

The electrode potential of TCNQ salts of Cd, Pb, and Cu have been determined by Sharp.[107] The paper by Nogami *et al.*[108] investigating crown ether complexes of TCNQ salts contains the Ba salt. An unusual preparative route to Pb(TCNQ) has been found by Dick *et al.*,[115] who prepared the compound from hexamethyldilead and TCNQ.

X-ray structure analyses of salts of TCNQ with simple metal ions other than alkali metals seem not to have been reported. Yet there is an evaluation of the crystal structure of Ag(TCNQ) by lattice imaging using high-resolution electron microscopy.[116] The structure resembles that of the simple alkali TCNQ salts. The tetragonal unit cell ($a = 12.5$, $c = 7.0$ Å) makes the occurrence of TCNQ diads likely.

TABLE 8. Electronic Binding Energies in Some TCNQ Compounds[a]

Compound	N_{1s}	Approximate splitting[b]	O_{1s}	M_{2s}	$M_{2p(1/2, 3/2)}$	M_{3s}	M_{3p}	$M_{3d(3/2, 5/2)}$	$M_{4d(3/2, 5/2)}$	P_{2s}	P_{2p}
1. TCNQ(7th sublim.)	398.4	2.4									
TCNQ(recrys. from soln.)	398.3	2.5									
2. TCNQ-anthracene	398.2	2.6									
3. TCNQ-pyrene	398.1	2.4									
4. TCNQ-benzidene (no solvent molecule)	(398.4)										
5. TCNQ-benzidine (including CH_2Cl_2)	(397.9)										
6. Li(TCNQ)	397.8										
7. Na(TCNQ)	397.9			62.5							
8. K(TCNQ)	397.6				295.0, 292.3	32.4[c]	16.4				
9. Cs(TCNQ)	397.2							736.9, 722.9	77.0, 74.6		
10. (TMPD[d])(TCNQ)	(397.6, 398.9)										
11. (Ph_3PMe[e])(TCNQ)	(397.4)									190.0	132.4
12. $Cs_2(TCNQ)_3$	(397.3)							737.6, 722.8	76.0, 74.5		
13. $(TMPD)(TCNQ)_2$	(397.4, 398.4)										
14. $(Ph_3PMe)(TCNQ)_2$	(397.8)									189.6	132.5
15. Cu(TCNQ)	398.1				951.0, 931.2	121.8	74.8	2.2			
16. $Cu(TCNQ)_2$	(398.3)				950.4, 930.4	121.5	74.5	2.1			
17. $Fe(TCNQ)_2 \cdot 3H_2O$	398.0		531.3		723.0, 709.6		55.6				
18. $Co(TCNQ)_2 \cdot 3H_2O$	397.9		531.8		792.2, 781.2		61.4				
19. $Ni(TCNQ)_2 \cdot 3H_2O$	398.0		532.2		874.1, 856.5		68.7				

[a] Reproduced with permission from Reference 98.
[b] It is rather difficult to measure precisely the splitting for complexes 4 to 19, where the satellite peaks are very broad; no figure is given for the splitting.
[c] The satellite is observed at 14.7 eV away from K_{3s} peak.
[d] TMPD, tetramethyl-*p*-phenylenediamine.
[e] Ph_3PMe, triphenylmethylphosphonium.

TABLE 9. *Transition Energies (σ_i, cm^{-1}) and Oscillator Strength of Some Simple TCNQ Salts*[a]

Compound	σ_{CT}	f_{CT}	σ_{YI}	f_{YI}	σ_{YII}	f_{YII}
Li	9.8×10^3	0.26	16.2×10^3	0.25	26.6×10^3	0.56
NH_4	9.5	0.24	16.0	0.25	26.9	0.54
Rb-I	9.8	0.31	15.7	0.23	27.1	0.56
Na	11.0	0.25	16.5	0.41	27.7	1.0
Ph_3PCH_3	10.5	0.24	15.3	0.39	26.4	0.88
K	9.5	0.36	16.0	0.27	27.7	0.62
Ni	9.0	0.44	15.7	0.49	25.3^s, 26.7	1.0
Fe	8.6, 10.5^s	0.58	15.4	0.63	25.5, 26.9	1.3
Co	8.8, 10.5^s	0.52	15.4	0.78	25.0^s, 26.6	1.5
Mn	9.0^b	0.52	15.3	0.69	25.0^s, 25.8	1.7
Rb-II	7.4	0.39	16.7	0.33	28.1	1.5
Cs	7.2, 9.3^s	0.22	14.0^s, 16.0	0.30	27.5	0.86
Ba	10.0	0.17	13.8, 15.4	1.0	26.2	2.2
Mor	10.5	~0.1	13.8, 15.1, 16.1	0.89	27.0	2.1

[a] Reproduced with permission from Reference 79.
s, shoulder; b, broad

4. *7,7,8,8-Tetracyano-p-quinodimethane Salts of Metal Complexes and Organometallic Compounds*

4.1 Salts Containing Main Groups Metals

Besides a report[117] on $R_2Tl^+(TCNQ)^-$ (R=phenyl, *p*-tolyl, and *p*-ClC_6H_4), most publications discuss salts containing the group-IVa metals Sn and Pb. With Sn, the following compounds have been prepared[118–121]: $R_3Sn(TCNQ)$, with $R=CH_3$; *n*-C_3H_7; *n*-C_4H_9 (for $R=CH_3$ a brown and a blue form has been obtained[121]); $CH_3(C_5H_5)_2Sb(TCNQ)$; $[(C_6H_5)_3Sn]_2(TCNQ)$; $(CH_3)_2Sn(TCNQ)_2$; $(C_6H_5)_3SnSn(C_6H_5)_3(TCNQ)$; $(C_5H_5)_2Sn(TCNQ)$; and a series of TCNQ salts of bis (β-keto-enolato)tin listed in Reference 120. Mössbauer spectra of the last group of compounds indicate the presence of both Sn(II) and Sn(IV), and it has been proposed that chains are formed in which the TCNQ is coordinated via the CN groups to the axial positions of the Sn complexes.

Infrared spectra of the pale green and diamagnetic complex with $Pb(\eta\text{-}C_5H_5)_2$ indicate[122] the presence of $(TCNQ)^{2-}$.

4.2 TCNQ Salts or Adducts of Transition-Metal Complexes

A series of TCNQ salts of transition-metal complexes have been prepared by Melby *et al.*[2] The observed conductivities are given in parentheses: $[Cu(NH_3)_2](TCNQ)_2$, (1.4×10^{-3}); $[Cu(en)_2](TCNQ)_2$, (2.5×10^{-5}) (en $= H_2NCH_2CH_2NH_2$); [Cu(2,2′-dipyridylamine)]$(TCNQ)_2$, (2.5×10^{-2}); [M(1,10-phenanthroline)$_3$]$(TCNQ)_2 \cdot 6H_2O$, M = Fe (10^{-6}), Ni (10^{-8}); [Ni(1,10-phenanthroline)$_2$]$(TCNQ)_2 \cdot 6H_2O$, (5×10^{-9}); $[Cr_2(\text{acetate})_4](OH)$ $(TCNQ) \cdot 6H_2O$, $(10^{-9}$ ohm^{-1} cm$^{-1})$. The most striking compound of this series appears to be the [Cu(2,2′-dipyridylamine)] salt with a conductivity of 2.5×10^{-2} ohm^{-1} cm^{-1} and an activation energy of 0.06 eV. Table 10 summarizes the properties of the metal containing TCNQ salts synthesized by Melby *et al.*[2]

For the rest of this section we will now follow the periodic system by groups, starting with group V and ending with Cu and Au. Figure 11 shows the crystal structure of the TCNQ salt of the cluster [di-μ-chloro-(hexamethylbenzene)niobium]trimer, $[Nb_3Cl_6(C_6Me_6)_3](TCNQ)_2$.[123] In this compound, which has a single-crystal conductivity of about 10^{-3} ohm^{-1} cm^{-1} (and a powder conductivity a factor of 15 higher), $(TCNQ)^-$ dimers link the unusual paramagnetic Nb clusters to infinite zigzag chains. The $(TCNQ)^-$ within a dimer show ring–external bond overlap and are 3.10 Å apart. The room-temperature esr spectrum of a polycrystalline sample consists of a relatively broad absorption with a g factor of 1.996 and a peak-to-peak derivative linewidth of 36.95 Oe. The g value indicates that the odd electron is localized on the niobium cluster.

A good deal of effort has been devoted to TCNQ salts of bis(arene)-chromium cations. Simple salts cation$^+(TCNQ)^-$ and complex salts cation$^+(TCNQ)_2^-$ have been obtained.[124] The crystal structures of the 1 : 1 and the 1 : 2 salt of bis(toluene)chromium have both been determined.[125, 126] Both compounds consist of segregated stacks of the cations and the $(TCNQ)^-$ anions. The TCNQ stacks (with a full negative charge in the 1 : 1 and half a negative charge in the 1 : salt) are regular. In the 1 : 1 salt the interplanar distance in the TCNQ stack is 3.42 Å, the overlap pattern is of the ring–ring type. A projection of the structure along the a axis is shown in Figure 12. The interplanar distance in the TCNQ stack in the 1 : 2 salt is 3.29 Å, the overlap pattern is ring–external bond. Figure 13 shows a projection of the structure along the a axis. As expected, the conductivity of the complex salt (2 ohm^{-1} cm^{-1}, $E_a = 0.08$ eV) is much

TABLE 10. *TCNQ Salts Containing Metal and Complexed Metal Cations*[a]

Product[b]	Cation source	Yield, %	Decompn. range, °C	Resistivity, ohm cm[c]	Remarks
$Li^+TCNQ^{\dot{-}}$ [d]	LiI	98	>300	2×10^5	Purple microcrystals
$Na^+TCNQ^{\dot{-}}$ [d]	NaI	59	>300	3×10^4	Purple microcrystals
$K^+TCNQ^{\dot{-}}$ [d]	KI	70	>300	5×10^3	Red needles
$Cs^+TCNQ^{\dot{-}}$	CsCl	67		3×10^4 [e]	Purple crystals
$(Cs^+)_2(TCNQ^{\dot{-}})_2(TCNQ)$ [d]	CsI	58–82		9×10^4 [f]	Purple prisms
$Ba^{++}(TCNQ^{\dot{-}})_2$ [d]	BaI_2	70		8×10^5	Purple crystals
$Mn^{++}(TCNQ^{\dot{-}})_2 \cdot 3H_2O$	$MnCl_2$	98	~170	9×10^4	Blue powder
$Fe^{++}(TCNQ^{\dot{-}})_2 \cdot 3H_2O$	$FeSO_4$	95	~170	5×10^4	Blue powder
$Co^{++}(TCNQ^{\dot{-}})_2 \cdot 3H_2O$	$CoSO_4$	91	~170	9×10^4	Blue powder
$Ni^{++}(TCNQ^{\dot{-}})_2 \cdot 3H_2O$	$NiCl_2$	95	~170	9×10^4	Blue powder
$Cu^+TCNQ^{\dot{-}}$	CuI	54		2×10^2 [g]	Blue-black needles
$Cu^{++}(TCNQ^{\dot{-}})_2$	$CuSO_4$	80		2×10^2 [h]	Green powder
$Ag^+TCNQ^{\dot{-}}$ [i]	$AgNO_3$	98		8×10^5	Blue powder
$Ce^{+3}(TCNQ^{\dot{-}})_3 \cdot 6H_2O$ [i]	$Ce(NO_3)_3$	98			Blue powder
$Sm^{+3}(TCNQ^{\dot{-}})_3 \cdot 6H_2O$	$Sm(NO_3)_3$	92		2×10^4	Blue powder
$Pb^{++}(TCNQ^{\dot{-}})_2 \cdot 1.5H_2O$	$Pb(NO_3)_2$	99		2×10^5	Blue powder
$Cu(NH_3)_2^{++}(TCNQ^{\dot{-}})_2$	$Cu(NH_3)_4^{++}SO_4^{=}$	94		7×10^2	Green powder
$Cu(H_2NCH_2CH_2NH_2)_2^{++}(TCNQ^{\dot{-}})_2$	$Cu(H_2NCH_2CH_2NH_2)^{++}SO_4^{=}$	97		4×10^4	Blue powder
Cu(2,2′-dipyridylamine)$^{++}(TCNQ^{\dot{-}})_2$	$CuCl_2$ + 2,2′-dipyridylamine	80		40[j]	
$Fe(phen)_3^{++}(TCNQ^{\dot{-}})_2$ [j] $6H_2O$ [k]	$FeSO_4$ + 3phen.	95	260	10^6	
$Ni(phen)_3^{++}(TCNQ^{\dot{-}})_2 \cdot 6H_2O$	$NiCl_2$ + 3phen.	98		10^8	Green powder
$Ni(phen)_2^{++}(TCNQ^{\dot{-}})_2 \cdot 6H_2O$	$Ni(phen)_2Cl_2$ [l]	96		2×10^8	Green powder
$(Cr^{+3})_2(AcO^-)_4(OH^-)(TCNQ^{\dot{-}}) \cdot 6H_2O$	$Cr(OAc)_3$	30		10^9	Blue powder

[a] Reproduced from *J. Am. Chem. Soc.* **84**, 3374 (1962). © 1962, American Chemical Society.

[b] Unless otherwise noted the preparations were carried out by metathesis of the cation source with Li^+TCNQ^- in water.

[c] Determined on compactions at room temperature unless otherwise noted.

[d] Reaction of cation source with TCNQ in boiling CH_3CN.

[e] Activation energy 0.15 e.v.

[f] Activation energy 0.36 e.v.; single crystal measurement gave the values 500, 6×10^4 and 6×10^4 along three principle axis.

[g] Activation energy 0.13 e.v.

[h] Activation energy 0.12 e.v.

[i] Forms free TCNQ in light.

[j] Activation energy 0.06 e.v.

[k] phen represents 1,10-phenanthroline.

[l] Pfeiffer and Tapperman, Z. Anog. Allgem. Chem. 215, 273 (1933).

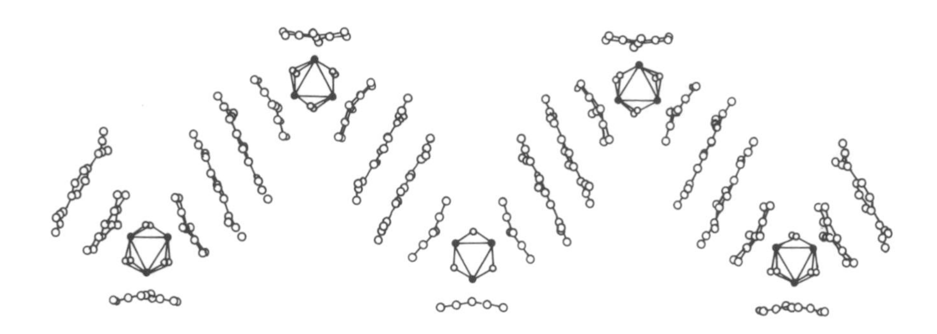

FIGURE 11. Crystal structure of $[Nb_3Cl_6(C_6Me_6)_3](TCNQ)_2$. [Reproduced from *J. Am. Chem. Soc.* **99**, 110 (1977), with permission. Copyright © 1977 American Chemical Society.]

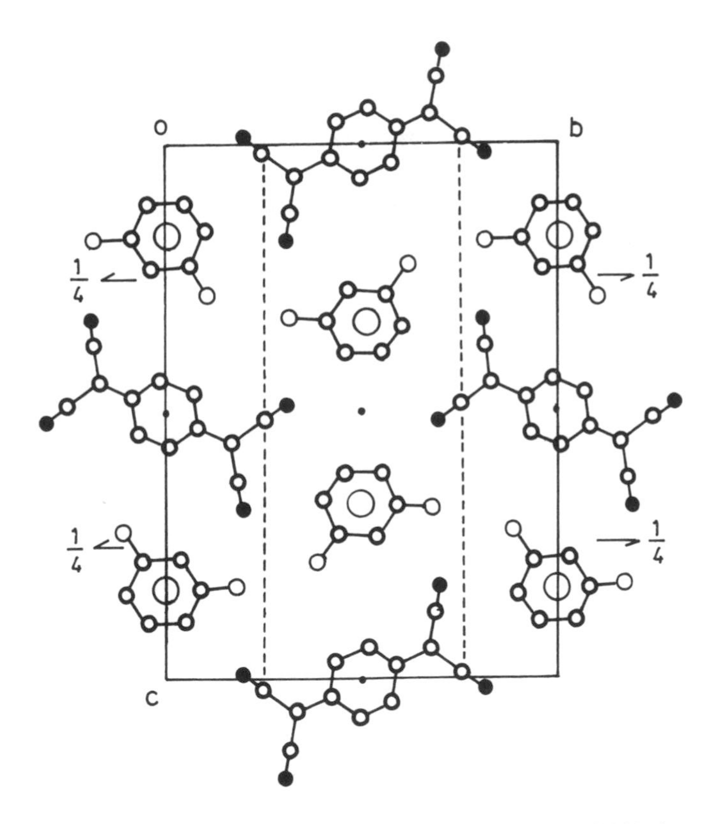

FIGURE 12. Projection of the structure of bis(toluene)chromium (TCNQ) along the *a* axis. (Reproduced from Reference 125, with permission.)

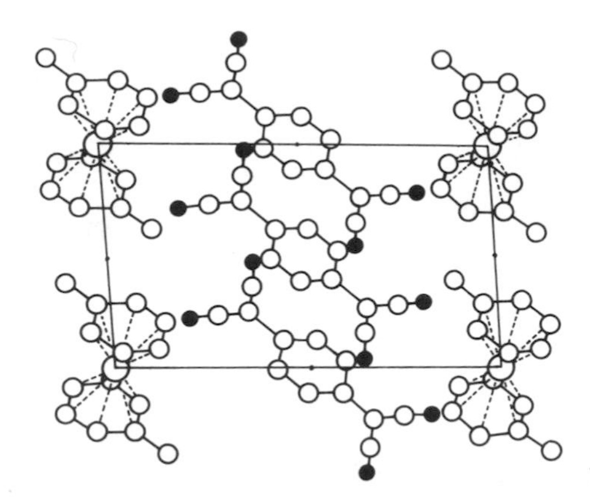

FIGURE 13. Projection of the structure of bis(toluene)chromium(TCNQ)$_2$ along the *a* axis. (Reproduced from Reference 126, with permission.)

higher than that of the simple salt, 5×10^{-6} ohm^{-1} cm^{-1}, $E_a = 0.36$ eV.[126] A structure determination[127] of the 1 : 2 salt at 153 K reveals that the TCNQ stacks remain regular and that the interplanar distance has shrunk to 3.16 Å. Measurements of the magnetic susceptibilities[128] showed that the values for the 1 : 1 salts bis(benzene)chromium(TCNQ) and bis(toluene) chromium(TCNQ) correspond to one electron per complex molecule. Electron-spin-resonance spectra suggest that the unpaired spins are connected solely with the cations. The temperature dependence of the susceptibility of bis(toluene)chromium(TCNQ) is reproduced in Figure 14. At high temperature (up to 100 K) the susceptibility follows the Curie Law; below

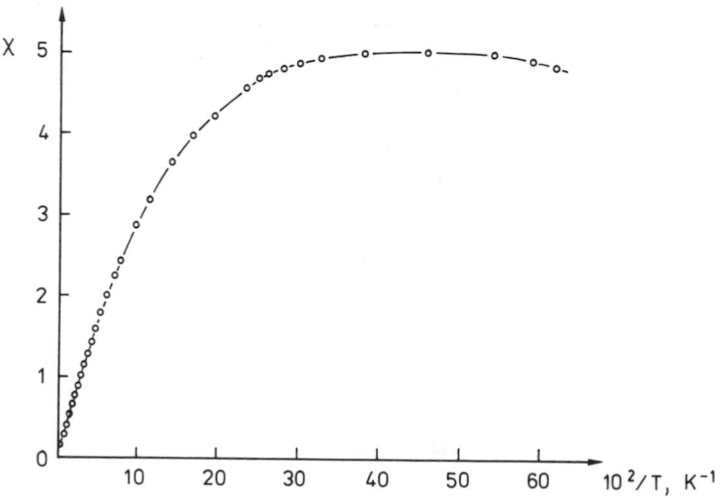

FIGURE 14. Temperature dependence of the susceptibility of bis(toluene)chromium (TCNQ). (Reproduced from Reference 128, with permission.)

this temperature it begins to deviate from the Curie law and passes through a maximum at 2.4 K. The temperature dependence of the susceptibility of the complex salts shows a small, nearly temperature-independent contribution, typical for highly conducting TCNQ salts. Nuclear-magnetic-resonance data[128] hint at the occurrence of two inequivalent groups of cations in bis(benzene)chromium(TCNQ), whereas in bis(toluene)chromium(TCNQ) all cations are presumably equivalent.

Ferrocenium and substituted ferrocenium salts of TCNQ found a pronounced interest in recent years. Decamethylferrocene, DMeFc, forms a simple (1 : 1) and a complex (1 : 2) salt with TCNQ. Conductivities[129] are of the order of 10^{-9} ohm^{-1} cm^{-1} for the simple salt and 0.1 ohm^{-1} cm^{-1} for the complex salt. Photoemission studies of these salts[129] show that the ionization threshold of the complex salt is about 0.8 eV below the Fermi energy, and the first peak is most likely due to the highest occupied state of $(TCNQ)^-$. More recent studies[130] revealed the existence of three decamethylferrocenium (DMeFc) salts with TCNQ: the fibrous violet $DMeFc(TCNQ)_2$, a green and metamagnetic (DMeFc)(TCNQ),[131] where the metamagnetic behavior is characterized by either antiferromagnetic or ferromagnetic properties depending on the magnitude of the external field (as illustrated in Figure 15), and a purple paramagnetic DMeFc(TCNQ).

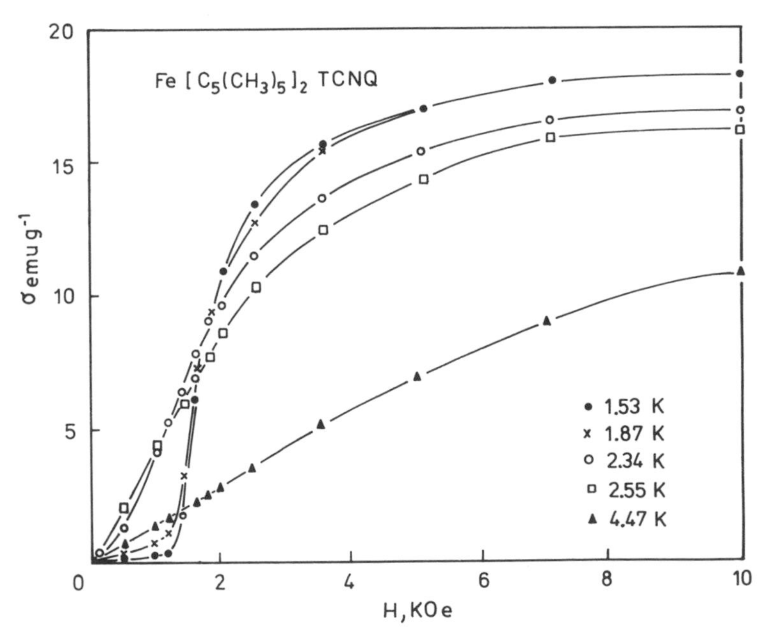

FIGURE 15. Isothermal plots of magnetic moment σ as a function of applied magnetic field H, for the metamagnetic compound $(DMeFc)^+(TCNQ)^-$. [Reprinted from *J. Am. Chem. Soc.* **101**, 2755 (1979), with permission. Copyright © 1979 American Chemical Society.]

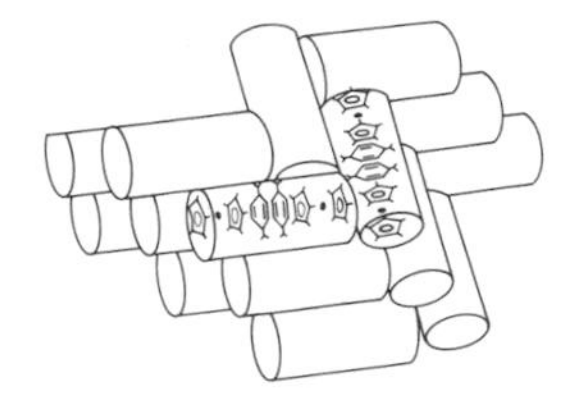

FIGURE 16. Schematic illustration of the crystal structure of $(DMeFc)_2^+(TCNQ)_2^-$. Each cylinder represents a dimer. [Reprinted from *J. Am. Chem. Soc.* **101**, 2756 (1979), with permission. Copyright © 1979 American Chemical Society.]

The structure[132] of the purple polymorph consists of isolated dimeric units DMeFc–TCNQ–TCNQ–DMeFc. A schematic illustration of the crystal structure is shown in Figure 16. The arrangement of the dimers in the crystal seems to be governed by packing criteria only. The TCNQ species carry a full negative charge each, as derived from the bond distances. This implies that iron is in a $S=1/2$ state, consistent with Mössbauer data. The TCNQ diads have interplanar distances of 3.147 Å and exhibit a slightly slipped (along the short axis) ring–ring overlap pattern.

The structure[133] of the green metamagnetic and air-sensitive DMeFc(TCNQ) consists of infinite mixed stacks DMeFc–TCNQ–DMeFc–TCNQ, as shown in Figure 17. The distances between cation and anion are 3.67(2) and 3.66(3) Å, compared with 3.55 Å in the purple dimeric phase.

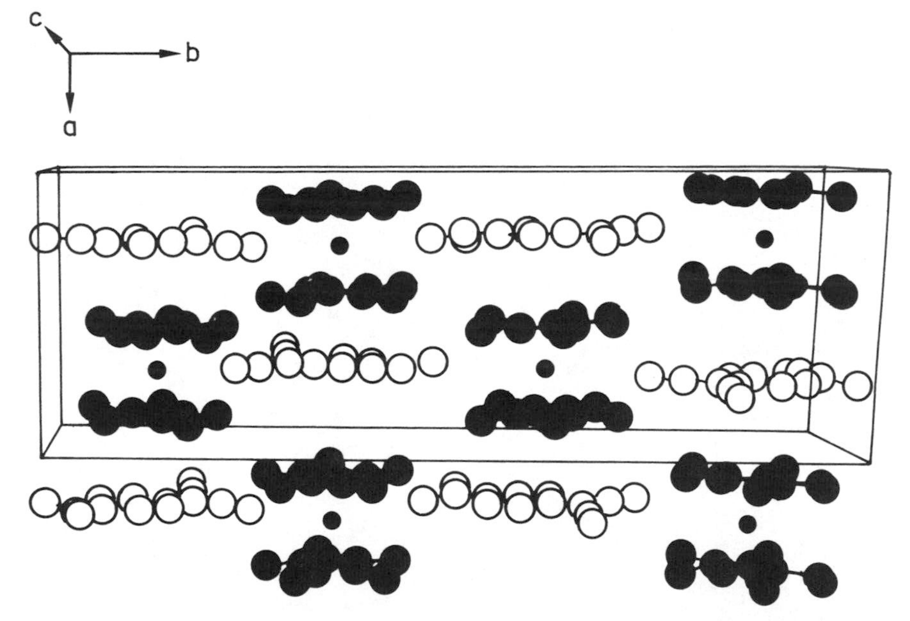

FIGURE 17. Unit cell of the green metamagnetic compound (DMeFc) (TCNQ). The $(DMeFc)^+$ cation is shaded. [Reprinted from *J. Am. Chem. Soc.* **101**, 7111 (1979), with permission. Copyright © 1979 American Chemical Society.]

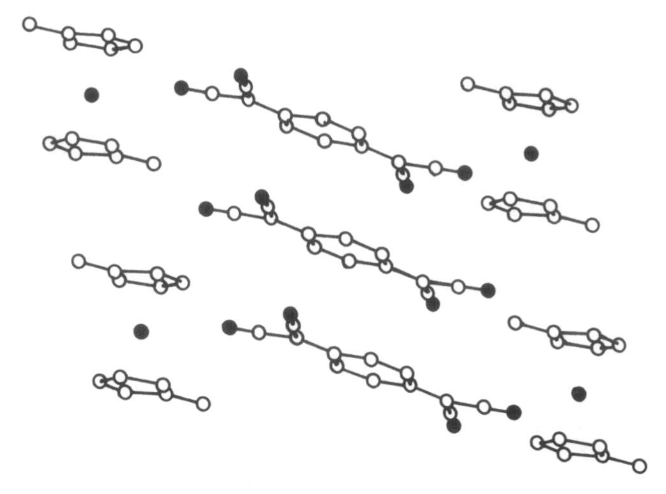

FIGURE 18. View of the structure of 1,1′-dimethylferrocenium$(TCNQ)_2$ along the *b* axis. (Reproduced from Reference 134, with permission.)

Some of the $(TCNQ)^-$ ions are replaced in a statistical manner by an oxidation product which contains a ketone group instead of a cyano group to form the α,α-dicyano-*p*-toluoyl-cyanide radical anion.

The complex, 1,1′-dimethylferrocenium$(TCNQ)_2$ crystallizes[134] with segregated stacks of cations and anions, as shown in Figure 18. Two adjacent TCNQ species of a stack are crystallographically different, and there seems to be a difference in the quinoid character and hence in the charge of the two species. The interplanar distances in the regular TCNQ stacks are 3.23 Å, the overlap is of the type ring–external bond. The magnetic susceptibility follows a Curie–Weiss law, $X = C/(T - \theta)$ with $\theta = -0.732°$ and $C = Ng^2\beta^2S(S + 1)/3k$ with $g = 2.837$. The authors state that the susceptibility is dominated by the 1,1′-dimethylferrocenium cation.

Reddish black crystals of $[Fe(bipy)_3](TCNQ)_2 \cdot 3H_2O$ (bipy = 2,2′-bipyridyl) have been described by Sano *et al.*[109]

Two products have been isolated from the reaction of TCNQ with Co(II) Schiff base complexes[135]: $[Co(acacen)(py)_2](TCNQ)$ and $[Co(acacen)(py)_2]_2(TCNQ)$, acacen = *N,N′*-ethylenebis(acetylacetoneimine), py = pyridine. The 1 : 1 compound is paramagnetic ($1.71\mu_B$) and the electronic spectrum indicates the presence of $(TCNQ)^-$. The 2 : 1 compound is diamagnetic and does not contain the $(TCNQ)^-$ anion. Air-sensitive $Co(bipy)_2$ (with Co in the oxidation state zero) has been prepared by metal atom synthesis and reacted with TCNQ to yield $[Co(bipy)_2](TCNQ)_2$.[136] It has been suggested from redox potentials that $[Co(bipy)_2]$ and $[Ni(bipy)_2]$ should give charge-transfer complexes with TCNQ.[137]

The $[Ni(dipy)_3](TCNQ)_2 \cdot H_2O$ molecules form greenish-black crystals.[109] The electronic spectra of the charge-transfer complexes of TCNQ with *o*-phenylendiamine (opd) and substituted *o*-phenylendiamine complexes of Ni have been reported.[138] These compounds have conductivities in the range 10^{-5}–10^{-9} ohm^{-1} cm^{-1} and activation energies of 0.65–0.86 eV. The compounds contain $(TCNQ)^-$ in the solid state (derived from ir spectra), the solution in acetone is characterized by the equilibrium.

$$Ni(opd)_2 + TCNQ \rightleftharpoons [Ni(opd)_2]^+(TCNQ)^-$$

The salt $[Ni(opd)_2]_2^+(TCNQ)^-$ and the analogous Pd compound[139] have conductivities of the order 10^{-4}–10^{-5} ohm^{-1} cm^{-1}.

The adducts of TCNQ with bis(1,2-benzoquinonedioximato) nickel and palladium,[140] $M(HBQD)_2$, contain neutral TCNQ and metal complex molecules. The crystal structure in principal, consists of segregated stacks of the metal complex and of TCNQ, but the repeat distances within a stack equal the lengths of the *c* axis [7.328(2) Å for the Pd compound] of the triclinic crystals and are rather large. The cyano group of the TCNQ stacks partially stick into the adjacent $M(HBQD)_2$ stacks. The solid-state electronic spectra of another TCNQ compound of a Pd complex, $Pd(ox)_2TCNQ$, ox = 8-quinolinoline, have been reported.[141] The crystal structure[142] of $[Pd(CNCH_3)_4](TCNQ)_4 \cdot 2NCCH_3$ contains stacks of tetrameric TCNQ groups flanked by square planar metal complex units which show no electronic interactions with each other or with TCNQ molecules (Figure 19). The center of a tetrameric (TCNQ) i.e. $(TCNQ)_4$, group coincides with a crystallographic inversion center, and the four TCNQ species may be denoted BAA′B′. The A and A′ molecules have an interplanar distance of 3.32 Å and exhibit ring–external bond overlap. The species A and B are 3.29 Å apart, their planes include a dihedral angle of 7.4°. The

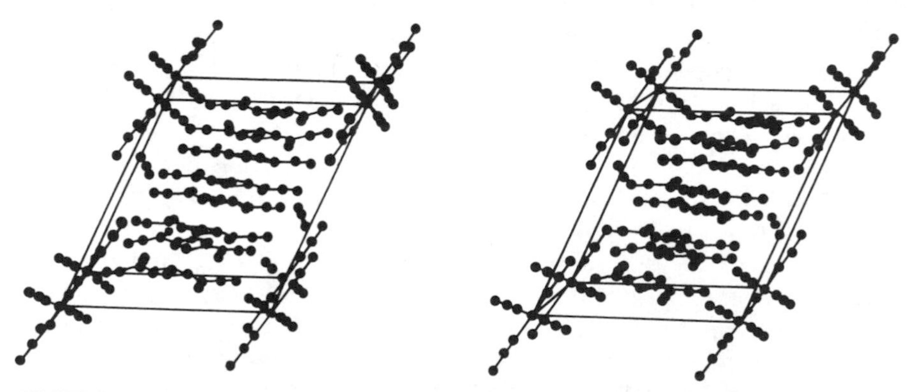

FIGURE 19. Stereoscopic packing diagram for $[Pd(CNCH_3)_4](TCNQ)_4 \cdot 2NCCH_3$. [Reproduced from *J. Am. Chem. Soc.* **98**, 5173 (1976), with permission. Copyright © 1976 American Chemical Society.]

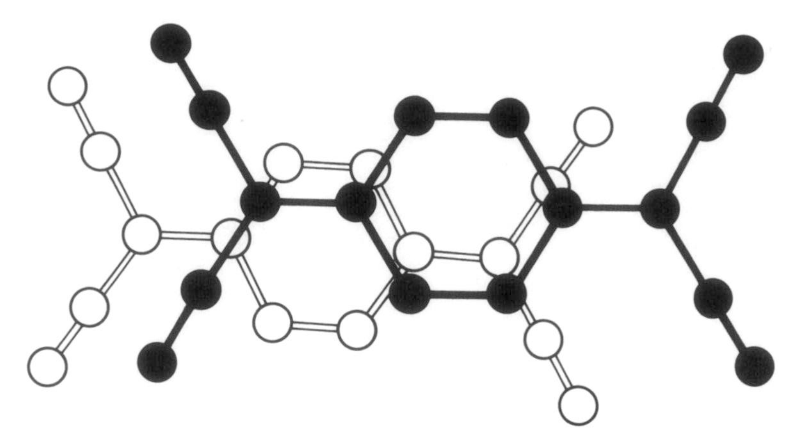

FIGURE 20. Overlap of two molecules A and B of the tetrameric $(TCNQ)_4$ group in $[Pd(CNCH_3)_4](TCNQ)_4 \cdot 2NCCH_3$. [Reproduced from *J. Am. Chem. Soc.* **98**, 5173 (1976), with permission. Copyright © 1976 American Chemical Society.]

somewhat unusual type of overlap is shown in Figure 20. At the junction between tetramers an interaction of a tetramer with two other adjacent ones can be assumed. The conductivity of the compound is 5.4×10^{-3} ohm^{-1} cm^{-1} along the needle axis and 1.3×10^{-2} ohm^{-1} cm^{-1} for powders.

A number of TCNQ salts or adducts with Pt complexes have been investigated. The charge-transfer interaction of TCNQ with *trans*-bis(trialkylphosphine)dialkynylplatinum(II) and related compounds has been studied with ir and visible spectra in solution, and $[(R_3P)_2Pt(C\equiv CH)_2](TCNQ)$ and $[(R_3P)_2Pt(C\equiv CCH_3)_2](TCNQ)$, with R=methyl or ethyl, have been isolated as stable, purple-black crystals.[143] These adducts show a characteristic absorption band in the visible region [506–555 nm (2.45–2.23 eV, 19 800–18 000 cm^{-1}) in nitromethane]. Polarized reflectane spectra of[144] $[Pt(S_2C_3H_3)_2]TCNQ$ show peaks at 16 660 cm^{-1} (2.07 eV, 600 nm) attributed to a charge-transfer excitation, 20 000 and 25 000 cm^{-1} ascribed to local excitations on $(TCNQ)^{-1}$, and 28 570 cm^{-1} (3.54 eV, 350 nm) interpreted as a metal-to-ligand transition. Mayerle[145] has determined the crystal structure of a compound with that composition. It consists of mixed stacks of TCNQ and the bis(propene-3-thione-1-thiolate)platinum(II) complex molecules. The interplanar distance is 3.42 Å and the bond lengths of the TCNQ indicate only small charge transfer.

Keller and co-workers[146] have prepared $[Pt(NH_3)_4](TCNQ)_2$ and a series of compounds $[Pt(a\text{-}a)_2](TCNQ)_x$, where a-a denotes a bidentate amine ligand, and x has values from 2 to 4. The observed conductivities range from 7×10^{-2} to 10^{-8} ohm^{-1} cm^{-1} and are, as usual, higher for the

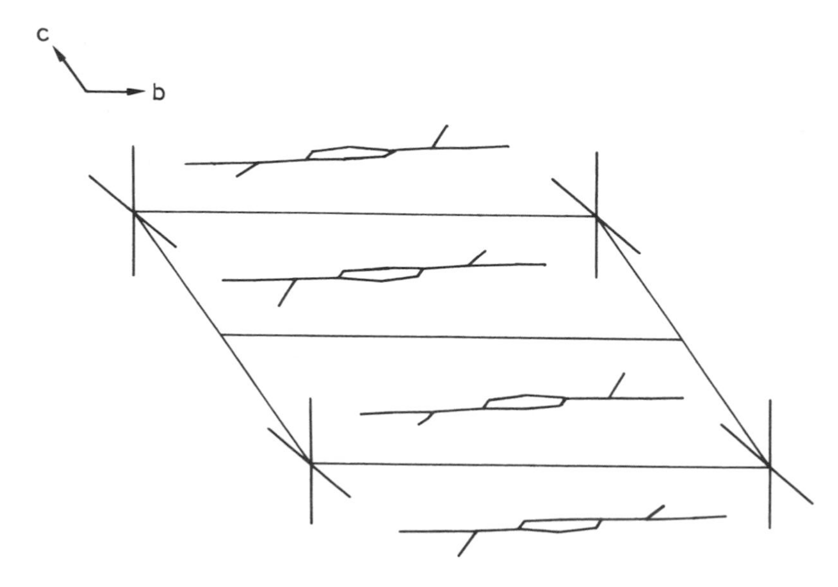

FIGURE 21. Projection of the structure of $[Pt(NH_3)_4](TCNQ)_2$.

complex salts than for the simple ones. (Simple, in this case, means 1 : 2, as the Pt complex molecules have a double positive charge.) The crystal structures of three of these salts have been determined: In $[Pt(NH_3)_4](TCNQ)_2$,[147] the TCNQ forms dimers with ring–ring overlap and interplanar distances of 3.14 Å. The dimers are aligned to stacks, but with no effective overlap between dimers. So it is most appropriate to regard the dimers as isolated. The cations flank the anion stacks with no electronic interaction between each other or to the anions (Figure 21), and resembling in that aspect the structure of $[Pd(CNCH_3)_4](TCNQ)_4 \cdot 2NCCH_3$.[142] (Compare Figure 18.) Electron-spin-resonance spectra of polycrystalline samples as well as of single crystals show one exchange-narrowed line of about 5G width. The paramagnetism does not follow a Curie law and increases with rising temperature with an activation energy of about 0.05 eV.

The crystal structure[148] of the complex salt $[Pt(bipy)_2](TCNQ)_3$ also contains noninteracting metal complex cations which flank stacks of $(TCNQ)_3^{2-}$ triads. The triads are centrosymmetric, the interplanar distances within a triad (defined as the mean deviations of the atoms of the quinoid part of the outer molecules to the plane through the quinoid part of the central molecule) are 3.23 Å, and between triads 3.33 Å. The overlap pattern is ring–external bond within a triad and slightly slipped ring–ring between triads. This arrangement is nearly identical with that in $Cs_2(TCNQ)_3$.[14] A projection of the structure is shown in Figure 22, details of the TCNQ

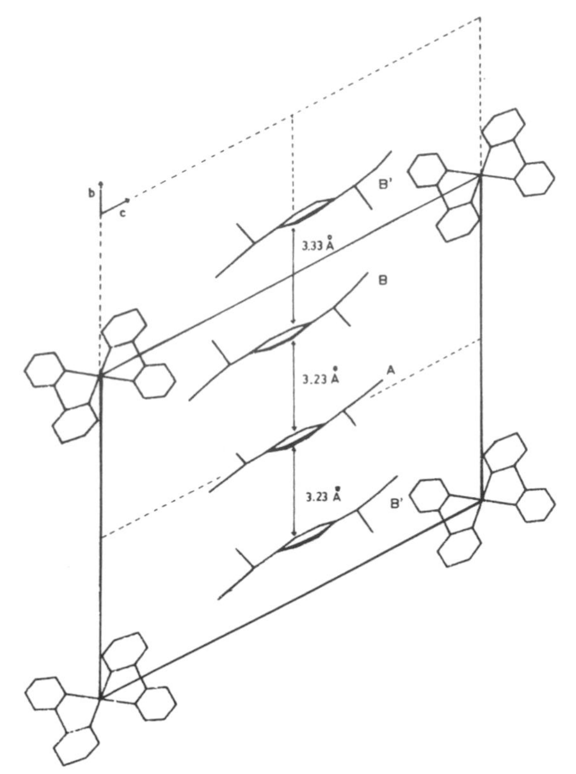

FIGURE 22. Projection of the structure of $[Pt(bipy)_2](TCNQ)_3$.

arrangement in Figure 23. The paramagnetism of the compound increases with increasing temperature with an activation energy of about 0.15 eV. The angular dependence of the electron-spin-resonance linewidth is shown in Figure 24, that of the g-value in Figure 25.

The complex $[Pt(dipy)_2](TCNQ)_2$ has a very unusual crystal structure[149] by the occurrence of σ-bonded $(TCNQ)_2^{2-}$ dimers. (There is just one other crystal structure known where dimerization of TCNQ by σ-bond formation has been observed: the 2 : 1 salt of N-ethylphenazinium and dimerized TCNQ.[150] It might be interesting in this context that TCNQ dimerizes by forming of a σ bond and adds acetone, if an acetonic solution is treated with a base like sodium benzoate.[151]) Figure 26 shows a sketch of the dimeric molecule with bond distances and angles. The radical character has disappeared by the σ-bond formation, and consequently the compound is nearly diamagnetic; an esr signal can barely be seen at room temperature.

Anisotropy measurements yielded the zero-field splitting parameters $D = 0.01\ \text{cm}^{-1}$, $E = 0.0014\ \text{cm}^{-1}$. The intensity of the fine-structure lines increases with increasing temperature with an activation energy of 0.25 eV.

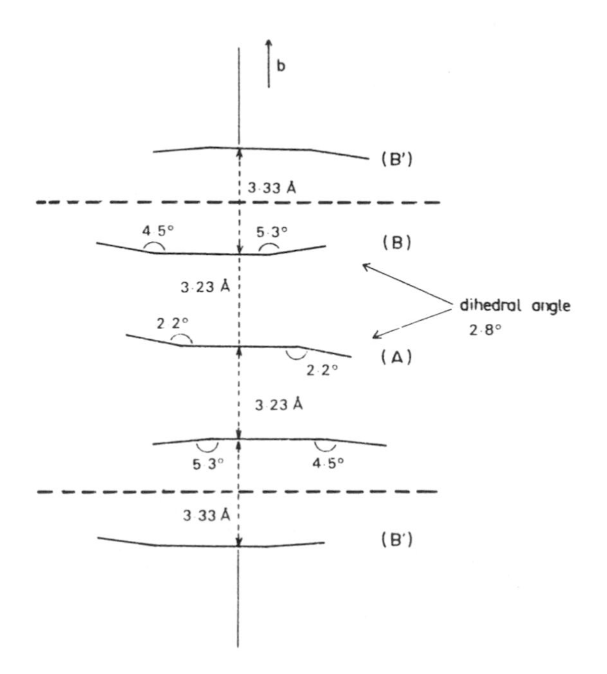

FIGURE 23. Details of the TCNQ arrangement in $[Pt(bipy)_2](TCNQ)_3$.

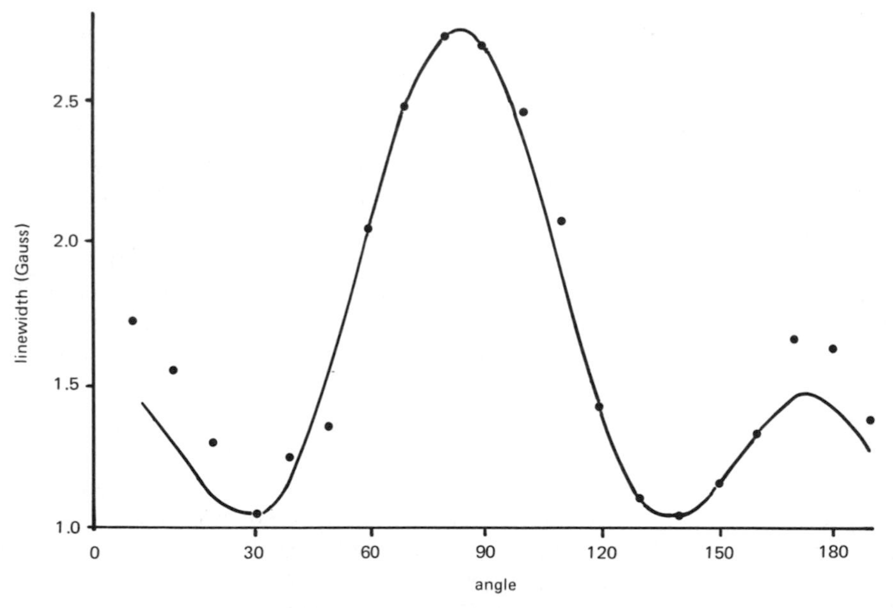

FIGURE 24. Angular dependence of the ESR linewidth in $[Pt(bipy)_2]^{2+}(TCNQ)_3^{2-}$ for rotation in the *bc* plane. 0^0 is the *c* axis $\parallel H_0$. The stacking axis is *b*.

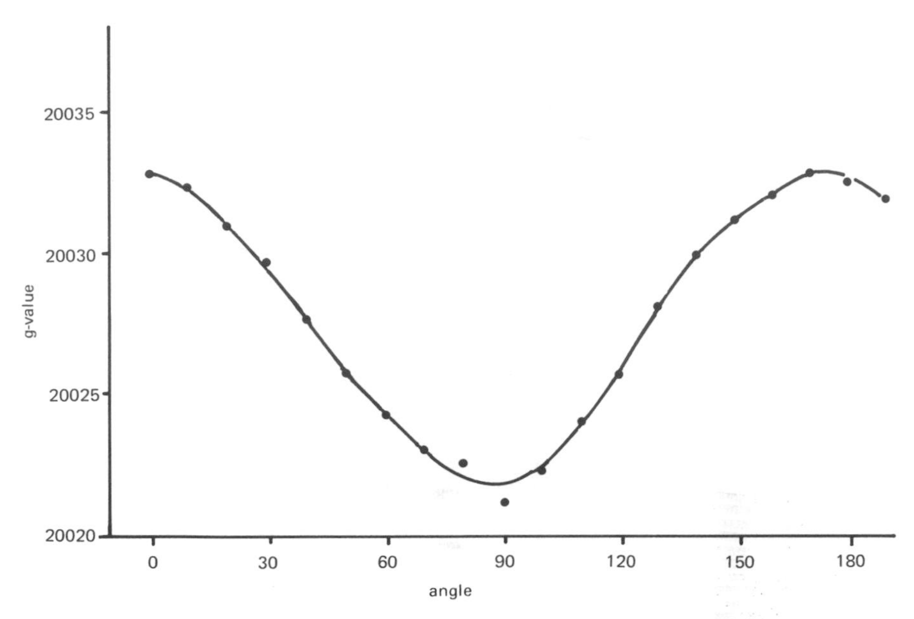

FIGURE 25. Angular dependence of the g value in $[Pt(bipy)_2]^{2+}(TCNQ)_3^{2-}$. Rotation as in Figure 24.

The fine-structure lines start to broaden at about 348 K, and at 360 K the lines suddenly coincide to a new single line, the intensity of which is three orders of magnitude larger. This phase transition from a nearly diamagnetic to a paramagnetic modification is accompanied by a change in color from grey-green to grass-green for powder samples.[146, 149]

A copper complex of the composition $[Cu(bipy)_3](TCNQ)_3Cl \cdot 4H_2O$ was reported by Sano *et al.*[109] to have a powder conductivity of 4×10^{-3} ohm^{-1} cm^{-1}. The structure of the 1 : 1 complex bis(8-hydroxyquinolinato)copper(II)(TCNQ) has been determined by Williams

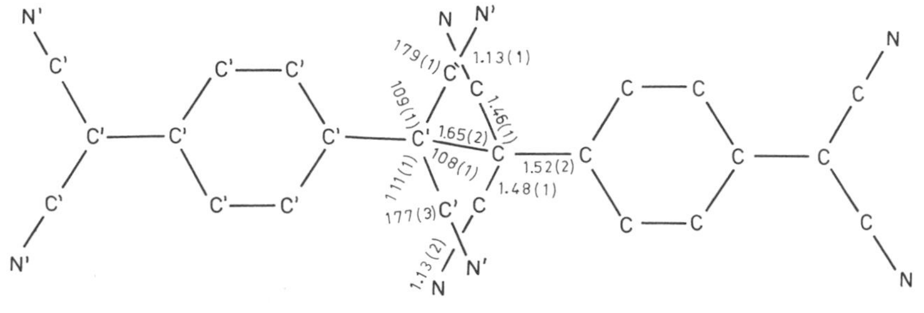

FIGURE 26. Dimeric $(TCNQ)_2^{2-}$ molecule in $[Pt(bipy)_2](TCNQ)_2$ with bond distances (Å) and angles (°).

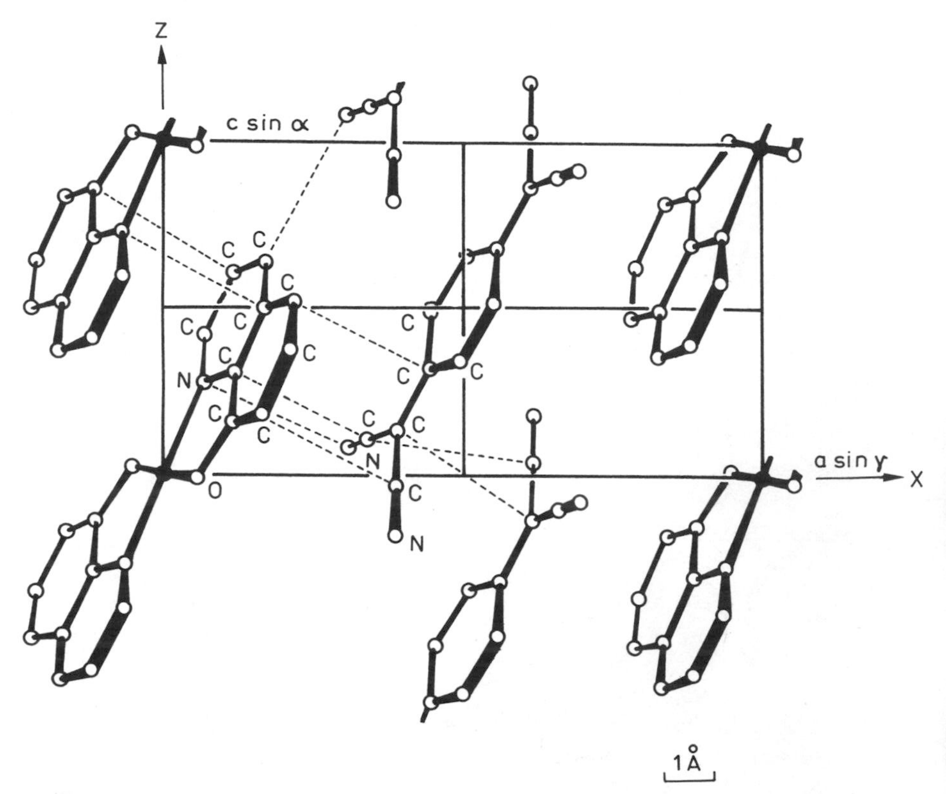

FIGURE 27. Structure of bis(8-hydroxyquinolinato)copper(II)(TCNQ) projected along the *b* axis. (Reproduced from Reference 152, with permission.)

and Wallwork.[152] It may be described as interpenetrating segregated stacks, but the overlap patterns suggest a description as mixed stacks with a large relative shift of adjacent molecules. This is indicated in Figures 27 and 28.

Cassoux and Gleizes[158] have prepared a 2 : 1 molecular adduct between *N*,*N*′-(1,2-phenylene)bis(salicylaldiminato)copper(II) and TCNQ, $[Cu(salphen)]_2(TCNQ)$. The molecular structure of the copper complex is shown in Figure 29. The structure consists of segregated stacks of the copper complex with interplanar distances of 3.24 and 3.30 Å, the interplanar distances in the TCNQ stack being twice as large. Hence there is no TCNQ–TCNQ interaction. The mode of overlap in the copper complex stack is shown in Figure 30. The charge transfer to the TCNQ stack is practically zero. The powder conductivity of unground samples is about 10^{-6} ohm cm^{-1}, and decreases on grinding, probably due to a change in the crystal structure. The esr spectrum is an axial spectrum with two g values ($g_{\parallel} = 2.180$ and $g_{\perp} = 2.043$); the magnetic susceptibility at room

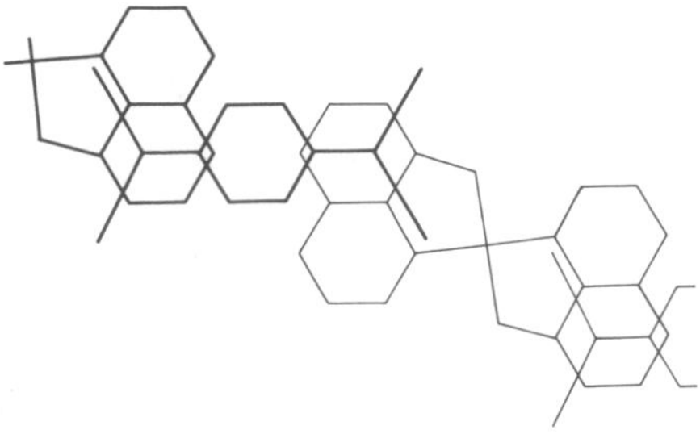

FIGURE 28. Molecular overlap in bis(8-hydroxyquinolinato)copper(II)(TCNQ) as viewed in a direction perpendicular to the mean molecular planes. (Reproduced from Reference 152, with permission.)

temperature, 2.227×10^{-3} cm^3/mol corresponds to an effective magnetic moment of $1.78\mu_B$. These data indicate that there are no significant interactions between the copper ions.

Yumoto *et al.*[153] have prepared and investigated a series of TCNQ salts of bis(dialkaldithiocarbamato)gold(III) complexes. The preparation of the simple salts was carried out by metathesis of the corresponding bis(dialkyldithiocarbamato)gold(III)chloride with Li(TCNQ). Complex salts were obtained by reaction of the simple salts with neutral TCNQ. Table 11 summarizes the compounds prepared in this way. The simple salts of the composition $[Au(S_2CNR_2)_2]^+(TCNQ)^-$, with $R = CH_3$ through n-C_4H_9, n-C_6H_{13}, n-C_8H_{17}, and benzyl ($=H_2C$–C_6H_5) have conductivities ranging from 3×10^{-5} to 5×10^{-11} ohm^{-1} cm^{-1} and activation energies from 0.30 to 0.97 eV. The complex salt $[Au(S_2CN(H_2C\text{-}C_6H_5)_2)_2]^+(TCNQ)_2^-$ and the same compound containing acetonitrile have conductivities (activation energies) of 1.5×10^{-3} ohm^{-1} cm^{-1} (0.26 eV) and 6×10^{-2} ohm^{-1} cm^{-1} (0.13 eV), respectively. The electronic spectra of the simple salts show only local excitations of $(TCNQ)^-$ [at 12000 (1.49 eV) and

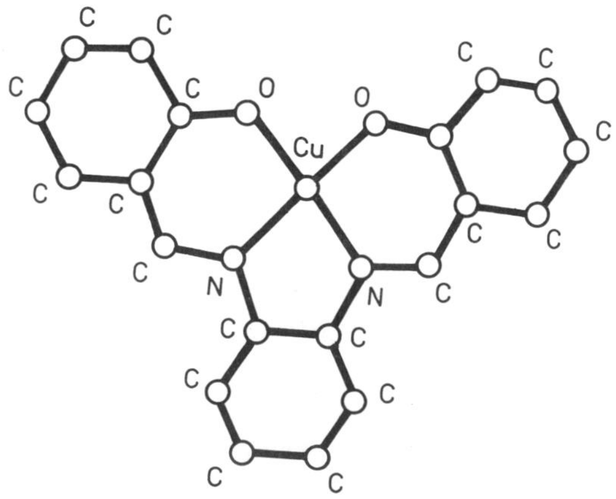

FIGURE 29. Molecular structure of the Cu(salphen) complex molecule in $[Cu(salphen)]_2(TCNQ)$. [Reprinted from *Inorg. Chem.* **19**, 665 (1980), with permission. Copyright © 1980 American Chemical Society.]

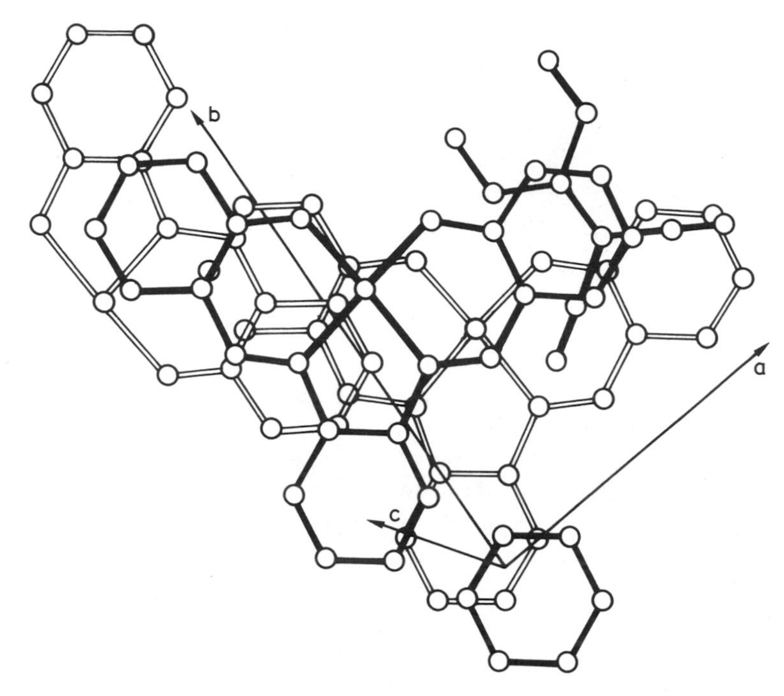

FIGURE 30. Overlap within the Cu(salphen) stack in $[Cu(salphen)]_2(TCNQ)$. [Reprinted from *Inorg. Chem.* **19**, 665 (1980), with permission. Copyright © 1980 American Chemical Society.]

24 000 cm^{-1} (2.98 eV)], those of the complex salts have peaks at 11 000 (1.36), 12 500 (1.55), and 22 000 cm^{-1} (2.73 eV). A summary of electrical and magnetic properties is given in Table 12. Structural data supported by x-ray evidence are not given but the authors conclude from the physical properties that compounds 1, 2, 3, and 5 (see Table 11 for the meaning of the

TABLE 11. Properties of Bis(dialkyldithiocarbamato)gold(III) TCNQ Salts[a]

No.	Salt	Color	Mp(dec) °C
1	$[Au(S_2CNMe_2)_2]^+TCNQ^{\overline{\cdot}}$	Dark-green needles	212–213
2	$[Au(S_2CNEt_2)_2]^+TCNQ^{\overline{\cdot}}$	Dark-green needles	201–202
3	$[Au(S_2CN(n\text{-}Pr)_2)_2]^+TCNQ^{\overline{\cdot}}$	Dark-green needles	174–177
4	$[Au(S_2CN(n\text{-}Bu)_2)_2]^+TCNQ^{\overline{\cdot}}$	Violet needles	148–150
5	$[Au(S_2CN(n\text{-}C_6H_{13})_2)_2]^+TCNQ^{\overline{\cdot}}$	Dark-green needles	86–87
6	$[Au(S_2CN(n\text{-}C_8H_{17})_2)_2]^+TCNQ^{\overline{\cdot}}$	Black microcrystals	55–57
7	$[Au(S_2CN(CH_2C_6H_5)_2)_2]^+TCNQ^{\overline{\cdot}}$	Black microcrystals	186–188
8	$[Au(S_2CN(CH_2C_6H_5)_2)_2]^+(TCNQ)_2^{\overline{\cdot}}$	Dark-blue needles	211–213
9	$[Au(S_2CN(CH_2C_6H_5)_2)_2]^+(TCNQ)_2^{\overline{\cdot}} \cdot (MeCN)$	Black needles	212–214

[a] Reproduced with permission from Reference 153.

TABLE 12. Resistivity ρ, Activation Energy E_a, and Magnetic Susceptibility χ_M of Bis(dialkyldithiocarbamato) Gold(III)(TCNQ) Salts

Salt	$\rho_{25°C}$(ohm cm)	E_a (eV)	χ_M (emu mol^{-1})
Simple salt			
1[b]	8.9×10^{8}	0.50	7.1×10^{-4}
2	2.2×10^{10}	0.58	9.8×10^{-4}
3	1.4×10^{9}	0.83	9.2×10^{-4}
4	3.0×10^{7}	0.43	-3.7×10^{-4}
5	5.4×10^{7}	0.55	8.2×10^{-4}
6	1.1×10^{7}	0.97	-5.2×10^{-4}
7	2.9×10^{4}	0.30	9.3×10^{-4}
Complex salt			
8	6.6×10^{2}	0.26	9.8×10^{-4}
9	1.6×10^{2}	0.13	4.0×10^{-4}

[a] Reproduced with permission from Reference 153.
[b] See Table 11 for molecular formulas.

numbers) contain mixed stacks, and that $(TCNQ)_2^{2-}$ dimers exist in compounds 4 and 6. A segregated stacking of TCNQ is suggested for compound 7. No structural conclusions are made for compounds 8 and 9, but the electronic spectra are interpreted in terms of a resonance between $(TCNQ)^-$ and $(TCNQ)^0$.

5. *Applications*

Metal TCNQ salts may find application as ion-selective electrodes (selective for the metal that is a component of the salt).[154] Thin films of Cu(TCNQ) between layers of Cu metal and Al metal may be used as switching and memory devices.[155] The films can be prepared by reacting the copper substrate with a TCNQ solution. The analogous reaction occurs with silver substrates. A sort of review of possible applications of TCNQ salts, e.g., for junction devices, in electrochemical cells, solid electrolyte capacitors (including thin-film capacitors[156]), and transducers has been given by Yoshimura and Murakami.[157]

6. *Summary*

The TCNQ structural arrangements of the compounds discussed in this Chapter are summarized in Table 13. Table 14 contains the lattice parameters of the transition-metal TCNQ compounds in the order in which they were mentioned above.

None of the TCNQ compounds discussed above has metallic properties, but most of them are semiconductors. The optimum room-temperature

TABLE 13. *Possible TCNQ Arrangements*[a]

Segregated stacks

Regular stacks	Reference
$Na(TCNQ)$[c] at 353 K	7
$K(TCNQ)$[c] at 395 K	9
$Rb(TCNQ)$(phase II)[b]	10
$Rb(TCNQ)$(phase III)[b]	11
$NH_4(TCNQ)$(phase I) at 311 K	15
$NH_4(TCNQ)$(phase II)[b]	15
$[(H_3CC_6H_5)_2Cr](TCNQ)_2$[b]	125
$[(H_3CC_6H_5)_2Cr](TCNQ)_2$[b]	126
$[(H_3CC_5H_4)_2Fe](TCNQ)_2$[b]	134
$[Pd(HBQD)_2](TCNQ)$[e]	140
[*N*,*N*′-(1,2-phenylene)bis (salicylaldiminato) copper(II)]$_2$(TCNQ)[e]	158

Stacks of diads	Reference
$Li(TCNQ)$	5
$Na(TCNQ)$[b]	6
$K(TCNQ)$[b]	9
$Rb(TCNQ)$(phase I)[b,c] at 113 K	5
$NH_4(TCNQ)$(phase I)	15
$Ag(TCNQ)$	116

Stacks of triads	Reference
$Rb_2(TCNQ)_3$[b,c]	12
$Rb_2(TCNQ)_3$[b,c] at 113 K	13
$Cs_2(TCNQ)_3$[b,c]	14
$[Pt(bipy)_2](TCNQ)_3$[b,c]	148

Stacks of tetrads	Reference
$[Pd(CNMe)_4](TCNQ_4) \cdot 2NCMe$[b,d,d]	142

Mixed stacks	Reference
$[(C_5(CH_3)_5)_2Fe](TCNQ)$	133
$[Pt(S_2C_3H_3)_2](TCNQ)$	145
$[Cu(C_{10}H_6NO)_2](TCNQ)$	152
$\{Nb_3Cl_6[C_6(CH_3)_6]_3\}(TCNQ)_2$[f]	123

Isolated units	Reference
$[(C_5(CH_3)_5)_2Fe]TCNQ$[c] (purple; TCNQ dimers) (Unit: [Fe]–TCNQ–TCNQ–[Fe])	132
$[Pt(NH_3)_4](TCNQ)_2$[c] (stacks of dimers, but without overlap between dimers)	147
$[Pt(bipy)_2](TCNQ)_2$ (σ-bonded TCNQ dimers)	149

[a] Room temperature measurements are quoted if no temperature is indicated.
[b] Ring–external bond overlap.
[c] Ring–ring overlap.
[d] Irregular overlap. When oligomeric units are present, the first letter gives the overlap within the unit, the last one between the units.
[e] No interactions along the TCNQ stacks because of large interplanar distances.
[f] Zigzag instead of linear chains.

TABLE 14. Crystallographic Data of Transition-Metal TCNQ Compounds

$[Nb_3Cl_6(C_6Me_6)_3](TCNQ)_2$ (Ref. 123)	
monoclinic, $P2_1/m$	
$a = 8.838(3)$ Å	
$b = 27.201(7)$ Å	$\beta = 90.96(2)°$
$c = 12.868(3)$ Å	
$Z = 2$, $\rho_{calc} = 1.49$ g/cm^3	
$[(MeC_6H_5)_2Cr](TCNQ)$ (Ref. 125)	
monoclinic, $P2_1/n$	
$a = 7.00(2)$ Å	
$b = 15.45(3)$ Å	$\beta = 97(1)°$
$c = 20.50(6)$ Å	
$Z = 4$, $\rho_{calc} = 1.34$ g/cm^3	
$[(MeC_6H_5)_2Cr](TCNQ)_2$ (Ref. 126)	
triclinic, $P\bar{1}$,	
$a = 8.25(2)$ Å	$\alpha = 94.7(5)°$
$b = 7.76(2)$ Å	$\beta = 92.3(5)°$
$c = 13.77(3)$ Å	$\gamma = 112.5(5)°$
$Z = 1$, $\rho_{calc} = 1.32$ g/cm^3	
$[Fe(C_5Me_5)_2]_2(TCNQ)_2$ purple paramagnetic polymorph (Ref. 132)	
monoclinic, $P2_1/c$	
$a = 9.7076(12)$ Å	
$b = 12.2113(17)$ Å	$\beta = 95.012(2)°$
$c = 23.5849(36)$ Å	
$Z = 4$, $\rho_{calc} = 1.265$ g/cm^3	
$[Fe(C_5Me_5)_2](TCNQ)$ green metamagnetic polymorph (Ref. 133)	
monoclinic, $P2_1/n$	
$a = 10.840(5)$ Å	
$b = 30.999(13)$ Å	$\beta = 99.20(3)°$
$c = 8.628(3)$ Å	
$Z = 4$, $\rho_{calc} = 1.238$ g/cm^3	
$[(MeC_5H_4)_2Fe](TCNQ)_2$ (Ref. 134)	
triclinic, $P\bar{1}$	
$a = 7.660(2)$ Å	$\alpha = 83.83(2)°$
$b = 7.530(2)$ Å	$\beta = 110.42(2)°$
$c = 14.083(4)$ Å	$\gamma = 94.48(2)°$
$Z = 1$, $\rho_{calc} = 1.370$ g/cm^3	
$[Pd(HBQD)_2](TCNQ)$ (Ref. 140)	
triclinic, $P\bar{1}$	
$a = 10.251(3)$ Å	$\alpha = 105.54(3)°$
$b = 7.843(3)$ Å	$\beta = 83.63(3)°$
$c = 7.328(2)$ Å	$\gamma = 92.28(3)°$
$Z = 1$, $\rho_{calc} = 1.72$ g/cm^3	

(continued overleaf)

TABLE 14 (*continued*)

$[Pd(CNCH_3)_4](TCNQ)_4 \cdot 2NCCH_3$ (Ref. 142)
triclinic, $P\bar{1}$

$a = 7.730(2)$ Å	$\alpha = 65.08(2)°$
$b = 14.978(7)$ Å	$\beta = 81.29(2)°$
$c = 14.389(4)$ Å	$\gamma = 73.46(2)°$
$Z = 1$, $\rho_{calc} = 1.34$ g/cm^3	

$[Pt(S_2C_3H_3)_2](TCNQ)$ (Ref. 144)
triclinic, $P\bar{1}$

$a = 8.744(5)$ Å	$\alpha = 92.10(2)°$
$b = 9.488(2)$ Å	$\beta = 102.73(2)°$
$c = 6.156(3)$ Å	$\gamma = 105.19(3)°$
$Z = 1$, $\rho_{calc} = 2.10$ g/cm^3	

$[Pt(NH_3)_4](TCNQ)_2$ (Ref. 147)
triclinic, $P\bar{1}$

$a = 7.20(1)$ Å	$\alpha = 127.07(7)°$
$b = 14.21(2)$ Å	$\beta = 122.71(8)°$
$c = 8.93(1)$ Å	$\gamma = 66.52(9)°$
$Z = 1$, $\rho_{calc} = 1.85$ g/cm^3	

$[Pt(bipy)_2](TCNQ)_3$ (Ref. 148)
triclinic, $P\bar{1}$

$a = 7.979(5)$ Å	$\alpha = 117.13(3)°$
$b = 10.567(6)$ Å	$\beta = 97.19(4)°$
$c = 17.203(7)$ Å	$\gamma = 111.28(5)°$
$Z = 1$, $\rho_{calc} = 2.01$ g/cm^3	

$[Pt(bipy)_2](TCNQ)_2$ (Ref. 149)
triclinic, $P\bar{1}$

$a = 7.891(7)$ Å	$\alpha = 130.01(5)°$
$b = 14.247(2)$ Å	$\beta = 120.06(6)°$
$c = 12.253(2)$ Å	$\gamma = 61.96(6)°$
$Z = 1$, $\rho_{calc} = 1.65$ g/cm^3	

Bis(8-hydroxyquinolinato)copper(II)(TCNQ) (Ref. 152)
triclinic, $P\bar{1}$

$a = 12.00(1)$ Å	$\alpha = 112.5(5)°$
$b = 7.54(1)$ Å	$\beta = 88.8(5)°$
$c = 7.12(1)$ Å	$\gamma = 96.8(5)°$
$Z = 1$, $\rho_{calc} = 1.608$ g/cm^3	

$[Cu(salphen)]_2(TCNQ)$ (Ref. 158)
triclinic, $P\bar{1}$

$a = 11.970(4)$ Å	$\alpha = 103.44(2)°$
$b = 12.567(4)$ Å	$\beta = 92.35(2)°$
$c = 7.020(2)$ Å	$\gamma = 92.48(2)°$
$Z = 1$, $\rho_{calc} = 1.555$ g/cm^3	

conductivities are of the order 10^{-1} ohm^{-1} cm^{-1}. The structural conditions for high conductivity and segregated regular stacks are fulfilled in the high-temperature forms of the simple alkali TCNQ salts, but the complete charge transfer prevents metallic behavior. Semiconductor to semiconductor transitions are observed in all of the alkali TCNQ salts, at least under pressure. The low-temperature phases of the simple alkali TCNQ salts are characterized by diadic TCNQ units.

It is a general and expected observation that the conductivities are higher in the compounds with regular stacks than in those with alternating stacks. Another condition for high conductivity, partial charge transfer, is obeyed in the so-called complex salts, which formally contain $(TCNQ)^-$ and $(TCNQ)^0$. But they hardly ever consist of *regular* stacks of the TCNQ species, and they do not exhibit metallic properties. The only complex salts where regular TCNQ stacks have been observed are the 1 : 2 salts bis(toluene)chromium$(TCNQ)_2$ and 1,1′-dimethylferrocenium$(TCNQ)_2$, which belong to the most conducting ones, but are not metallic either. The reason for this could be a charge localization that manifests itself in a difference in the quinoid character of adjacent TCNQ species, as it was argued in the iron compound.

Some effort has been devoted to the preparation of compounds with segregated stacks of a planar transition-metal complex and TCNQ, but such a compound could not yet be synthesized. Mixed stacks may be obtained, as in $[Pt(S_2C_3H_3)_2](TCNQ)$, or structures where the TCNQ species form stacks flanked by noninteracting transition-metal complexes. Only the bis(arene)chromium(TCNQ) and the ferrocenium(TCNQ) salts tend to crystallize in segregated stacks, but mixed stacks or oligomeric units may also be obtained. Although the TCNQ salts or adducts of simple and complex metal cations do not show metallic behavior or outstanding conductivities, the investigation of transport and other physical properties has contributed a good deal to the understanding of low-dimensional electronic interactions.

ACKNOWLEDGMENT

I thank J. S. Miller for supplying valuable references, and the copyright holders for their permission to use the materials reproduced in this Chapter.

Notation

Latin

a	Crystallographic a axis
a-a	Bidentate amine ligand
acacen	*N,N′*-ethylenebis(acetylacetoneimine)

b	Crystallographic b axis
bipy	2,2′-bipyridyl
c	Crystallographic c axis
C_p	Molar specific heat at constant pressure
D	Zero-field splitting parameter
DMeFc	Decamethylferrocenium
E	Zero-field splitting parameter
E_a	Activation energy
esr	Electron spin resonance
g	g factor in magnetic measurements
HBQD	*o*-benzoquinonedioximato
ir	Infrared
kK	10^3 cm^{-1}
M	Metal
nqr	Nuclear quadrupole resonance
opd	*o*-phenylenediamine
py	Pyridine
R	Crystallographic reliability factor
R	Alkyl or aryl substituent
R–B	Ring–external double bond overlap
R–R	Ring–ring overlap
salphen	*N,N*′-(1,2-phenylene)bis(salicylaldiminato)
TCNQ	7,7,8,8-tetracyano-*p*-quinodimethane; IUPAC name: 2,2′-(2,5-cyclohexadien-1,4-diyliden)bis(propandinitrile)
TTF	2,2′,5,5′-tetrathiafulvalene [$=\Delta^{2,2'}$-bis(1,3-dithiole)]
X	Halogen

Greek

α	Unit cell angle; absorption band designation
β	Unit cell angle; absorption band designation; some constant
γ	Unit cell angle
ΔH	Enthalphy difference
μ_B	Bohr magneton
ρ	Electrical resistivity
ρ_{calc}	Calculated crystallographic density
σ	Electrical conductivity
σ_{bl}	Standard deviation in bond lengths
χ_M	Molar magnetic susceptibility
$\parallel$	Parallel to chain direction
$\perp$	Perpendicular to the chain direction

References‡

1. Acker, D. S., and Hertler, W. R., *J. Am. Chem. Soc.* **84**, 3370 (1962).
2. Melby, L. R., Harder, R. J., Hertler, W. R., Mahler, W., Benson, R. E., and Mochel, W. E., *J. Am. Chem. Soc.* **84**, 3374 (1962); (a) Siedle, A. R., *ibid.* **97**, 5931 (1975); (b) Kathirgamanathan, P., and Rosseinsky, D. R., *J. Chem. Soc. Chem. Commun.*, 839 (1980).

‡ Citation of a Chemical Abstracts reference means that the original publication was not available to the author.

3. Sakai, N., Shirotani, I., and Minomura, S., *Bull. Chem. Soc. Jap.* **45**, 3321 (1972).
4. Pott, G. T., and Kommandeur, J., *Mol. Phys.* **13**, 373 (1967). *Chem. Abstr.* **68**, 109, 782m (1968).
5. Hoekstra, A., Spoelder, T., and Vos, A., *Acta Crystallogr. B* **28**, 14 (1972).
6. Konno, M., and Saito, Y., *Acta Crystallogr. B* **30**, 1294 (1974).
7. Konno, M., and Saito, Y., *Acta Crystallogr. B* **31**, 2007 (1975).
8. Richard, P., Zanghi, J.-C., and Guédon, J.-F., *Acta Crystallogr. B* **34**, 788 (1978).
9. Konno, M., Ishii, T., and Saito, Y., *Acta Crystallogr. B* **33**, 763 (1977).
10. Shirotani, I., and Kobayashi, H., *Bull. Chem. Soc. Jap.* **46**, 2595 (1973).
11. van Bodegom, B., de Boer, J. L., and Vos, A., *Acta Crystallogr. B* **33**, 602 (1977).
12. van der Wal, R. J., and van Bodegom, B., *Acta Crystallogr. B* **34**, 1700 (1978).
13. van der Wal, R. J., and van Bodegom, B., *Acta Crystallogr. B* **35**, 2003 (1979).
14. Fritchie, Jr., C. J., and Arthur Jr., P., *Acta Crystallogr.* **21**, 139 (1966).
15. Kobayashi, H., *Acta Crystallogr. B* **34**, 2818 (1978).
16. Flandrois, S., and Chasseau, D., *Acta Crystallogr. B* **33**, 2744 (1977).
17. Girlando, A., Bozio, R., and Pecile, C., *Chem. Phys. Lett.* **25**, 409 (1974).
18. Iida, Y., *Bull. Chem. Soc. Jap.* **42**, 637 (1969).
19. Metzger, R. M., *J. Chem. Phys.* **63**, 5090 (1975).
20. Afifi, H. H., Abdel-Kerim, F. M., Aly, H. F., and Shabaka, A. A., *Z. Naturforsch. A* **33**, 344 (1978).
21. Sakai, N., Shirotani, I., and Minomura, S., *Bull. Chem. Soc. Jap.* **45**, 3314 (1972).
22. Vegter, J. G., and Kommandeur, J., *Mol. Cryst. Liq. Cryst.* **30**, 11 (1975).
23. Blanc, J. P., Cheminat, B., and Robert, H., *C.R. Acad. Sci. Ser. B* **273**, 147 (1971).
24. Murgich, J., *J. Chem. Phys.* **70**, 5354 (1979).
25. Bozio, R., and Pecile, C., *J. Chem. Phys.* **67**, 3864 (1977).
26. Iqbal, Z., Christoe, C. W., and Dawson, D. K., *J. Chem. Phys.* **63**, 4485 (1975).
27. Terauchi, H., Sakamoto, N., and Shirotani, I., *J. Chem. Phys.* **64**, 437 (1976).
28. Terauchi, H., *Phys. Rev. B* **17**, 2446 (1978).
29. Vegter, J. G., Hibma, T., and Kommandeur, J., *Chem. Phys. Lett.* **3**, 427 (1979) [*Chem. Abstr.* **71**, 64, 787k (1969)].
30. Kommandeur, J., in *Low Dimensional Cooperative Phenomena*, H. J. Keller, Ed., Plenum Press, New York (1975), pp. 65–88; and in *Physics and Chemistry of Low-Dimensional Solids*, L. Alcacer, Ed., Reidel Publishing, Boston, (1980), pp. 197–212.
31. Shirotani, I., and Sakai, N., *J. Solid State Chem.* **18**, 17 (1976).
32. Lépine, Y., Caillé, A., and Larochelle, V., *Phys. Rev. B* **18**, 3585 (1978).
33. Siemons, W. J., Bierstedt, P. E., and Kepler, R. G., *J. Chem. Phys.* **39**, 3523 (1963).
34. Khanna, S. K., Bright, A. A., Garito, A. F., and Heeger, A. J., *Phys. Rev. B* **10**, 2139 (1974).
35. Kepler, R. G., *J. Chem. Phys.* **39**, 3528 (1963).
36. Holczer, K., Grüner, G., Miljak, M., and Cooper, J., *Solid State Commun.* **24**, 97 (1977).
37. LeBlanc, Jr., O. H., *J. Chem. Phys.* **42**, 4307 (1965).
38. Blakemore, J. S., Lane, J. E., and Woodbury, D. A., *Phys. Rev. B* **18**, 6797 (1978).
39. Blythe, A. R., Boon, M. R., and Wright, P. G., *Discuss. Faraday Soc.* **51**, 110 (1971) [*Chem. Abstr.* **75**, 145, 270w (1971)].
40. Nurullaev, Y. G., Kocharli, K. S., and Vlasova, R. M., *Ser. Fiz. Math. Nauk*, 108 (1976) [*Chem. Abstr.* **87**, 207, 588q (1977)].
41. Vlasova, R. M., Smirnov, I. A., Sochava, L. S., and Sherle, A. I., *Fiz. Tverd. Tela* **10**, 2990 (1968) [*Chem. Abstr.* **70**, 15, 301t (1969)].
42. Kondow, T., Mizutani, U., and Massalski, T. B., *Phys. Status Solidi B* **81**, 157 (1977) [*Chem. Abstr.* **87**, 30, 003s (1977)].
43. Etemad, S., Garito, A. F., and Heeger, A. J., *Phys. Lett. A* **40**, 45 (1972).
44. Siratori, K., Kondow, T., and Tasaki, A., *J. Phys. Chem. Solids* **39**, 225 (1978).

45. Chesnut, D. B., and Arthur, Jr., P., *J. Chem. Phys.* **36**, 2969 (1962).
46. Jones, M. T., and Chesnut, D. B., *J. Chem. Phys.* **38**, 1311 (1963).
47. Faucher, J. P., and Robert, H., *C.R. Acad. Sci. Ser. B* **270**, 174 (1970).
48. Hibma, T., Dupuis, P., and Kommandeur, J., *Chem. Phys. Lett.* **15**, 17 (1972) [*Chem. Abstr.* **77**, 107, 369u (1972)].
49. Vegter, J. G., Kommandeur, J., and Fedders, P. A., *Phys. Rev. B* **7**, 2929 (1973).
50. Hibma, T., and Kommandeur, J., *Solid State Commun.* **17**, 259 (1975).
51. Hibma, T., and Kommandeur, J., *Phys. Rev. B* **12**, 2608 (1975).
52. Hibma, T., Sawatzky, G. A., and Kommandeur, J., *Phys. Rev. B* **15**, 3959 (1977).
53. Schwerdtfeger, C. F., *Solid State Commun.* **23**, 621 (1977).
54. Flandrois, S., and Boissonade, J., *Chem. Phys. Lett.* **58**, 596 (1978).
55. Hibma, T., Sawatzky, G. A., and Kommandeur, J., *Chem. Phys. Lett.* **23**, 21 (1973).
56. Alizon, J., Berthet, G., Blanc, J. P., Gallice, J., and Robert, H., *Phys. Status Solidi B* **74**, 217 (1976) [*Chem. Abstr.* **84**, 171, 845t (1976)].
57. Silverstein, A. J., and Soos, Z. G., *Chem. Phys. Lett.* **39**, 525 (1976).
58. Suhai, S., *Solid State Commun.* **21**, 117 (1977).
59. Berthet, G., Blanc, J. P., Gallice, J., Robert, H., Thibaud, C., Fabre, J. M., and Giral, L., *Solid State Commun.* **22**, 251 (1977).
60. Stösser, R., and Siegmund, M., *J. Prakt. Chem.* **319**, 827 (1977).
61. Murgich, J., and Pissanetzky, S., *Chem. Phys. Lett.* **18**, 420 (1973).
62. Gillis, P., Alizon, J., Berthet, G., Blanc, J. P., Gallice, J., Robert, H., and Résibois, P., *Physica B* **90**, 154 (1977).
63. Alizon, J., Blanc, J. P., Gallice, J., and Robert, H., *C.R. Acad. Sci. Ser. B* **276**, 623 (1973).
64. Alizon, J., Berthet, G., Blanc, J. P., Gallice, J., and Robert, H., *Mol. Cryst. Liq. Cryst.* **32**, 245 (1976).
65. Kondow, T., and Sakata, T., *Phys. Status Solidi A* **6**, 551 (1971) [*Chem. Abstr.* **75**, 135, 337j (1971)].
66. Bozio, R., Girlando, A., and Pecile, C., *J. Chem. Soc. Faraday Trans. II*, 1237 (1975).
67. Anderson, G. R., and Devlin, J. P., *J. Phys. Chem.* **79**, 1100 (1975).
68. Bozio, R., Zanon, I., Girlando, A., and Pecile, C., *J. Chem. Soc. Faraday Trans. II*, 235 (1978).
69. Wozniak, W. T., Depasquali, G., and Klein, M. V., *Chem. Phys. Lett.* **40**, 93 (1976).
70. Chi, C.-K., and Nixon, E. R., *Spectrochim. Acta A* **31**, 1739 (1975) [*Chem. Abstr.* **84**, 16, 417m (1976)].
71. Tanner, D. B., Jacobsen, C. S., Bright, A. A., and Heeger, A. J., *Phys. Rev. B* **16**, 3283 (1977).
72. Kuzmany, H., Kundu, B., and Stolz, H. J., in *Proceedings of the International Conference on Lattice Dynamics*, M. Balkanski, Ed., Flammarion, Paris (1978), p. 584 [*Chem. Abstr.* **90**, 113, 495s (1979)].
73. Vegter, J. G., and Kommandeur, J., *Phys. Rev. B* **9**, 5150 (1974).
74. Torrance, J. B., Scott, B. A., and Kaufmann, F. B., *Solid State Commun.* **17**, 1369 (1975).
75. Boyd, R. H., and Phillips, W. D., *J. Chem. Phys.* **43**, 2927 (1965).
76. Iida, Y., *Bull. Chem. Soc. Jap.* **42**, 71 (1969).
77. Vlasova, R. M., Gutman, A. I., Rozenshtein, L. D., and Kartenko, N. F., *Phys. Status Solidi B* **47**, 435 (1971) [*Chem. Abstr.* **75**, 156, 653z (1971)].
78. Hiroma, S., Kuroda, H., and Akamatu, H., *Bull. Chem. Soc. Jap.* **44**, 9 (1971).
79. Oohashi, Y., and Sakata, T., *Bull. Chem. Soc. Jap.* **46**, 3330 (1973).
80. Oohashi, Y., and Sakata, T., *Bull. Chem. Soc. Jap.* **48**, 1725 (1975).
81. Tanaka, J., Tanaka M., Kawai, T., Takabe, T., and Maki, O., *Bull. Chem. Soc. Jap.* **49**, 2358 (1976).
82. Papavassiliou, G. C., and Spanou, S. S., *J. Chem. Soc. Faraday Trans. II* **73**, 1425 (1977).

83. Papavassiliou, G. C., *J. Chem. Soc. Faraday Trans. II* **74**, 1446 (1978).
84. Iida, Y., *Bull. Chem. Soc. Jap.* **50**, 1445 (1977).
85. Michaud, M., Carlone, C., Hota, K., and Zauhar, J., *Chem. Phys.* **36**, 79 (1979).
86. Iida, Y., *Bull. Chem. Soc. Jap.* **51**, 434 (1978).
87. Soos, Z. G., and Klein, D. J., *J. Chem. Phys.* **55**, 3284 (1971).
88. Kral, K., *Czech. J. Phys. B* **26**, 660 (1976) [*Chem. Abstr.* **85**, 114, 042a (1976)].
89. Gasser, W., and Hoefling, R., *Phys. Status Solidi B* **92**, 91 (1979) [*Chem. Abstr.* **90**, 160, 837q (1979)].
90. Maas Jr., E. T., *Mater. Res. Bull.* **9**, 815 (1974) [*Chem. Abstr.* **81**, 120, 055r (1974)].
91. Khatkale, M. S., and Devlin, J. P., *J. Chem. Phys.* **70**, 1851 (1979).
92. Vlasova, R. M., Gutman, A. I., Kartenko, N. F., Semkin, V. N., Titov, V. V., Khursandova, S. M., and Sherle, A. I., *Izv. Akad. Nauk. SSSR, Ser. Khim.*, 102 (1975) [*Chem. Abstr.* **83**, 19, 120z (1975)].
93. Seki, K., Kamura, Y., Shirotani, I., and Inokuchi, H., *Chem. Phys. Lett.* **35**, 513 (1975).
94. Nielsen, P., Epstein, A. J., and Sandman, D. J., *Solid State Commun.* **15**, 53 (1974).
95. Alcacer, L., Saramago, B. J. V., and Soares, A. M. P., *Phys. Status Solidi A* **20**, K151 (1973) [*Chem. Abstr.* **80**, 65, 263j (1974)].
96. Lin, S. F., Spicer, W. E., and Schechtman, B. H., *Phys. Rev. B* **12**, 4184 (1975).
97. Borodko, Y. G., Kaplunov, M. G., Moravskaya, T. M., and Pokhodnya, K. I., *Phys. Status Solidi B* K141 (1977) [*Chem. Abstr.* **87**, 209, 208q (1977)].
98. Ikemoto, I., Thomas, J. M., and Kuroda, H., *Faraday Discuss. Chem. Soc.* **54**, 208 (1972).
99. Woo, J. C., *Sae Mulli* **16**, 207 (1976) [*Chem. Abstr.* **90**, 65, 088n (1979)].
100. Shirotani, I., Kajiwara, T., Inokuchi, H., and Akimoto, S., *Bull. Chem. Soc. Jap.* **42**, 366 (1969).
101. Shirotani, I., Onodera, A., and Sakai, N., *Bull. Chem. Soc. Jap.* **48**, 167 (1975).
102. Fujii, G., Shirotani, I., and Nagano, H., *Bull. Chem. Soc. Jap.* **50**, 1726 (1977).
103. Berlin, A. A., Brodzeli, M. I., Gurtsiev, S. I., Eligulashvili, I. A., Kertsman, E. L., Kuzina, V. V., and Sherle, A. I., *Zh. Fiz. Khim.* **51**, 2928 (1977) [*Chem. Abstr.* **88**, 30, 815d (1978)].
104. Shirotani, I., Sakai, N., Inokuchi, H., and Minomura, S., *Bull. Chem. Soc. Jap.* **42**, 2087 (1969).
105. Sakai, N., Shirotani, I., and Minomura, S., *Bull. Chem. Soc. Jap.* **43**, 57 (1970).
106. Sherle, A. I., Mezhikovskii, S. M., Kuzina, V. V., and Berlin, A. A., *Izv. Akad. Nauk. SSSR, Ser. Khim.*, 554 (1974) [*Chem. Abstr.* **81**, 13, 164d (1974)].
107. Sharp, M., *Anal. Chim. Acta* **85**, 17 (1976) [*Chem. Abstr.* **86**, 80, 764r (1977)].
108. Nogami, T., Morinaga, N., Kanda, Y., and Mikawa, H., *Chem. Lett.*, 111 (1979).
109. Sano, M., Ohta, T., and Akamatu, H., *Bull. Chem. Soc. Jap.* **41**, 2204 (1968).
110. Ikemoto, I., Thomas, J. M., and Kuroda, H., *Bull. Chem. Soc. Jap.* **46**, 2237 (1973).
111. Thompson, R. C., Gujral, V. K., Wagner, H. J., and Schwerdtfeger, C. F., *Phys. Status Solidi A* **53**, 181 (1979) [*Chem. Abstr.* **91**, 150, 240w (1979)].
112. Thompson, R. C., Hoyano, Y., and Schwerdtfeger, C. F., *Solid State Commun.* **23**, 633 (1977).
113. Chain, E. E., Kevill, D. N., Kimball, C. W., and Weber, L. W., *J. Phys. Chem. Solids* **37**, 817 (1976).
114. Siedle, A. R., Candela, G. A., and Finnegan, T. F., *Inorg. Chim. Acta* **35**, 125 (1979).
115. Dick, A. W. S., Holliday, A. K., and Puddephatt, R. J., *J. Organomet. Chem.* **96**, C41 (1975).
116. Uyeda, N., Kobayashi, T., Ishizuka, K., and Fujiyoshi, Y., *Nature* **285**, 95 (1980).
117. Syutkina, O. P., Panov, E. M., and Kocheshkov, K. A., *Zh. Obshch. Khim.* **49**, 651 (1979) [*Chem. Abstr.* **91**, 20, 575y (1979)].
118. Richards, J. A., and Harrison, P. G., *J. Organomet. Chem.* **64**, C3 (1974).
119. Cornwell, A. B., Harrison, P. G., and Richards, J. A., *J. Organomet. Chem.* **67**, C43 (1974).

120. Cornwell, A. B., Cornwell, C. A., and Harrison, P. G., *J. Chem. Soc. Dalton*, 1612 (1976).
121. Cornwell, A. B., Harrison, P. G., and Richards, J. A., *J. Organomet. Chem.* **140**, 273 (1977).
122. Holliday, A. K., Makin, P. H., and Puddephatt, R. J., *J. Chem. Soc. Dalton*, 228 (1979).
123. Goldberg, S. Z., Spivack, B., Stanley, G., Eisenberg, R., Braitsch, D. M., Miller, J. S., and Abkovitz, M., *J. Am. Chem. Soc.* **99**, 110 (1977).
124. Yagubskii, E. B., Khidekel, M. L., Shchegolev, I. F., Buravov, L. I., Gribov, B. G., and Makova, M. K., *Izv. Akad. Nauk. SSSR, Ser. Khim.*, 2124 (1968) [*Chem. Abstr.* **70**, 29, 029x (1969)].
125. Shibaeva, R. P., Atovmyan, L. O., and Rozenberg, L. P., *J. Chem. Soc. Chem. Commun.*, 649 (1969).
126. Shibaeva, R. P., Atovmyan, L. O., and Orfanova, M. N., *J. Chem. Soc. Chem. Commun.*, 1494 (1969).
127. Shibaeva, R. P., Atovmyan, L. O., and Ponomarev, V. I., *Zh. Strukt. Khim.* **16**, 860 (1975) [*J. Struct. Chem.* **16**, 792 (1975)].
128. Zvarykina, A. V., Karimov, Y. S., Lyubovskii, R. B., Makova, M. K., Khidekel, M. L., Shchegolev, I. F., and Yagubskii, E. B., *Mol. Cryst. Liq. Cryst.* **11**, 217 (1970).
129. Ritsko, J. J., Nielsen, P., and Miller, J. S., *J. Chem. Phys.* **67**, 687 (1977).
130. Miller, J. S., Ries, Jr., A. H., and Candela, G. A., in *Quasi-One-Dimensional Conductors II*, S. Barisić, A. Bjelis, J. R. Cooper, and B. Leontić, Eds., Springer, Berlin (1979), pp. 313–321.
131. Candela, G. A., Swartzendruber, L. J., Miller, J. S., and Rice, M. J., *J. Am. Chem. Soc.* **101**, 2755 (1979).
132. Reis, Jr., A. H., Preston, L. D., Williams, J. M., Peterson, S. W., Candela, G. A., Swartzendruber, L. J., and Miller, J. S., *J. Am. Chem. Soc.* **101**, 2756 (1979).
133. Miller, J. S., Reis, Jr., A. H., Gebert, E., Ritsko, J. J., Salaneck, W. R., Kovnat, L., Cape, T. W., and Van Duyne, R. P., *J. Am. Chem. Soc.* **101**, 7111 (1979).
134. Wilson, S. R., Corvan, P. J., Seiders, R. P., Hodgson, D. J., Brookhart, M., Hatfield, W. E., Miller, J. S., Reis, Jr., A. H., Rogan, P. K., Gebert, E., and Epstein, A. J., in *Molecular Metals*, W. E. Hatfield, Ed., Plenum Press, New York (1979), pp. 407–414.
135. Clarkson, S. G., Lane, B. C., and Basolo, F., *Inorg. Chem.* **11**, 662 (1972).
136. Groshens, T. J., and Klabunde, K. J., *Proc. N.D. Acad. Sci.* **33**, 50 (1979).
137. Henne, B. J., and Bartak, D. E., *Proc. N.D. Acad. Sci.* **33**, 67 (1979).
138. Bäbler, F., and von Zelewsky, A., *Helv. Chim. Acta* **60**, 2723 (1977).
139. Miles, M. G., and Wilson, J. D., *Inorg. Chem.* **14**, 2357 (1975).
140. Keller, H. J., Leichert, I., Mégnamisi-Bélombé, M., Nöthe, D., and Weiss, J., *Z. Anorg. Allg. Chem.* **429**, 231 (1977).
141. Koizumi, S., and Iida, Y., *Bull. Chem. Soc. Jap.* **46**, 629 (1973).
142. Goldberg, S. Z., Eisenberg, R., Miller, J. S., and Epstein, A. J., *J. Am. Chem. Soc.* **98**, 5173 (1976).
143. Masai, H., Sonogashira, K., and Hagihara, N., *J. Organomet. Chem.* **34**, 397 (1972).
144. Philpott, M. R., and Brillante, A., *Mol. Cryst. Liq. Cryst.* **50**, 163 (1979).
145. Mayerle, J. J., *Inorg. Chem.* **16**, 916 (1977).
146. Endres, H., Keller, H. J., Moroni, W., and Nöthe, D., *Z. Naturforsch. B* **31**, 1322 (1976).
147. Endres, H., Keller, H. J., Moroni, W., Nöthe, D., and Dong, V., *Acta Crystallogr. B* **34**, 1703 (1978).
148. Endres, H., Keller, H. J., Moroni, W., Nöthe, D., and Dong, V. *Acta Crystallogr. B* **34**, 1823 (1978).
149. Dong, V., Endres, H., Keller, H. J., Moroni, W., and Nöthe, D., *Acta Crystallogr. B* **33**, 2428 (1977).
150. Morosin, B., Plastas, H. J., Coleman, L. B., and Stewart, J. M., *Acta Crystallogr. B* **34**, 540 (1978).

151. Farcasiu, M., and Russell, C. S., *J. Org. Chem.* **41**, 571 (1976).
152. Williams, R. M., and Wallwork, S. C., *Acta Crystallogr.* **23**, 448 (1967).
153. Yumoto, Y., Kato, F., and Tanaka, T., *Bull. Chem. Soc. Jap.* **52**, 1072 (1979).
154. Sharp, M., and Johansson, G., *Anal. Chim. Acta* **54**, 13 (1971) [*Chem. Abstr.* **74**, 82, 418g (1971)].
155. Potember, R. S., Poehler, T. O., and Cowan, D. O., *Appl. Phys. Lett.* **34**, 405 (1979).
156. Yoshimura, S., Ito, Y., and Murakami, M., *Ger. Offen.* **2**, 313, 211 (1974) [*Chem. Abstr.* **81**, 31, 137b (1974)].
157. Yoshimura, S., and Murakami, M., in *Synthesis and Properties of Low-Dimensional Materials*, J. S. Miller, and A. J. Epstein, Eds., The New York Academy of Sciences, New York (1978), pp. 269–292.
158. Cassoux, P., and Gleizes, A., *Inorg. Chem.* **19**, 665 (1980).

6

Linear Chain 1,2-Dithiolene Complexes

L. Alcácer and H. Novais

1. Introduction

This chapter focuses on linear chain materials formed by the association of 1,2-dithiolene metal complex molecules with organic or inorganic donors and cations. Formally, 1,2-dithiolene complexes are obtained from the combination of an appropriate metal atom with the ligand

$$\begin{array}{c} R\diagdown\;\;\diagup S^- \\ C \\ \| \\ C \\ R\diagup\;\;\diagdown S^- \end{array}$$

in which R = H, CF_3, CN, C_6H_5, CH_3, and C_2H_5. (The RC=CR unit may be part of a benzene ring.)

The ligand in which R = CN was originally obtained by Bähr and Schleitzer in 1955[1] and is usually designated maleonitriledithiolate, mnt. Although these authors had prepared a palladium complex of this ligand, they could not completely identify its structure. Those complexes received little attention until 1962, when the $[M(mnt)_2]^z$ square planar complexes with M = Ni, Cu, Pd, Pt, and Au and $z = 1-$ and $2-$ were prepared and studied in detail.[2,3] Since then, these complexes have been thoroughly investigated and many others have been prepared, including those with R = CF_3, H, CH_3, C_6H_5, etc.

Two or three ligands can be associated with the metal atom giving different structures. Square planar complexes with two ligands are obtained for M = Ni, Cu, Zn, Pd, Pt, and Au, and in some cases, M = Fe and Co,

L. Alcácer and H. Novais • Serviço de Química-Física Estado Sólido, LNETI, Sacavém, Portugal.

the molecules being then associated in dimers. Tris-maleonitriledithiolene complexes $[M(mnt)_3]^z$ have been obtained for M = V, Cr, Mn, Ge, Mo, Rh, In, W, Re, and Tl, with structures that are intermediate between the trigonal prismatic and the distorted octahedral.

We will only consider the bis(1,2-dithiolene) square planar complexes, collectively referred to by $[M(S_2C_2R_2)_2]^z$ since only they can form linear chain compounds.

One important and general feature of these complexes is the variety of oxidation states they can present, as well as the fact that they can easily be converted into each other. The $z = 0, 1-$, and $2-$ species are common and even the $3-$ and $4-$ ionic species have been detected. Cation complexes are essentially never observed.

The generic term bis(1,2-dithiolene) refers to all ligands of the *cis*-1,2-disubstituted ethylene-1,2-dithiol type and their arene-1,2-dithiol analogs, when complexed with transition metals in their several possible charged forms. By doing so, it avoids ascribing formal oxidation states, namely, to the metal complexes whose ligands become intermediate between dithiodiketones and *cis*-1,2-ethylene-1,2-dithiolate dianions.

The most interesting linear chain complexes of this series with respect to their electrical and magnetic properties are those in which the organic or inorganic donors or cations are stacked and interact strongly. The $[M(S_2C_2R_2)_2]^z$ molecules when stacked can interact through metal–metal or metal–sulfur contacts, but in most cases intermolecular interactions are weak, sometimes giving, however, interesting magnetic properties due to one-dimensional exchange. There is evidence that the highest occupied molecular orbital, in the ground state, is highly delocalized and has very low electronic density on the metal atom but does have considerable spatial extent perpendicular to the plane, which can account for intermolecular interactions along the chain.

We shall review the chemistry of the ligand and its complexes, as well as describe and discuss the most interesting crystal structures and the electrical and magnetic properties.

2. *Chemistry of the 1,2-Dithiolene Ligand and Its Complexes*

The synthesis of square planar metal complexes of $[M(S_2C_2R_2)_2]^z$ type and their ligands are reviewed. Thorough surveys on metal bis(1,2-dithiolenes) were published by McCleverty,[4] Schrauzer,[5] Hoyer *et al.*,[6] and Eisenberg;[7] the later one focuses on structural systematics.

Chemical methods leading to linear chain complexes of this class may be generally classified in the following types:

(a) Methatesis, MT:

$$\text{cat}^+\text{an}^- + A^+[M(S_2C_2R_2)_2]^{1-} \longrightarrow R^+[M(S_2C_2R_2)_2]^{1-} + A^+\text{an}^-$$

where cat^+an^- is a convenient soluble salt of the cation, cat^+, to be introduced (an^- = anion).

(b) Charge transfer (CT) when there is a direct electron transfer between two molecules. Some complexes formed from tetrathiafulvalene or tetrathiotetracene as donors and metal bis(1,2-ditholenes) as acceptors are examples of this type.

(c) Redox reactions (RD) when a previous oxidation–reduction step is necessary for the formation of the complex as, for instance, in the preparation of some perylene derivatives.

(d) Electrochemical techniques (EL).

To obtain single crystals of these complexes following the (a), (b), and (c) methods described above, several techniques are normally used. They usually consist in slow diffusion of the components in a U-shaped tube, using an appropriate solvent; or evaporation of part of the solvent followed by slow cooling. Vapor-phase preparations have also been successful in some cases.

The electrochemical techniques normally consist of oxidizing the donor molecule in the presence of the anion, at a constant current of a few microamperes, for a period of about one week.

Compounds derived from metal bis(1,2-dithiolenes) by coordinative adduct formation, binuclear bridging, or ligand substitution are not included, since they have not as yet shown any particular interesting electrical or magnetic properties of one-dimensional character.

All linear chain complexes found in the literature have been included, with the exception of those whose counterions are of the large spherical type—namely, tetraalkylammonium, tetraphenylphosphonium, and the like, or are simple metal ions, unless they have shown unusual properties. Electrical conductivities and other physical data for most of the salts not covered here, as well as those for simple neutral 1,2-dithiolene transition-metal complexes have been previously reviewed and tabulated by Miller and Epstein.[8]

The compounds are arranged in Tables 1–6 by $[M(S_2C_2R_2)_2]^z$ type, giving for each one, the stoichiometry, method of preparation, and electrical conductivities in compressed pellet or in a single crystal. Activation energies, E_a, are also included whenever available. Literature references are mainly those where the chemical preparation of the compounds and their electrical conductivities can be found, though some have been added that include further physical data, in particular, the crystal structure.

2.1. Aliphatic 1,2-Dithiolene Complexes: $[M(S_2C_2R_2)_2]^z$ Type

2.1.1. Metal Bis(cis-ethylene-1,2-dithiolene) (R = H; z = 0 and 1−; M = Co, Ni, Cu, Pd, and Pt)

The ligand *cis*-ethylenedithiolate, edt, has been prepared by the method of Schroth and Peschel[9] which was slightly modified later.[10] Although $Ni(edt)_2$ had been prepared directly from the ligand in a low yield[11] it is most conveniently obtained through iodine oxidation of the monoanion.[12–14] The cobalt, copper, palladium, and platinum monoanionic complexes have also been prepared.[13,14] The palladium and platinum neutral complexes were also obtained, but in much lower yields than the nickel analog.[14]

The linear chain complexes derived from this ligand are listed in Table 1.

2.1.2. Metal Bis(cis-1,2-ditrifluoromethylethylene-1,2-dithiolene) ($R = CF_3$, z = 0, 1−, and 2−; M = Fe, Co, Ni, Cu, Pd, Pt and Au)

The synthesis of bis(trifluoromethyldithiethene),[21] $[S_2C_2(CF_3)_2]^{2-}$, has allowed the preparation of the complexes of this series. The first to be reported was that of molybdenum obtained by King through the reaction of $[S_2C_2(CF_3)_2]^{2-}$ with molybdenum hexacarbonyl.[22] The following one was

TABLE 1. Complexes with Metal Bis(cis-ethylene-1,2-dithiolene), $M(edt)_2$

Donor or cation	Transition metal	Stoichiometry	Reaction type	Conductivity		Reference
				σ_{RT} ohm^{-1} cm^{-1}	E_a (eV)	
Qn	Ni	1:1	MT	1.3×10^{-7}	0.26	14
MeQn	Ni	1:1	MT	5.0×10^{-10}	0.44	14
MeQn	Pd	1:1	MT	1.8×10^{-9}	0.42	14
MeQn	Pt	1:1	MT	$\sim 10^{-7}$	0.38	14
MePy	Ni	1:1	MT	3.3×10^{-10}	1.29	14
MePy	Pt	1:1	MT	8.7×10^{-9}	1.00	14
Mp	Ni	1:1	MT	1.9×10^{-8}	0.48	14
NMP	Ni	1:1	MT	5×10^{-4}		15
$Ni(opd)_2$	Ni	1:1	MT	1.7×10^{-3}	0.24	15
$Pd(opd)_2$	Ni	1:1	MT	5×10^{-4}		15
$Pd(opd)_2$	Pd	1:1	MT	2.3×10^{-5}		15
$Ni(dmopd)_2$	Ni	1:1	MT	6.2×10^{-7}		15
TTF	Ni	2:3	CT	—		16
TTF	Ni	2:1	CT	7.4×10^{-3} [a]	0.23	16, 18
TTT	Ni	1.2:1	CT	30[a]		17, 19
TMTTF	Ni	1:1	CT	$\sim 10^{-6}$		20

[a] Single crystal.

TABLE 2. Complexes with Metal Bis(cis-1,2-ditrifluoromethylethylene-1,2-dithiolene), $M(S_2C_2(CF_3)_2)_2$

Donor or cation	Transition metal	Stoichio-metry	Reaction type	Conductivity σ (ohm^{-1} cm^{-1})	Reference
Per	Ni	1:1	CT	10^{-5} [a]	29, 30
Per	Pt	1:1	CT	5×10^{-9}	29, 30
Pyr	Ni	1:1	CT	10^{-5} [a]	29, 30
Pyr	Pt	1:1	CT	2×10^{-6}	29, 30
AZ	Pt	2:1	CT	3×10^{-3}	30
Trp	Ni	1:1	MT	—	31
KOS	Ni	2:1	MT	—	32
CNpyMe	Ni	2:1	MT	—	32
PTZ	Ni	1:1	CT, MT	$<10^{-8}$	15, 33, 34
POZ	Ni	1:1	CT, MT	2.6×10^{-4}	15, 33, 34
TTF	Ni, Cu, Pt, Au	1:1	MT	$<10^{-9}$	16, 35, 36, 37
TTT	Ni	1:1	CT	$<10^{-10}$	38
MeQn	Pd	1:1	MT	—	25
TTN	Ni	1:1	—	—	39
		Complex with Metal Bis(1,2-ditrifluoromethylethylene-1,2-diselenenolene), $M(Se_2C_2(CF_3)_2)_2$			
TTF	Cu	1:1	—	—	40

[a] Single crystal.

the nickel complex, obtained by the reaction of $[S_2C_2(CF_3)_2]^{2-}$ with nickel tetracarbonyl.[3, 23, 24] This complex is readily reduced in several basic solvents, such as ketones, amides, nitriles, amines, and dimethylsulfoxide (DMSO), but it is stable in nonpolar solvents like hydrocarbons or dichloromethane. Further reduction to the dianion is achieved by *p*-phenylenediamine in DMSO. The iron, cobalt, platinum, and palladium complexes were prepared using, respectively, iron pentacarbonyl, dicobalt octacarbonyl, and tris(triphenylphosphine)platinum or palladium,[25] as neutral or monoanionic complexes, except the last one which has only been obtained as the monoanionic species. A different and more convenient synthetic route to these anionic complexes, using halotriphenylphosphine metal compounds, was reported for cobalt, nickel, palladium, copper, and gold.[26] Iron and cobalt complexes were also obtained in the dianionic form.[27]

Selenium analogs of the type $[M(Se_2C_2(CF_3)_2)_2]^z$, where $z = 0$, $1-$, and $2-$, and M = Fe, Co, Ni, and Cu, have been prepared from bis(trifluoromethyl-1,2-diselenetene, $Se_2C_2(CF_3)_2$.[28] These 1,2-deselenetene complexes are generally less stable than the corresponding 1,2-dithiolene complexes.

The linear chain compounds derived from these complexes are listed in Table 2.

2.1.3. *Metal Bis(cis-1,2-dicyanoethylene-1,2-dithiolene) (R = CN; z = 1− and 2−; M = Fe, Co, Ni, Cu, Zn, Pd, Pt and Au)*

The ligand *cis*-1,2-dicyanoethylene-1,2-dithiolate, more commonly known as maleonitriledithiolate, mnt, had been originally prepared by Bähr and Schleitzer[1] from sodium cyanide and carbon disulfide to give sodium cyanodithioformate which then dimerized with loss of sulfur. This technique was later slightly modified and simplified in the work-up procedure[24,41,42,43] and by the use of acetone instead of *N,N*-dimethylformamide, DMF.[44] The tetraalkylammonium salts in the reduced form are in general readily prepared.[24,45,46] The monoanionic complexes are obtained by oxidation of the precedent, either with iodine in the cases of nickel and platinum, using DMSO as solvent,[23,24,46] and using sulfolane in the case of palladium[47] or with $Ni[S_2C_2(CF_3)_2]_2$ for copper and palladium.[23,24] The gold complex can be obtained directly in the monoanionic form[23] although it has been also obtained in the reduced 2-form.[48,49] Bis complexes of cobalt and iron appear as dimers[23,46] and are prepared more conveniently by a modified technique.[43,50]

Linear chain complexes prepared from $[M(mnt)_2]^{1-}$ and perylene were obtained either through prior iodine oxidation of perylene, Per, or by electrolysis.[63,64,99] A very convenient electrolytic technique,[51] giving generally good crystals, is accomplished by maintaining the current intensity at a constant value of about 10 μA, during the entire electrolysis period of approximately one week. Most of the other linear chain complexes derived from $[M(mnt)_2]^z$ have been obtained by metathesis. All complexes of this series are listed in Table 3.

TABLE 3. Complexes with Metal Bis(cis-1,2-dicyanoethylene-1,2-dithiolene), $M(mnt)_2$

Donor or cation	Transition metal	Stoichiometry	Reaction type	Conductivity σ (ohm^{-1} cm^{-1})	Conductivity E_a (eV)	Reference
TMPD	Ni, Ni + Cu	2:1	MT	—		52, 53
KOS	Co, Ni, Cu, Zn	2:1	MT	—		32, 54
KOS	Fe, Co	2:2	MT	—		32
KOS	Ni	1:1	MT	—		32
pyMe	Co, Ni	2:1	MT	—		32
pyMe	Fe	2:2	MT	—		32
CNpyR	Ni	2:1	MT	—		32
CNpyMe	Co, Zn	2:1	MT	—		32
TAE	Pt	1:1	MT	10^{-11}		30
TTF	Ni	1:1	CT, MT	$1.6 \times 10^{-3}/5 \times 10^{-4}$	0.15	15, 20
TTF	Cu	1:1	CT, MT	2.3×10^{-5}		55
TTF	Pd	1:1	CT, MT	2.1×10^{-4}		55

(continued)

TABLE 3 (continued)

Donor or cation	Transition metal	Stoichio-metry	Reaction type	Conductivity σ (ohm^{-1} cm^{-1})	Conductivity E_a (eV)	Reference
TTF	Pt	1:1	CT, MT	1.4×10^{-4}		55, 56
TTF	Fe	2:2	MT	2×10^{-4}		15
TTF	Co	2:2	MT	4×10^{-4}	0.24	15
TTF	Ni	2:1	MT	1.3×10^{-5}		15, 56
TTF	Cu	2:1	MT	5×10^{-7}		15, 56
TTF	Pd	2:1	MT	1.7×10^{-6}		15, 56
TTF	Pt	2:1	MT	1.7×10^{-4}		30, 56
TTT	Pd	2:1	MT	2×10^{-4}		30
TTT	Pt	2:1	MT	0.33		30
TTT	Ni	1:1	MT	3.1×10^{-3}		15, 30
TTT	Pt	1:1	MT	1.25×10^{-2}		15, 30
TCNTTF	Ni	1:1	MT	3.3×10^{-8}		15, 57
$Ni(opd)_2$	Ni	1:1	MT	1.2×10^{-6}	0.24	15
$Ni(opd)_2$	Ni	2:1	MT	7.9×10^{-6}		15
$Ni(dmopd)_2$	Ni	1:1	MT	3.8×10^{-4}	0.33	15
$Ni(dmopd)_2$	Ni	2:1	MT	$<10^{-7}$		15
CV	Ni	1:1	MT	3×10^{-9}		15
An_3C	Ni	1:1	MT	3×10^{-9}		15
MG	Ni	1:1	MT	3×10^{-9}		15
NMP	Ni	1:1	MT	$2 \times 10^{-8}/10^{-10}$		15, 58
NMP	Ni	2:1	MT	—		59
Thian	Ni	1:1	MT	3.3×10^{-5}		15
WB	Ni	1:1	MT	2.1×10^{-6}		15
MB	Ni	1:1	MT	6×10^{-7}	0.28	60, 61, 62
MB	Ni	2:1	MT	10^{-9}	0.41	61, 62
MB	Fe	2:2	MT	2×10^{-11}	0.54	60, 61, 62
$Fe(HMB)_2$	Ni	1:1	MT	10^{-8}	0.52	60
Per	Ni	2:1	RD, EL	50[a]		63
Per	Cu	2:1	RD, EL	6[a]		63
Per	Pd	2:1	RD, EL	7×10^{-2}[a]		63
Per	Pt	2:1	RD, EL	280[a]		64, 99
NH_4	Ni	1:1	MT	5×10^{-2}		65, 114
NH_4	Pd	1:1	MT	1.1		65
Na	Ni	1:1	MT	$4 \times 10^{-4}/0.25$		65
Na	Pd	1:1	MT	0.4		65
$Ag(PPh_3)_2$	Ni	2:1	MT	—		66, 67
$Ag(PTol_3)_2$	Ni	2:1	MT	—		66
$Cu(PPh_3)_2$	Ni	2:1	MT	—		66
$Cu(PPh_3)_3(Ac)$	Ni	2:1	MT	—		66
$Cu(PPh_3)(Py)$	Ni	2:1	MT	—		66
$(Ph_3Z)_2Av$	Au (Z = P, As)	1:1	MT	—		68
TMTTF	Ni, Pd, Pt	2:1	MT	10^{-8}		20
TMpy	Cu	2:1	MT	2.1×10^{-12}	0.9(<64°C)	69
Tpy	Cu	2:1	MT	6.5×10^{-12}	1.3(<85°C)	69
Pol	Cu	1:1	—	2.5×10^{-9}	0.58	69
Ag(TMA	Cu	1:1	MT	2.9×10^{-9}	0.84	69
Ni(TMA)	Cu	1:1	MT	3.2×10^{-4}	0.27	69

[a] Single crystal.

2.1.4. Metal Bis(cis-stilbene-1,2-dithiolene) ($R = C_6H_5$; $z = 0, 1-$, and $2-$; M = Fe, Co, Ni, Pd, and Pt)

The first complex of this series to be prepared was the neutral bis(*cis*-stilbenedithiolate)nickel, $Ni(S_2C_2Ph_2)_2$, also named bis(dithiobenzyl)nickel, from the reaction of nickel sulfide with diphenylacetylene in toluene.[70, 71] A more general method for the preparation of these complexes was developed afterward starting from benzoin and P_4S_{10}, followed by reaction of the formed thiophosphoric ester of dithiobenzoin with adequate metal salts.[72–74] Iron and cobalt dimer complexes were also prepared conforming to this synthesis.[75, 76] Reduction to the monoanion takes place with *p*-phenylenediamine in DMSO,[3, 23] a reaction that can be reversed by iodine in dichloromethane. Reduction of the neutral nickel complex to the dianion was achieved with $NaBH_4$ in methanol.[90]

The only linear chain complex of this series reported in the literature was (4-acetyloxy-1-ethylpyridinium) $[Ni(S_2C_2Ph_2)_2]$.[32]

2.1.5. Metal Bis(cis-1,2-dialkylethylene-1,2-dithiolene) (R = Alkyl; $z = 0$ and $1-$; M = Ni, Pd, and Pt)

The synthesis of 1,2-dialkylethylene-1,2-dithiolenes is achieved by a method analogous to that used for the diphenyl derivative, except that the adequate acyloin must be used in the reaction with P_4S_{10}.[72, 74, 77] Only two linear chain complexes of this type have been reported (Table 4). The selenium analogs, neutral and dianionic bis(dimethyldiselenolene)-nickel, were synthesized from dimethyl-1,3-diselenole-2-thione.[78] Although it was affirmed that the method is a general one for the preparation of other substituted bis(1,2-diselenolene) complexes, only those have been characterized. It was also said that charge-transfer complexes with donors like tetrathiafulvalene could be prepared, but they have not been described.

2.1.6. Metal Bis(tetramethylene-ethylene-1,2-dithiolene) (R = Tetramethylene; $z = 0$ and $1-$; M = Ni)

The neutral complex, formally the bis(cyclohexene-1,2-dithiolene)nickel was prepared by the reaction of 2-hydroxycyclohexanone thiophosphoric ester with nickel chloride. The work-up procedure includes ether extraction and chromatographic purification (mp 185°C).[20] The monoanionic species was obtained through potassium ethoxide ring opening of tetramethylene-1,3-dithiole-2-thione to give the dipotassium salt of cyclohexene-1,2-dithiol,

TABLE 4. Complexes with Metal Bis(cis-1,2-dimethylethylene-1,2-dithiolene), $M(Medt)_2$

Donor or cation	Transition metal	Stoichiometry	Reaction type	Conductivity σ (ohm^{-1} cm^{-1})	Reference
MBT	Pt	1:1	CT	5×10^{-7}	30
TTT	Pt	3:1	CT	2×10^{-5}	30

which was made to react with nickel chloride and tetrabutylammonium bromide.[20]

No linear chain compounds were reported to be prepared with these complexes.

2.2. *Aromatic 1,2-Dithiolene Complexes: $[M(S_2Ar)_2]^z$ Type; $z = 1-$ and $2-$*

Known bis(arene-1,2-dithiolenes) include the following aromatic ligands

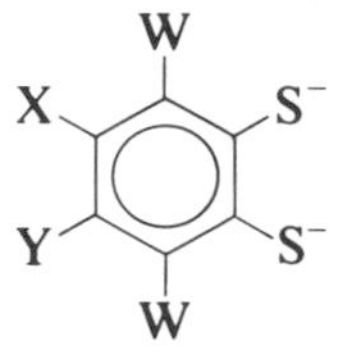

$$W = X = Y = H \text{ (bdt); } CH_3 \text{ (tmdt); } F \text{ (tfdt); } Cl \text{ (tcdt)}$$

$$W = X = H;\ Y = CH_3 \text{ (tdt)}$$

$$W = H;\ X = Y = CH_3 \text{ (xdt)}$$

The ligand benzene-1,2-dithiol, classically obtained by reduction of 1,2-benzenesulphonyl chloride with Zn powder[79] is more conveniently prepared from 1,2-dibromobenzene through 1,2-bis(butylthio)benzene.[80] This method is also adequate for the preparation of *o*-xylene-4,5-dithiol and 3,4,5,6-tetramethylbenzene-1,2-dithiol, starting from the suitably substituted *o*-dibromobenzenes.[81] Benzene-1,2-dithiol has also been synthesized from *o*-aminotiophenol[82] and more recently from anthranilic acid.[83] 3,4,5,6-Tetrachlorobenzene-1,2-dithiol was prepared from hexachlorobenzene[81,117] and 3,4,5,6-tetrafluorobenzene-1,2-dithiol from 1,2,3,4-tetrafluorobenzene.[84] Toluene-3,4-dithiol is commercially available since it is an analytical reagent.

TABLE 5. Complexes with Metal Bis(benzene-1,2-dithiolene), $M(bdt)_2$

Donor or cation	Transition metal	Stoichiometry	Reaction type	Conductivity σ (ohm^{-1} cm^{-1})	Reference
$Ni(opd)_2$	Ni	1:1	MT	4×10^{-4}	15
TTF	Ni	2:1	EL	10^{-2}	20, 88, 89
TTF	Ni	1:1	MT	10^{-2}	20
TMTTF	Ni, Cu	1:1	MT, EL	10^{-2}	20, 88, 89
DEDMTTF	Ni	1:1	EL	—	88, 89

TABLE 6. Complexes with Metal Bis(4-methylbenzene-1,2-dithiolene), $M(tdt)_2$

Donor or cation	Transition metal	Stoichiometry	Reaction type	Conductivity		Reference
				σ (ohm^{-1} cm^{-1})	E_a (eV)	
NMP	Ni	1:1	MT	10^{-7}		15
$Ni(opd)_2$	Ni	1:1	MT	2.5×10^{-3}	0.17	15
$Ni(opd)_2$	Fe	2:2	MT	3.5×10^{-5}		15
$Ni(opd)_2$	Co	2:2	MT	1.4×10^{-4}		15
$Ni(dmopd)_2$	Ni	1:1	MT	7.1×10^{-9}	0.33	15
$Ni(dmopd)_2$	Fe	2:2	MT	3.9×10^{-9}		15
$Ni(dmopd)_2$	Co	2:2	MT	8.7×10^{-10}		15

Bis(arene-1,2-dithiolene)metal complexes of these ligands are obtained in the 1− and 2− forms, the monoanionic salts being usually the most stable. The following have been prepared:

$M(bdt)_2$: M = Co, Ni, Cu Ref. 81; Pt, Pd Ref. 20

$M(tdt)_2$: M = Fe Ref. 27; Co, Ni, Cu, Pt, Au Refs. 85, 86; Pd Ref. 20

$M(xdt)_2$: M = Co, Ni, Cu Ref. 81; Au Ref. 87

$M(tmdt)_2$: M = Co, Ni, Cu Ref. 81; Au Ref. 87

$M(tcdt)_2$: M = Fe Refs. 76, 117; Co, Ni, Cu, Au Ref. 81

$M(tfdt)_2$: M = Cr, Mn, Fe, Co, Ni, Cu Ref. 84

Linear chain complexes prepared with $M(bdt)_2$ and with $M(tdt)_2$ are listed in Tables 5 and 6.

3. *Molecular and Electronic Structures of Metal 1,2-Dithiolene Complex Molecules*

3.1. *Molecular Structures*

The 1,2-dithiolene complex molecules that form linear chain compounds are square planar and result from the combination of two ligands

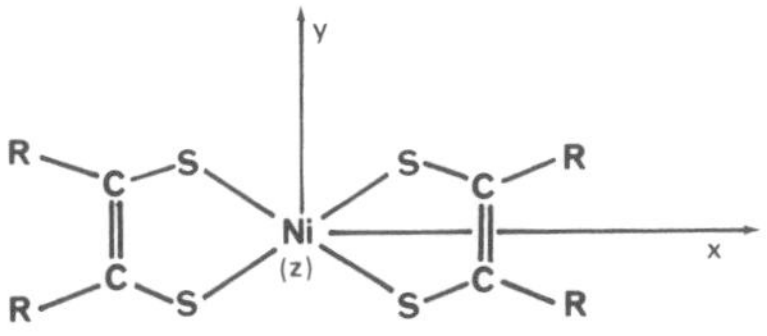

FIGURE 1. Coordinate axis assumed in the description of the electronic structures of $Ni(S_2C_2R_2)_2$.

TABLE 7. Comparison of Assigned Observed Frequencies for Some Complexes of the Type $[Ni(S_2C_2R_2)_2]^z$ in cm^{-1} [a]

$[Ni(S_2C_2R_2)_2]^z$		ν(C=C)	ν(Ni–S), B_{2u}	ν(C–S)	δ(S–Ni–S)	ν(Ni–S), B_{3u}
R = H	$z = 0$	1350 s, 1358 sh	428 m	798 m	237 m	398 vw
	$z = 1-$	1435 s	415.5 m	790 sh	233 w	385 sh
$R = C_6H_5$	$z = 0$	1365 s	408 vw, 475 m		202 w, 130 m	454 w
	$z = 1-$	1475 s	406 vw, 465 w		130 m	428 vw
	$z = 2-$	1533 s	401 s, 450 w		207 m, 128 s	418 sh
$R = CF_3$	$z = 0$	1425 m	465 m		211 w	425 w
	$z = 1-$	1485 s	449 m		210 w	415 w
	$z = 2-$	1534 s	422 w		208 m	394 w
R = CN	$z = 1-$	1435 sh, m[b]				
	$z = 2-$	1485 s				

[a] ν, stretching; δ, in-plane bending; s, strong; m, medium; w, weak; vw, very weak; sh, shoulder.
[b] sh, for the (Et4N) salt; m, for the sodium salt.

with a metal atom (Figure 1). They therefore have D_{2h} symmetry. The infrared spectra of these complexes have been studied in great detail by Schläpfer and Nakamoto[91] and by Schrauzer.[5] An important point that resulted from these studies is the dependence of the frequencies of some modes on the net charge of the complex which is illustrated in Table 7.

Some of the complexes present dimeric structures, $[M(S_2C_2R_2)_2]_2^z$, M = Fe and Co; $R = CF_3$ ($z = 0$, $1-$, and $2-$); or R = CN ($z = 2-$).[27, 119, 120] The dimerization occurs through metal–sulfur bonds to give a square pyramidal coordination about the metal:

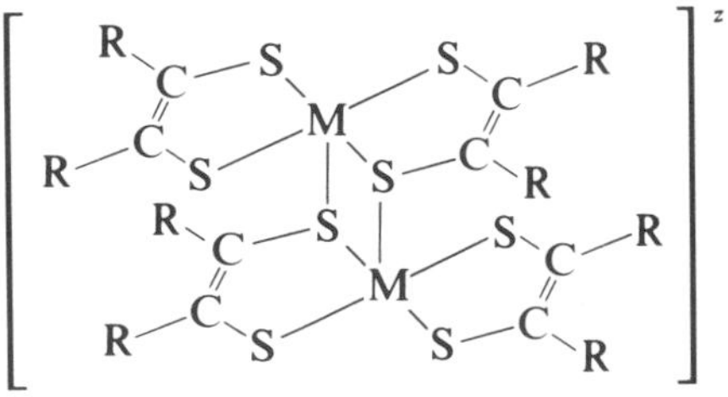

Other dimeric structures[121,122] present direct metal–metal bonding as in $[M(S_2C_2H_2)_2]_2$, M = Pd and Pt:

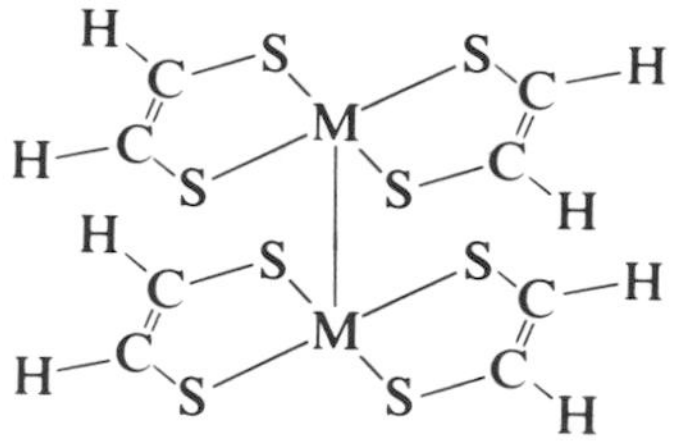

We will now briefly discuss the electronic structures of 1,2-dithiolene complexes and their potentialities to form linear chains.

3.2. Electronic Structures

There have been three major attempts to understand and explain the electronic structures of the planar bis(1,2-dithiolene)metal complexes.[11,92,93] Two of the calculations provided essentially the same answer to the assignment of the ground state of the mnt complexes which for the $[Ni(mnt)_2]^{1-}$ species is $^2B_{2g}$. This is basically the assignment of Schrauzer[11] also confirmed by an electron spin resonance, esr, study of the ^{33}S hyperfine splittings in $[Ni(mnt)_2]^{1-}$.[94] The unpaired electron resides in an orbital of d_{xz} symmetry. Many more contributions to the study and understanding of the electronic structures of bis(1,2-dithiolene)metal complexes have been made.[4,5]

In spite of some still existing controversy,[95] we present in Figure 2 Schrauzer's scheme of the electronic structure of bis(1,2-dithiolene)metal complexes. From that description, and from all other studies of the electronic structures of bis(1,2-dithiolene)metal complexes, it is clear that the monoanionic and the neutral complexes have highly delocalized ground states. The molecular orbitals of those complexes receive important contributions from the ligand. In the case of the neutral $Ni(edt)_2$ complex, the lowest unoccupied molecular orbital (LUMO) has $3b_{2g}$ symmetry (Figure 3) and the metal orbitals contribution (d_{xz}) is according to Schrauzer[11] of the order of 18%.

X-ray photoelectron spectroscopy[96] confirms the early assumption on delocalization and contribution to the molecular orbitals of the complexes. It has been pointed out that the charge on the metal remains essentially constant in the 0, 1−, and 2− series, and the addition of electronic charge gained on reduction resides mainly on the ligands and, in particular, on the sulfur atoms.

The magnetic properties and the esr spectra of the 1,2-dithiolene complexes are, in general, consistent with the above picture of the electronic structure; namely, with Schrauzer's molecular orbitals scheme. In fact, the Ni, Pd, and Pt complexes with formula $[M(S_2C_2R_2)_2]^{1-}$ (R = H, CN, CF_3,

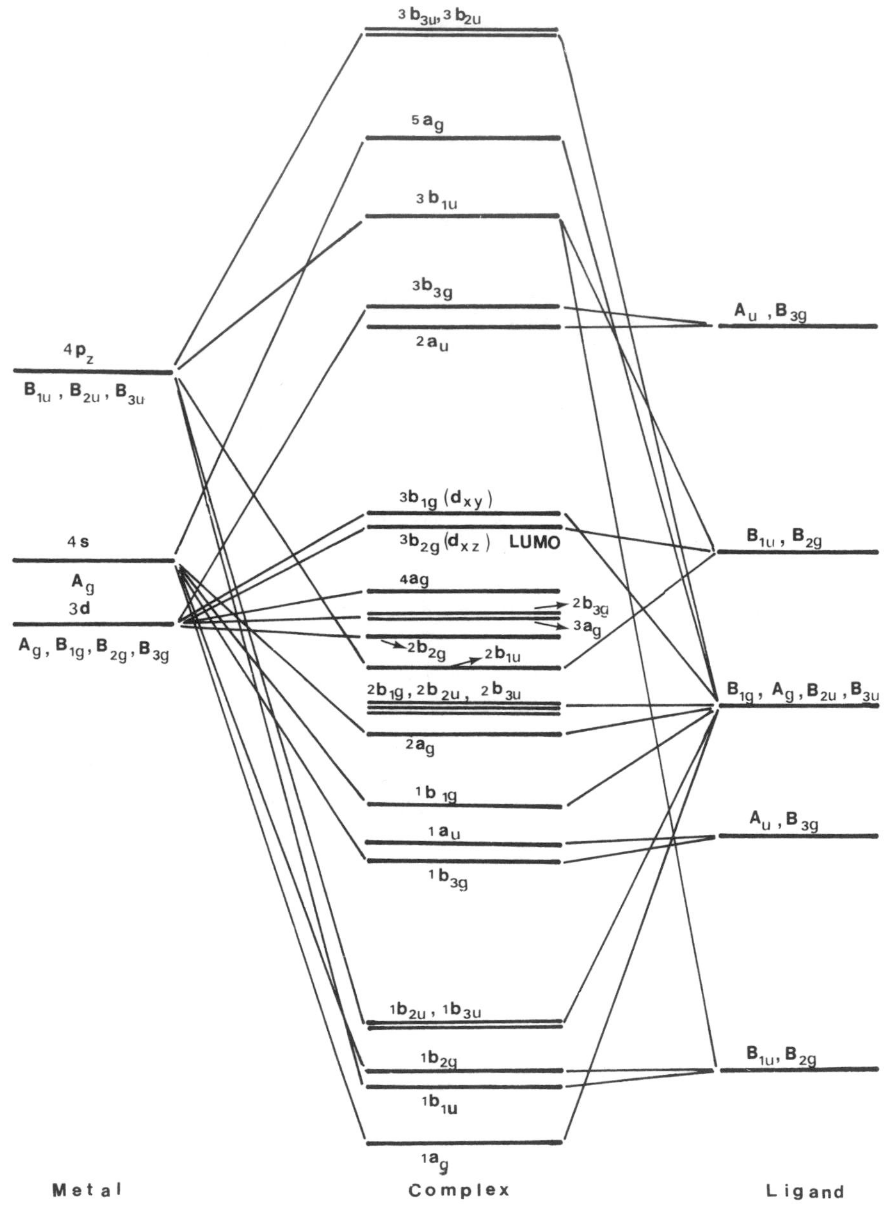

FIGURE 2. Schematic representation of the energy levels in $Ni(S_2C_2R_2)_2$.

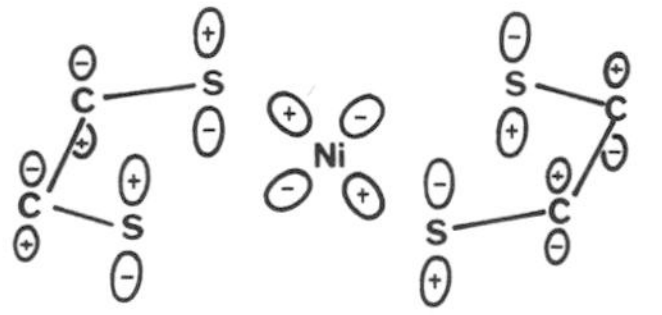

FIGURE 3. The $3b_{2g}$ molecular orbital in $Ni(S_2C_2R_2)_2$.

C_6H_5, and $C_6H_4CH_3$) are paramagnetic with a doublet ground state ($S = 1/2$), showing in the solid state lower effective magnetic moments than expected due to exchange interactions through metal–sulfur bonding. In the formalism of the electronic configuration of the metal oxidation state, these complexes would be M(III), d^7 complexes; but in this case, the formalism is obviously inadequate since they are electron delocalized. The corresponding dianions, currently described as M(II), d^8 complexes, are diamagnetic, and the trianion of the nickel complex proved to be paramagnetic, as expected.

The dianions of the copper and gold complexes have been thoroughly studied by esr.[100, 123] For $[Au(mnt)_2]^{2-}$ the data[123] are consistent with a ground-state hole configuration ($b_{1g}^2 a_g^1$) corresponding to the valence electronic configuration ($b_{2g}^2 a_g^1$). b_{1g} is the normally half-filled orbital in d^9 complexes and a_g, lying below b_{1g}, is a primarily ligand orbital with a metal 6*s* component (~15%) and small admixtures of metal 5*d* components. Such a scheme requires that the b_{1g} orbital be raised in energy above the a_g orbital. This is in agreement with the fact that the metal *d* orbitals are, in general, raised in energy in the heavier transition metals, whereas the ligand-based orbitals are relatively unaffected.

The electronic structure, orbital occupancy, electronic configuration, and metal oxidation state as well as the magnetic and esr properties of the most common nonassociated 1,2-dithiolene complexes are summarized in Table 8.

The bis(1,2-dithiolene)metal complexes offer a great variety of electron-rich oxidation states. The 0, 1−, and 2− species are common, and even the 3− and the 4− species have been detected in some cases.[97] Oxidation–reduction data on bis(1,2-dithiolene)metal complexes have been extensively tabulated in the literature.[4, 30, 77, 98]

Several attempts have been made to correlate redox potentials, electron affinities, ionization potentials, polarizabilities, and steric factors with the tendency to give conducting donor–acceptor complexes.[30, 98] The most serious attempts that have been made concern the formation of segregated stacks. It has been suggested that segregated stacks and high conductivities can be expected when the oxidation potential of the donor, E_D, is close to the reduction potential of the acceptor, E_A, such that $|E_A - E_D| < 0.25$ V. In such cases, partial charge transfer can occur, resulting in unfilled conduction bands that would lead to high conductivities. In the case of the bis(1,2-

TABLE 8. Electronic Configuration, Metal Oxidation State, and Magnetic and ESR Properties for $[M(S_2C_2R_2)]^z$ Complexes

M	z	Number of d electrons	Assigned ground state of complex	μ_{eff} (μ_B)	$\langle g \rangle$	g_1	g_2	g_3	A_1	A_2	A_3	Reference
									($cm^{-1} \times 10^4$)			
Ni(III)[a]	1−	d^7 [e]	$\cdots 4a_g^2\, 3b_{2_g}^1 = {}^2B_{2_g}$	1.83[c]	2.0633	1.998	2.042	2.160[f]	(−)15	(+)2.9	<2	4,93
Ni(III)[b]	1−	d^7 [e]		1.85	2.0618	1.996	2.043	2.139[f]				4
Pd(III)[a]	1−	d^7		Diamagnetic[d]	2.0238	1.956	2.056	2.065[f]	(+)9.0	(+)6.3	(+)5.5	4
Pd(III)[b]	1−	d^7		1.73	2.0238	1.955	2.049	2.065[f]				4
Pt(III)[a]	1−	d^7		1.05	2.042	1.825	2.067	2.221				4
Pt(III)[b]	1−	d^7		1.73	2.039	1.823	2.074	2.210				4
Ni, Pd, Pt(II)	2−	d^8		Diamagnetic								
Ni(I)[a]	3−	d^9			2.063	2.062	2.081	2.2055	(−)53	(±)5.8	(±)5.8	97
Pd(I)[a]	3−	d^9				2.034	2.042	2.116	(+)44	(+)29	(+)26	97
Co(III)[a]	1−	d^6	spin–triplet	2.81								4
Co(II)[a]	2−	d^7	$\cdots 4a_g^2\, 3b_{2_g}^1 = {}^2B_{2_g}$	2.16	2.255	1.977	2.025	2.798[c, f]	23	28	50[c, f]	4, 93
Ph(II)[a]	2−	d^7		1.91	2.24	1.936	2.019	2.447[c, f]	<4	7.5	<4[c, f]	4, 93
Cu, Au(III)	1−	d^8		Diamagnetic								
Cu(II)[a]	2−	d^9	$\cdots 3b_{2_g}^2\, 3b_{1_g}^1 = {}^2B_1$	1.78	2.0458	2.086	2.026	2.026[c, f]	(−)162	(−)39	(−)39[c, f]	93, 97
Au(II)[a]	2−	d^9	$\cdots 3b_{2_g}^2\, a_g^1 = {}^2A_g$	1.85	2.009	2.0160	1.9777	2.0059[c, f]	±41.5	±39.2	±40.7[c, f]	123

[a] R = CN.
[b] R = CF_3.
[c] In magnetically dilute sample.
[d] In solid.
[e] The formalism of the electron configuration of the metal oxidation state is inadequate since the complexes are spin delocalized.
[f] $g_1 = g_z$, $g_2 = g_x$, $g_3 = g_y$.

dithiolene)metal complexes, due to the great variety of structures, it seems quite difficult with the data available and the limited number of reasonably good conductors to attempt any simple correlation.

Any serious work to attempt such correlation should take into account the energies and symmetries of the lowest unoccupied molecular orbitals (LUMO) and of the highest occupied molecular orbital (HOMO) of acceptor and donor, respectively, as well as redox potentials, electron affinities, ionization potentials, and steric factors. An interesting example of those difficulties are the $(Per)_2[M(mnt)_2]$ and $(Per)[M(S_2C_2(CF_3)_2)_2]$ complexes (per = perylene).[29, 99] In the first series of complexes, segregated stacks as well as high conductivities are observed, whereas in the $(Per)[M(S_2C_2(CF_3)_2)_2]$ there are alternating stacks and poor conductivities. The donor in both cases is perylene and the acceptor is expected to have similar LUMO orbitals and symmetries. Probably the reason for such different behavior is the presence of the CF_3 bulky group in the second case, therefore a steric factor.

From the electronic structures of $[M(S_2C_2R_2)_2]^z$ molecules, it is then clear that the potentials do exist for the formation of linear chain complexes and for intermolecular interactions along a chain, as well as for the combination in dimers with metal–sulfur or metal–metal bonding. This would be possible through overlap of the out-of-plane molecular orbitals, which usually have metal d_{xz} and sulfur p_z components.

The redox potentials certainly play a role in the formation of linear chain complexes which can be fully charge transferred when the oxidation potential of the donor, E_D, is very different from the reduction potential of the acceptor, E_A, or simply associate as neutral species when they are very similar. For example, $(TTF)[Ni(S_2C_2(CF_3)_2)_2]$ (TTF = tetrathiafulvalene), with $E_A = +0.92$ V versus SCE[4] and $E_D = +0.33$ V versus SCE[30] is a fully charge-transferred D^+A^- salt,[16] while TTF can combine with $Ni(edt)_2$, $E_A = +0.09$ V versus SCE,[4] in $(TTF)_2[Ni(edt)_2]_3$ which is not charge transferred[16] and is actually a disordered solid solution of the very similar neutral donor and acceptor molecules (see Section 4.1.9). These same molecules, however, do form a charge-transfer complex with a different structure in $(TTF)_2[Ni(edt)_2]$[16] (Section 4.1.6).

Although there is the possibility for intermolecular interactions along a chain of $[M(S_2C_2R_2)_2]^z$ molecules, there is no clear evidence for truly one-dimensional metal 1,2-dithiolene chains which would be responsible for high electrical conductivities. The only example in which this situation could arise is the compound $(NH_4)[Ni(mnt)_2]\cdot H_2O$ which exhibits a single-crystal conductivity of about 5×10^{-2} ohm^{-1} cm^{-1} at room temperature.[65] However, the conductivity is not very high and there is insufficient information on that compound to clearly decide on the nature of the interactions involved.

4. Linear Chains of Metal 1,2-Dithiolene Complex Molecules

4.1. Structural Data

We will review the structural data on the molecular complexes in which there is evidence of one-dimensional character and/or which are relevant for their electrical and magnetic properties. There is a great variety of known structures, from regular segregated stacks to alternate stacks. In the structures with segregated stacks of $[M(S_2C_2R_2)_2]^z$ molecules, there is the possibility of intermolecular interactions along the chain through overlapping of the out-of-plane $^3b_{2g}$ molecular orbitals.

The ability to form stacks is certainly related to this possibility but it is often limited by the nature of the cations or donor molecules which, in many cases, form stacks themselves or combine in an alternate arrangement. In general, the formation of segregated or alternate stacks give rise to cooperative electrical and magnetic behavior.

We shall now describe examples of the most typical structures and relate them to the physical properties in Section 4.2.

4.1.1. Segregated Regular Stacks—Two Stacks

The crystal structure $(Per)_2[M(mnt)_2]$, M = Pt and Au (Per = perylene), consists of two segregated regular stacks, one of $[M(mnt)_2]^{1-}$ molecules and another of $(Per)^{1/2+}$ molecules, both parallel to the *b* axis. Each $[M(mnt)_2]^{1-}$ column is surrounded by six columns of perylene. Interchain coupling is weak. For the platinum compound, for which the structure determination is now completed,[99] adjacent $[Pt(mnt)_2]^{1-}$ molecules

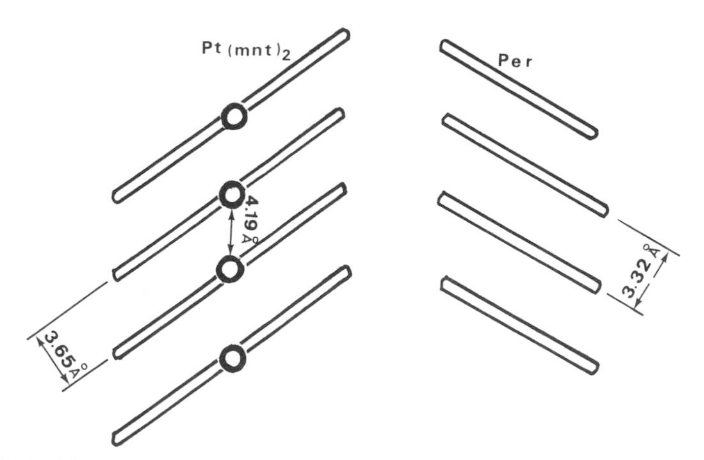

FIGURE 4. Schematic representation of the molecular arrangement in the structure of $(Per)_2[Pt(mnt)_2]$.

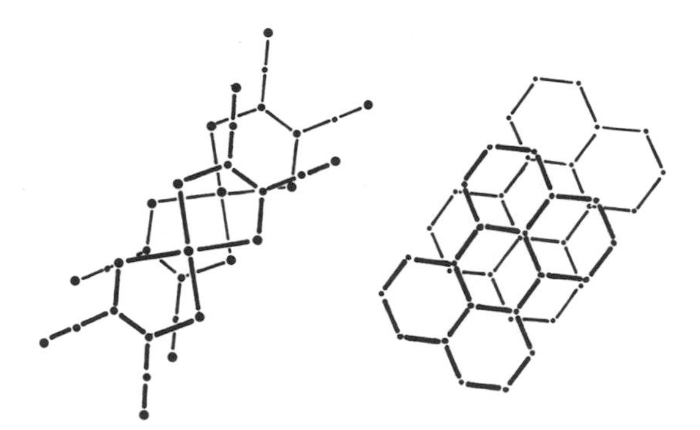

FIGURE 5. Overlap of the perylene and $Pt(mnt)_2$ molecules in $(Per)_2[Pt(mnt)_2]$.

are separated by 3.65 Å, and the Pt–Pt distance is 4.19 Å. The perylene column presents a very good graphitelike mode of overlap at a short distance of 3.32 Å, which is good evidence for strong interactions. The molecular arrangement as well as the actual overlap of perylene and $[M(mnt)_2]^{1-}$ molecules are schematically represented in Figures 4 and 5.

4.1.2. *Segregated Regular Stacks—One Stack of* $[M(S_2C_2R_2)]^{2-}$ *Ions*

$(Bu_4N)_2[M(mnt)_2]$, M = Ni and Cu, and $(Et_4N)_2[M(mnt)_2]$, M = Ni and Cu, are simple structures consisting of segregated stacks of $[M(mnt)_2]^{2-}$ ions regularly spaced, but in which the distance between two neighboring dianions is large and only very weak interactions of spin-exchange type can take place when species are paramagnetic.[100–103] In the case of $(Bu_4N)_2[Cu(mnt)_2]$,[100] the distance between planes is 6.91 Å and the distance between copper atoms in two adjacent planes is 9.4 Å. For the nickel analog, the distance between neighboring Ni atoms is 9.84 Å. This arrangement is shown schematically in Figure 6.

4.1.3. *Segregated Stacks—Dimeric and Tetrameric Units of* $[M(S_2C_2R_2)_2]^{1-}$ *Ions*

For the $(Et_4N)[Ni(mnt)_2]$ structure,[101] the anions are stacked in columns whose axes are parallel to *c*. Those columns are surrounded by cations. The $[Ni(mnt)_2]^{1-}$ ions are weakly connected with each other and form dimeric units within the columns. Overlapping of nickel and sulfur atoms is especially large. The average interplanar spacing within the dimeric unit is ~3.5 Å (Figure 7).

The $(Ph_3MeP)[M(mnt)_2]$ complexes, M = Ni and Pt, have the same type of structure[104] consisting of isolated stacks with dimerization within each stack. There is a metal–sulfur bond of 3.41 Å.

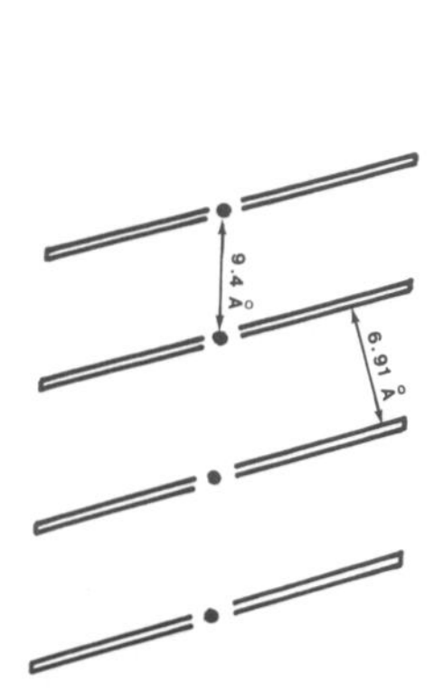

FIGURE 6. Schematic representation of the arrangement of $[Cu(mnt)_2]^{2-}$ in the crystal structure of $(Bu_4N)_2[Cu(mnt)_2]$.

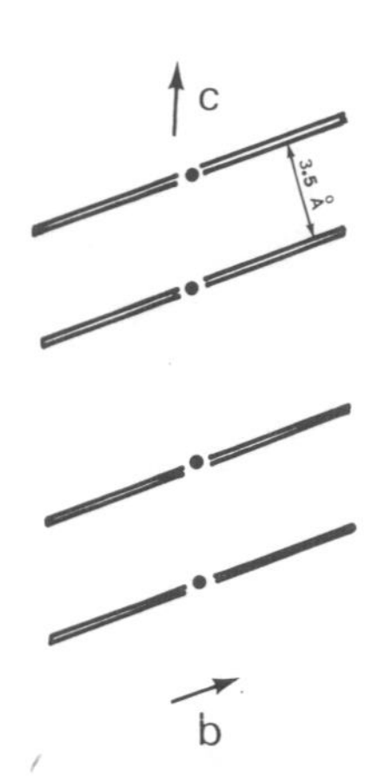

FIGURE 7. Schematic representation of the molecular arrangement in the crystal structure of $(Et_4N)[Ni(mnt)_2]$.

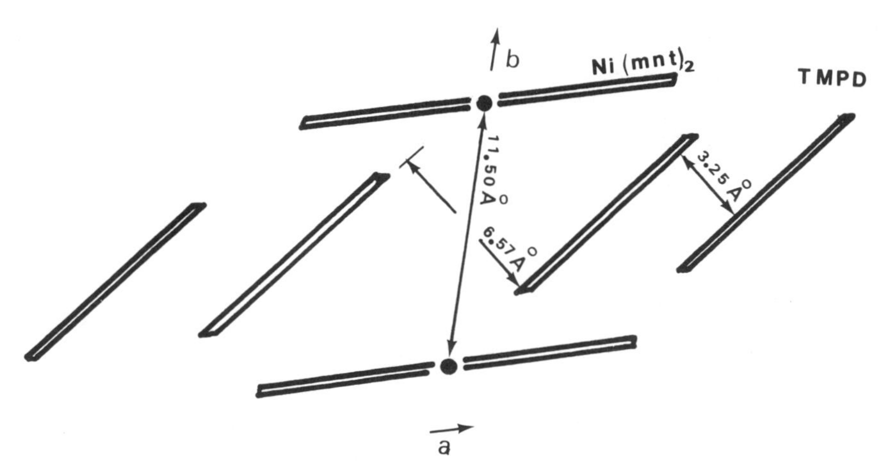

FIGURE 8. Schematic representation of the molecular arrangement in the crystal structure of $(TMPD)_2[Ni(mnt)_2]$.

In $(Bu_4N)[Cu(mnt)_2]$,[101, 105] the columns are composed of tetrameric units of $[Cu(mnt)_2]^{1-}$. The average interplanar spacing within each unit is 3.6 Å.

4.1.4. Segregated Stacks—One Regular Stack of $[M(S_2C_2R_2)_2]^{2-}$ Ions

The structure[52] of $(TMPD)_2[Ni(mnt)_2]$ (TMPD = *N,N,N′,N′*-tetramethyl-*p*-phenylenediamine) consists of alternating flat sheets of $[Ni(mnt)_2]^{2-}$ regularly spaced along the *c* axis and stacked sheets of

$(TMPD)^+$ (Figure 8). The organic free radicals are stacked in dimerized units along the *a* direction with an intradimer separation of only 3.25 Å, the shortest interatomic separation being 3.65 Å. In contrast, for adjacent radicals in different dimer units, the plane-to-plane separation is 6.57 Å and the shortest interatomic distance is 6.97 Å.

4.1.5. Segregated Regular Stacks—One Stack of Cations

The crystal structure[17, 19] of $(TTT)_{1.2}[Ni(edt)_2]$ (TTT = tetrathiotetracene) consists of a one-dimensional chain of stacked $TTT^{5/6+}$ units separated by $[Ni(edt)_2]^{1-}$ ions arranged in two separated orthorhombic subcells. In Figure 9, we schematically represent the molecular arrangement of the two subcells normal to the *ac* plane, showing coincidence of the $Ni(edt)_2$

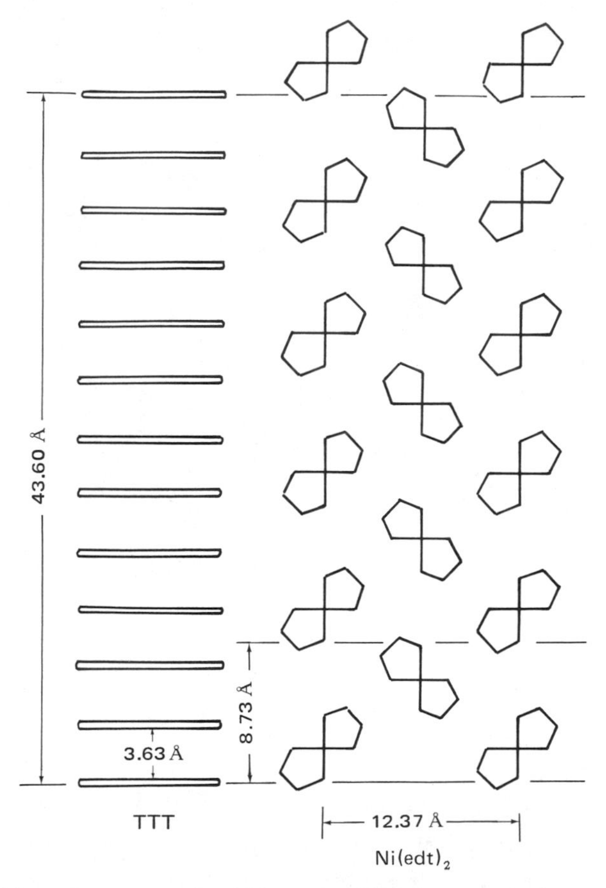

FIGURE 9. Schematic representation of the molecular arrangement in the crystal structure of $(TTT)_{1.2}[Ni(edt)_2]$.

and TTT subcells at $5c_{Ni} = 12c_{TTT}$. The TTT molecules are stacked along the c axis with uniform spacing $c_{TTT} = 3.63$ Å and are almost parallel to the ab plane. The $[Ni(edt)_2]^{1-}$ molecules are parallel to the ac plane. There is no stacking of $[Ni(edt)_2]^{1-}$ molecules.

4.1.6. Segregated Stacks—One Stack of Cations

The structure[16, 124] of $(TTF)_2[Ni(edt)_2]$ (TTF = tetrathiafulvalene) consists of three crystallographically distinct types of TTF molecules occurring in stripes parallel to (100) and alternating with stripes of $Ni(edt)_2$ molecules (Figure 10a). Two of the crystallographically different TTF molecules form columnar stacks along b, types I and II, the later being rotated by about 60° with respect to the first (Figure 10b). Types II and II′ are fully

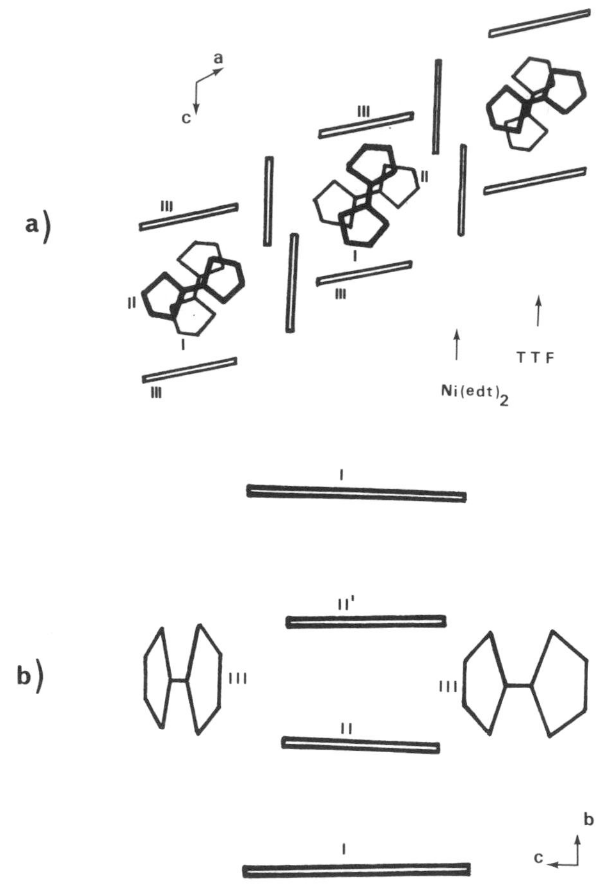

FIGURE 10. Schematic representation of the molecular arrangement in the crystal structure of $(TTF)_2[Ni(edt)_2]$.

eclipsed with a short sulfur–sulfur distance of 3.48 Å, allowing strong intermolecular interactions. The third crystallographically different TTF molecules (type III) have their molecular planes oriented perpendicular to both TTF molecules arranged in stacks and the $Ni(edt)_2$ molecules. Two of the sulfur atoms in these TTF type-III units make close contact (3.5 Å) with sulfur atoms in type-II TTF molecules, on both sides, bridging the stacks together along the *c* direction.

4.1.7. Alternate Stacks—Regular Spacing

The compounds $(TTF)[M(S_2C_2(CF_3)_2)_2]$, M = Pt, Cu, and Au, occur as fully charge-transferred 1:1, D^+A^- salts.[36,37,106] The donor and acceptor molecules form alternating stacks with equal spacing (Figure 11). Each of these compounds undergoes first-order phase transitions, between room temperature and 200 K. These transitions are accompanied by an ordering of the CF_3 substituents on the $[M(S_2C_2(CF_3)_2)_2]^{1-}$ units which are rotationally disordered in the room-temperature structure. For the copper complex, a lattice dimerization was observed to occur below 12 K, corresponding to the pairing of $(TTF)^+$ along a diagonal axis c_p, which proved to be the true one-dimensional chain axis.[112,118]

The molecular packing of the $(Per)[Ni(S_2C_2(CF_3)_2)_2]$ complex[29] consists of true alternate stacks with regular spacing. The overlap between the perylene and the $Ni(S_2C_2(CF_3)_2)_2$ molecules is unsymmetrical, there being a tilting of the molecular planes with respect to the stacking axis of 32°. The maximum interplanar spacing is 3.54 Å.

4.1.8. Mixed Stacks—Trimers

In $(NMP)_2[Ni(mnt)_2]$ (NMP = *N*-methylphenazinium), planar $[Ni(mnt)_2]^{2-}$ ions are sandwiched between slightly bent (4°) $(NMP)^+$ ions.[59] The average interplanar separation is 3.48 Å. The trimers are stacked along the *a* axis with a 3.35-Å interplanar separation between adjacent units. This results in a very weak electronic overlap (Figure 12).

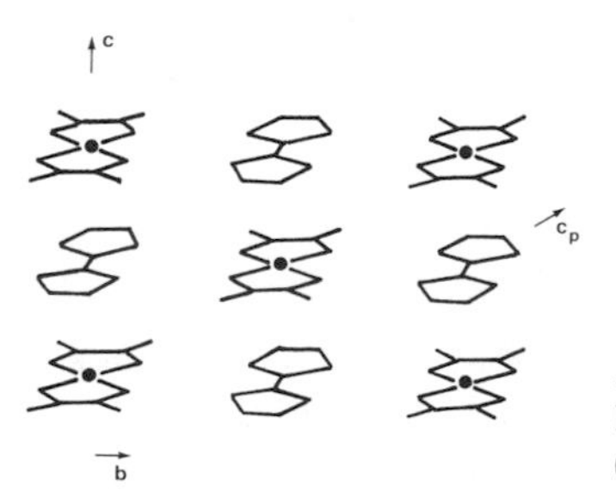

FIGURE 11. Schematic representation of the molecular arrangement in the crystal structure of $(TTF)[Cu(S_2C_2(CF_3)_2)_2]$.

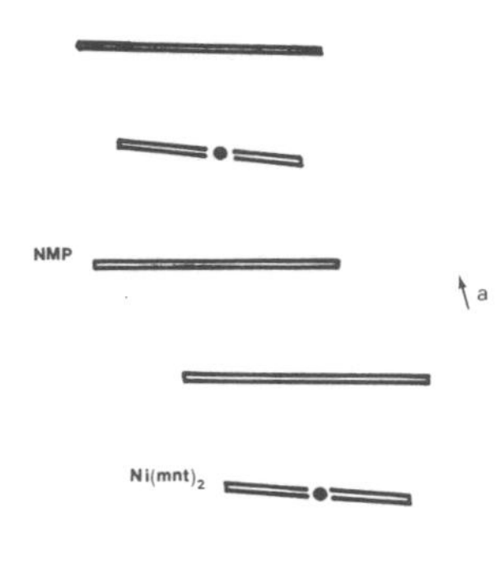

FIGURE 12. Schematic representation of the molecular arrangement in the crystal structure of $(NMP)_2[Ni(mnt)_2]$.

4.1.9. Disordered Solid Solution

The compound[16, 125] $(TTF)_2[Ni(edt)_2]_3$, which is not charge transferred, corresponds to a disordered solid solution of neutral molecules of TTF and $Ni(edt)_2$. The unit cell (Figure 13) contains two sites, each one accommodating either a TTF or a $Ni(edt)_2$ molecule in a proportion corresponding to the stoichiometry. This structure is a consequence of the close similarity of the two molecules. The sulfur and outer carbon positions are in virtual superposition for both molecules, the difference corresponding to a replacement of a C=C pair in TTF by a nickel atom in $Ni(edt)_2$. The x-ray diffraction pattern contains weak satellite reflections which result from a sinusoidal modulation corresponding to 10 unit cells. There is also evidence of displacive modulations in a direction perpendicular to the main modulation.

4.2. Magnetic Properties

Only a limited number of bis(1,2-dithiolene)metal complexes are known to exhibit highly anisotropic character in their electrical or magnetic properties—weak magnetic interactions being the most common. The most

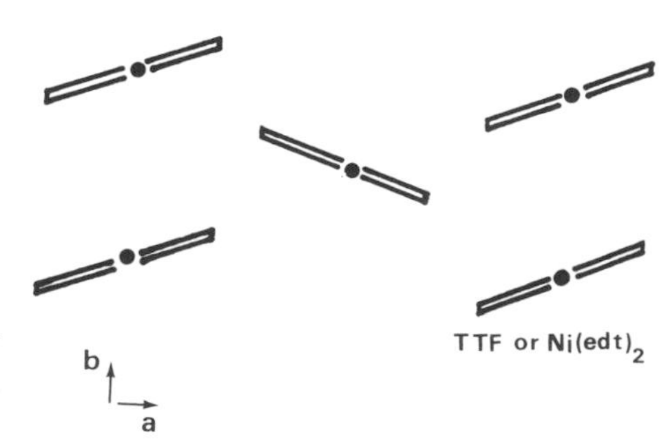

FIGURE 13. Schematic representation of the molecular arrangement in the crystal structure of $(TTF)_2[Ni(edt)_2]_3$.

interesting case of magnetic behavior observed for $[M(S_2C_2R_2)_2]^z$ complexes is the one shown by $(TTF)[M(S_2C_2(CF_3)_2)_2]$ (M = Cu and Au). These isostructural compounds, with alternating regular stacks, are the first examples of materials where a spin–Peierls transition (the magnetic analog of an electronic Peierls transition) occurs.[35, 36, 40, 106–108, 118]

Both compounds have a phase transition, at 250 K for M = Cu and at 200 K for M = Au. Throughout this transition, the basic crystal structure remains unaltered, but there are abrupt changes in several lattice parameters and there is a discontinuity on the magnetic susceptibility. Below this transition, the magnetic susceptibility of both materials is characteristic of one-dimensional antiferromagnetically coupled systems of the Heisenberg type, with $J/k_B = 77$ K and $g = 1.97$ for M = Cu, and $J/k_B = 68$ K and $g = 2.05$ for M = Au. The coupling constant J is defined through the Hamiltonian $H_{ij} = JS_i \cdot S_j$, and g is the Landé factor. At 12 K, for the copper complex and 2.1 K for the gold analog, a second-order transition occurs and the magnetic susceptibility starts freezing down continuously towards zero at 0 K; i.e., to a singlet ground state with a magnetic gap. This freezing down of the susceptibility has been interpreted in terms of a spin–Peierls distortion corresponding to a spin-lattice dimerization. Curiously enough, this lattice dimerization occurs along the diagonal c_p axis as described in Section 4.1.7, and indeed corresponds to the easiest pairing of spins localized on $(TTF)^+$.[118]

Nuclear relaxation studies[111] for these complexes reinforce the interpretation given for the static susceptibility. Above the transition, the compounds are accurately described as $S = 1/2$ uniform Heisenberg chains. The continuous opening of the energy gap at the spin–Peierls transition first observed through the susceptibility measurements shows up clearly in the nmr results.

In $(TTF)[Cu(Se_2C_2(CF_3)_2)_2]$, which is isostructural with the sulfur analog, the susceptibility versus temperature behavior is similar to that of the sulfur compound,[40] including the position of the susceptibility maximum, which provides a measure of the magnetic coupling parameter J. In this case the transition temperature is 6 K.

The complex $(TTF)[Pt(S_2C_2(CF_3)_2)_2]$ follows the Curie–Weiss law, ($\theta = 16$ K) for dominant ferromagnetic interactions from 270 to about 40 K. At lower temperatures, this behavior changes and there is a peak at 12 K, indicating antiferromagnetic coupling.[109] The effective magnetic moment μ_{eff}, is 2.24 μ_B, where μ_B is the Bohr magneton. This is consistent with two $S = 1/2$ spins per formula. In this platinum analog, which is isostructural with the copper and gold compounds but where the monoanion $[Pt(S_2C_2(CF_3)_2]^{1-}$ is also paramagnetic, the combined results of the magnetic susceptibility and specific heat were analyzed using a simple model[110] that ignores differences between the two types of $S = 1/2$ spin carriers and involves a system of ferromagnetic chains treated "exactly" with interchain antiferromagnetic interaction evaluated in a mean-field approximation. In this platinum complex, above an apparent ordering transition at 8 K, the susceptibility is well described by the model irrespective of whether the ferromagnetic exchange is Heisenberg, Ising, or their intermediate. Comparison with specific-heat calculations for the Heisenberg, Ising, and intermediate cases points to an intermediate anisotropic exchange close to the Heisenberg limit.

The nickel analog has a different crystal structure.[106] The ions of this compound are tilted in a way suggesting more interchain coupling. This compound has a μ_{eff} of $2.27\mu_B$, but exhibits dominant antiferromagnetic interactions ($\theta = -18$ K) down to 40 K. At lower temperatures its Curie constant is reduced to that of a single $S = 1/2$ spin with a ferromagnetic intercept ($\theta = 5$ K). For some samples of this material, presumably of higher purity, an abrupt antiferromagnetic transition occurs near 4 K. Below the transition, the magnetization isotherms are metamagnetic, with considerable field histeresis, suggesting magnetoelastic behavior.

Heisenberg-type interactions have also been observed for $(Bu_4N)_2[Cu(mnt)_2]$. As pointed out above, in this complex the $[Cu(mnt)_2]^{2-}$ ions crystallize in stacks with crystallographic equivalent sites, and the stacks are isolated by the cations. The susceptibility is Curie–Weiss with $\theta = +0.5 \pm 1$ K, indicating extremely weak exchange interactions between anions.[100,103] Analysis of esr data demonstrates that the one-dimensional exchange taking place along the c axis is weak, $J = 0.0107$ $cm^{-1} = 107$ G at 4.2 K. The esr spectrum is quite complicated because of the exchange in the presence of nuclear hyperfine interactions of comparable magnitude.

The complex $(TMPD)_2[Ni(mnt)_2]$ possesses paramagnetic $(TMPD)^+$ and diamagnetic $[Ni(mnt)_2]^{2-}$. The $(TMPD)^+$ ions are stacked in an alternating one-dimensional array of exchange—coupled molecules. This compound gives a good example of triplet-exciton motion of the Frenkel type.[52] The esr studies demonstrate that the spectra are of the triplet-exciton type, characteristic of an alternating chain of spins. Exciton creation energy is $\Delta E = 0.240$ eV, and the zero-field splitting parameters D and E of the spin Hamiltonian are $|D| = 215 \pm 1$ G, $|E| = 28 \pm 1$ G. The intensity of the esr signal follows the relation

$$I \propto (1/T)[\tfrac{1}{3}\exp(\Delta E/k_B T) + 1]^{-1}$$

where ΔE is the energy for the singlet–triplet separation (exciton-creation energy). At room temperature, the esr spectrum consists of a single line that broadens as the temperature is lowered, i.e., as the exchange rate slows down. At approximately 230 K, the line splits into two lines that narrow as the temperature is lowered. At low temperatures, the linewidth of the individual doublet lines reaches a plateau which varies from 4 to 8 G depending on crystal orientation. The absence of hyperfine splitting indicates that interaction between the dimers is not zero.

On grounds of structure and electrical conductivity, a very interesting material is $(TTT)_{1.2}[Ni(edt)_2]$, whose structure was discussed above. The magnetic behavior, however, is not very exciting. It follows the Curie–Weiss law for one unpaired electron per formula unit ($\theta = -5.5$ K) characteristic of antiferromagnetic interactions between localized spins.[17,19,113] It appears as if only $[Ni(edt)_2]^{1-}$ contributes to the susceptibility. It would be expected to have a small, almost temperature-independent susceptibility due to the TTT chain, but this contribution was not accounted for. The esr spectrum is clearly due to the $[Ni(edt)_2]^{1-}$ species with a weak exchange interaction between the two different orientations of the ions in the unit cell.

$(TTF)_2[Ni(edt)_2]$ follows a Curie law with $\mu_{eff} \simeq 1.67\mu_B$/formula corresponding to one unpaired spin in the $[M(S_2C_2R_2)_2]^{1-}$ ion.[16] No unpaired spins are observed in TTF, a fact supported by the crystal structure.[18] Actually, the columnar stacking of TTF molecules along the *b* axis allows for π orbital overlap within the stacks, the odd electrons being paired due to the association of $[(TTF)_2]^+$ units.

The $(Pyr)[Ni(S_2C_2(CF_3)_2)_2]$ and $(Per)[Ni(S_2C_2(CF_3)_2)_2]$ complexes are nonionic and therefore diamagnetic—impurities being responsible for a weak paramagnetism exhibited by both complexes.[29] The series of $(Per)_2[M(mnt)_2]$ also exhibit unusual magnetic properties.[63] The most interesting case is the one where M = Pd. The esr data and static susceptibility are explained by exchange interactions between the $(Per)^+$ ions and the $[Pd(mnt)_2]^{1-}$ paramagnetic ions, giving rise to averaged *g* values. From the data, it was possible to separate the contribution to the susceptibility due to the perylene chain, which is weak and almost temperature independent, from the Curie–Weiss type susceptibility due to the anionic chain corresponding to 0.12 spins per formula unit.

For the platinum analog, whose structure is described above and which has metallic behavior at room temperature,[64] the susceptibility follows an apparent Curie–Weiss law with $\theta \simeq -30$ K. The data can be better fitted to a Curie–Weiss law if we subtract a small contribution due to the perylene chain, as in the palladium complex. In some crystals of this compound, a transition in the susceptibility has been observed at approximately 10 K. This could be due to a spin–Peierls distortion. At the same temperature the conductivity also exhibits a metal–insulator transition.

Other compounds of the $[M(S_2C_2R_2)_2]^z$ series have less interesting magnetic behavior which is easily explained on grounds of the magnetic nature of the species in the formula unit.

4.3. Electrical Properties

Very few bis(1,2-dithiolene)metal complexes have electrical conductivities above 10^{-6} ohm^{-1} cm^{-1}. The highest conductivities are found in the series $(Per)_2[M(mnt)_2]$, M = Ni, Cu, Pd, Pt and Au. The nickel and the copper compounds are semiconductors and have low activation energies[63] (Table 3). The platinum, palladium, and gold compounds behave as metals at room temperature.[64] For the platinum compound, the conductivity increases as the temperature is lowered until approximately 10 K when a metal–insulator transition seems to occur. From the crystal structure [99] it is clear that the perylene chain is responsible for the high conductivity and for the metallic behavior.

A conductivity of 30 ohm^{-1} cm^{-1} at room temperature is reported for $(TTT)_{1.2}Ni(edt)_2$ whose behavior is similar to that of $K_2Pt(CN)_4Br_{0.3}\cdot 3H_2O$.[19, 113] The conductivity is thermally activated with an activation energy on the order of 0.11 eV in the vicinity of 150 K, showing a very broad peak in the plot of $d(\ln R)/d(1/T)$ vs T, centered at this temperature. Such a peak may indicate the onset of interchain correlation or of extrinsic semiconducting behavior due to defects or impurities, which would decrease the activation energy. It has been suggested that a Peierls distortion of period $12c_{TTT}$ (Figure 7), present at all temperatures associated with the commensurate periodic crystal field of the same periodicity, is responsible for the conductivity behavior.

A series of reasonably good conductors have been reported recently[65] in which it is suggested that charge transport occurs along the $Ni(mnt)_2$ anionic chain. One of the compounds, $(NH_4)[Ni(mnt)_2]\cdot H_2O$ has a uniform segregated stack arrangement[114] with a Ni–Ni separation of 3.92 Å and exhibits an activated conductivity with a value of 5×10^{-2} ohm^{-1} cm^{-1} in single-crystal measurements at room temperature. The compound has a broad phase transition with lattice dimerization occurring between 200 and 150 K, which is observed in magnetic susceptibility and electrical resistivity.

The complex $(TTF)_2[Ni(edt)_2]$ also has reasonably high conductivities along two of the crystallographic directions and small activation energies (semiconductor).[16, 18] At room temperature the conductivity along b is 7.4 $\times 10^{-3}$ ohm^{-1} cm^{-1} and 2.7×10^{-4} ohm^{-1} cm^{-1} along c. In both directions the activition energy is of the order of 0.23 eV. Along the a direction the conductivity is at least one order of magnitude lower. This anisotropy is certainly related to the two-dimensional character of the arrangement of the TTF molecules.

Conductivity data for most $M(S_2C_2R_2)_2$ derived complexes are collected in Tables 1–6.

5. Conclusions

The unique properties of 1,2-dithiolene complexes, particularly the variety of oxidation states and the possibility of reversible one-electron transfer reactions, as well as the wide range of possibilities of the chemistry available, make these compounds of great interest for chemists, solid-state physicists, and even scientists concerned with materials science and technology. Recently, some interesting materials have been produced, such as the nickel bis-1,2-dithiolene polymer[115]:

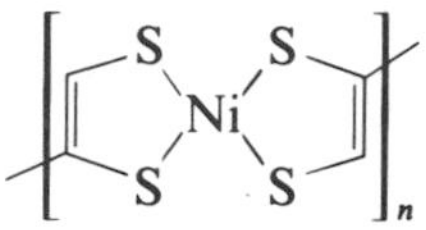

and the oligomers

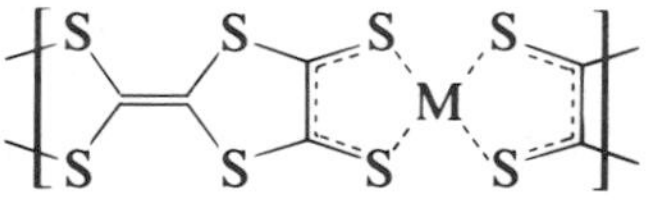

which have conductivities of the order of 30 ohm^{-1} cm^{-1} for M = Ni, and lower values for the other analogs.[116]

Some of these 1,2-dithiolene derivatives have an intense, broad electronic absorption around 900 nm, which is near the output frequency of the GaAs laser. Because of this property these compounds are expected to be useful as *Q* switches.

Notation

A	Acceptor
Ac	Acetone
an^-	Anion
An_3C	Tris(*p*-methoxyphenyl)carbenium
Ar	Aryl
Az	*N*,*N*-dimethylbenzothiazoloneazine
bdt	Benzene-1,2-dithiolate
Bu	Butyl
CNpyMe	4-Cyano-1-methylpyridinium
CNpyR	4-Cyano-1-alkylpyridinium (R = Me, Et, Bu)
CT	Charge transfer
CV	Crystal violet
cat^+	Cation
D	Donor
D	Zero-field splitting parameter

DEDMTTF	Diethyldimethyltetrathiafulvalene
DMF	*N*,*N*-dimethylformamide
dmopd	4,5-Dimethyl-*o*-phenylenediimine
DMSO	Dimethylsulfoxide
E	Zero-field splitting parameter
ΔE	Exciton creation energy
E_a	Activation energy
E_A	Reduction potential of the acceptor
E_D	Oxidation potential of the donor
edt	*cis*-Ethylenedithiolate
EL	Electrochemical
esr	Electron spin resonance
Et	Ethyl
g	Landé factor
HMB	Hexamethylbenze
HOMO	Highest occupied molecular orbital
J	Coupling constant
k_B	Boltzmann constant
KCP	Potassium tetracyanoplatinate
KOS	4-Acetyloxy-1-ethylpyridinium
LUMO	Lowest unnocupied molecular orbital
M	Transition metal
MB	Methylene blue
MBT	Bis(3-methylbenzothiazolinylidene-2)
Medt	*cis*-1,2-Dimethylethylene-1,2-dithiolate
MePy	*N*-methylpyridinium
MeQn	*N*-methylquinolinium
MG	Malachite green
μ_B	Bohr magneton
μ_{eff}	Effective magnetic moment
mnt	Maleonitriledithiolate
MP	Morpholine
MT	Methatesis
NMP	*N*-methylphenazinium
nmr	Nuclear magnetic resonance
opd	*o*-phenylenediimine
Per	Perylene
Phdt	*cis*-1,2-Diphenylethylene-1,2-dithiolate
Pol	TRpy-terephthalaldeyd polymer
POZ	Phenoxazine
PTZ	Phenothiazine
py	Pyridine
pyMe	*N*-methylpyridinium
Pyr	Pyrene
Qn	Quinoline
R	Organic substituent group, alyphatic or aromatic
R	Electrical resistance
RD	Redox reactions
RT	Room temperature
S	Spin quantum number
SCE	Saturated calomel electrode

σ	Electrical conductivity
TAE	1,1′,3,3′-diethylene-2,2-bibenzimidazolidene
tcdt	Tetrachlorobenzene-1,2-dithiolate
TCNTTF	Tetracyanotetrathiafulvalene
tdt	4-Methylbenzene-1,2-dithiolate
θ	Weiss constant
tfd	*cis*-Bis(trifluoromethyl)ethylene-1,2-dithiolate
tfdt	Tetrafluorobenzene-1,2-dithiolate
Thian	Thianthrene
TMA	Tetramethylammonium
tmdt	Tetramethylbenzene-1,2-dithiolate
TMPD	*N*,*N*,*N*′,*N*′-Tetramethyl-*p*-phenylenediamine
TMpy	1,2,4,6-Tetramethylpyridinium
TMTTF	Tetramethyltetrathiafulvalene
Tol	Tolyl
Trp	Cycloheptatrienylium(tropylium)
TRpy	1,2,6-Trimethylpyridinium
TTF	Tetrathiafulvalene
TTN	Tetrathionaphtalene
TTT	Tetrathiotetracene
xdt	4,5-Dimethylbenzene-1,2-dithiolate
WB	$(TMPD)ClO_4$

References

1. (a) Bähr, G., and Schleitzer, H., *Chem. Ber.* **88**, 1771 (1955); (b) Bähr, G., and Schleitzer, H., *ibid.* **90**, 438 (1957).
2. Gray, H. B., Williams, R., Bernal, I., and Billig, E., *J. Am. Chem. Soc.* **84**, 3596 (1962).
3. Davison, A., Edelstein, N., Holm, R. H., and Maki, A. H., *J. Am. Chem. Soc.* **85**, 2029 (1963).
4. (a) McCleverty, J. A., in *Progress in Inorganic Chemistry*, Vol. 10, F. A. Cotton, Ed., Interscience, New York (1968), pp. 49–221; (b) McCleverty, J. A., in *MTP International Review of Science, Inorganic Chemistry, Series One*, Vol. 2, C. C. Addison and D. B. Sowerby, Eds., Butterworths, London (1972), pp. 301–327.
5. (a) Schrauzer, G. N., in *Transition Metal Chemistry*, Vol. 4, R. C. Carlin, Ed., Marcel Dekker, New York (1968), pp. 299–335; (b) Schrauzer, G. N., *Acc. Chem. Res.* **2**, 72 (1969).
6. Hoyer, E., Dietzsch, W., and Schroth, W., *Z. Chem.* **11**, 41 (1971).
7. Eisenberg, R., in *Progress in Inorganic Chemistry*, Vol. 12, S. J. Lippard, Ed., Interscience, New York (1970), pp. 295–369.
8. Miller, J. S., and Epstein, A. J., in *Progress in Inorganic Chemistry*, Vol. 20, S. J. Lippard, Ed., Interscience, New York (1976), pp. 1–151.
9. Schroth, W., and Peschel, J., *Chimia* **18**, 171 (1964).
10. King, R. B., and Eggers, C. A., *Inorg. Chem.* **7**, 340 (1968).
11. Schrauzer, G. N., and Mayweg, V. P., *J. Am. Chem. Soc.* **87**, 3585 (1965).
12. Hoyer, E., and Schroth, W., *Chem. Indus.*, 652 (1965).
13. Hoyer, E., Dietzsch, W., Hennig, H., and Schroth, W., *Chem. Ber.* **102**, 603 (1969).
14. Browall, K. W., and Interrante, L. V., *J. Coord. Chem.* **3**, 27 (1973).
15. Miles, M. G., and Wilson, J. D., *Inorg. Chem.* **14**, 2357 (1975).

16. Interrante, L. V., Browall, K. W., Hart, Jr., H. R., Jacobs, I. S., Watkins, G. D., and Wee, S. H., *J. Am. Chem. Soc.* **97**, 889 (1975).
17. Interrante, L. V., Bray, J. W., Hart, Jr., H. R., Kasper, J. S., Piacente, P. A., and Watkins, G. D., *J. Am. Chem. Soc.* **99**, 3523 (1977).
18. Kasper, J. S., Interrante, L. V., and Secaur, C. A., *J. Am. Chem. Soc.* **97**, 890 (1975).
19. Bray, J. W., Hart, Jr., H. R., Interrante, L. V., Jacobs, I. S., Kasper, J. S., Piacente, P. A., and Watkins, G. D., *Phys. Rev. B* **16**, 1359 (1977).
20. Getten, C., These, Université des Sciences et Techniques du Languedoc, Académie de Montpellier, 1979.
21. (a) Krespan, C. G., McKusick, B. C., and Cairns, T. L., *J. Am. Chem. Soc.* **82**, 1515 (1960); (b) Krespan, C. G., *ibid.* **83**, 3434 (1961).
22. King, R. B., *Inorg. Chem.* **2**, 641 (1963).
23. Davison, A., Edelstein, N., Holm, R. H., and Maki, A. H., *Inorg. Chem.* **2**, 1227 (1963).
24. Davison, A., and Holm, R. H., *Inorg. Synth.* **10**, 19 (1967).
25. Davison, A., Edelstein, N., Holm, R. H., and Maki, A. H., *Inorg. Chem.* **3**, 814 (1964).
26. Davison, A., Howe, D. V., and Shawl, E. T., *Inorg. Chem.* **6**, 458 (1967).
27. Balch, A. L., Dance, I. G., and Holm, R. H., *J. Am. Chem. Soc.* **90**, 1139 (1968).
28. Davison, A., and Shawl, E. T., *Inorg. Chem.* **9**, 1820 (1970).
29. Schmitt, R. D., Wing, R. M., and Maki, A. H., *J. Am. Chem. Soc.* **91**, 4394 (1969).
30. Wheland, R. C., and Gillson, J. L., *J. Am. Chem. Soc.* **98**, 3916 (1976).
31. Wing, R. M., and Schlupp, R. L., *Inorg. Chem.* **9**, 471 (1970).
32. Dance, I. G., and Solstad, P. J., *J. Am. Chem. Soc.* **95**, 7256 (1973).
33. Geiger, Jr., W. E., and Maki, A. H., *J. Phys. Chem.* **75**, 2387 (1971).
34. Singhabhandhu, A., Robinson, P. D., Fang, J. H., and Geiger, Jr., W. E., *Inorg. Chem.* **14**, 318 (1975).
35. Bray, J. W., Hart, Jr., H. R., Interrante, L. V., Jacobs, I. S., Kasper, J. S., Watkins, G. D., Lee, S. H., and Bonner, J. C., *Phys. Rev. Lett.* **35**, 744 (1975).
36. Jacobs, I. S., Bray, J. W., Hart, Jr., H. R., Interrante, L. V., Kasper, J. S., Watkins, G. D., Prober, D. E., and Bonner, J. C., *Phys. Rev. B* **14**, 3036 (1976).
37. Kasper, J. S., and Interrante, L. V., *Acta Crystallogr. B* **32**, 2914 (1976).
38. Geiger, Jr., W. E., *J. Phys. Chem.* **77**, 1862 (1973).
39. Wudl, F., in *The Physics and Chemistry of Low Dimensional Solids*, L. J. Alcácer, Ed., Reidel, Dordrecht (1980), pp. 265–279.
40. Interrante, L. V., Bray, J. W., Hart, Jr., H. R., Jacobs, I. S., Kasper, J. S., and Piacente, P. A., in *Lecture Notes in Physics: Quasi-One-Dimensional Conductors II*, S. Barisic, A. Bjelis, J. R. Cooper, and B. Leontic, Eds., Springer-Verlag, Berlin (1979), pp. 55–68.
41. Locke, J., and McKleverty, J. A., *Inorg. Chem.* **5**, 1157 (1966).
42. Stiefel, E. I., Bennett, L. E., Dori, Z., Crawford, T. H., Simo, C., and Gray, H. B., *Inorg. Chem.* **9**, 281 (1970).
43. Bray, J., Locke, J., McCleverty, J. A., and Coucouvanis, D., *Inorg. Synth.* **13**, 187 (1971).
44. Ciganek, E., and Krespan, C. G., *J. Org. Chem.* **33**, 541 (1968).
45. Billig, E., Williams, R., Bernal, I., Waters, J. H., and Bray, H. B., *Inorg. Chem.* **3**, 663 (1964).
46. Weiher, J. F., Melby, L. R., and Benson, R. E., *J. Am. Chem. Soc.* **86**, 4329 (1964).
47. Novais, H. M., unpublished.
48. Waters, J. H., and Gray, H. B., *J. Am. Chem. Soc.* **87**, 3534 (1965).
49. Waters, J. H., Bergendahl, T. J., and Lewis, S. R., *Chem. Commun.*, 834 (1971).
50. Dance, I. G., and Miller, T. R., *Inorg. Chem.* **13**, 525 (1974).
51. Bechgaard, K., private communication.
52. Hove, M. J., Hoffman, B. M., and Ibers, J. A., *J. Chem. Phys.* **56**, 3490 (1972).
53. Inoue, M., Kuramoto, H., and Nakamura, D., *J. Magn. Reson.* **33**, 409 (1979).

54. Dance, I. G., Solstad, P. J., and Calabrese, J. C., *Inorg. Chem.* **12**, 2161 (1973).
55. Wudl, F., Ho, C. H., and Nagel, A., *Chem. Commun.*, 923 (1973).
56. Wudl, F., *J. Am. Chem. Soc.* **97**, 1962 (1975).
57. Miles, M. G., Wilson, J. D., Dahm, D. J., and Wagenknecht, J. H., *Chem. Commun.*, 751 (1974).
58. Miller, J. S., and Epstein, A. J., *J. Coord. Chem.* **8**, 191 (1979).
59. Endres, H., Keller, H. J., Moroni, W., and Nöthe, D., *Acta Crystallogr. B* **35**, 353 (1979).
60. Rosseinsky, D. R., Kite, K., Malpas, R. E., and Hann, R. A., *J. Electroanal. Chem.* **68**, 120 (1976).
61. Rosseinsky, D. R., and Malpas, R. E., *J. Electroanal. Chem.* **89**, 433 (1978).
62. Rosseinsky, D. R., and Malpas, R. E., *J. Chem. Soc. Dalton*, 749 (1979).
63. (a) Alcácer, L. J., and Maki, A. H., *J. Phys. Chem.* **78**, 215 (1974); (b) Alcácer, L. J., and Maki, A. H., *ibid.* **80**, 1912 (1976).
64. Alcácer, L. J., Novais, H. M., and Pedroso, F. P., in *Molecular Metals*, W. E. Hatfield, Ed., Plenum Press, New York (1979), pp. 415–418.
65. Perez-Albuerne, E. A., Isett, L. C., and Haller, R. K., *Chem. Commun.*, 417 (1977).
66. Caffery, M. L., and Coucouvanis, D., *J. Inorg. Nucl. Chem.* **37**, 2081 (1975).
67. Coucouvanis, D., Baenziger, N. C., and Johnson, S. M., *Inorg. Chem.* **13**, 1191 (1974).
68. Bergendahl, T. J., and Waters, J. H., *Inorg. Chem.* **14**, 2556 (1975).
69. Manecke, G., and Wöhrle, D., *Makrom. Chem.* **116**, 36 (1968).
70. Schrauzer, G. N., and Mayweg, V. P., *J. Am. Chem. Soc.* **84**, 3221 (1962).
71. Schrauzer, G. N., and Mayweg, V. P., *Z. Naturforsch. B* **19**, 192 (1964).
72. Schrauzer, G. N., and Mayweg, V. P., *J. Am. Chem. Soc.* **87**, 1483 (1965).
73. Schrauzer, G. N., Mayweg, V. P., and Heinrich, W., *Inorg. Chem.* **4**, 1615 (1965).
74. (a) Schrauzer, G. N., Mayweg, V. P., Finck, H. W., Müller-Westerhoff, U., and Heinrich, W., *Angew. Chem.* (Int. Ed.) **3**, 381 (1964); (b) Schrauzer, G. N., and Mayweg, V. P. *Angew. Chem.* (Int. Ed.) **3**, 639 (1964).
75. Schrauzer, G. N., Mayweg, V. P., Finck, H. W., and Heinrich, W., *J. Am. Chem. Soc.* **88**, 4604 (1966).
76. McCleverty, J. A., Atherton, N. M., Locke, J., Wharton, E. J., and Winscom, C. J., *J. Am. Chem. Soc.* **89**, 6082 (1966).
77. Olson, D. C., Mayweg, V. P., and Schrauzer, G. N., *J. Am. Chem. Soc.* **88**, 4876 (1966).
78. Engler, E. M., Patel, V. V., and Schumaker, R. R., *IBM Tech. Discl. Bull.* **20**, 2858 (1977).
79. Hurtley, W. R. H., and Smiles, S., *J. Chem. Soc.* **128**, 1821 (1962).
80. Ferretti, A., *Org. Synth. Coll.* **5**, 419 (1973).
81. Baker-Hawkes, M. J., Billig, E., and Gray, H. B., *J. Am. Chem. Soc.* **88**, 4870 (1966).
82. Hünig, S., and Fleckenstein, E., *Liebigs Ann. Chem.* **738**, 192 (1970).
83. Degani, I., and Fochi, R., *Synthesis*, 471 (1976).
84. Callaghan, A., Layton, A. J., and Nyholm, R. S., *Chem. Commun.*, 399 (1969).
85. Gray, H. B., and Billig, E., *J. Am. Chem. Soc.* **85**, 2019 (1963).
86. Williams, R., Billig, E., and Gray, H. B., *J. Am. Chem. Soc.* **88**, 43 (1966).
87. Jenkins II, J. J., *Diss. Abstr. Int. B* **38**, 5364 (1978).
88. Calas, P., Fabre, J. M., Khalife-el-Saleh, M., Mas, A., Torreilles, E., and Giral, L., *Tetrahedron Lett.*, 4475 (1975).
89. Calas, P., Fabre, J. M., Mas, A., Khalife-el-Saleh, M., Torreilles, E., Cot, L., and Giral, L., *Mol. Cryst. Liq. Cryst.* **32**, 151 (1976).
90. Schrauzer, G. N., and Rabinowitz, H. N., *J. Am. Chem. Soc.* **90**, 4297 (1968).
91. Schläpfer, C. W., and Nakamoto, K., *Inorg. Chem.* **14**, 1338 (1975).
92. Shupack, S. I., Billig, E., Clark, R. J. H., Williams, R., and Gray, H. B., *J. Am. Chem. Soc.* **86**, 4594 (1964).
93. Maki, A. H., Edelstein, N., Davison, A., and Holm, R. H., *J. Am. Chem. Soc.* **86**, 4580 (1964).

94. Schmitt, R. D., and Maki, A. H., *J. Am. Chem. Soc.* **90**, 2288 (1968).
95. Clark, R. J. H., and Turtle, P. C., *J. Chem. Soc. Dalton*, 2142 (1977).
96. Grim, S. O., Matienzo, L. J., and Swartz, Jr., W. E., *Inorg. Chem.* **13**, 447 (1974).
97. Geiger, Jr., W. E., Allen, C. S., Mines, T. E., and Senftleber, F. C., *Inorg. Chem.* **16**, 2003 (1977).
98. Rosa, E. J., and Schrauzer, G. N., *J. Phys. Chem.* **73**, 3132 (1969).
99. Alcácer, L. J., Novais, H. M., Pedroso, F. P., Flandrois, S., Coulon, C., Chasseau, D., and Gaultier, J., *Solid State Commun.* **35**, 945 (1980).
100. Plumlee, K. W., Hoffman, B. M., Ibers, J. A., and Soos, Z. G., *J. Chem. Phys.* **63**, 1926 (1975).
101. Kobayashi, A., and Sasaki, Y., *Bull. Chem. Soc. Jap.* **50**, 2650 (1977).
102. Plumlee, K. W., Hoffman, B. M., Ratajack, M. T., and Kannewurf, C. R., *Solid State Commun.* **15**, 1651 (1974).
103. Plumlee, K. W., *Diss. Abstr. Int. B* **36**, 6196 (1976).
104. Fritchie, C. J., *Acta Crystallogr.* **20**, 107 (1966).
105. Forrester, J. D., Zalkin, A., and Templeton, D. H., *Inorg. Chem.* **3**, 1500 (1964).
106. Jacobs, I. S., Hart, Jr., H. R., Interrante, L. V., Bray, J. W., Kasper, J. S., Watkins, G. D., Prober, D. E., Wolf, W. P., and Bonner, J. C., *Physica B* **86–88**, 655 (1977).
107. Bray, J. W., Hart, Jr., H. R., Interrante, L. V., Jacobs, I. S., Kasper, J. S., Watkins, G. D., Wee, S. H., and Bonner, J. C., *AIP Conf. Proc.* **29**, 504 (1976).
108. Bloch, D., Voiron, J., Bonner, J. C., Bray, J. W., Jacobs, I. S., and Interrante, L. V., *Phys. Rev. Lett.* **44**, 294 (1980).
109. Jacobs, I. S., Interrante, L. V., and Hart, Jr., H. R., *AIP Conf. Proc.* **24**, 355 (1974).
110. Bonner, J. C., Wei, T. S., Hart, Jr., H. R., Interrante, L. V., Jacobs, I. S., Kasper, J. S., Watkins, G. D., and Blöte, H. W. J., *J. Appl. Phys.* **49**, 1321 (1978).
111. Smith, L. S., Ehrenfreund, E., Heeger, A. T., Interrante, L. V., Bray, J. W., Hart, Jr., H. R., and Jacobs, I. S., *Solid State Commun.* **19**, 377 (1976).
112. Moncton, D. E., Birgeneau, R. J., Interrante, L. V., and Wudl, F., *Phys. Rev. Lett.* **39**, 507 (1977).
113. Interrante, L. V., Bray, J. W., Hart, Jr., H. R., Kasper, J. S., Piacente, P. A., and Watkins, G. D., *Ann. N.Y. Acad. Sci.* **313**, 407 (1978).
114. Isett, L. C., Rosso, D. M., and Bottger, G. L., *Phys. Rev. B* **22**, 4739 (1980).
115. Andersen, J. R., Patel, V. V., and Engler, E. M., *Tetrahedron Lett.*, 239 (1978).
116. Rivera, N. M., Engler, E. M., and Schumaker, R. R., *Chem. Commun.*, 184 (1979).
117. Wharton, E. J., and McCleverty, J. A., *J. Chem. Soc. A*, 2258 (1969).
118. Kasper, J. S., and Moncton, D. E., *Phys. Rev. B* **20**, 2341 (1979).
119. Enemark, J. H., and Lipscomb, W. N., *Inorg. Chem.* **4**, 1729 (1965).
120. Hamilton, W. C., and Bernal, I., *Inorg. Chem.* **6**, 2003 (1967).
121. Browall, K. W., Interrante, L. V., and Kasper, J. S., *J. Am. Chem. Soc.* **93**, 6289 (1971).
122. Browall, K. W., Bursh, T., Interrante, L. V., and Kasper, J. S., *Inorg. Chem.* **11**, 1800 (1972).
123. Schlupp, R. L., and Maki, A. H., *Inorg. Chem.* **13**, 44 (1974).
124. Interrante, L. V., *Adv. Chem. Ser.* **150**, 1 (1976).
125. Interrante, L. V., and Kasper, J. S., *AIP Conf. Proc.* **53**, 205 (1979).

7

The Spin–Peierls Transition

James W. Bray, Leonard V. Interrante, Israel S. Jacobs, and Jill C. Bonner

1. *Introduction*

1.1. *Introductory Remarks*

The spin–Peierls transition is an unusual kind of magnetoelastic transition occurring (at least in in its simplest form) in a very limited number of quasi-one-dimensional insulating systems. It was predicted theoretically almost two decades ago, in the general context of physical chemistry and the properties of certain organic free radicals in particular. The first clear-cut experimental realization was not discovered until 1975.[1,2] The discovery was made in the wake of a high level of interest which developed among physicists in the properties of quasi-one-dimensional organic conductors (organic metals), of which (tetrathiafulvalene)(7,7,8,8-tetracyano-*p*-quinodimethane), (TTF)(TCNQ), remains the best-known example. Typical programs proceeded by synthesis followed by conductivity measurements, and low-conductivity samples were routinely discarded. The program followed at General Electric involved combining organic donors (e.g., TTF) with metallo-organic planar acceptors called metal bisdithiolene (BDT) complexes, MBDT, where M is a metal atom, such as Cu, Au, or Pt. In the case of (TTF)(CuBDT), the system was found to be an excellent insulator; however, magnetization studies suggested low-dimensional behavior (quasi-one-dimensional behavior) with curious properties.

Most magnetic systems enter a long-range ordered phase, commonly ferromagnetic or antiferromagnetic, as the temperature is lowered through a

James W. Bray, Leonard V. Interrante, and Israel S. Jacobs • General Electric Corporate Research and Development, P.O. Box 8, Schenectady, New York 12301. *Jill C. Bonner* • Department of Physics, University of Rhode Island, Kingston, Rhode Island 02881.

transition temperature. However, in the case of an ideal one-dimensional magnetic model with short-range interactions, ordering can occur only at $T = 0$ K. Weak interchain interactions are necessary to induce a nonzero transition temperature. Such systems are termed quasi-one-dimensional. The ordering temperature is related to the ratio of interchain to intrachain coupling and is therefore relatively low. This ratio is, in fact, a measure of the degree of one-dimensional character of the material.

Quasi-one-dimensional magnetic ordering occurs in a model of exchange-coupled spins on a rigid lattice. If allowance is made for the possibility of an elastic distortion of the lattice, a new type of ordering can occur at low temperatures which is magnetoelastic. This ordering is now called a spin–Peierls transition in antiferromagnetic materials, since it has strong similarities to the Peierls transition in a quasi-one-dimensional metal. However, spin–Peierls systems are insulating at all temperatures: no metallic phase is involved.

A spin–Peierls (SP) system can be qualitatively described as follows. Consider the substance to consist of an assembly of quantum spin chains described by a spin-1/2 Heisenberg Hamiltonian with nearest neighbor only exchange coupling of antiferromagnetic type. These chains are stacked parallel to one another, and interchain magnetic coupling will be neglected. However, since the exchange energy of the spin chains is a function of separation between adjacent lattice sites, an elastic distortion of the lattice will influence the spin Hamiltonian of the chains. This effect can be represented by adding to the one-dimensional Heisenberg Hamiltonian on a rigid lattice a term representing spin–phonon coupling, where the phonons are simply the wave-vector, k, space Fourier transform of the real-space

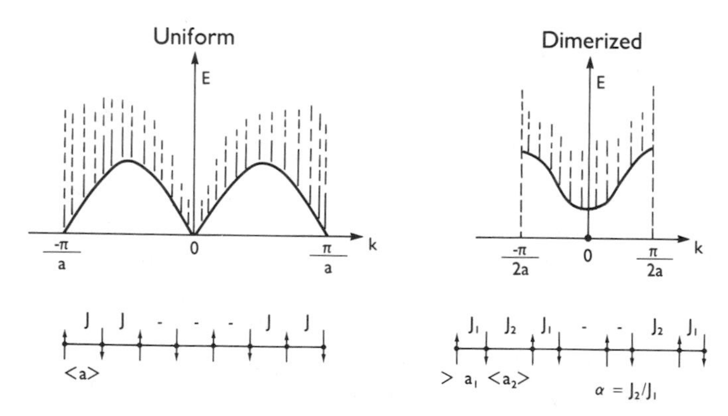

FIGURE 1. Schematic representation of the conformation and elementary excitations vs wave vector k for a uniform Heisenberg AF chain and an alternating chain ($J_2/J_1 < 1$). For the latter, the heavy dot at $k = 0$ indicates the ground state. Also the unit cell is doubled ($a_1 + a_2 = 2a$), which halves the zone-boundary wave vector.

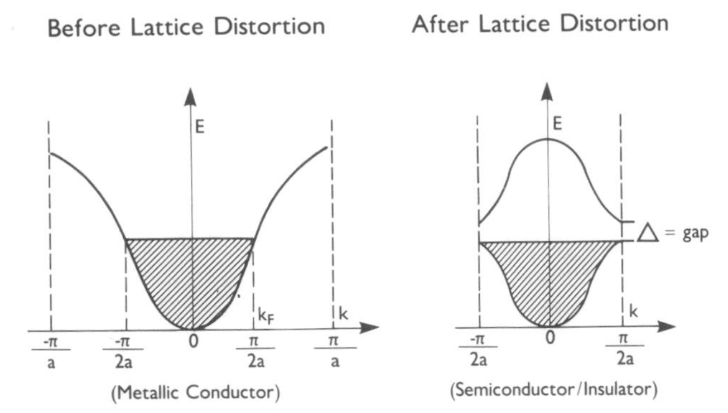

FIGURE 2. Schematic representation of the Peierls transition, comparing the energy dispersion $E(k)$ for a half-filled band of a uniform chain with that for a dimerized chain. The electronic energy is lowered as the system undergoes a metal-to-insulator transition.

lattice vibrations. Since the phonons have a three-dimensional character, interchain spin–phonon coupling is automatically included, and to this the finite transition temperature may be attributed. Full details are to be found in Section 2.

Since the question of finite temperature ordering is nontrivial, it is worth noting that an important paper by Pytte[3] treats one-dimensional *Ising* chains coupled to three-dimensional phonons. An exact transformation takes this system into a fully three-dimensional Ising pseudospin model, thus demonstrating without doubt the existence of a ($T > 0$ K) transition. Such a transformation does not work with Heisenberg spin chains, but the point is made.

In the above picture, as the temperature is lowered, the uniform spin chain undergoes a transition (at a critical temperature T_c) to a dimerized state [Figure 1]. This can be visualized as a state in which neighboring pairs of ions are displaced a small amount from their "uniform locations," alternately closer and further apart. The transition results from an inherent instability in quantum Heisenberg (or XY) systems, where dimerization produces a decrease in magnetic free energy which outweighs the increase in lattice free energy due to the distortion. The situation strongly resembles that in a one-dimensional conductor, where Peierls[4] was able to demonstrate that a half-filled conduction band is unstable against a dimerization of the lattice ions which introduces a gap at the Fermi energy (Figure 2). This result is a filled valence band, an empty conduction band, and an overall lowering of the energy of the system. Clearly, the Peierls process transforms a conductor into an insulator (semiconductor), and hence it is a metal–insulator transition.

In the case of the experimental SP systems, the dimerization proceeds continuously and progressively below T_c to a maximum value (which depends on the particular system) at $T = 0$ K. Hence the SP transitions in zero magnetic field observed experimentally are second order. A spin–Peierls distortion is such that its onset precludes additional magnetic ordering at lower temperatures; i.e., the spin excitations remain paramagnetic down to absolute zero. The order parameter is therefore given by the degree of lattice distortion, or equivalently the magnitude of the magnetic gap. Assuming that interchain spin–spin coupling is not negligible, one may view the situation as a competition between spin–phonon and spin–spin coupling. In the familiar magnetically ordered systems, the rigid lattice approximation is good. The spin–spin coupling mechanism therefore dominates the spin–phonon effects, whereas the reverse is true in the case of SP systems.

Theories of the cooperative ordering of purely magnetic insulating systems are at this point well characterized and understood through a variety of exact analytical and approximate techniques in statistical mechanics. This is much less the case with regard to magnetoelastic phenomena. In particular, for the SP systems, the *mixed dimensionality*, i.e., one-dimensional spin chains coupled to a three-dimensional lattice, makes a comprehensive microscopic treatment still more difficult. Even the popular and versatile renormalization group approach cannot, at present, be successfully formulated for the entire SP problem. The theoretical difficulties may be understood in terms of the three major ingredients of the problem: uniform Heisenberg chains (above T_c), alternating Heisenberg chains (below T_c), and spin–phonon coupling. An analytical Bethe Ansatz approach has been formulated for the problem of the generalized uniform spin chain,[5] but no finite temperature solutions have been forthcoming in the Heisenberg limit. At present, the most reliable information on the behavior of Heisenberg spin chains comes from numerical studies.[6] For the alternating Heisenberg spin chain there is not even a Bethe Ansatz formulation, and all information is based on approximate techniques.[7–9] The salient feature of a spin-1/2 alternating chain is the existence of an energy gap between the singlet ground state and the lowest-lying band of (triplet) excited states for all nonzero alternation (dimerization), which is sketched in Figure 1. The presence of this gap is a quantum feature, since a classical Heisenberg alternating antiferromagnetic chain shows no gap. Such a gap for the quantum case is essential both for spin–Peierls theory and spin–exciton theory. An important theory of the SP transition due to Pytte,[10] which has been employed in modified form to fit experiment and theory for[1] (TTF)-(CuBDT) and methyethylmorpholinium$(TCNQ)_2$,[11] $MEM(TCNQ)_2$, is based on a Hartree–Fock method due to Bulaevskii, which gives a remarkably good qualitative account of spin-1/2 uniform[12] and alternating[7] Heisenberg antiferromagnetics.[13] A test of the degree of accuracy of the Hartree–Fock approach is the comparison of the Bulaevskii antiferromagnetic susceptibilities as a function of alternation with more accurate finite chain extrapolation studies.[14] For example, while the overall behavioral trend is remarkably similar, the Bulaevskii susceptibility maximum is approximately 10% higher than numerical results for all degrees of alternation. A Luttinger model approach of Cross and Fisher[15] was employed to treat the spin interactions more accurately, but this approach calculates only the functional forms of the thermodynamic quantities accurately, and it therefore appears somewhat difficult to make direct contact with experiment. The susceptibility χ of an antiferromagnetic alternating Heisenberg chain is shown in Figure 3a and has the remarkable property of being independent of crystalline orientation (because of the spin–space isotropy of the Heisenberg Hamiltonian). Since for SP systems, χ vanishes as $T \to 0$ K for both single crystal and powder specimens, such systems are readily distinguished from Heisenberg-like antiferromagnetically ordered systems, where the single-crystal susceptibility vanishes for at most one crystal orientation and the powder susceptibility remains nonzero as $T \to 0$ K.[16] The contrasting types of ordering are illustrated in Figure 3b and 3c, respectively.

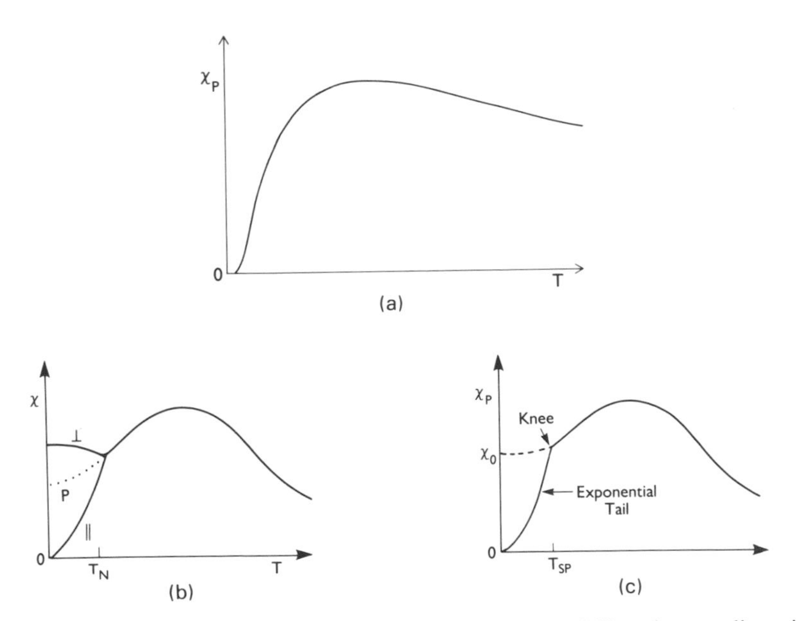

FIGURE 3. (a) Schematic of powder (and single-crystal) susceptibility of a one-dimensional antiferromagnetic alternating Heisenberg chain with $\alpha = 0.8$ ($H = 0$). (b) Powder and oriented crystal susceptibilities of a quasi-one-dimensional Heisenberg-like antiferromagnetic system ($H = 0$). The temperature of long-range ordering is T_N, the Néel point. (c) Powder (and single-crystal) susceptibility of a quasi-one-dimensional spin–Peierls antiferromagnetic system ($H = 0$). The transition from uniform chain to alternating chain commences at T_{SP} ($\equiv T_c$ in text).

At this point it may be valuable to outline in more detail similarities and differences between two major approaches to this problem[1,2,10,15] and throw light on the problem of spin–phonon coupling. As already noted, a simple approach to the SP problem leads to a spin–spin term, a spin–phonon term, and a phonon term. Use of the well-known Jordan–Wigner transform converts the one-dimensional spin–spin term into quadratic and quartic pseudofermion interaction terms. Similar treatment of the spin–phonon term produces, in addition, terms quadratic and quartic in pseudofermion operators coupled to phonon operators. Since an exact solution of the three-dimensional lattice phonons is out of the question, a random phase approximation (RPA) is used. Such an approximation to the spin–phonon coupling is intrinsically mean-field in character. In the case of the four-fermion terms, Cross and Fisher[15] have attempted an "exact" treatment of the salient aspects of the problem using Luttinger model techniques, whereas Pytte–Bray–Bulaevskii have used a Hartree–Fock-type approximation on the quartic fermion terms to decouple them into quadratic terms. Such an approach is also mean-field. Hence, while the Cross–Fisher approach is mean-field in the spin–phonon coupling, the Pytte–Bray–Bulaevskii approach is mean-field for both the spin–spin and spin–phonon terms.

Spin–Peierls systems are therefore treated in a mean-field approximation (see Section 2.2 for details). As argued by Bulaevskii *et al.*,[17] a treatment involving no special features of the lattice yields very small SP transition temperatures, expected to be well below the temperature at which interchain ordering to a three-dimensional magnetically ordered state is likely to take place. However, most substances in which the SP transition have been observed, namely, (TTF)(CuBDT), (TTF)(AuBDT), (TTF)-(CuBDSe), and MEM$(TCNQ)_2$, exhibit an important additional feature. This is a precursive three-dimensional softening of the phonon corresponding to the dimerization wavevector, which reduces the phonon frequency to about an order of magnitude below its usual value. This softening phenomenon appears to be associated with a structural transition in these substances at a temperature much higher than T_c. For example, in (TTF)-(CuBDT) the structural transition occurs at about 240 K, whereas T_c is only 12 K. The structural transition and the soft mode have been observed experimentally in x-ray diffuse scattering studies.[18,19] The pre-existing soft mode greatly facilitates spin–phonon ordering, and a SP transition sets in at a relatively high temperature; i.e., spin–phonon coupling wins the battle with regular magnetic ordering, at least in zero applied magnetic field. In addition, the three-dimensional (rather than one-dimensional) character of the soft phonon, which corresponds essentially to lattice planes perpendicular to the chains moving in unison,[15] goes a long way toward validating a mean-field approach to the problem. The Ginzburg critical region, where deviations from mean-field behavior may be expected, has been estimated[15] to be $0.04T_c$, i.e., relatively small.

As noted above, theoretical treatments of the SP transition usually start by employing a well-known transformation from spin operators to pseudofermion operators (Jordan–Wigner transformation). This transforms the SP problem in a one-dimensional insulator into a mathematical form very similar to the regular electronic Peierls (RP) problem (a metal–insulator transition), which helps explain the similarity of the associated instabilities mentioned above. However, mean-field solutions agree with experiment much less well in the RP case than in the SP case, and this is presumably due to the precursive soft mode in the SP systems as well as the differing energy scales of the two cases.

Similarities between RP and SP systems make it natural to consider the possibility of a generalized Peierls transition, such that the electronic Peierls (RP) and spin–Peierls (SP) transitions correspond, respectively, to the conducting and insulating limits of such a generalized distortion process (see Section 2.5). We therefore suggest that a study of the SP transition may throw light on the RP transition—directly via the Jordan–Wigner transformation, or indirectly via generalized Peierls transition reasoning. Such knowledge would be of valuable assistance in designing desired physical properties into organic conductors. In this connection, of the four existing experimental SP systems mentioned above, $MEM(TCNQ)_2$ is particularly interesting. It displays an RP-like transition near room temperature and an SP transition at a much lower temperature ($\sim$ 18 K).

Since SP systems are magnetoelastic, the "natural" thermodynamic parameters with which to examine the phase behavior are temperature, applied magnetic field, and pressure. The effects of magnetic field on the one hand and pressure on the other are particularly interesting. It follows from the Jordan–Wigner transformation that magnetic field in the SP system maps onto chemical potential (Fermi energy) in the corresponding (spinless) fermion system. Interesting phenomena such as commensurate–incommensurate transitions have been predicted to occur in electronic systems[20] as the chemical potential is varied. Since, however, the chemical potential is the degree of filling of the band and therefore is determined directly by the material parameters, it cannot be varied at will. In a SP system, on the other hand, the applied magnetic field is a continuously tunable parameter, which allows experimental investigation of its analog, Fermi energy. Theoretical expectations[17,21,22] have been borne out by recent high-field magnetic measurements. As the field increases, the Fermi level progressively decreases, favoring other types of lattice distortion, namely, incommensurate or higher-order commensurate phases. However, the commensurate dimerized phase is not immediately destroyed,[23] but persists up to a critical field and temperature called H_c and $T_c(H_c)$. This is because lattice Umklapp energies initially keep the lattice distortion "pinned" at the dimerization wave vector, until finally overcome by the magnetic energy. Experimental evidence for a new high-field phase now exists for (TTF)(CuBDT),[24–26] (TTF)(AuBDT),[27] and MEM$(TCNQ)_2$,[28] above ~ 115 kOe, ~ 17 kOe, and ~ 190 kOe, respectively. It appears, however, that microscopic probes such as neutron scattering are required to determine the precise nature of the high-field phase behavior.

The effects of pressure on SP systems is a very interesting area[29] for which only preliminary experimental results are available at this point.[25] Pressure experiments have played a valuable role in studies of quasi-one-dimensional conductors, as evidenced by recent initial experiments[30, 31] on a conductor that undergoes a metal–insulator transition at atmospheric pressure. Application of 6 kbars of pressure suppresses the metal–insulator transition and allows the appearance of a long-sought superconducting transition at ~ 1 K. This has given a new lease of life to the search for superconductivity in one-dimensional organic metals, and a zero-pressure superconducting organic has now been discovered.[32] Clearly an effect of applied pressure is to "stiffen" the lattice, i.e., to make it more rigid. In the case of SP indulators, early experiments on (TTF)(CuBDT) have revealed effects of pressure in terms of an enhanced (uniform) exchange constant and a significant reduction in T_c. An exciting possibility inherent in the pressure experiments is that of "scaling down" the entire H-T phase diagram by reducing T_c and thereby rendering high-field phenomena more accessible (in field and temperature) to structure-revealing neutron scattering experiments.

A final point to consider is the possible role of interchain spin–spin coupling in determining the general phase behavior. The possibility has not been considered in the extant[15, 17, 21, 22] theories of field-dependent behavior. However, as is argued in Section 2.4, the effect of an applied field is to enhance the tendency toward three-dimensional antiferromagnetic ordering (by tending to increase the Néel temperature), whereas the spin–phonon-induced SP ordering tendency is diminished (T_c decreases). Hence one may speculate that in a sufficiently large field, spin–spin interchain coupling will dominate spin–phonon interchain coupling, and SP behavior will be suppressed. In a similar vein, the effect of pressure is also to enhance spin–spin ordering (by increasing the interchain J), while SP ordering becomes less favorable.[25, 29] Clearly a rich variety of phase behavior may be expected from SP systems under the influence of field and pressure.

1.2. Background

McConnell and co-workers[33] appear to have been the first (in 1962) to make application of the dimerization instability of a spin-1/2 magnetic linear chain in the context of paramagnetic resonance of solid free radicals.

They drew upon several background sources and were especially influenced by the work of Peierls[4] in 1954 on the instability of a one-dimensional metallic lattice. Peierls' work stimulated Fröhlich and Kuper[34, 35] to investigate a one-dimensional metallic model for possible relevance to superconductivity. A second background source cited is the study of the alternation of bond lengths in long conjugated chain molecules. The work of Ooshika[36] and of Longuet-Higgins and Salem[37] are to be noted in this regard, with credit for seminal ideas traced back to Platt[38] and earlier work by Kuhn.[39]

The approach of McConnell *et al.* was essentially qualitative. The first quantitative treatment, to our knowledge, was that of Chesnut.[40] Chesnut first introduced a model akin to that discussed in Section 1.1, namely, one containing a spin–spin coupling term, a site-separation-dependent exchange expanded as far as the linear term, and an elastic lattice. He obtained a mean-field, second-order transition from a uniformly spaced lattice to one of alternating character. Essentially this behavior is retained in the calculations that followed. In fact, in the historical development of the spin–Peierls model, more detailed theories have generally elaborated upon their predecessors without negating their essential features. All extant theories have at least some mean-field character, which is necessary to induce a finite temperature phase transition in an otherwise one-dimensional model. This mean-field character enters into the calculation in various ways, for example, as a long-range constraint,[40] via the random phase approximation,[10, 15] or by a Hartree–Fock approximation.[15] In fact, the basic view of SP systems as consisting of identical uniform chains above the transition and identical alternating chains below ignores distortion fluctuations and is therefore inherently mean-field. A variety of subsequent theories are discussed in detail in Section 2.

Finally, we note that another review of spin–Peierls transitions has recently appeared by Buzdin and Bulaevskii[41] in the Russian literature. It is less comprehensive than this work but is a useful complement.

2. *Theory of the Spin–Peierls Transition*

2.1. *Physical Ideas and Formulations of the Problem*

The model spin–Peierls (SP) system consists of a parallel array of one-dimensional antiferromagnetic (AF) chains that are coupled into a three-dimensional lattice. The AF interaction is typically Heisenberg or XY in character and the spin value $|\mathbf{S}| = 1/2$, for reasons which we explore in this section. We include no interchain magnetic coupling in the ideal model.

The model Hamiltonian may thus be written

$$H = \sum_{j} J(j, j+1)(\mathbf{S}_j \cdot \mathbf{S}_{j+1}) + \sum_{q,\alpha} \omega_0(\mathbf{q}, \alpha) b^{\dagger}_{q\alpha} b_{q\alpha} \tag{1}$$

where the sum over lattice sites j includes nearest intrachain neighbors only, $b_{q\alpha}^{\dagger}$ ($b_{q\alpha}$) is the creation (destruction) operator for three-dimensional phonons with wave vector $\mathbf{q}$ on branch α, and ω_0 is the unrenormalized phonon energy. Since the exchange energy $J(j, j+1)$ is a function of the three-dimensional spatial separation of sites j and $j+1$, the one-dimensional spin interactions depend on the three-dimensional motion of the lattice sites. Thus we are treating a one-dimensional magnetic system coupled to a three-dimensional phonon system. The spin vector $\mathbf{S}$ may be $2-n$ (XY) or $3-n$ (Heisenberg), but we will emphasize the Heisenberg case since it is most likely to be found in reality. (We take $\hbar = k_B = 1$).

Before proceeding with a solution of the Hamiltonian, it is useful to ask the question: Why is it energetically favorable for a one-dimensional AF chain to dimerize? The answer lies in the nature of the excitation spectrum and in the concept of quantum fluctuations. The excitation spectrum[42] of an infinite one-dimensional AF Heisenberg chain is depicted in Figure 1. The salient point is that the excitation spectrum is degenerate with the ground state at $q = 0, \pm\pi/a$. This degeneracy brings excited states infinitely close to the ground state (of which the Néel state is a component). Therefore, quantum zero-point fluctuations of the chain will populate the low-lying excited states. This implies that the observed $T = 0$ K state is a composite of the singlet ground state and triplet excited states. The consequences are that the Néel state is not a true eigenstate of the Hamiltonian, there is no long-range order at $T = 0$ K, and the magnetic susceptibility at $T = 0$ K is nonzero.

If the chain is dimerized, a gap develops in the excitation spectrum which lifts the above mentioned degeneracy of the ground and excited states (Figure 1). The zero-point fluctuations can now no longer populate the excited states, and the net magnetic energy is lowered. This is the essence of the spin–Peierls transition. This is the only change of magnetic energy if the exchange J is a linear function of intrachain lattice spacing, because then dimerization leaves the total exchange energy unchanged. If J is a more rapid function of intrachain spacing, then exchange energy can also favor dimerization.

We note that the above arguments hold equally well for XY AF chains, because their excitation spectrum also exhibits the requisite degeneracy with the ground state. An Ising or classical AF chain cannot show the spin–Peierls effect. This is because an Ising chain effectively has an energy gap between the ground and excited states, and a classical chain has no zero-point energy, and therefore its magnetic energy is independent of chain dimerization. Similarly, quantum fluctuations decrease very rapidly as $|\mathbf{S}|$ increases to values greater than 1/2, and so one would not expect high-spin materials to show a SP transition. Also, finite chains develop in their excitation spectrum a gap, whose magnitude scales as N^{-1}, where N is the

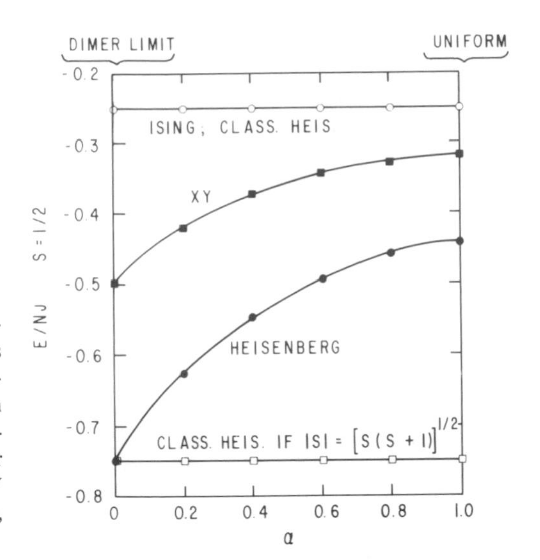

FIGURE 4. Ground-state free energies of alternating magnetic chains as a function of the amount of dimerization, $\alpha = J_2/J_1$, in Eq. (7) with $J_1 + J_2 = 2J$, and $J_1, J_2 > 0$. Calculations are for the Heisenberg (after Duffy and Barr, Reference 8), the XY (after Pincus, Reference 47), the Ising, and classical models of an AF chain.

number of sites. Therefore, we expect the reduction of chain length below some value to interfere with the SP phase. Once again, however, an appropriately rapid dependence of J on intrachain spacing can cause any chain to dimerize in order to gain net exchange energy.[3] See Figure 4 for ground-state magnetic energies versus dimerization amplitude for several interaction types in the linear J case.

In solving the Hamiltonian, Eq. (1), the theorist is faced with a number of choices and approximations, for the Hamiltonian cannot be solved exactly. By far the most common initial tactic that authors[1,2,10,15,20,43] have used is to convert the spin operators to pseudofermion operators via the Jordan–Wigner transformation.[44] One defines

$$\Psi_j = (-2)^{j-1} S_1^z S_2^z \cdots S_j^- \tag{2}$$

where

$$S^{\pm} = S^x \pm iS^y$$

and

$$S_j^z = 1/2 - \Psi_j^\dagger \Psi_j$$

Substitution of Eq. (2) into Eq. (1) yields

$$\begin{aligned} H = \sum_j 1/2J(j, j+1) \\ [\Psi_j^\dagger \Psi_{j+1} + \Psi_{j+1}^\dagger \Psi_j - 2\Psi_j^\dagger \Psi_j + 2\Psi_j^\dagger \Psi_j \Psi_{j+1}^\dagger \Psi_{j+1} + 1/2] \\ + \sum \omega_0(\boldsymbol{q}, \alpha) b_{q\alpha}^\dagger b_{q\alpha} \end{aligned} \tag{3}$$

Here the first two terms in the *sum* over j come from the XY terms of $\mathbf{S}_j \cdot \mathbf{S}_{j+1}$, and the last three terms come from the Ising term.

An important approximation involves the treatment of the spin–phonon coupling. By far the most common approach of workers has been to expand $J(j, j+1)$ only to first order in intersite spacing, thereby obtaining a spin–phonon coupling term.[10]

$$J(j+1) = J + \sum_j [\mathbf{u}(j) - \mathbf{u}(j+1)] \cdot \nabla_j J(j, j+1) + \cdots \tag{4}$$

Here the $\mathbf{u}$ are the lattice displacement operators in three dimensions. It is important that the phonons are three-dimensional, because the one-dimensional spin system by itself cannot undergo a phase transition at finite temperature.[45] The treatment of the three-dimensionality of the lattice is therefore crucial to the ultimate form of the transition. This has been discussed in some detail by Cross and Fisher.[15] An exact treatment of the three-dimensional lattice is impossible, however, and most workers have effectively treated the chains as decoupled (mean-field, random phase approximation) and have used their three-dimensional nature as the implicit means of suppressing the one-dimensional fluctuations of a single chain, which would otherwise drive T_c to zero.

Proceeding with this "linear J" approximation, we Fourier transform $\mathbf{u}$

$$\mathbf{u}(j) = (mN)^{-1/2} \sum_{\alpha q} \mathbf{e}(\alpha\mathbf{q}) e^{i\mathbf{q} \cdot \mathbf{R}_j} Q(\alpha\mathbf{q})$$

where

$$Q(\alpha\mathbf{q}) = [2\omega_0(\alpha\mathbf{q})]^{-1/2}(b_{q\alpha} + b^{\dagger}_{-q\alpha}) \tag{5}$$

and where $\mathbf{e}(\alpha\mathbf{q})$ is the phonon polarization vector, m is the mass of the magnetic lattice site, and N is the number of chain sites.

Using Eqs. (5) and (4) and Fourier transforming ($\Psi_k = N^{-1/2} \sum_j e^{ikj}\Psi_j$), Eq. (3) becomes[10]

$$\begin{aligned} H = &\sum \varepsilon_k \Psi_k^{\dagger}\Psi_k + N^{-1} \sum_{k_1+k_2=k_3+k_4} v(k_2 - k_3)\Psi_{k_1}^{\dagger}\Psi_{k_2}^{\dagger}\Psi_{k_3}\Psi_{k_4} \\ &+ N^{-1/2} \sum_{kq\alpha} g_1(k\mathbf{q}\alpha)\Psi_k^{\dagger}\Psi_{k-q}(b_{q\alpha} + b^{\dagger}_{-q\alpha}) \\ &+ N^{-1} \sum_{k_1+k_2=k_3+k_4-q} g_2(k_2 - k_3, \mathbf{q}\alpha)\Psi_{k_1}^{\dagger}\Psi_{k_2}^{\dagger}\Psi_{k_3}\Psi_{k_4}(b_{q\alpha} + b^{\dagger}_{-q\alpha}) \\ &+ \sum \omega_0(\mathbf{q}\alpha) b_{q\alpha}^{\dagger} b_{q\alpha} \end{aligned} \tag{6}$$

where

$$\varepsilon_k = J(\cos ka - 1); \qquad v(k) = J \cos ka$$

$$g_1(k\mathbf{q}\alpha) = \tfrac{1}{2} g(\mathbf{q}\alpha)(1 - e^{-iqa})(e^{ika} - 1)(1 - e^{i(q-k)a})$$

$$g_2(k\mathbf{q}\alpha) = g(\mathbf{q}\alpha)e^{ika}(1 - e^{-iqa})$$

$$g(\mathbf{q}\alpha) = [2m\omega_0(\mathbf{q}\alpha)]^{-1/2}\mathbf{e}(\mathbf{q}\alpha) \cdot \nabla J(j, j+1)$$

Equation (6) is exact except for the linearization of $J(j, j+1)$. It may be argued that this approximation is inadequate for a given real system, but within the treatment presented above, retention of higher-order terms in the expansion of J obviously leads to even more intractable (phonon–phonon interaction) terms. For a given real material, the linear J approximation should be scrutinized on the basis of the character of the intersite orbital interactions and, ultimately, by the accuracy of the resulting predictions.

At this point we note that many authors[46–53] have elected to begin with a different Hamiltonian than Eq. (1), viz.,

$$H' = \sum_{j=1}^{N-2} (J_1 \mathbf{S}_{j,1} \cdot \mathbf{S}_{j,2} + J_2 \mathbf{S}_{j,2} \cdot \mathbf{S}_{j+1,1}) + \text{phonons} \tag{7}$$

This is obviously the Hamiltonian for a dimerized AF chain (Figure 1), where J_1 and J_2 represent the intra- and inter-dimer exchange interactions and j labels the dimer. J_1 and J_2 are functions of intersite spacing exactly as in Eq. (1), and Eq. (7) simply represents the same chains below the SP transition temperature T_c. The treatment of Eq. (7) is developed essentially the same way as Eq. (1), except that two sets of pseudofermion operators are defined—one for the interdimer and one for intradimer interactions.[7] The spatial dependence of J_1 and J_2, as noted before, has usually been treated in the linear approximation to provide a spin–phonon coupling, and the same remarks are valid for this case. However, it is easier in this formulation to treat other functional dependences of J on intersite spacing, and a few authors have done this for exponential and power-law dependences.[51–53]

Since Eqs. (1) and (7) represent the same physical system but in different temperature regimes [$J_1 = J_2 = J(j, j+1)$ at T_c], their proper solution should give identical results. The problem is that for the mean-field methods most often used, their solutions are not the same. This will be further discussed in the Section 2.2. Fortunately, their mean-field solutions are not wildly different. Opinions have been registered both ways[15, 46] about which formulation is the superior one to treat the SP problem, but there is *a priori* no rigorous answer to this—it depends on the approximations employed.

Some generalizations of the SP Hamiltonian not explicit in Eqs. (1) and (7) can be made. The one most often employed by authors is to allow anisotropy of J ($J^x \neq J^y \neq J^z$). This is fairly easily accomplished by multiplying the three vector components of $\mathbf{S} \cdot \mathbf{S}$ by different variable parameters and calculating the effects of changing these parameters on the solution equations. If $S_j^z S_{j+1}^z$ is eliminated, the XY model is obtained. Many authors[46–48, 50, 53] have elected to work with the XY model because it is much more easily solved, since the troublesome four-fermion terms in Eq. (6) come only from the Ising term. The effects of other anisotropies (anisotropic XY,[50, 53] anisotropic Heisenberg[43]) on the SP state have been considered by a few authors. Another generalization of the problem, only touched upon so far,[54, 55] is the inclusion of next-nearest-neighbor (or even more distant) exchange interactions.

We have defined the SP transition rigorously as a magnetically driven spin–lattice dimerization of antiferromagnetic chains. We note that there are theoretical studies of related dimerizations which are also caused by spin–lattice coupling. For instance, Ising chains with lattice-induced alternating ferromagnetic and antiferromagnetic intrachain exchange imbedded in a three-dimensional lattice may undergo either first- or second-order dimerization transitions for sufficiently strong spin–lattice coupling.[3, 56] Furthermore, this is true for both classical n-component spin systems[56] and $S = 1, 3/2$ Ising systems.[57] None of these results are applicable to the true SP transition as we have defined it.

2.2. *Mean-Field Results*

Most SP treatments have been totally mean-field (MF), and all SP treatments have employed at least some MF approximations. As mentioned in Section 2.1, the justification for MF treatments must come from the properties of the three-dimensional phonon subsystem. Since real SP systems may exhibit significant lattice anisotropies in addition to the nearly one-dimensional magnetic interactions, the assumption of the applicability of MF methods certainly should be carefully scrutinized in any given case. There will certainly be a noticeable deviation from MF behavior sufficiently close to T_c. Some real systems have particular lattice characteristics (e.g., precursive soft modes) that support MF approximations, but this will be discussed in Section 3, where actual materials are described.

One of the most common MF approximations is to use a Hartree–Fock (HF) approximation on Eq. (6). This is identical to the Hartree approximation since the fermions are spinless, and spin is required to produce the Fock term. All four-fermion terms are converted to two-fermion terms:

$$\langle \Psi_k^\dagger \Psi_q \rangle = n_k \delta_{kq} \tag{8}$$

where $n_k = [\exp(\beta E_k) + 1]^{-1}$, $(\beta \equiv 1/k_B T)$, and Eq. (6) becomes[10]

$$H = \sum_k E_k \Psi_k^\dagger \Psi_k + \sum g(\alpha \mathbf{q} k) \Psi_k^\dagger \Psi_{k-q}(b_{\alpha q} + b_{\alpha - q}^\dagger) + \sum \omega_0(\mathbf{q}\alpha) b_{q\alpha}^\dagger b_{q\alpha} \tag{9}$$

where

$$E_k \equiv pJ \cos ka$$

$$g(\alpha \mathbf{q} k) \equiv ipg(\alpha \mathbf{q})\{\sin ka - \sin[(k-q)a]\}$$

$$p = 1 - 2N^{-1} \sum_k n_k \cos ka$$

The renormalization factor p arises from the S^z spin interactions and is approximately constant at $p = 1.64$ for temperatures $T \ll J$. In the notation we are using, the boldface wave vectors are three-dimensional, and the ones in italic type are the components along the chain.

Equation (9) is identical to the Fröhlich Hamiltonian commonly used to model the regular Peierls transition except that spin is absent. It is most commonly solved by applying another MF approximation, the random phase approximation (RPA), to the fermion–phonon coupling term. This gives an expression for the renormalized phonon frequency ω:

$$\omega^2 = \omega_0^2(\alpha\mathbf{q}) + N^{-1} \sum g(\alpha\mathbf{q}k) \times [g(\alpha\mathbf{q}, k-q)n_k - g^*(\alpha\mathbf{q}k)n_{k-q}]/(\omega - E_{k-q} + E_k) \qquad (10)$$

which leads at $q = \pi/a$ and near T_c to

$$\omega^2 = \omega_0^2(T/T_c - 1)/\ln(0.83pJ/T_c)$$

At T_c, $\omega(q = 2k_F) = 0$, where $k_F = \pi/2a$ is the Fermi wave vector. This leads to the gap equation:

$$\omega_0^2(\mathbf{q}, q = 2k_F) = \frac{g^2}{J^2}\frac{2}{\pi}\int_{-Jp}^{Jp} dE\, \frac{[(Jp)^2 - E^2]^{1/2}}{E} \tanh(\tfrac{1}{2}\beta_c E) \qquad (11)$$

Equation (11) relates T_c to the parameters of the SP problem. To obtain the usual Bardeen–Cooper–Schrieffer (BCS) form, we must make the weak-coupling assumption $T_c \ll J$. Then

$$T_c = 0.83pJ \exp(-1/\lambda) \qquad (12)$$

where

$$\lambda \equiv 4g^2p/\omega_0^2 \pi J$$

$$g \equiv g(\alpha\mathbf{q}, q = 2k_F)$$

$$\omega_0 \equiv \omega_0(\alpha\mathbf{q}, q = 2k_F)$$

When weak coupling is not valid, Eq. (11) must be solved numerically for T_c. Of course, if coupling is so strong that the resulting lattice distortion is very large, then the assumptions of linearized J dependence and harmonic phonons must concomitantly be questioned.

The magnetic gap $\Delta(T)$ is also given in weak coupling by the BCS formula

$$\Delta(T = 0) \equiv \Delta(0) = 1.765\ T_c \qquad (13)$$

and $\Delta(T)$ follows the temperature dependence of the standard BCS energy gap.[58] Below T_c, the lattice is dimerized and two unequal and alternating J's are produced:

$$J_{1,2} \equiv J(1 \pm \delta(T)) \qquad (14)$$

Other related MF results are

$$\delta = 2g\langle Q\rangle/J \tag{15}$$

where

$$\langle Q\rangle = \langle b_{q\alpha} + b^{\dagger}_{-q\alpha}\rangle \delta_{q,\,2k_F}/(2\omega_0)^{1/2}$$

The angular brackets denote a thermal average over the condensed phonon modes. Furthermore,

$$\Delta = 2gp\langle Q\rangle \tag{16}$$

and therefore

$$\delta(T) = \Delta(T)/pJ \tag{17}$$

The ground-state magnetic energy as a function of dimerization is found to be

$$E \propto -\delta^2 \ln^2 \delta \tag{18}$$

The specific heat of the magnetic subsystem is predicted to have a BCS-like jump at T_c of magnitude

$$\Delta C = 0.92\ k_B T_c/J \tag{19}$$

per formula unit.

In the above MF treatment, the entire effect of the Ising term is subsumed in p. Therefore, the results of Eqs. (9)–(17) also give the results for the XY model if pJ and pg are replaced by J and g, respectively. The resulting MF XY solution involves no HF approximation but does still require the RPA approximation. One may undertake to solve the general anisotropic Heisenberg–XY model within the same approximations, and the results[43] show quantitative changes of the relations, Eqs. (9)–(19), with first-order SP transitions possible as a function of anisotropy.

As stated in Section 2.1, beginning the MF treatment with the Hamiltonian of the dimerized state [Eq. (7)] leads to quantitative (but not severe) differences in the results. The "below T_c" MF treatment has many similarities to the above-described "above T_c" one, but two sets of pseudo-fermion operators are used for the inter- and intra-dimer interactions.[7] A canonical transformation is performed to diagonalize the XY part of the Hamiltonian *including* the fermion–phonon coupling, and a HF factorization of the new four-fermion terms is performed.[48] Thus HF averages $\langle\Psi_k^{\dagger}\Psi_{k-q}\rangle$ are kept in proportion to gQ. We reproduce here the results of this MF treatment of Eq. (7):

$$\begin{aligned} H' = {} & \sum_k (\omega_k^{\alpha} n_k^{\alpha} + \omega_k^{\beta} n_k^{\beta}) \\ & + \frac{1}{2} \sum_{k_1 k_2} [V(k_1 k_2 k_2 k_1)(n_{k_1}^{\alpha} n_{k_2}^{\alpha} + n_{k_1}^{\beta} n_{k_2}^{\beta} + n_{k_1}^{\alpha} n_{k_2}^{\beta} + n_{k_1}^{\beta} n_{k_2}^{\alpha}) \\ & + V(k_1 k_2 k_1 k_2)(-n_{k_1}^{\alpha} n_{k_2}^{\alpha} - n_{k_1}^{\beta} n_{k_2}^{\beta} + n_{k_1}^{\alpha} n_{k_2}^{\beta} + n_{k_1}^{\beta} n_{k_2}^{\alpha})] \end{aligned} \tag{20}$$

where

$$n_k^{\alpha,\beta} = [1 + \exp(\varepsilon_k^{\alpha,\beta}/T)]^{-1}$$

and

$$\varepsilon_k^{\alpha,\beta} = \omega_k^{\alpha,\beta} + \tfrac{1}{2}(J_1 + J_2) \sum_k (n_k^\alpha + n_k^\beta)$$
$$\pm \frac{1}{N} \sum_{k_1} V(k_1, k)(n_{k_1}^\beta - n_{k_1}^\alpha) \tag{21}$$

where

$$\omega_k^{\alpha,\beta} = -\tfrac{1}{2}(J_1 + J_2) \pm \tfrac{1}{2}(J_1^2 + J_2^2 + 2J_1J_2 \cos k)^{1/2}$$

and

$$k = 2\pi n/N \qquad \text{with } n = 1, 2, \ldots, N$$

Furthermore we have

$$V(k_1, k_2) = \tfrac{1}{2}J_1 \cos 2(\phi_{k_1} - \phi_{k_2})$$
$$+ \tfrac{1}{2}J_2 \cos(k_2 - k_1 - 2\phi_{k_1} + 2\phi_{k_2}) \tag{22}$$

where

$$\phi_k = -\tfrac{1}{2} \tan^{-1}[\gamma \sin k/(1 + \gamma \cos k)]$$
$$\gamma \equiv J_2/J_1$$

Since the spin–phonon dimerization is "built-in" to this approach, no RPA of spin–phonon coupling is used. The gap equation is obtained by minimizing the free energy with respect to the dimerization parameter δ:

$$b\delta = \int_0^{\pi/2} \frac{dk}{J} \left(\frac{\partial J_k}{\partial \delta}\right) \tanh(\tfrac{1}{2}\beta\varepsilon_k^\alpha), \tag{23}$$

where

$$J_k \equiv J[1 - (1 - \delta^2)\sin^2 k]^{1/2}$$
$$b \equiv \frac{2\lambda'}{1 + (1 + 8\lambda'/\pi)^{1/2}}$$
$$\lambda' \equiv 4p/\lambda$$

In the corresponding XY case,[48] Eq. (23) is the same except ε_k^α and b are replaced by J_k and λ', respectively. The weak-coupling result for T_c in the XY case is

$$T_c = 0.83\, Je^{-\lambda'}. \tag{24}$$

However, Eq. (23) is not of the BCS form [as Eq. (11)] and leads to results like

$$\Delta(0) = J\delta(0)[1 + 4(\ln 2)/\pi - 2(\ln \delta)/\pi]. \tag{25}$$

No analytical expression for T_c can be extracted in this Heisenberg case, although Eqs. (23) and (24) suggest that $T_c \sim \exp(-1/\lambda^{1/2})$ for weak coupling, in contrast both to the XY case [Eq. (24)] and to the XY and Heisenberg cases in the MF treatment above T_c [Eq. (12)]. This difference is accentuated by numerical solutions[46, 52] of Eq. (23), which show that the SP transition in this MF Heisenberg treatment is always first order, for all values of the parameters.

A few other variations of this below-T_c MF calculation have been performed. The XY model with anisotropic exchange has been treated[50, 53] with a result that anisotropy can cause the transition to go first order and also to be suppressed near the Ising limit. It is interesting to note that Dubois and Carton[50] derive a strong-coupling expression for T_c in the XY model which has a linear dependence on λ:

$$T_c = \pi J\lambda/8p \qquad \text{for } T_c \gg J/2 \tag{26}$$

Nonlinear (exponential) variation of J has also been treated[51, 53] with the consequence that the first-order nature of the transition is either produced or enhanced. This agrees with the reasoning of Section 2.1 that strong dependence of J on intersite spacing enhances dimerization. In general, all of these modifications of the basic SP model will lead to varied and non-BCS-like functional dependences of the system parameters [$\Delta(T)$, $\delta(T)$, T_c, etc.].

2.3. *Beyond Mean-Field Solutions*

If one accepts that linearizing J is a good approximation for a given system, then the mean-field method of solution of the Hamiltonian [Eq. (1) or (7)] is the most questionable approximation involved in reaching the solutions presented in Section 2.2. The magnetic subsystem is one-dimensional and these materials can exhibit lattice anisotropies, all of which can lead to fluctuations that are not included in MF treatments. However, the fact is that the best examples of SP systems have so far been fairly well described by the MF equations. Therefore, there has not been strong impetus for improving the solutions beyond MF. The reasons for this MF behavior lie in characteristics of the three-dimensional phonon spectra which cause all the chains to respond together. This soft-mode character will be described in Section 3.

Nevertheless, it is desirable to go beyond MF solutions to determine what artifacts the MF approximations have introduced, to determine whether new behavior can be predicted, and to provide the best possible models for comparison with experiment. The most successful step in this direction so far has been made by Cross and Fisher (CF),[15] who begin with Eq. (6) and use a boson algebra to avoid the HF approximation on the four-fermion terms. Their treatment (and all other treatments beginning above T_c) also uses the RPA on the spin–phonon coupling term. We note that the approach beginning below T_c [Eq. (7)] may be less easily improved by a non-MF treatment, since it assumes (i.e., builds in) a lattice dimerization that would not occur in a fully non-MF treatment of a truly one-dimensional chain.

The CF treatment takes advantage of the close similarity of the pseudofermion representation of the one-dimensional Heisenberg chain [Eq. (6)] to the exactly soluble Luttinger–Tomonaga[59] model, as originally developed by Luther and Peschel.[60] The use of the word "exact" to describe this treatment or its solutions is unfortunate, since several approximations are necessary to the treatment. These include the linearization of ε_k [see Eq. (6)].

$$\varepsilon_k \simeq Ja(|k| - k_F) \tag{27}$$

the approximation of $v(k)$,

$$v(k) \simeq \begin{cases} J, & k \sim 0 \\ -J, & k \sim 2k_F, \end{cases} \tag{28}$$

the use of a momentum cutoff which is taken to be approximately equal to the inverse chain spacing (a^{-1}), and the use of the weak-coupling approximation ($T \ll J$). Nevertheless, this approach is known to give improved results in the treatment of the fermion interaction terms (four-fermion terms), especially with regard to the functional dependences of the correlation functions. The approach does not give correct amplitudes, however, and CF have improved this aspect of their results by substituting known exact values for the spin wave velocity and the critical exponent of the correlation functions. The final result should be an improvement over the MF solutions. For reference, we show the equivalent form taken by the first two terms of Eq. (6) in this Bose–operator–algebra representation:

$$\begin{aligned} H_0 + H_1 = {} & Ja\frac{2\pi}{L}\sum_{k>0}[\rho_1(k)\rho_1(-k) + \rho_2(k)\rho_2(-k)] \\ & + \frac{J}{N}\sum_k [\rho_1(k)\rho_1(-k) + \rho_2(k)\rho_2(-k) + 4\rho_1(k)\rho_2(-k)] \end{aligned} \tag{29}$$

where

$$\rho_1(q) = \sum_k \Psi^\dagger_{1k+q}\Psi_{1k}$$

$$\rho_2(q) = \sum_k \Psi^\dagger_{2k+q}\Psi_{2k}$$

The subscripts 1 and 2 above denote electrons with wave vector $k \sim k_F$ and $k \sim -k_F$, respectively. The second sum of Eq. (29) may be diagonalized by a canonical transformation, and the treatment of the spin–phonon interaction terms proceeds similarly.

In general, the CF results support the approximate accuracy of thermodynamic properties calculated in MF. However, there are some notable differences in their solutions. For T_c, they obtain a linear functional dependence on λ:

$$T_c = 0.8J\lambda_{\mathrm{CF}} \tag{30}$$

where $\lambda_{\mathrm{CF}} \equiv \lambda/p = 4g^2/\pi J\omega_0^2$ [cf. Eq. (12)]. This result implies a smaller λ than would be obtained in MF for a given T_c, and this in turn implies a weaker spin–phonon coupling, which might make unnecessary any corrections to the one-dimensional magnetic properties due to phonon coupling. Also, they find an increase of the rate of phonon softening above T_c:

$$\omega^2(2k_F, T) = \omega_0^2(T/T_c - 1) \tag{31}$$

[cf. Eq. (10)]. They find that the MF order parameter should be scaled by a factor $(J/T_c)^{1/2}$ and that the lattice distortion for a given λ and T/T_c is reduced from MF by the factor $(T_c/J)^{1/2}$. Finally, they find at $T = 0$ K the dependence of the ground-state energy of the spin system E_0 and the excitation gap Δ on the distortion parameter δ to be

$$E_0 \propto \delta^{4/3}$$

$$\Delta \propto \delta^{2/3}, \tag{32}$$

[cf. Eq. (17)]. This result is very close to that of Duffy and Barr,[8] who derive their results by extrapolation from numerical calculations of finite dimerized chains. Reference 61 contains a detailed comparison of $E_0(\delta)$ for several calculation methods.

Solitons have gained considerable visibility in the literature of one-dimensional systems in the last few years. Recently, they have been applied to SP systems as well.[62–64] Soliton is a general name applied to any solitary wave solution of a nonlinear equation. Since they are nonlinear excitations, they are not found in solutions that are based on linear response theory, such as those we have discussed so far. These excitations are very necessary to obtain a complete picture of the behavior of a system if their formation energy E_s is less than or equal to the other energies under consideration. In a Heisenberg or XY AF chain, a soliton may be visualized as a local change of the order of alternation of the spins (a π-phase soliton) or, in the dimerized phase, as a local change in the degree of dimerization (an amplitude soliton) accompanying the phase soliton.

Nakano and Fukuyama[62] extend the boson treatment of CF to study solitons. They transform the boson variables into phase variables to produce a sine–Gordon equation, which they solve for the soliton creation energy. After this transformation the SP Hamiltonian becomes

$$H = \int dx \left[\tilde{A}(\nabla\theta_+)^2 + CP^2(x) - B\,\frac{u(x)}{u_0}\cos\theta_+ + 2\,cu^2(x)/a \right] \tag{33}$$

where

$$\tilde{A} \equiv Ja/8, \qquad B \equiv J\delta/a, \qquad C \equiv \pi^2 Ja/2$$

$$c = \omega_0^2 m, \qquad u_0 = \delta/2\nabla J$$

$$\theta_\pm(x) = \frac{2\pi i}{L} \sum_{q\neq 0} \frac{1}{q}\,[\rho_1(q) \pm \rho_2(q)]\exp(-\tfrac{1}{2}\alpha\,|\,q\,| - iqx)$$

$$P(x) \equiv \frac{1}{4\pi}\,\nabla\theta_-(x)$$

Here $u(x)$ is the spatial variation of the dimerization magnitude u_0. The first two terms come from the transformation of Eq. (29), the third from the transformation of the spin–phonon term in the dimerized state, and the fourth from the elastic distortion energy. The nonlinearity obviously arises from the third term. The soliton creation energy E_s is found to be

$$E_s \simeq 0.29\ \Delta \tag{34}$$

which is less than the gap energy. One might expect such excitations to be observable in experiments.

Grabowski *et al.*[63] have examined solitons in a half-filled fermion band, which of course is closely related to the SP model. They, too, derive an E_s less than the linear excitation gap (by a factor π^{-1}). Their nonlinear Hamiltonian is also of the sine–Gordon type, coupling both amplitude and phase, but is actually derived within the MF Hartree–Fock approximation. It therefore differs in details from Eq. (33). However, Horovitz[64] has also produced a non-MF ("exact") solution of this problem which supports the conclusions of Reference 63.

The only other calculation to date which goes beyond the MF approach is a many-body valence bond calculation of Klein and García-Bach.[65] Their results appear accurate for ground-state energies and predict ground-state bond alternation (dimerization) for $|\mathbf{S}| = 1/2$ Heisenberg AF chains. Interestingly, they also suggest dimerization for the $|\mathbf{S}| = 1$ case, which has not been treated before. However, this approach does not include the lattice effects and yields little information about how other model parameters (e.g., spin–lattice coupling, temperature) affect the results.

2.4. Effects of External Probes: Magnetic Field, Pressure, and NMR

Since the spin–Peierls transition is a phenomenon of lattice-coupled antiferromagnetic chains, one expects magnetic fields to have important and interesting effects upon the transition. This is indeed the case. The magnetic field H couples to the system via a Zeeman term, which must be added to the Hamiltonian, Eq. (1):

$$H_M \equiv -\tilde{g}\mu_B H \sum_j S_j^z \tag{35}$$

where $\tilde{g}$ is the gyromagnetic ratio $\simeq 2$ and μ_B is the Bohr magneton. In the pseudofermion representation [Eq. (2)] this may be written

$$H_M = -2\mu_B H \sum_j \Psi_j^\dagger \Psi_j \tag{36}$$

$$H'_M = -2\mu_B H \sum_k (n_k^\alpha + n_k^\beta) \tag{37}$$

Here H'_M is the equivalent result below T_c in the dimerized state [Eq. (7)] and is the required addition to Eq. (20). Inspection of Eq. (6) shows that the addition of Eq. (36) merely modifies ε_k:

$$\varepsilon_k = J(\cos ka - 1 - h) \tag{38}$$

where $h \equiv 2\mu_B H/J$. Equation (37) is equally easily incorporated into the below-T_c formalism. Equation (38) shows that the effect of H is to modify the Fermi level (band-filling) of the pseudofermions. We shall present the resulting MF and non-MF solutions, but first we discuss qualitatively some important expected effects of magnetic fields.

Very large magnetic fields may have drastic effects on an AF chain. For instance, H may flip all spins to the field direction and make the chain effectively ferromagnetic. To accomplish this at $T = 0$ K, the field energy must exceed the AF exchange energy of a spin antiparallel to the field. This requires[16] $H \geq 2S(J_1 + J_1^2/J_2)/\tilde{g}\mu_B$, which is about 1 MG for spin-1/2 and $J_1 \sim J_2 \sim J = 70$ K. At lower but usually substantial fields, a dimerized AF chain may be rendered magnetic at $T = 0$ K by applying a field greater in energy than the magnetic gap;[16] viz., $H \gtrsim \Delta/\tilde{g}\mu_B$, which is about 150 kG for $\Delta = 20$ K. Our treatment in this chapter will not be concerned with effects of this type, but the concerned experimenter should be aware of them. For smaller fields, the spin wave spectrum of the AF chain is modified, and these modified excitations coupled to the lattice produce the effects of concern here.

A SP system and a quasi-one-dimensional magnetic system which orders in a three-dimensional AF state at low temperatures show distinctively different responses to a magnetic field. In a SP system, $T_c(H)$ is a monotonically decreasing function of H, whereas in an ordered quasi-one-

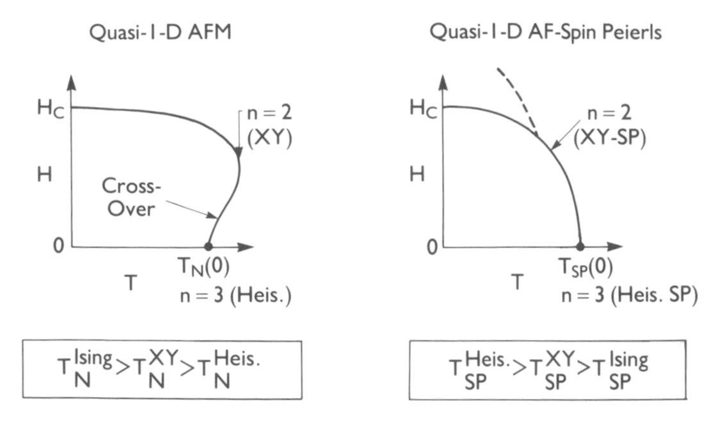

FIGURE 5. Qualitative comparison of the effect of magnetic field H on the transition temperature T_{SP} ($\equiv T_c$ in text) for a spin–Peierls system and T_N for an antiferromagnetic chain system that undergoes three-dimensional magnetic ordering. The dashed line on the spin–Peierls diagram denotes the occurrence of a high-field magnetic phase (see text).

dimensional magnetic system, the three-dimensional ordering temperature $T_N(H)$ first increases as H increases and then decreases, giving a maximum at $H \neq 0$ (see Figure 5).[66] This phenomenon can be understood in terms of the differing role of quantum fluctuations in the two systems.[23] The effect of H is effectively to decrease the number of spin components n by one [e.g., Heisenberg ($n = 3$) becomes XY ($n = 2$)]. Magnetic quantum fluctuations decrease as n decreases. These fluctuations decrease intrachain spin coherence and thereby depress T_N.[67] Therefore, H initially increases T_N by decreasing fluctuations until this tendency is counteracted by the increasing Zeeman energy, which reduces T_N to zero. In contrast, in a SP system it is the quantum spin fluctuations which provide the driving force for the SP transition, when J depends linearly on intrachain spacing (see Section 2.1). Since H reduces these fluctuations, it reduces the energy available to form the SP phase, and this effect and the Zeeman energy combine to decrease $T_c(H)$ for all H. We note that thermal fluctuations[68] also increase with n (or with any increase of degrees of freedom) and also reduce intrachain coherence length. For this reason, a quasi-one-dimensional system of classical Heisenberg chains, which cannot have a SP transition for linear J, can easily undergo a three-dimensional AF ordering that exhibits the described initial increase of $T_N(H)$.[68]

We first present solutions for $T_c(H)$ and other consequences of H within the MF approximations. The consequence of Eq. (38) in the MF Hamiltonian, Eq. (9), is simple;[12] E_k is modified

$$E_k = pJ \cos ka + \tilde{g}\mu_B H - 2sJ \tag{39}$$

where $s = 1/2 - N^{-1} \sum_j S_j^z$ is the average z component of the spin and is determined self-consistently [like p, Eq. (9)] by

$$s = \tfrac{1}{2} - N^{-1} \sum_k [1 + \exp(E_k/T)]^{-1} \tag{40}$$

Equation (39) shows that H moves the band filling away from half-filled and therefore produces an effective chemical potential (Fermi level) μ, measured from the half-filled-band point:

$$\mu(H)/J = 2s - h \tag{41}$$

As H is increased from zero, the Fermi level μ moves away from $\mu = 0$. For $\mu_B H \ll J$,[21]

$$\mu(H) \simeq -1.44\ \mu_B H \tag{42}$$

The corresponding Fermi wave vector, k_F, is also decreased:

$$ak_F(H) = \arccos(1.44\ \mu_B H/pJ)$$

or,

$$k_F(0) - k_F(H) = 1.44\ \mu_B H/pJa \qquad \text{for } \mu_B H \ll J \tag{43}$$

Since k_F is changed by H, the magnetic energy now has a minimum at a new wave vector, $2k_F(H)$. However, the lattice Umklapp energy associated with the reciprocal lattice vector $Q = \pi/a = 2k_F(0)$ causes the lattice to distort at π/a until H becomes large enough for the reduction in magnetic energy to exceed the Umklapp energy. This Umklapp energy is also called commensurability energy and is present for all values of $2k_F(H)$ that are commensurate with the lattice. However, the lowest-order commensurability is $k_F = \pi/ja$, where $j = 2$, and the effect is by far strongest here. In general, the wave vector q at which the system distorts in a field H and the transition temperature T_c are determined by calculating the q and T for which the soft phonon energy $\omega(q)$ first goes to zero in the (RPA) response equation

$$\omega^2(q) = \omega_0^2(q) + \Pi(q) \tag{44}$$

where $\Pi(q)$ is the relevant linear response function of the spin system [see, e.g., Eq. (10)].

The response of the regular Peierls transition to changes of chemical potential has been calculated by several authors, and Bray[21] and Bulaevskii *et al.*[17] used these calculations and Eq. (41) to calculate MF results for $T_c(H)$. In the regime of H where the distortion wave vector q is fixed at π/a, $T_c(H)$ is determined by numerical solution of Eq. (42) and

$$\ln(T_c(H)/T_{c0}) + \mathrm{Re}[\Psi(\tfrac{1}{2} + i\mu(H)/2\pi T_c(H)) - \Psi(\tfrac{1}{2})] = 0$$
$$T_{c0} \equiv T_c(H = 0) \tag{45}$$

where Ψ is the digamma function. For small $\mu_B H/k_B T_{c0}$, one obtains the following expansion[17, 21]:

$$\frac{T_c(H) - T_{c0}}{T_{c0}} = -0.44\left(\frac{\mu_B H}{k_B T_{c0}}\right)^2 - 0{\cdot}2\left(\frac{\mu_B H}{k_B T_{c0}}\right)^4 \tag{46}$$

When H exceeds a critical value H_c, the q that minimizes Eq. (44) is no longer π/a. Above H_c, q rapidly moves away from π/a either to $2k_F(H)$ [commensurate–incommensurate (C–I) transition] or to some new commensurate value where it sticks for some range of H [commensurate–commensurate (C–C) transition]. The details of the actual phonon spectrum can be very important in determining the nature of this transition and the value of H_c, and more will be said of this when we examine real systems in Section 3. When the phonon spectrum causes no modification, the MF result for H_c is

$$H_c \simeq 0.75\ k_B T_{c0}/\mu_B$$

$$T_c(H_c) \simeq 0.54\ T_{c0}\,. \tag{47}$$

The transition at H_c has been speculated to be first order.[21, 22, 63, 64] Cross[22] reached this conclusion by comparing energies of soliton states and states with a single Fourier component. The usual soliton picture calls for the nucleation of small regions (solitons) of phase II ($H > H_c$) in phase I ($H < H_c$) for H near H_c. The number of solitons increases continuously if they repel each other as H increases near H_c, and a second-order transition to phase II results. However, if the solitons attract one another, then the nucleation of one causes an avalanche of soliton production and the transition must be first order,[22, 63] akin to the single-Fourier-component situation.

If the transition at H_c is of the C–C type, then the possibility exists that many commensurate phases can be present above H_c, each with some commensurate q and each realized in some range of H. The situation would then involve a cascade of transitions (probably first order) as H increases above H_c. This situation has been called a "Devil's staircase"[69] for an infinity of such transitions and a "harmless staircase"[70] for a finite number.

We note that MF results for the XY model in a field have also been calculated.[17, 20, 71] In Reference 71, the wave vector q is frozen at π/a, allowing only the dimerized and normal SP phases to exist. This results in some quantitative and qualitative changes (besides the usual scaling from Heisenberg to XY MF results), such as the appearance of a tricritical point separating first- and second-order uniform-to-SP transitions at low T. Reference 20 emphasizes an important caveat in the analogy between the SP system in a field and the regular Peierls system with changing chemical potential: the fermion number is *not* conserved in the SP system, but is in the regular Peierls system. This causes some qualitative differences in behavior.[20]

Non-MF solutions have been calculated by Cross[22] using the boson algebra approach[15] to solve Eq. (44). The results are qualitatively the same as the MF equations for the most part, with some quantitative differences. We review the results corresponding to Eqs. (43), (46), and (47), respectively:

$$k_F(0) - k_F(H) = 1.27\ \mu_B H/Ja \tag{48}$$

$$\frac{T_c(H) - T_{c0}}{T_{c0}} = -0.36\left(\frac{\mu_B H}{k_B T_{c0}}\right)^2 \tag{49}$$

$$H_c \simeq 0.69\ k_B T_{c0}/\mu_B$$

$$T_c(H_c) \simeq 0.77\ T_{c0} \tag{50}$$

In addition, the result for H_c in the case of a frozen q (only uniform and dimerized states allowed) is

$$H_c \simeq 0.87\ k_B T_{c0}/\mu_B. \tag{51}$$

As noted in Section 2.3, Cross uses the exact susceptibility and spin wave velocity when calculating the above results. If one does this in the MF results, then the coefficient of the H^2 term in Eq. (46) becomes smaller than the value 0.36 of Eq. (49). We note that another difference with stated MF results is that in MF the uniform-to-SP phase boundary above H_c is stated to go rapidly toward $T = 0$ K as H increases,[17] although this result may stem from inadequate calculation. Cross[22] notes that as q moves away from π/a, λ_{CF} decreases by a factor of 2 due to disappearance of the Umklapp term, and therefore that $T_c(H)/T_{c0}$ approaches 1/2 at high fields on this phase boundary.

Finally, using nonlinear solutions within the boson-algebra approach,[15] Nakano and Fukuyama[62] note that the soliton creation energy E_s decreases linearly with applied H. They therefore state that the field at which E_s goes to zero is H_c, the C–I transition field. Their result is

$$H_c \simeq (\pi/28)\lambda_{CF} J/\mu_B. \tag{52}$$

Using Eq. (30) this becomes

$$H_c \simeq 0.14\ k_B T_{c0}/\mu_B \tag{53}$$

which is considerably smaller than Eqs. (47) and (50).

The effects of pressure P on the SP transition have been calculated[29] using both the MF and Cross–Fisher (CF) results. To first order, pressure may effect T_c by altering the exchange J, spin–phonon coupling g, or the bare $2k_F$ phonon frequency ω_0. Within the assumption of linear variation of J on intrachain spacing, g is not affected directly by P. An interesting difference in the MF and CF results is that T_c depends on J only in the MF formulation, and hence the pressure-induced increase of J (and susceptibility) has no effect on T_c in the CF formulation. Hence, the only source of change of T_c with P in the CF formulation (and the main source in the MF approach) is the variation of ω_0 with P. This can be quite complicated depending on details of the actual phonon spectrum. An obvious fact is that ω_0 must be anharmonic if $\omega_0(P)$ is not to be constant. We shall see in Section 3 that anharmonicity is almost certainly the case in several real systems. Beyond this, the functional form of $T_c(P)$ depends on assumptions made for $\omega_0(P)$. Calculated results[29] for a strongly anharmonic (very soft phonon) case are, in the MF case,

$$\ln[T_c(P)/T_c(0)] = \ln(1 + \alpha P) - \lambda^{-1}(1 + \alpha P)(1 - \gamma P)^{-\xi} + \lambda^{-1} \qquad (54)$$

where $\alpha \equiv m^{1/2}gA/Jc$, $\gamma \equiv A/ca$, A is the cross-sectional area of a chain, and $c = m\omega_0^2$ is the chain force constant. The corresponding CF result is

$$T_c(P)/T_c(0) = (1 - \gamma P)^{\xi} \qquad (55)$$

The power ξ is 4 in the extremely soft phonon limit and may take values between 0 and 4 (0 is the harmonic limit). The MF formulation also leads to a nonzero result in the harmonic ω_0 limit:

$$T_c(P)/T_c(0) = (1 + \alpha P)\exp(-\alpha\lambda^{-1}P). \qquad (56)$$

The reader is referred to Reference 29 for calculations of force constants, Debye energies, and related observables.

Nuclear magnetic resonance (nmr) of an appropriate atomic nucleus that is coupled to the $S = 1/2$ electron on the chain site provides a useful probe of SP dynamics. The nuclear spin–lattice relaxation rate, T_1^{-1}, is the result of interaction between the nuclear spin of interest and the moment of the site electron. The nuclear spin may flip (relax) when accompanied by an electron spin flip. The electron spin flip is part of the spin wave excitation spectrum which characterizes the one-dimensional, $S = 1/2$ AF chain. At high temperatures in the uniform AF chain (above T_c), the spin wave correlation length is short, and T_1^{-1} may be dominated by the one-dimensional diffusive character of the long-wavelength spin excitations (they may not be diffusive[72]) and is frequency dependent.[73] This frequency dependence becomes less apparent at low temperatures as the diffusive character lessens due to increasing correlation length. At low temperature (but above T_c), T_1^{-1} also becomes temperature independent as expected for an $S = 1/2$, one-dimensional AF chain with well-defined excitations.[74]

Below T_c, a gap develops in the dimerizing AF spin-wave spectrum, and T_1^{-1} should decrease since the main relaxation mechanism of the nuclear spin is being destroyed. A MF calculation of T_1^{-1} has been performed for the Heisenberg chain below T_c in terms of nuclear relaxation by scattering of three triplet excitons, which are thermally excited across the SP gap.[75] Diederix *et al.*[76] and Groen *et al.*[77] have studied the spin dynamics of an alternating chain with fixed alternation.

2.5. Relationship to Regular Peierls Transition

The regular electronic Peierls transition (RP) has been amply described in the literature.[4, 78, 79] In brief, it is a metal-to-insulator transition in a quasi-one-dimensional conductor caused by a lattice distortion with wave vector $2k_F$ (k_F = Fermi wave vector), which opens an energy gap at the Fermi level. We have already pointed out that the RP transition can be modeled by the Fröhlich Hamiltonian, which is identical to the transformed XY (or MF Heisenberg) SP Hamiltonian, Eq. (9), except that spin is present. In this section, we shall use the Hubbard model[80] to show a further connection between the SP and RP transitions. Indeed, we believe that these transitions can be thought of as two aspects of a generalized Peierls transition.

The Hubbard Hamiltonian in one-dimension can be written

$$\sum_{\substack{ij\\ \sigma}} t a_{i\sigma}^{\dagger} a_{j\sigma} + U \sum_{i} n_{i\uparrow} n_{i\downarrow} = H_H \tag{57}$$

$$n_{j\sigma} \equiv a_{j\sigma}^{\dagger} a_{j\sigma}$$

where t is the nearest-neighbor hopping integral and U is the Coulomb repulsion if two electrons reside on the same site. The electron creation (destruction) operator is $a^{\dagger}$ (a). The spin index is σ, and the sum is over nearest-neighbor sites i, j. t is a function of intrachain lattice spacing and may be expanded once in lattice spacing to yield an electron–phonon coupling term, exactly as we did for J in Eq. (4) to get spin–phonon coupling. Doing this, Fourier transforming Eq. (57) for $U = 0$ and one electron per site, and adding a harmonic phonon term yields:

$$H_H = \sum_{k} \varepsilon(k) a_k^{\dagger} a_k + \sum g'(k, q) a_k^{\dagger} a_{k+q} (b_q^{\dagger} + b_{-q}) + \sum \omega_0(\mathbf{q}) b_q^{\dagger} b_q \tag{58}$$

where

$$\varepsilon(k) \equiv 2t \cos ka$$

$$g'(k, q) \equiv \frac{\partial t}{\partial x} \left(\frac{\hbar}{2m\omega_0} \right)^{1/2} (1 - e^{+iqa})(e^{ika} - 1)(1 - e^{-i(k+q)a})$$

The similarity to Eq. (9) is apparent, except that here the spin index is absorbed into the wave-vector indices and t replaces the role of J. In the RPA approximation, Eq. (58) yields a formula for the RP transition temperature:

$$T_{\mathrm{RP}} = 4.56\, t \exp(-1/\lambda') \tag{59}$$

where

$$\lambda' \equiv N\,|\,g'(2k_F)\,|^2/\hbar\omega_0\,\pi t$$

Since typically t for conducting systems is a couple of orders of magnitude larger than J for magnetic systems, one expects T_{RP} to be much greater than T_c. In other words, there is more energy associated with electron translation (t) than with spin exchange (J), and consequently the RP transition has more energy at its disposal than the SP transition.

If we now allow U to increase through positive values in a half-filled band, the single-band picture associated with the RP transition breaks down. This is because the translating electron has different energies when it is on an occupied or unoccupied site, translational invariance therefore breaks down, and k^{-1} space no longer has the same simple interpretation for the dispersion relations, density functions, etc. For $U \ll t$, the U term can be treated as a perturbation on the $U = 0$ procedure we have just presented. For $U \sim t$, we are unable to write a simple expression for a Peierls transition like Eq. (59), and one is reduced to numerical methods to derive properties of the Hubbard model. As U becomes larger than t, the electrons tend to localize on each site for a half-filled band in order to minimize Coulomb repulsion energy. This localization leads to a so-called $4k_F$ instability, so named because its spatial periodicity is half that of the RP transition.[81, 82] This $4k_F$ instability can be viewed as a kind of Wigner crystallization which becomes complete for $U \gg t$. Its expected effect is to destroy the translational freedom of the electron system and cause a metal-to-insulator transition, which is called a Mott transition.[83]

For $U \gg t$, it is well known[84] than an expansion of Eq. (57) in t/U will yield the Heisenberg Hamiltonian [Eq. (1)], as the first term, where J is given by $4t^2/U$. This Hamiltonian then yields the SP instability, which occurs at the $2k_F$ wave vector.

We arrive then at the following overview of the Peierls transition in the Hubbard model for a half-filled band. For $U = 0$, the charge and spin degrees of freedom are unseparated and the RP transition opens equal gaps for both the charge and magnetic excitations by means of a lattice distortion at $2k_F$. As U increases, a new instability appears at $4k_F$ which is chiefly associated with electron localization and therefore with reduction of the charge degrees of freedom (i.e., electron translation). The $2k_F$ instability remains but becomes more associated with the spin degrees of freedom as U increases, gradually evolving into a SP transition for $U \gg t$. For $U \neq 0$, the gaps in the charge and spin excitation spectra will not be equal.

The presence of $U > 0$ does tend to separate the charge and spin degrees of freedom, but this separation is not complete until $U \gg t$. Therefore, the "generalized" Peierls instabilities at $2k_F$ and $4k_F$ should have both charge and spin character for $U \neq 0$, ∞. This has been demonstrated by numerical calculations of the $2k_F$ and $4k_F$ correlation functions.[85]

For a real system, an instability at $2k_F$ or $4k_F$ may reduce the electrical conductivity σ, the magnetic susceptibility χ, or both. If both are reduced to zero by the $2k_F$ instability with no $4k_F$ instability, one suspects the RP transition ($U \ll t$). If the material is an insulator (electrons localized) and only χ is reduced to zero by the $2k_F$ instability, one suspects the SP transition ($U \gg t$). If both σ and χ are affected but unequally at $2k_F$ or $4k_F$, one suspects an intermediate case. It is important to bear in mind that the energy associated with the charge degrees of freedom (t) is much larger than that associated with the spin degrees of freedom (J). Therefore, a small reduction of σ can involve more electronic energy than the total reduction of χ. Thus we suspect that the $2k_F$ transition in materials is like[86] (K)(TCNQ) results from the intermediate case; the susceptibility χ is completely lost below the transition, but the high transition temperature (and therefore high effective J) probably results from the concomitant loss of some charge (translational) freedom (i.e., σ is reduced[87]).

For a non-half-filled Hubbard band, the above reasoning must be extended somewhat. Localization of the electrons for this case requires longer-range Coulomb interactions (e.g., nearest-neighbor repulsion for a quarter-filled band), and the expression for an effective J, when the Coulomb repulsion greatly exceeds t,[88] will be modified. Other than this, the same arguments as made previously should hold except that now the magnitudes of the longer-range Coulomb interactions are also relevant variables in addition to U.

We note that other authors[89–91] have tried to couple the SP and RP phenomena by starting from a Hamiltonian that includes both the Fröhlich electron–phonon coupling [Eq. (58)] and the Hubbard U term. These attempts have all been MF and are basically inadequate to obtain proper results for one-dimensional electronic systems. (Recall that only the phonon system is three-dimensional). Their solutions show various dimerized and AF ground states but should not be considered definitive. Much more accurate calculations of the ground state of interacting one-dimensional electron systems have been produced by renormalization group and boson algebra methods,[79] but these do not include the electron–phonon coupling with similar accuracy. Lépine and Caillé,[92] again in MF, have compared the spin ($U \gg t$) and electronic ($U = 0$) free energies of a dimerized chain and find the spin contribution very important for the dimerization of large U systems. Kondo[93] finds that U enhances the spin-density-wave and SP instability, also by MF techniques. All these MF results are suspect, at least with respect to their quantitative aspects.

Finally, it is interesting to note that the spin degrees of freedom appear to have important roles even in some highly conducting quasi-one-dimensional systems. The organic superconductors, bis(tetramethyltetraselenafulvalenium) hexafluorophosphate, $(Me_4TSeF)_2PF_6$, and some isostructural analogs, give evidence of having a spin-density-wave transition at atmospheric pressure [and at $T \simeq 12$ K for $(Me_4TSeF)_2PF_6$],[32, 94–96] but at this writing more clarification of this situation is required. The presence of spin density waves is evidence for important correlation effects coexisting with high conductivity (translational freedom) in these materials in a complex way, but probably not unrelated to the discussions in this section.

3. Experimental Systems

3. Tetrathiafulvalenium Bis-Dithiolene Metal Compounds

3.1.1. Introduction

Numerous π-donor–acceptor compounds have been prepared that combine the organic π-donor, tetrathiafulvalene (TTF) (Figure 6a), with a series of bis-dithiolene (BDT) metal complexes (MBDT) (Figure 6b).[97] Several members of this group of compounds, where the bis-dithiolene complex is of the type $MX_4C_4(CF_3)_4$ (X = S, Se), exhibit magnetic properties that are indicative of long-range cooperative interactions in the solid state.[2, 98–100] Among these compounds are an example of a quasi-one-dimensional ferromagnetic system[98] as well as one that exhibits metamagnetic behavior.[99] Three of these compounds [$(TTF)(MX_4C_4(CF_3)_4$; M = Cu, X = S, Se; M = Au, X = S] exhibit the characteristics of a quasi-one-dimensional Heisenberg antiferromagnetic spin system at temperatures above ~ 12 K but undergo a second-order transition at lower temperatures which has been characterized as a spin–Peierls transition.

3.1.2. Structural Considerations

The $(TTF)[MX_4C_4(CF_3)_4]$ (M = Cu, Au; X = S, Se) compounds are isostructural with each other and with the corresponding M = Pt, X = S compound, for which a complete three-dimensional crystal-structure determination has been reported.[101] This structure can be viewed in terms of a

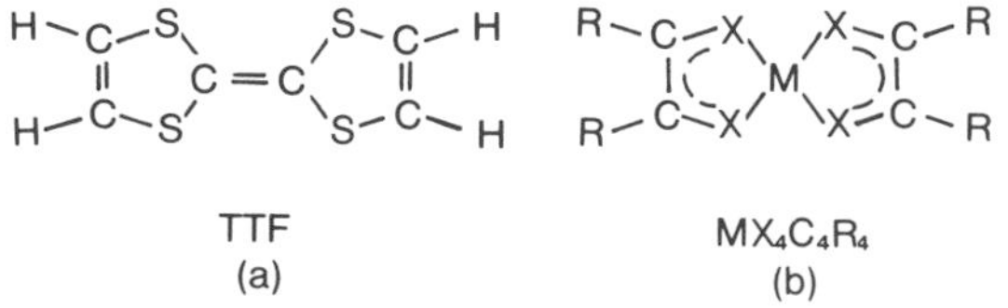

TTF
(a)

$MX_4C_4R_4$
(b)

FIGURE 6. Molecular structures for (a) tetrathiafulvalene, and (b) the bis-ethylenedithiolene and bis-ethylenediselenene metal complexes.

face-centered cell (space group $F\bar{1}$; $Z = 4$; $a_F = 23.34$, $b_F = 13.10$, $c_F = 7.82$, $\alpha = 90.6$, $\beta = 104.4$, $\gamma = 92.0$) in which $(TTF)^+$ and $[MX_4C_4(CF_3)_4^-$ ions alternate along all three axes with a stacking of the units along c_F.

An alternative description of the structure is in terms of a primitive triclinic cell ($P\bar{1}$) whose lattice vectors (a_P, b_P, and c_P) are related to those of the face-centered ($F\bar{1}$) cell by means of the following transformation[18]:

$$\begin{matrix} \mathbf{a}_F \\ \mathbf{b}_F \\ \mathbf{c}_F \end{matrix} = \begin{bmatrix} -1 & -2 & -1 \\ -1 & 0 & 1 \\ 1 & 0 & 1 \end{bmatrix} \begin{matrix} \mathbf{a}_P \\ \mathbf{b}_P \\ \mathbf{c}_P \end{matrix}$$

The relationship of these two cells is shown in Figure 7 for the (010) plane of the primitive cell which is also the (100) plane of the face-centered cell.

This "sodium-chloride-like" structural arrangement, as well as the magnetic properties, indicates that these are fully charge-transferred π-donor–acceptor compounds comprised of $(TTF)^+$ and $[MX_4C_4(CF_3)_4]^-$ ions. The relatively large separation between the ions in this structure, along with alternation of the donors and acceptors, apparently results in a

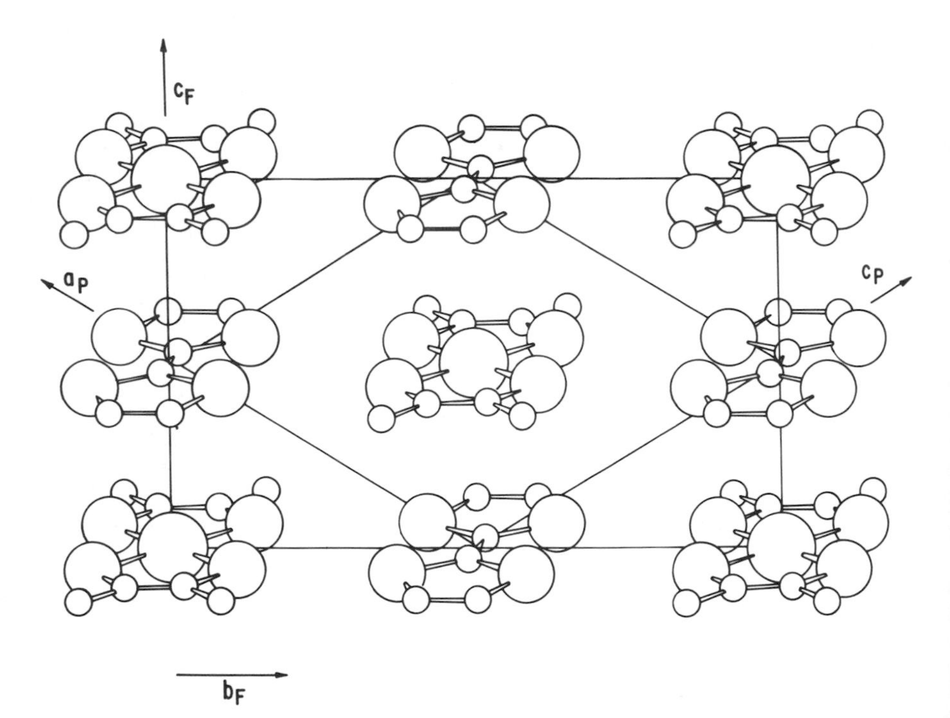

FIGURE 7. Arrangement of TTF and $MX_4C_4(CF_3)_4$ units in the (100) plane of the $F\bar{1}$ cell and (010) plane of the $P\bar{1}$ cell. The largest circle represents the metal M at the center of a $MX_4C_4(CF_3)_4$ unit. The F atoms of the CF_3 groups and the H atoms of TTF have been omitted for clarity of presentation.

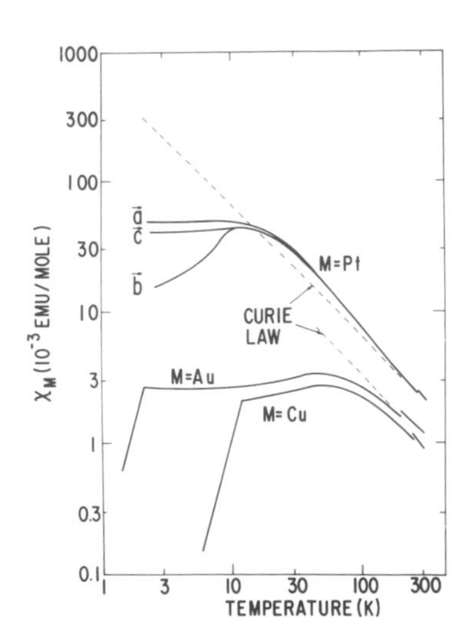

FIGURE 8. χ vs T for the (TTF)[$MS_4C_4(CF_3)_4$] complexes.

highly localized electronic structure, as is suggested by the extremely low electrical conductivity observed for these compounds.[2]

In the case of the M = Pt, X = S compound, both the $(TTF)^+$ and $[PtX_4C_4(CF_3)_4]^{\dot{-}}$ ions carry an unpaired electron spin, and a direct exchange interaction along the stacking direction c_F is possible. The magnetic susceptibility behavior of this compound is in fact consistent with a description that involves one-dimensional ferromagnetic chains which couple together antiferromagnetically below ~ 12 K.[98] On the other hand, when M = Cu or Au, the corresponding $[MX_4C_4(CF_3)_4]^-$ ions are spin paired, and the only spin carrier is the $(TTF)^+$ ion. As will be described in Section 3.1.3 in more detail, this has a pronounced effect on the magnetic exchange interactions and consequently the magnetic susceptibility behavior. These differences are illustrated in the log χ vs log T diagram in Figure 8, in which the change in number of unpaired spins per (TTF)[$MX_4C_4(CF_3)_4$] formula unit is reflected in the respective Curie law constants for the M = Pt and M = Cu, Au derivatives (intercepts of dashed lines in Figure 8).

In all of these (TTF)[$MX_4C_4(CF_3)_4$] (M = Pt, Cu, Au) compounds, there is evidence from both the magnetic susceptibility and the x-ray structural studies of a first-order structural transformation between room temperature and 200 K. As is shown in Figure 8, this structural transition results in an abrupt change in the magnetic susceptibility which is most obvious in the case of the M = Cu and Au derivatives but is also observable in the M = Pt case.

Under certain conditions single crystals of the (TTF)[$MX_4C_4(CF_3)_4$] compounds can be taken through this transition without a large increase in the mosaic spread and studied by means of x-ray diffraction and other

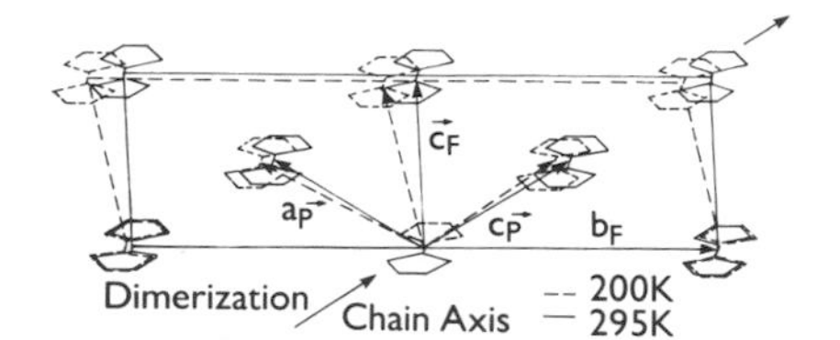

FIGURE 9. Projection of the TTF units in $(TTF)[CuS_4C_4(CF_3)_4]$ on the α_p-c_P or c_F-b_F plane, showing the change in crystal structure which occurs at ~ 240 K.

methods. Such a single-crystal x-ray diffraction study has been performed on $(TTF)[CuS_4C_4(CF_3)_4]$ both above and at various temperatures below this transition, resulting in a detailed physical picture of the crystal structure before and after the transition.[102] The transition is reflected mainly in changes within the $(010)_P$ plane, as shown in Figure 9. Notably, the essential equality of a_P and c_P at 297 K is destroyed, and c_P becomes 1.75 Å (or 20%) shorter than a_P at 20 K. As will be discussed in more detail later, this change is apparently important in determining the directionality of the magnetic interactions in the solid state and, in particular, in establishing the c_P axis as the one-dimensional magnetic chain axis. The "high-temperature transition" in this compound also results in an ordering of the trifluoromethyl substituents on the $[CuS_4C_4(CF_3)_4]^-$ units, which are rotationally disordered in the room-temperature structure. This causes a decrease in the volume of the unit cell and presumably provides the driving force for the transition.

Although a detailed study of the M = Au and Pt compounds below the first-order transition has not been carried out, preliminary x-ray precession studies on the Au compound[103] and the abrupt drop in magnetic susceptibility observed for both compounds indicates that a similar change in structure occurs in these cases but that the magnitude of the change in lattice parameter (Δa) may be quite different, with $\Delta a_{Au} \geq \Delta a_{Cu}$ and with Δa_{Pt} probably considerably less.

3.1.3. *Magnetic Susceptibility and Electron Paramagnetic Resonance Measurements*[2]

The magnitude of the spin susceptibility at room temperature for each of the $(TTF)[MX_4C_4(CF_3)_4]$ (M = Cu, Au; X = S, Se) compounds is consistent with the presence of one $\tilde{g} \simeq 2$, $S = 1/2$ spin per formula unit, as expected for the completely charge-transferred $(TTF)^+[MX_4C_4(CF_3)_4]^-$ formulation. The identification of the $(TTF)^+$ ion as the principal spin carrier in this system is confirmed by electron paramagnetic resonance (epr) studies of the $(TTF)[MS_4C_4(CF_3)_4]$ (M = Cu, Au) compounds, which show principal $\tilde{g}$ values and $\tilde{g}$-tensor orientation in the crystal in detailed agreement with the known structure and previous eps results for $(TTF)^+$ derivatives. The integrated intensity of this epr signal mirrors the temperature dependence of the static susceptibility throughout the entire temperature range and shows directly the continuous quenching of the $(TTF)^+$ spins below the spin–Peierls transition temperature. Moreover, the approximate constancy and isotropy of the $\tilde{g}$ values as a function of temperature provide evidence for the Heisenberg character of the exchange interactions.

The temperature dependence of the static susceptibility of all three compounds below the first-order transition and down to the spin–Peierls transition are fit accurately by a Bonner–Fisher calculation[6] of a $S = 1/2$, one-dimensional uniform Heisenberg antiferromagnetic (AF) chain. The respective values of J/k_B and $\tilde{g}$ obtained from these fits are 77 K and 1.97 for M = Cu, X = S and 68 K and 2.05 for M = Au, X = S. These $\tilde{g}$ values are, within experimental error, in agreement with those obtained from the epr measurements. (The principal $\tilde{g}$ values for the M = Cu, X = S compound range from 2.0016 to 2.0151 by epr.[2]) For the M = Cu, X = Se compound, $J/k_B = 88$ K and the $\tilde{g}$ values are not precisely determined. Below 12 K for M = Cu, X = S, 6 K for M = Cu, X = Se, and 2 K for M = Au, X = S, the spin susceptibility decreases sharply, independent of the orientation of the crystals in the applied field, and falls to essentially zero within a few degrees.

The temperature dependence of the susceptibility in this region has been found to conform in good qualitative detail to that expected for a system undergoing the progressive, temperature-dependent dimerization of spins anticipated for the spin–Peierils transition. The results of such an analysis for the $(TTF)[CuS_4C_4(CF_3)_4]$ compound is illustrated in Figure 10. The solid line below 12 K in this figure was calculated using the Bulaevskii model for the χ of an alternating chain system[7] assuming a temperature-dependent alternation with BCS functional form, where the alternation at $T = 0$ K $[\delta(0)]$ is taken as an adjustable parameter. The Bulaevskii susceptibility was matched to the actual susceptibility at T_c. The resultant fit to

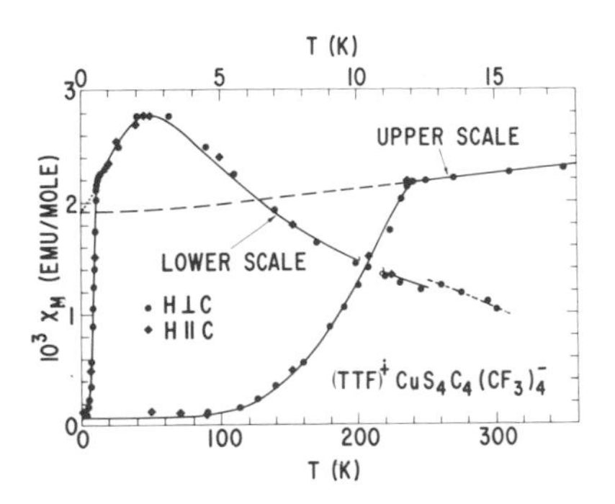

FIGURE 10. Magnetic susceptibility of $(TTF)[CuS_4C_4(CF_3)_4]$ along two directions. Solid lines are calculated from a spin–Peierls theory which contains AF chains with uniform exchange above 12 K and temperature-dependent alternating exchange below.

the experimental data is quite good, whereas the assumption of a temperature-independent alternation, as would be expected in the case of a first-order transition, fails to reproduce the shape of the χ vs T curve. Using the spin–Peierls theory and the experimental values of T_c and J/k_B, $\delta(0)$ is calculated and compared with that obtained from the curve-fitting procedure. The discrepancy between the predicted (0.167) and observed (0.127) value is small and may be due to the basic limitations of the purely mean-field approach used in the theoretical calculation. A similarly good correspondence with the mean-field spin–Peierls model is obtained in the case of the other two members of this group of compounds.

In summary, the magnetic data obtained for these three compounds are consistent with a one-dimensional Heisenberg, $S = 1/2$, AF description for the spin system and indicates a second-order transition to a singlet ground state at low temperature due to a progressive spin–lattice dimerization.

3.1.4. Other Data on the $(TTF)[MX_4C_4(CF_3)_4]$ System

Mean-field spin–Peierls theory yields a prediction of both the form and magnitude of the specific-heat change at the transition. A typical mean-field BCS-like jump in the magnetic specific heat is anticipated of magnitude $\sim 0.1R$ for the M = Cu, X = S salt and $\sim 0.02R$ for the corresponding Au derivative.[1,2] Experimental results, employing both dc measurements on polycrystalline samples and ac measurements on a single crystal, have confirmed these expectations in detail, at least for the Cu compound, where the specific-heat anomaly at $T_c = 12$ K is quite close in both form and magnitude to that predicted.[104] In the case of the M = Au compound, the transition initially appeared to be less mean-field-like,[104] possibly because of crystal imperfections. However, more recent experiments[105] suggest that its specific-heat anomaly is rather like that of the Cu compound. For both of these systems, a linear term in the specific heat was observed with coefficients that are in good agreement with expectations for one-dimensional Heisenberg, S = 1/2, AF chains.

Further support for the view of these systems as uniform one-dimensional Heisenberg, $S = 1/2$, AF chain systems above T_c has been obtained from temperature-dependent proton nmr studies carried out for the M = Cu and Au, X = S compounds.[106] These measurements give values for T_1^{-1}, the spin–lattice relaxation rate, which are in reasonable quantitative agreement with that anticipated theoretically[75] and which show the temperature and frequency dependence characteristic of a $S = 1/2$, one-dimensional AF spin system, as well as the progressive formation of the gap in the magnetic energy spectrum below T_c.

Another approach to evaluating the quasi-one-dimensional character of the system is found in higher-temperature nmr measurements of T_1 as a function of frequency.[107] The frequency dependence of T_1 reflects the spin dynamics. The goal of this work was to obtain interchain as well as intrachain exchange couplings for (TTF)(CuBDT) on both sides of the 240 K structural transition. The authors conclude that below 240 K the intrachain exchange is 20 to 40 times greater than the interchain one, while above this transition the ratio is only 5 to 10, reflecting a poor one-dimensional character in that range. However, there are some problems with the magnitude of intrachain exchange obtained in this analysis compared to that found from static susceptibility, so one hesitates to pursue calculations of quantities, such as the "virtual" Néel temperature, which are derived from interchain magnetic exchange. The question of effects ascribable to interchain exchange is considered again in Section 3.4, but it remains elusive.

Direct observation of the anticipated lattice dimerization below T_c has been obtained in the case of (TTF)[$CuS_4C_4(CF_3)_4$] by means of x-ray diffraction measurements.[18] Below T_c new superlattice reflections develop which require a doubling of the a_P and c_P cell dimensions. The intensity of these new peaks builds up progressively below T_c as the square of the BCS gap function in excellent agreement with the expected mean-field character of the transition. A detailed examination of the intensities of the new x-ray reflections has been carried out and provides a microscopic picture of the nature of the lattice dimerization.[19] The basic feature is a translation of the $(TTF)^+$ units along c_P which is accompanied by a movement of the $[CuS_4C_4(CF_3)_4]^-$ in an orthogonal direction so as to fill the space left by the $(TTF)^+$ units (see Figure 11). The maximum excursion of these ions is on the order of 0.04 Å, in approximate agreement with the prediction of the mean-field theory.[1] These results clearly establish the existence of a progressive lattice dimerization below T_c and, by virtue of the observed direction of dimerization, lead to the identification of the one-dimensional AF exchange direction as the c_P-axis direction in the crystal. The "dimensional AF chains" in this system are therefore comprised of $(TTF)^+$ units that presumably interact by a direct exchange mechanism along the direction of their closest approach, viz. c_P.

A diffuse x-ray study of the (TTF)[$CuS_4C_4(CF_3)_4$] compound above T_c indicates the persistence, up to at least 225 K, of enhanced scattering at the

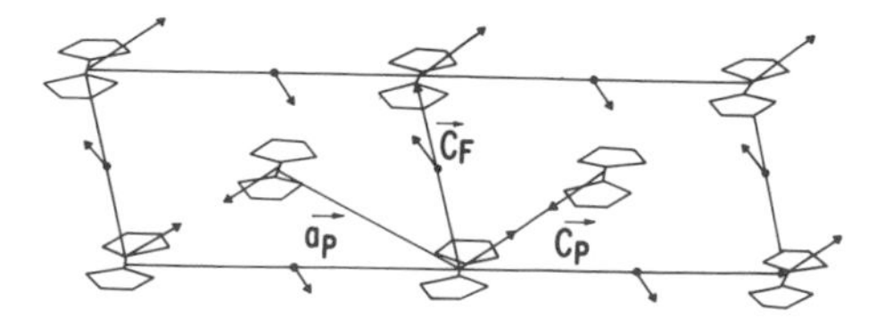

FIGURE 11. Directions of displacements for the $(TTF)^+$ and $[CuS_4C_4(CF_3)_4]^-$ units (only the center of mass of the latter species is shown here, for clarity) below the spin–Peierls transition temperature.

position of the new Bragg peaks seen below T_c.[18] This scattering was found to be approximately isotropic in k-space, indicating that the implied precursive lattice softening is fully three-dimensional, in contrast to the presumably one-dimensional magnetic fluctuations. These results show that a lattice mode of very low energy ($\sim$ 1 meV) exists in this structure in the absence of substantial magnetic correlations. The apparently fortuitous existence of this "soft mode" at the wave vector ($2k_F$) appropriate for the lattice dimerization clearly facilitates the occurrence of the spin–Peierls transition and, in fact, may be an important prerequisite for its observation.[18] As mentioned in Section 2.2, this soft mode causes the chains to act cooperatively at $2k_F$ and therefore is probably responsible for the quite mean-field-like behavior.

3.2. (*Methylethylmorpholinium*)(*Bis-7,7,8,8-Tetracyano-p-quinodimethanide*), *MEM* $(TCNQ)_2$

The other system for which good evidence for the occurrence of a spin–Peierls transition has been obtained is a 2 : 1 complex salt of 7,7,8,8-tetracyano-*p*-quinodimethane (TCNQ) with the methylethylmorpholinium cation $(MEM)^+$, i.e., $MEM(TCNQ)_2$.[11] The planar TCNQ units in this compound are stacked to form one-dimensional chains (see Figure 12). Above $\sim$ 335 K these chains are almost uniform and the system is a good electrical conductor [$\sigma(> 340\ K) \sim 15$–$30\ ohm^{-1}\ cm^{-1}$]. A phase transition occurs at this temperature which dimerizes the TCNQ units in the chain and decreases the conductivity by about four orders of magnitude. As is evidenced by both magnetic susceptibility and specific heat, there is another transition at $\sim$ 20 K. Preliminary x-ray measurements below this temperature have shown that the unit cell doubles along the chain direction, leading to a tetramerized structure.

Between 335 and 20 K the magnetic susceptibility can be fit quite well[11] by the Bonner–Fisher calculation for a $S = 1/2$, one-dimensional Heisenberg AF chain system using a $\tilde{g}$ value of 2.003, as determined from epr, and an exchange interaction J ($H = J \sum_i S_i S_{i+1}$) of 106 K. Below 20 K the susceptibility dips sharply below the Bonner–Fisher curve in a manner much like that observed for the $(TTF)[MX_4C_4(CF_3)_4]$ (M = Cu, Au; X = S, Se) compounds, extrapolating to zero at 0 K.

The analysis of the susceptibility[11] change below 20 K follows essentially the same approach used in the case of the $(TTF)[MX_4C_4(CF_3)_4]$ compounds, and yields a good fit to the mean-field spin–Peierls model with $T_c = 17.7$ K and an alternation parameter at 0 K, $\delta(0) = 0.161$. The corresponding magnetic gap at $T = 0$ K derived from the fit was $2\Delta(0) = 56$ K, yielding $\Delta(0)/T_c = 1.58$, while the mean-field theory predicts a value of 1.76.

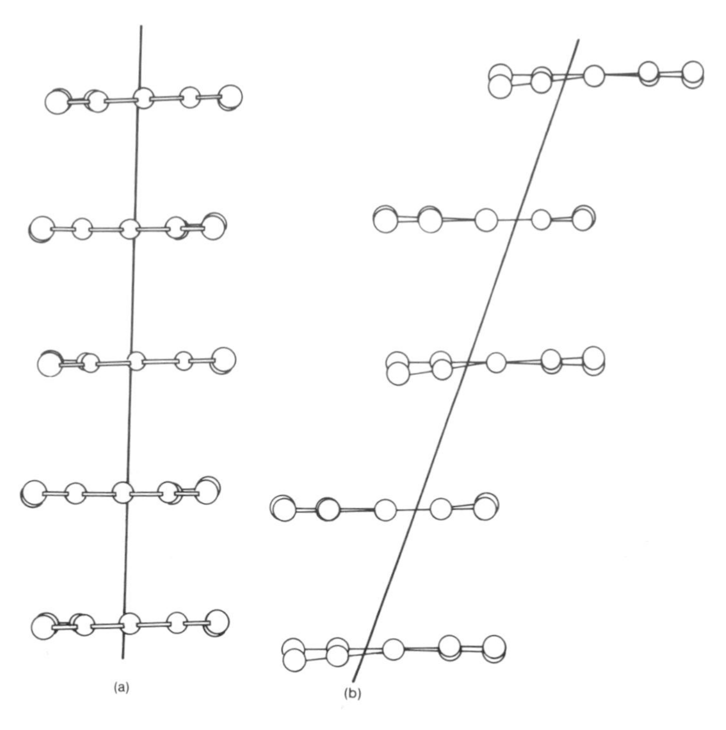

FIGURE 12. The TCNQ chains in MEM$(TCNQ)_2$ as viewed along their longest axis at (a) 346 K and (b) 113 K.

The Knight shift,[11] which probes the electron spin density at the position of the TCNQ protons, has also been determined for this compound as a function of temperature from proton magnetic resonance experiments. The measured Knight shift, in fact, mirrors the static susceptibility throughout the range of measurements including the region around the 20 K transition. The spin susceptibility,[108] measured by epr, also agrees with the static susceptibility and the Knight shift data. These measurements confirm the expectation that the sharp drop in susceptibility observed below this temperature is indeed reflecting a pairing of the spins which reside on the $(TCNQ)_2^{\dot{-}}$ dimer units. Finally, specific-heat measurements have been carried out on a collection of small crystals of MEM$(TCNQ)_2$. An anomaly was observed near 18 K whose form and magnitude are in reasonable agreement with that predicted by the mean-field spin–Peierls theory.

The similarity of MEM$(TCNQ)_2$ to the bis-dithiolene materials appears to extend beyond spin–Peierls behavior to some lattice characteristics. For example, there is evidence[108, 109] for a precursive soft mode at the dimerization wave vector in MEM$(TCNQ)_2$. Presumably this mode plays a similar role of promoting mean-field behavior at T_c. It is also noteworthy that the upper crystal-phase transition (~ 335 K) in MEM(TCNQ)$_2$ is also severe and first order.[108,110]

3.3. *Other Possible Spin–Peierls Systems*

Over the last twenty years a variety of other compounds have been suggested as examples of what we now refer to as spin–Peierls behavior. One of the earliest of these was Würster's blue perchlorate [i.e., *N,N,N′,N′*-tetramethyl-*p*-diaminobenzene (TMPD) perchlorate], which exhibits a phase transition at 186 K, below which the magnetic susceptibility drops sharply, and the formerly uniform chain of $(TMPD)^+$ radical cations becomes strongly dimerized.[111–113] In this case, however, the transition is clearly first order in nature, and the dimerization is essentially temperature independent below the transition. Moreover, it is accompanied by an ordering of the formerly rotationally disordered perchlorate anions,[114] indicating that the driving force for the transition is provided by a crystal-structure instability in which the magnetic system plays no important role.

Several members of the alkali TCNQ salts undergo phase transitions that result in a dimerization of uniform chains with concomitant loss of magnetic susceptibility and that have been speculated to be of the spin–Peierls type.[115] The transitions are usually first order but may in some cases be second order. One of these materials, (K)(TCNQ), has been treated by two groups in terms of a modified spin–Peierls transition which begins with a displacive, first-order transition at $T_c = 395$ K. Lépine *et al.*[86b] fit the experimental susceptibility results for $T < T_c$ using the usual Pytte-type mean-field equations which they modify for (K)(TCNQ) by using an exchange $J(T)$ that is temperature dependent due to thermal lattice expansion. They then truncate the equations at T_c by the displacive structural phase transition to get first-order behavior. Takaoka and Motizuki[86a] obtain a reasonable fit using the equations of a mean-field approach which yields a first-order spin–Peierls transition.

As first discussed in Section 2.5, the transition temperature in (K)(TCNQ) is very high ($T_c = 395$ K) as is the inferred value of $J \simeq 1800$ K, and the material is not an electrical insulator above T_c.[87] Therefore, the transition might better be considered to be of the generalized Peierls type, wherein both the charge and spin degrees of freedom of the electrons contribute to the transition in the sense that some amount of each is lost at T_c. The contribution of the energy associated with loss of part of the translational (charge) freedom of the electrons (i.e., decrease of conductivity[87]) might then account for the high values of T_c and J. Similar considerations apply to other alkali TCNQ's (or indeed to any other candidates) with high T_c; e.g., (Na)(TCNQ) ($T_c = 348$ K) and (Rb)(TCNQ) ($T_c = 381$ K). These high transition temperatures also make it extremely difficult to establish the existence of a uniform one-dimensional AF chain system above T_c (e.g., by susceptibility), a prerequisite for spin–Peierls behavior.

Other one-to-one TCNQ-derivative compounds for which spin–Peierls transitions have been suggested are hexamethylenetetrathiafulvalene 2,3,5,6-tetrafluoro-7,7,8,8-tetracyano-*p*-quinodimethane (HMTTF)($TCNQF_4$)[116] and its isostructural selenium analog,[117] (HMTSeF)($TCNQF_4$). These materials are fully charge transferred and so have half-filled bands, which have severely restricted electron mobility if the Coulomb repulsion (U) is large, as it seems to be. This suggests that the chain dimerization transitions are magnetically driven, i.e., of the spin–Peierls type. However, as discussed in the previous paragraph, the high transition temperatures [e.g., $T_c \simeq 205$ K for (TMTSeF)($TCNQF_4$)] suggest that the generalized Peierls transition might be a more apt description of the systems. The salt (NH_4)(TCNQ) has been thought to be a spin–Peierls system,[118] and the magnitude of its T_c (~ 28 K) makes this more reasonable, but the complexity of its possible crystal structures makes more data desirable.

The material (DBTTF)($TCNQCl_2$) [$\Delta^{2,2'}$-bi-*d*-benzo-1,3-dithiol 2,5-dichloro-7,7,8,8-tetracyano-*p*-quinodimethane] appears to be analogous to MEM$(TCNQ)_2$ and a good candidate for spin–Peierls behavior.[119] It has two phase transitions, one at 180 K which reduces the conductivity severely but not the susceptibility, and one at $T_c \simeq 38$ K which reduces the susceptibility in spin–Peierls fashion. The susceptibility can be approximately fit with the exchange $J \simeq 160$ K and the spin–Peierls dimerization parameter $\delta(0) = 0.25$. The values of T_c and J are low enough that the driving force for the transition at T_c may come from the spin degrees of freedom (magnetic interactions) alone. The analogy of this material with (MEM)$(TCNQ)_2$ suggests a quarter-filled band with a dimerization transition at 180 K, preparing the chains for the tetramerization by magnetic interactions at T_c.

The material (diethylmorpholinium)$(TCNQ)_2$,[120] DEM$(TCNQ)_2$, might be expected to be an analog to MEM$(TCNQ)_2$, but these closely related materials have different crystal structures. Like MEM$(TCNQ)_2$, DEM$(TCNQ)_2$ appears to be fully charge transferred but has two inequivalent TCNQ stacks. One of these sets of stacks does appear to behave like the TCNQ stacks in the MEM compound compound in that its spin susceptibility decreases rapidly below $T \simeq 22$ K. These epr results have provoked the interesting suggestion of a SP transition at 22 K in half the TCNQ stacks of this material.[120]

Another interesting claimed spin–Peierls system is stressed[121] or doped[122,123] vanadium(IV) oxide(VO_2). The effect of doping or stress is to introduce a new crystal phase which contains zig-zag chains of V^{4+} ions with localized spins.[124] At $T_C \simeq 220$–320 K, depending on doping, a first-

or second-order transition (depending on doping or stress) leads to progressive dimerization of the chains, as expected for a spin–Peierls transition. With this high value of T_c, the same caveats as discussed in the previous paragraphs apply (generalized Peierls behavior, one-dimensional AF behavior unestablished above T_c). In addition, the complicated crystal environment causes one to worry about crystal-structure instabilities unrelated to magnetism, and indeed others have occasionally interpreted the crystal phases differently.[125] Likewise, the complicated mixed-stack donor–acceptor material tetraethyltetrathiafulvalene(TCNQ)$_2$, $Zt_4TF(TCNQ)_2$, which has also been analyzed as a spin–Peierls system[126] ($T_c \simeq 170$ K), is subject to similar caveats as vanadium(IV) oxide.

Finally, we note that spin–Peierls effects, i.e., spin density waves, may play a role in the phase transitions of quasi-one-dimensional conductors. This statement is essentially the recognition that the spin degrees of freedom are always present and can participate in phase transitions in various ways. As mentioned in Section 2.5, $(Me_4TSeF)_2PF_6$, a quasi-one-dimensional organic superconductor, appears to have important spin interactions at its phase transitions.[94–96]

3.4. High-Magnetic-Field Experiments

3.4.1. Introduction

We presented in Section 2.4 a theoretical consideration of what might be expected for spin–Peierls systems in high magnetic fields. A brief summary of the expectations follows. The transition temperature $T_c(H)$ is depressed with increasing field, the amount of the depression going initially as H^2. This behavior defines a line of second-order transitions as T_c decreases until a predicted special point $[H_c, T_c(H_c)]$ is reached. Beyond this point there are various speculations about the expected properties. In many cases the extension below $T_c(H_c)$ of the boundary of the dimerized phase is a line of first-order transitions. For a fixed-periodicity model, the special point $[H_c, T_c(H_c)]$ is a tricritical point marking the change-over in the order of the transitions, and all transitions are from the dimerized to the uniform state. In other models, some more exotic phases have been suggested; e.g., an incommensurate high-field state in which the spin periodicity does not coincide with the lattice (C–I transition), or a new commensurate state that persists for some field range (C–C transition), or the possibility of many commensurate phases attained in the cascade of (first-order) transitions called Devil's staircase or harmless staircase. For these exotic phases, the H-T diagram obviously has additional boundary lines.

Experiments have been reported for (TTF)(CuBDT) with neutron diffraction on crystals to 80 kOe,[23] and with magnetization measurements on powders to ~200 kOe,[23–25] and for MEM(TCNQ)$_2$ with magnetization[28] on powders to 200 kOe. Research on (TTF)(AuBDT) powders is in progress[27] to 40 kOe between 1.1 K and 4.2 K. These studies are interesting in themselves, and intercomparison of the results from the different compounds casts light on questions raised in the individual investigations.

3.4.2. *(TTF)(CuBDT)*

Initial neutron experiments[23] were carried out at the Brookhaven National Laboratory High Flux Beam Reactor on a small crystal using a triple-axis spectrometer set for elastic scattering, with a superconducting magnet producing fields up to 80 kOe. Upon making scans through the dimerization Bragg peak $(1.5, 0, 0.5)_P$ (of the primitive unit cell) at 5 K with zero field and with 78 kOe, the results of Figure 13 were obtained. There is no magnetic-field dependence of the wave vector associated with the dimerization, but the intensity reduction is a real effect. The latter was investigated further by measuring the intensity of this peak as a function of temperature at four different field values, as shown in Figure 14. Approximate transition temperatures $T_c(H)$ were obtained by the extrapolations shown, yielding a depression ΔT_c proportional to H^2.

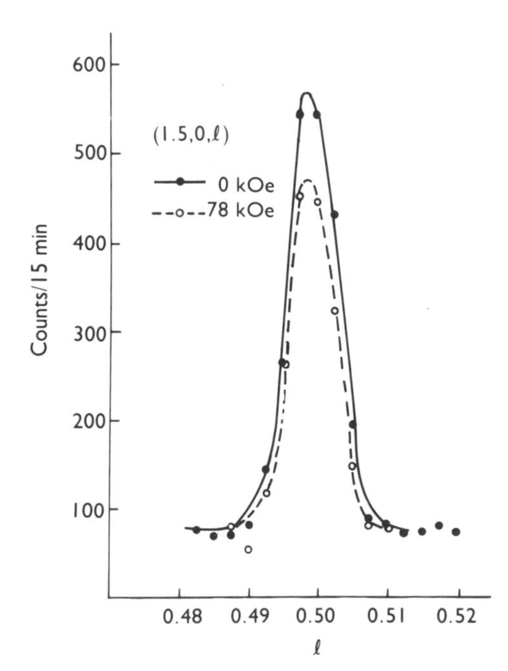

FIGURE 13. Neutron scattering scans through the (1.5, 0, 0.5) dimerization Bragg peak of (TTF)[$CuS_4C_4(CF_3)_4$] at zero field and at 78 kOe, the highest obtainable field ($T = 5$ K).

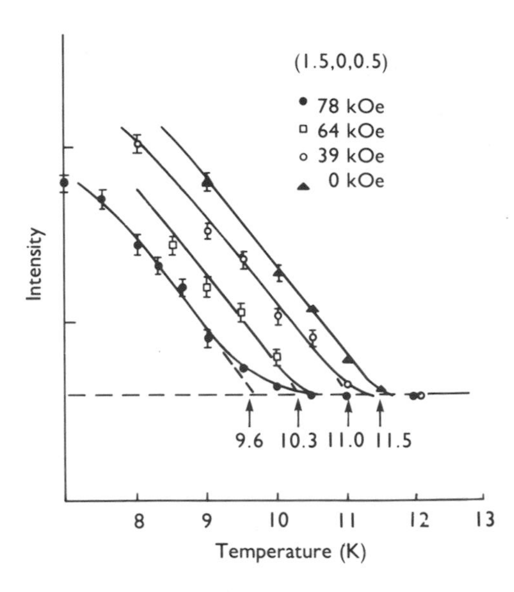

FIGURE 14. Temperature dependence of the neutron scattering intensity of the (1.5, 0, 0.5) Bragg peak of $(TTF)[CuS_4C_4(CF_3)_4]$ at a series of different applied magnetic fields.

Two separate series of high-field magnetization measurements were carried out on powders at the Service National des Champs Intenses, CNRS, Grenoble. In each case the field was produced in a water-cooled Bitter-type solenoid and the magnetization was measured with a sample motion magnetometer operating in a uniform field.

In the first series of experiments,[23] the maximum field was 155 kOe but the accuracy of the experiments was poor owing to the small amount of sample, as well as to a possible inadvertent dilution during a prior experiment. Nevertheless, the isofield M vs T curves to 100 kOe showed a depression of the temperature of the "knee," the feature taken to indicate the transition. The isofield curve at 150 kOe appeared to suggest a complete return to the uniform chain. (This "result" was refuted by later experiments.[24]) The magnetization isotherms, M vs H, progressed with decreasing temperature from being linear above T_c, to a nonlinear S-shape below T_c, gradually sharpening at lower temperatures to present a more nearly abrupt rise in magnetization at a critical field. The inflection points of the isotherms were taken to indicate a phase transition. The collection of transition points from the two types of curves defined a simple phase diagram in rough overall agreement with theory (cf. Figure 5), including a crude correspondence between the highest experimental transition field, 125 kOe, and the mean-field theoretical result for H_c from Eq. (47) equivalent to 134 kOe.

A follow-up series of experiments, with a newly prepared and larger sample, added considerably to knowledge of the phase behavior. Initial

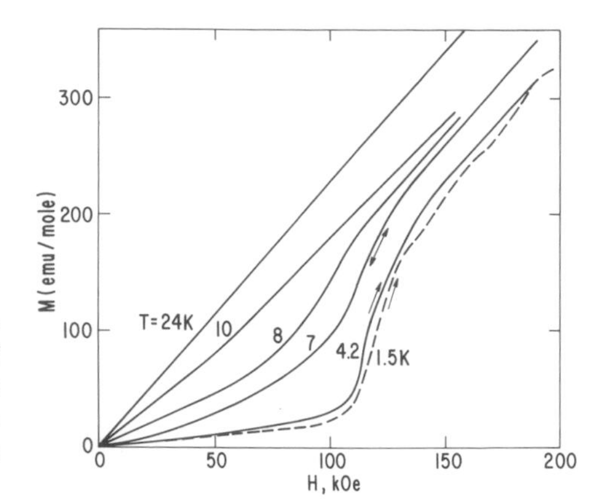

FIGURE 15. Isothermal curves of magnetization vs field for powder samples of (TTF)[$CuS_4C_4(CF_3)_4$]. For the two lowest temperatures, only the rising-field branch is shown; the other curves are reversible.

measurements[24] again went to 155 kOe, but some later runs were carried to 200 kOe.[25] Figure 15 shows a series of magnetization isotherms. These curves were reversible for $T \gtrsim 5.5$ K but showed a marked hysteresis at lower temperatures. An example of this behaviour is shown in Figure 16. Included is an inset both displaying the temperature dependence of the width of the hysteresis, which is seen to decrease in a well-characterized manner and to vanish at about 5.5 K, and delineating a region of first-order transitions anticipated in some theoretical treatments.[21, 22, 63, 64, 71, 127] The isotherm at 1.5 K in Figure 16 is peculiar in that it is sheared in field (by much more than can be accounted for by demagnetization effects). It is also curved so that it extends over at least 50 kOe, as demonstrated by the

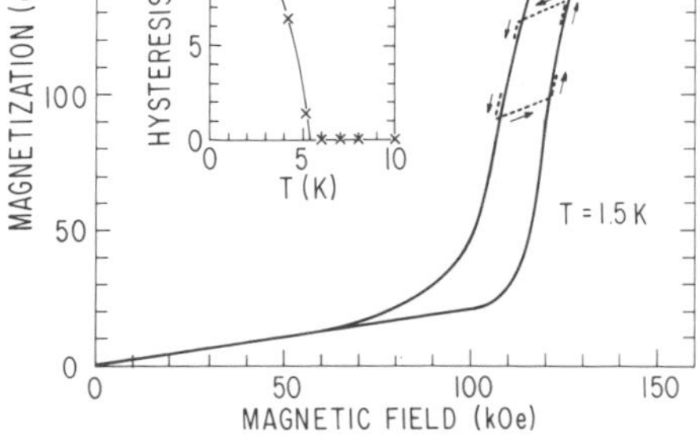

FIGURE 16. Magnetization vs field at 1.5 K for powder samples of (TTF)[$CuS_4C_4(CF_3)_4$]. Inset shows width of hysteresis loop vs temperature. Similar behavior persists when H_{max} is raised to 200 kOe.

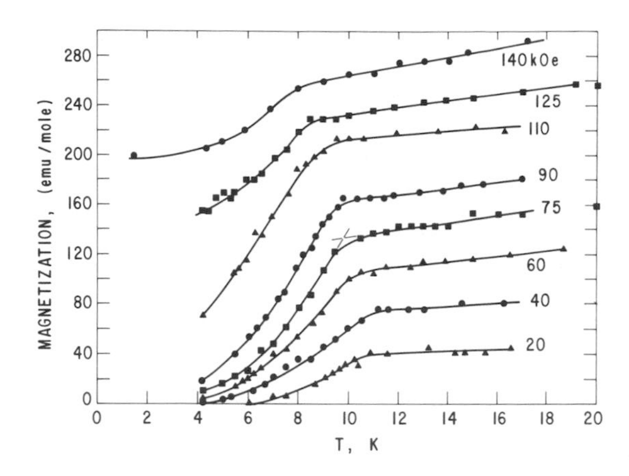

FIGURE 17. Magnetization vs temperature in selected fields for powder samples of (TTF)[$CuS_4C_4(CF_3)_4$]. An example of the mean field "knee" construction is shown for $H = 75$ kOe.

persistence of hysteresis (failure to close to a reversible portion) to the upper limit of measurement, and this continued to be the case when the limit was raised to 200 kOe (the hysteretic region then exceeds 100 kOe). We return to a discussion of this behavior below.

A second major feature of this follow-up series of experiments is revealed in the family of selected isofield curves in Figure 17. The "knees" (selected above as transition points) appear to persist to the highest fields. Concurrently, the magnetization for fields of $H \gtrsim 125$ kOe appears headed toward distinctly nonzero values as temperature decreases toward 0 K [$M(T = 0 \text{ K}) > 0$]. These results strongly suggest the existence of one or more new phases in (H, T) space. Figure 18 shows an overview in the form of a composite M, H, T plot taken (and smoothed) from many isofield and isothermal curves. Below $T = 5.5$ K, the averages of rising- and falling-field magnetization measures are presented. The lines of solid circles (knees) and crosses (inflection points) are a guide to the eye suggesting an approximate H-T phase diagram.

A note of caution is needed in the matter of phase-boundary indicators. The work in progress[27] on (TTF)(AuBDT) uses ac susceptibility in a superposed steady field as a probe, and this technique reveals more detailed structure in both isothermal and isofield sweeps. In particular, there is more basis for using peak susceptibilities than the mean-field "knee" construction as transition indicators. This has a small but nontrivial effect on the shape of the phase boundary at lower fields and higher temperatures, but is rather more important in the high-field region bounding the suggested new phase(s).

of the transition at slightly higher temperatures, e.g., by looking for time effects in magnetization after a rapid change of field, do not support the notion that there are first-order transitions for $T > 5.5$ K. The original work speculated on the nature of the boundary lines in the region of 9 K and the possibility of a multicritical point, M. However, in view of the present uncertainty regarding the exact location of the upper transition boundary, additional remarks would be inappropriate.

3.4.3. *Methylethylmorpholinium Bis(7,7,8,8-Tetracyano-p-Quinodimethanide), MEM(TCNQ)*$_2$

Magnetization measurements on a small powder sample of this compound[28] were carried out at CNRS–Grenoble with the same equipment as for (TTF)(CuBDT), going to 200 kOe. The experimental procedures were also similar, with qualitatively similar results. Owing to the rather higher $T_c(0)$, about 50% more field was needed to map out an equivalent area of H-T space. [See Section 2.4, especially Eq. (47)].

Figure 20 shows the results of the isofield measurements, M vs T, up to 197 kOe, and Figure 21 displays the isotherms, M vs H, for temperatures

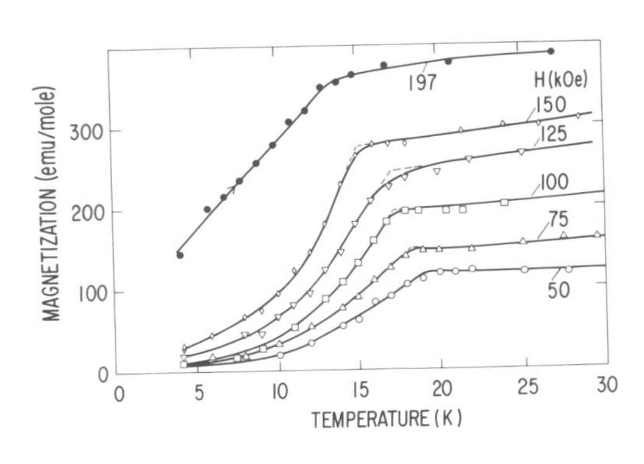

FIGURE 20. Isofield magnetization curves vs temperature for powder samples of MEM(TCNQ)$_2$. The 197-kOe measurements were preceded by cooling the sample to 4.2 K in 175 kOe.

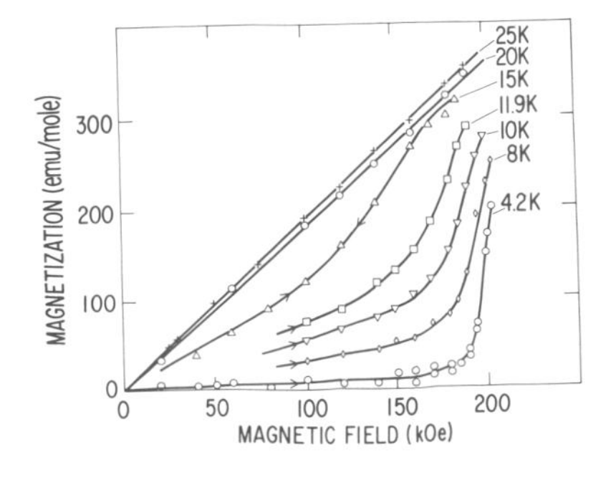

FIGURE 21. Isothermal magnetization curves vs field for powder MEM(TCNQ)$_2$ up to 200 kOe. For the isotherms below 12 K, only the increasing-field branch is shown.

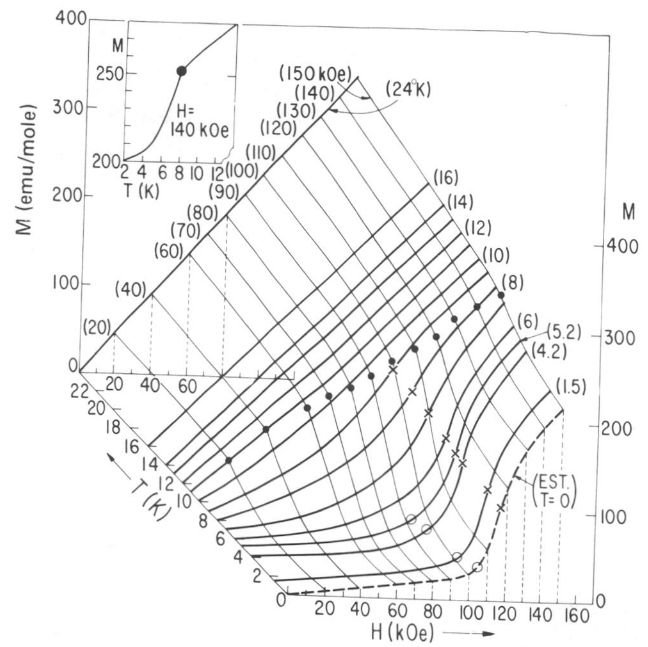

FIGURE 18. Composite *M*, *H*, *T* plot of high-field magnetization data for powder (TTF)[$CuS_4C_4(CF_3)_4$]. The crosses are isotherm inflection points, the solid circles are isofield "knees" (example in inset), and the open circles represent the onset of hysteresis.

Accurately reevaluated transitions are not available for (TTF) (CuBDT), so we show in Figure 19 the results obtained with the coarser techniques, with the reservation that the temperatures indicated are likely to be higher than will be found by other techniques. In this figure solid circles and crosses have the same significance as in Figure 18. The point B marks the end of the hysteretic transitions. Attempts to establish the order

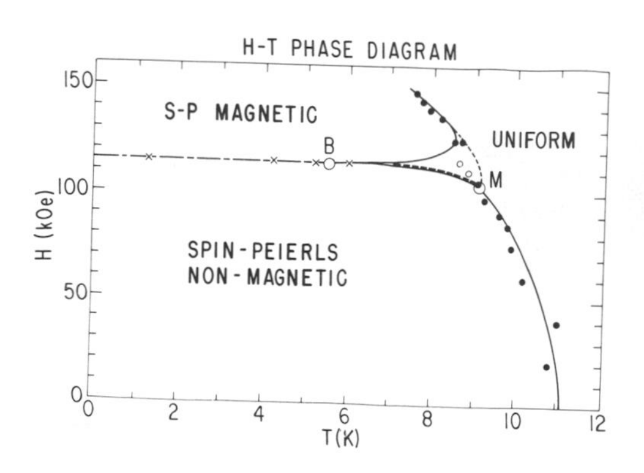

FIGURE 19. Spin–Peierls *H-T* phase diagram for (TTF)[$CuS_4C_4(CF_3)_4$] powder. The boundary separating the SP–nonmagnetic and SP–magnetic phases may possibly be described by the dashed lines or by the solid ones.

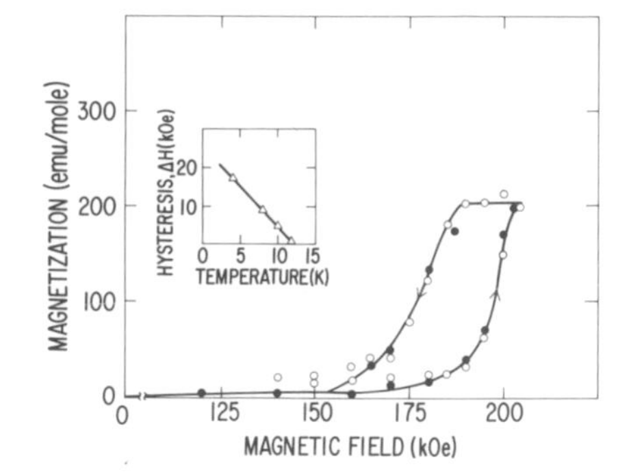

FIGURE 22. Isothermal magnetization vs field up to 203 kOe at 4.2 K for powder MEM(TCNQ)$_2$. Insert: width of the hysteresis loop vs temperature.

from 4.2 to 25 K. For $T < 12$ K, only the branch measured with increasing field is shown. The rough parallels with results for (TTF)(CuBDT) are evident.

In Figure 22, we give the magnetization results at 4.2 K for increasing and then decreasing values of the magnetic field. There is a nonmagnetic to SP-magnetic transition that appears to be centered at 190 kOe. The large hysteresis starts around 160 kOe and is certainly incomplete at 203 kOe, producing an almost square hysteresis loop, a truncated version of the behavior seen in (TTF)(CuBDT). The hysteresis region extends to 12 K (see inset, Figure 22), a greater fraction of $T_c(0)$ than in the former compound. We take the average of the inflection points from increasing and decreasing fields as the transition field and combine these with the mean-field "knees" from Figure 21 to construct a phase diagram shown in Figure 23. The transition field is $\sim$192 kOe at 4.2 K and 189 kOe at 12 K. The inflection point has not been observed at 1.5 K, due to the experimental high-field limitations. The upper limit temperature of the hysteresis region is very close to the intersection of the SP-magnetic to nonmagnetic transition line with the SP (nonmagnetic) to uniform transition line.

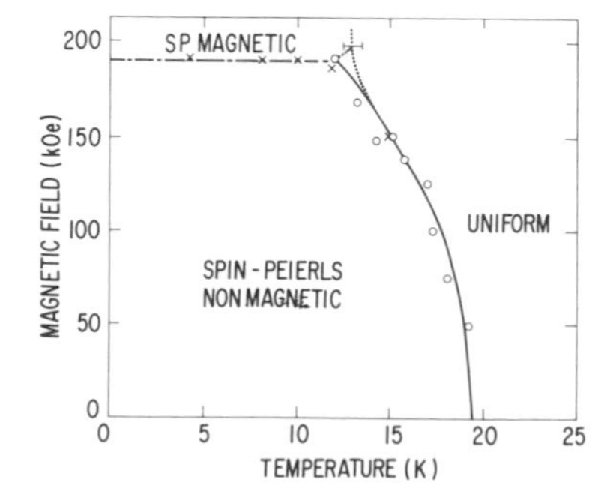

FIGURE 23. Spin–Peierls H-T phase diagram for MEM(TCNQ)$_2$ powder. The circles are from the knees of isofield curves, while the crosses indicate transition fields estimated from isotherms. The dotted line curves for $H > 190$ kOe are speculative; they represent two distinct and mutually exclusive possibilities.

The considerable hysteresis also makes it more complicated to evaluate the isofield magnetization curves at high fields because their trajectory depends on the detailed H, T prehistory. There is, however, an indication in the 197-kOe curve of a knee suggesting the start of a boundary for the SP-magnetic phase, somewhat similar to that seen in (TTF)(CuBDT). The uncertainties about its precise location are even greater than in the case of the previously described compound.

3.4.4. *(TTF)(AuBDT)*

High-field studies of susceptibility, magnetization, and specific heat are in progress[27,105] on powders of this member of the family at the Kamerlingh Onnes Laboratory, Leiden. The significantly lower temperature of its spin–Peierls transition, $T_c(0) \simeq 2$ K, offers considerable experimental advantages in that much less magnetic field is needed to probe an equivalent region of H-T space and temperature measurements (in field) and other techniques become easier.

Initial results show many features qualitatively similar to those measured on (TTF)(CuBDT) and MEM$(TCNQ)_2$. Some of these are included in Section 3.4.5 below. More significantly, the improved detail is enriching our perception of previous results and permitting a deeper insight into the phase behavior.

3.4.5. *Discussion*

The results summarized in the published phase diagrams (Figures 19 and 23) have many of the qualitative features of the theoretical predictions. At the outset, if any doubt remains, the considerable field dependence of the behavior permits one to discard models of the zero-field transition as a structural transition not involving spin—phonon coupling.

At the first level of comparison, our focus falls on the second-order transition line, $T_c(H)$, which decreases initially as H^2. This line is predicted to extend to 0.54 $T_c(0)$ or 0.77 $T_c(0)$ in the existing models.[17,21,22] The interpretation of the experimental indicators of T_c had initially suggested coefficients of H^2 about twice as large as calculated in the models. A more critical evaluation of these indicators now suggests that the agreement between experiment and theory in this region is much better. It is instructive to gather the experimental results for intercomparison as well as for comparison with theory by using a normalized phase diagram, Figure 24. The normalized scales are $H/T_c(0)$ and $T_c(H)/T_c(0)$. Detailed normalization is sensitive to the choice of $T_c(0)$, and locating the transitions at the peak of $(\partial M/\partial T)_H$ is preferable to the mean-field knee construction adopted earlier. For this comparison we have chosen $T_c(0)$ at 10.3 K for (TTF)(CuBDT) and

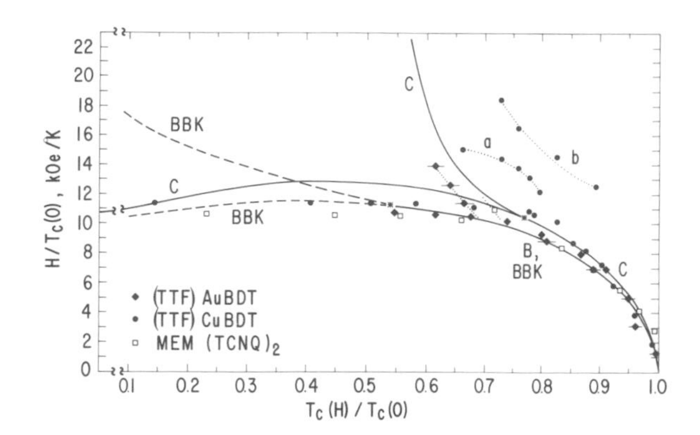

FIGURE 24. Composite *H-T* normalized phase diagram for (TTF)(AuBDT), (TTF)(CuBDT), and $MEM(TCNQ)_2$ powders. The normalized scales are $H/T_c(0)$ and $T_c(H)/T_c(0)$. Where possible, transitions are located by peak values of $(\partial M/\partial T)_H$ or $(\partial M/\partial H)_T$. For normalization, the values chosen for $T_c(0)$ are 2.03, 10.3, and 18.0 K for the compounds as listed. The designations a and b represent mean-field estimates for the high-field phase boundary from two separate experiments (5 MW and 10 MW solenoid power). The solid and dashed curves represent theory from Cross (Reference 22), C; Bray (Reference 21), B; and Bulaevskii *et al.* (Reference 17), BBK. The * locates the special point $(H_c/T_c(0), T_c(H_c)/T_c(0))$ of the various theories.

at 18.0 K for $MEM(TCNQ)_2$. Initial results for (TTF)(AuBDT) $[T_c(0) = 2.03\text{ K}]$[27] are also shown. The superposition as a quasi-universal curve, including the field scale, is very satisfactory. The experimental results in the vicinity of the second-order transition are in rather general accord with theory and should not be interpreted as distinguishing between the mean-field and Cross-Fisher theories, in view of the described accuracy.

As shown in Section 2.4, at low values of $T/T_c(0)$, the field H_c has special meaning, marking a transition out of the nonmagnetic (dimerized) spin–Peierls state. In the Hartree theory of Bray[21] and Bulaevskii *et al.*,[17] its value is $H_c = 11.2\ T_c(0)$ where H_c is in kOe and T_c is in K, and in the non-mean-field theory of Cross,[22] its value is $H_c = 10.3\ T_c(0)$. The results for all three compounds lie in the range $(10.5 \pm 0.6)T_c(0)$, with much of the uncertainty resting on the choice of $T_c(0)$. Inasmuch as the intrachain (uniform chain) exchange J/k_B ranges from 68 to 103 K, while the spin–Peierls transition temperature (spin–phonon coupling) ranges over nearly a factor of 10, this critical field scaling is a significant affirmation of the theoretical approaches.

At the next level of comparison, we consider the presence of the high-field phase and the occurrence of hysteresis. Although the latter seems often to be associated with isothermal entry into the high-field phase, it is interesting to consider them separately. As already noted, the evidence for the presence of a high-field phase gets stronger as the experiments become "easier," i.e., require less field. Concurrently, the location of the phase boundary is probably most accurate for the (TTF)(AuBDT), for which differential susceptibility peaks (vs T) were used, and less so for the other two compounds, for which a mean-field knee-like graphical construction was used. The latter boundaries are undoubtedly too high in temperature by perhaps some 10–20%. Even without much adjustment, the high-field phase line for the different compounds joins the second-order phase line at reduced temperature values $T/T_c(0)$ between about 0.65 and 0.8. Again, considering the range of $T_c(0)$ and J/k_B values, this near (and likely to improve) superposition strongly suggests that the high-field phase has its origin in models inspired mainly by the physics of the spin–Peierls system. It is equally important that the theoretical models locate this junction at $T/T_c(0)$ values of 0.54 and 0.77.[22] Although there must be some *inter*chain spin–spin coupling in these compounds, and one could rightly expect to find evidence for it somewhere in the phase diagram, the results so far do not exhibit a recognizable influence thereof. Whether the high-field phase is indeed one of the theoretically suggested incommensurate or higher-order commensurate phases must await results of more microscopically probing experiments on suitable samples. It is probably significant that the positions of the existing, incomplete, and possibly imprecise high-field phase lines $[T_c(H > H_c)]$ are closer to Cross' estimate than to that of Bulaevskii *et al.*

We consider now the presence of hysteresis in the dc magnetization isotherms and speculate about its origins. It dominates the entry into the high-field phase region for $MEM(TCNQ)_2$, continuing up to a value $T/T_c(0) \simeq 0.67$, very close to the value of 0.69 at which the high-field phase line for (TTF)(AuBDT) joins the second-order line. On the other hand, for (TTF)(CuBDT) hysteresis extends only up to $T/T_c(0) \simeq 0.5$, and it has not been seen in (TTF)(AuBDT) in measurements that descend only to $T/T_c(0) = 0.54$. It may be significant that for this series $k_B T_c(0)/J$ descends as 0.17, 0.14, and 0.03, as a measure of the spin–phonon coupling. While the presence of dc magnetization hysteresis is proof of the first-order nature of a transition, its absence does not prove the contrary. It may simply be a question of the time scale of observation. More experiments are needed to resolve the question of the order of the transition all across this boundary between the dimerized SP phase and the high-field phase.

The origin of the considerable hysteresis has been the subject of some speculation,[24, 128] which is made difficult by the fact that all the experiments have been with powders. That fact promotes models in which elastic or magnetostrictive strain occurs along selected axes and impedes or distributes the transition field (e.g., by geometrical as well as physical considerations). The remarkable spread of field [over 100 kOe in (TTF)(CuBDT)] over which hysteresis continues, along with the near isotropy of epr $\tilde{g}$ factors $[(\Delta\tilde{g}/\tilde{g}) < 0.5\%]$, are facts that are offered to refute those models. An intrinsic alternative to such explanations for the global hysteresis may be found in the cascade of higher-order commensurate phases known as the Devil's staircase or the harmless staircase (Section 2.4). The steps are presumably too small to be resolved at our lowest temperatures, or they may be partly smeared. Finally, one may speculate that the high-field region has commensurate phase(s) at low temperature and an incommensurate phase closer to the boundary for transition to the uniform chain phase. This could pose thermodynamic complications, and such speculation needs some experimental facts for support.

3.5. Pressure Experiments

A brief theoretical discussion of the factors governing the effects of pressure on the SP transition appears in Section 2.4, with more details in Reference 29. Although only a few results exist now, there is an interesting potential in such experiments that is likely to stimulate future exploration.

Initial results[25] of work by Bloch and Voiron, measuring magnetization to 60 kOe as a function of temperature and of pressure to 6 kbars (0.6 GPa) on (TTF)(CuBDT) are shown in Figure 25. A decrease in T_c of about 1.1 K is observed and an increase of J (intrachain) can be inferred. The data are too limited to test the several functional forms developed from different theoretical models and assumptions. One may note that some soft-mode (ferroelectric) systems show remarkably strong pressure dependences of their transition temperatures[129] owing to a delicate balance between competing interactions. Although spin–Peierls systems are profoundly influenced by the presence of a soft mode, the pressure dependence shown in Figure 25 is not that steep.

A particularly attractive feature of pressure experiments is that, by depressing T_c, they permit a broader exploration of normalized phase space (H/T_c, T/T_c). This could be quite useful in probing the nature of the

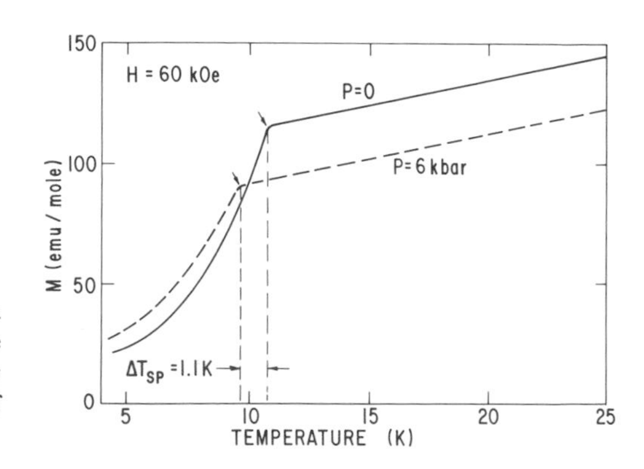

FIGURE 25. Experimental data showing the depression of the SP transition temperature by 6 kbars pressure for powder of $(TTF)[CuS_4C_4(CF_3)_4]$.

high-field phase, equivalent in a sense to choosing a compound with lower T_c. A second consequence of this depression or shrinking of the phase diagram (unnormalized) is that it increases the opportunity of seeing some manifestation of *inter*chain spin–spin coupling. Indeed, pressure may also serve to increase the magnitude of the interchain magnetic coupling. Certainly, if the pressure effects serve to suppress completely the spin–Peierls nonmagnetic phase, the spin–spin effects must take over, providing a three-dimensional antiferromagnetically ordered state. With these prospects, further results will be awaited with much interest.

4. Conclusions and Summary

In this chapter, we have surveyed the properties of a novel class of low-dimensional (quasi-one-dimensional) donor–acceptor complexes in the light of a variety of experimental studies and also in relation to current theoretical ideas. The four experimental compounds emphasized here, namely, (TTF)(CuBDT), (TTF)(AuBDT), (TTF)(CuBDSe), and $MEM(TCNQ)_2$, show an unusual magnetoelastic transition (spin–Peierls transition), which is an insulating analog of the famous Peierls transition in a one-dimensional conductor. The Peierls transition is a transition from a metallic, conducting state to a low-temperature insulating (semiconducting) phase where the chain sites are dimerized in a charge-density-wave state. In the spin–Peierls (SP) system, at high temperatures the system shows the characteristic magnetic behavior of an assembly of essentially noninteracting uniform Heisenberg spin-1/2 antiferromagnetic chains. Below the transition, the chains progressively develop an alternating or dimerized character.

Since experimental investigation of spin–Peierls systems began in 1975, only a few examples of spin–Peierls systems have been discovered. The vast majority of quasi-one-dimensional insulating systems order magnetically and are well described in a rigid-lattice approximation. Usually spin–phonon coupling is so weak that the ordering is dominated by interchange spin–spin coupling. In spin–Peierls systems, the lattice plays an important role in the sense that the transition arises from spin–phonon coupling along the magnetic chains. Magnetoelastic (SP) ordering appears to be intimately associated with a pre-existing soft mode in the correct position in reciprocal space for dimerization. It should be noted, however, that a pre-existing soft mode may be a necessary but not sufficient condition for a SP transition. An additional member of the (TTF)(MBDT) family, (TTF)(PtBDT), presumably also has a pre-existing soft mode (since it is isostructural with the M = Cu, Au compounds and shows a similar first-order transition at high temperature), but the ultimate low-temperature ordering remains magnetic in character. Of course in (TTF)(PtBDT), the intrachain exchange interactions are ferromagnetic, presumably because there are spins on both the $(TTF)^+$ and $(PtBDT)^{\dot{-}}$ ions, whereas in the SP BDT compounds only the $(TTF)^+$ bear spins. There is some degree of interest in reexamining various quasi-one-dimensional organic materials synthesized in the quest for good organic conductors and discarded upon discovery of poor conductivity properties. Additional examples of SP systems may be discovered this way.

Claims have been made for spin–Peierls-type behavior in a number of systems other than the four mentioned above. We have reviewed some of these in Section 3.3. Others include systems that are not good insulators, but instead are semiconductors or conductors. These may be examples of a generalized Peierls transition, wherein both charge translation (regular Peierls) and spin freedom (spin–Peierls) are involved (restricted) at the transition. This concept is discussed in Sections 2.5 and 3.3. There can be no doubt, with the discovery of the quasi-one-dimensional superconductors like $(Me_4TSF)_2PF_6$, that spin correlations (spin density waves) can be important even in the conductors.

Theoretical investigations show that simple canonical spin–Peierls transitions, where the exchange varies linearly with intermolecular spacing, can occur only for quantum magnetic interactions. Ising chains and classical chains are excluded. Whether SP transitions occur for higher spin values, $1/2 < S < \infty$, remains to be investigated. The question turns on whether the magnetic free energy below some temperature is lowered by dimerization more than the lattice energy is increased. In practice, however, it may well be that the only type of SP system realizable experimentally is $S = 1/2$ with very good Heisenberg exchange, i.e., with $\tilde{g}$ factors close to the free-electron values. In the case of the commonly studied magnetic insulating systems, the magnetic character arises from a metal ion such as Cu, Mn, Fe, or Co. To avoid confusion, it is important to emphasize that in the case of all metal–bisdithiolenes, the magnetic character is not derived from the metal atom but is due to a free spin associated with the entire metallo-organic ion.

Experimental investigations of most of the SP systems considered in this chapter have now become quite extensive. In summary, in zero (or very low) field, susceptibility, specific heat, epr, nmr, x-ray, and neutron scattering studies have been performed on (TTF)(CuBDT). The nmr, x-ray, and neutron scattering experiments have not been done on (TTF)(AuBDT). For (TTF)(CuBDSe), existing experimental data include only susceptibility studies. Unfortunately, some experimental work on some of these bisdithiolenes is made very difficult by their severe high-temperature structural distortion, which tends to shatter large single crystals. $MEM(TCNQ)_2$ is a very interesting substance which shows a $4k_F$ transition near room temperature in addition to a low-temperature SP transition. It may be easier to grow large single crystals of this compound. Existing experimental studies on $MEM(TCNQ)_2$ include zero-field susceptibility, specific heat, and nmr.

In nonzero magnetic field, neutron scattering studies (on the behavior of the dimerization Bragg peak) have been carried out on (TTF)(CuBDT) up to ~80 kOe. Magnetization experiments have been done up to 200 kOe at Grenoble. In the case of (TTF)(AuBDT), extensive susceptibility, magnetization, and specific-heat experiments are in progress for fields up to 25–30 kOe, at the Kamerlingh Onnes Laboratory, Leiden. For $MEM(TCNQ)_2$, magnetization studies up to 200 kOe have been done at Grenoble.

Since the "rational" thermodynamic parameters for a magnetoelastic system are magnetic field, pressure, and temperature, it would be valuable to map out the detailed global phase behavior as a function of all three parameters experimentally. In addition to the field experiments just noted, preliminary pressure experiments have been undertaken on both (TTF)-(CuBDT) and $MEM(TCNQ)_2$ at Grenoble. The $MEM(TCNQ)_2$ complex appears particularly interesting in this connection, but much more work is needed.

In the case of phase behavior in a field, theoretical studies have indicated several interesting possibilities such as commensurate–incommensurate transitions or commensurate to higher-order commensurate transitions (possibly even including Devil's-staircase-like behavior). The possibility exists also of entering an antiferromagnetically ordered phase. The current theories of SP high-field phase behavior are due to Bray, Bulaevskii *et al.*, and Cross and Fisher, extended by Cross. Both Bulaevskii *et al.* and Cross favor an incommensurate high-field phase (in addition to a low-field dimerized phase). Present experimental data is generally consistent with these theories in a rather gratifying way, but it is not at this point sufficiently extensive and precise to be definitive. Hence alternative theories involving fixed-periodicity systems, transitions to magnetically ordered phases, or a soliton picture, cannot be completely ruled out.

Understanding of the SP phenomenon continues to develop as more experimental results are obtained. For example, evidence for the presence of a new high-field phase is now strong, but absolute confirmation of its existence and, particularly, determination of its nature is still lacking. What is needed is a microscopic probe like neutron scattering. For the much studied case of (TTF)(CuBDT), however, the applied magnetic field required to probe the high-field phase region is in excess of 120 kOe. No existing neutron scattering facility can operate at such high fields. The (TTF)-(AuBDT) or (TTF)(CuBDSe) complex with spin–Peierls transition temperatures of 2 and 6 K, respectively, [cf 12 K for (TTF)(CuBDT)] offer more attractive possibilities for neutron exploration, since the entire phase diagram theoretically should and experimentally appears to "scale" with T_c. Early experiments on (TTF)(AuBDT) with single crystals were unsuccessful. It appears that (TTF)(AuBDT) crystals are more sensitive to the effects of strain than (TTF)(CuBDT), and the crystals shatter while being cooled through the severe distortive transition at 200 K. Future neutron experiments on the Au compound are planned at Grenoble, using special techniques intended to avoid this shattering problem.

At this point we summarize the extent to which theoretical and experimental knowledge of spin–Peierls systems is on a solid footing. In zero field, susceptibility and specific-heat experimental studies can be fitted by theoretical calculations based on a simple mean-field model with an accuracy that is surprisingly good. More specifically, the susceptibility can be fitted above the transition to a uniform Heisenberg chain model and below the transition to a progressively dimerizing Heisenberg chain model. The specific heat displays a jump at T_c whose magnitude is in good agreement with mean-field predictions. Perhaps even more fundamental, the existence and progressive nature of the dimerization below T_c has been confirmed by x-ray studies. In nonzero field, the transition temperature has been found to decrease with increasing field, as predicted theoretically. Further, recent detailed experimental analysis has shown that the shape of the phase boundary for the low-field dimerized phase for three SP systems, (TTF)(CuBDT), (TTF)(AuBDT), and MEM$(TCNQ)_2$, agrees, within experimental uncertainty, with *either* of the two leading calculations. (The phase boundaries predicted by Bray, Bulaevskii *et al.*, or Cross do not differ very much except at the highest fields and lowest temperatures. Since both approaches treat the spin–phonon coupling in RPA, the relatively small deviations between them must result from the differences in treatment of the magnetic interactions.) In the same three systems, a special point has been detected in the H-T plane $[H_c, T_c(H_c)]$ whose location is in rough agreement with both sets of theoretical predictions. These results are all the more remarkable because of the very different lattice structures of the bisdithiolenes on the one hand and MEM$(TCNQ)_2$ on the other (mixed stack structure versus segregated stack structure). There appears to be a rather notable "universality" of behavior among the different SP systems. More precisely, this is an experimental demonstration of the theoretical prediction that a major portion of the H-T phase diagram (at least) is a universal function of $H/T_c(0)$ vs $T_c(H)/T_c(0)$.

Features of interest for which complete understanding is not at present available include the observation and extent of magnetization hysteresis across the boundary between SP–nonmagnetic and SP–magnetic phases; the precise nature of the phase at high fields; the behavior in general with pressure; and the extent and possible role in the phase behavior of spin–spin interchain coupling.

It appears possible to make some judgments about the degree of validity of mean-field theory for this problem. Some inconsistency appears in the estimation of the zero-temperature dimerization parameter $\delta(0)$ between a direct theoretical estimate based on knowledge of just J and T_c and one based on a fitting procedure to the susceptibility below T_c. Newly available susceptibility data as a function of temperature in nonzero fields for (TTF) (AuBDT) reveal a peak in the susceptibility at $T_c(H)$ rather than a mean-field knee and suggest a modification of the criterion used to locate $T_c(H)$ (inflection points rather than "knee" in the magnetization curves). General theoretical considerations suggest an anomaly at T_c in the zero- and nonzero-field specific heat having the form of a finite cusp rather than a step. A small "tail" on the cusp on the high-temperature side indicates deviations from ideal (simple) mean-field behavior. The extent of the tail should be governed by the Ginzburg criterion, predicted theoretically to be 0.04 T_c. Experimental observations on (TTF)(CuBDT) and (TTF)(AuBDT), while preliminary and in need of further refinement and analysis, seem to be in agreement with a tail of this magnitude. The relative smallness of the Ginzburg criterion indicates that overall mean-field theory should give a good account of the phase behavior of SP systems, and all experimental investigations so far are in agreement with this.

It is hoped that a detailed study of SP systems may yield, in addition to valuable new knowledge about quasi-one-dimensional magnetoelastic systems, information on other quasi-one-dimensional organics which may not be insulators but show varying degrees of conductivity. An interesting example might be the organic conductor $(Me_4TSeF)_2PF_6$, which shows spin-density-wave ordering at atmospheric pressure, as do the SP systems, and which shows superconducting behavior at higher pressure.

Acknowledgments

In preparing this chapter we have benefited from helpful and stimulating discussions with many of our colleagues and collaborators. In particular, we wish to thank D. Bloch, L. J. de Jongh, and J. A. Northby for discussions of the experimental phase diagrams and their significance, J. S. Kasper for continual guidance on structural aspects, M. Mohan for special assistance with the introductory overview, and W. P. Wolf for discussions of critical phenomena. In addition, we gratefully acknowledge the contributions of H. R. Hart, Jr., J. A. Northby, L. J. de Jongh, F. Greidanus, and H. A. Groenendijk, who have kindly permitted us to use some of their results prior to publication.

This work was supported in part at General Electric by the Air Force Office of Scientific Research (AFSC), United States Air Force, under Contract F49620-79-C-0051. The United States Government is authorized to reproduce and distribute reprints for governmental purposes, notwithstanding any copyright notation hereon. This work was supported in part at the University of Rhode Island by National Science Foundation Grant DMR-80-10819.

Notation

Latin

A	Cross-sectional area of a chain
a	Intrachain lattice spacing
$a^\dagger$ (a)	Electron creation (destruction) operator
AF	Antiferromagnetic
$b^\dagger$ (b)	Phonon creation (destruction) operator
BDT	Bis-dithiolene
BDSe	Bis-disfelenene
c	Intrachain force constant
DBTTF	$\Delta^{2,2}$-bi-*d*-benzo-1,3-dithiolene
DEM	Diethylmorpholinium
$\mathbf{e}$	Phonon polarization vector
E_s	Soliton creation energy
g, g'	Bare spin–phonon coupling constants
$\tilde{g}$	Electron gyromagnetic ratio
H	Magnetic field
H_c	Critical field between nonmagnetic and magnetic spin–Peierls phases
$\hbar$	Planck's constant/2π
HMTTF	Hexamethylenetetrathiafulvalene
J	Intrachain exchange interaction
k_B	Boltzmann constant
k_F	Fermi wavevector
L	Chain length
M	Magnetization
m	Mass of magnetic lattice site
MEM	Methylethylmorpholinium
MF	Mean-field
N	Number of chain sites
P	Pressure
RP	Electronic Peierls
$\mathbf{S}$	Spin operator
SP	Spin–Peierls
T	Temperature
t	Intrachain nearest-neighbor transfer energy
T_1	Nuclear spin–lattice relaxation rate
T_c	Spin–Peierls transition temperature
TCNQ	7,7,8,8-Tetracyano-*p*-quinodimethane
$TCNQCl_2$	2,5-Dichloro-7,7,8,8-tetracyano-*p*-quinodimethane
$TCNQF_4$	2,3,5,6-Tetrafluoro-7,7,8,8-tetracyano-*p*-quinodimethane
TMPD	*N*,*N*,*N'*,*N'*-Tetramethyl-*p*-diaminobenzene
TMTSeF	Tetramethyltetraselenafulvalene
TETTF	Tetraethyltetrathiafulvalene
T_N	Néel temperature
T_{RP}	Electronic Peierls transition temperature
TTF	Tetrathiafulvalene
U	On-site Coulomb interaction

Greek

χ	Magnetic susceptibility
Δ	Magnetic energy gap
δ	Dimensionless dimerization order parameter
λ, λ'	Dimensionless spin–phonon coupling constants
μ	Chemical potential
μ_B	Bohr magneton
Ψ	Pseudofermion operator
ω_0	Phonon energy

References

1. Bray, J. W., Hart, H. R., Interrante, L. V., Jacobs, I. S., Kasper, J. S., Watkins, G. D., Wee, S. H., and Bonner, J. C., *Phys. Rev. Lett.* **35**, 744 (1975).
2. Jacobs, I. S., Bray, J. W., Hart, H. R., Interrante, L. V., Kasper, J. S., Watkins, G. D., Prober, D. E., and Bonner, J. C., *Phys. Rev. B* **14**, 3036 (1976).
3. Pytte, E., *Phys. Rev. B* **10**, 2039 (1974).
4. Peierls, R. E., *Quantum Theory of Solids*, Oxford University Press, London (1955), p. 108.
5. Yang, C. N., and Yang, C. P., *Phys. Rev.* **150**, 321 (1966) and succeeding series of papers in the *Physical Review*; Gaudin, M., *Phys. Rev. Lett.* **26**, 1301 (1971); Johnson, J. D., and McCoy, B. M., *Phys. Rev. A* **6**, 1613 (1972).
6. Bonner, J. C., and Fisher, M. E., *Phys. Rev.* **135**, A640 (1964).
7. Bulaevskii, L. N., *Sov. Phys.-JETP* **17**, 684 (1963).
8. Duffy, Jr., W., and Barr, K. P., *Phys. Rev.* **165**, 647 (1968).
9. Bonner, J. C., and Blöte, H. W. J., *Phys. Rev.* **B1** (1982).
10. Pytte, E., *Phys. Rev. B* **10**, 4637 (1974).
11. Huizinga, S., Kommandeur, J., Sawatzky, G. A., Thole, B. T., Kopinga, K., de Jonge, W. J., and Roos, J., *Phys. Rev. B* **19**, 4723 (1979), and references therein.
12. Bulaevskii, L. N., *Sov. Phys.-JETP* **16**, 685 (1963).
13. But not for the corresponding ferromagnets. See Reference 12.
14. Bonner, J. C., Blöte, H. W. J., Bray, J. W., and Jacobs, I. S., *J. Appl. Phys.* **50**, 1810 (1979).
15. Cross, M. C., and Fisher, D. S., *Phys. Rev. B* **19**, 402 (1979).
16. Bonner, J. C., Blöte, H. W. J., and Johnson, J. D., *J. Appl. Phys.* **50**, 7379 (1979).
17. Bulaevskii, L. N., Buzdin, A. I., and Khomskii, D. I., *Solid State Commun.* **27**, 5 (1978).
18. Moncton, D. E., Birgeneau, R. J., Interrante, L. V., and Wudl, F., *Phys. Rev. Lett.* **39**, 507 (1977).
19. Kasper, J. S., and Moncton, D. E., *Phys. Rev. B* **20**, 2341 (1979).
20. Kotani, A., and Harada, I., *J. Phys. Soc. Jap.* **49**, 535 (1980).
21. Bray, J. W., *Solid State Commun.* **26**, 771 (1978).
22. Cross, M. C., *Phys. Rev. B* **20**, 4606 (1979).
23. Bray, J. W., Interrante, L. V., Jacobs, I. S., Bloch, D., Moncton, D. E., Shirane, G., and Bonner, J. C., *Phys. Rev. B* **20**, 2067 (1979).
24. Bloch, D., Voiron, J., Bonner, J. C., Bray, J. W., Jacobs, I. S., and Interrante, L. V., *Phys. Rev. Lett.* **44**, 294 (1980).
25. Jacobs, I. S., Bray, J. W., Interrante, L. V., Bloch, D., Voiron, J., and Bonner, J. C., *Physics In One Dimension*, J. Bernasconi and T. Schneider, Eds., Springer, New York (1981), p. 173.
26. Jacobs, I. S., Bray, J. W., Hart, Jr., H. R., Interrante, L. V., Kasper, J. S., Bloch, D., Voiron, J., Bonner, J. C., Moncton, D. E., and Shirane, G., *J. Magn. Magn. Mater.* **15–18**, 332 (1980).

27. Northby, J. A., Groenendijk, H. A., and de Jongh, L. J., private communication.
28. Bloch, D., Voiron, J., Bray, J. W., Jacobs, I. S., Bonner, J. C., and Kommandeur, J., *Phys. Lett. A* **82**, 21 (1981).
29. Bray, J. W., *Solid State Commun.* **35**, 853 (1980).
30. Jérome, D., Mazaud, A., Ribault, M., and Bechgaard, K., *J. Phys. Lett.* **41**, 95 (1980).
31. Ribault, M., Benedek, G., Jérome, D., and Bechgaard, K., *J. Phys. Lett.* **41**, 397 (1980).
32. Bechgaard, K., Carneiro, K., Olsen, M., Rasmussen, F. B., and Jacobsen, C. S., *Phys. Rev. Lett.* **46**, 852 (1981).
33. McConnell, H. M., and Lynden-Bell, R., *J. Chem. Phys.* **36**, 2393 (1962); Thomas, D. D., Keller, H., and McConnell, H. M., *ibid.* **39**, 2321 (1962).
34. Fröhlich, H., *Proc. R. Soc. Lond. A* **223**, 296 (1954).
35. Kuper, C. G., *Proc. R. Soc. Lond. A* **227**, 214 (1955).
36. Ooshika, Y., *J. Phys. Soc. Jap.* **12**, 1238, 1246 (1957); **14**, 747 (1959).
37. Longuet-Higgins, H. C., and Salem, L., *Proc. R. Soc. Lond. A* **251**, 172 (1959).
38. Platt, J. R., *J. Chem. Phys.* **25**, 80 (1956).
39. Kuhn, H., *J. Chem. Phys.* **17**, 1198 (1949).
40. Chesnut, D. B., *J. Chem. Phys.* **45**, 4677 (1966).
41. Buzdin, A. I., and Bulaevskii, L. N., *Usp. Fiz. Nauk* **131**, 495 (1980); *Sov. Phys. Usp.* **23**, 409 (1980).
42. des Cloizeaux, J., and Pearson, J. J., *Phys. Rev.* **128**, 2131 (1962).
43. Holz, A., Penson, K. A., and Bennemann, K. H., *Phys. Rev. B* **16**, 3999 (1977).
44. Jordan, P., and Wigner, E., *Z. Phys.* **47**, 631 (1928).
45. Landau, L. D., and Lifshitz, I. M., *Statistical Physics*, Pergamon, New York (1958), p. 482.
46. Lépine, Y., Tannous, C., and Caillé, A., *Phys. Rev. B* **20**, 3753 (1979).
47. Pincus, P., *Solid State Commun.* **9**, 1971 (1971).
48. Beni, G., and Pincus, P., *J. Chem. Phys.* **57**, 3531 (1972).
49. Beni, G., *J. Chem. Phys.* **58**, 3200 (1973).
50. DuBois, J. Y., and Carton, J. P., *J. Phys.* **35**, 371 (1974).
51. Lépine, Y., and Caillé, A., *J. Chem. Phys.* **67**, 5598 (1977).
52. Takaoka, Y., and Motizuki, K., *J. Phys. Soc. Jap.* **47**, 1752 (1979).
53. Lépine, Y., and Caillé, A., *J. Chem. Phys.* **71**, 3728 (1979).
54. Oguchi, A., and Tsuchida, Y., *Prog. Theor. Phys.* **56**, 1976 (1976).
55. Mijatović, M., and Milošević, S., *J. Magn. Magn. Mater.* **15–18**, 1029 (1980).
56. Penson, K. A., Holz, A., and Bennemann, K. H., *Phys. Rev. B* **13**, 433 (1976).
57. Mijatović, M., Milošević, S., and Urumov, V., *J. Magn. Magn. Mater.* **23**, 79 (1981).
58. Rickayzen, G., *Theory of Superconductivity*, John Wiley, New York (1965), p. 443.
59. Luttinger, J. M., *J. Math. Phys.* **4**, 1154 (1963); Mattis, D. C., and Lieb, E. H., *ibid.* **6**, 304 (1965).
60. Luther, A., and Peschel, I., *Phys. Rev. B* **12**, 3908 (1975).
61. Fields, J. N., Blöte, H. W. J., and Bonner, J. C., *J. Appl. Phys.* **50**, 1808 (1979).
62. Nakano, T., and Fukuyama, H., *J. Phys. Soc. Jap.* **49**, 1679 (1980).
63. Grabowski, M., Subbaswamy, K. R., and Horovitz, B., *Solid State Commun.* **34**, 911 (1980); Horovitz, B., *Solid State Commun.* **34**, 61 (1980).
64. Horovitz, B., *Phys. Rev. Lett.* **46**, 742 (1981).
65. Klein, D. J., and García-Bach, M. A., *Phys. Rev. B* **19**, 877 (1979).
66. de Jonge, W. J. M., Hijmans, J. P. A. M., Boersma, F., Schouten, J. C., and Kopinga, K., *Phys. Rev. B* **17**, 2922 (1978).
67. Imry, Y., Pincus, P., and Scalapino, D., *Phys. Rev. B* **12**, 1978 (1975).
68. Villain, J., and Loveluck, J. M., *J. Phys. Lett. (Paris)* **38**, L77 (1977).
69. Von Boehm, J., and Bak, P., *Phys. Rev. Lett.* **42**, 122 (1979).
70. Villain, J., and Gordon, M. B., *J. Phys. C* **13**, 3117 (1980).

71. Tannous, C., and Caillé, A., *Can. J. Phys.* **57**, 508 (1979).
72. Mohan, M., *Phys. Rev. B* **21**, 1264 (1980); see also erratum, *ibid.* **23**, 433 (1981).
73. Ajiro, Y., Nakajima, Y., Turukawa, Y., and Kiriyama, H., *J. Phys. Soc. Jap.* **44**, 420 (1978).
74. Ehrenfreund, E., Rybaczewski, E. F., Garito, A. F., Heeger, A. J., and Pincus, P., *Phys. Rev. B* **7**, 421 (1973).
75. Ehrenfreund, E., and Smith, L. S., *Phys. Rev. B* **16**, 1870 (1977).
76. Diederix, K. M., Groen, J. P., Klaasen, T. O., and Poulis, N. J., *Physica* **96B**, 41 (1979).
77. Groen, J. P., Klaassen, T. O., Poulis, W. J., Müller, G., Thomas, H., and Beck, H., *Phys. Rev. B* **22**, 5369 (1980).
78. Allender, D. W., Bray, J. W., and Bardeen, J., *Phys. Rev. B* **9**, 119 (1974).
79. Emery, V. J., *Highly Conducting One-Dimensional Solids*, J. T. Devreese, R. P. Evrard, and V. E. van Doren, Eds., Plenum, New York (1979), p. 247.
80. Hubbard, J., *Proc. R. Soc. Lond. A* **276**, 238 (1963); **281**, 401 (1964).
81. Torrance, J. B., *Phys. Rev. B* **19**, 3099 (1978).
82. Emery, V. J., *Phys. Rev. Lett.* **37**, 107 (1976).
83. Mott, N. F., *Metal-Insulator Transitions*, Taylor and Francis, Ltd., London (1974), Chapter 4.
84. Klein, D. J., and Seitz, W. A., *Phys. Rev. B* **8**, 2236 (1973).
85. Chui, S. T., and Bray, J. W., *Phys. Rev. B* **18**, 2426 (1978); **21**, 1380 (1980); Bray, J. W., and Chui, S. T., *Phys. Rev. B* **19**, 4876 (1979).
86. (a) Takaoka, Y., and Motizuki, K., *J. Phys. Soc. Jap.* **47**, 1752 (1979); (b) Lépine, Y., Caillé, A., and Larochelle, V., *Phys. Rev. B* **18**, 3585 (1978).
87. Afify, H. H., Abdel-Kerim, F. M., Aly, H. F., and Shabaka, A. A., *Z. Naturforsch. A* **33**, 344 (1978).
88. Chui, S. T., unpublished.
89. Egri, I., *Solid State Commun.* **17**, 441 (1975); *Z. Phys. B* **23**, 381 (1976); *Solid State Commun.* **22**, 281 (1977).
90. Mertsching, J., *Phys. Status Solidi B* **87**, 599 (1978).
91. Robaszkiewicz, S., *Phys. Status Solidi B* **70**, K51 (1975).
92. Lépine, Y., and Caillé, A., *Solid State Commun.* **28**, 655 (1978).
93. Kondo, J., *Physica* **98B**, 176 (1980).
94. Walsh, W. M., Wudl, F., Thomas, G. A., Nalewajek, D., Hauser, J. J., Lee, P. A., and Poehler, T., *Phys. Rev. Lett.* **45**, 829 (1980).
95. Scott, J. C., Pedersen, H. J., and Bechgaard, K., *Phys. Rev. Lett.* **45**, 2125 (1980).
96. Andrieux, A., Jérome, D., and Bechgaard, K., *J. Phys. Lett.* (*Paris*) **42**, L87 (1981).
97. See Alcucer, L., this volume; Wudl, F., Ho., C. H., and Nagel, A., *J. Chem. Soc. Chem. Commun.*, 923 (1973); Wudl, F., *J. Am. Chem. Soc.* **97**, 1962 (1975); Interrante, L. V., Browall, K. W., Hart, Jr., H. R., Jacobs, I. S., Watkins, G. D., and Wee, S. H., *J. Am. Chem. Soc.* **97**, 889 (1975); Kasper, J. S., Interrante, L. V., and Secaur, C. A., *J. Am. Chem. Soc.* **97**, 890 (1975); Calas, P., Fabre, J. M., Khalife-El-Saleh, M., Mas, A., Torreilles, E., and Giral, L., *Tetrahedron Lett.*, 4475 (1975); Miles, M. G., and Wilson, J. D., *Inorg. Chem.* **14**, 2357 (1975); Wheland, R. C., and Gillson, J. L., *J. Am. Chem. Soc.* **98**, 3916 (1976); Interrante, L. V., Bray, J. W., Hart, Jr., H. R., Kasper, J. S., and Piacente, P. A., *Ann. N.Y. Acad. Sci.* **313**, 407 (1978); Interrante, L. V., and Kasper, J. S., *AIP Conf. Proc.* **53**, 205 (1979).
98. Bonner, J. C., Wei, T. S., Hart, Jr., H. R., Interrante, L. V., Jacobs, I. S., Kasper, J. S., Watkins, G. D., and Blote, H. W. J., *J. Appl. Phys.* **49**, 1321 (1978).
99. Jacobs, I. S., Hart, Jr., H. R., Interrante, L. V., Bray, J. W., Kasper, J. S., Watkins, G. D., Prober, D. E., Wolf, W. P., and Bonner, J. C., *Physica* **86–88B**, 655 (1977).

100. Interrante, L. V., Bray, J. W., Hart, Jr., H. R., Jacobs, I. S., Kasper, J. S., Piacente, P. A., and Bonner, J. C., *Proceedings of the International Conference on Quasi-One-Dimensional Conductors, Lecture Notes in Physics 96*, Vol. II, Springer-Verlag, New York (1979), p. 55.
101. Kasper, J. S., and Interrante, L. V., *Acta Crystallogr. B* **32**, 2914 (1976).
102. Delker, G. E., Ph.D. thesis, University of Illinois, 1976.
103. Kasper, J. S., private communication.
104. Wei, T. S., Heeger, A. J., Salamon, M. B., and Delker, G. E., *Solid State Commun.* **21**, 595 (1977).
105. de Jongh, L. J., Greidanus, F., and Northby, J. A., private communication.
106. Smith, L. S., Ehrenfreund, E., Heeger, A. J., Interrante, L. V., Bray, J. W., Hart, Jr., H. R., and Jacobs, I. S., *Solid State Commun.* **19**, 377 (1976).
107. Jeandey, C., and Nechtschein, M., *J. Magn. Magn. Mater.* **15–18**, 1053 (1980).
108. Huizinga, S., $2k_F$ and $4k_F$, Thesis, University of Gronigen, Gronigen, The Netherlands (unpublished).
109. van Bodegom, B., Larsen, B. C., and Mook, H. A., *Phys. Rev. B* **24**, 1520 (1981).
110. Lacoe, R. C., Gruner, G., and Chaikin, P. M., *Solid State Commun.* **36**, 599 (1980).
111. Terauchi, H., Sakamoto, N., and Shirotani, I., *J. Chem. Phys.* **64**, 437 (1976).
112. Chihara, H., Nakamura, M., and Seki, S., *Bull. Chem. Soc. Jap.* **38**, 1776 (1965).
113. Soos, Z. G., and Hughes, R. G., *J. Chem. Phys.* **46**, 253 (1967).
114. DeBoer, J. L., and Vos, A., *Acta Crystallogr. B* **28**, 835 (1972).
115. Torrance, J. B., *Accts. Chem. Res.* **12**, 79 (1979); *Ann. N.Y. Acad. Sci.* **313**, 210 (1978).
116. Tomkiewicz, Y., Torrance, J. B., Bechgaard, K., and Mayerle, J. J., *Bull. Am. Phys. Soc.* **24**, 232 (1979); Torrance, J. B., Mayerle, J. J., and Bechgaard, K., *Phys. Rev. B* **22**, 4960 (1980).
117. Hawley, M. E., Bryden, W. A., Bloch, A. N., Cowan, D. O., Poehler, T. O., and Stokes, J. P., *Bull. Am. Phys. Soc.* **24**, 232 (1979).
118. Kobayashi, H., *Acta Crystallogr. B* **34**, 2818 (1978).
119. Jacobsen, C. S., Pedersen, H. J., Mortensen, K., and Bechgaard, K., *J. Phys. C* **13**, 3411 (1980).
120. Schwerdtfeger, C. F., Wagner, H. J., and Sawatzky, G. A., *Solid State Commun.* **35**, 7 (1980).
121. Pouget, J. P., Launois, H., D'Haenens, J. P., Merenda, P., and Rice, T. M., *Phys. Rev. Lett.* **35**, 873 (1975).
122. D'Haenens, J. P., Kaplan, D., and Merenda, P., *J. Phys. C* **8**, 2267 (1975).
123. Villeneuve, G., Drillon, M., Launay, J. C., Marquestaut, E., and Hagenmuller, P., *Solid State Commun.* **17**, 657 (1975).
124. Pouget, J. P., Launois, H., Rice, T. M., Dernier, P., Gossard, A., Villeneuve, G., and Hagenmuller, P., *Phys. Rev. B* **10**, 1801 (1974).
125. Reyes, J. M., Segel, S. L., and Sayer, M., *Can. J. Phys.* **54**, 1 (1976).
126. Flandrois, S., Delhaes, P., Amiell, J., Brun, G., Torreilles, E., Fabre, J. M., and Giral, L., *Phys. Lett. A* **66**, 244 (1978).
127. Kosevich, A. M., and Khokhlov, V. I., *Solid State Commun.* **11**, 461 (1972).
128. Bonner, J. C., Bray, J. W., and Jacobs, I. S., *Bull. Am. Phys. Soc.* **25**, 339 (1980).
129. Samara, G. A., *High-Pressure and Low-Temperature Physics*, C. W. Chu and J. A. Woolam, Eds., Plenum, New York (1978), p. 255.

8

Polypyrrole: An Electrochemical Approach to Conducting Polymers

A. F. Diaz and K. K. Kanazawa

1. Introduction

Until recent times, almost all of the known organic polymers were essentially electrically insulating, with room-temperature conductivities of 10^{-10} ohm^{-1} cm^{-1} or less. The desirability of having low-density, flexible, processible conductors provided the impetus for finding ways of enhancing the electronic conductivity of polymers. The electrical and optical properties of these materials depend on the electronic structure and basically on the chemical nature of the repeating unit. The general requirements of the electronic structure in these polymers were recognized and described many years ago.[1] The electronic conductivity is proportional to both the density and the drift mobility of the carriers. The carrier drift mobility is defined as the ratio of the drift velocity to the electric field and reflects the ease with which carriers are propagated. Enhancing the electrical conductivity of polymers then requires an increase in the carrier mobility and the density of charge carriers. The particular importance of the delocalized π-electrons to form energy bands of high-mobility carriers was stressed early on[2] and a large number of polymers were considered as having this characteristic.[3] These same considerations have been discussed from a somewhat broader perspective that include polymeric charge-transfer complexes and organometallics.[4] The dramatic conductivity enhancements reported recently[5] in polyacetylene when treated with strong oxidants have spurred a resurgence of interest in these systems including several articles reviewing the electronic structure of these polymers.[6–8]

A. F. Diaz and K. K. Kanazawa • IBM Research Laboratory, San José, California 95193.

In searching for new polymeric materials with useful conducting properties, it is important to build an intuitive understanding of the relationship between the properties and the chemical structure of the polymers. Only very recently, however, have different polymer systems exhibiting greatly enhanced electronic conduction been reported. These include poly-*p*-phenylene,[9] polypyrrole,[10, 11] polyaniline,[12–14] poly-*p*-phenylene sulfide,[15–17] poly(2,5-thienylene),[18] fluoroaluminum, and fluorogallium phthalocyanine.[19] While the number of polymers becoming available for comparative studies is growing, the actual number is still limited and their chemical structures are significantly different, so that it is presently difficult to make useful quantitative generalizations.

Polypyrrole is a particularly important addition to these conducting polymers for a number of reasons. Polypyrrole is thermally stable and can be easily doped without employing hazardous oxidants. This is the first electrically conducting polymer film to be prepared electrochemically,[10, 20] and we have now made evident the importance of the electrochemical approach for the preparation of conducting polymer films.[21, 22] The electrochemical approach has the advantage that the properties of the film can be changed by simply varying the electrolysis conditions (e.g., electrode potential, current density, solvent, electrolyte) in a controlled way. While the polymerization reaction proceeds with the formation of the polymer in the oxidized (conducting) form, the film can be prepared for removal in either its conducting or insulating form. This approach also permits the preparation of derivatized polypyrrole polymers by polymerizing the appropriately substituted pyrrole. It is with approaches of this type which permits the polymers to be modified in a systematic way that correlations between the chemical structure and the electrical and optical properties of these materials can begin to be established. The sections that follow present the results of recent studies on the polypyrrole films and stress the importance of the electrochemical approach for the preparation and study of these materials. This chapter is not intended as an extensive review of the present state of conducting organic polymers.

2. Pyrrole Black

The electrochemical preparation of polypyrrole, "pyrrole black," was first reported thirteen years ago.[20] Pyrrole black was produced as an insoluble, brittle, powdery film on a platinum electrode by the electrooxidation of pyrrole in aqueous sulfuric acid. The film had a conductivity of 8 ohm^{-1} cm^{-1} and a high number of free spins that generated a strong electron spin resonance, epr, signal with a g value of 2.0026. From the elemental analysis, one gets the formula $C_{4.0}H_{3.4}N_{0.99}S_{0.15}O_{0.92}$, suggesting the presence of 0.15 sulfate dianions per pyrrole unit. This relationship implies that there is one positive charge for every 3–4 pyrrole units. Although a more complete characterization of this material was not available, it must be similar to the pyrrole black powders generated when pyrrole is oxidized chemically, e.g., with H_2O_2.[23] The formula for this material obtained from the elemental analysis is $C_{4.0-4.5}H_{3.0-3.4}N_{1.0}O_{1.0-1.5}$ which indicates that the material consists of linked pyrrole units. The anion in this material could be formate, in which case it would be present in the ratio of one formate ion for every two pyrrole units. Oxidation degradation of this polymer yields mainly pyrrole-2,5-dicarboxylic acid, indicating that the pyrrole rings are linked via the α positions.

3. *Electrochemical Preparation of Polypyrrole*

The electrochemical oxidation of pyrrole to produce conducting polypyrrole films is of considerable interest in our laboratories, and the generality of this approach for the preparation of conducting films is only now being appreciated.[21, 22] This preparative procedure has several attractive features. First, the polymer in its conducting form is produced directly from the monomer without the need to use strong chemical oxidants. A much wider selection of anions for use as counterions becomes available since the anion in the film is provided directly by the electrolyte. This is in contrast with the conventional approaches for generating conducting polymers where the neutral polymer structure is first synthesized and subsequently treated with a strong oxidant in order to produce the conducting form of the organic polymer.[5] With this latter approach the number of anions that can be used is more limited since it must be generated from the chemical oxidant.[24] For example, in the oxidation of polyacetylene with bromine the resulting bromide counterion is produced from the oxidant.

The electrosynthesis of these films proceeds via the oxidation of pyrrole at the platinum electrode to produce an unstable π-radical cation which then reacts with the neighboring pyrrole species [Eq. (1)].

Pyrrole (N–H) ⟶ pyrrole radical cation (+•, N–H) ⟶ (a) + pyrrole (N–H) → Polymer film; (b) Nucleophile → Soluble product (1)

The cyclic voltammogram of these solutions shows an irreversible peak for the oxidation of pyrrole at +1.2 V (E_{pa}) versus the sodium chloride calomel reference electrode (SSCE).[25] The mechanism of the overall reaction for the formation of the fully aromatized product is very complicated and involves a series of oxidation and deprotonation steps [Eq. (2)].

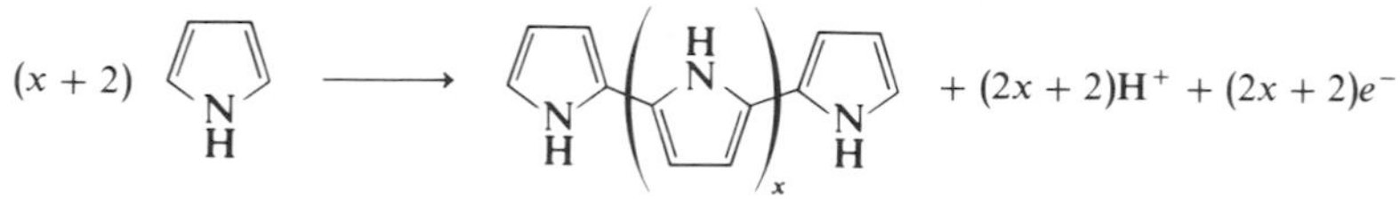

$$(x+2)\ \text{pyrrole} \longrightarrow \text{Py}-(\text{Py})_x-\text{Py} + (2x+2)H^+ + (2x+2)e^- \quad (2)$$

It is not clear whether the reaction involves the coupling of two pyrrole radical cations[26] or of a neutral and a radical cation species. However, the reaction occurs only when the potential is sufficiently anodic to oxidize pyrrole, and at these voltages the polymer is in the oxidized form. Thus it seems reasonable that the reaction proceeds via a radical cation combination process. The coupling reaction occurs primarily in the α position of the pyrrole ring to produce a linear polymer. Thus as may be expected, polymeric films are not produced when one of the α positions on the pyrrole is blocked by a substituent,[25] when there are large groups on the nitrogen that sterically hinder the molecule,[25] or when there are nucleophiles in the solution that compete for the pyrrole radical cation intermediate.[27]

This reaction has electrochemical stoichiometry where two electrons/pyrrole ring are involved in the formation of the polymer. In fact we measure 2.2–2.4 electrons/pyrrole ring; however, the extra charge (0.2–0.4) consumed during the reaction results from the concurrent oxidation of the resulting polymer which has a lower oxidation potential ($E^0 = -0.2$ V) than the monomer [Eq. (3)].

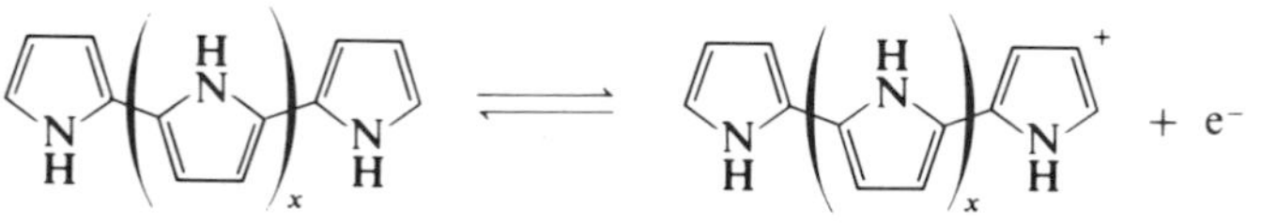

$$\text{Py}-(\text{Py})_x-\text{Py} \rightleftharpoons [\text{Py}-(\text{Py})_x-\text{Py}]^+ + e^- \quad (3)$$

The electrochemical stoichiometry of this reaction precludes the presence of a polymerization mechanism involving a cationic propagation step. In this regard, the reaction is different from the electrochemical polymerization reactions normally found in the literature where the reaction is initiated at the electrode surface but the polymer is formed in a chain propagation step that occurs in the bulk of the solution. Because the film remains on the electrode surface, the conducting nature of the film is important for the continuation of the reaction. The stoichiometry of this reaction also provides a real convenience in the preparation of these films since the desired thicknesses can be controlled by monitoring the current density of the reaction.

In practice, polypyrrole films are prepared by the electro-oxidation of pyrrole in a one-compartment cell equipped with a platinum working electrode, gold wire counter electrode, and a sodium chloride calomel reference electrode (SSCE). In a typical preparation, an acetonitrile solution containing 0.1 *M* tetraethylammonium tetrafluoroborate plus ca. 0.02 *M* pyrrole is employed.[10] The oxygen in the solution is swept out with an inert gas prior to the electrolysis. In practice, a wide variety of solvents and electrolytes

FIGURE 1. Scanning electron micrographs of polypyrrole tetrafluoroborate grown in (a) anhydrous acetonitrile and (b) acetonitrile containing 1% H_2O.

can be used as long as the electrical resistance of the solution is not high and the nucleophilicity does not interfere with the polymerization reaction. These conditions can be accomplished by selecting solutions where the electrolyte is highly dissociated and which are slightly acidic. As will be described below, the physical properties of the resulting films are sensitive to changes in the reaction conditions. For convenience, thin films ($<0.5\ \mu m$ thick) are usually prepared potentiostatically using a potential of 0.8 V. These films were left attached to the electrode surface, rinsed, and used for the electrochemical characterizations. The thicker films used as free-standing films for most of the physical characterizations are prepared galvanostatically using a current density of 0.22 mA/cm^2. These films were grown to thicknesses of 10–200 μm so that the films could be physically separated from the electrode using a scalpel. The thinner free-standing films used for infrared and optical studies separated themselves from the electrode on immersion in hot water. The free-standing films were then washed with acetonitrile in a Soxhlet extractor for several hours.

There is one aspect about the preparative procedure which is fairly unique and should be mentioned here. That is, the topography of the growing surface changes dramatically when the solvent is changed from moist to anhydrous acetonitrile. The growing surface is the face of the film which is in contact with the solution. Films grown in anhydrous acetonitrile have a rough surface with dendritelike structures as seen in Figure 1a. These structures must be due to the presence of highly conducting centers of preferred growth which survive when the Lewis base characteristics of the solution is low. When the solution contains 1% H_2O or other hydroxylic solvents, the growing surface becomes much smoother, although still rather uneven (Figure 1b). Under the latter conditions, the chemical nature of the surface is probably more homogeneous. The face that was attached to the electrode typically appears quite smooth when the film is removed from the metal electrode.

4. *Free-Standing Films*

Some of the organic polymers that can be made conducting are shown in Figure 2. Included in this figure is the inorganic polymer, poly(sulfurnitride), $(SN)_x$, which is intrinsically conducting[28] and indeed is superconducting below 0.3 K.[29] The conductivity scale in the figure is only intended to serve as a guide. This listing is not complete by any means and is intended to be suggestive of the kinds of polymer structures and of the range of conductivities encountered. Polypyrrole shares the feature with the other polymers of having an extended conjugation of π electrons. Except for the radical polymers, poly(sulfurnitride) and polyacene quinone radical

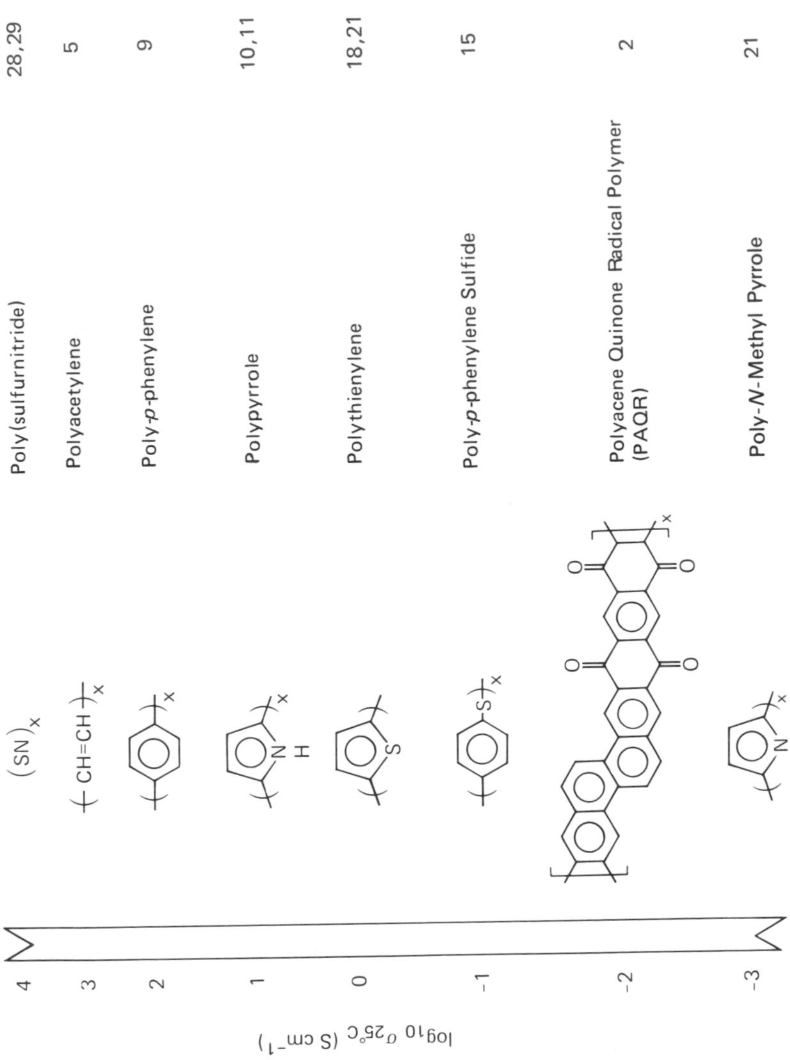

$\log_{10} \sigma_{25°C}$ (S cm^{-1})	Structure	Polymer	Reference
4	$(SN)_x$	Poly(sulfurnitride)	28,29
3	$\left(CH{=}CH \right)_x$	Polyacetylene	5
2		Poly-*p*-phenylene	9
1		Polypyrrole	10,11
0		Polythienylene	18,21
-1		Poly-*p*-phenylene Sulfide	15
-2		Polyacene Quinone Radical Polymer (PAQR)	2
-3		Poly-*N*-Methyl Pyrrole	21

FIGURE 2. Some conjugated polymers showing enhanced electronic conductivity.

polymer (PAQR), the other polymers are conducting by virtue of the oxidation or reduction of the polymer π system. As discussed earlier, it is not sufficient that high mobility states exist for the carriers. They must also be accessible to population by carriers. The oxidation or reduction of the polymers then serves to populate these states with holes or electrons, respectively. In most of the polymer systems, the oxidation process is accomplished by exposing the polymer to strong chemical oxidants. In the case of the electrochemically derived polypyrrole films, the film is conducting as grown and contains electrolyte anions that balance the cationic charge on the polymer.

This section deals with the properties of those films that are physically removed from the electrode and then studied. The electrical transport behaviors of the thicker free-standing films show a marked similarity to those of conducting polyacetylene. These transport characteristics are detailed in the following and leave little doubt that the transport is electronic and metallic in nature. The term "metallic" is being used in the sense that the carrier density is only weakly temperature dependent—the temperature dependence of the electronic behaviors being dominated by scattering.

The electrical conductivity of polypyrrole fluoroborate films were checked using both four-probe and Van der Pauw[30] methods with no significant differences. The temperature, T, dependence of the conductivity is shown in Figure 3.[31] The abcissa was chosen to be $T^{-1/4}$ primarily because the conductivity-temperature behavior for a number of polypyrrole films prepared under a variety of conditions showed a straight-line behavior when plotted on a semilog plot of this type. The physical significance of the $T^{-1/4}$ behavior is currently unknown. In any event, the conductivity is seen

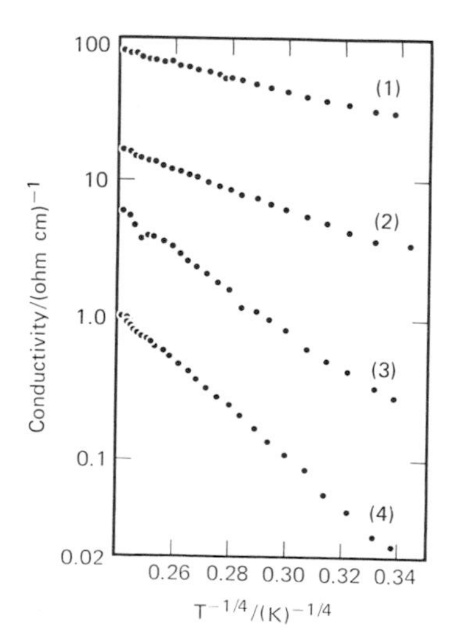

FIGURE 3. Plot of log conductivity vs $T^{-1/4}$ for several polypyrrole tetrafluoroborate samples having different conductivities.

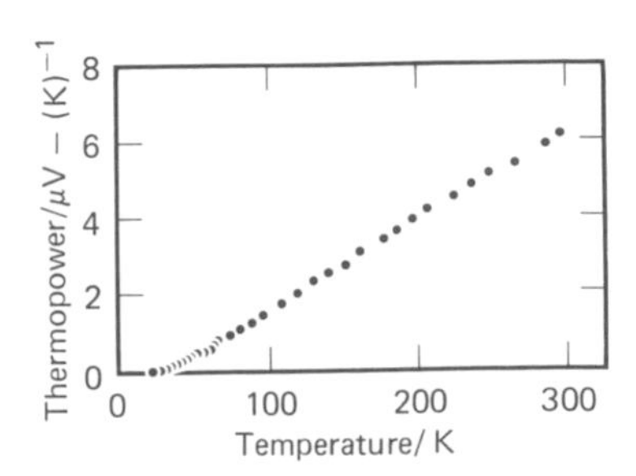

FIGURE 4. Thermoelectric power of polypyrrole tetrafluoroborate and its temperature dependence.

to increase with increasing temperature, unlike the behavior that would be expected for a simple metal. While it is not likely that the variable range hopping picture of Mott[32] is appropriate at these high temperatures, it is likely that some localization effect is being made. This temperature dependence then does not necessarily reflect the absence of metallic conduction. The ac conductivity at room temperature was also measured at frequencies up to 100 kHz and showed no frequency dependence.

The thermoelectric power of polypyrrole tetrafluoroborate films and its temperature dependence was measured and its behavior is shown in Figure 4.[31] The thermoelectric power is positive, implying conduction by holes. This observation is consistent with the notion of conduction in an oxidized polymer film. Moreover, the magnitude of the room-temperature thermoelectric power of $+8\ \mu V\ K^{-1}$ is compatible with a metal-like population of states. The sign and linearity of the thermoelectric power with temperature is consistent with the interpretation of these behaviors as that of a *p*-type metal.

Further evidence for the metallic nature of the conduction was obtained from a measurement of the thermal conductivity of polypyrrole tetrafluoroborate.[31] Using a modification of the technique of Bideau *et al.*,[33] the thermal conductivity K was determined to be 9×10^{-3} cal/s cm K. This value is intermediate between that of metals and insulating polymers and indicates that the carriers contribute substantially to the thermal transport. No significant ionic contribution to the conductivity could be detected. The resistance of one sample remained unchanged after the passage of more than a hundred times the amount of charge involved in the polymerization of the film.

Hall measurements were performed using a double ac technique.[34] An anomalous behavior was observed, where the Hall constant is negative, implying electronic, or *n*-type conduction. This type of behavior is not uncommon in amorphous systems and is observed also with polyphthalocyanine.[35] While this anomaly has not been definitely resolved as yet, a present view of its origin invokes localized concepts.[36] This data in addition to the $T^{-1/4}$ behavior of the conductivity might also suggest some degree of localization of the extended carrier states.

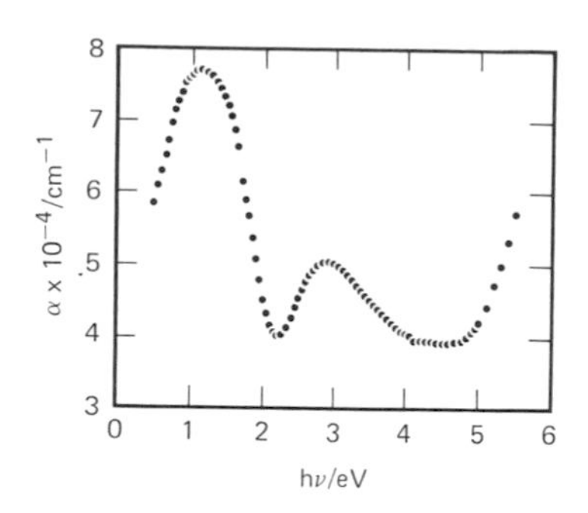

FIGURE 5. Absorption coefficient of polypyrrole tetrafluoroborate from 0.5 to 5 eV.

The appearance of strong absorptions or large reflections below a threshold energy in the electromagnetic spectrum is frequently used as evidence for metallic behavior. The absorption constant for thin free-standing films of polypyrrole tetrafluoroborate is shown in Figure 5 as a function of the incident energy.[31] The spectrum is reminiscent of those observed with polyacetylene[37] and poly-*p*-phenylene sulfide.[15,16] The higher-energy absorption is presumably an interband transition while the lower-energy absorption, which appears in the infrared region and is rather symmetrical, is associated with the polymer conductivity. In the case of polyacetylene, the lower-energy absorption appears when the polymer is oxidized. The detailed nature of the transitions giving rise to this low-energy absorption is not known at this time.

The freshly prepared films have a dark brown coloration, however, upon exposure to air (oxygen), the appearance becomes bluish-black. No other visible color changes are observed in the appearance of the films even after they are washed by extraction. As mentioned above, the topography of the growing surface of the polypyrrole tetrafluoroborate film will vary with the nature of the solvent.[21] The edge-on view of the film indicates that the film is not fibrillar, as are[29] $(SN)_x$ and $(CH)_x$.[38] It is very poorly crystalline and space filling.[31] The films have a density of 1.48 g/cm^3, determined by a mixed solvent technique, which is fairly insensitive to the nature of the anion.[27,31] The films are flexible, although extended washing over a 10-h period resulted in brittle films. It was found that an elongation of less than 1% was sufficient to break the films.

The conducting films are insoluble in the common solvents. This is a disadvantage for their characterization since it eliminates a number of polymer characterization methods that require solubility. This property, however, may be an advantage for technical applications. The films are thermally stable. The thermogravimetric scans show only a slow weight loss increasing to 10% at 300°C, with markedly larger rates of weight loss above that temperature as shown in Figure 6.[39] The electrical conductivity of the film is reversible to temperature cycling for temperatures below 150°C.[39]

The elemental analyses account for 89–97% of the material and provide the formula $C_{4.0}H_{3.5}N_{0.94-0.99}B_{0.41-0.52}F_{0.7-1.0}$. The films are composed of approximately 70% pyrrole polymer and 30% anion by weight. The analyses for boron and fluorine vary from sample to sample and are poorly reproducible. Even the samples of known tetrafluoroborate salts produce poor results for boron and fluorine. Thus the ratio of pyrrole units to fluoroborate anion (ideally tetrafluoroborate) is approximately 3 : 1. From these analyses it is concluded that the conducting films are oxidized and contain approximately 0.3 cationic charges per pyrrole unit. This estimate agrees with the degree of oxidation of the film calculated from the cyclic voltammetric studies of thin films left intact on the platinum electrode. Analysis of the surface region by ESCA clearly shows signals for boron and fluorine in the oxidized films. These signals, which must be due to the anion, are not present, however, when the film is made neutral by switching the electrode potential to cathodic values. These experiments are discussed below. The infrared and Raman spectra of polypyrrole show broad bands that could correspond to the pyrrole moiety. The detailed structure of the polymer is very difficult to ascertain. X-ray analysis revealed no peak in the spectrum, suggesting an amorphous polymer.[31] Electron diffraction in transmission electron microscopy, however, showed some very diffused diffraction rings.[40] The spacings inferred from these rings suggest a graphitelike structure.

The results of the electrical measurements in combination with the proposed chemical structure are consistent with the interpretation that the conductivity enhancement results from the generation of holelike states in the oxidation reaction of the extended π system of the polymer.

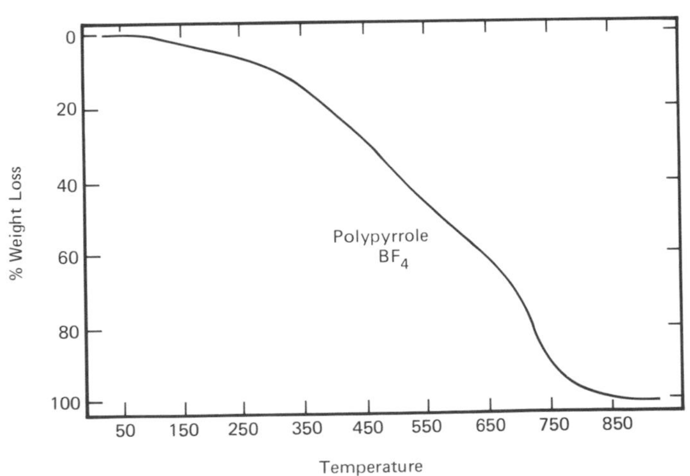

FIGURE 6. Plot of weight loss vs temperature for polypyrrole tetrafluoroborate. (From Reference 39.)

5. Electrochemical Properties of Thin Films

A complete study of conducting polymers should include electrochemistry especially since many of the applications of these materials may lie in their use as electrodes. This fact has been appreciated in certain laboratories; for example, studies are reported on the electrode behavior of $(SN)_x$ in aqueous solutions,[41,42] the use of $(CH)_x$ in a photocell,[43,44] and on the electrochemical doping of $(CH)_x$.[45] With polypyrrole we determined for the first time the redox potential of a conducting polymer and demonstrated that the polymer can be driven repeatedly between the conducting (oxidized) and nonconducting (neutral) state.[46] A unique feature of this redox reaction is that it involves electron transfer out of the extended π-electron system, which is an integral part of the polymer backbone structure [Eq.(1)]. In this regard, the polymer is inherently electroactive and the reaction is different from those described in the literature involving polymer films where the polymer backbone is saturated and not electroactive but instead contains covalently attached pendant groups that are electroactive.[47–51] The redox reaction is accompanied by a color change where the film changes from light yellow when neutral to brown-black when oxidized. The absorption spectra of the film in the visible region is shown in Figure 7.

Electrochemical switching experiments require the use of thin films, approximately 0.02–0.1 μm, with good electrical contact to the electrode, typically platinum. The electrochemical preparation of polypyrrole is particularly convenient for this study because films with the desired thickness can be produced directly on the electrode by monitoring the current density

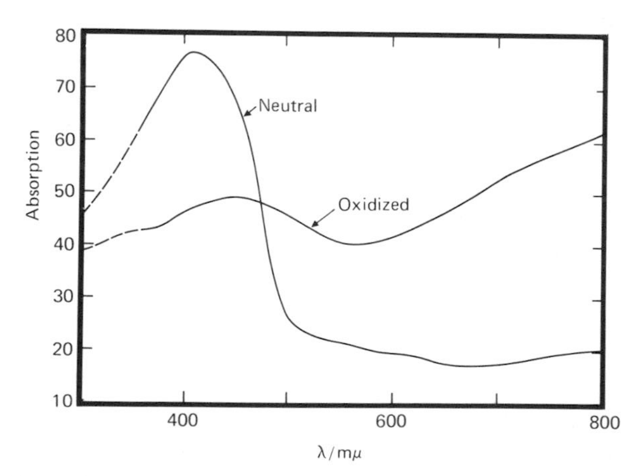

FIGURE 7. Absorption spectra (×100) of a polypyrrole film (neutral and oxidized) on a transparent SnO_2 electrode in acetonitrile containing Et_4NBF_4.

and with good electrical contact. Films prepared for these measurements are rinsed in acetonitrile and mounted in an electrochemical cell containing Et_4NBF_4/CH_3CN. In the case of the polypyrrole films, the complete washing and mounting procedure was performed in the absence of air because the kinetics of the redox reaction are sensitive to oxygen. The cyclic voltammogram for the polypyrrole film which results when the potential applied to the film is varied linearly from -0.5 V to 0.6 V to -1.2 V and back to -0.5 V is seen in Figure 8. The plot shows a peak for the oxidation of the film at approximately -0.1 V on the anodic sweep and the corresponding reduction peak at approximately -0.3 V in the cathodic sweep. The peak heights of the oxidation (i_{pa}) and the reduction (i_{pc}) waves scale linearly with the sweep rate and each wave appears symmetrical about the peak potential, E_p. The voltammograms do not change when the solutions are stirred. These observations generally indicate that the electroactive material is localized on the electrode surface and is not involved in a diffusion process. The symmetry of the peak is more easily seen with the oxidation wave, which is better defined and can be better corrected for the difference in the capacitance currents appearing on opposite sides of the wave. The peaks are not perfectly symmetrical, however, and some tailing can be seen after the peak. This tailing is probably produced by the slow rate of diffusion of the counterions in the film. The large capacitive currents on the anodic side of the peaks reflect the change in the effective surface area of the platinum/polymer electrode. The peak for the reduction reaction is more difficult to interpret. The integrated current under the oxidation peak remains constant and is, e.g., 0.69 mC/cm^2 for a film prepared using 8 mC/cm^2. The relative amount of current involved in the preparation and the switching reaction of the polymer can be compared directly considering that the apparent n value for the film-forming reaction is 2.2 F/mole where 2 F/mole is involved in the polymerization reaction and the extra 0.2 F/mole is consumed in the partial oxidation of the film. The redox reaction of a film prepared using 8 mC/cm^2 involves 0.7 mC/cm^2 [$8 \times (0.2/2.2)$ mC/cm^2], which agrees well with the value 0.69 mC/cm^2 obtained from the cyclic voltammetry experiments.

There are two characteristics of the cyclic voltammogram that are very important, the broadness of the peaks and the different E_p values for the oxidation and the reduction waves. Both of these characteristics indicate that the redox reaction is not a Nernstian process.[52–54] The broadness of the peaks may be due to interactions between the electroactive sites and/or electrochemical nonequivalence of the sites. Even the anodic peaks which are more narrow have peak widths at half-heights of approximately 130 mV, which is greater than the expected 90-mV value for a surface-localized reaction with n equal to 1 and with Nernstian behavior.[52] A peak width of 130 mV for a film 200 Å thick is not unusually large when considering that

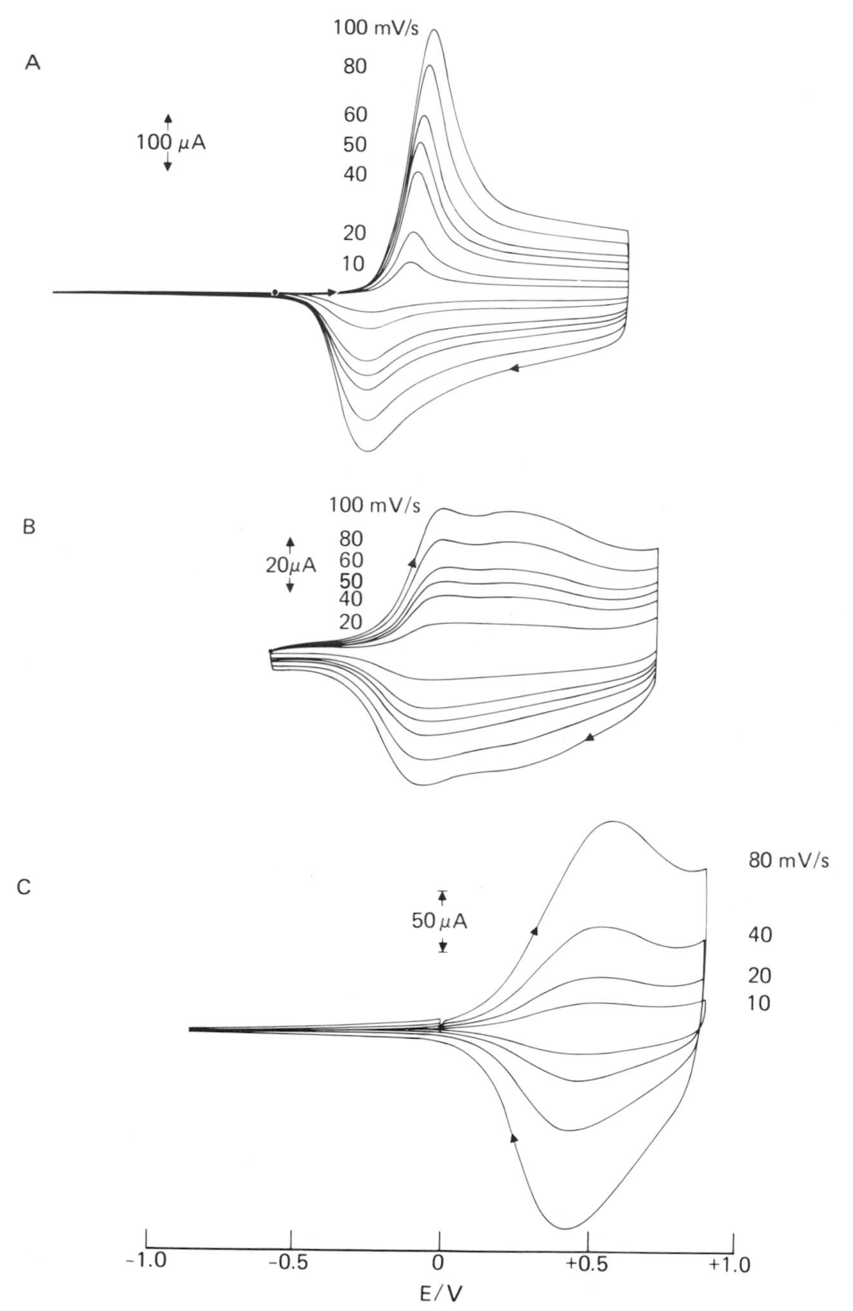

FIGURE 8. Cyclic voltammograms of thin polymer films on platinum measured in Et_4NBF_4/CH_3CN using a SSCE reference electrode. A, polypyrrole; B, polypyrrole-poly-*N*-methylpyrrole; C, poly-*N*-methylpyrrole.

peak widths of this magnitude have been reported for other polymeric films that are much thinner.[47, 50, 51, 55, 56] Thus, the interactions between the redox sites in the polypyrrole film must be less severe than in those cases involving pendent molecular redox centers, and the difference probably reflects the fact that the redox reaction of polypyrrole involves an extended π-electron system.

The nonequivalence of E_{pa} (-0.1 V) and E_{pc} (-0.3 V), where E_{pc} has a less anodic value, suggests that the redox reaction produces conformational changes in the polymer. The π-electronic structure of the neutral polymer must consist of segments produced by twists and bends along the polymer chain. The length of these conjugated segments become longer as the oxidation level of the polymer becomes higher and should produce a correspondingly less anodic E_p value. Such a correlation between the E_p values and the length of the π segment was shown to exist for a series of aromatic oligomers of pyrrole, thiophene, and benzene.[57] Slow electron transfer or counterion mobility which accompanies the redox reaction may also contribute to the nonequivalence of the E_p values.

The redox reaction of the polymer is accompanied by movement of ions in and out of the film in order to compensate the cationic nature of the oxidized polymer. The dependence of the redox reaction on ion mobility is evident by the lack of complete symmetry of the peaks about the E_p value in Figure 8. The ESCA results, in fact, show that the film contains fluoroborate only when it is in the oxidized form, but not when neutral.[58] As seen in Figure 9, the shape of the voltammogram depends on the nature of the electrolyte used. As observed with Et_4NBF_4, the i_{pa} values vary linearly with the sweep rate, and the peak areas remain constant with every electrolyte. The difference in the voltammograms produced by the various ions must be due to differences in the rates of ion diffusion in the film. The influence of the salt does not depend on the nature of the anion alone but on the cation–anion combination. Thus in these electrolytic conditions, ion aggregation (pairs, triads, etc.) must be very important in the film. The preliminary results of a spectroelectrochemical study of the redox reaction of these films provide a value for the ion diffusion rate in the film of 10^{-10} cm^2/s.[59] Considering that the nature of the electrolyte can change the peak positions and produce multiple peaks, plus the fact that E_{pa} does not equal E_{pc}, caution must be used in interpreting these voltammograms since the films correspond to approximately 100 monolayers of electroactive sites and the available theory for electrochemical processes, which include the interactions between the redox centers, are for one or several monolayers at the electrode.[55, 56]

The electroactive behavior observed with the polypyrrole films is not unique to this material but is a characteristic of many of the π-conjugated polymers that are conducting in the oxidized state. This is seen in Table 1

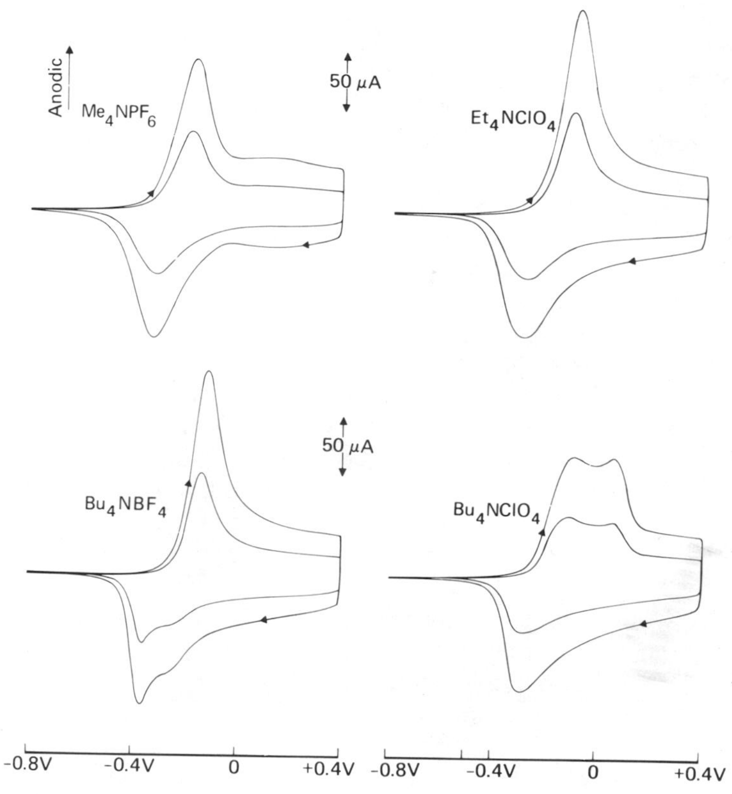

FIGURE 9. Cyclic voltammograms of (Pt)polypyrrole in CH_3CN containing various electrolyte salts.

which lists the switching potentials for the films of various unsaturated polymers. All of the values were experimentally measured except the value for poly-*p*-phenylene, which was estimated from a $h\nu_{\min}$-E_{pa} relationship involving the lower-molecular-weight oligomers.[57] The oxidation potential for poly-*p*-phenylene is much higher than any of the heterocyclic polymers. This result is not unreasonable since we find that the relative ease of oxidation of these polymers parallels the relative ease of oxidation of the corresponding monomers, and the E_{pa} value for benzene is more anodic than the values for thiophene and pyrrole.[57] It is sufficiently high that it will be more difficult to promote the switching behavior of this polymer even if a procedure becomes available for preparing thin polymeric films on the electrode. The redox potentials of polypyrrole and polyaniline are uniquely low compared to the other values. Hydrogen bonding in these materials could give rise to these low values since it would provide additional stabilization to the π cation in the oxidized polymer. The nature of the hydrogen bond

TABLE 1. Switching Potentials of Some Conducting Polymer Films

Polymer	Switching potential (mV)[a]	Conductivity, (25 C)(ohm^{-1} cm^{-1})	Degree of oxidation	Reference
 polypyrrole	−200	1–100	0.25–0.30	46
 polythiophene	700	10^{-3}–10^{-1}	0.1	21
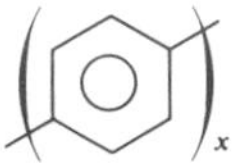 poly(p-phenylene)	1600 (estimated)			21, 57
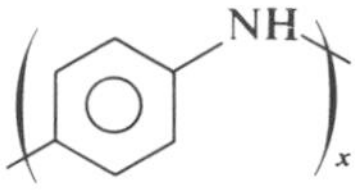 polyaniline	100[b]	10^{-2}	(1)	14
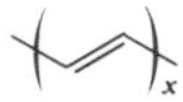 polyacetylene	700	10^2–10^3	0.01	73

[a] Measured in CH_3CN vs SSCE.
[b] Measured in 0.1M H_2SO_4 (aq).

would be sensitive to the oxidation level of the polymer and could produce a protonation–deprotonation reaction which accompanies the polymer redox reaction [Eqs. (4) and (5)].

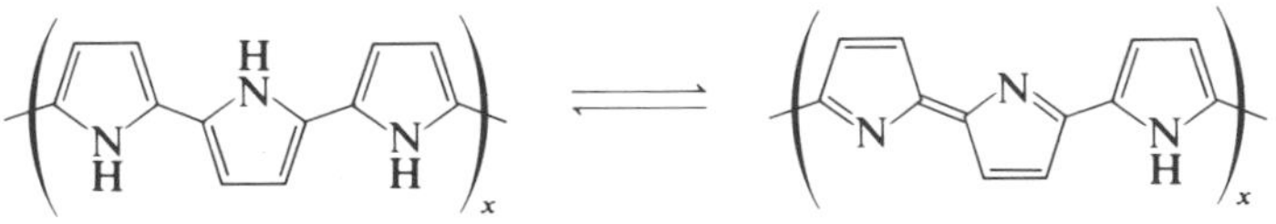

(4)

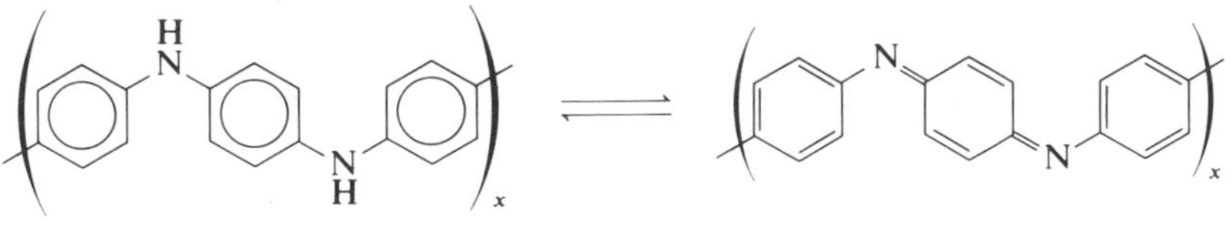

(5)

6. *N*-Substituted Polypyrroles

The redox potentials of the *N*-substituted polypyrroles are considerably more positive than the value for polypyrrole [Eq. (6)].[60]

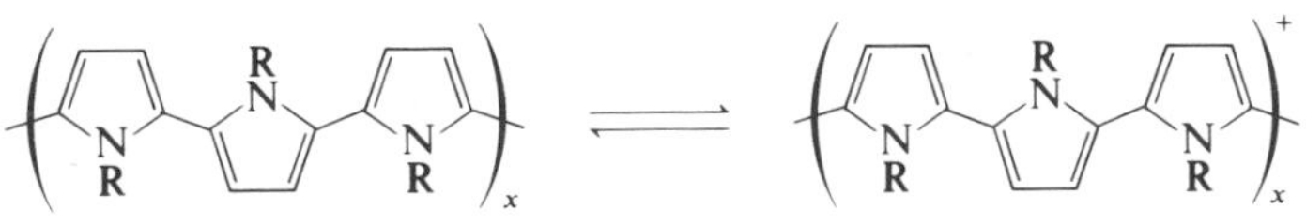

(6)

TABLE 2. Summary of Data for Poly-N-Alkylpyrrole Polymers

	Thin films	Thick films		
N–R	E^0 (mV)[a]	Degree of oxidation	Density (g/cm^3)	Conductivity (ohm^{-1} cm^{-1})
H	−200	0.25–0.30	1.48	30–100
Methyl	450	0.23–0.29	1.46	10^{-3}
		<0.05	1.33	10^{-6}
Ethyl	450	0.20	1.36	2×10^{-3}
n-Propyl	500	0.20	1.28	10^{-3}
n-Butyl	640	0.11	1.24	10^{-4}
i-Butyl	600	0.08	1.25	2×10^{-5}
Phenyl	650	0.15	1.42	10^{-3}

[a] Pt vs SSCE in Et_4NBF_4/CH_3CN.

These polymers are more difficult to oxidize, and the redox reactions of the polymers are correspondingly less sensitive to air. The reactions of these compounds are also accompanied by a color change from yellow (neutral polymer) to dark brown (oxidized polymer). The presence of these closely related alkyl substituents is expected to produce regular changes in the properties of the polymer. As seen in Table 2 the E^0 values for the *N*-substituted polypyrroles are much more positive than the value for polypyrrole, and they show a gradual positive shift as the size of the substituent increases. These shifts are probably due to the steric effects of the substituent which could disturb the ring–ring planarity along the polymer chain and thus destabilize the cationic form of the polymers, plus the reduction of any stabilizing effects resulting from chain–chain interactions.

The thicker polymeric films needed for measurement of the physical properties become progressively more difficult to prepare as the size of the alkyl group increases. These films are produced with a rough and wrinkled appearance and in lower yields as the formation of soluble products which darken the reaction solution becomes more prominent. With very large alkyl groups as with *N*-*t*-butylpyrrole and *N*-cyclohexylpyrrole, continuous films are not produced on the electrode, only a powdery brown-black deposit. Thus the steric effects of the *N*-substituent are very important in the electrochemical preparation of these films. Even though these polymers with large alkyl groups can not be prepared electrochemically, as free-standing films, it may still be possible to prepare them by a coupling reaction analogous to the preparation of polythiophene[18] and polyphenylene.[9]

The films are space filling and continuous like polypyrrole. The results of the elemental analyses indicate that the films consist of linearly coupled pyrrole units that remain unsaturated and retain the substituent plus contain the anion. Since we have not been able to form polymers with α-substituted pyrroles,[25] these polymers probably are α,α′-coupled as in the case of the parent polypyrrole unit.[23] The degree of oxidation, the density, and conductivity of these films also varies with the size of the substituent. As seen in Table 2, the degree of oxidation of the films decreases somewhat regularly from approximately 0.25–0.29 for the pyrrole and *N*-methylpyrrole polymers to 0.08 for the *i*-butylpyrrole polymer. The density of the films also varies as is expected since both the chemical nature and the degree of oxidation of the polymer is changing. The density of the films is expected to change with variations in the degree of oxidation, since it produces changes in the polymer/anion proportion in the film. The variation in the density resulting from this change can be estimated by comparing the change in the density of poly-*N*-methylpyrrole, which is 1.46 when the degree of oxidation is 0.26 and 1.33 when it is <0.05.[58] The range in the density values for the list of substituted polymers is greater than these two values, suggesting that the chemical nature of the polymer must have a strong influence on the packing structure of the film.

The conductivity of the films also decreases as the size of the alkyl group increases. The decrease in the conductivity (over six orders of magnitude) of the films must be one of the important factors underlying the observed difficulty in the preparation of these films. This change in the conductivity is accompanied by an increase in the thermoelectric power. The poly-*N*-methyl-pyrrole and poly-*N*-isobutylpyrrole films show room-temperature thermoelectric powers of $+50$ and $+80$ μV/K, respectively, which are much higher than the value, $+8$ μV/K, for polypyrrole.[60] Here again, the polypyrrole film stands apart from the *N*-substituted pyrrole polymers both with regard to its uniquely high conductivity and its distinct thermoelectric power value. The free-standing films (in the oxidized form) are stable to air except for the polybutylpyrroles which are slightly air sensitive and show a decrease in their conductivity by a factor of 5 after standing in air for approximately three days.

The *N*-phenyl group produces an effect that is catagorically different from that of the other substituents. It produces the largest effect on the switching potential of the thin films as may be expected considering the inductive effect of the group. Poly-*N*-phenylpyrrole is only about half as oxidized as polypyrrole and has a conductivity and density like poly-*N*-methylpyrrole even though the phenyl group is much larger than the methyl group. Thus the electronic effects of the phenyl π-electron system must be opposing the steric effect of the group. Poly-*N*-phenylpyrrole is of particular interest because it provides the possibility of introducing highly polar groups into the polymer structure as substituents on the phenyl ring.

An obvious variation of the electrochemical preparation procedure is the copolymerization of pyrrole/substituted-pyrrole mixtures.[21] The cyclic voltammetric analysis of films prepared by electropolymerizing pyrrole/*N*-methylpyrrole mixtures indicate that both monomers are incorporated into the film (Figure 8) and that the fraction of pyrrole in the film is greater than in the monomer mixture.[21, 22] This result indicates that the polymerization reaction has some selectivity. The conductivities and thermoelectric power values of the free-standing films are intermediate between that of the pyrrole and *N*-methylpyrrole polymers, and show a smooth variation with the composition of the monomer mixture used in the polymerization.[21, 22, 39]

Polypyrrole also offers the possibility of varying its properties by introducing substituents in the β positions of the pyrrole ring. For example, thin films of poly-β-methylpyrrole has a switching potential similar to polypyrrole, while the free-standing film has a room-temperature conductivity of 4 ohm^{-1} cm^{-1} which is intermediate between polypyrrole and poly-*N*-methylpyrrole.[61] Substituents placed in these positions are expected to produce less of a steric perturbation along the polymer chain compared to the *N*-substituents. Therefore it may be possible to introduce large groups in the β positions.

Besides polypyrrole, studies of chemically derivatized conducting polymers have also been reported with polyacetylene. The conductivity of this material is also reduced, and the air sensitivity is enhanced when the polymer is derivatized by replacing one of the hydrogens in the ethylene group by a methyl[62] or when it is doubly substituted as in poly(hepta-1,6-diyne).[63]

7. *Anion Variation*

The tetrafluoroborate anion makes up a significant fraction (25–30%) of the polypyrrole films and accumulates in the film during the preparative process in order to provide overall neutrality to the film. As previously stated, the anion comes from the electrolyte present in the solution and is not generated by some chemical oxidant, e.g., Br_2 or AsF_5, which would be used to oxidize the polymer.[24] This situation arises because the polymer is oxidized electrochemically during the electropolymerization process. This is a very important characteristic of the electropolymerization preparation since it permits us to vary the nature of the film by simply changing the anion of the electrolyte and more importantly to incorporate anions which cannot be generated from the oxidant, e.g., sulfonate and carboxylate anions.

We have already seen in Figure 9 that the redox kinetics of the polypyrrole films depend very much on the nature of the anion. The anion also produces a variation of 10^5 in the conductivity of the films (Table 3).[21] This is somewhat unexpected since neither the density nor the degree of oxidation of the polymer in the film show much variation with the change of the

TABLE 3. Polypyrrole Films with Various Anions

Anion	Degree of oxidation	Density (g/cm^3)	Conductivity (25 C)(ohm^{-1} cm^{-1})
BF_4^-	0.26–0.30	1.48	30–100
PF_6^-	0.24	1.48	100
AsF_6^-	0.33	1.47	100
ClO_4^-	0.30	1.51	100–200
HSO_4^-	0.30	1.58	0.3
$CF_3SO_3^-$	0.31	1.48	0.3–1
$CH_3C_6H_4SO_3^-$	0.32	1.37	20–100
CF_3COO^-	0.33	1.45	12
$HC_2O_4^-$	0.40	1.46	10^{-3}–10^{-2}

anion.[64] The degree of oxidation of the film is a thermodynamic state and depends on the nature of the polymer and, surprisingly, not on the polymer–anion combination of the film. The wide variation in the conductivity of the films listed in Table 3 must be due primarily to the nature of the anion, and not the degree of oxidation of the polymer since the latter remains fairly constant for the majority of the films listed. On the other hand, the trend in the conductivity values relates directly to the relative nucleophilicity of the various anions. This correlation suggests that the change in the conductivity could result from some form of anion-induced localization of the radical cation in the polymer π-electron structure. Finally, the electrochemical preparation procedure of polypyrrole films provides a very convenient means for producing films with the desired conducting properties by simply varying the electrolyte.

8. *Applications*

A discussion of the technical applications of these films is of course a speculative venture. In order to minimize this characteristic from our discussion, we will review briefly the results of some studies that relate to the development of applications for these materials, rather than list all the possible applications. A study based on one possible application involves coating *n*-GaAs photoanodes with a protective film of polypyrrole for use in a solar cell.[65] This film is very effective in preventing the dissolution of GaAs into the electrolyte, while permitting electron exchange. The film remains electroactive on this surface and switches at a less anodic potential than when on platinum in acetonitrile solution. The photoanode output characteristics of the electrode with the protective film are comparable to the bare electrode. In this system the adhesion of the film to the GaAs is presently poor and the film lifts off under illumination in aqueous solution in approximately 15 min. Polypyrrole films have also been used as protective films for *n*-type crystalline and polycrystalline silicon photoelectrodes.[66] The coated electrodes show typical *n*-type behavior and have photoanode output characteristics that are comparable to the uncoated electrode. The coated electrodes are very stable and show no evidence of poor film adhesion. For example, they function well up to 122 h whereas the uncoated electrode decomposes in less than 30 s.

The ability to switch these films between the neutral and the oxidized state and correspondingly to switch them between the nonconducting and conducting form[46, 58] in an electrochemical environment also suggests other applications. For example, an electrode surface (on platinum) for organic reaction in aprotic solvents which would be selective to a certain range of redox potentials is possible. In this application, the film used as an anode is conducting and will drive oxidation reactions. The film used as a cathode is poorly conducting and will inhibit the reduction of organic compounds, thus eliminating the use of a divided cell. Because the neutral film in the cathode is porous to protons, this application would require the presence of protons or small ions, e.g., Li^+, in the electrolytic solution. The color changes observed during the redox reaction give rise to the use of electroactive polymers for passive display devices.[67] These films will switch with Coulombic reversibility and as fast since the films of polymers containing electroactive pendent groups have been characterized electrochemically for use in this application.[51]

Thicker films of polypyrrole may find use as inexpensive electrodes for general electrochemical use.[68] In aprotic solvents, these films work well in the potential range −1.6–1.0 V versus SSCE. The possibility of attaching electroactive centers on the surface region of the polypyrrole films provides a new approach for the preparation of chemically modified electrodes.[69] The studies of chemically modified surfaces have stressed the attachment of redox couples on metal oxide and semiconductor electrode surfaces for use in the areas of chemical selectivity, electrocatalysis, and photocorrosion.[47, 69, 70] Redox centers can also be attached to the surface region of polypyrrole by first nitrating it, reducing the nitro groups electrochemically and then acylating the resulting group with the desired reagent.[71] There is also the possibility of using a polypyrrole in battery applications as has been demonstrated with $(CH)_x$.[72]

As stated before, polypyrrole and other conducting polymers will probably find many of their applications in the area of electrochemistry. As regards the free-standing films, they may find use in the applications identified for the other conducting polymers. Polypyrrole has a very important advantage over the other conducting polymers being studied currently because it is stable in ambient conditions for several months and can be chemically modified.[21, 60]

ACKNOWLEDGMENTS

The authors wish to thank our co-workers who have contributed significantly to the development of this electrochemical approach for the preparation and study of conducting polymers: J. I. Castillo, J. M. Vasquez, J. A. Logan, M. Salmon, and G. P. Gardini, who first suggested the potential of polypyrrole. We also wish to thank B. Gallegos and K. Bryan for their assistance in the preparation of this manuscript.

References

1. See for example, Okamoto, Y., and Brenner, W., *Organic Semiconductors*, Reinhold Publishing, London (1964), Chapter 7.
2. Pohl, H. A., *J. Polym. Sci. C* **17**, 13 (1967).
3. Kanda, S., and Pohl, H. A., *Organic Semiconducting Polymers*, J. E. Katon, Ed., Marcel Dekker, New York (1968).
4. Goodings, E. P., *Chem. Rev.* **5**, 95 (1976).
5. Chiang, C. K., Fincher, Jr., C. R., Park, Y. W., Heeger, A. J., Shirakawa, H., Louis, E. J., Gau, S. C., and MacDiarmid, A. G., *Phys. Rev. Lett.* **39**, 1098 (1977).
6. Whangbo, M-H., Hoffmann, R., and Woodward, R. B., *Proc. R. Soc. Lond. A* **366**, 23 (1979).
7. Fabish, T. J., *Crit. Rev. Solid State Mater. Sci.* **8**, 383 (1979).
8. Grant, P. M., and Batra, J. P., *Syn. Metals* **1**, 193 (1979/80).
9. Ivory, P. M., Miller, G. G., Sowa, J. M., Shacklette, L. W., Chance, R. R., and Baughman, R. H., *J. Chem. Phys.* **71**, 1506 (1979).
10. Diaz, A. F., Kanazawa, K. K., and Gardini, G. P., *J. Chem. Soc. Chem. Commun.*, 635 (1979).
11. Kanazawa, K. K., Diaz, A. F., Geiss, R. H., Gill, W. D., Kwak, J. F., Logan, J. A., Rabolt, J. F., and Street, G. B., *J. Chem. Soc. Chem. Commun.*, 854 (1979).
12. Jozefowicz, M., Yu, L. T., Perichon, J., and Buvet, R., *J. Polym. Sci. C* **22**, 1187 (1969).
13. Langer, J., *Solid State Commun.* **26**, 839 (1978).
14. Diaz, A., and Logan, J. A., *J. Electroanal. Chem.* **111**, 111 (1980).
15. Clarke, T. C., Kanazawa, K. K., Lee, V. Y., Rabolt, J. F., Reynolds, J. R., and Street, G. B., *J. Polym. Sci.*, to be published.
16. Rabolt, J. F., Clarke, T. C., Kanazawa, K. K., Reynolds, J. R., and Street, G. B., *J. Chem. Soc. Chem. Commun.*, 347 (1980).
17. Chance, R. R., Shacklette, L. W., Miller, G. G., Ivory, D. M., Sowa, J. M., Elsenbaumer, R. L., and Baughman, R. H., *J. Chem. Soc. Chem. Commun.*, 348 (1980).
18. Yamamoto, T., Sanechika, K., and Yamamoto, A., *J. Polym. Sci. Polym. Lett. Ed.* **18**, 9 (1980).
19. Kuznesot, P. M., Wynne, K. J., Nohr, R. S., and Kenny, M. E., *J. Chem. Soc. Chem. Commun.*, 121 (1980).
20. Dall'Olio, A., Dascola, Y., Varacca, V., and Bocchi, V., *Compt. Pend. C* **267**, 433 (1968).
21. Diaz, A., *Chem. Scrip.* **17**, 145 (1981).
22. Diaz, A. F., Kanazawa, K. K., Castillo, J. I., and Logan, J. A., American Chemical Society, 180th Meeting, Conductive Polymers Symposium, Las Vegas, Nevada, August 1980, Abstract Number ORPL 159.
23. Gardini, G. P., *Adv. Heterocyclic Chem.* **15**, 67 (1973).
24. Gau, S-C., Milliken, J., Pron, A., MacDiarmid, A. C., and Heeger, A. J., *J. Chem. Soc. Chem. Commun.*, 662 (1979).

25. Diaz, A. F., Martinez, A., Kanazawa, K. K., and Salmon, M., *J. Electroanal. Chem.* **130**, 181 (1981).
26. This mechanism has been proposed for the formation of aniline blacks. See Baizer, M. M., *Organic Electrochemistry*, Marcel Dekker, New York (1973), p. 515.
27. Diaz, A., and Salmon, M., unpublished results.
28. Walatka, V. V., Lakes, M. M., and Perlstein, J. H., *Phys. Rev. Lett.* **31**, 1139 (1973).
29. Street, G. B., and Greene, R. L., *IBM J. Res. Dev.* **21**, 99 (1977).
30. Hornstra, J., and Van der Pauw, L. J., *J. Electronics Control* **7**, 169 (1959).
31. Kanazawa, K. K., Diaz, A. F., Gill, W. D., Grant, P. B., Street, G. B., Gardini, G. P., and Kwak, J. F., *Syn. Metals* **1**, 329 (1980).
32. Mott, N. F., *Philos. Mag.* **19**, 835 (1969).
33. Bideau, D., Troadec, J. P., Meury, J. L., Rosse, G., and Quan, Dang Tran, *Rev. Phys. Appl.* **31**, 415 (1978).
34. Seeger, K., and Gill W. D., IBM Research Rep. RJ 2339 (1979).
35. Burnay, S. G., and Pohl, H. A., *J. Non Cryst. Solids* **30**, 221 (1978).
36. See, for example, Friedman, L., and Pollak, M., *Philos. Mag. B* **38**, 173 (1978), and references therein.
37. Fincher, Jr., C. R., Peebles, D. L., Heeger, A. J., Droy, M. A., Matsumura, Y., MacDiarmid, A. G., Shirakawa, H., and Ikeda, S., *Solid State Commun.* **27**, 489 (1978).
38. Ito, T., Shirakawa, H., and Ikado, S., *J. Polym. Sci. Polym. Chem. Ed.* **12**, 11 (1974).
39. Kanazawa, K. K., Diaz, A. F., Krounbi, M. T., and Street, G. B., *J. Syn. Metals* **4**, 119 (1981).
40. Geiss, R. H., *Proceedings from the 38th Meeting of Electron Microscopy Society of America*, Claitor's Publishing Division, Baton Rouge, 1980, p. 238.
41. Nowak, R. J., Mark, Jr., H. B., MacDiarmid, A. G., and Weber, D., *J. Chem. Soc. Chem. Commun.*, 9 (1977).
42. Nowak, R. J., Kutner, W., and Mark, Jr., H. B., *J. Electrochem. Soc.*, 125 (1978).
43. MacDiarmid, A. G., Heeger, A. J., Kletter, M. J., Nigney, P. J., Park, Y. W., Pron, A., and Waerner, T., International Conference on Low Dimensional Synthetic Metals, Denmark, August 1980, Abstract 9.3.
44. Chen, S. N., Heeger, A. J., Kiss, Z., MacDiarmid, A. G., Gan, S. C., and Peebles, D. L., *Appl. Phys. Lett.* **36**, 96 (1980).
45. Nigrey, P. J., MacDiarmid, A. C., and Heeger, A. J., *J. Chem. Soc. Chem. Commun.*, 594 (1979).
46. Diaz, A. F., and Castillo, J. I., *J. Chem. Soc. Chem. Commun.*, 397 (1980).
47. Wrighton, M. S., Austin, R. G., Bocarsly, A. B., Bolts, J. M., Haas, O., Legg, K. P., Nadjo, L., and Palazzotto, M., *J. Electroanal. Chem.* **87**, 429 (1978).
48. Merz, A., and Bard, A. J., *J. Am. Chem. Soc.* **100**, 3222 (1978).
49. Oyama, N., and Anson, F., *J. Electrochem. Soc.* **127**, 640 (1980).
50. Kerr, J. B., Miller, L. L., and Van de Mark, M. R., *J. Am. Chem. Soc.* **102**, 3383 (1980).
51. Kaufman, F. B., Schroeder, A. H., Engler, E. M., and Patel, V. V., *Appl. Phys. Lett.* **36**, 422 (1980).
52. Hubbard, A. T., and Anson, F. C., *Anal. Chem.* **58**, (1966).
53. Laviron, E., *Bull. Soc. Chem. France*, 3717 (1967).
54. Laviron, E., *J. Electroanal. Chem.* **39**, 1 (1972).
55. Brown, A. P., and Anson, F. C., *Anal. Chem.* **49**, 1589 (1977).
56. Smith, D. F., William, K., Kuo, K., and Murray, R. W., *J. Electroanal. Chem.* **95**, 217 (1979).
57. Diaz, A. F., Crowley, J. I., Bargon, J., Gardini, G. P., and Torrance, J. B., *J. Electroanal. Chem.* **121**, 355 (1981).
58. Diaz, A. F., Castillo, J. I., Logan, J. A., and Lee, W. Y., *J. Electroanal. Chem.* **129**, 115 (1981).
59. Genies, E. M., Diaz, A. F., and Laviron, E., unpublished results.

60. Diaz, A., Castillo, J., Kanazawa, K. K., Logan, J. A., Salmon, M., and Fajardo, O. **133**, 237 (1981).
61. Salmon, M., Diaz, A. F., Waltman, R., and Bargon, J., unpublished results.
62. Chien, J. C. W., and Karasz, F. E., The Electrochemical Society, 157th Meeting, May 11–16, 1980, Abstract 56.
63. Gibson, H. W., Bailey, F. C., Epstein, A. J., Rommelmann, H., and Pochan, J. M., *J. Chem. Soc. Chem. Commun.*, 426 (1980).
64. Diaz, A., Salmon, M., Logan, J. A. and Krounbi, M., unpublished results.
65. Noufi, R., Tench, D., and Warren, L. F., *J. Electrochem. Soc.* **127**, 2310 (1980).
66. Noufi, R., Frank, A. J., and Nozik, A. J., *J. Am. Chem. Soc.* **103**, 1849 (1981).
67. Diaz, A., U.S. Patent Application, 1980.
68. Diaz, A. F., Vasquez, J. M., and Martinez, A., *IBM J. Res. Develop.* **25**, 42 (1981).
69. Murray, R. W., *Acc. Chem. Res.* **13**, 135 (1980).
70. Snell, K. D., and Keenan, A. G., *Chem. Soc. Rev.* **8**, 259 (1979).
71. Diaz, A. F., Lee, W. Y., Logan, A., and Green, D. C., *J. Electroanal. Chem.* **108**, 377 (1980).
72. MacDiarmid, A. G., and Heeger, A. J., The Electrochemical Society, 157th Meeting, Organic Metals and Semiconductors Symposium, St. Louis, Mo., May 1980, Abstract 55.
73. Diaz, A. F., and Clarke, T. C., *J. Electroanal. Chem.* **111**, 115 (1980).

9

Compendium of Synthetic Procedures for One-Dimensional Substances

Paul J. Nigrey

1. Introduction

In recent years, the area of one-dimensional materials has attracted considerable interest in the international research community and, due to the unique nature of these materials, a broad range of scientists, including chemists, experimental physicists, and theoretical physicists, have been joined together in an attempt to understand the chemical and physical properties of these materials. Since a close collaborative effort between chemists and physicists is required, a simplified understanding of the synthetic techniques is useful. The purpose of this chapter, therefore, is to enable the novice (chemist or physicist) to have at his disposal a collection of synthetic procedures for the preparation of representative one-dimensional materials. These procedures were chosen, by the author, from the sometimes large selection of literature articles, in such a way as to permit the facile synthesis of the desired materials. The literature citations reflect the author's desire to provide only specific synthetic procedures while at the same time allowing a survey of other possible methods.

2. Organic Materials

2.1. Electron-Donor Molecules

2.1.1. Nitrogen Heterocyclic Compounds

The nitrogen heterocyclic compounds used as donor molecules (see Table 1) are the quaternary salts formed by the action of organic halides or

Paul J. Nigrey • Laboratory for Research on the Structure of Matter, University of Pennsylvania, Philadelphia, Pennsylvania 19104. Dr. Nigrey's present address is: Betz Laboratories, Inc., Trevose, Pennsylvania 19047.

TABLE 1.

Compound	Reference[a]
Quinolinium iodide	3
Quinolinium = **1**(R = H)	
N-Methylquinolinium iodide	4
N-Methylquinolinium = **1**[R = CH_3(Me)]	
Phenazinium iodide	5
Phenazinium = **2**(R = H)	
N-Methylphenazinium methylsulfate [$(NMP)^+(MeOSO_3^-)$]	6
N-Methylphenazium = **2**(R = Me)	
N-Ethylphenazinium ethylsulfate	6
N-Ethylphenazinium = **2**(R = Et)	
Acridinium iodide	7
Acridinium = **3**(R = H)	
N-Methylacridinium = **3**(R = Me)	8
Acradizinium bromide	9
Acradizinium = **4**	
(*N*,*N*′-diethyl-4,4′-bipyridylium)diiodide	10
N,*N*′-diethyl-4,4′-bipyridylium = DEBP	
[1,2-di(*N*-ethyl-4-pyridinum)]ethylene dibromide	10
1,2-di(*N*-ethyl-4-pyridinium)ethylene = DEPE	
N,*N*,*N*′,*N*′-tetramethyl-*p*-phenylenediamine (TMPD) = **5**	11

[a] Reference in which synthetic procedure can be found.

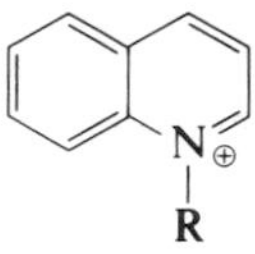

1 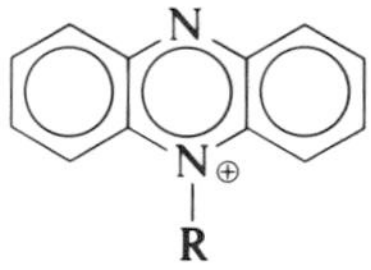 **2** 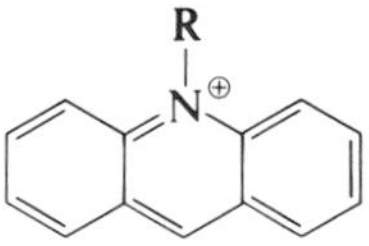 **3** 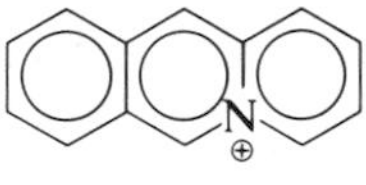 **4** 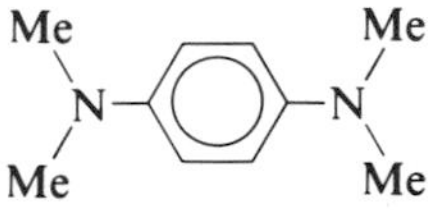 **5**

hydrohalic acids on the heterocyclic base, e.g., quinoline, phenazine, and acridine. This type of a reaction, sometimes called the Menschutkin reaction,[1] may be considered a displacement of the halogen as a halide ion by the heterocyclic base acting as a nucleophilic agent. A solvent such as nitrobenzene may be used to temper the reaction or access elevated temperatures[2] to obtain practical speeds of quaternization.

2.1.2. *Sulfur Heterocyclic Compounds*

A number of synthetic methods[14] exist for the preparation of tetrathiafulvalene (TTF) (see Table 2) and its derivatives. By far the most commonly used methods are deprotonation of the 1,3-dithiolium ions (method I) and the desulfurization of 2-thioxo-1,3-dithioles using phosphite reagents (method II).

Dithionaphthalene (DTN) (see Table 2) was first isolated from the reaction of naphthalene and sulfur vapors in a hot iron tube,[25] while Price and Smiles[26] reported its preparation starting with 1-aminonaphthalene-8-sulfonic acid. More recently Meinwald *et al.*[27] reported its preparation from the low-temperture reaction of 1,8-dilithionaphthalene with sublimed sulfur.

Tetrathiotetracene (TTT) (see Table 2) was first prepared by Marschalk and Stumm[30] from the reaction of naphthacene and sulfur in trichlorobenzene. Perez-Albuerne[31] slightly modified the original procedure to give higher yields of TTT with minimized side reactions.

Dithiapyranylidenes (DTPs) (see Table 2) are of interest since they represent a class of organosulfur donors that are similar to TTF by virtue of their symmetry (D_{2h}), reversible formation of dications in solution, iso-π-electronic structure, and ability to form ion-radical salts of both 1 : 1 and 1 : 2 stoichiometry with 7,7,8,8-tetracyano-*p*-quinodimethane (TCNQ). Dithiapyranylidenes can typically be prepared by reductive coupling or induced to eliminate a chalcogen moeity by thermal or photochemical means. Sandman *et al.*[32] has recently succeeded in preparing a number of dithiapyranylidenes using a novel single-step procedure utilizing an acyclic precursor containing no sulfur atoms.

2.1.3. *Selenium Heterocyclic Compounds*

While TTF can readily be prepared by the deprotonation of the 1,3-dithiolium salts, the analogous reactions with the corresponding selenium compounds fail.[14] It appears that tetraselenofulvalene (TSeF) (see Table 3) and its derivatives must be prepared by treatment with either triphenylphosphine or trialkyl phosphite.[33,34]

2.1.4. *Sulfur-Selenium Heterocyclic Compounds*

The mixed sulfur-selenium compounds shown in Table 4 are prepared by methods similar to those described in Section 2.1.2 and 2.1.3.

TABLE 2.

Compound	References
Tetrathiafulvalene (TTF), **6**($R_1 = R_2 = R_3 = R_4 = H$)	12, 14
Symmetrically substituted TTF	
deuterio TTF, **6**($R_1 = R_2 = R_3 = R_4 = D$), method I	13, 15
dimethyl TTF, **6**($R_1 = R_4 = Me$; $R_2 = R_3 = H$), method I	16
tetramethyl TTF, **6**($R_1 = R_2 = R_3 = R_4 = Me$), method I	17
tetraethyl TTF, **6**($R_1 = R_2 = R_3 = R_4 = Et$), method I	18
hexamethylene TTF, **7**, method I	19, 20
octamethylene TTF, **8**, method I	19, 20
tetracyano TTF, **6**($R_1 = R_2 = R_3 = R_4 = CN$), method II	21, 22
Unsymmetrically substituted TTF	
dimethyl TTF, **6**($R_1 = R_2 = Me$; $R_3 = R_4 = H$) method I	16
benzo TTF, **9**	23
methyl TTF, **6**($R_1 = Me$; $R_2 = R_3 = R_4 = H$), method I	24
benzo, dimethyl TTF, **10**	23
dimethyl, diethyl TTF, **6**($R_1 = R_2 = Me$; $R_3 = R_4 = H$), method I	18
Dithionaphthalene (DTN), **11**	25, 26, 27
Tetrathionaphthalene (TTN), **12**	28
Dithiotetracene (DTT), **13**	29
Tetrathiotetracene (TTT), **14**	30, 31
Dithiapyranylidene (DTP)	32

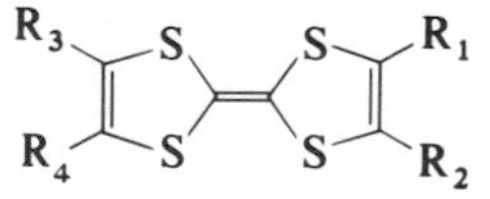

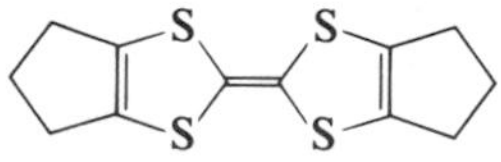

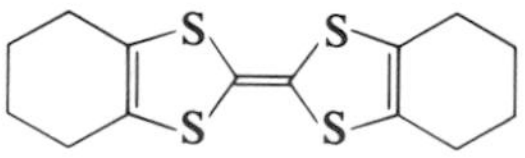

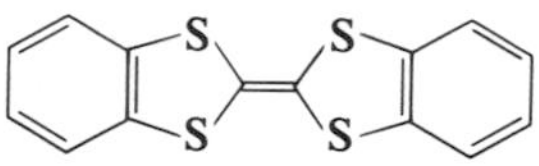

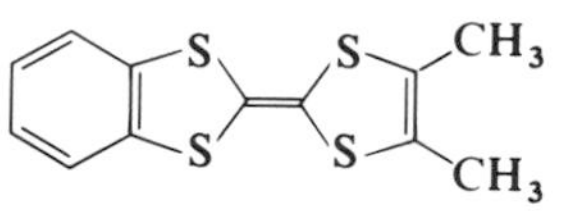

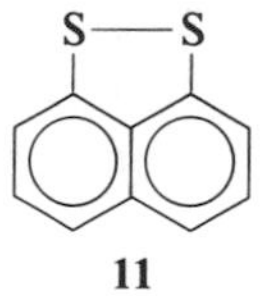

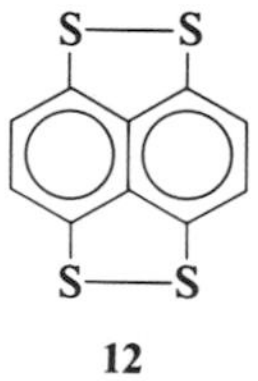

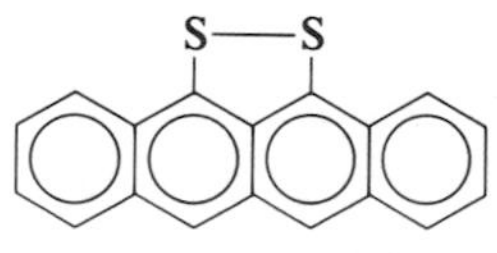

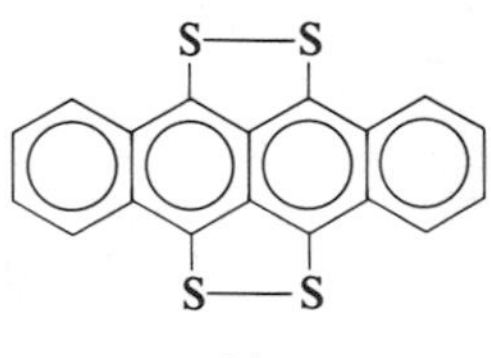

TABLE 3.

Compound	References
Tetraselenofulvalene (TSeF), **6**(S = Se)	14, 33, 34
Symmetrically substituted TSeF	
tetramethyl TSeF, **6**S = Se; $R_1 = R_2 = R_3 = R_4 =$ Me)	35
hexamethylene TSeF, **7**(S = Se)	36, 37
tetramethylseleno TSeF	38
Diselenonaphthalene (DSeN), **11**(S = Se)	27
Tetraselenotetracene (TSeT), **14**(S = Se)	31, 39

TABLE 4.

Compound	References
Dithiodiselenofulvalene (DSDSeF), **15**($R_1 = R_2 = R_3 = R_4 =$ H)	40
Symmetrically substituted DSDSeF	
dimethyl DSDSeF	41
hexamethylene DSDSeF	37, 42
octamethylene DSDSeF	43
Naphtho (1,8 cd)-1,2-selenathiole	27

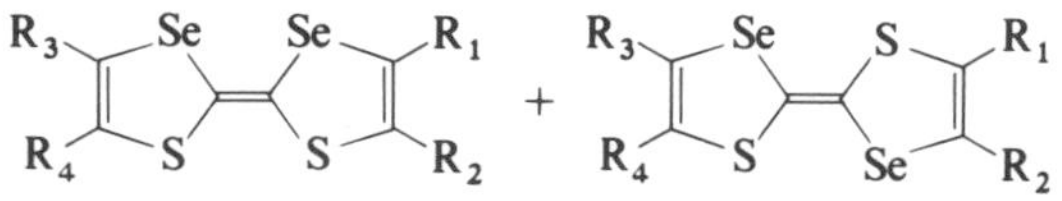

15

2.2. *Electron-Acceptor Molecules*

7,7,8,8-tetracyano-*p*-quinodimethane (TCNQ) (see Table 5) was first synthesized by Acker and Hertler by the condensation of 1,4-cyclohexanedione with malononitrile to give 1,4-bis(dicyanomethylene)-cyclohexane which can easily be oxidized with bromine or *N*-bromosuccinimide to give TCNQ.[45] This material has been shown to be a powerful electron-acceptor reacting with a wide variety of donor molecules. Substituted TCNQ can be prepared by starting with the corresponding p-xylene dihalide and treating it with sodium cyanide to give p-xylene dicyanide. Treatment of the dicyanide with sodium methoxide and cyanogen halide affords the tetracyanodiacetate which upon basic hydrolysis and acidification gives dihydro TCNQ. Oxidation of the dihydro TCNQ with bromine yields the corresponding substituted TCNQ.[45]

TABLE 5.

Compound	Reference
Tetracyanoethylene, $(NC)_2C = C(CN)_2$	46
7,7,8,8-tetracyano-*p*-quinodimethane(TCNQ), **16**($R_1 = R_2 = R_3 = R_4 = H$)	44
TCNQ-d_4, **16**($R_1 = R_2 = R_3 = R_4 = D$)	47
methyl TCNQ, **16**($R_1 = Me$; $R_2 = R_3 = R_4 = H$) MeTCNQ	48
dimethyl TCNQ, **16**($R_1 = R_4 = CH_3$; $R_2 = R_3 = H$)	48
substituted TCNQ, **16**($R_1 = R_3 = H$; $R_2 = Br$; $R_4 = Me$)	46
monofluoro TCNQ, **16**($R_1 = F$; $R_2 = R_3 = R_4 = H$)	49
difluoro TCNQ, **16**($R_1 = R_4 = F$; $R_2 = R_3 = H$)	50
11,11,12,12-tetracyano-2,6-naphtho-*p*-quinodimethane (TNAP), **17**	51
Hexacyanobutadiene $(NC)_2C[(C(CN)]_2C(CN)_2$	52
Hexacyanotrimethylenecyclopropanediide	53
Tetracyanoquinoquinazolinoquinazoline	54
Metal dithiolene	55

16 **17**

2.3. *Organic Charge-Transfer Complexes*

2.3.1. *Simple Salts*

The simple salts are defined as compounds in which the donor to acceptor stoichiometry is 1 : 1. These compounds (see Table 6) are typically prepared by adding hot equimolar solutions of the donor to TCNQ in hot, dry solvents such as acetonitrile, tetrahydrofuran, or dichloromethane (method I) or by metathesis when hot solutions of a donor salt is combined with TCNQ salts in solvents such as acetonitrile, acetone, or ethanol (method II). The later method relies on the donor and acceptor moiety combining in a ratio that reflects their charge. However, in order to evaluate their solid-state properties by conductivity measurements and structural determinations, these simple salts must be prepared in the form of single crystals. There exist numerous techniques[56] to obtain single crystals of the simple salts of which diffusion growth and electrochemical methods have been most commonly used.

TABLE 6.

Compound	Reference
K(TCNQ)	57
(NMP)(TCNQ), method I	6
(HMP)(TCNQ), **18**	58
(TTF)(TCNQ-d_8), method I	59
(TSeF)(TCNQ), method I	60
(TTF)(TCNQR_2)	43
(TTN)(TCNQ), method I	28
(TTT)(TCNQ), method II	61
(DTP)(TCNQ), method I	62
(TSeF)(TCNQR_2), method I	63

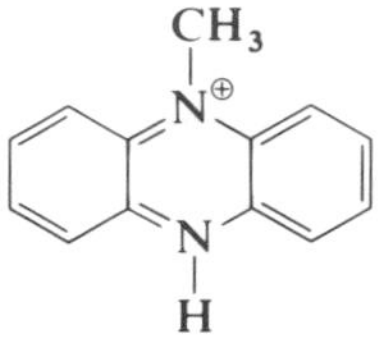

CH_3 N$^{\oplus}$ N H

18

2.3.2. Complex Salts

The complex salts are compounds in which the donor-to-acceptor stoichiometry is primarily 1 : 2, although 2 : 3 stoichiometries have also been observed. These materials (see Table 7) are typically prepared by the addition of neutral TCNQ to a solution of the simple salt (method I) or by the addition of hot solutions containing excess ethyl, aryl, or *N*-heterocyclic ammonium iodides to hot solutions of neutral TCNQ (method II). Another procedure (method III) involves the use of the acid, H_2(TCNQ), as both a proton and electron source and is particularly suitable for the preparation of trisubstituted ammonium complex salts starting from free amines.[57]

TABLE 7.

Compound	Reference
(triethylammonium)$(TCNQ)_2$, methods I, II, and III	57
(quinolinium)$(TCNQ)_2$, methods II and III	57
(acridinium)$(TCNQ)_2$, method II	64
(TTT)$(TCNQ)_2$	61
(DEPE)$(TCNQ)_4$, method II	65
$(TTF)_2(TCNQR_2)$, method I	66
$(NMe_3H)(I_3)_{1/3}$ (TCNQ)	70

TABLE 8.

Compound	Reference
$(TSeF)_x(TTF)_{1-x}(TCNQ)$	67
$(TTF)(TCNQ)_x(MeTCNQ)_{1-x}$	68
$(TSeF)(TCNQ)_x(MeTCNQ)_{1-x}$	69

Single crystals of the complex salts have been prepared principally by the slow cooling of saturated solutions, usually acetonitrile, of the complex salts.

2.3.3. Simple Alloy Salts

These materials (see Table 8) are simple salts in which a small fraction of the donor or acceptor molecules have been replaced by an isostructural donor or acceptor molecule. The overall donor to acceptor stoichiometry is 1 : 1. Single crystals are usually prepared by the slow cooling of hot saturated solutions of donor and acceptor in acetonitrile.

2.3.4. Complex Alloy Salt

The only example of such a substance, $(NMP^+)_x(Phen)_{1-x}(TCNQ)_x$-$(TCNQ)_{1-x}$ (Phen = phenazine)[71] is prepared by addition of phenazine and TCNQ to solutions containing the simple or complex salt.

2.3.5. Mixed-Valence Salts

Mixed valence charge-transfer salts (see Table 9) are systems in which the donor molecules have been oxidized by halogen or pseudohalogen acceptors. The oxidation state of the donors in these materials is nonintegral and, usually, $\rho < 1$. In systems where the donor species is TTF, the preparation is carried out by three different chemical methods: Method I—titration of TTF with solutions of halogen in acetonitrile and CCl_4;[72]

TABLE 9.

Compound	References
$(TTF)X_p$ (X = Cl, Br, BF_4)	13, 72–74, 76
$(DSeN)I_{1.82}$	77
$(TTT)I_{1.5}$	78, 79
$(TSeT)I_{0.5}$	80
$(TMSeF)X_{0.5}$ (X = PF_6^-, AsF_6^-, SbF_6^-, BF_4^-, NO_3^-)	81

method II—electrochemical oxidation of TTF to predetermined values of ρ followed by the addition of halide;[73] method III—photochemical oxidation of TTF in CCl_4 or CBr_4 solutions by uv light.[74] In systems where the donor species is TTT, the mixed-valence salts can also be prepared by co-sublimation of donor and acceptor species.[75]

3. *Inorganic Materials*

3.1. *Metal Dithiolenes*

Metal dithiolenes or dithiolates, (see Table 10) are most generally prepared by one of three synthetic routes. In the first route, the reaction of an acyloin or benzoin with phosphorus pentasulfide in refluxing xylene or dioxane, followed by the addition of an aqueous or alcoholic solution of the metal salt and heating yields the neutral dithiolene complexes.[82] The second procedure involves the reaction of the dianionic 1,2-dithiolates with the respective metal salts.[83] The third major synthetic route utilizes the reaction of low-valent metal carbonyl or phosphine compounds with bistrifuloromethyldithiete to give the dithiolene complexes.[84] Although the neutral material and a large number of complexes[85] are known, only a few of these complexes exhibit conductivity in the metallic regime. These highly conducting complexes are generally prepared by the direct combination of the neutral components in organic solvents and by metathesis reactions.[86]

3.2. *Partially Oxidized Metal-Chain Systems*

3.2.1. *Nonstoichiometric Anion Containing Tetracyanoplatinates*

Of all the highly conducting inorganic one-dimensional systems, $K_2Pt(CN)_4Br_{0.3} \cdot 3H_2O$ (KCP-Br) (see Table 11) has been exhaustively studied and indeed has provided an understanding of other low-

TABLE 10.

Compound	Reference
$TTT_{1.2}[Ni(S_2C_2H_2)_2]$ **19**(M = Ni; R = H)	87
$(TTF)_2[Ni(S_2C_2H_2)_2]$ **19**(M = Ni; R = H)	88
$NH_4[Pd(S_2C_2R_2)_2]$ **19**(M = Pd; R = CN)	89

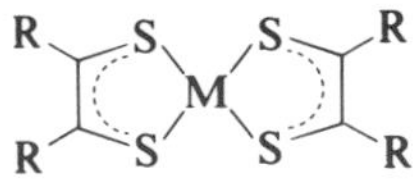

19

TABLE 11.

Compound	References
$K_2Pt(CN)_4Br_{0.3}\cdot 3H_2O$ (KCP–Br)	90–95
$K_2Pt(CN)_4Cl_{0.3}\cdot 3H_2O$ (KCP–Cl)	96
$K_2Pt(CN)_4Cl_{0.5}Br_{0.5}\,xH_2O$	97
$K_2Pt(CN)_4(FHF)_{0.3}\cdot 3H_2O$	96, 98
$Rb_2Pt(CN)_4Cl_{0.3}\cdot 3H_2O$	96, 98
$Rb_2Pt(CN)_4(N_3)_{0.25}\cdot xH_2O$	96
$Rb_2Pt(CN)_4(FHF)_{0.29}\cdot 1.7H_2O$	96, 98, 99
$Rb_2Pt(CN)_4(FHF)_{0.38}$	96, 98, 100–102
$Cs_2Pt(CN)_4Cl_{0.3}$	96, 98
$Cs_2Pt(CN)_4F_{0.19}$	96, 98
$Cs_2Pt(CN)_4(N_3)_{0.25}\cdot 0.5H_2O$	96
$Cs_2Pt(CN)_4(FHF)_{0.39}$	96, 98
$Cs_2Pt(CN)_4(FHF)_{0.23}$	96, 98, 104
$(NH_4)(H_3O)_{0.17}Pt(CN)_4Cl_{0.42}\cdot 2.8H_2O$	96
$[C(NH_2)_3]_2Pt(CN)_4Br_{0.25}\cdot H_2O$	103
$[C(NH_2)_3]_2Pt(CN)_4(FHF)_{0.26}\cdot xH_2O$	96, 98

dimensional systems. This material has been prepared by chemical[90] and electrochemical oxidation[91] of the tetracyanoplatinate (II) ion or by cocrystallizing the Pt(II) and Pt(IV) complexes.[92] Although the above procedures result in crystals of sufficient size, specific procedures exist for the preparation of large, high-purity, single crystals. These crystal-growth procedures are slow evaporation techniques,[93] saturated solutions containing growth modifiers,[94] and diffusion growth.[95] In any of the above techniques, great care must be taken to prevent the loss of water of crystallization due to improper humidity conditions or by improper heating. In order to assure high-quality crystals, ultrapure starting materials should be utilized in the preparation.

TABLE 12.

Compound	References
$K_{1.75}Pt(CN)_4\cdot 1.5H_2O$	91, 105, 106
$Rb_{1.75}Pt(CN)_4\cdot 2H_2O$	96, 98, 104
$Cs_{1.25}Pt(CN)_4\cdot xH_2O$	96, 98
$Rb_3Pt(CN)_4[(O_3SO)_2H]_{0.49}\cdot H_2O$	96, 98, 107
$K_2Pt(O_2C_2O_2)_2\cdot 2H_2O$	91, 108
$Ni(bqd)_2I_{0.5}$, bqd = oxidized 1,2-benzoquinodioxime	109, 110a
$Pd(hdpg)_2I_y$, hdpg = oxidized diphenylglyoxime, $y < 1.15$	110
$Ir(CO)_3X$, (X = Cl, Br, I)	111
$Ir(CO)_2Cl_2^{z-}$	107, 111b,c, 112

TABLE 13.

Compound	References
$Hg_{3-\delta}AsF_6$	113, 114
$Hg_{3-\delta}SbF_6$	114, 115

3.2.2. *Nonstoichiometric Cation Containing Tetracyanoplatinates*

Nonstoichiometric cation-containing substances, shown in Table 12, are prepared by methods similar to those discussed in Section 3.2.1.

3.3. *Polymercury Cation Chain Compounds*

$Hg_{3-\delta}AsF_6$, an extremely air- and moisture-sensitive material (see Table 13), is prepared by reacting mercury with a slight excess of arsenic pentafluoride in liquid sulfur dioxide[113] or by reacting with $Hg_3(AsF_6)_2$[114].

4. *Polymeric Materials*

4.1. *Poly(sulfurnitride) [Polythiazyl], $(SN)_x$*

$(SN)_x$ can be prepared[116] by the solid-state polymerization of S_2N_2 in the form of high-purity single crystals. The later, S_2N_2, is prepared by passing heated vapors of S_4N_4 over heated silver wool.[117] Epitaxial films[118] of $(SN)_x$ can also be prepared by the controlled sublimation of crystalling $(SN)_x$ onto an appropriate substrate. Polythiazyl halides[119–123] [e.g., $(SnBr_{0.4})_x$, Refs. 119, 120] can be prepared by the reaction of halogen vapors with analytically pure $(SN)_x$ or directly from S_4N_4 to give the highly conducting polythiazyl halides, $(SNX_y)_x$.

4.2. *Polyacetylene, $(CH)_x$*[124]

$(CH)_x$ free-standing films may be prepared by wetting the inside walls of a glass reactor vessel with a Ziegler catalyst[125] solution of triethylaluminum and tetra-*n*-butoxy titanium in toluene and then immediately admitting purified acetylene gas at a pressure of several centimeters to ca. 1 atm. The cohesive film grows on all surfaces that had been wetted by the catalyst solution. If the polymerization had been carried out at ca. $-78°C$, the film is formed almost completely as the *cis*-isomer. The *trans*-isomer can similarly be grown using a temperature of 150°C (decane solvent) to carry out polymerization or by thermal isomerization at 200°C

TABLE 14.

Compound	Reference
$(CHI_{0.28})_x$, method I	126
$[CH(H_2O)_{0.30}(HClO_4)_{0.13}]_x$, method II	127
$[CH(SO_3F)_{2y}]_x$, method II	128
$[CH(SbF_6)_{0.05}]_x$, method II	127
$[CH(ClO_4)_{0.065}]_x$, method IV	129

of the *cis*-isomer. When the film is carefully washed with pentane, analytically pure $(CH)_x$ is obtained. $(CH)_x$ can be doped to *p*- or *n*-type semiconductors or metals (see Table 14) by treating the silvery films with a known vapor pressure of volatile dopants such as iodine, arsenic pentafluoride, or perchloric acid (method I), with a solution of the dopant such as sodium naphthalide in THF or nitrosonium hexafluoroantimonate in a nitromethane/methylene chloride mixture (method II), with a liquid sodium/potassium alloy at room temperature (method III), or electrochemically (method IV) to give highly conducting golden-silvery films of $(CHX_y)_x$.

4.3. Polypyrroles

Polypyrrole films[130] can be synthesized galvanostatically on a platinum surface in a two-electrode cell containing 0.1 *M* tetraethylammonium tetrafluoroborate and 0.06 *M* pyrrole in 99% aqueous acetonitrile.

References

1. Menschutkin, B. N., *Z. Phys. Chem.* **6**, 45 (1890).
2. Nigrey, P. J., and Garito, A. F., *J. Chem. Eng. Data* **22**, 451 (1977).
3. Trowbridge, P. F., *J. Am. Chem. Soc.* **21**, 67 (1899).
4. La Coste, W., *Chem. Ber.* **15**, 192 (1882).
5. Claus, A., *Chem. Ber.* **8**, 600 (1875).
6. Melby, L. R., *Can. J. Chem.* **43**, 1448 (1965).
7. Graebe, C., and Caro, M., *Liebig's Ann. der Chemie* **158**, 265 (1871).
8. Freund, M., and Bode, G., *Chem. Ber.* **42**, 1746 (1904).
9. Bradsher, C. K., and Beavers, L. E., *J. Am. Chem. Soc.* **77**, 4812 (1955).
10. Rembaum, A., Hermann, A. M., Stewart, F. E., and Gutmann, F., *J. Phys. Chem.* **73**, 513 (1969).
11. Hünig, S., Quast, M., Brenninger, W., and Frankenfeld, E., *Org. Syn.* **49**, 107 (1969).
12. Narita, M., and Pittman, C. V., Jr., *Synthesis*, 489 (1976).
13. Wudl, F., Smith, G. M., and Hufnagel, E. J., *J. Chem. Soc. Chem. Commun.*, 1453 (1970); Wudl, F., and Kaplan, M. L., *Inorg. Synthesis* **19**, 27 (1979).
14. Melby, L. R., Harzler, H. D., and Sheppard, W. A., *J. Org. Chem.* **39**, 2456 (1974).
15. Dolphin, D., Pegg, W., and Wirz, P., *Can. J. Chem.* **52**, 4078 (1974).

16. Wudl, F., Kruger, A. A., Kaplan, M. L., and Hutton, R. S., *J. Org. Chem.* **42**, 768 (1977).
17. Ferraris, J. P., Poehler, T. O., Bloch, A. N., and Cowan, D. O., *Tetrahedron Lett.*, 2553 (1973).
18. Mas, A., Fabre, J.-M., Toreilles, E., Giral, L., and Brun, G., *Tetrahedron Lett.*, 2579 (1977).
19. Gemmer, R. V., Cowan, D. O., Poehler, T. O., Bloch, A. N., Pyle, R. E., and Banks, R. H., *J. Org. Chem.* **40**, 3544 (1975).
20. Spencer, H. K., Cava, M. P., Yamagishi, F. G., and Garito, A. F., *J. Org. Chem.* **41**, 730 (1976).
21. Miles, M. G., Wilson, J. D., Dahm, D. J., and Wagenknecht, J. H., *J. Chem. Soc. Chem. Commun.*, 751 (1974).
22. Yoneda, S., Kawase, T., Inabe, M., and Yoshida, Z., *J. Org. Chem.* **43**, 595 (1978).
23. Gonnella, N. C., and Cava, M. P., *J. Org. Chem.* **43**, 369 (1978).
24. Green, D. C., *J. Org. Chem.* **44**, 1476 (1979).
25. Lanfry, M., *C. R. Hebd. Seances Acad. Sci.* **152**, 92 (1911).
26. Price, W. B., and Smiles, S., *J. Chem. Soc.*, 2372 (1928).
27. Meinwald, J., Dauplaise, D., Wudl, F., and Hauser, J. J., *J. Am. Chem. Soc.* **99**, 255 (1977).
28. Wudl, F., Schafer, D. E., and Miller, B., *J. Am. Chem. Soc.* **98**, 252 (1976).
29. Nigrey, P. J., and Garito, A. F., *J. Chem. Eng. Data* **23**, 183 (1978).
30. Marschalk, C., and Stumm, C., *Bull. Soc. Chim. France,* 418 (1948).
31. Perez-Albuerne, E. A., U. S. Patent 3,723,417, March 27, 1973.
32. Sandman, D. J., Holmes, T. J., and Warner, D. E., *J. Org. Chem.* **44**, 880 (1979).
33. Engler, E. M., and Patel, V. V., *J. Am. Chem. Soc.* **96**, 7376 (1974).
34. Lakshmikantham, M. V., and Cava, M. P., *J. Org. Chem.* **41**, 882 (1976).
35. Bechgaard, K., Cowan, D. O., and Bloch, A. N., *J. Chem. Soc., Chem. Commun.*, 937 (1974).
36. Bloch, A. N., Cowan, D. O., Bechgaard, K., Pyle, R. E., Banks, R. H., and Poehler, T. O., *Phys. Rev. Lett.* **34**, 1561 (1975).
37. Shu, P., Bloch, A. N., Carruthers, T. F., and Cowan, D. O., *J. Chem. Soc. Chem. Commun.*, 505 (1977).
38. Engler, E. M., and Green, D. C., *J. Chem. Soc., Chem. Commun.*, 148 (1976).
39. Goodings, E. P., Mitchard, D. A., and Owen, G., *J. Chem. Soc. Perkin Trans. I,* 1310 (1972).
40. Engler, E. M., and Patel, V. V., *J. Chem. Soc. Chem. Commun.,* 671 (1975).
41. Engler, E. M., and Patel, V. V., *Tetrahedron Lett.*, 423 (1976).
42. Spencer, H. K., Lakshmikantham, M. V., Cava, M. P., and Garito, A. F., *J. Chem. Soc. Chem. Commun.*, 807 (1975).
43. Wheland, R. C., and Gillson, J. L., *J. Am. Chem. Soc.* **98**, 3916 (1976).
44. Acker, D. S., and Hertler, W. R., *J. Am. Chem. Soc.* **84**, 3370 (1962).
45. Wheland, R. C., and Martin, E. L., *J. Org. Chem.* **40**, 3101 (1975).
46. Cairns, T. L., Carbon, R. A., Coffman, D. D., Engelhardt, V. A., Heckert, R. E., Little, E. L., McGeer, E. G., McKusuck, B. C., Middleton, W. J., Scribner, R. M., Theobald, C. W., and Winberg, H. E., *J. Am. Chem. Soc.* **80**, 2775 (1958).
47. Lunelli, B., and Pecile, C., *Gazz. Chim. Ital.* **99**, 496 (1969).
48. Diekmann, J., Hertler, W. R., and Benson, R. E., *J. Org. Chem.* **28**, 2719 (1963).
49. Ferraris, J. P., and Saito, G., *J. Chem. Soc. Chem. Commun.*, 992 (1978).
50. Saito, G., and Ferraris, J. P., *J. Chem. Soc. Chem. Commun.*, 1027 (1979).
51. Sandman, D. J., and Garito, A. F., *J. Org. Chem.* **39**, 1165 (1974).
52. Webster, O. W., *J. Am. Chem. Soc.* **86**, 2898 (1964).
53. Fukunaga, T., *J. Am. Chem. Soc.* **98**, 610 (1976).
54. Wudl, F., Kaplan, M. L., Teo, B. K., and Marshall, J., *J. Org. Chem.* **42**, 1666 (1977).

55. (a) Alcacer, L., in *Extended Linear Chain Compounds*, Vol. 3, J. Miller, Ed., Plenum Press, New York (1982), (b) Burns, R. P., and Mcauliffe, C. A., in *Advances in Inorganic Chemistry and Radiochemistry*, Vol. 22, H. J. Emeléus and A. G. Sharpe, Eds., Academic Press, New York (1979), pp. 303–348.
56. Andersen, J. R., and Engler, E. M., *Ann. N. Y. Acad. Sci.* **313**, 293 (1978).
57. Melby, L. R., Harder, R. J., Hertler, W. R., Mahler, W., Benson, R. E., and Mochel, W. E., *J. Am. Chem. Soc.* **84**, 3374 (1962).
58. Soos, Z. G., Keller, H. J., Moron, W., and Noethe, D., *J. Am. Chem. Soc.* **99**, 5040 (1977).
59. Nigrey, P. J., *J. Cryst. Growth* **40**, 265 (1977).
60. Engler, E. M., Scott, B. A., Etemad, S., Penney, T., and Patel, V. V., *J. Am. Chem. Soc.* **99**, 5909 (1977).
61. Delhaés, P., Flandrois, S., Keryer, G., and Dupuis, P., *Mater. Res. Bull.* **10**, 825 (1975).
62. Darocha, B. F., Titus, D. D., and Sandman, D. J., *Acta Crystallogr. B.* **35**, 2445 (1979).
63. Andersen, J. R., Craven, R. A., Weidenborner, J. E., and Engler, E. M., *J. Chem. Soc. Chem. Commun.*, 526 (1977).
64. Ehrenfreund, E., and Nigrey, P. J., *Phys. Rev. B* **21**, 48 (1980).
65. Ashwell, G. J., Eley, D. D., Wallwork, S. C., and Willis, M. R., *Proc. R. Soc. Lond. A* **343**, 461 (1975).
66. Wheland, R. C., *J. Am. Chem. Soc.* **99**, 291 (1977).
67. Engler, E. M., Scott, B. A., Etemad, S., Penney, T. and Patel, V. V., *J. Am. Chem. Soc.* **99**, 5909 (1977).
68. Tomkiewicz, Y., Craven, R. A., Schultz, T. D., Engler, E. M., and Taranko, A. R., *Phys. Rev. B* **15**, 1011 (1977).
69. Engler, E. M., Craven, R. A., Tomkiewicz, Y., Scott, B. A., Bechgaard, K., and Andersen, J. R., *J. Chem. Soc. Chem. Commun.*, 337 (1976).
70. Abkowitz, M. A., Epstein, A. J., Griffiths, C. H., Miller, J. S., and Slade, M. L., *J. Am. Chem. Soc.* **99**, 5304 (1977).
71. Miller, J. S., and Epstein, A. J., *J. Am. Chem. Soc.* **100**, 1639 (1978).
72. Scott, B. A., LaPlaca, S. J., Torrance, J. B., Silverman, B. D., and Welber, B., *J. Am. Chem. Soc.* **99**, 6631 (1977).
73. Kaufman, F. B., Engler, E. M., Green, D. C., and Chambers, J. Q., *J. Am. Chem. Soc.* **98**, 1596 (1976).
74. Scott, B. A., Kaufman, F. B., and Engler, E. M., *J. Am. Chem. Soc.* **98**, 4342 (1976).
75. Hilti, B., and Mayer, C. W., *Helv. Chim. Acta* **61**, 501, 1462 (1978).
76. Wudl, F., *J. Am. Chem. Soc.* **97**, 1962 (1975).
77. Dauplaise, D., Meinwald, J., Scott, J. C., Temkin, H., and Clardy, J., *Ann. N. Y. Acad. Sci.* **313**, 382 (1978).
78. Kaminskii, V. F., Khidekel, M. L., Lyubovskii, R. B., Shchegolev, I. F., Shibaeva, R. P., Yagubskii, E. B., Zvarykina, A. V., and Zuereva, G. L., *Phys. Status Solidi A* **44**, 77 (1977).
79. Isett, L. C., and Perez-Albuerne, E. A., *Solid State Commun.*, **21**, 433 (1977).
80. Delhaes, P., Coulon, C., Manceau, J. P., Flandrois, S., Hilti, B., and Mayer, C. W., in *Molecular Metals*, Vol. VI: 1, W. E. Hatfield, Ed., Plenum Press, New York (1978), pp. 59–66.
81. Bechgaard, K., Jacobsen, C. S., Mortensen, K., Pedersen, H. J., and Thorup, N., *Solid State Commun.* **33**, 1119 (1980).
82. Van der Put, P. J., and Schilperoord, A. A., *Inorg. Chem.* **13**, 2476 (1974).
83. Gray, H. B., Williams, R., Bernaland, I., and Billig, E., *J. Am. Chem. Soc.* **84**, 4756 (1962).
84. King, R. B., *Inorg. Chem.* **2**, 641 (1963).
85. Miller, J. S., and Epstein, A. J., in *Progress in Inorganic Chemistry*, Vol. 20, S. J. Lippard, Ed., John Wiley, New York (1976) pp. 1–151.
86. Interrante, L. V., Bray, J. W., Hart, H. R., Jr., and Watkins, G. D., *Ann. N. Y. Acad. Sci.* **313**, 407 (1978).

87. Interrante, L. V., Bray, J. W., Hart, H. R., Jr., Kasper, J. S., Piacente, P. A., and Watkins, G. D., *J. Am. Chem. Soc.* **99**, 3523 (1977).
88. Interrante, L. V., Browall, K. W., Hart, H. M., Jr., Jacobs, I. S., Watkins, G. D., and Wee, S. H., *J. Am. Chem. Soc.* **97**, 889 (1975).
89. Perez-Albuerne, E. A., Isett, L. C., and Haller, R. K., *J. Chem. Soc. Chem. Commun.*, 417 (1977).
90. Krogmann, K., and Hauser, H.-D., *Z. Anorg. Allg. Chem.* **358**, 67 (1968); Abys, J. A., Enright, N. P., Gerdes, H. M., Hall, T. L., and Williams, J. M., *Inorg. Synthesis* **19**, 1 (1979).
91. Miller, J. S., *Science* **194**, 189 (1976); Miller, J. S., *Inorg. Synthesis,* **19**, 13 (1979).
92. Bernasconi, J., Breusch, P., Kuse, D., and Zeller, H. R., *J. Phys. Chem. Solids* **35**, 145 (1974).
93. Zeller, H. R., and Beck, A., *J. Phys. Chem. Solids* **35**, 77 (1974).
94. Jaklevic, R. C., and Bedford, C. D., *Mater. Res. Bull.* **9**, 289 (1974).
95. Guggenheim, H. J., and Bahnck, D., *J. Cryst. Growth* **26**, 29 (1974).
96. Williams, J. M., Schultz, A. J., Underhill, A. E., and Carnerio, K., in *Extended Linear Chain Compounds*, Vol. 1, J. S. Miller, Ed., Plenum Press, New York (1982).
97. Miller, J. S., and Weagly, R. J., *Inorg. Chem.* **16**, 2965 (1977).
98. Williams, J. M., and Schultz, A. J., *NATO Conf. Ser. VI* **1**, 337 (1980).
99. Stearly, K. L., and Williams, J. M., *Inorg. Syn.* **20**, 24 (1980).
100. Williams, J. M., and Schultz, A. J., *Ann. N. Y. Acad. Sci.* **313**, 509 (1978).
101. Williams, J. M., Gerrity, D. P., and Schultz, A. J., *J. Am. Chem. Soc.* **99**, 1668 (1977).
102. Coffey, C., and Williams, J. M., *Inorg. Syn* **20**, 25 (1980).
103. Cornish, T. F., and Williams, J. M., *Inorg. Syn* **19**, 10 (1979).
104. Koch, T. R., Abys, S. A., and Williams, J. M., *Inorg. Syn.* **19**, 9 (1979).
105. Levy, L. A., *J. Chem. Soc.,* 1081 (1912); Terrey, H., *J. Chem. Soc.,* 202 (1928).
106. Koch, T. R., Enright, N. P., and Williams, J. M., *Inorg. Syn.* **19**, 8 (1979).
107. Lee, G. C., Stearley, K. L., and Williams, J. M., *Inorg. Syn.* **20**, 29 (1980).
108. Underhill, A. E., Watkins, A. E., Williams, J. M., and Carnerio, K., in *Extended Linear Chain Compounds*, Vol. 1, J. S. Miller, Ed., Plenum Press, New York (1982).
109. (a) Megnamisi-Belombe, M., *Ann. N. Y. Acad. Sci.* **313**, 633 (1978); (b) Miller, J. S., and Griffiths, C. M., *J. Am. Chem. Soc.* **99**, 749 (1977).
110. (a) Peterson, J. L., Schramm, C. S., Stojakovic, D. R., Hoffman, B. M., and Marks, T. J., *J. Am. Chem. Soc.* **99**, 286 (1977); (b) Marks, T. J., and Kalinck, D. W., in *Extended Linear Chain Compounds*, Vol. 1, Plenum Press, J. S. Miller, Ed., New York (1982).
111. (a) Fischer, E. O., and Brenner, K. S., *Z. Naturforsch.* B 17, 774 (1962); (b) Reis, Jr., A. H., in *Extended Linear Chain Compounds*, Vol. 1, J. S. Miller, Ed. Plenum Press, New York (1982); (c) Ginsberg, A. P., Koepke, J. W., and Sprinkle, C. R., *Inorg. Syn.* **19**, 18 (1979); (d) Reis, Jr., A. H., Hagley, V. S., and Peterson, S. W., *J. Am. Chem. Soc.* **99**, 4184 (1977).
112. Ginsberg, A. P., Koepke, J. W., Hauser, J. J., West, K. W., DiSalvo, F. J., Sprinkle, C. R., and Cohen, R. L., *Inorg. Chem.* **15**, 514 (1976).
113. Miro, N. D., MacDiarmid, A. G., Heeger, A. J., Garito, A. F., Chiang, C. K., Schultz, A. J., and Williams, J. E., *J. Inorg. Nucl. Chem.* **40**, 1351 (1978).
114. Cutforth, B. D., and Gillespie, R. J., *Inorg. Syn.* **19**, 22 (1979).
115. Cutforth, B. D., Datars, W. R., van Schyndel, A., and Gillespie, R. J., *Solid State Commun.* **21**, 377 (1977).
116. Labes, M. M., Love, P., and Nichols, L. F., *Chem. Rev.* **79**, 1 (1979).
117. Mikulski, C. M., Russo, P. J., Saran, M. S., MacDiarmid, A. G., Garito, A. F., and Heeger, A. J., *J. Am. Chem. Soc.* **97**, 6385 (1975).
118. Bright, A. A., Cohen, M. J., Garito, A. F., Heeger, A. J., Mikulski, C. M., and MacDiarmid, A. G., *Appl. Phys. Lett.* **26**, 612 (1975).

119. Ahktar, M., Chiang, C. K., Heeger, A. J., Milliken, J., and MacDiarmid, A. G., *Inorg. Chem.* **17**, 1539 (1978).
120. Chiang, C. K., Cohen, M. J., Peebles, D. L., Heeger, A. J., Akhtar, M., Kleppinger, J., MacDiarmid, A. G., Milliken, J., and Moran, M. J., *Solid State Commun.* **23**, 607 (1977).
121. Street, G. B., Bingham, R. L., Crowley, J. I., and Kuyper, J., *J. Chem. Soc. Chem. Commun.*, 464 (1977).
122. Akhtar, M., Chiang, C. K., Heeger, A. J., and MacDiarmid, A. G., *J. Chem. Soc. Chem. Commun.*, 846 (1977).
123. Milliken, J., Ph.D. Thesis, University of Pennsylvania, Philadelphia, Pennsylvania, 1980.
124. MacDiarmid, A. G., and Heeger, A. J., *Syn. Metals* **1**, 101 (1979/80). in press.
125. Ito, T., Shirakawa, H., and Ikeda, S., *J. Polym. Sci. Polym. Chem. Ed.* **12**, 11 (1974).
126. Chiang, C. K., Druy, M. A., Gau, S. C., Heeger, A. J., Louis, E. J., MacDiarmid, A. G., Park, Y. W., and Shirakawa, H., *J. Am. Chem. Soc.* **100**, 1013 (1978).
127. Gau, S. C., Milliken, J., Pron, A., MacDiarmid, A. G., and Heeger, A. J., *J. Chem. Soc. Chem. Commun.*, 663 (1979).
128. Anderson, L. R., Pez, G. P., and Hsu, S. L., *J. Chem. Soc. Chem. Commun.*, 1066 (1978).
129. Nigrey, P. J., Heeger, A. J., and MacDiarmid, A. G., *J. Chem. Soc. Chem. Commun.*, 594 (1979).
130. Diaz, A. F., Kanazawa, K. K., and Gardini, G. P., *J. Chem. Soc. Chem. Commun.*, 635 (1979); Diaz, A. F., and Kanazawa, K. K., this volume.
131. Michalczyk, M. J., Vidusek, D. A., and Williams, J. M., *Inorg. Syn.* **20**, 26 (1980).
132. Gerrety, D. P., and Williams, J. M., *Inorg. Syn.* **20**, 28 (1980).
133. Cleave, M. J., and Griffith, W. P., *J. Chem. Soc. A*, 2788 (1970).
134. Besinger, R., Vidusek, D. A., Gerrity, D. P., and Williams, J. M., *Inorg. Syn.* **20**, 20 (1980).

10

Structural, Magnetic, and Charge-Transport Properties of Stacked Metal Chelate Complexes

Brian M. Hoffman, Jens Martinsen, Laurel J. Pace, and James A. Ibers

1. Introduction

Intense effort devoted to the synthesis and physical characterization of crystals containing stacked, planar, transition-metal chelate complexes has yielded many materials of widely divergent properties. Some are primarily of structural interest. Others have unusual magnetic properties; for example, the first well-documented instance of a spin Peierls transition (a solid state analogue of the Jahn–Teller effect) was observed in such a material.[1] Yet others exhibit high or unusual electrical conductivities, and the greatest effort is currently being expended toward the understanding and control of charge-transport properties in molecular crystals. Here compounds prepared from metallomacrocycles are of particular interest and challenge. For example, with phthalocyaninatonickel iodide, Ni(pc)I,† a high, metallike conductivity along the stacking axis is achieved,[2,3] whereas bis(diphenylglyoximato)nickel iodide, Ni(dpg)I,[4] has the same structural motif but exhibits much lower conductivity. To understand and control such divergent physical properties in related solid state structures has engendered considerable research, particularly since the possibility exists of tailor-making compounds with specific transport properties through well-defined chemical modifications.

† Abbreviations for ligands and compounds will be found in Table 1. Table 1 also contains a list of symbols used in ensuing tables.

Brian M. Hoffman, Jens Martinsen, Laurel J. Pace, and James A. Ibers • Department of Chemistry and Materials Research Center, Northwestern University, Evanston, Illinois 60201.

TABLE 1. Abbreviations

a. Chemical abbreviations[a]

aona	*N*-Isopropyl-2-oxy-1-naphthylidenaminato
BA	Bromanil
bipy	2,2′-Bipyridyl
bqd	1,2-Benzoquinonedioximato
BTF	Benzotrifuroxan
CA	Chloranil
dbtaa	Dibenzo[b,i]-1,4,8,11-tetraaza[14]annulenato
DDQ	2,3-Dichloro-5,6-dicyano-1,4-benzoquinone
dhg	1,2-Dihydroglyoximato
disn	Diiminosuccinonitrilo
dmg	1,2-Dimethylglyoximato
DNBF	Dinitrobenzofuroxan
dpg	1,2-Diphenylglyoximato
dta	Dithioaceto
dtc	*N*,*N*-Di-*n*-butylthiocarbamato
edt	Ethylene-1,2-dithiolato
en	Ethylenediamine
etio	2,7,12,17-Tetramethyl-3,8,13,18-tetraethylporphyrinato
HMTSF	Hexamethylenetetraselenafulvalene
KCP(Br)	$K_2[Pt(CN)_4]Br_{0.3} \cdot 3H_2O$
IA	Iodanil
meg	1-Methyl-2-ethylglyoximato
mnt	Maleonitriledithiolato
NMP^+	*N*-Methylphenazinium
nmsim	Bis-*N*-methylsalicylaldiminato
ntbsim	*N*-*t*-Butylsalicylideneiminato
oep	2,3,7,8,12,13,17,18-Octaethylporphyrinato
omtbp	1,4,5,8,9,12,13,16-Octamethyltetrabenzporphyrinato
opd	*o*-Phenylenediiminato
ox	Oxalato
PA	Picryl azide
pc	Phthalocyaninato
POZ	Phenoxazine
$pQBr_2Cl_2$	2,3-Dibromo-5,6-dichloro-1,4-benzoquinone
$pQBr_3Cl$	2,3,5-Tribromo-6-chloro-1,4-benzoquinone
$pQ(N_3)_2Cl_2$	2,5-Diazido-3,6-dichloro-1,4-benzoquinone
$pQ(N_3)_2Br_2$	2,5-Diazido-3,6-dibromo-1,4-benzoquinone
ptt	Propene-3-thione-1-thiolato
PTZ	Phenothiazine
qnl	8-Hydroxyquinolinato
salphen	*N*,*N*′-1,2-Phenylene-bis(salicylaldiminato)
taab	Tetrabenz[b,f,j,n][1,5,9,13]tetraazacyclohexadecine
tatbp	Triazatetrabenzporphyrinato
tatma	1,4,8,11-Tetraaza-5,7,12,14-tetramethyl[14]annulenato
tbp	Tetrabenzporphyrinato
TCB	Tetracyanobenzene
tcdt	1,2,3,4-Tetrachlorobenzene-5,6-dithiolato
TCNQ	7,7,8,8-Tetracyano-*p*-quinodimethane
tfd	*Cis*-1,2-bis(trifluoromethylethylene)-1,2-dithiolato
TFZ	Benzotrifurazan
tmdbtaa	*o*-Tetramethyldibenzo[b,i]-1,4,8,11-tetraaza[14]annulenato
tmp	5,10,15,20-Tetramethylporphyrinato
TMPD	*N*,*N*,*N*′,*N*′-Tetramethyl-*p*-phenylenediamine

(*continued*)

TABLE 1 (continued)

TMTTF	Tetramethyltetrathiafulvalene
TNB	1,3,5-Trinitrobenzene
TNF	2,4,7-Trinitro-9-fluorenone
tpp	5,10,15,20-Tetraphenylporphyrinato
TRNF	2,4,5,7-Tetranitro-9-fluorenone
TRP^+	Tropylium
TTF	Tetrathiafulvalene
TTT	Tetrathiatetacene

b. Table abbreviations

A	Acceptor
AF	Antiferromagnet
AR	Absorption peak reference
BF	Bonner–Fisher
C	Conductivity reference
CL	Curie law temperature dependence
CWL	Curie–Weiss law temperature dependence
d( )	Distances in Å
D	Donor
DC	Dichroism
DI	Dimer
HAF	Heisenberg antiferromagnet
J	Exchange constant
M	Magnetism reference
MM	Metal-to-metal
MOM	Metal-over-metal stacking arrangement
MT	Magnetic transition
NP	Not parallel; angle away from parallel follows
P	Powder results
PP	Plane-to-plane
PS	Pauli susceptibility; nearly temperature independent
S	Structure reference
SC	Single-crystal results
sem	Semiconductor
SO	Symmetric overlap
SP	Spin-Peierls distortion
SS	Slipped stack
ST	Singlet–triplet temperature dependence
TAP	Temperature-activated paramagnetism
Type A	Metallic conductivity ($\sigma \propto T^{-\alpha}$) below room temperature, reaching a broad maximum and then falling off in an activated manner
Type B	Metallic conductivity ($\sigma \propto T^{-\alpha}$) below room temperature, then undergoes a sharp and discontinuous transition to a semiconducting or insulating state
T_m	Temperature at which type-A conductivity reaches a maximum
T_{MI}	Temperature of metal insulator transition for type-B conductivity
U	Uniform stacks
UO	Unsymmetric overlap
YM	Young's modulus reference
Γ	EPR peak-to-peak linewidth
Δ	Activation energy for semiconductor conductivity
σ	Conductivity
χ	Magnetic susceptibility
χ^P	Paramagnetic susceptibility

[a] We have adopted the convention of using lower-case letters for ligands and upper-case letters for molecules.

Despite the many excellent descriptions of stacked molecular solids,[5–14] it has been roughly five years since the last comprehensive review of systems based on metal complexes.[5] This review attempts to document all such materials comprehensively within the few well-defined limitations forced on us by the wealth of information now available in certain areas. We have chosen to ignore the ionic platinum systems (e.g., tetracyanoplatinates, oxalates, and squarates), which are amply reviewed elsewhere.[6–10] Instead we limit the discussion to crystals that are generally comprised of two components, one of which may be viewed as an electron donor, D, or its cation, D^{+p}, the other an acceptor, A or anion, A^{-q}. At least one component must be a planar transition-metal chelate complex; the other may be another complex, or a purely organic molecule or ion. Materials that have not been isolated as single crystals and subjected to an x-ray crystal structure determination are included only when it is illustrative to do so.

In bringing together information on such divergent materials, we have found it is useful to employ a variant of a classification scheme for organic donor–acceptor complexes.[15, 16] This is illustrated both by Figure 1 and by the headings in this chapter. In the crystals discussed here, either one or both components tend to form stacks, and the first major division is structural: crystals either contain stacks that integrate the D and A units, or the D and A units can be segregated with D or A stacking, or both. We define a stack of transition-metal complexes as having parallel units with a metal–metal separation of less than 5 Å and a stack of organic molecules as having an interplanar separation of less than 4 Å. The second major division is physical. In an integrated stack crystal the metal–ligand complex is either neutral or ionic; in a segregated stack crystal it either has an integral

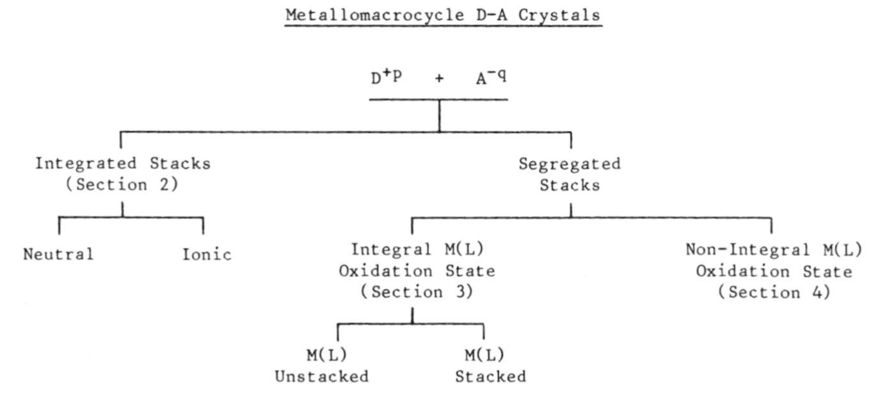

FIGURE 1. Scheme for classification of materials prepared from a donor or donor cation, D^{+p}, and acceptor or acceptor cation, A^{-q}, (p, $q = 0, 1, \ldots$) with one or both components being a planar transition-metal complex.

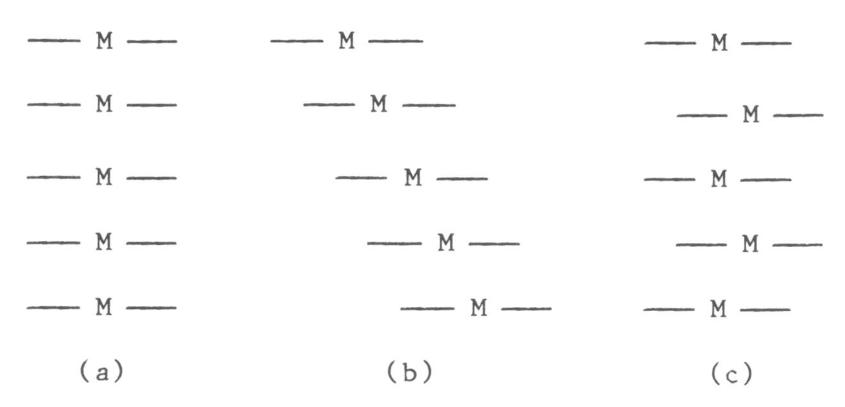

FIGURE 2. Some structural motifs for the parallel stacking of planar metallomacrocycles. Motif (a), which is the most important, is often called the "metal-over-metal" arrangement. Motif (b) is often called the "slipped-stack" arrangement. Many other motifs are possible, (c) being but one example.

oxidation state or a nonintegral oxidation state. The nonintegral oxidation state is sometimes referred to as "partial oxidation," "mixed valence," or "incomplete charge transfer," and although this state has generally been effected with oxidizing agents, in principle it could equally well be achieved by partial reduction. The classification scheme is convenient for it allows us to group together diverse materials that show similar properties. But it is not unambiguous, as the distinction between integrated and segregated stacks can be arbitrary.[17] Moreover, the many variations in stacking arrangements are not addressed in this classification scheme. To take the single example of the parallel stacking of planar metallomacrocycles, a number of structural motifs are possible (Figure 2). By far the most important is the "metal-over-metal" arrangement illustrated in Figure 2a.

The classification scheme is particularly effective in arranging crystals by their charge transport properties. Integrated stack crystals (Section 2) and integral oxidation state segregated stack crystals (Section 3) are invariably semiconductors, with typical room temperature conductivities being $\sigma \lesssim 10^{-3}\ \Omega^{-1}\ \text{cm}^{-1}$. In contrast, many of the nonintegral oxidation state, segregated stack crystals (Section 4) have high room temperature conductivities, $\sigma \sim 20\text{–}500\ \Omega^{-1}\ \text{cm}^{-1}$, which increase with decreasing temperature in a metallike fashion.

Why is this last class of compounds unique in displaying such effective charge transport? It appears that a molecular crystal based on a metal complex (or an organic molecule) can be highly conducting only if it meets two criteria. The first is that the metal–ligand molecules normally must be in crystallographically similar environments and be planar so that they can array in close communication. A crystal satisfying this criterion will exhibit the intermolecular interactions necessary to generate a conductive pathway. However, all of the categories in Figure 1 potentially may be comprised of crystals containing stacked, planar metal complexes. It appears that a second criterion, that of a nonintegral oxidation state, must be satisfied if effective charge transport is to occur.[16,18,19] In order to appreciate the importance of a nonintegral oxidation state, first consider the conduction process in a one-dimensional array of closed-shell molecules. In order to transport charge, it is necessary first to separate positive and negative charges along a chain, and this requires a large activation energy:

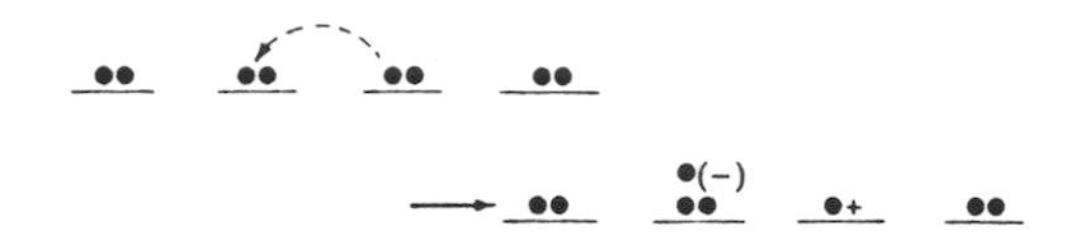

An equivalent picture is obtained from a simple band-theory argument. Such a crystal is predicted to be a semiconductor (or insulator), for it has a filled valence band and carrier generation would require excitation across a band gap.

However, an apparent discrepancy arises when one considers a crystal prepared from open-shell molecules. In the simplest band theory of a regular one-dimensional chain of atoms or molecules, occurrence of an odd number of electrons per site would give rise to a partially filled band and metallic conductivity.[5] Nevertheless, crystals of organic free radicals and of open-shell transition-metal complexes are for the most part no better conductors than are their closed-shell counterparts,[20] and the reason is the same. In order to transport charge in a crystal with one odd electron per site, it is necessary, at least transiently, to perturb the equilibrium electron distribution and create states with, say, two electrons on one site, no electrons on another:

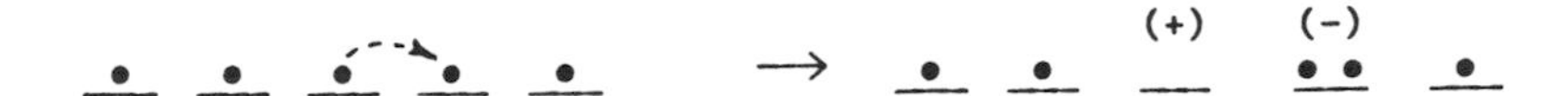

We accept the view that this charge separation process, whose Coulombic repulsion energy is commonly called U, is almost always energetically unfavorable for the low density of "conduction" electrons in a molecular crystal.[6,16,18,19] In short, a molecular crystal is effectively on the low-density side of a Mott transition and the electrons "crystallize" out on lattice sites and are unable to transport charge.[21]

But what happens if the molecules are in a nonintegral oxidation state? In such crystals there is no need to change the number of sites with zero, one, or two electrons during charge transport. For example, with closed-shell molecules that have been partially oxidized we might schematically envision charge transport by a limiting process in which preformed "holes" move between isoenergetic configurations:

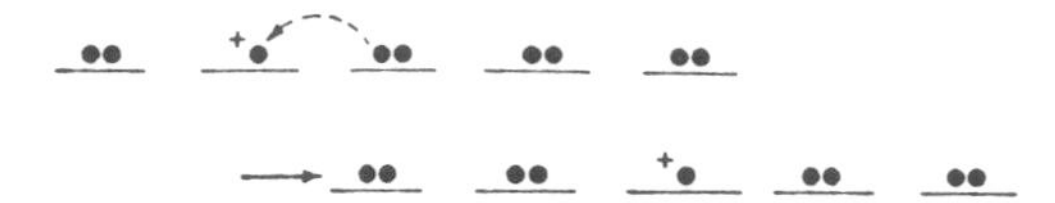

The on-site repulsion energy does not change in this process and thus a large value of U is not incompatible with good conductivity.

This picture gives a useful representation of charge transport in the "atomic limit," where correlation effects dominate and there is a finite matrix element for electron transfer (t), but a large value of U, such that the ratio $4t/U \rightarrow 0$.[22, 23] The best molecular conductors appear to be representatives of the complicated, mathematically challenging regime in which delocalization and electron-correlation effects are of comparable magnitude: $4t/U \sim 1$.[11–14] Nevertheless, all the highly conductive transition-metal complexes are of the partially oxidized class, and evidence to date indicates that a nonintegral oxidation state is necessary if a molecular crystal is to conduct well.

However, for a crystal to contain regular stacks of metallomacrocycles in a nonintegral oxidation state is apparently not sufficient to ensure high conductivity. Various bis(glyoximato) complexes appear to possess the necessary requirements of structure and valency, and yet are semiconductors with very modest room temperature conductivity (Section 4). We return to this conundrum.

Consideration of the results collected in this review will make it apparent that chemists recently have developed entirely new classes of materials, composed in part of metallomacrocycles, whose physical properties potentially may be controlled by chemical modification. In some cases these physical properties, for example, metallike conductivity in the stacking direction, rival those shown by the "organic metals."[6, 8, 11] The field of stacked metallomacrocycles is an exceedingly active one. We anticipate that as the relation of physical to chemical properties becomes better understood, new inexpensive materials with commercially useful physical properties will be synthesized.

2. *Integrated Stack Crystals*

2.1. *General Features*

Many compounds exhibit integrated stacks of planar transition-metal complexes and planar π organics. These systems may be classified either as neutral charge-transfer crystals[24–44] (Tables 2 and 3) or as ionic charge-

TABLE 2. Structure and Properties of Neutral 1:1 Charge-Transfer Compounds

Compound	Structure; description of stacking	Optical charge-transfer transition (cm^{-1})	Conductivity (Ω^{-1} cm^{-1})	References
[Ni(tmp)][TCNQ]	DA (U) along b; UO d(PP) = 3.43	9100 11100	SC: $\sigma < 10^{-5}$	44
[Pd(bqd)$_2$][TCNQ]	DA (U) along c; UO d(DA) = 3.4			35, 36
[Ni(bqd)$_2$][TCNQ]	DA (U) along c; UO d(DA) = 3.4			35, 36
[Pt(ptt)$_2$][TCNQ]	DA (U) along a; UO d(DA) = 3.42			37
[perylene][Ni(tfd)$_2$]	DA (U) along c; UO Angle of DA to $c = 32°$	4000	SC: $\sigma_c = 2.6 \times 10^{-3}$ $\sigma_{c \times a*} = 6.5 \times 10^{-5}$ $\sigma_{a*} = 1.4 \times 10^{-5}$ $\Delta_{c \times a*} = 0.32$ eV $\Delta_{a*} = 0.41$ eV P: $\sigma = 6.0 \times 10^{-6}$ (sem)	33
[perylene][Pt(tfd)$_2$]			P: $\sigma = 5 \times 10^{-9}$ (sem)	34
[pyrene][Ni(tfd)$_2$]	DA (U) along c; SO Small angle to c axis d(PP) = 3.54	8500 doublet	SC: $\sigma_c = 0.09$ P: $\sigma = 9.9 \times 10^{-9}$ (sem) $\Delta = 0.32$ eV	33
[pyrene][Pt(tfd)$_2$]			P: $\sigma = 2 \times 10^{-6}$ (sem)	34
[Cu(ntbsim)$_2$][TNB]	DA (pairs) along c d(DA) = 3.35 Cu(ntbsim)$_2$ nonplanar			39
[Ni(ntbsim)$_2$][TNB]	DA (pairs) along c d(DA) = 3.35 Ni(ntbsim)$_2$ nonplanar			39
[Co(ntbsim)$_2$][TNB]	DA (pairs) along c d(DA) = 3.11 Co(ntbsim)$_2$ nonplanar			39

$[Pd(qnl)_2][CA]$	DA(NP) along *a* axis; UO Angle = 15.5° *d*(Pd–Cl) = 3.44		24, 28, 30
$[Pd(qnl)_2][BA]$		14 600	28
$[Pd(qnl)_2][IA]$		14 300	28
$[Pd(qnl)_2][BTF]$		19 700	24
$[Pd(qnl)_2][DNBF]$		15 400	24
$[Pd(qnl)_2][DDQ]$		12 100	24
$[Pd(qnl)_2][pQBr_2Cl_2]$		14 800	28
$[Pd(qnl)_2][pQBr_3Cl]$		14, 800	28
$[Pd(qnl)_2][pQ(N_3)_2Cl_2]$		14 700	27
$[Pd(qnl)_2][pQ(N_3)_2Br_2]$		14 800	27
$[Pd(qnl)_2][TCNQ]$		11 600	25
$[Pd(qnl)_2][TCB]$	DA (U) along *a* axis; UO *d*(PP) = 3.99	17 600	24, 29
$[Pd(qnl)_2][TNF]$		17 000	24
$[Cu(qnl)_2][CA]$		15 900	28
$[Cu(qnl)_2][BA]$		15 900	28
$[Cu(qnl)_2][IA]$		15 700	28
$[Cu(qnl)_2][pQBr_2Cl_2]$		15 700	28
$[Cu(qnl)_2][pQBr_3Cl]$		16 000	27
$[Cu(qnl)_2][pQ(N_3)_2Cl_2]$		16 000	27
$[Cu(qnl)_2][p(N_3)_2Br_2]$		16 000	25, 27
$[Cu(qnl)_2][TCNQ]$	DA along *a* axis; UO *d*(PP) = 3.2 *d*(DA) interplanar = 3.23 *d*(DD) and *d*(AA) interplanar = 3.41–3.47	12 100	28
$[Cu(qnl)_2][TFZ]$		22 700	24
$[Cu(qnl)_2][BTF]$	DA along *a* axis; UO	21 500	24, 26
$[Cu(qnl)_2][DNBF]$		15 700	24
$[Cu(qnl)_2][TNF]$		18 200	24
$[Cu(qnl)_2][DDQ]$		12 900	24

TABLE 3. Structure and Properties of Neutral 1 : 2 and 2 : 1 Charge-Transfer Compounds

Compound	Structure; description of stacking	Optical charge-transfer transition (cm^{-1})	Conductivity (Ω^{-1} cm^{-1}	Reference
$[Cu(salphen)]_2[TCNQ]$	DDA along *c*; UO *d*(DD) = 3.19		P: $\sigma \approx 10^{-8}$ (sem) on grinding, blue-black crystals turn green $\sigma = 10^{-6}$–10^{-7}	40
$[Cu(qnl)_2][PA]_2$	D_A^A along *c*; UO d(DA) = 3.4 (average) = 3.1 (closest)	19600 ⊥ polarized		S 31 24
$[Cu(qnl)_2][TNB]_2$		20700		24
$[Cu(qnl)_2][TCB]_2$	D_A^A; UO *d*(PP) ≈ 3.4 *d*(DA) = 3.07 (shortest)	17900		32
$[Cu(nmsim)_2][TNB]_2$	D_A^A along *c*; UO *d*(PP) ≈ 3.4 *d*(DA) = 3.2 (shortest)			38
$[Co(anoa)_2][TCNQ]_2$	D_A^A along *c*; UO *d*(DA) < 3.5			41
$[Ni(etio)][TRNF]_2$	ADA trimers along *b*; *d*(Ni–TNF) = 3.33			42
$[Pd(dta)_2][Pd_2(dta)_4]$	DA along *a* *d*(Pd–Pd) = 2.754 (dimer–dimer) = 3.399 (monomer–dimer)			43

transfer salts[1,45–64] (Tables 4 and 5).

The nature of the ground state typically has been assigned by an analysis of the optical properties. To a good approximation the optical absorption spectrum of such a compound is given by the sum of the spectra of the component molecules[65] (with an additional band or set of bands, the so-called charge-transfer bands, occurring between 5 and 30 kK). The charge of the components typically is assigned by noting whether a better fit of the overall spectrum is obtained using spectra of the neutral donors and acceptors or of their ions, but recently other methods have become available.[66] In all known cases in this section, both subunits can be assigned an integral oxidation state and the resultant solids are either insulators or semiconductors.

$M(qnl)_2$

Most of the neutral DA crystals have been prepared from an uncharged transition-metal complex and a benzenoid or quinoid organic acceptor. One class of these donors involves the complex $[M(qnl)_2]$, where M = Cu or Pd.[17,24–32] There is one example where a metal-containing subunit acts as the acceptor: $M(tfd)_2$ reacts with the organic donors perylene and pyrene to afford a complex with neutral DA stacks.[33,34]

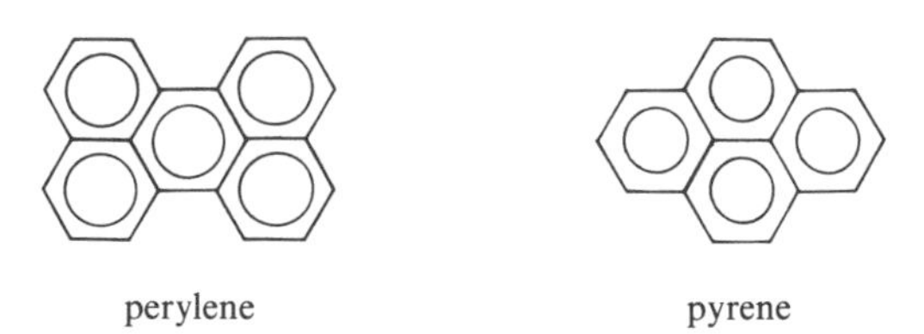

perylene pyrene

Although most compounds prepared from neutral parent molecules are neutral DA complexes, in some cases an ionic charge-transfer salt results. In addition, compounds prepared from D^{+p} and A^{-q} subunits have thus far been found to be ionic. Many of the ionic crystals involve A^{-q} units that are variants on metal *bis*(ethylene)1,2-dithiolenes, namely, $[M(tfd)_2]^-$ and $[M(mnt)_2]^{-2}$. A variety of planar organic cations have been used to prepare such solids: TTF^+,[1,45] POZ^+,[56,57] PTZ^+,[56,57] and NMP^+.[59] A number of charge-transfer salts have been prepared with nonplanar metal-containing cations, for example, ferrocenium salts with TCNQ,[67] but the only one to involve a planar metal-containing cation and organic anion is

TABLE 4. Structure and Properties of Ionic 1:1 Charge-Transfer Salts

Compound	Structure; description of stacking	Magnetism	Optical charge-transfer transition (cm^{-1})	Conductivity (Ω^{-1} cm^{-1})	References
$[TTF]^+[Cu(tfd)_2]^-$	DA(U) along *a*, *b*, and *c* $d(PP) = 3.90$ (*c*) CF_3 rotationally disordered Order at 240 K	TTF^+: $S = 1/2$ $g \sim 2.00$ (EPR) HAF (298–12 K) $J/k_b = 77$ K SP, $T < 12$ K (see text)		SC: $\sigma < 10^{-9}$	S 51, 52, M 1, 45, 46, 48, 54, 55
$[TTF]^+[Au(tfd)_2)_2]^-$	DA(U) along *a*, *b*, and *c* $d(PP) = 3.93$ (*c*) CF_3 rotationally disordered Order at 260 K	TTF^+: $S = 1/2$ $g \sim 2.00$ (EPR) HAF (298–2.1 K) $J/k_b = 68$ K SP, $T < 2.1$ K (see text)		SC: $\sigma < 10^{-9}$	72, 1, 45, 46, 48, 54, 55, 72
$[TTF]^+[Pt(tfd)_2]^-$	DA(U) along *a*, *b*, and *c* $d(PP) = 3.19(c)$ CF_3 rotationally disordered Order at 270 K	TTF^+: $S = 1/2$ $g \sim 2.00$ (EPR) $[Pt(tfd)_2]^-$: $S = 1/2$ χ greater than CL (300–40 K) Intermediate AF ordering along *b*, $T \leq 40$ K (see text)		SC: $\sigma < 10^{-9}$	S 47, M 45, 50
$[TTF]^+[Ni(tfd)_2]^-$	DADA along *a*, *c*; repeat unit along *b* every three chains	TTF^+: $S = 1/2$ $g \sim 2.00$ (EPR) $[Ni(tfd)_2]^-$: $S = 1/2$ Ferrimagnetic (see text)		SC: $\sigma < 10^{-9}$	45, 64,

$[PTZ]^+[Ni(tfd)_2]^-$	DAADDA along *b*; $d(DA) = 3.36$ $d(AA) = 3.83$ $d(DD) = 3.4–3.9$	PTZ^+: $S = 1/2$ $[Ni(tfd)_2]^-$: $S = 1/2$ EPR: average of D^+ and A^- $g = 2.03$ $\Gamma = 250$ G TAP $\chi^p = 3.7 \times 10^{-4}$ emu/mol @ 295 K $= 1.6 \times 10^{-4}$ emu/mol @ 77 K	7700	P: $\sigma \leq 10^{-8}$	S 57 C, M 56
$[POZ]^+[Ni(tfd)_2]^-$	DA along *b* $d(PP) = 3.66$	POZ^+: $S = 1/2$ $[Ni(tfd)_2]^-$: $S = 1/2$ EPR: average of D^+ and A^- $g = 2.03$ $\Gamma = 250$ G CWL intensity $\chi^p = 2.08 \times 10^{-3}$ emu/mol	8000 Weak unassigned band at 720 nm	P: $\sigma \leq 10^{-8}$	S 57 C, M 56
$[TRP]^+[Ni(tfd)_2]^-$	DA(NP); UO Angle 17° $d(PP) = 3.9$	$[Ni(tfd)_2]^-$: $S = 1/2$ $g_{xx} = 2.137$ $g_{yy} = 2.044$ $g_{zz} = 1.996$ $\chi^p = 1.4 \times 10^{-3}$ emu/mol CWL			63
$[Au(dmg)_2][AuCl_2]$	DA along *c*; MOM $d(MM) = 3.26$ Å		CD band at 550 nm shows no Au–Au interaction	P: $\sigma = 10^{-8}$ $= 3.7 \times 10^{-4}$ at 590 kbar	S 60 61, 62

TABLE 5. *Structure and Properties of Ionic 2 : 1 and 1 : 2 Charge-Transfer Salts*

Compound	Structure; description of stacking	Magnetism	Conductivity (Ω^{-1} cm^{-1})	References
$[NMP]_2[Ni(mnt)_2]$	DAD trimers along *a* d(DA) = 3.48 d(DD) = 3.35 Cations slightly bent Good overlap in trimers		P: $\sigma = 6 \times 10^{-3}$ (sem)	59
$[Pt(bipy)_2][TCNQ]_2$	D^A_A along *a* axis d(DA) = 3.3 Two TCNQ moities appear to be σ bonded	Weak signal at RT Biradical or Frenkel excitons Large crystals: triplet or center-line impurity $D = 0.01$ cm^{-1} $E = 0.0014$ cm^{-1} TAP $E_{act} = 0.25$ eV At 87 K increase in paramagnetism $\mu = 2.35$ BM or $2e^-$/unit σ bond break?		58

$[Pt(bipy)_2][TCNQ]_2$.[58] Two other compounds are also reported in which both the anion and cation units are metal complexes.[43, 60]

R = CF_3 : $M(tfd)_2$
R = CN: $M(mnt)_2$
R = H: $M(edt)_2$

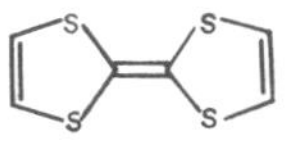

TTF

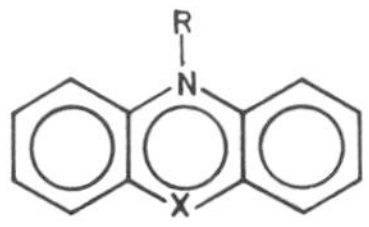

R = H, X = O: POZ^+
R = H, X = S: PTZ
R = CH, X = N: NMP^+

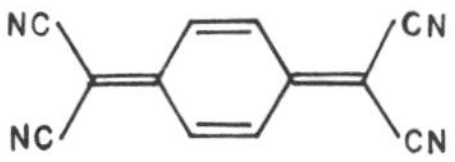

TCNQ

The most common stoichiometry in integrated-stack materials is 1 : 1 ($D^{+\delta} : A^{-\delta}$), although compounds with 1 : 2 or 2 : 1 stoichiometry are known. The 1 : 1 solids are listed in Tables 2 and 4. They are composed of stacks of alternating, parallel donor and acceptor molecules in which the DA arrangement ranges from "eclipsed," where the molecules lie directly on top of each other and overlap maximally, to "slipped," where the overlap is diminished. The "slipped" arrangement is illustrated by $[perylene][Ni(tfd)_2]$[33] and $[Pd(bqd)_2][TCNQ]$[35] (Figure 3). Donor–acceptor interplanar spacings range from 3.1 Å, which is less than the van der Waals' contact distance of ~3.4 Å, and thus is indicative appreciable orbital overlap, to 3.9 Å. Several of the compounds with spacings greater than 3.5 Å have contact distances for nonhydrogen atoms of less than 3.4 Å.

Compounds of stoichiometry 2 : 1 and 1 : 2 exhibit two different stacking arrangements (Tables 3 and 5). The YXYYXY arrangement, represented by $[NMP]_2[Ni(mnt)_2]$,[59] consists of collinear X and Y units that orient plane-to-plane along the stacking axis (Figure 4). These compounds often show only weak Y–Y interaction [d(Y–Y) > 3.8 Å] in comparison with the X–Y interaction [d(X–Y) < 3.4 Å], and are thus best described as weakly interacting YXY trimers. The system $[Cu(qnl)_2][PA]_2$[31] exemplifies the

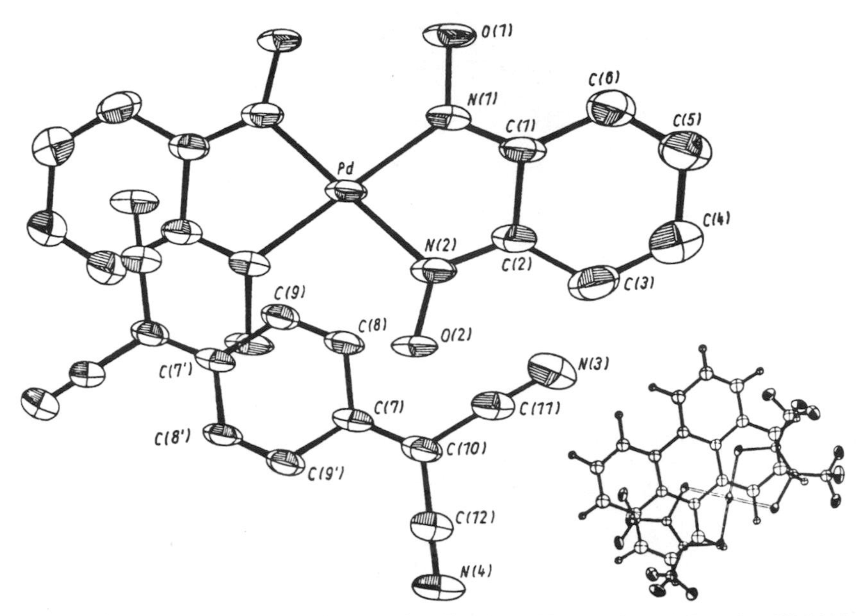

FIGURE 3. Donor–acceptor projections for $[Pd(bqd)_2][TCNQ]$ and $[perylene][Ni(tfd)_2]$ (inset) showing the unsymmetrical molecular overlaps resulting from the slipped stacking arrangement [References 35 and 33 (inset)].

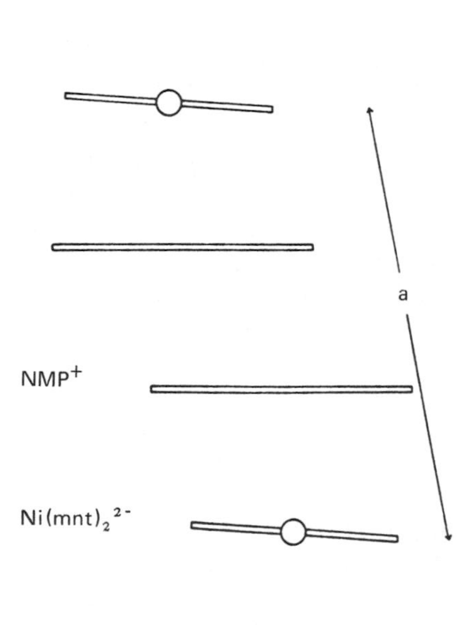

FIGURE 4. Schematic representation of the DAD trimer stacking arrangement found for $[NMP]_2[Ni(mnt)_2]$ (Reference 59).

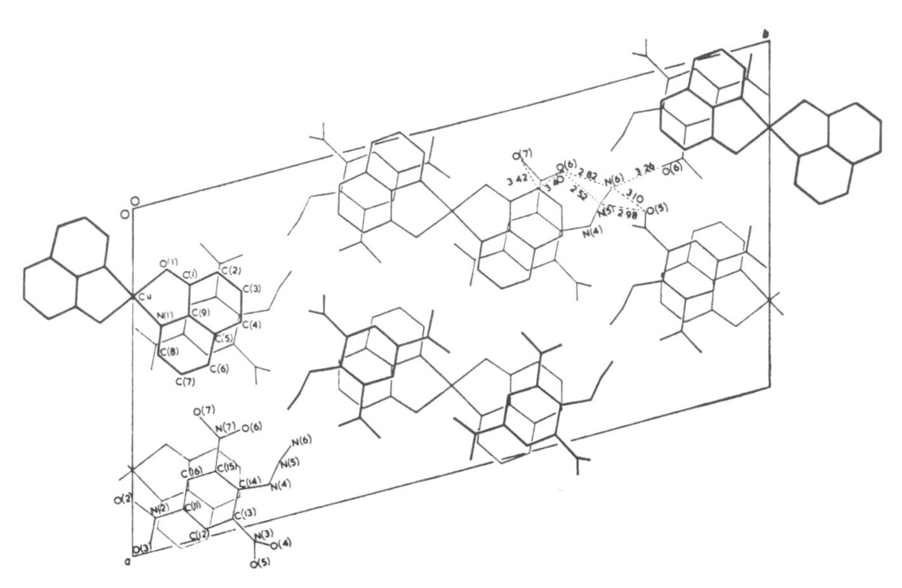

FIGURE 5. Projection down *c* axis for $[Cu(qnl)_2][PA]_2$ showing picryl azide molecules overlapping with only the organic portion of the metal complex (Reference 31).

stacking arrangement, X_Y^Y, which results when the Y subunits orient roughly edge-to-edge with each other and face-to-face with the X subunits (Figure 5). This pattern generally occurs when the X unit is large and the Y units are small, as in the $M(qnl)_2$ [31, 32] systems. The metal seems merely to hold the two 8-quinolinol ligands together; each acceptor interacts individually with only one ligand, or one-half of the X unit. The X–Y separation ranges from 3.2 to 3.5 Å, with varying eclipsed to slipped overlap.

The strength of the charge-transfer interaction between subunits along a neutral stack depends on the degree of intermolecular π-orbital overlap, and is thus dependent on both the relative orientation of adjacent component molecules within the stack and on the donor–acceptor spacing. The role that charge-transfer stabilization plays in determining the structures of such compounds has been probed theoretically.[68] Calculations indicate that charge-transfer interactions have little effect on the plane-to-plane spacing; rather this distance is probably governed by dipole–dipole and van der Waals' interactions. However, it is found that the interaction energy is very sensitive to the donor–acceptor orientation. Thus, in the absence of sizable orientational contributions from charge–dipole, London dispersion, and hydrogen-bonding interactions, subunits will adopt an orientation that maximizes charge-transfer stabilization. The principle of maximum charge-transfer stabilization is useful in predicting the molecular orientations in many organic DA complexes, but has been less successful in systems with metal-containing subunits because of the presence of large charge–dipole and hydrogen-bonding interactions.

In a majority of neutral compounds both subunits are closed shell and therefore diamagnetic. Nevertheless, in some a weak intrinsic EPR signal is observed whose intensity is temperature insensitive or decreases with decreasing temperature.[56] In these compounds the ground state is diamagnetic, but there is a thermally accessible ionic paramagnetic state. An adequate theoretical treatment of such systems has become available only recently.[69, 70] Several neutral compounds contain a paramagnetic Cu^{II} ion, but the magnetic behavior shows no evidence of magnetic coupling between Cu sites, even in $[Cu(salphen)]_2[TCNQ]$,[40] where the D–D spacing is 3.19 Å. Ionic complexes frequently exhibit EPR signals that are interpreted in terms of exchange interactions.[1]

2.2. *Metal Complex D^{+p} or A^{-q}, Organic Molecule A^{-q} or D^{+p}*

2.2.1. *Neutral Compounds* ($p = q = 0$)

A majority of the neutral compounds have $M(qnl)_2$ [17, 24–32] as one of the subunits. The disparity in size between this large donor and the smaller acceptors often results in unsymmetrical DA interactions, with the acceptors overlapping with only one of the organic ligands[26] (Figure 6). Although the energy of the optical charge-transfer band in these crystals is sensitive to the nature or electron affinity of the acceptor, it is also sensitive to the ionization potential of the metal, with the band in the Pd system at lower energy than the band in the Cu system. Schiff base complexes, such as Cu(salphen),[40] as well as metal glyoximates, such as the $M(bqd)_2$ complexes,[35, 36] also react with benzenoid and quinoid acceptors to form neutral charge-transfer compounds and in these compounds the DA interactions are also found to be unsymmetrical.

2.2.2. *Ionic Compounds* ($p, q > 0$)

Compounds of the type $[TTF][M(tfd)_2]$, where M = Cu, Au, Ni, and Pt, exhibit the most interesting magnetic properties[1, 45] of the systems considered in this section and consequently will be discussed in some detail. All four compounds show a similar dominant EPR signal at room temperature which has been attributed to TTF^+ radicals.[48] Magnetic susceptibility measurements on the four systems have established that the M = Cu and Au subunits are diamagnetic while the M = Ni and Pt subunits are paramagnetic. Thus, there is complete charge transfer between the donor and the acceptor, and the compounds are best formulated as $[TTF]^+[M(tfd)_2]^-$. On the basis of complete structures for the Cu [48, 49]

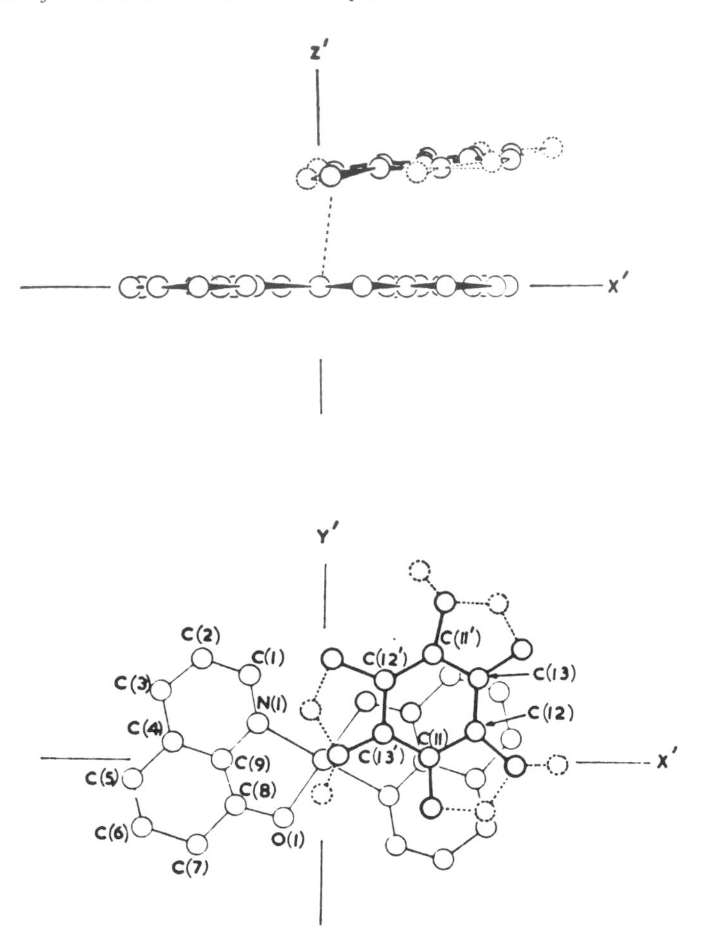

FIGURE 6. Projections for $[Cu(qnl)_2][BTF]$ (top) parallel to and (bottom) perpendicular to the metal complex plane demonstrating the unsymmetrical molecular overlap often found in 1:1 $[Cu(qnl)_2]$ complexes (Reference 26).

and Pt[47] compounds and precession data for the Au material,[48] the three solids appear to be isostructural. The crystals assume a rock-salt packing arrangement, with alternating D and A units along each of the three nearly orthogonal crystal axes in an $F\bar{1}$ cell (Figure 7). The molecular planes of both subunits lie roughly parallel to the (001) plane, thereby defining the *c* axis as the stacking axis. The intermolecular spacing along this axis is large ($\sim$3.9 Å), suggestive of highly localized electronic states; this has been borne out by the low observed electronic conductivity ($\sigma < 10^{-9}$ Ω^{-1} cm^{-1}).[45] The crystals may also be viewed as being composed of segregated slipped stacks. By choosing three vectors from the origin of the $F\bar{1}$ cell to

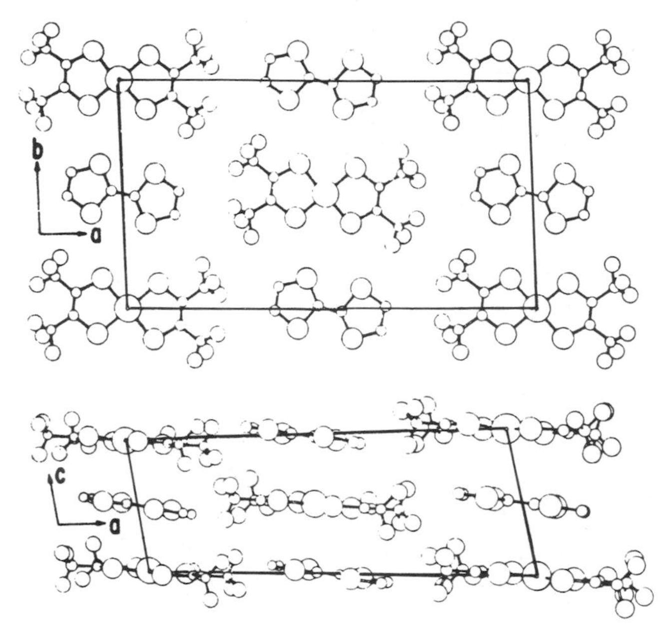

FIGURE 7. View of the structure of $[TTF]^+[M(tfd)_2]^-$ (M = Cu, Au, and Pt) for an $F\bar{1}$ cell in the (001) and (010) planes (Reference 46).

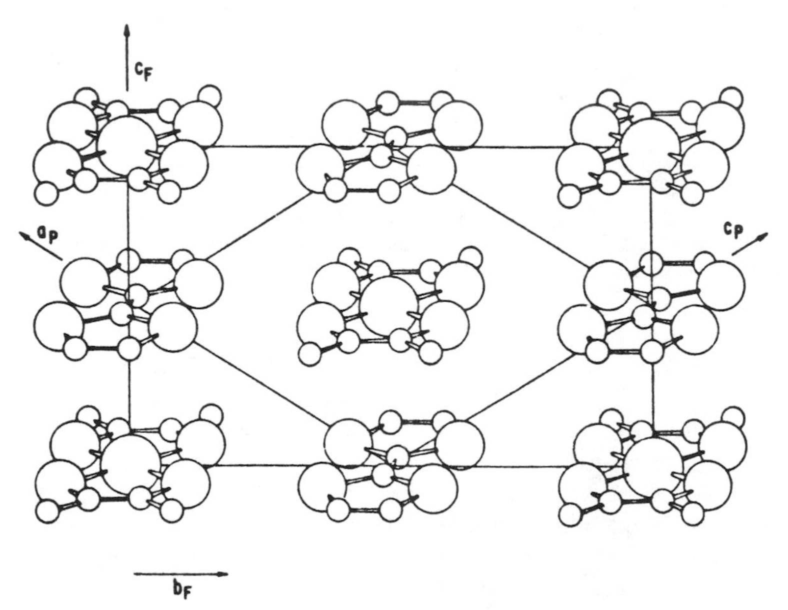

FIGURE 8. View of the structure of $[TTF]^+[M(tfd)_2]^-$ (M = Cu, Au, and Pt) in the (100) plane showing the relationship between axes in the $F\bar{1}$ cell and in the $P\bar{1}$ cell (Reference 1).

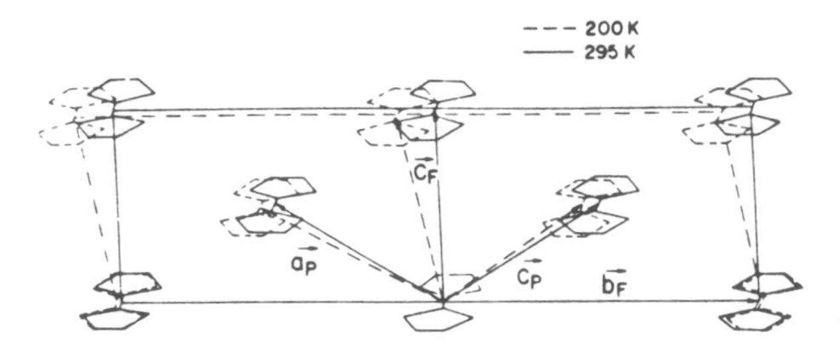

FIGURE 9. Illustration of the first-order structural change in $[TTF]^+[Cu(tfd)_2]^-$ associated with the ordering at 240 K of the rotationally disordered $-CF_3$ groups. Only the TTF units are shown for clarity (Reference 52).

the face centers, one may define a $P\bar{1}$ cell (Figure 8).[1] Each of these vectors, labeled a_p, b_p, and c_p, may be viewed as lying along segregated stack axes, although the long interplanar spacings (> 7.5 Å) suggest that direct D–D or A–A π-orbital overlap is very small within such stacks.

Each of these compounds is observed to undergo a first-order phase transition between room temperature and 200 K. This transition, which has been fully characterized only for the Cu compound,[49] results in the ordering of the rotationally disordered $-CF_3$ substituents, and is probably driven by the accompanying decrease of the unit cell volume. Although the basic site symmetry remains the same as that found at room temperature, there are significant changes in some of the unit cell parameters (Figure 9). For example, TTF units along the a_p axis move further apart, from 7.80 to 8.47 Å, while along the c_p axis the TTF–TTF spacing decreases from 7.53 to 6.86 Å. This results in a chain of regularly spaced TTF units along the c_p axis. Preliminary x-ray work indicates that there is a similarly large or perhaps even larger change in the structural parameters of the Au compound at ~200 K, but that the structural change in the Pt compound at 270 K is relatively small.[1]

The magnetic susceptibility, given in Figure 10 for M = Cu and Au, has been measured by both static and EPR methods and is found to be isotropic at all temperatures. The first-order transition described above can be seen as a break in the curve at the transition temperature. The data for the Cu system above 12 K have been fit accurately by a Bonner–Fisher calculation[71] for an $S = 1/2$ one-dimensional Heisenberg antiferromagnetic spin system, and are described by the Hamiltonian $\hat{H}_{ex} = J \sum_i \hat{S}_i \cdot \hat{S}_{i+1}$, with $J/k_B = 77$ K.[48] A similar analysis of the data for the Au system at temperatures above 2.1 K yields a value of $J/k_B = 68$ K. Below 12 K and 2.1 K for Cu and Au, respectively, the susceptibility shows a sharp, field independent decrease indicative of a progressive transition to a singlet state via a second-order phase transition.

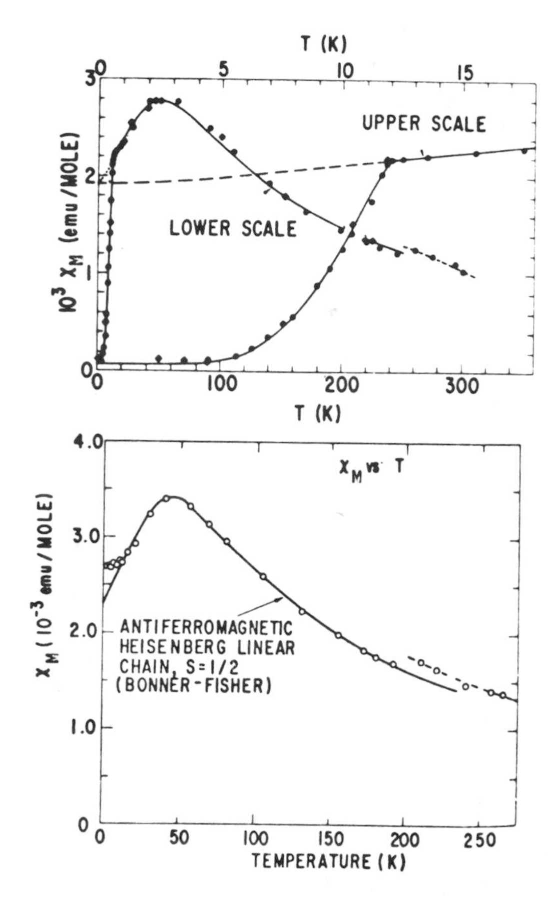

FIGURE 10. Temperature dependence of the magnetic susceptibility for $[TTF]^+[Cu(tfd)_2]^-$ (top) and $[TTF]^+[Au(tfd)_2]^-$ (bottom). Solid lines are based on a Bonner–Fisher calculation for antiferromagnetic chains with a uniform exchange above the transition and a spin Peierls theory with alternating exchange below (Reference 48).

Theoretical calculations on $S = 1/2$ antiferromagnetic spin systems have for some time predicted a fundamental instability in a regular chain of spins in favor of the formation of an alternating chain with a singlet ground state.[72–79] This transition is called a spin Peierls transition and is the magnetic analogue of the electronically driven Peierls distortion predicted for one-dimensional metals.[5] Although other systems have initially been reported to exhibit this phenomenon,[79] $[TTF][Cu(tfd)_2]$ and $[TTF][Au(tfd)_2]$ are the first systems where confirmatory experiments have borne out the interpretation. In both systems the susceptibility is shown to follow the prediction of a regular one-dimensional chain (solid line, Figure 10) above the transition, and below to go to zero in an activated manner, as predicted for a progressive dimerization.

Specific heat measurements[55] and nuclear relaxation studies[53,54] on the Cu and Au compounds have also confirmed the nature of the transition. But perhaps the best evidence for the spin Peierls nature of the transition

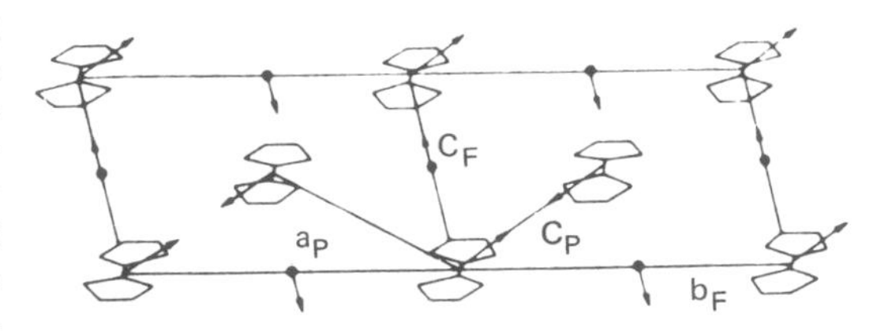

FIGURE 11. Illustration of the molecular displacements occurring in $[TTF]^+[Cu(tfd)_2]^-$ below the spin Peierls transition temperature of 12 K in the a_p–c_p plane. Only the $[TTF]^+$ units are shown for clarity; however, the translation of the center of mass of the $[Cu(tfd)_2]^-$ units is indicated (Reference 51).

comes from x-ray scattering measurements taken on the Cu sample at varying temperatures.[51, 52] New superlattice reflections that require a doubling of the a_p and c_p cell dimensions appear below T_c. The TTF units dimerize via translations of 0.036 Å along the c_p axis (Figure 11) with a net decrease in the TTF–TTF spacing of 0.072 Å. The magnitude of the translation is in reasonable agreement with theory. In addition, the TTF units rotate slightly, and there is a movement of the $Cu(tfd)_2$ units by 0.015 Å in an orthogonal direction to fill in the space vacated by the TTF molecules. In particular, the existence of a progressive dimerization at temperatures below the transition establishes that the transition is consistent with a spin Peierls distortion, and also defines the c_p axis direction as the one-dimensional antiferromagnetic exchange direction.

An additional feature of the temperature-dependent x-ray scattering is the persistence above T_c of intensity at the superlattice positions.[51] This is consistent with a soft phonon mode at a wave vector commensurate with the changes that occur on dimerization. This low-frequency lattice mode may be a requirement for the observation of a spin Peierls transition.

The Pt system differs from the other two in that the $Pt(tfd)_2$ unit is paramagnetic and contributes to the susceptibility.[50] The temperature dependent susceptibility (Figure 12) is isotropic and weakly ferromagnetic in the temperature region of 40–270 K. Below 12 K the susceptibility becomes anisotropic and shows some degree of antiferromagnetic coupling, generally indicative of long-range three-dimensional ordering. The specific heat data, however, are not consistent with long-range order. Rather, the results have been interpreted as evidence for intermediate range antiferromagnetic ordering between ferromagnetically coupled D^+ and A^- chains. If the coupling within D^+ and A^- chains is taken to be equal, if the intrachain ferromagnetic interactions are treated exactly, and if the interchain antiferromagnetic interactions are treated with a mean field approximation, then exchange constants of $J_F/k_B = -65$ K and $J_{AF}/k_B = 11$ K may be determined from the susceptibility and specific heat data. The exchange is intermediate between the Heisenberg and Ising models. Solid solutions of the composition $[TTF][Pt_{1-x}Au_x(tfd)_2]$ with $0 \leq x \leq 0.2$ have also been studied.[45]

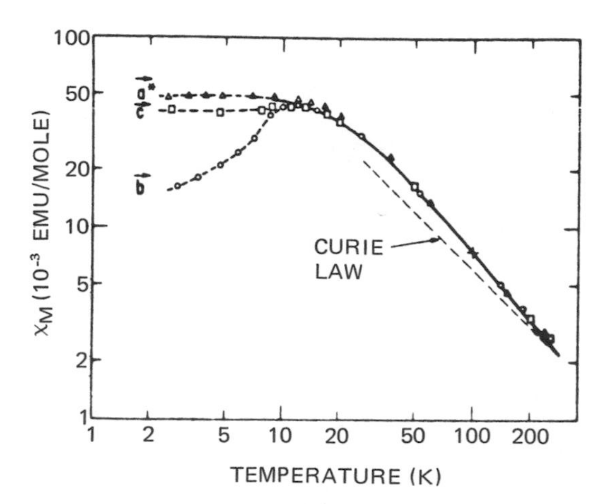

FIGURE 12. Temperature dependence of the magnetic susceptibility for $[TTF]^+[Pt(tfd)_2]^-$ along the a^*, b, and c axes. Solid lines are based on a model for intermediate antiferromagnetic ordering (Reference 50).

These show a maximum susceptibility at lower temperatures presumably because the ferromagnetic chains have been interrupted (Figure 13).

The structure of the Ni compound differs from that of the other three systems in that the repeat unit along b is every three chains.[45, 64] In addition, there is tilting of the subunits that offers the possibility of greater interchain interactions. The magnetic susceptibility behavior has been interpreted in terms of linear antiferromagnetic trimers that are weakly coupled ferromagnetically along the c axis to form a "trimer ladder." Some samples show no ordering down to 1.5 K, while others of higher purity show an

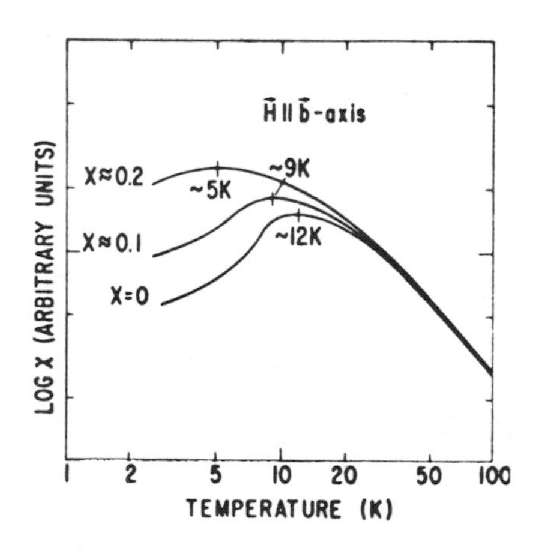

FIGURE 13. Temperature dependence of the magnetic susceptibility for the solid solutions $[TTF]^+[Pt_{1-x}Au_x(tfd)_2]^-$ showing breakdown of antiferromagnetic order with increasing x (Reference 45).

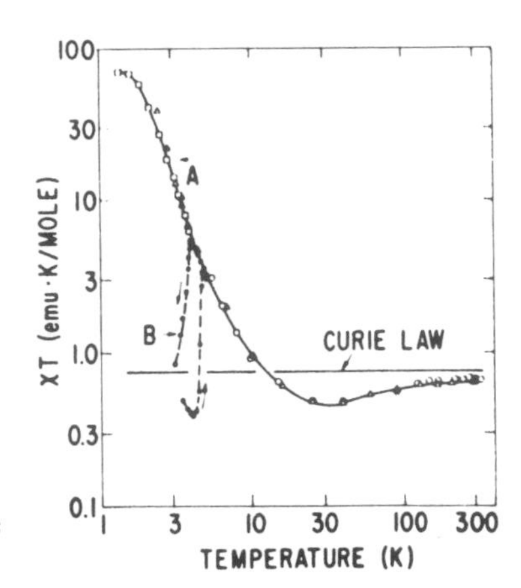

FIGURE 14. Temperature dependence of the magnetic susceptibility for $[TTF]^+[Ni(tfd)_2]^-$ for several samples, some of which exhibit a hysteretic transition at ~5 K (Reference 45).

abrupt, first-order antiferromagnetic transition with thermal hysteresis near 4 K (Figure 14). Below this transition the magnetic isotherms are metamagnetic with field hysteresis, suggestive of magnetoelastic behavior.

The systems $[PTZ][Ni(tfd)_2]$ and $[POZ][Ni(tfd)_2]$ reveal how a seemingly small modification in the donor molecule can cause a dramatic change in solid state properties.[56] Although the donor molecules have a similar molecular structure, the stacking in the PTZ crystals is in a DAADDAAD pattern, while the stacking in the POZ crystals is in a DADA pattern.[57] Both solids have been found by optical spectroscopy to be ionic at room temperature, and both show a single-line EPR signal at $g = 2.03$ with a linewidth of 250 G. This g value is roughly the average of the isotropic g values of the donor and acceptor radicals, an indication that there is an exchange interaction that is large compared with the Zeeman energies of the two sites. The bulk susceptibility for the POZ complex corresponds to two unpaired spins per ion pair, and shows a Curie law temperature dependence. The ground state is thus a paramagnetic ionic state with negligible interaction between spins. The PTZ complex is only weakly paramagnetic at room temperature, and its static and EPR susceptibilities decrease with decreasing temperature. This behavior indicates strong D^+A^- interactions that lead to a spin-paired singlet ground state with thermally accessible triplet excitations.

2.3. *Mixed-Metal Complexes*

The complex $[Au(dmg)_2][AuCl_2]$, one of two systems in Table 3 in which both the D and A units are metal containing, exhibits chains of alternating Au^{III}–Au^{I} ions, with a metal–metal spacing of 3.26 Å.[60] This spacing is similar to the metal–metal distance found in some Ni^{II}, Pd^{II}, and

Pt^{II} systems where weak metal–metal interactions are found.[80] However, dichroism[61] and pressure-dependent conductivity studies[62] have shown the absence of any appreciable metal–metal interactions between Au^{I} and Au^{III} ions. The second system involves one of the crystal forms of $Pd(dta)_2$.[43] The structure reveals stacks of alternating $Pd(dta)_2$ monomers and $Pd_2(dta)_4$ dimers (Figure 15). The monomer–dimer metal–metal spacing is 3.399 Å, while the Pd–Pd distance within the dimer is 2.754 Å. The latter distance is shorter than the Pd–Pd distance in $Pd(edt)_2$ (2.79 Å), which has been taken as an example of direct M–M bonding. Moreover, the Pd–Pd distance is 0.14 Å shorter than that between the centers of the S_4 planes, another indication of Pd–Pd interactions. Such interactions are consistent with the existence of dimers in solution and in the vapor phase. Single-crystal polarized spectra show properties that can be related to the one-dimensional nature of these interactions, and to the different Pd–Pd separations in a stack.

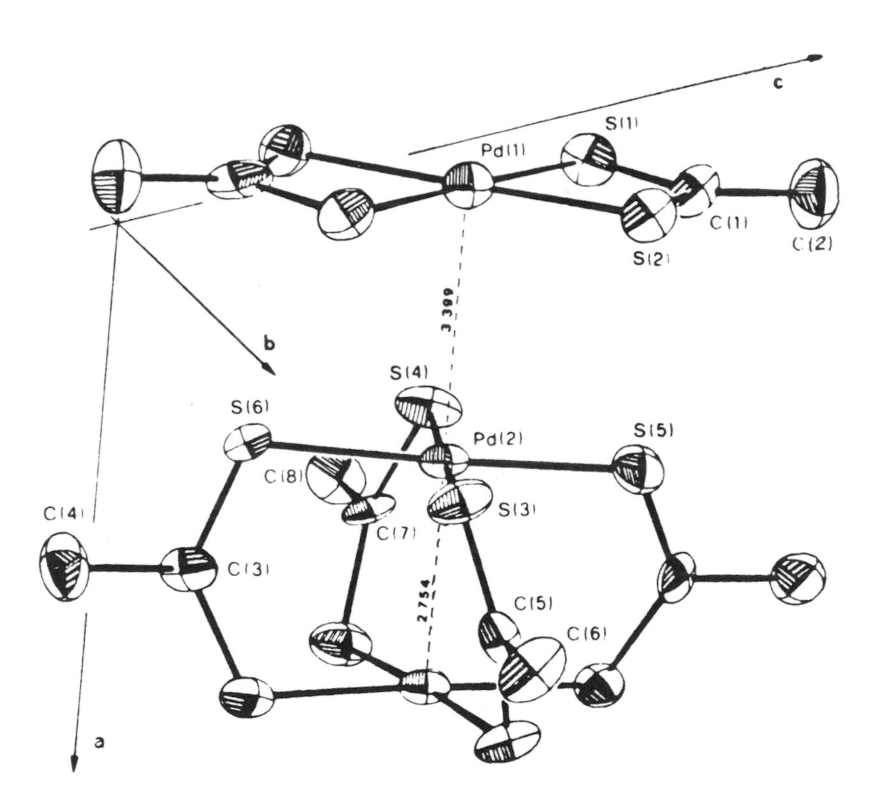

FIGURE 15. Illustration of $[Pd(dta)][Pd_2(dta)_4]$ unit showing molecular geometry and the Pd–Pd distances (Reference 43).

3. Segregated Stack Crystals, Integral Oxidation State Metal Complexes

3.1. General Features

All crystals in this class are either insulators or semiconductors, and with a single exception (discussed in Section 3.3), they are ionic charge-transfer salts. The cations are generally either planar aromatics, such as $TMPD^{\dot{+}}$ and $TTT^{\dot{+}}$ (Section 3.2) or "onium" counterions, such as Et_4N^+ and $MePh_3P^+$ (Section 3.3), although some cationic metal complexes are also known. For a majority of the systems, the anionic components are bis(1,2-dithiolene) metal complexes. The molecular properties of dithiolenes have been studied and reviewed extensively.[81–84] However, our interest is in solid-state interactions, and many of these solids (particularly those listed in Table 7) have not been thoroughly characterized.

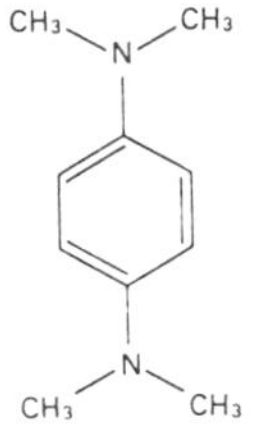

TMPD

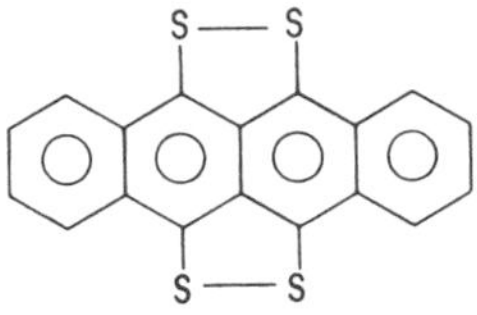

TTT

There are two general structural motifs for segregated stack, integral oxidation state donor–acceptor compounds involving transition-metal complexes. The first, discussed in Section 3.2, is one in which the transition-metal complex does not stack and the counterion, which in all examples to date has been a planar aromatic organic system, does stack.[85–91] Compounds in this category, along with pertinent structural and physical properties, are included in Table 6. The second type of system, discussed in Section 3.3, is one in which the transition-metal complex does stack and the counterion, which may also be a transition-metal complex, usually is located in columns around the stacks of transition-metal complexes.[92–110] These compounds are listed in Table 7. We include a discussion of the $[perylene]_2[M(mnt)_2]$ system[111–113] (Section 3.4) because it has been postulated, on the basis of magnetic measurements, that the $perylene^+$ cations are stacked. Although the physical properties of this system have been investigated extensively (Table 8), it is not possible to classify this system further since no crystal structures are known.

TABLE 6. *Structure and Properties of Segregated Stack, Integral Oxidation State Systems with the Metal Complex Not Stacked*

Compound	Structure; description of stacking[a]	Magnetism	Conductivity (Ω^{-1} cm^{-1})	References
$[TMPD]_2[Ni(mnt)_2]$	Stacks of $(TMPD^+)_2$ dimers d(intra) = 3.25 d(inter) = 6.57 Symmetry $\bar{1}$[b]	EPR: $g_{xx} = 2.0035$ $g_{yy} = 2.0033$ $g_{zz} = 2.0029$ Triplet exciton spectrum $\Delta E = 0.24$ eV		85
$[Pt(bipy)_2][TCNQ]_3$	Stacks of $(TCNQ)_3^{2-}$ trimers d(intra) = 3.23 d(inter) = 3.33 Symmetry $\bar{1}$	EPR: $g_\perp = 2.0022 < g_\parallel$ $\Gamma \propto [3 \cos^2 \theta - 1]^2$ $\ln(\chi T)$ vs. $1/T$ is linear $J = 0.15$ eV		86
$[TTF]_2[Ni(edt)_2]$	$(TTF^+)_2$ dimers connected by TTF° units; (see text) Symmetry m	EPR: $g_x = 2.1217$ $g_y = 2.045$ $g_z = 1.9967$ χ: CL ($\theta < 1$ K) $\mu_{eff} = 1.67$ BM	SC: $\sigma_b = 7.4 \times 10^{-3}$ (sem) $\sigma_c = 2.7 \times 10^{-4}$ $\Delta = 0.23(2)$ eV	S 87 C, M 88
$[TTT]_{1.2}[Ni(edt)_2]$[c]	TTT in uniform stacks with nonintegral oxidation states Symmetry 2	EPR: $g_x = 2.088$ $g_y = 2.073$ $g_z = 1.997$ $\Gamma = 100$ G at 298 K $= 30$ G at 77 K χ: AF ($\theta = -5.5$ K) $Ni(edt)_2$—CWL TTT—small contribution	SC: $\sigma = 30$ (sem) $\Delta = 0.11$ eV	89, 90, 91

[a] In all cases, the transition-metal complex is isolated from the stacks of the organic anion or cation.
[b] Here and in succeeding tables crystallographic symmetry imposed on the metal complex.
[c] Temperature-independent thermopower.

TABLE 7. Structure and Properties of Segregated Stack, Integral Oxidation State Systems with the Metal Complex Stacked

Compound	Structure; description of stacking[a]	Magnetism	Conductivity (Ω^{-1} cm^{-1})	References
$[n\text{-}Bu_4N][Cu(mnt)_2]$	DI d(Cu–Cu′) = 4.03[b] 4.43 d(Cu–S) = 2.17 d(Cu′–S) = 3.64	Diamagnetic		S 92 M 93
$[n\text{-}Bu_4N][Ni(mnt)_2]$[c]	Isomorphous with M = Cu (powder data)	$\mu_{eff} = 1.54$ BM		S 93 M 93, 94, 95
$[n\text{-}Bu_4N][Fe(mnt)_2]$	DI Fe: square pyramidal; d(Fe–Fe′) = 3.08 5.37 d(Fe–S) = 2.23 d(Fe′–S) = 2.46	$\mu_{eff} \sim 1.62$ BM		S 96 M 97, 98
$[Et_4N][Ni(mnt)_2]$	DI d(Ni–Ni′) = 4.14 4.31 d(Ni–S) = 2.15 d(Ni′–S) = 3.52	$\mu_{eff} \sim 1.0$ BM ST, $J = 620$ cm^{-1}	SC: $\sigma = 2.9 \times 10^{-6}$ (sem) $\Delta = 0.35$ eV P: $\sigma = 1.0 \times 10^{-8}$ (sem)	S, C 99 C 100 M 93, 97, 101
$[Et_4N][Pd(mnt)_2]$		Diamagnetic $J > 1000$ cm^{-1}	P: $\sigma = 2.5 \times 10^{-8}$ (sem)	C 100 M 93, 97, 101
$[Et_4N][Pt(mnt)_2]$		$\mu_{eff} = 1.1$ BM ST, $J = 350$ cm^{-1}		93, 97, 101

(continued overleaf)

TABLE 7 (*continued*)

Compound	Structure; description of stacking[a]	Magnetism	Conductivity (Ω^{-1} cm^{-1})	References
$[Et_4N][Au(mnt)_2]$	Isomorphous with M = Pt (powder data)	Diamagnetic		93
$[MePh_3P][Ni(mnt)_2]$	DI d(Ni–Ni′) = 3.47 3.62 d(Ni–S) = 2.15 d(Ni′–S) = 3.59	μ_{eff} = 0.79 BM ST, J = 490 cm^{-1}		S 102 M 97
$[n\text{-}Bu_4N][Co(tcdt)_2]$	DI (isolated) Co: square pyramidal d(Co–Co′) = 3.10 10.6 d(Co–S) = 2.19 d(Co′–S) = 2.40 Not isomorphous with analogous M = Ni, Cu, Au systems (powder data)	Diamagnetic		103, 104
$[Ph_3PC1][Au(tfd)_2]$	DI d(Au–Au′) = 4.66 = 7.61 d(Au–S) = 2.29 d(Au′–S) = 3.96			105

$[Pt(disn)_2][PtCl_2(PhCN)_2]$[d]	Both species in separate columns held together by H bonds $d(Pt\text{–}Pt') = 4.48$ Symmetry $2/m$	Diamagnetic (impurities present)	P: $\sigma < 10^{-9}$	106
$[Ni(tatma)]_2[Ni(edt)_2]$	Stacks of $[Ni(tatma)^+]_2$ dimers separated by $Ni(tatma)^{\circ}$ units, $Ni(edt)_2$ not stacked	CL	SC: $\sigma = 6 \times 10^{-5}$ (sem) $\Delta = 0.23$ eV	107
$[Cu(en)_2][Pt(ox)_2]$	Zig-zag chains of $Pt(ox)_2^{2-}$ Isolated $Cu(en)_2^{2+}$ $d(Pt\text{–}Pt') = 3.55$ $= 3.86$			108
$[Au(dtc)_2][AuBr_2]$	MOM Adjacent cations staggered $d(Au\text{–}Au') = 4.06$ Symmetry $\bar{1}$	Diamagnetic		109
$[Au(dtc)_2]Br$	MOM Adjacent cations eclipsed $d(Au\text{–}Au') = 4.97$ Symmetry $\bar{1}$			110

[a] Distances not supplied in the original paper were calculated using the published atomic coordinates.
[b] M–M′ distances are the intra- and interdimer distances, M–S is the average intramolecular bond length, and M′–S is the shortest intermolecular MS contact.
[c] EPR: 1% M = Ni in crystal of M = Cu: $g_x = 2.160$, $g_y = 2.042$, $g_z = 1.998$.
[d] X-ray PES spectrum: 74.2 eV (Pt^{II}).

TABLE 8. *Properties of Structurally Uncharacterized Systems*[a]

Compound: $[perylene]_2[M(mnt)_2]$	Magnetism	Conductivity (Ω^{-1} cm^{-1})	References
M = Ni	EPR: $g = 2.012$ $\Gamma = 600$ G χ: CWL $\theta = -60$ K $\mu = 2.2$ BM	SC: $\sigma = 50$ (sem) $\Delta = 0.102$ eV	C, M 111 M 112
M = Pd	EPR: $g_1 = 2.0290$ $g_2 = 2.0230$ $g_3 = 1.9849$ χ: $Pd(mnt)_2^- =$ CWL $[Perylene]^+ =$ TAP ($\Delta E_p \sim 0.0167$ eV)	P: $\sigma = 0.07$ (sem) $\Delta = 0.168$ eV	C 111 M 112
M = Pt	χ: CWL $\theta \sim 30$ K $\mu = 1.87$ BM	SC: $\sigma = 125$–280 P: $\sigma = 6$	M, C 113
M = Cu	EPR: $g = 2.020$ $\Gamma = 68$ G at 298 K $= 55$ G at 77 K	SC: $\sigma = 6$ (sem) $\Delta = 0.128$ eV	C, M 111 M 112

[a] Perylene is formally in a nonintegral (+ 1/2) oxidation state.

3.2. *Metal Complex Not Stacked*

The four systems listed in Table 6 consist of stacks of planar organic molecules surrounded by transition-metal complexes that are not closely associated with each other. All but the $[Pt(bipy)_2][TCNQ]_3$ system contain planar organic cations and metal dithiolene anions. The properties of these compounds vary considerably, so each system will be described briefly.

The first compound in this class to be studied extensively was $[TMPD]_2[Ni(mnt)_2]$.[85] The transition-metal component of $[TMPD]_2[Ni(mnt)_2]$ is diamagnetic and therefore does not contribute to the magnetic properties. The structure consists of stacks of $[TMPD^+]_2$ dimers which are surrounded by isolated $Ni(mnt)_2^{2-}$ anions (Figure 16). Adjacent $[TMPD^+]_2$ dimers are well separated with an interplanar distance of 6.57 Å between dimers as compared with 3.25 Å within the dimer. The coupling of spins within the dimer unit results in a singlet ground state with a low-lying triplet excited state (a triplet exciton). There is also a very weak coupling interaction between dimers, which allows slow exciton motion along the $[TMPD^+]_2$ chain.

The transition-metal component of $[Pt(bipy)_2][TCNQ]_3$ is also diamagnetic. The structure[86] consists of stacks of trimers, $[TCNQ]_3^{2-}$, sur-

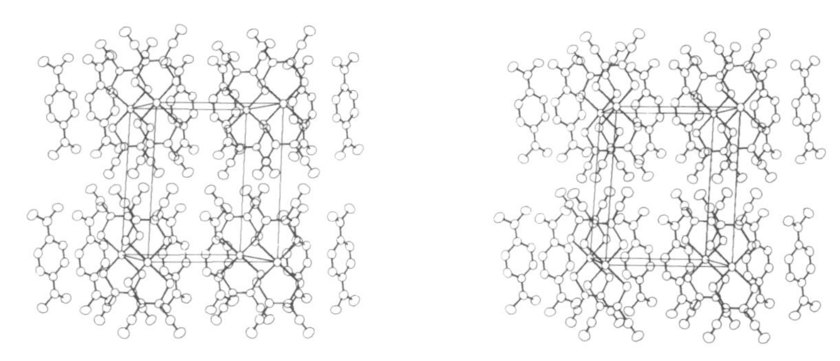

FIGURE 16. A stereoscopic view of the unit cell of $[TMPD]_2[Ni(mnt)_2]$ (Reference 85).

rounded by noninteracting $Pt(bipy)_2^{2+}$ cations. Adjacent trimers are closely spaced with an interplanar separation of only 3.33 Å as compared with 3.23 Å within the trimer. The magnetic properties of this system have been studied much less extensively than those of $[TMPD]_2[Ni(mnt)_2]$. Single-crystal EPR spectra do show, however, that spins are interacting along the TCNQ stacks. Structurally, this compound is very different from the 1 : 2 compound $[Pt(bipy)_2][TCNQ]_2$ [58] (Section 2) which forms D_A^A stacks with the two TCNQ moieties σ-bonded to each other.

In each of the remaining compounds listed in Table 6, only the transition-metal complex contributes to the magnetic properties. The system $[TTF]_2[Ni(edt)_2]$ has a magnetic moment of 1.67 BM, which corresponds to one unpaired electron per formula unit.[88] On the basis of solid state optical and EPR spectra, this unpaired electron has been assigned to the $Ni(edt)_2^-$ anion. The absence of any unpaired electrons associated with the TTF^+ cation can be explained by strong electron pairing within $[TTF^+]_2$ dimers. These dimers are located along the *b* axis and are connected along the *b* and *c* axes by neutral TTF molecules.[87] The molecular planes of the $Ni(edt)_2$ units are parallel to the *bc* plane, as shown in Figure 17. The TTF units in the dimer are eclipsed with intermolecular S–S contacts of 3.48 Å. Given the overall composition of the unit cell, this compound is most instructively formulated as $[TTF^\circ]_2[TTF^+]_2[Ni(edt)_2^-]_2$.

The $[TTT]_{1.2}[Ni(edt)_2]$[89–91] system is similar to the $[TTF]_2[Ni(edt)_2]$ system in that only the unpaired electron from the $Ni(edt)_2^-$ anion contributes to the susceptibility. However, here the TTT units are uniformly spaced in the stack, and there are no crystallographically distinguishable TTT° and TTT^+ entities. This suggests that the electrons are delocalized along the stack; the most accurate formulation for this complex is then $[TTT]_{1.2}^{+5/6}[Ni(edt)_2]^-$, with the organic donor ion in a nonintegral oxidation state and the transition-metal complex in an integral oxidation state. The susceptibility of the unstacked anions is Curie–Weiss, and that of the stacked anions is quenched.

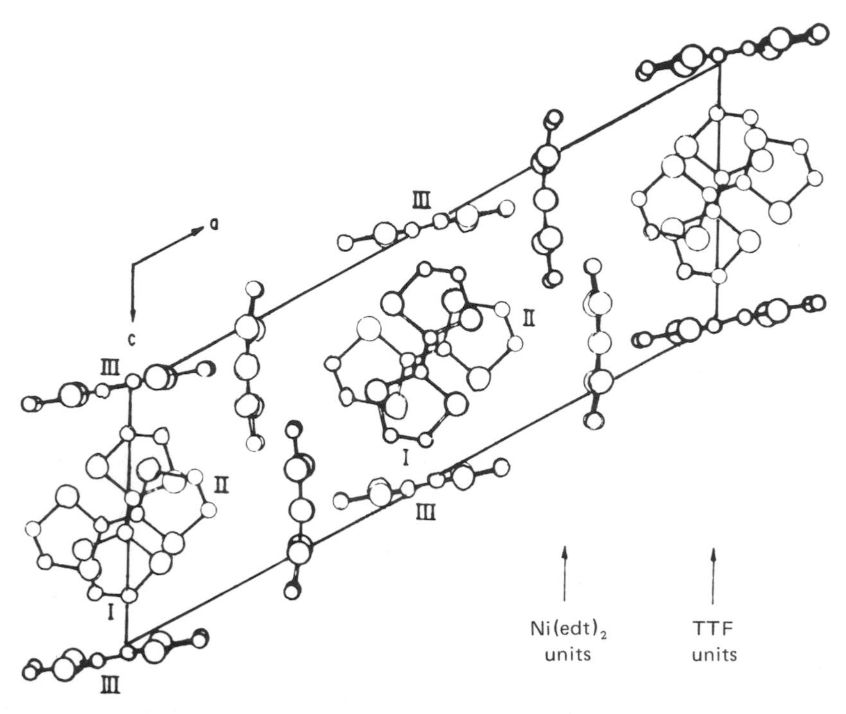

FIGURE 17. The unit cell of $[TTF]_2[Ni(edt)_2]$ viewed down [010]. The molecules centered at $y = 1/2$ are darkened; except for molecules labeled II, all other molecules have their centers at $y = 0$. Three different types of TTF units are identified by Roman numerals (Reference 87).

Electrical conductivity measurements of the $[TTF]_2[Ni(edt)_2]$ and $[TTT]_{1.2}[Ni(edt)_2]$ systems show them to be semiconductors with conductivities of $\sim 10^{-3}$ and 30 Ω^{-1} cm^{-1}, respectively. In both compounds, the direction of highest conductivity is along the stack of organic molecules. Since the $[TTT]_{1.2}[Ni(edt)_2]$ system contains chains of partially oxidized organic molecules that appear to be uniformly spaced along the stacking axis, an even higher conductivity might be expected, as discussed in Section 1. The low room temperature conductivity, compared with other partially oxidized TTT salts, and its temperature dependence, which is typical of a semiconductor and the quenched susceptibility of the cation are presumed to be a consequence of the postulated Peierls distortion.

3.3. *Metal Complex Stacked*

Most of the compounds listed in this class (Table 7) are tetraalkylammonium (R_4N^+) salts of metal dithiolene complexes. These salts generally exist in a 2 : 1 or 1 : 1 stoichiometry. In the 2 : 1 salts, the transition-metal components are isolated from each other (d(M–M > 5 Å). However, the existence of large metal–metal separations does not necessarily prevent interaction between the transition-metal components or the occurrence of interesting properties. Particularly interesting magnetic behavior is displayed by $[n\text{-}Bu_4N]_2[Cu(mnt)_2]$,[114] which consists of regular, slipped stacks of $Cu(mnt)_2^{2-}$ anions with a Cu–Cu distance of 9.40 Å and a closest intermolecular S---S contact of 7.20 Å. Based on a theoretical description involving weak one-dimensional exchange, the well-resolved EPR spectrum (Figure 18a) of this compound has been simulated (Figure 18b,c). This analysis permits the direct determination of an unusually small exchange coupling, $J_0 = 0.01\ \text{cm}^{-1}$. The intermolecular spacing in the $[Et_4N]_2[Cu(mnt)_2]$ system[115,116] is smaller (d(Cu–Cu) = 7.60 Å) than that in $[n\text{-}Bu_4N]_2[Cu(mnt)_2]$, but the exchange coupling, though much larger, is still weak ($J_0 = 2\ \text{cm}^{-1}$).

All of the 1 : 1 metal dithiolene systems listed in Table 7 are composed of a phosphonium or ammonium cation and a metal^{III} bis(dithiolene) monoanion. These compounds are structurally very similar, consisting of slipped stacks of metal dithiolene anions that are associated in pairs; the stacks are surrounded by noninteracting cations. A view of a representative unit cell

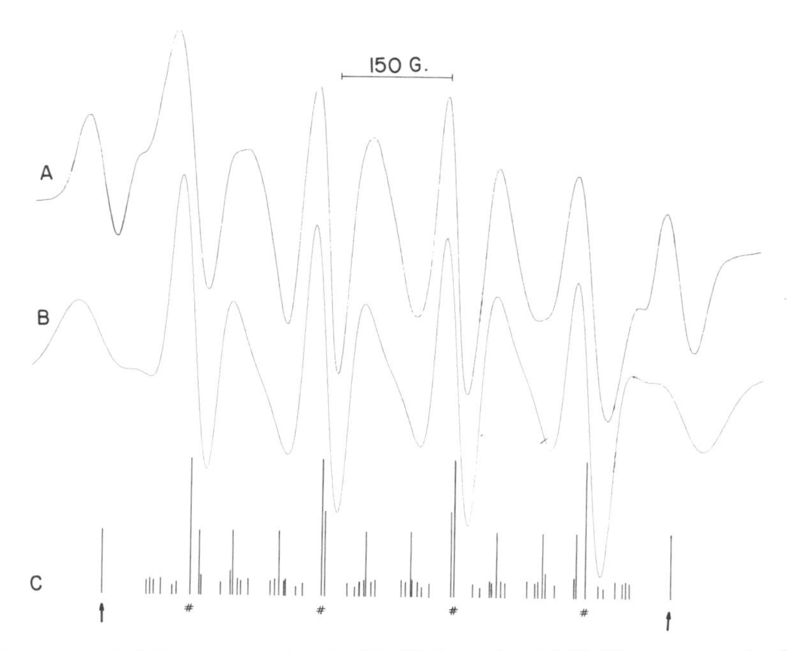

FIGURE 18. (a) EPR spectrum of $[n\text{-}Bu_4N]_2[Cu(mnt)_2]$ at 4.2 K. The computer simulation (b) and stick diagram (c) employ line positions that were calculated assuming that the one-dimensional exchange along the $Cu(mnt)_2^{2-}$ stack is weak ($J = 0.011\ \text{cm}^{-1}$), being smaller than the parallel copper hyperfine coupling ($0.014\ \text{cm}^{-1}$) (Reference 114).

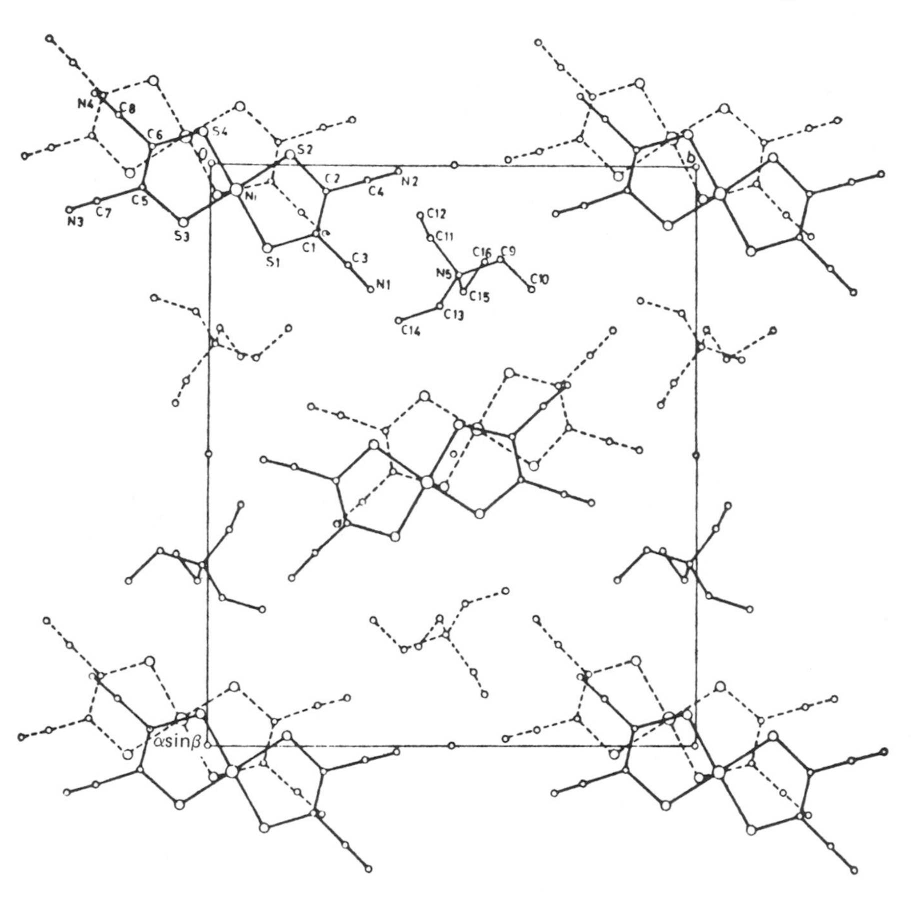

FIGURE 19. A projection of the structure of the 1:1 salt $[Et_4N][Ni(mnt)_2]$ as viewed along the *c* axis (Reference 99).

for this type of system is shown in Figure 19. Only $[n\text{-}Bu_4N][Cu(mnt)_2]$[92] displays a slightly different association of anion pairs in the stack (Figure 20). To avoid classifying compounds incorrectly we have included in Table 7 only those compounds for which a full x-ray crystal structure or x-ray powder data and corroborative information from magnetic studies (e.g., evidence for short-range magnetic coupling) have been reported. The well-known hazard of using x-ray powder patterns to determine whether compounds of similar overall composition have identical crystal structures arose yet again when the original claim[93] that $[n\text{-}Bu_4N][Co(mnt)_2]$ and its Cu analogue were isomorphous had to be withdrawn on the basis of differences in intensities from single crystals.[117]

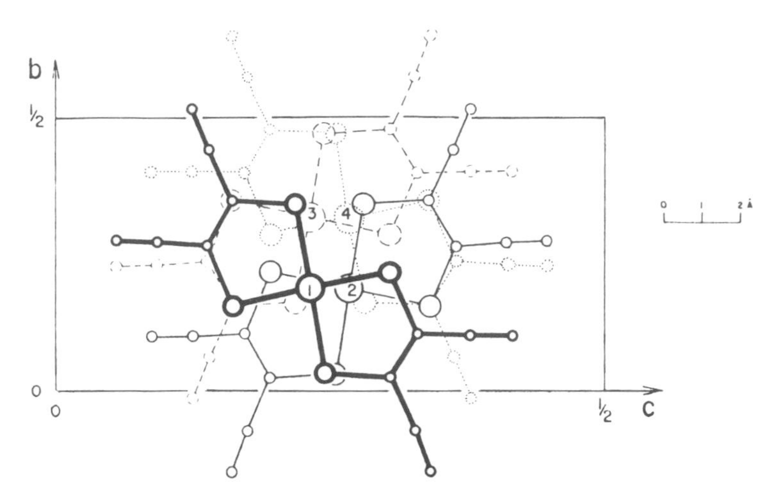

FIGURE 20. A projection down the *a* axis of [*n*-Bu_4N][$Cu(mnt)_2$] showing the stacking of the [$Cu(mnt)_2$]$^-$ anions as viewed down a stack (Reference 92).

As has been noted previously,[81] the structural parameters of metal dithiolene complexes are relatively insensitive to the overall charge on the complex and, consequently, the monoanionic dithiolene complexes generally retain the mmm (D_{2h}) molecular symmetry found in both neutral and dianionic dithiolene complexes. The only notable structural trend is a lengthening of the M–S bond as the overall charge on the transition-metal complex increases.

The transition metal in the bis(dithiolene) monoanions is formally trivalent. The complexes with Au^{III} and Cu^{III} are diamagnetic, those with Ni^{III}, Pd^{III}, Pt^{III}, and Fe^{III} exhibit a doublet ground state, and that with Co^{III} exhibits singlet–triplet behavior. A reduction in the magnetic moments of the Ni, Pd, and Pt systems is observed in the solid state[93, 101] and has been attributed to a pairwise spin–spin interaction where the $S = 1/2$ transition-metal complexes are antiferromagnetically coupled into singlet ground state pairs in the solid. There is excellent agreement between the experimentally determined temperature dependence of the magnetic susceptibility of these systems and that predicted for a system with a singlet ground state and a thermally accessible triplet state (Figure 21). For the $M(mnt)_2^-$ systems measured,[97] the exchange coupling constant, J, is around 500 cm^{-1}, although the spin exchange can be so substantial that the compound is diamagnetic in the solid state, as for the [Et_4N][$Pd(mnt)_2$] system where J is calculated to be greater than 1000 cm^{-1}. A similar pairwise interaction is probably responsible for the magnetic behavior of [*n*-Bu_4N][$Co(tcdt)_2$], which exhibits a full triplet ground state in weakly coordinating solvents

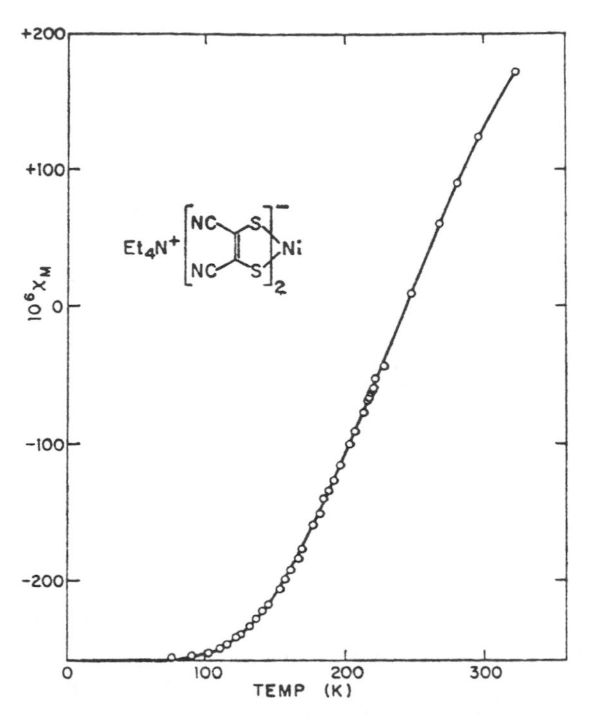

FIGURE 21. Temperature dependence of the magnetic susceptibility for $[Et_4N][Ni(mnt)_2]$. The circles are experimental points and the solid curve is the theoretical dependence calculated assuming singlet–triplet behavior with $J = 620$ cm^{-1} (Reference 97).

(e.g., $\mu = 3.18$ BM in tetrahydrofuran)[103] but is diamagnetic in the solid state.† A particularly puzzling feature of these systems is that in most of them J is small enough to allow an appreciable population of paramagnetic excited states, and yet they do not exhibit EPR spectra attributable to the constituent spins at any temperature.[118] The absence of an EPR spectrum might arise from rapid spin relaxation, but such relaxation must be associated with properties of the stacked solid, for EPR signals of the monomeric anions are readily observed in solutions[93, 98] or glasses.[98]

The metal–metal separation in these systems is generally too large to allow direct spin correlation through the metal atoms. Consequently, this correlation is postulated[97] to occur through an intermediate sulfur atom by way of a M–S---M pathway (Figure 22) that is accessible because the metal

† An alternative explanation for the magnetic behavior of $[n\text{-}Bu_4N][Co(tcdt)_2]$ is that the complex has a spin–singlet ground state in the solid, whereas in solution, weak solvent interactions change the ground state to a spin triplet.[103] However, the existence of $Co(tcdt)_2^-$ dimers in the solid is consistent with the explanation that pairwise interactions cause the observed magnetic behavior.

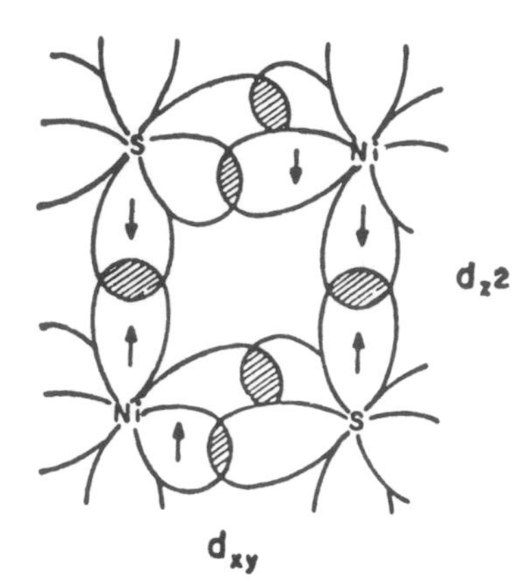

FIGURE 22. Schematic representation of proposed spin correlation mechanism in $[Ni(mnt)_2]^-$ systems (Reference 97).

dithiolene anions dimerize by forming two M---S linkages[117] (Figure 23). The dimerization can be very weak, as is apparent in the structures of $[Et_4N][Ni(mnt)_2]$[99] and $[MePh_3P][Ni(mnt)_2]$[102] where there is less than 0.2-Å difference between the distance within and between dimers. This dimerization also can be very strong as in $[n\text{-}Bu_4N][Fe(mnt)_2]$,[96] $[n\text{-}Bu_4N][Co(tcdt)_2]$,[104] and $[Ph_3PCl][Au(tfd)_2]$[105]; in each of these systems the distance between the dimers is much greater than the distance within them. In $[n\text{-}Bu_4N][Co(tcdt)_2]$ the distance between dimers is so large (> 10 Å) that the overall structure resembles that of the 2 : 1 R_4N^+ salts of metal dithiolene systems with the isolated dimer units in the 1 : 1 salts in place of the isolated monomer units in the 2 : 1 salts. In both the Co and Fe systems, the linkage between dimers is so strong that the intermolecular M---S distance is nearly equal to the intramolecular M–S distance, and the transition metal is best described as possessing a square-pyramidal coordination geometry. The effect of dimerization, whether weak or strong,

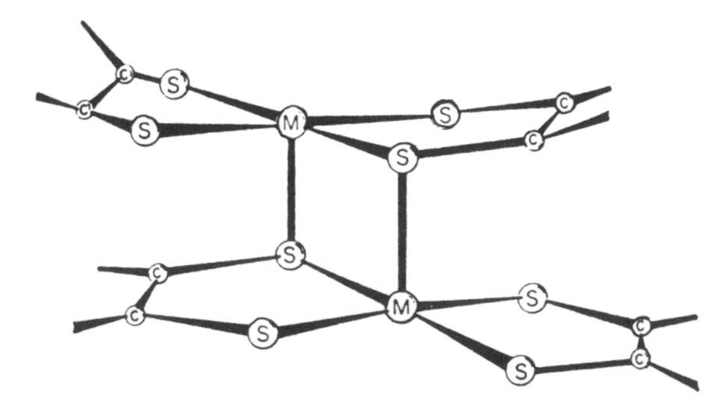

FIGURE 23. Idealized structure showing the two M---S linkages that form upon dimerization of two dithiolene complexes (Reference 117).

on the metal dithiolene complex is limited to minor deviations of the MS_4 unit from planarity.[81] In the more tightly bound dimers,[96,104] the metal atom is displaced by about 0.3 Å out of the plane of the four sulfur atoms.

Electrical conductivity measurements have been made on a few of these metal dithiolene systems and show that they are, at best, poor semiconductors, with room temperature conductivities of less than $10^{-6}\ \Omega^{-1}\ \mathrm{cm}^{-1}$. This is not surprising as the molecules are in an integral oxidation

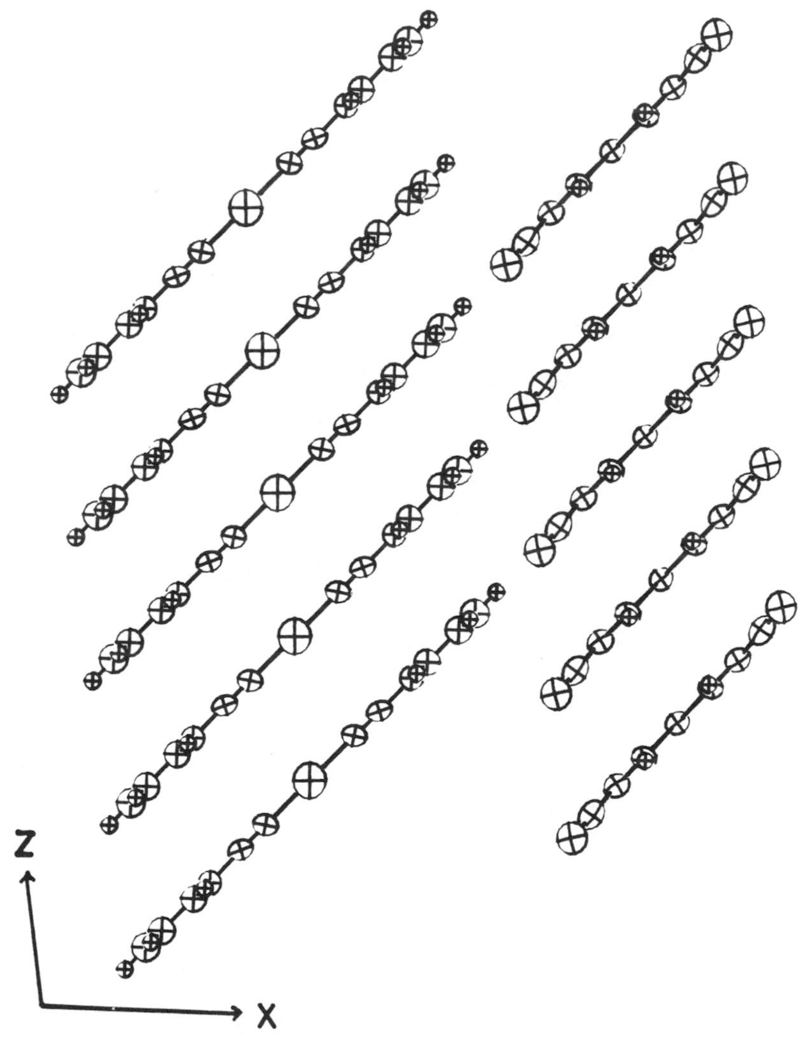

FIGURE 24. A perspective view of $[Pt(disn)_2][PtCl_2(PhCN)_2]$ looking along the *b* axis. The column on the left contains the $PtCl_2(PhCN)_2$ molecules, and the column on the right contains the $Pt(disn)_2$ molecules (Reference 106).

state. Moreover, the existence of dimers destroys the crystallographic uniformity of the stacks and, as the magnetic studies have shown, the unpaired electrons tend to couple within the dimers rather than delocalize along the stacks.

Of the five compounds in Table 7 that do not contain a metal dithiolene system, the most interesting structurally is $[Pt(disn)_2][PtCl_2(PhCN)_2]$.[106] The compound, which is made by cocrystallizing $Pt(disn)_2$ and $PtCl_2(PhCN)_2$, consists of segregated stacks of the component molecules with an interplanar distance of ~ 3.40 Å. A view of the structure of $[Pt(disn)_2][PtCl_2(PhCN)_2]$ along the *b* axis is shown in Figure 24. Both Pt atoms are still in the Pt^{II} state as measured by x-ray photoelectron spectroscopy. Moreover, it is unlikely that the disn ligand alone has been reduced since $M(disn)_2$ systems have much lower electron affinities than the closely related $M(mnt)_2$ systems.[119] Thus, $[Pt(disn)_2][PtCl_2(PhCN)_2]$ is the only example of a 1 : 1 DA compound consisting of neutral component molecules that does not crystallize in stacks of alternating donor and acceptor molecules (Section 2). Stabilization of the segregated stacking arrangement can be attributed to intermolecular hydrogen bonding between the imine hydrogen atoms and the chlorine atoms on the $PtCl_2(PhCN)_2$ molecule. Although uniform, segregated stacks are present in $[Pt(disn)_2][PtCl_2(PhCN)_2]$, the components are not partially oxidized, and consequently, the electrical conductivity is low ($< 10^{-9}\ \Omega^{-1}\ cm^{-1}$ for a pressed pellet).

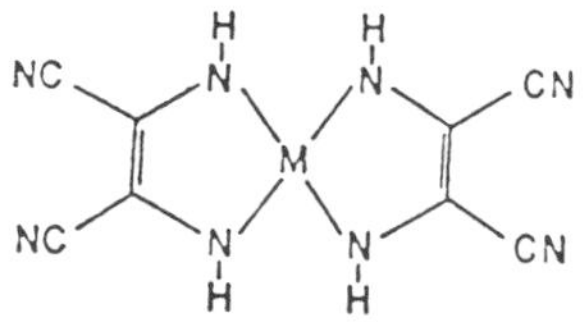

$M(disn)_2$

Another interesting compound in this class is $[Ni(tatma)]_2[Ni(edt)_2]$,[107] which is structurally similar to $[TTF]_2[Ni(edt)_2]$[87] (Section 3.2) and contains stacks of dimers, $[Ni(tatma)^+]_2$, connected by neutral Ni(tatma) molecules. These stacks are similarly surrounded by the unstacked, paramagnetic $[Ni(edt)_2]^-$ anions. The compound $[Ni(tatma)]_2[TCNQ]$ has also been prepared, but powder diffraction data indicate that it has a different structure.

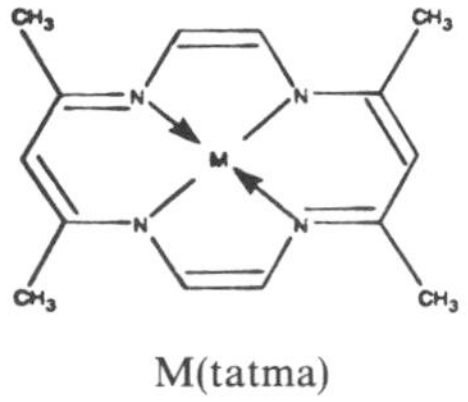

M(tatma)

3.4. Structurally Uncharacterized Systems

In the [perylene]$_2$[M(mnt)$_2$] system (M = Ni,[111,112] Pd,[112] Pt,[113] or Cu[111,112]) (Table 8) both the perylene cation and the M(mnt)$_2$ anion contribute to the susceptibility. On the basis of the single-crystal EPR spectra of the Pd system, a Curie–Weiss contribution to the susceptibility was attributed to the Pd(mnt)$_2^-$ anions, while a thermally activated (ΔE_p = 0.0167 eV) contribution was attributed to the perylene cations. The Pt system exhibits magnetic behavior similar to that of the Pd system, and the EPR spectra of the Ni and Cu systems are not sufficiently resolved to allow a detailed analysis. The degree of charge transfer from the perylene molecule to the M(mnt)$_2$ complex has not been quantitated. Although no crystal structure is known, it seems certain from the EPR results that segregated stacks of perylene molecules in some form must be present. This is in marked contrast to the [perylene][Ni(tfd)$_2$][33] system (Section 2) in which the component molecules are neutral and crystallize in integrated DA stacks. Single crystals of [perylene]$_2$[Pt(mnt)$_2$] exhibit the highest conductivity of these systems (125–280 Ω^{-1} cm^{-1}), but measurements of the temperature dependence of the conductivity were not reproducible so it has not been established whether this compound is a conductor or semiconductor.

4. Segregated Stack Crystals, Nonintegral Oxidation State Metal Complexes

As noted in the Introduction, effective charge transport in crystals of metallomacrocycles is limited to this category. With but one exception to date,[120,121] all such metallomacrocycles contain the metal surrounded by an essentially square-planar array of four coordinating nitrogen atoms. The most fruitful approach to the preparation of highly conducting molecular solids based on metallomacrocycle building blocks has used complexes that may be viewed as variants on the metalloporphyrin skeleton.[2,3,122–138] The chemical flexibility of the porphyrinlike metallomacrocycles[139–141] provides an unusual opportunity to vary purposely and rationally the electronic properties of the molecular building block, and therefore of the resulting solid, through choice of the basic ligand, of peripheral substituents, and of incorporated metal. There are three principal variants to the porphyrin skeleton[142] that have been examined. The tetrabenzporphyrin (tbp) skeleton can be viewed as having an expanded aromatic π-electron system obtained by fusing benzene rings onto the β-carbon atoms of the porphyrin pyrrole rings. Phthalocyanine (pc) may be imagined to arise from four methine$\rightarrow$aza bridge substitutions in tbp. In addition, an "intermediate" complex, triazetetrabenzporphyrin (tatbp), which can be obtained conceptually from tbp by three methine → aza bridge substitutions, has also been studied.

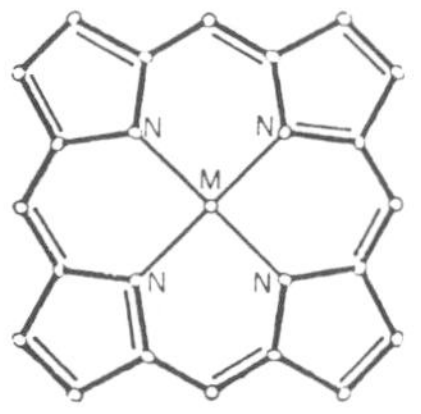

M(porphyrin)

M(tbn) 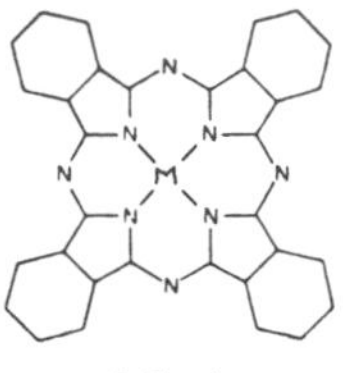 M(pc)

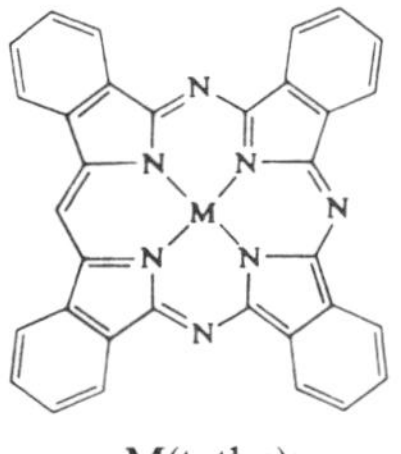

M(tatbp)

Thus, this class of complexes contains two sizes of ring systems, and within each size the chemical and physical properties can be modulated by choice of bridge atom. Moreover, the basic structures are the foundation for an enormous range of complexes: almost every metal ion in the Periodic Table can be inserted into the central hole of these doughnut-shaped structures, and the periphery of the macrocycles is also susceptible to substitution. Such substitutions, which are well known for the small-ring porphyrin system,[139, 140] lead to such complexes as M(oep),[140] and M(tmp),[143] and the octathioalkyl tetraazaporphyrins,[144] and may also be applied to the large-ring, tetrabenzporphyrin (tbp) system to give such complexes as M(omtbp)[140] and analogously substituted phthalocyanines. In short, porphyrinic complexes form the basis of a large class of new materials which exhibit high conductivities as well as other novel properties and which are particularly useful in the study of the relationship between molecular structure and solid state transport properties.

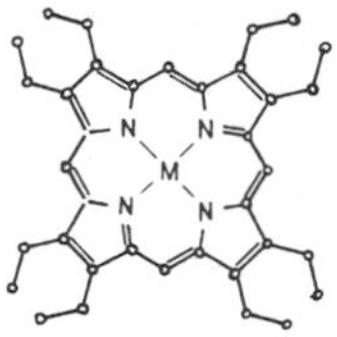

M(oep)

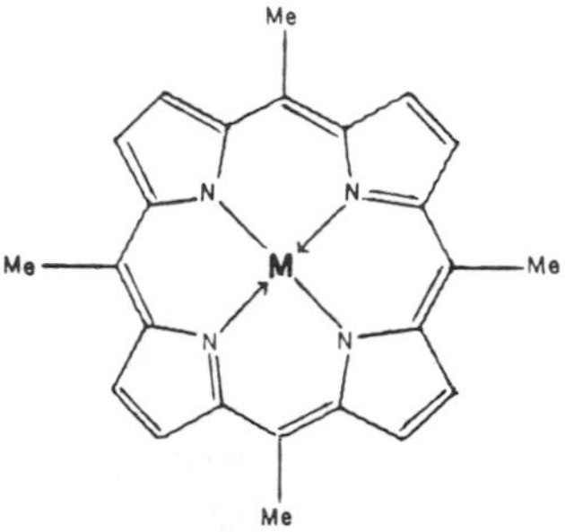

M(tmp)

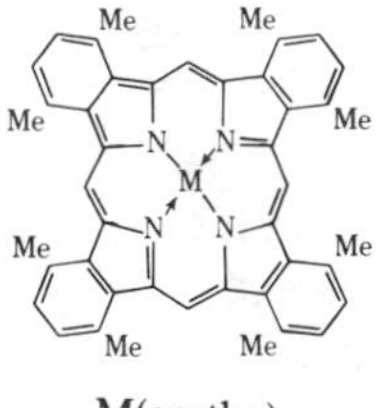

M(omtbp)

A second series of compounds, based on the bis(α,β-dionedioximates) of various transition metals, especially the Ni triad, has been extensively studied over a longer period.[4, 145–171]

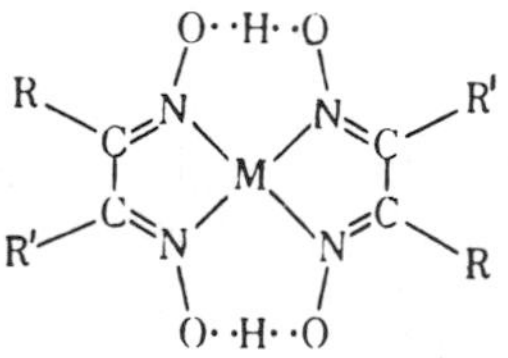

R′=H : $M(dhg)_2$

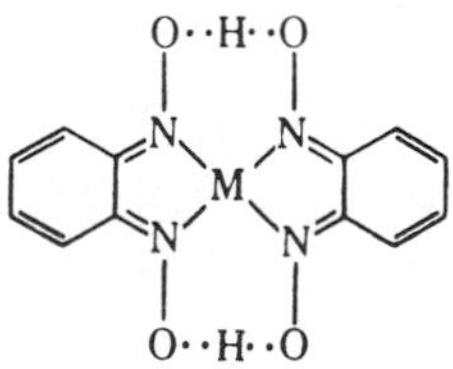

$M(tqd)_2$

This series was the first to be subjected to partial oxidation through the use of iodine.[171] The integral oxidation state parent complexes of Ni, Pd, and Pt are insulators, while the partially oxidized derivatives are semiconductors. These complexes are of particular interest within the context of this review because of the diversity of crystal structures they adopt and because they serve as a contrast to related systems, the porphyrins and phthalocyanines, that display more interesting physical properties. As we have noted, one condition for metallike conductivity in the stacking direction is a structural motif that involves stacked columns of closely interacting, planar metallomolecules (Figure 2a). An interesting puzzle is to understand those factors that result in the bis(α,β-dionedioximates) of various metals occasionally adopting this motif, yet failing to exhibit high conductivities.

Finally, the first reports[120, 121] of a conducting sulfur chelate, involving a partially oxidized metal dithiolate, have appeared, as have reports of materials based on dibenzotetraaza[14]annulenes,[172–175] further analogues of the porphyrins:

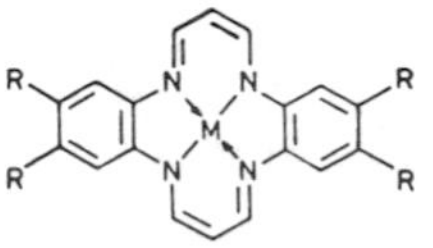

R=H : M(dbta)

We first give an overview of the general features of these various compounds and the means by which they have been characterized (Section 4.1). The individual classes of compounds are then described, beginning with the dionedioximates (Section 4.2) followed by the most important class, the porphyrinic complexes (Section 4.3), and then by the dithiolates and porphyrinlike analogues (Section 4.4).

4.1. General Features and Methods of Characterization

Although the generation of compounds in which the metal component is in a nonintegral oxidation state can be accomplished in principle by either reduction or oxidation, all highly conducting materials reported to date have been prepared by halogen oxidation of the neutral parent molecule. A major reason that partially reduced materials have not been much studied is that, unlike the oxidized materials, they are generally air sensitive. Molecular iodine has been an especially advantageous oxidant because of the stability of polyiodide anions (I_3^-, I_5^-) in nonpolar environments and because such ions are readily accommodated in channels in the various structures. Molecular bromine may be similarly useful, but it has not been

employed frequently. Other means of oxidation are, of course, applicable. All of the iodine-oxidized systems discussed here crystallize with stacks of metallomacrocycles around which are four parallel channels containing linear chains of iodine (Figure 25) with one exception: In the $Pd(bqd)_2I_{0.5}$ $\cdot 1/2S$ system (S = aromatic solvent), only two of the channels contain linear chains of iodine while two contain solvent.[162] In all cases metallomacrocycles are stacked metal-over-metal such that the mean molecular plane is perpendicular to the stacking (*c*) axis (Figure 2a). They alternate in an ABAB pattern along *c*, where molecule B is related to molecule A by symmetry but is rotated about the stacking axis by either ~40° in the porphyrin and phthalocyanine systems, ~65° in the bqd systems, or 90° in the dpg systems.

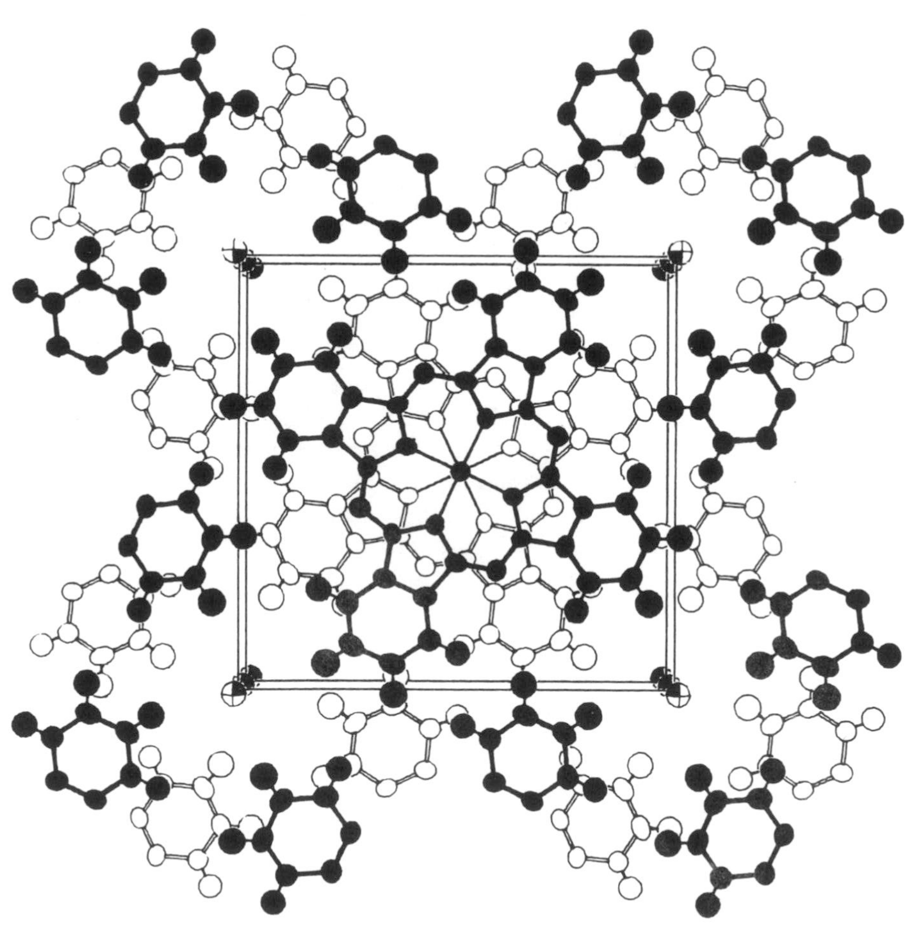

FIGURE 25. View down the *c* axis of the unit cell of Ni(pc)I (Reference 2). The similar view for Ni(tbp)I (Reference 132) is indistinguishable.

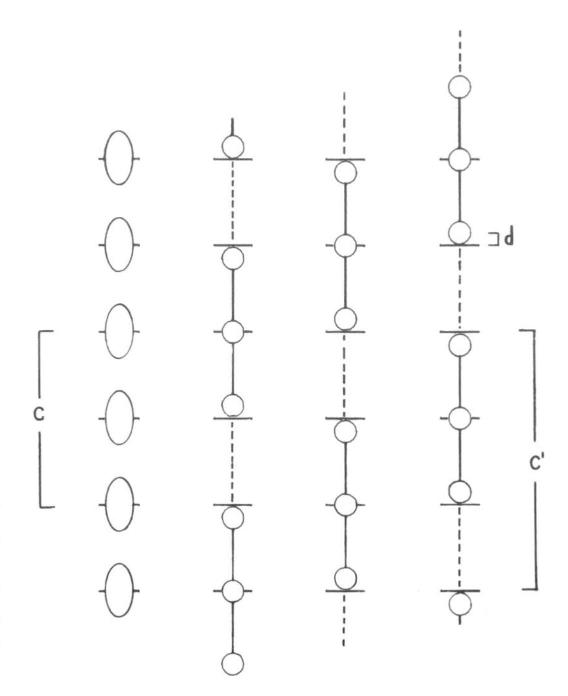

FIGURE 26. Disorder model for the iodine chains in Ni(pc)I. Average positions are shown on the left; the three distinct supercells each containing three such sites are shown with the same scale on the right (Reference 2).

The iodine species is usually disordered, as indicated by the presence of diffuse x-ray scattering in planes perpendicular to the stacking direction.[4,164,176] In the Ni(pc)I,[2] Ni(tbp)I,[132] and $Ni(bqd)_2I_{0.5}$ [164] systems, it has been possible to index all of the diffuse lines on the basis of a superlattice spacing of 9.5–9.8 Å, which is typical for a triiodide superlattice. The diffuse x-ray intensities for Ni(pc)I and $Ni(bqd)_2I_{0.5}$ have been collected and analyzed, and for each a model consisting of ordered chains of symmetrical triiodide units, which are disordered (translated) with respect to their neighbors, is consistent with the data (Figure 26). Often the intensity of a given diffuse line is slightly modulated, which indicates that there is a degree of three-dimensional order, but this feature of the diffuse scattering has not been investigated. For several compounds the diffuse x-ray scattering is too complex to be indexed on the basis of a single superlattice spacing. In Ni(tmp)I,[128,135] for example, most, but not all, of the diffuse lines can be indexed with a triiodide superlattice spacing of 9.8 Å, while in $Ni(omtbp)I_{1.08}$,[125] a 19.46-Å superlattice spacing only approximates the positions of the diffuse lines. There are no reports of a superlattice spacing that is typical of the I_5^- ion.

Convenient spectroscopic methods of identifying iodine-containing species have been developed in order to allow rapid characterization of newly prepared $M(L)I_x$ materials.[177,178] Resonance Raman spectroscopy

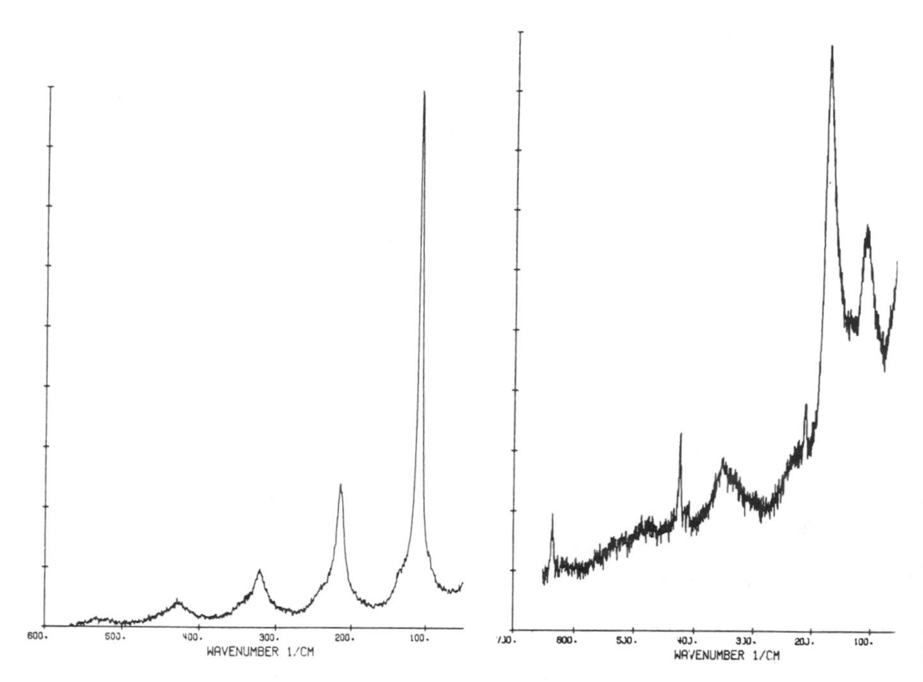

FIGURE 27. Representative resonance Raman spectra of partially oxidized M(L) crystals containing I_3^- (left, a) and I_5^- (right, b) (Reference 123).

offers a simple means of characterizing polyiodide anions. For example, Figure 27a displays the resonance Raman spectra characteristic of most of the $M(L)I_x$ systems reported to date. One may observe the intense, totally symmetric stretching fundamental (107 cm^{-1}) of a chain of symmetric I_3^- ions, along with the characteristic overtone progression. These materials give no evidence of I_5^- ($\nu \simeq 170$ cm^{-1}) or of I_2. On the other hand, the $M(pc)I_x$, $x \sim 3$, systems exhibit the scattering of I_2 coordinated to a Lewis base such as I_3^-,[122] and the $M(dpg)I_x$[4] and the $M(oep)I_x$[134] systems exhibit only scattering from the linear I_5^- ion (Figure 27b). Raman spectroscopy, of course, cannot be used to test for the presence of I^-. However, ^{129}I Mössbauer spectroscopy can be used to quantify the amounts of I^-, as well as to support the Raman assignment of polyiodide species. For example, the Mössbauer spectrum of $Ni(pc)I_{2.16}$ was shown to be consistent with the presence of I_3^- anions, and to limit to less than 3% and 5% the possible amounts of I^- and I_2, respectively.[2]

An understanding of charge transport in a molecular crystal must begin with knowledge of a number of characteristics of the material, such as composition and structure, but knowledge of the ionicity, or degree of partial oxidation, is also essential. In many systems the ionicity is difficult to obtain.[6, 8, 11] However, the techniques applicable to halogen-oxidized donors eliminate such difficulties. A knowledge of the composition of an $M(L)I_x$ crystal and characterization of the anionic species by resonance Raman and Mössbauer spectroscopies and x-ray diffraction techniques makes it possible to determine the ionicity directly. For example, a crystal of composition M(L)I in which the iodine-containing species is shown to be I_5^- ions is immediately characterizable as $(M(L))^{+0.2}(I_5^-)_{0.2}$. However, the use of spectroscopic methods without corroboration can be misleading. Reaction of the dication $[Pd(taab)]^{+2}$ with molecular iodine produces a compound formulated by chemical analysis and resonance Raman spectroscopy as the nonintegral oxidation state compound $Pd(taab)^{+2.7}(I_3^-)_{2.7}$.[179, 180] However, later the crystal structure showed the compound to be $[Pd(taab)][I_8]$, an integral oxidation state system containing the unusual I_8^{-2} polyiodide dianion.[181]

Describing the cation state generated from a neutral, $M^{II}(L)$ metallomacrocyle is more complex than describing an oxidized organic compound. One must consider the site of oxidation, since both the central metal and the aromatic π system of the macrocycle can be redox active.[182–184] For some metals, such as Mn, Fe, Co, and Ag, the first oxidation normally occurs at the metal ion and the process can be written

$$M^{II}(L) \longrightarrow M^{III}(L) + e^-$$

However, for metals without a readily accessible trivalent state, such as Mg and Zn, the highest occupied orbital of the complex is a π-molecular orbital of L, and the first oxidation is ligand centered:

$$M^{II}(L) \longrightarrow M^{II}(L^+) + e^-$$

Neutral molecules with an anion bound to a trivalent metal also behave in this manner.[129, 130]

There are instances, however, where the metal- and ligand-oxidized states of an individual complex are roughly degenerate and are interconvertible. The oxidized form of Ni(tpp) has been shown to exhibit an equilibrium between the $Ni^{II}(tpp^+)$ and $Ni^{III}(tpp)$ forms, with the former exhibiting a typically free-radical-like EPR signal and the latter displaying an EPR spectrum characteristic of a low-spin, d^7, Ni^{III} ion.[184] Similarly, Ni(pc) can form either a ligand- or metal-oxidized species, depending on conditions.[182, 185] As we discuss below, such borderline cases become particularly interesting when they occur in one-dimensional solids.

EPR measurements provide the means of ascertaining the nature of the carriers in a partial oxidized stack of M(L) complexes. It is straightforward to construct a formal description for the general case of a carrier in an arbitrary $M(L)I_x$ crystal.[125, 132] Consider a crystal of composition

$[M(L)]^{\rho+}(I_3^-)_\rho$ in which the hole state might display characteristics both of a planar M^{III} and of a π-cation radical species, with perhaps an additional small spin delocalization onto iodine. The carrier spin in this crystal can be described by a phenomenological wavefunction:

$$|\psi> = \gamma|[M^{II}(L)^+]_{\rho-1}(I_3^-)\rangle + \beta|[M^{III}(L)]_{\rho-1}(I_3^-)\rangle + \alpha|[M(L)]_{\rho-1}(I_3\cdot)\rangle \quad (1)$$

The normalization condition gives $\gamma^2 + \beta^2 + \alpha^2 = 1$, but typically $\alpha^2 \ll 1$, so $\gamma^2 + \beta^2 \cong 1$. If the $[M(L)]^+$ state is a quantum mechanical d–π mixture, then β^2/γ^2 gives the ratio of d and ligand contributions. Alternatively, in the limiting case where intrastack intermolecular interactions among the metal d and among the ligand π orbitals are large compared with d–π mixing, then β^2/γ^2 represents the relative susceptibility of d and π bands. This corresponds to an intrastack, two-band model analogous to the two bands, one on the donor and one on the acceptor stacks, that underlie the g-factor decomposition developed for the (TTF)(TCNQ) class of organic molecular metals.[6, 11, 186] Through use of Eq. (1), the g-tensor components for an $M(L)I_x$ compound are found to be[125, 132]

$$g_i = \gamma^2 g_i(M^{II}(L)^+) + \beta^2 g_i(M^{III}(L)) + \alpha^2 g_i(I_3\cdot) \quad (2)$$

where g_i represents g_x, g_y, or g_z, and reference g values for the isolated component species are to be used.

Analysis of the observed temperature-dependent g tensor allows one to obtain the phenomenological coefficients and through them to deduce the properties of the carrier spin states. For example, the observed g tensors of $Ni(omtbp)I_{1.08}$ [125] and Ni(pc)I[2] are free-radicallike and temperature independent. They were analyzed in terms of a ligand-centered oxidation, but with a small spin density (<0.5%) on the iodine, arising from back charge transfer from I_3^- to the macrocycle stack. However, the g values of Ni(tbp)I [132] and Ni(tatbp)I [128, 133] deviate appreciably from the free-electron value and indicate that the carrier spins have a substantial metal d character.

4.2. The α,β-Dionedioximates

The bis(α,β-dionedioximates) of various transition metals, especially dmg, dpg, and bqd complexes of the Ni triad, have been extensively studied.[4, 145–171] Table 9 lists those glyoximate systems for which complete structural data are available. Table 10 lists the corresponding data for the $M(bqd)_2$ systems. These complexes adopt two basic structural motifs. The first, typified by α-$Pd(bqd)_2$,[158] consists of planar units stacked in columns with the M–M vector parallel to the stacking axis (Figure 2a). The repeat in such a motif is usually two units, with a rotation of approximately 45°

between successive units. The metal–metal distances are short (3.15–3.40 Å), the packing is somewhat inefficient, leading to relatively low density and the existence of channels in the stacking direction, and even the unoxidized parents show evidence of anisotropic properties.[157] The channels serve not only as a potential residence for an oxidant, such as a halogen, but also for

TABLE 9. Structure and Properties of Metal Glyoximates

Compound	Structure; description of stacking	Conductivity (Ω^{-1} cm^{-1})	References
$Pd(dhg)_2$	SS *d*(Pd–Pd) = 3.558(3)		S 145
$Ni(dhg)_2$	SS *d*(Ni–Ni) = 4.196(1) Symmetry $\bar{1}$		S 146 Also see 147
$Pt(dhg)_2$	SS *d*(Pt–Pt) = 3.504(1) Symmetry $\bar{1}$		S 148
$Ni(dmg)_2$	MOM *d*(Ni–Ni) = 3.245 Symmetry 2/*m*	$\sigma = 1.0 \times 10^{-10}$ (sem) $\Delta = 0.57$ eV	S 149, 150 C 151
$Ni(meg)_2$	SS *d*(Ni–Ni) = 4.747(1) Symmetry $\bar{1}$		S 152 (earlier 153)
$Pd(dmg)_2$	MOM *d*(Pd–Pd) = 3.25 Symmetry 2/*m*	$\sigma = 3.3 \times 10^{-10}$ (sem)	S 150 C 151
$Pt(dmg)_2$	MOM *d*(Pt–Pt) = 3.23(1) Symmetry 2/*m*	$\sigma < 10^{-12}$	S 154 S 151
$Ni(dpg)_2I$	MOM Ruffled units *d*(Ni–Ni) = 3.223(2) Symmetry 222 I_5^-	$\sigma = 2.7 \times 10^{-3}$ (dc) (sem) 3.0×10^{-3} (ac) $\Delta = 0.19$ eV	S 4 C 4, 169, 170
$Pd(dhg)_2I$	MOM *d*(Pd–Pd) = 3.244 Symmetry *mmm*		S 168
$Pd(dpg)_2I$	*d*(Pd–Pd) = 3.26 (P) I_5^-	$\sigma = 2.0$–8.0×10^{-5} (sem) $\Delta = 0.388$ eV	4, 169, 170
$Ni(dpg)_2Br$	*d*(Ni–Ni) = 3.36 (P) Br_5^-	$\sigma = 3.8$–9.1×10^{-4} (sem) $\Delta = 0.327$ eV	169, 170
$Pd(dpg)_2Br_{1.1}$	*d*(Pd–Pd) = 3.28 (P) Br_5^-	$\sigma = 8.0$–15×10^{-5} (sem) $\Delta = 0.208$ eV	169
$[Pt(dpg)_2][ClO_4]$	MOM Ruffled units *d*(Pt–Pt) = 3.259(4) Symmetry 222		S 155

TABLE 10. Structure and Properties of Metal Benzoquinonedioximates

Compound	Structure; description of stacking	Conductivity (Ω^{-1} cm^{-1})	Other	References
$Ni(bqd)_2$	SS *d*(Ni–Ni) = 3.856(3) Symmetry $\bar{1}$		Young's modulus: $0.40(10) \times 10^{11}$ dyn/cm^2	S 156, YM 157
$Pd(bqd)_2$ α form	MOM *d*(Pd–Pd) = 3.202(1) Symmetry 2/*m*		Young's modulus: $1.7(9) \times 10^{11}$ dyn/cm^2 Absorption peak: 1.8 eV	S 158 A, YM 157
$Pd(bqd)_2$ β form	SS *d*(Pd–Pd) = 3.774(2) Symmetry I		Young's modulus: $0.67(14) \times 10^{11}$ dyn/cm^2	S 159 A, YM 157
$Pt(bqd)_2$	MOM *d*(Pt–Pt) = 3.173(1) Symmetry 2/*m*	$\sigma = 10^{-3}$ (sem) $\Delta = 0.25$ eV	Young's modulus: $3.6(1.5) \times 10^{11}$ dyn/cm^2 Absorption peak: 0.7 eV	S 160 A, YM 157 C 161
$Ni(bqd)_2I_{0.018}$	MOM *d*(Ni–Ni) = 3.180(1) Symmetry 2/*m*	$\sigma < 9 \times 10^{-9}$		S, C 162

$Ni(bqd)_2I_{0.52} \cdot 0.32C_6H_5CH_3$	MOM Symmetry $2/m$ I_3^- ions and solvent in channels d(Ni–Ni) = 3.153(3)	$\sigma = 1.8 \times 10^{-6}$ (dc) 1.1×10^{-7} (ac) $\Delta = 0.54(8)$ eV	Absorption peak: 1.6 eV	S 162, 163, 164 A 157 C 162
$Pd(bqd)_2I_{0.50} \cdot 0.52o\text{-}C_6H_4Cl_2$	MOM Symmetry $2/m$ I_3^- ions and solvent in channels d(Pd–Pd) = 3.184(3)	$\sigma = 5.6 \times 10^{-3}$ (dc) 4.5×10^{-3} (ac) $\Delta = 0.22(3)$ eV	Young's modulus: $1.2(3) \times 10^{11}$ dyn/cm^2 Absorption peak: 1.7 eV	S, C 162 A, YM 157
$Pt(bqd)_2 \cdot 1/2AgClO_4$	Two half, nearly planar units in asymmetric unit Symmetry $\bar{1}$ SS d(Pt–Pt) = 3.386(3) Ag^+, ClO_4^- ions fill channels			S 165
$Pt(bqd)_2I_2$ Modification I	Octahedral Pt^{IV} No Pt–Pt interaction Symmetry $\bar{1}$			S 166
$Pt(bqd)_2I_2$ Modification II	Octahedral Pt^{IV} No Pt–Pt interaction			S 166

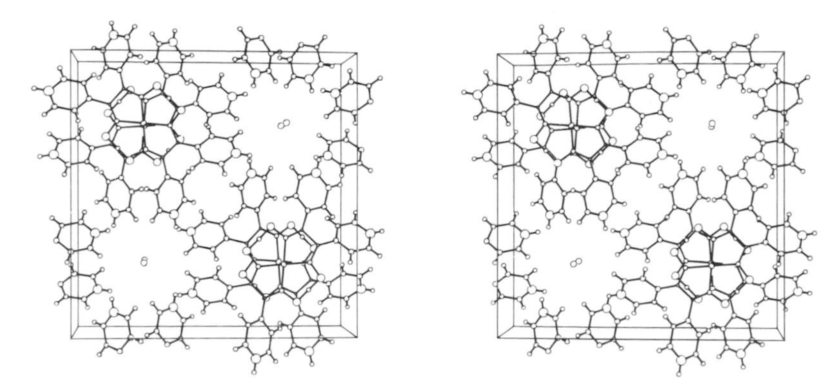

FIGURE 28. A stereoview of the unit cell of $Ni(dpg)_2I$. The *a* axis is horizontal to the right, the *b* axis is vertical from bottom to top, and the *c* axis is toward the reader. The vibrational ellipsoids are drawn at the 50% level, except hydrogen atoms, which are drawn arbitrarily small (Reference 4).

solvent molecules (occasionally undetected).[187] Figures 28 and 29 show two views of the $Ni(dpg)_2I$ structure.[4] The channels are clearly evident. But other arrangements of channels are possible, as shown in Figure 30, where the metal complexes maintain this same motif. The second motif (Figure 2b), typified by $Ni(bqd)_2$[156] (Figure 31), involves the placement of planar units in a slipped-stack arrangement so that the M–M vector is not parallel to the stacking direction. The result is a longer M–M distance, a more efficient packing with higher density, no channels, and often a herringbone

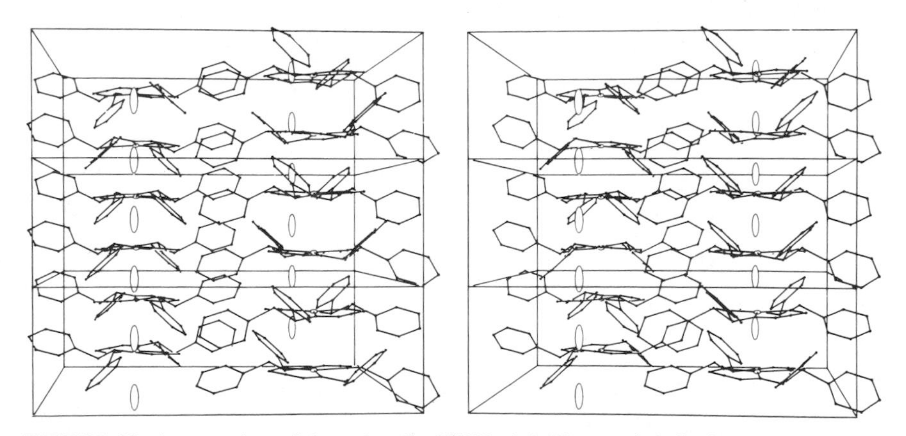

FIGURE 29. A stereoview of the unit cell of $Ni(dpg)_2I$. The *a* axis is horizontal to the right, and *b* axis is away from the reader, and the *c* axis is vertical from bottom to top. The vibrational ellipsoids of the Ni and I atoms are drawn at the 50% level. All other atoms are drawn arbitrarily small (Reference 4).

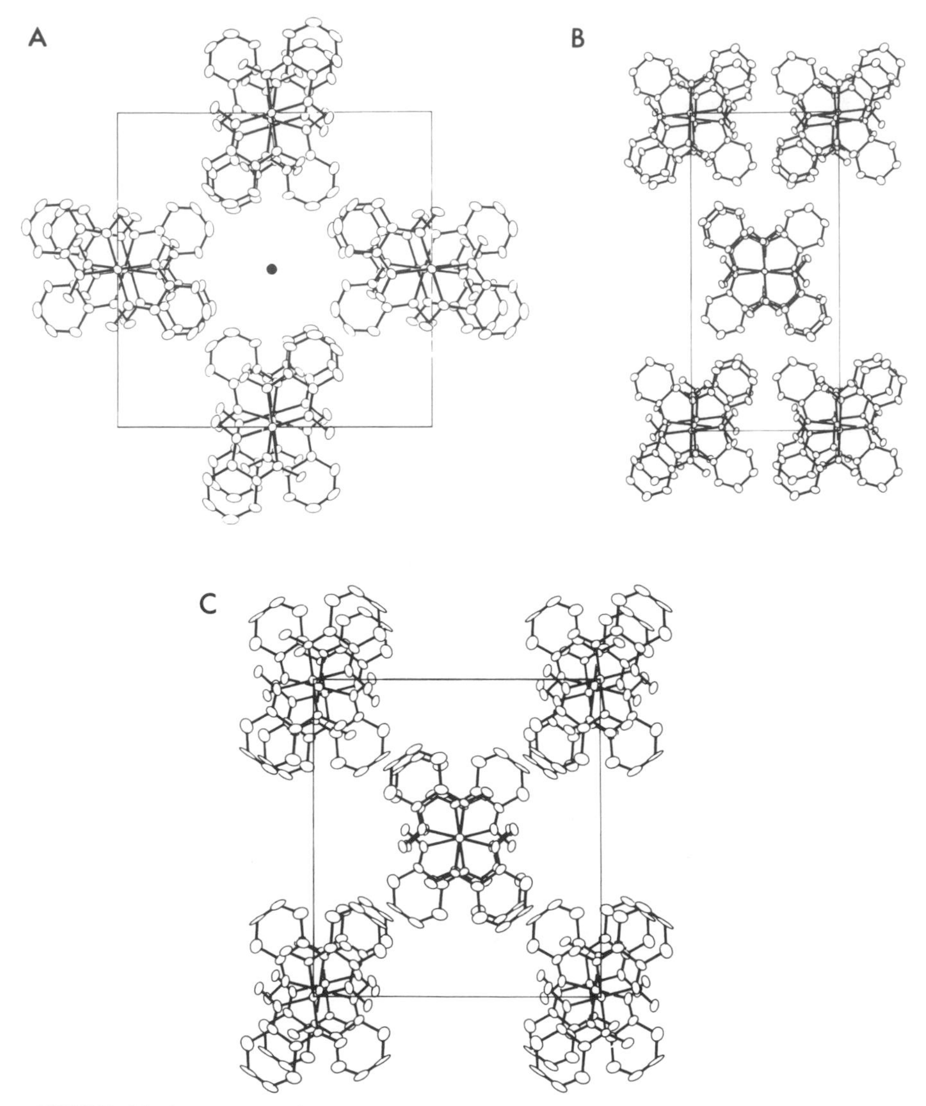

FIGURE 30. Comparison of metal bisbenzoquinonedioximate crystal structures viewed along the stacking direction. A is $Ni(bqd)_2I_{0.5}$ and $Pd(bqd)_2I_{0.5}$ (Reference 162), B is $Pd(bqd)_2$ (Reference 158), and C is $Ni(bqd)_2I_{0.018}$ (Reference 162).

arrangement. Clearly the tendency to form metal–metal interactions favors the first motif, while intermolecular van der Waals' and packing forces favor the second motif.

In addition, polymeric complexes of $M(dpg)_2$ and $M(dmg)_2$, M = Fe or Co, have been prepared with pyrazine as a bridging ligand. The structure consists of pyrazine–M–pyrazine–M · · · infinite chains with dmg or dpg units perpendicular to the chains.[188]

Endres et al.[189] have attempted to define those factors that influence the motif through the synthesis of a large number of bis(α,β-dionedioximates) of the Ni triad. Although they carried out no structure determinations of the resultant compounds, they did determine unit cell dimensions and crystal systems. Based on the correspondence of these cells with those of known structures (Tables 9 and 10) they attempted to deduce the motif present and in the case of motif 1 the M–M distance from the unit cell dimensions. Such an approach involves risks, as similar unit cells do not necessarily guarantee similar structures. They conclude that the first motif, involving direct metal–metal interactions, is favored for the Ni triad when strongly electron withdrawing ligands of low bulk are employed. But probably more subtle factors are involved. Thus, while $Ni(bqd)_2$ [156] (Figure 31) adopts motif 2, $Ni(bqd)_2I_{0.018}$ [162] (Figure 30) adopts motif 1; it is by no means clear that the small amount of iodine present occupies a specific position in the structure. Similarly, $Pd(bqd)_2$ exists in two modifications as does $Ni(meg)_2$. But when $Ni(meg)_2$ is recrystallized from $CHCl_3$, only the modification adopting the second motif is obtained.[190] Clearly the packing forces in these systems are deserving of further theoretical consideration.

The question of structural motif aside, these complexes are of interest

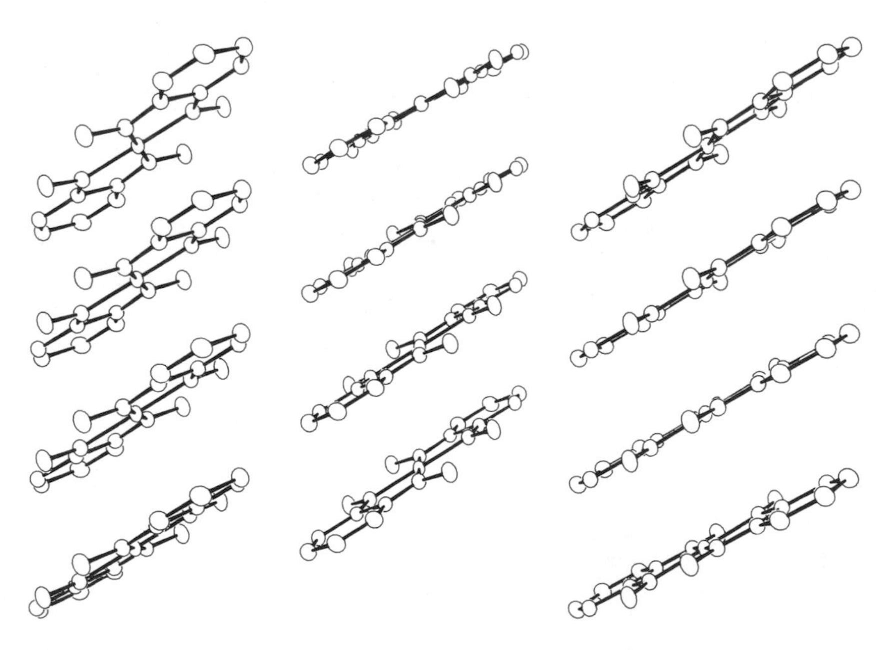

FIGURE 31. Packing diagram of $Ni(bqd)_2$ plotted from the data of Reference 156. Vibrational ellipsoids are drawn at the 50% level (Reference 162).

owing to the changes in physical properties that occur on partial oxidation. The report by Edelman[171] that $Ni(dpg)_2$ could be oxidized by I_2 or Br_2 to afford $Ni(dpg)_2X$ ($X=Br$ or I) appears to be the first in what is now an extensive group of reports on the oxidation of the bis(α,β-dionedioximates) of the Ni triad as well as of planar metallomacrocycles in general. Edelman logically guessed that $Ni(dpg)_2X$ was a compound of Ni^{III}, a relatively unusual, but by no means inaccessible state of Ni. However, several lines of evidence, then and later, did not support the presence of Ni^{III} in these systems. An x-ray photoelectron spectroscopic study of $Ni(dpg)_2I$[4] showed no evidence of a trapped valence or of any appreciable change in the charge on the metal upon oxidation. Although early EPR studies suggested a ligand oxidation to yield a free radical,[167] later studies on purer materials found no EPR signal at all.[191] Foust and Soderberg,[167] on the basis of a limited crystallographic study, deduced that the $M(dpg)_2X$ systems ($M=Ni$ or Pd; $X=I$ or Br) adopted structural motif 1 (Figure 2a), with M–M distances of 3.26–3.36 Å, depending upon M and X, and with channels in which the halogen atoms (nature unknown) were located. Such an arrangement was consistent with the known nature of the starch–iodine complex.[192] A complete three-dimensional structure determination of $Ni(dpg)_2I$, in combination with spectroscopic studies, has established that this compound contains mainly, if not exclusively, I_5^- ions in the channels and hence that the Ni is in the formal oxidation state of 2.20 on the assumption that the ligand is not oxidized. In addition, this study established that the electrical conductivity in the Ni–Ni direction is about 10^5 times that in the parent complex. Thus partial oxidation converts an insulator to a semiconductor possessing highly anisotropic conductivity properties.

The site of partial oxidation in these halogenated compounds remains a mystery, owing in part to the paucity of magnetic data. Miller has suggested an indirect method of determining whether the oxidation is metal- or ligand-centered based on changes in the relative orientation of adjacent molecules upon halogenation.[193] Oxidation out of a metal orbital, because of its a_g symmetry, should not affect the relative orientation, whereas oxidation out of a ligand orbital should result in a rotation to allow for better ligand–ligand overlap between adjacent molecules and thus a stabilization of the partially occupied molecular orbital. On the basis of this argument and the known crystal structures of parents and oxidized products, $Ni(dpg)_2I$ and $Pd(dpg)_2I$ are metal oxidized, while $Ni(bqd)_2I_{0.52}$ is ligand oxidized. Clearly, this argument ignores crystal packing considerations and may be incorrect. A simple rationalization of the low conductivity displayed by these materials would be that in every case oxidation is metal centered, and that the M–M distances are too great to support metallike conductivity. However, this would require a reinterpretation of the photoelectron spectroscopic data.[4]

TABLE 11. Structural Data for Metal Glyoximate Systems

Compound	M–M (Å)	M–N(1) (Å)	M–N(2) (Å)	N(1)–M–N(2) (deg.)	References
$Pd(dhg)_2$	3.558(3)	1.957(7)	1.953(9)	80.3(4)	145
		1.991(8)	1.958(7)	80.3(4)	
$Ni(dhg)_2$	4.196(1)	1.868(4)	1.880(4)	82.2(2)	146
$Pt(dhg)_2$	3.504(1)	2.013(14)	1.968(14)	78.0(7)	148
$Ni(dmg)_2$	3.245	1.85(2)	1.85(2)		149, 150
$Ni(meg)_2$	4.747(1)	1.861(4)	1.862(4)	82.5(2)	152, 153
$Pd(dmg)_2$	3.25	1.99(2)	1.93(2)		150
$Pt(dmg)_2$	3.23(1)	1.93(4)	1.95(4)	83	154
$Ni(dpg)_2I$	3.223(2)	1.868(15)		82(1)	4
$Pd(dhg)_2I$	3.244(1)	1.989(19)		79.3(10)	168
$[Pt(dpg)_2][ClO_4]$	3.259(4)	1.95(2)		77(1)	155

Are there other manifestations of partial oxidation? The answer is not clear. One can surely note that the few structures known of the partially oxidized materials show structural motif 1, even though some of the parents show motif 2. (In this sense, which is somewhat illusory, partial oxidation brings about a "reduction" in the M–M distance.) But there is no direct evidence that partial oxidation reduces the M–M distance in strictly comparable cases, where parent and oxidized product have the same motif. In particular, note that the Pt–Pt distance of 3.173 Å in $Pt(bqd)_2$ is slightly shorter than the Pd–Pd distance of 3.184 Å in $Pd(bqd)_2I_{0.50}$, despite the fact that the Pt radius is larger than the Pd radius. Tables 11 and 12 tabulate some metrical details on these molecules. Although the data are generally of disappointing accuracy, the effects of oxidation state of the

TABLE 12. Structural Data for Metal Benzoquinonedioximate Systems

Compound	M–M (Å)	M–N(1) (Å)	M–N(2) (Å)	N(1)–M–N(2) (deg.)	References
$Ni(bqd)_2I_{0.018}$	3.180(2)	1.858(10)	1.940(10)	83.0(5)	162
$Ni(bqd)_2I_{0.52} \cdot 0.32C_6H_5CH_3$	3.153(3)	1.91(2)	1.90(2)	84(1)	162, 163
$Ni(bqd)_2$	3.856(3)	1.868(4)	1.860(5)	83.6(2)	156
$Pd(bqd)_2I_{0.5} \cdot 0.52C_6H_4Cl_2$	3.184(3)	1.996(7)	1.955(7)	80.8(3)	162
$Pd(bqd)_2$ (α form)	3.202(1)	2.00(2)	1.95(2)	81(1)	158
$Pd(bqd)_2$ (β form)	3.774(2)	1.963(8)	1.967(8)	79.5(4)	159
$Pt(bqd)_2$	3.173(1)	1.99(1)	1.99(1)	80.0(7)	160
$Pt(bqd)_2 \cdot 1/2AgClO_4$	3.386(3)	1.99(2)	1.98(2)	77(1)	165
		1.95(2)	1.99(2)	77(1)	
$Pt(bqd)_2I_2$ modification I	—	2.00(1)	2.01(1)	80.0(6)	166
$Pt(bqd)_2I_2$ modification II		2.011(6)	1.987(6)	79.4(3)	166
		2.024(6)	1.980(6)	78.9(3)	

metal (ranging in some instances from +2 to +4) are not evident. The only discernible trend is increasing M–N distances as one descends from Ni to Pd to Pt. Referring again to the conductivity data, we find that the nature of the metal apparently is unimportant. On the other hand, when we examine the porphyrin and phthalocyanine systems we will find upon oxidation by iodine a dramatic increase in conductivity—conductivity that in some instances is metallike. Clearly the ligand can have an extremely important effect on the transport properties, although such effects are not obvious within the bis(α,β-dionedioximate) compounds of the Ni triad.

4.3. The Porphyrins and Derivatives

4.3.1. Molecular Architecture

The porphyrinic complexes are all essentially planar. However, the simple porphyrin and tbp macrocycles are somewhat flexible, and can adopt a ruffled, or saddle-shaped distortion through twisting at the methine carbon atoms. These deformations cause almost negligible changes in electronic structure, but the conformational mobility offers an added element of subunit variability.

This conformational mobility is exemplified in the structures of Ni(oep) (Figure 32). In the triclinic form of Ni(oep) [194] the molecules are planar and centrosymmetric (Figure 32a) and are stacked inclined relative to the stacking axis in a herringbone pattern similar to that adopted by most of the metallophthalocyanines. By change of solvent, Ni(oep) can be made to crystallize in a tetragonal form[195] in which the molecule has crystallographically imposed $\bar{4}$ symmetry, and the packing is entirely different from that found for the triclinic form. The pyrrole rings in the tetragonal form of Ni(oep) retain their planarity, but the molecule as a whole is nonplanar

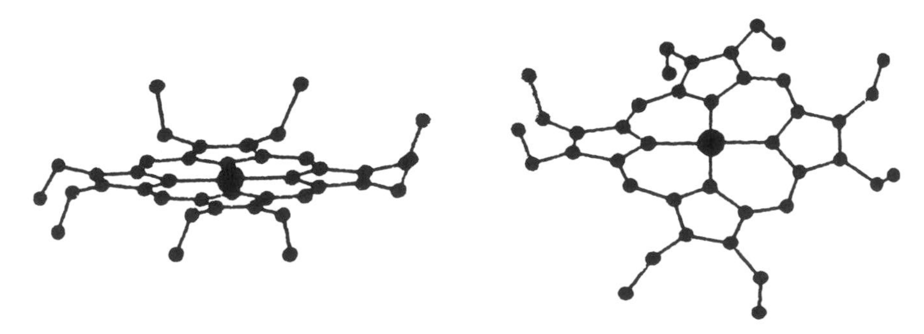

FIGURE 32. The Ni(oep) molecule in (left) triclinic Ni(oep) where the porphyrin ring is planar and (right) tetragonal Ni(oep) where the porphyrin ring is puckered (References 194).

because the pyrrole rings are tilted at an angle of 16.4° with respect to the mean molecular plane (Figure 32b). The interplanar angle between adjacent pyrrole rings is calculated† to be ~23° in this ruffled form of Ni(oep) compared with only 2.1° in planar Ni(oep). As a result of the puckering of the Ni(oep) molecule, the β-carbon atoms of the pyrrole rings lie as much as 0.5 Å above the mean porphyrin plane, and both the Ni–N and Ni–C_m distances decrease from values of 1.958(2) and 3.381(3) Å, respectively, found for the planar Ni(oep) [194] molecule to 1.929(3) and 3.355(4) Å, respectively, found for the puckered Ni(oep) [195] molecule. The existence of two conformations of Ni(oep) in crystals without molecules of solvation implies that the energy to deform the porphyrin ring from a planar to a ruffled conformation is not large. A similar mobility is exhibited by the tbp system, as evidenced by the crystal structures obtained for iodine-oxidized compounds (*vide infra*).

In contrast, the aza bridges of M(pc) compounds provide rigid connections and the molecules tend more toward planarity. Most four-coordinate metallophthalocyanines have α- and β-polymorphic crystalline forms with the β-polymorph being the more stable and more completely characterized.[141] This packing motif is very different from the metal-over-metal stacks (Figure 2a) adopted by the partially oxidized Ni(pc)I_x complex (Figure 25). Both forms consist of M(pc) slipped stacks (Figure 2b) in which the M(pc) molecule is planar and has a crystallographically imposed center of symmetry. The intramolecular distances and angles of the metallophthalocyanine molecules in the α and β crystalline modifications are very similar, the major differences being in the crystal packing and in particular in the inclination of the M(pc) molecule relative to the stacking axis. The β polymorphs of metal-free "H_2," Fe^{II}, Ni^{II}, Mn^{II}, Co^{II}, Cu^{II}, and Zn^{II} phthalocyanines are very similar structurally,‡ and a good description of the minor differences in the crystal packing in these systems has been reported by Mason et al.[197] The M(pc) molecules in all of these systems stack in a characteristic herringbone pattern as shown in Figure 33. The angle between the perpendicular to the MN_4 plane and the stacking (*b*) axis ranges from 45.7° in H_2pc [198] to 48.4° in Zn(pc).[199] As a result of the inclination of the M(pc) molecule relative to the stacking axis, the central metal atom of one M(pc) molecule is close enough to an azamethine nitrogen atom in the M(pc) molecules above and below it so that two weak

† Meyer[195] reported the crystal structure of tetragonal Ni(oep) and defines the angle between adjacent pyrrole rings as twice the angle between the plane of one pyrrole ring and the mean molecular plane. We prefer to define this angle as the dihedral angle between the planes of two adjacent pyrrole rings and have calculated this angle using the atomic coordinates given in Reference 195.

‡ The Ni(pc) molecule is not included in any comparisons since the structure of Ni(pc) completed in 1937,[196] is not as accurate as the structures of the other M(pc) systems.

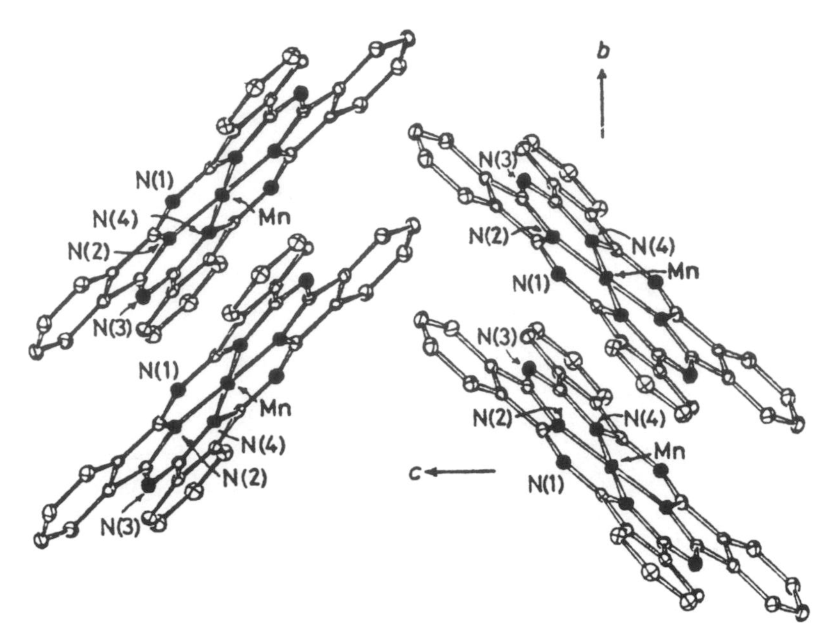

FIGURE 33. A view of the herringbone stacking pattern found in β-Mn(pc), a typical β-phthalocyanine (Reference 197).

intermolecular M · · · N bonds are formed, giving the central metal atom a distorted octahedral geometry. Figure 34 shows the intermolecular overlap that makes possible the formation of this weak bond. The intermolecular M · · · N bond lengths range from 3.15 Å in Mn(pc) [197] to 3.28 Å in Cu(pc).[200] The isoindole moieties of the pc ring retain their planarity although the azamethine nitrogen atom involved in the intermolecular M · · · N bond deviates slightly from the mean molecular plane in the direction of the metal atom. The stabilization of the β polymorph relative to the α polymorph is attributed to the formation of this intermolecular M · · · N bond. A metal such as Pt^{II} is less likely to be stabilized by this octahedral coordination, and indeed the β polymorph of Pt(pc) has not yet been isolated. Instead, α and γ polymorphs of Pt(pc) are stable, and both have been structurally characterized.[201, 202] In these systems, the angles between the perpendicular to the PtN_4 plane and the stacking axis are 25.3° and 30.8° for the α and γ polymorphs, respectively, as compared with the ~47° angle found in the β polymorphs described above. As a result, the intermolecular distance between the Pt atom and the closest azamethine nitrogen atom in the α and γ polymorphs is increased (relative to the β polymorphs) to 3.8 Å.

Of the metallophthalocyanines that crystallize in forms other than the α, β, and γ polymorphs, the most notable are Pb(pc) [203] and Ga(pc)F.[130]

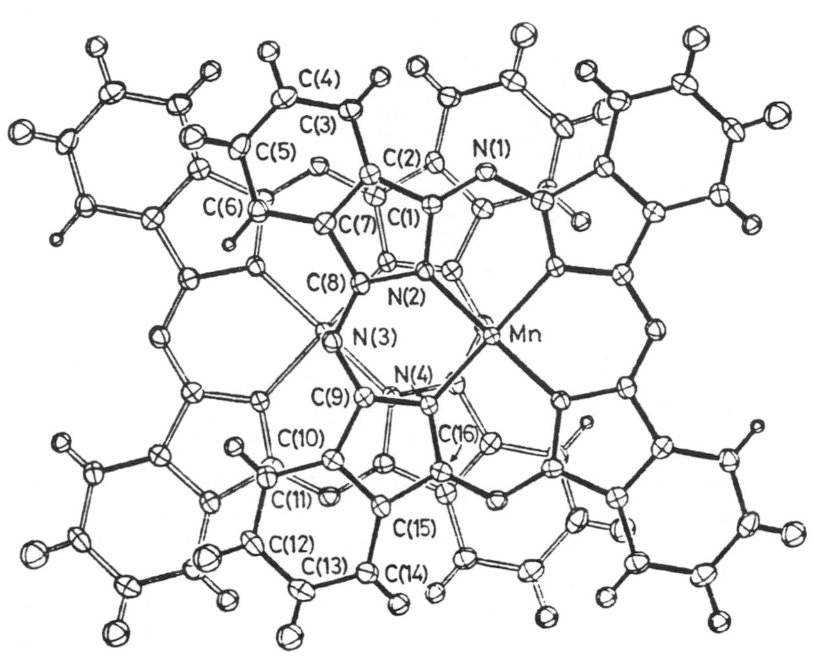

FIGURE 34. Normal projection of two parallel molecules in β-Mn(pc) (Reference 197).

The Pb(pc) molecules are stacked metal-over-metal, as shown in Figure 35, and the Pb–Pb spacing is 3.73 Å. The phthalocyanine ring deviates markedly from planarity, although the separate isoindole moieties retain their planarity. In Ga(pc)F the Ga atoms are symmetrically bridged by F, with a Ga–F distance of 3.92 Å; the pc rings are eclipsed, rather than staggered.[130]

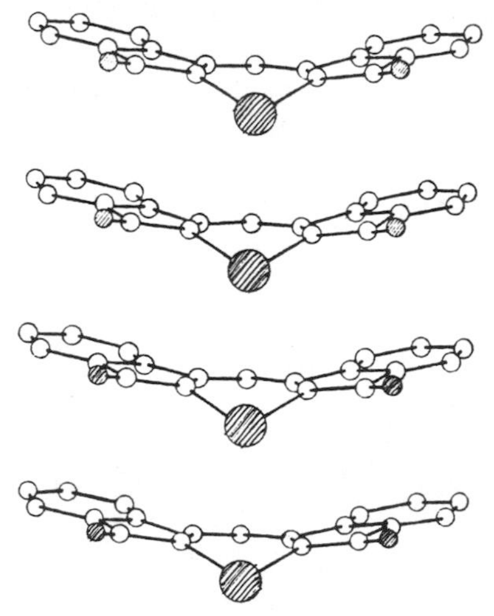

FIGURE 35. A view of the MOM P6(PC) stacks, showing half-molecules (Reference 203).

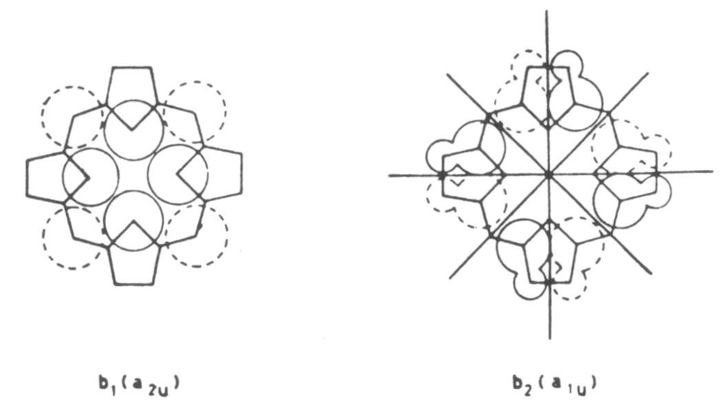

FIGURE 36. Porphyrin top-filled molecular orbitals. The orbital coefficients are proportional to the size of the circles. Solid or dashed circles indicate sign. Symmetry nodes are drawn in heavy lines (Reference 204).

4.3.2. Electronic Properties

In order to understand the solid state properties of the porphyrinic conductors, one must consider their electronic structure,[204] first as the isolated molecules and then as they interact in a one-dimensional stack.[133] The two highest-occupied ligand molecular orbitals of these complexes are of symmetry a_{2u} and a_{1u}, the former having large densities at the metal and nitrogen atoms, the latter having nodes passing through these atoms (Figure 36).[204] In the large-ring system, which we will denote generically as (tbp), calculations indicate that the a_{1u} orbital is at higher energy than a_{2u}, but that the size of the energy gap decreases with the number of aza nitrogen (=N–) bridges (Figure 37). Thus, ring oxidation of one of these

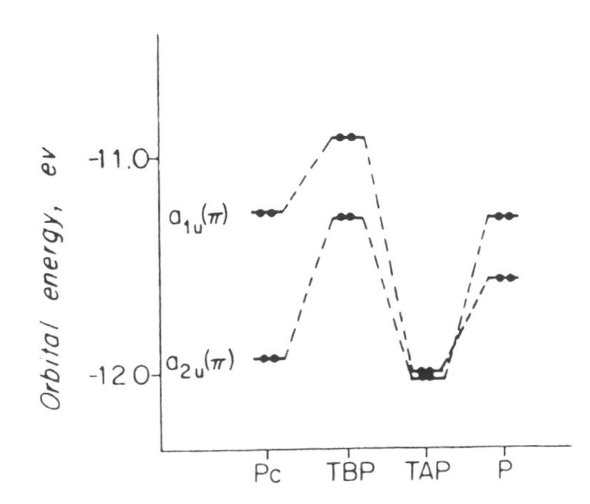

FIGURE 37. Top-filled π-molecular orbitals of Zn(pc), Zn(tbp), Zn(tap), and zinc porphine [Zn(p)] (redrawn from Reference 204).

metallomacrocycles is expected to yield the A_{1u} cation radical state. However, the appropriate choice of metal and peripheral substituents can actually alter the symmetry of the cation state formed by ring oxidation of the small, porphyrinato ring.

The metal-over-metal stacking arrangement typically adopted by partially oxidized M(L) complexes (Figures 2a and 25) is ideal for intermolecular interactions between the highest occupied ligand π orbitals and between the occupied metal d orbitals. In addition, there are classes of molecular orbitals that, though symmetry forbidden to interact in an isolated molecule, can hybridize through the influence of intermolecular interactions.[132, 133] In such M(L) stacks, neither intra- nor intermolecular interactions are of proper symmetry to mix the a_{1u} and a_{2u} ligand π-molecular orbitals, and mixing between the a_{1u} and d_{z^2} orbitals remains forbidden as well. However, the a_{2u} and d_{z^2} orbitals can hybridize. Thus, as a function of the π- and d-orbital energy level scheme for the individual M(L) molecules, the interaction within a crystal in a nonintegral oxidation state can lead to one of a variety of cases. A ligand-centered carrier can arise from oxidation from a band that is composed solely of the macrocycle

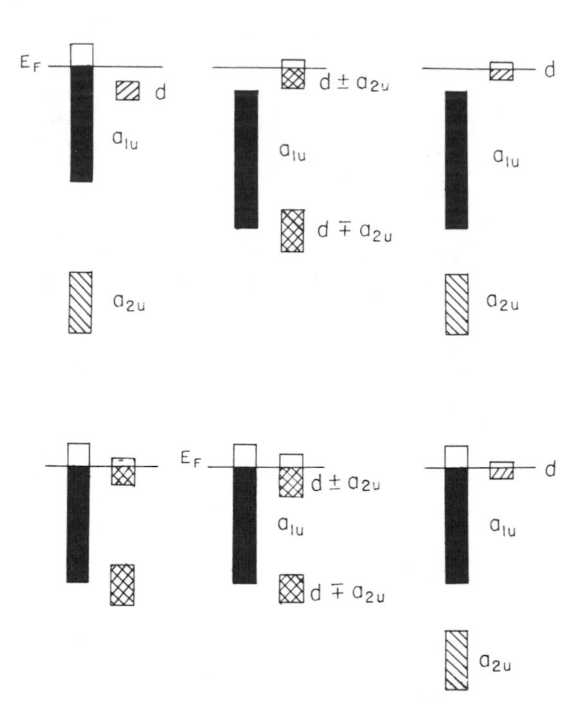

FIGURE 38. Some possible band structures for single-band conduction in partially oxidized M(L) stacks.

a_{1u} molecular orbital, while in the opposite limit oxidation can be purely metal centered. As an intermediate case, oxidation might occur from a d–π hybridized band (Figure 38). Even more interesting is the possibility of oxidation from two independent bands, one of π character, the second of either d–π or pure d character. As discussed below, such crystals exist[132, 133] and exhibit a novel, doubly mixed valence state. In addition to the intermolecular mixed valency along a M(L) stack, there is a second form of mixed valency within an individual oxidized complex, the $[M(L)]^+$ cation itself having characteristics intermediate between those of $[M^{II}(L)^+]$ and $[M^{III}(L)]$.

4.3.3. Large-Ring Metallomacrocycles

Studies of the compounds prepared by iodination of large-ring metallomacrocycles graphically demonstrate the chemical flexibility inherent in the use of such complexes to build molecular conductors. The first indication that M(pc) compounds could form the basis for a wide class of molecular conductors was the isolation of polycrystalline $M(pc)I_x$ powders.[122] Oxidation of "H_2," Fe, Co, Ni, Cu, Zn, and Pt phthalocyanines by molecular iodine results in materials with a range of compositions, $0.5 < x < 4$. For $x \lesssim 3$, resonance Raman (Figure 27) and ^{129}I Mössbauer spectroscopy showed the only form of iodine to be I_3^-. Pressed-pellet conductivity measurement showed semiconducting behavior but with the room temperature conductivity increased by over ten orders of magnitude from that of the parent complex. The activation energies in several cases (Ni, Cu, Pt) were so low as to lead to the suggestion that single crystals would have high, metallic ($d\sigma/dT < 0$) conductivities.

This result prompted a substantial interest in preparing single crystals of I_2-oxidized, large-ring metallomacrocycles. The first published report was the isolation of single crystals of a pair of materials obtained by the partial oxidation of Ni(omtbp) with iodine[124] and this was followed shortly by a report of the preparation of Ni(pc)I single crystals.[3] Both early reports have been followed by detailed single-crystal studies of structural, magnetic, and charge-transport properties of these compounds[2, 125] and the characterization of numerous related systems. Table 13 lists the key features of the currently known materials. Individual compounds will be described in a fashion most convenient for comparisons.

4.3.3.1. Ni(omtbp).[124–126] The two $Ni(omtbp)I_x$ materials isolated following oxidation with iodine have very different stoichiometries, $x = 1.08 \pm 0.01$ and 2.9 ± 0.3, and represent the only case to date in which a porphyrinic metallomacrocycle has yielded crystals at different levels of partial oxidation. The crystal structure for the $x = 1.08$ [125] material exhi-

TABLE 13. *Properties of Large-Ring Porphyrinic Systems*

Compound	M	x	Magnetic properties		Conductivity (Ω^{-1} cm^{-1})	References
$M(pc)(I_3)_x$	Ni	0.33	SC: $g_{\parallel} = 2.0075$ $g_{\perp} = 2.0007$ $\Gamma_{\parallel} = 7.2$ G $\Gamma_{\perp} = 5.4$ G	$\chi^p = 1.2 \times 10^{-4}$ emu/mol (static) $n_{eff} = 0.14$ spins/macr. (static, EPR) PS	SC: $\sigma = 260–750$ (type B) $\sigma^{-1} \propto T^{1.9}$ $T_{MI} = 55$ K P: $\sigma = 0.7$ (sem) $\Delta = 0.036$ eV	C 3, 2 127
	Ni	1.27			P: $\sigma = 0.6$ (sem)	122
	"H_2"	0.33	SC: $g_{\parallel} = 2.0057$ $g_{\perp} = 2.0020$ $\Gamma_{\parallel} = 4.4$ G $\Gamma_{\perp} = 3.0$ G	$\chi^p = 1.1 \times 10^{-4}$ emu/mol (static) $n_{eff} = 0.13$ spins/macr. (static, EPR PS	SC: $\sigma = 200–450$ (type B) $T_{MI} = 36$ K	131 127
	Pd	0.33	SC: $g_{\parallel} = 2.0082$ $g_{\perp} = 1.9993$ $\Gamma_{\parallel} = 3.4$ G $\Gamma_{\perp} = 2.3$ G	$n_{eff} = 0.22$ spins/macr. (EPR)	SC: $\sigma = 180–340$ (type A) $T_m = 82$ K	131 127
	Pt	0.33	SC: $g_{\parallel} = 2.027$ $g_{\perp} = 1.995$ $\Gamma_{\parallel} = 13$ G $\Gamma_{\perp} = 14.5$ G	PS	SC: $\sigma = 140–200$ (type A) $T_m = 121$ K	131 127

	Cu	0.33	SC: $g_{\parallel} = 2.134$ $g_{\perp} = 2.033$ $\Gamma_{\parallel} = 50$ G $\Gamma_{\perp} = 38$ G	$\chi^{p} = 1.75 \times 10^{-3}$ emu/mol $n_{eff} = 0.69$ spins/macr. (EPR) $n_{eff} = 1.25$ spins/macr. (static) CL	SC: $\sigma = 180$–300 (type B) $T_{MI} = 60$–80 K	131
	Cu	0.57			P: $\sigma = 5$ (sem)	122
	Co	0.33	No signal		SC: $\sigma = 70$ (type A) $T_{m} = 280$ K	131
	Fe	0.91			P: $\sigma = 3 \times 10^{-3}$ (sem)	122
$M(pc)(Br_3)_x$	Ni	0.35	P: $g_{\parallel} = 2.0044$ $g_{\perp} = 2.000$ $\Gamma_{\perp} = 1.0$–1.5 G	$n_{eff} = 0.24$ spins/macr. (EPR) PS	SC: $\sigma = 295$ (type A) $T_{m} = 120$ K	131
$M(tbp)(I_3)_x$	"H_2"	0.33	SC: $g_{iso} = 2.0027$ $\Gamma_{iso} = 3.7$ G		P: $\sigma = 0.01$	131
	Ni	0.33	$g_{avg} = 2.024$ $\Gamma = 107$ G $g_{\parallel} = 2.060$ $g_{\perp} = 2.018$ $\Gamma_{\parallel} = 110$ G $\Gamma_{\perp} = 75$ G g and Γ T dependent (see text)	$\chi^{p} = 9.0 \times 10^{-5}$ emu/mol (static) $n_{eff} = 0.11$ spins/macr. (static, EPR) PS	SC: $\sigma = 150$–340 (type A) $T_{m} = 95$ K $\sigma \propto T^{-1}$ ($T > 95$) $\Delta = 110$ K ($T < 95$)	132 128

(continued overleaf)

TABLE 13 (*continued*)

Compound	M	x	Magnetic properties		Conductivity (Ω^{-1} cm^{-1})	References
$M(tatbp)(I_3)_x$	"H_2"	0.33	SC: $g_{\parallel} = 2.0038$ $g_{\perp} = 2.023$ $\Gamma_{\parallel} = 3.3$ G $\Gamma_{\perp} = 2.6$ G	$\chi^p = 1.2 \times 10^{-4}$ emu/mol (static) $n_{eff} = 0.13$ spins/macr. (static, EPR) PS	SC: $\sigma = 10$–70 (type A) $T_m = 200$ K type B $T_{MI} = 130$ K Hysteretic transition	133
	Ni	0.33	P: $g_{iso} = 2.011$ $\Gamma_{iso} = 33$ G g and Γ T dependent (see text)	$n_{eff} = 0.14$ spins/macr. (EPR) PS	SC: $\sigma = 110$–200 (type A) $T_m = 150$ K $\sigma \propto T^{-1}$ ($T > 150$) $\Delta = 0.095$ eV	133 128
$M(omtbp)(I_5)_x$	"H_2"	0.31	SC: $g_{iso} = 2.0028$ $\Gamma_{\parallel} = 0.73$ G $\Gamma_{\perp} = 1.12$ G Γ increases with decreasing temp.	$n_{eff} = 0.33$ spins/macr. (EPR)	SC: $\sigma = 0.1$–0.4 (sem) $\Delta = 0.095$ eV	131
$M(omtbp)(I_3)_x$	Ni	0.36	SC: $g_{\parallel} = 2.0102$ $g_{\perp} = 2.0033$ $\Gamma_{\parallel} = 5.23$ G $\Gamma_{\perp} = 3.51$ G	$\chi^p = 3.9 \times 10^{-4}$ emu/mol (static) $n_{eff} = 0.40$ spins/macr. (static, EPR) CL to 10 K 4–10 K MT	SC: $\sigma = 4$–16 (type A) $T_m = 300$ K	124 125 126

		0.97	SC: $g_{\parallel} = 2.0108$ $g_{\perp} = 2.0029$ $\Gamma_{\parallel} = 2.80$ G $\Gamma_{\perp} = 2.26$	$\chi^{p} = 1.1 \times 10^{-3}$ emu/mol (static) $n_{eff} = 0.90$ spins/macr. (static, EPR) BF $J < 10\ cm^{-1}$	SC: $\sigma = 1$–4 (type A) $T_m = 340$ K	124 125 126
	Cu	0.33			SC: $\sigma = 0.1$–0.3 (sem) $\Delta = 0.08$ eV	131
$(\text{-M(pc)-O})_n\text{I}_x$	Si	0.5 1.4 4.6		$\chi^{p} = 3.0$–5.0×10^{-4} emu/mol (static)	P: $\sigma = 2 \times 10^{-2}$ (0.5)[a] $= 2 \times 10^{-1}$ (1.4) $\Delta = 0.04$ eV	136
	Ge	1.80 1.90 1.94 2.0		$\chi^{P} = 3.0$–5.0×10^{-4} emu/mol PS	P: $\sigma = 3 \times 10^{-2}$ (1.82) $\Delta = 0.08$ eV $\sigma = 5 \times 10^{-2}$ (1.90) $\Delta = 0.06$ eV $\sigma = 6 \times 10^{-2}$ (1.94) $\Delta = 0.05$ eV $\sigma = 1 \times 10^{-1}$ (2.0) (sem)	136
	Sn	1.2 5.5		$\chi^{P} = 3.0$–5.0×10^{-4} emu/mol PS	P: $\sigma = 1 \times 10^{-6}$ (1.2) (sem) $\Delta = 0.68$ eV	136
$\text{-(M(pc)-O)}_n\text{-Br}_x$	Si	1.0			P: $\sigma = 6 \times 10^{-2}$ (sem)	136
$\text{-(M(}t\text{-Bu}_4\text{pc)-O)}_n\text{-I}_x$	Si	$0.5 \leq x \leq 3.9$			P: $\sigma = 2 \times 10^{-4}$ (0.5) P: $\sigma = 2 \times 10^{-3}$ (3.9) (sem)	137
$\text{-(M(pc)-F)}_n\text{-I}_x$	Al	$0.012 \leq x \leq 3.4$			P: $\sigma = 2.2 \times 10^{-4}$ (0.02) $= 4.5$ (2.7) (sem)	129, 130
	Ga	$0.048 \leq x \leq 2.1$			P: $\sigma = 3.5 \times 10^{-5}$ (0.048) $= 0.15$ (2.1) (sem) $\Delta = 0.04$ eV	129, 130

[a] Here and in succeeding tables the numbers in parentheses denote stoichiometries.

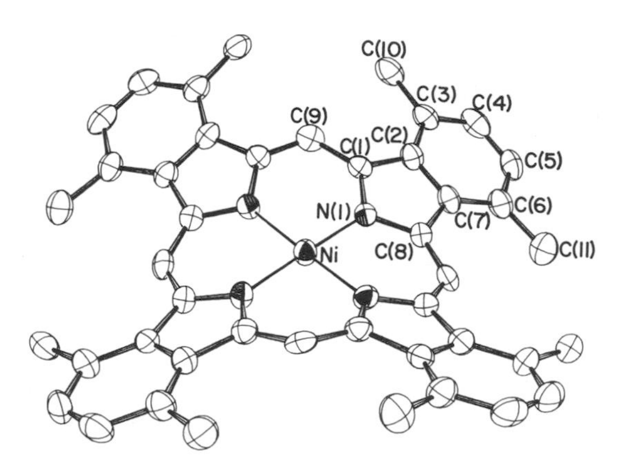

FIGURE 39. A perspective drawing of the Ni(omtbp) molecule. Hydrogen atoms have been ommitted (Reference 125).

bits columns of macrocycles, as illustrated in Figures 2a and 25. The iodine occurs as I_3^- chains which parallel the macrocyclic columns, and which exhibit an interesting one-dimensional disorder. The iodine disorder in Ni(omtbp)$I_{1.08}$ is much more severe than that found in the Ni(pc)I[2] or Ni(tbp)I[132] systems (see below). In order to solve the structure of Ni(omtbp)$I_{1.08}$, the disordered iodine atom had to be treated as a statistical distribution of electron density, with the fit for both cosine and triangle-shaped distributions being comparable.

The compound Ni(omtbp)$I_{1.08}$ thus exhibits the general topology for one-dimensional conductivity. However, in this compound the macrocycle is "puckered" by intramolecular steric hindrance among the exocyclic methyl groups; it is on a site of $\bar{4}$ symmetry and is severely ruffled (Figure 39). The isoindole moieties in the Ni(omtbp) molecule are essentially planar but are tilted at an angle of ~20° with respect to the mean molecular plane. This large deviation from planarity allows the molecule to avoid close contacts between the bulky methyl substituents on the fused benzene ring of the isoindole unit. As a result of this ruffling of the porphyrin ring, the Ni(omtbp) molecules cannot approach each other closely. The intrastack spacing, as measured by the distance between adjacent Ni atoms, is 3.778(5) Å [125] at 298 K, as compared with ~3.25 Å in other macrocyclic compounds in nonintegral oxidation states. Similarly, the shortest interatomic contacts between adjacent Ni(omtbp) units are also long, 3.46 Å compared with typical values of ~3.25 Å.[2, 132]

These materials probably offer the most telling example of the critical role that the nonintegral oxidation state plays in creating a conductive molecular crystal. The poor intermolecular overlap suggests that the electron transfer matrix element, t, is much smaller than in other conductive molecular crystals. Indeed, that the magnetic susceptibility of the $x = 1.08$ crystals is Curie-like (Figure 40) sets a limit on the electron transfer integral of $t < 3\ \text{cm}^{-1}$, while $U \sim 10^4\ \text{cm}^{-1}$.[125] Nevertheless, both Ni(omtbp)I_x materials exhibit moderately high, metallike conductivities at or above room temperature, thus demonstrating the overriding importance of the nonintegral oxidation state in obtaining ready charge transport in molecular crystals (Section 1).

These compounds also play an important role in correlating experimental and theoretical studies of charge transport in quasi-one-dimensional molecular crystals. The magnetic, optical, and electrical properties of quasi-one-dimensional molecular crystals are frequently discussed in terms of Hubbard models, which provide the simplest approximation for the interplay between band and correlation effects.[6,11,19] The latter are approximated by the on-site repulsion $U \geq 0$ defined above, while the former lead to noninteracting electrons in typical $-2|t| \cos kc$ bands for a one-dimensional crystal with lattice spacing c. Exact results are available only in the band limit ($|t| > 0$, $U = 0$)[5] and in the atomic limit ($U \gg 4|t|$) of strong correlations,[22,23] although the best one-dimensional organic conductors apparently fall in the difficult regime of intermediate correlations,

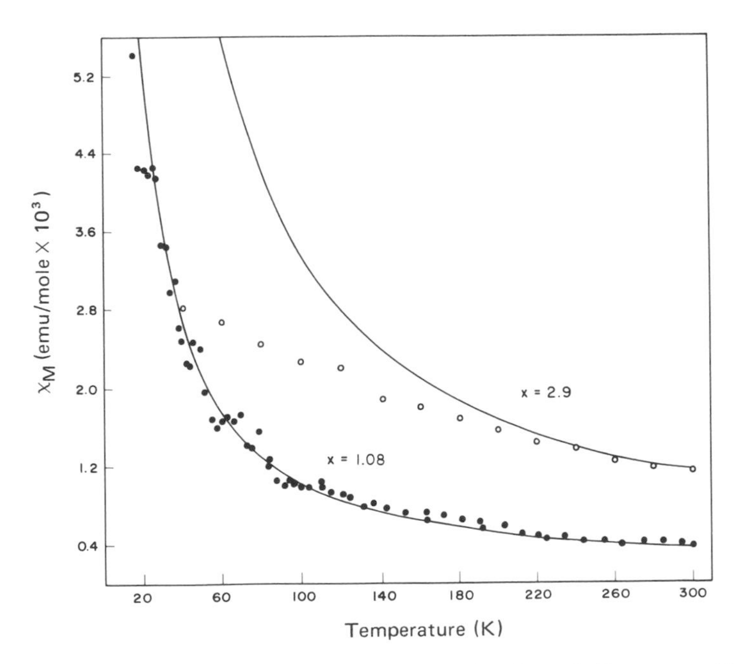

FIGURE 40. Temperature-dependent EPR intensities for Ni(omtbp)I_x systems. The lower data set is for a single crystal of Ni(omtbp)$(I_3)_{0.36}$ oriented with the field parallel to [001] and includes a least-squares fit of the data to a Curie law function. The upper set is the EPR intensity for powdered Ni(omtbp)$(I_3)_{0.97}$. The solid trace is a calculated, hypothetical Curie law response scaled to the ambient temperature susceptibility (Reference 125).

with $U \sim 4|t|$.[11] The analysis of the unusual properties of the Ni(omtbp)I_x complexes, namely, their metallike conductivity but Curie-like susceptibility, showed them to be the first experimental realization of the atomic limit for a partially filled band. Moreover, the joint analysis of the EPR spectra and conduction of the crystals showed that these properties can be understood in terms of paramagnetic $[Ni(omtbp)]^+$ polarons and correlated one-dimensional hopping.[126] The most important prediction of the polaron approach is a relationship between the conductivity, $\sigma(T)$, and EPR linewidth, $\Gamma(T)$. Both quantities are expected to be proportional to the polaron hopping frequency, $\omega_h(T)$. Analysis indicated that the ratio, $R = \Gamma(T)/\sigma(T)T^2$, should be independent of both temperature and hopping frequency whenever motional broadening dominates the linewidth. Such was actually observed, thus confirming the analysis (Figure 41). The value of R is reasonably constant for $T > 200$ K for several crystals whose conductivities each exhibit a slightly different temperature response; upward deviations at low temperature reflect other small ($\sim$0.5 G) contributions to Γ.

4.3.3.2. Ni(L), L = pc, tbp, and tatbp. Iodine oxidation of Ni(pc),[2,3,122] Ni(tbp),[128,132] and Ni(tatbp)[128,133,205] in each case leads to isolation of crystals with composition Ni(L)I. Full x-ray diffraction, resonance Raman, and ^{129}I Mössbauer studies show that the first two iodinated materials are isostructural and isoionic, consisting of columnar stacks of planar metallo-

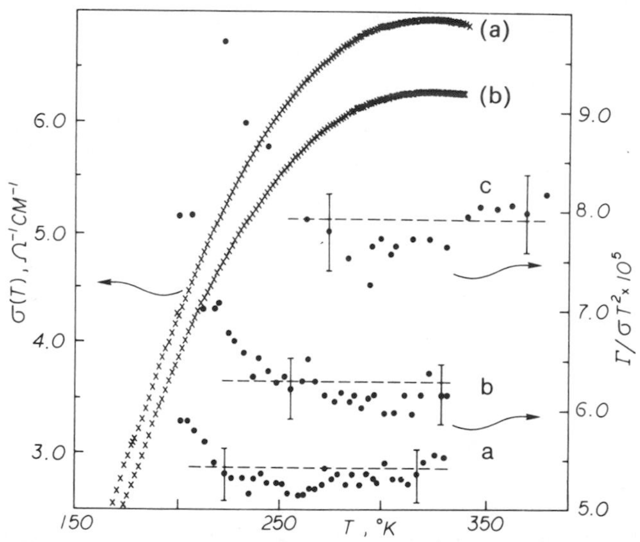

FIGURE 41. Comparison of EPR linewidths and conductivities for Ni(omtbp)$(I_3)_{0.36}$ crystals. Curves (a) and (b) are the conductivities $\sigma(T)$ of two crystals. Curves a and b give the ratio $R = \Gamma/\sigma T^2$ for the same crystals. Curve c gives R for $\sigma(T)$ and linewidth Γ taken from different crystals (Reference 126).

TABLE 14. Structural Data for Selected Nickel Porphyrin Systems

Compound	Ni–Ni (Å)	Ni–N (Å)	Molecular symmetry	References
Ni(pc)	4.71	1.83	$\bar{1}$	196
Ni(pc)I	3.244(2)	1.887(6)	4/*m*	2
Ni(tbp)I	3.217(5)[a] 3.26(2)	1.966(9)	4/*m*	132
Ni(omtbp)$I_{1.08}$	3.778(5)	1.953(5)	$\bar{4}$	125
triclinic Ni(oep)	7.617(2)	1.958(5)	$\bar{1}$	194
tetragonal Ni(oep)	8.228(6)	1.929(3)	$\bar{4}$	195
Ni(tmp)	5.648(6)[b]	1.959(10)	$\bar{1}$	143
[Ni(tmp)][TCNQ]	—	1.950(4)[a]	1	44
Ni(tmp)I	3.466(3)[a]	1.938(3)	$\bar{4}$	135

[a] 116 K.
[b] 123 K.

macrocycles with parallel chains of triiodide counterions (Figure 25). Furthermore, x-ray photographs of the Ni(tatbp)I [133] system are identical to those of Ni(pc)I and Ni(tbp)I, so all three of these systems appear to be structurally very similar. The electronic properties of the three Ni(L) parent molecules differ significantly, as discussed above. Since the Ni(L)I crystals exhibit no concomitant changes in ionicity or structure, they offer an opportunity to examine the dependence of solid state properties on purely electronic characteristics of the subunit in the absence of competing effects. The metallomacrocycle in each of these compounds is on a site of 4/*m* symmetry and consequently is crystallographically constrained to be planar. Thus, there is no impediment to strong intrastack interactions, and the metal–metal spacing along a column of Ni(pc)I is 3.244(2) Å at 298 K,[2] and the spacings for Ni(tbp)I are 3.26(2) Å at 298 K and 3.217(5) Å at 116 K.[132]

Structural parameters for Ni(pc), Ni(pc)I, Ni(tbp)I, and Ni(omtbp)$I_{1.08}$ are included in Table 14 along with similar information for small-ring porphyrinatonickel systems.[44, 135, 143, 194, 195] Even the major structural differences between the large-ring systems are small, and can be attributed to the expected differences between the phthalocyanine and tetrabenzporphyrin rings themselves. The larger central hole that is present in the tetrabenzporphyrin ring is reflected in larger Ni–N bond distances for the Ni(tbp)I and Ni(omtbp)$I_{1.08}$ systems [1.966(9) and 1.953(5) Å, respectively] relative to the Ni(pc) and Ni(pc)I systems (1.83 and 1.887(6) Å, respectively). Also, the C_a–C_m–C_a angles in Ni(tbp)I and Ni(omtbp)$I_{1.08}$ [124.6(11) and 124.7(7)°, respectively] are larger than the corresponding C_a–N_m–C_a angles in Ni(pc) and Ni(pc)I [117 and 119.2(6)°, respectively]. This contraction of the azamethine bridge angle in the Ni(pc) system is probably caused by the steric requirements of the lone-pair electrons on the azamethine nitrogen atom.[206]

In Ni(pc)I and Ni(tbp)I, the iodine superlattice spacing, c', is a multiple of the Bragg spacing, c ($c' = 3/2c$), so the superlattice is commensurate with the Bragg lattice and the repeat unit contains three Bragg sites. The disorder in these two systems is not severe and could be modeled satisfactorily (Figure 26).[2, 132, 176] No special treatment of the iodine atoms was necessary to solve the Bragg structures, and the only structural manifestation of the disorder, aside from diffuse x-ray scattering, is an increase in the thermal parameter of I along the chain direction.

The properties of both Ni(tbp)I and Ni(pc)I are characteristic of "molecular metals," but the details of the charge transport properties differ significantly and there are striking contrasts in the magnetic properties of the two materials. The properties of Ni(tatbp)I are not yet as well established but nonetheless can be compared instructively with those of the other two materials.

Each of the three materials shows a reduced and nearly temperature-dependent paramagnetic susceptibility, χ^S. Relating χ^S to the Pauli susceptibility of a one-dimensional tight-binding band[5, 12] permits one to estimate the intermolecular interactions within a stack, and in each case one obtains $|t| \sim 0.1$ eV.

Room temperature conductivities and, of more significance, mean-free paths for carrier motion along the stack are comparable with values reported for other one-dimensional "metals," including the organic charge-transfer salts[5, 207–209] (Table 15). Both Ni(pc)I and Ni(tbp)I show a metal-like increase in conductivity upon cooling from room temperature (Figure 42) but the resistivity of the former accurately varies with $T^{1.9}$, while that of the latter changes more gradually. The major difference in the temperature dependence of the conductivity for these compounds occurs at lower temperatures. The conductivity for Ni(tbp)I reaches a broad maximum before falling off rapidly. In contrast, the conductivity of Ni(pc)I

TABLE 15. Comparison of Mean Free Paths for Some Organic and Inorganic Molecular Conductors

Compound	RT conductivity (Ω^{-1} cm^{-1})	Mean free path (intermolecular spacings)	References
$(TTT)_2I_3$	1000–10 000	2.0–20	207
(HMTSF)(TCNQ)	1400–2200	1.6–2.5	208
(TTF)(TCNQ)	500–1000	0.4–0.8	209
$K_2Pt(CN)_3Br_{0.3} \cdot 3H_2O$	100	0.6	5
Ni(pc)I	260–750	1.0–2.3	2
Ni(tbp)I	180–330	0.7–1.3	132
Ni(tatbp)I	110–200	0.4–0.8	133
$Ni(omtbp)I_{1.08}$	16	0.04	125
Ni(tmp)I	40–280	0.07–0.51	135

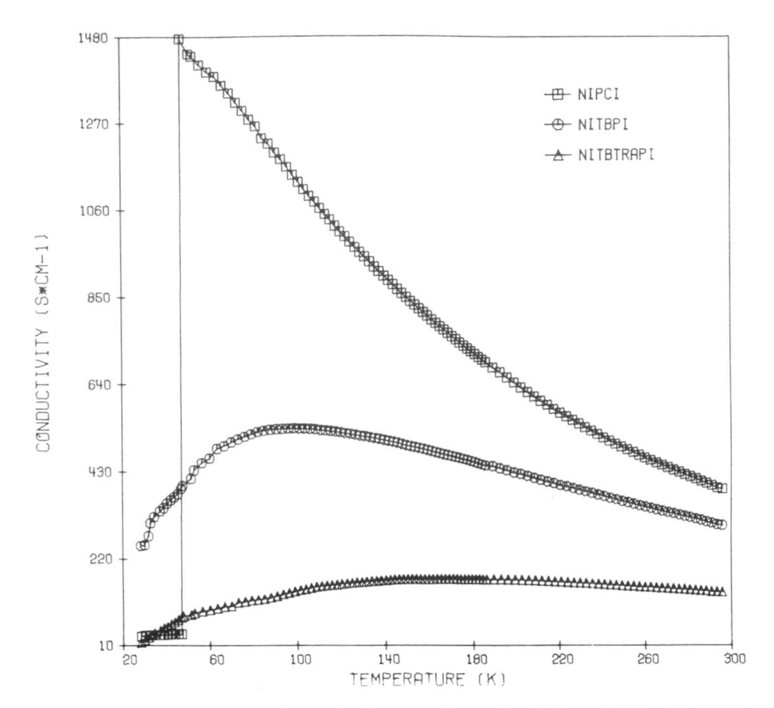

FIGURE 42. Temperature-dependent single-crystal conductivity of Ni(pc)I, Ni(tbp)I, and Ni(tatbp)I (References 2, 132, and 133).

increases with decreasing temperature until it reaches a maximum value of $\sim 1500\ \Omega^{-1}\ \text{cm}^{-1}$ at ~ 55 K. It appears to undergo a discontinuous transition to a semiconducting state with further cooling.† The characteristics of Ni(tatbp)I follow the temperature response of Ni(tbp)I, but are not simply intermediate between those of Ni(pc)I and Ni(tbp)I, although the room temperature conductivity is comparable.

Probably the most unusual result that emerges from this comparison concerns the nature of the charge carriers. In the study of Ni(pc)I, analysis of the g tensor showed that the metal retains its formal Ni^{II} oxidation state, playing a secondary role in the conduction process, and that the charge carriers are associated with delocalized π orbitals on the macrocycle. However, the carrier spin g values and linewidths (Γ) for Ni(tbp)I and Ni(tatbp)I are anomalous; both quantities are unusually large at room temperature [for Ni(tbp)I, $g_{\text{av}} = 2.03$, $\Gamma = 105$ G] and increase strongly as the temperature is lowered (Figure 43). Moreover, the EPR linewidths show an unprecedented increase with increasing spectrometer frequency.[128, 132, 205]

† Recent measurements indicate that Ni(pc)I indeed maintains a high conductivity ($\sigma > 10^3\ \Omega^{-1}\ \text{cm}^{-1}$) below 4.2 K (Martinsen, J., Greene, R. B., Palmer, S. M., and Hoffman, B. M., unpublished results).

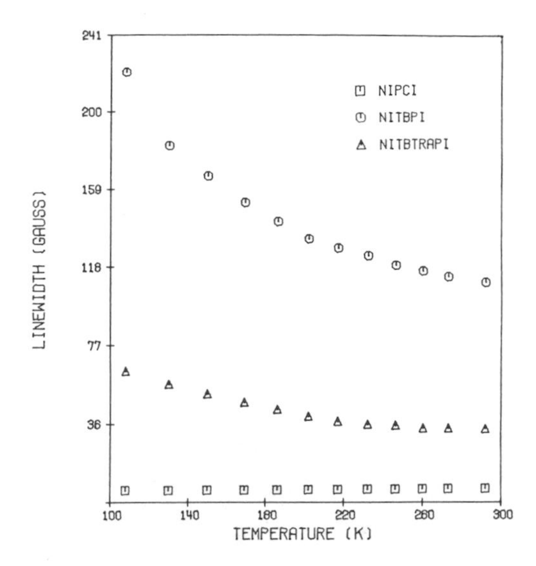

FIGURE 43. Temperature-dependent EPR linewidths of Ni(pc)I, Ni(tbp)I, and Ni(tatbp)I (References 2, 132, and 133).

This may be contrasted with frequency independence of the linewidth for Ni(pc)I, Ni(omtbp)I, and, for example, TTF–TCNQ[11]. These results have been interpreted to indicate that Ni(tbp)I and Ni(tatbp)I are the first partially oxidized complexes in which the charge carriers exhibit both metal and ligand properties, and that the electron "hole" created by iodine oxidation can "jump" between the metal and the macrocycle, as well as between one unit and its neighbors. In short, these complexes display a novel, doubly mixed valence state, the $[Ni(L)]^+$ cation itself exhibiting interconversion between the electronic tautomers, $[Ni^{II}(L)^+]$ and $[Ni^{III}(L)]$.[132] Based on this analysis, the frequency dependence of the linewidth requires that the carrier interchange between ligand and metal bands occur with a frequency, ω_h, that is comparable with the difference $\delta\omega$ between the resonance frequency for a pure metal d spin and a pure ligand π radical (i.e., at X band, $\delta\omega\beta(g_{Ni} - g_L)H_0/h \sim 10^9$–$10^{10}\ s^{-1}$).

The temperature-dependent conductivity of Ni(pc)I [and of H_2(pc)I,[131]; see below] shows a sharp, metal–semiconductor transition. However, very surprisingly, neither the Ni(pc)I static susceptibility nor the EPR linewidth shows any discontinuity in value or slope at T_{MI}.[2] Thus, the nature of the transition is unclear, and further work is in progress (see footnote on p. 533).

4.3.3.3. Incompletely Characterized Systems. Single crystals of a number of other iodinated phthalocyanines, tetrabenzporphyrins, and triazatetrabenzporphyrins have also been prepared,[127,131,133] as have crystals of Co(pc)Br (Table 13).[127,131] Typically, these compounds have the stoichiometry $M(L)I_{1.0}$, and it is likely that the structures are the same as for Ni(pc)I and Ni(tbp)I. All these compounds conduct well, but with a variety of behaviors (Figure 44). For example, H_2(pc)I appears to show a sharp metal–semiconductor transition as does Ni(pc)I, while most of the others show the rounded conductivity maximum of Ni(tbp)I. Yet, H_2(tatbp)I shows a rounded maximum, but a reversible and sharp but very hysteretic transition at a temperature below that where the maximum in the conductivity occurs. It may also be significant that crystals of Co(pc)I and Co(pc)Br exhibit very different conductivities. Detailed charge transport measurements on these materials will be of considerable interest.

The compounds Al(pc)F and Ga(pc)F form conductive powders, $(M^{III}(pc)F)Y_x$, $Y = I_3^-$, BF_4^-, PF_6^-, upon partial oxidation.[129, 130] The value of x varies from 0 to 1, but as with the M^{II} (pc) systems, the isolation of pure phases will be necessary for characterization of the stoichiometries. The observation of narrow EPR signals (≤ 1 G) at a g value of 2 shows the oxidation to be ligand centered (Section 4.1).

Phthalocyanine and porphyrin polymers of the form $(-M(L)-B)_n$ are well known, where L = pc or substituted pc, M = Si, Ge, or Sn, and the bridging group B may be O^{-2}, $(-O-CH_2-CH_2-O-)^{-2}$, or $(O-C_6H_4-O)^{-2}$.[210, 211] Treatment of phthalocyaninato polymers with iodine produces oxidized materials with a wide range of stoichiometries (Table 13).[136, 137] The iodine is in the form of I_3^-, except for high iodine

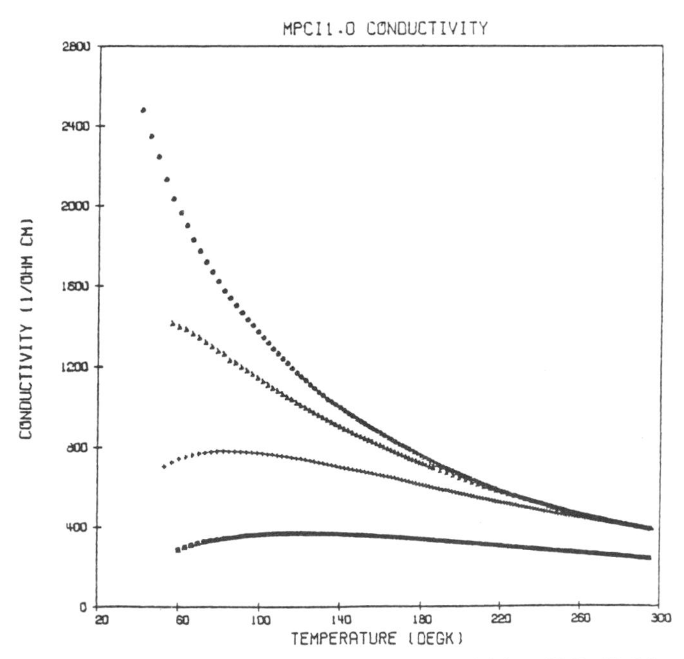

FIGURE 44. Temperature-dependent single-crystal conductivities of M(pc)I; M = H_2 (●), Ni (▲), Pd (+), Pt (■) (References 127 and 131).

content, where I_5^- is observed by resonance Raman spectroscopy. The materials are semiconductors, with room temperature conductivities as high as 5 Ω^{-1} cm^{-1} for a Si–O-linked powder. The conductivity decreases in the order $\sigma_{Si} > \sigma_{Ge} > \sigma_{Sn}$, in accord with the observation that there is a corresponding increase in the interplanar spacing. Further consideration of the temperature-dependent conductivities suggests that the siloxy and perhaps germyloxy materials could be intrinsically metallike in the chain direction.[136]

A synthetically novel approach to the preparation of porphyrin-based polymeric conductors involves the preparation of alkynyl-bridged systems, of the form $(M(L)\text{–}C{\equiv}C\text{–})_n$, where L = pc or various substituted pc and M = Si or Ge. Treatment with I_2 yields oxidized materials with conductivities in the range 10^{-4}–10^{-1} Ω^{-1} cm^{-1}.[138]

4.3.4. *Small-Ring Metallomacrocycles*

4.3.4.1. M(oep). The first report of high conductivity with a small-ring metallomacrocycle is found in the study of the iodine oxidation products of metallooctaethylporphyrins, M(oep). Iodine oxidation of Ni(oep), Cu(oep), and H_2(oep) yields polycrystalline materials with a range of stoichiometries[134] (Table 16). Resonance Raman spectroscopy indicates that the iodine occurs as the polyiodide I_5^- species, and thus the materials present an interesting contrast with the large-ring M(L)I_x compounds, where, to date, iodine occurs only as the I_3^- anion (Figure 27).

The Ni(oep)I_x and Cu(oep)I_x powders exhibit high electrical conductivities, which depend upon x and are approximately eight orders of magnitude greater than those of the parent porphyrin (Figure 45). The conductivity of H_2(oep)I_x is approximately three orders of magnitude greater than the unoxidized porphyrin, but far below that of the Ni and Cu complexes. Comparisons with the powder conductivities for other materials of this type indicate that single crystals of Ni(oep)I_x at room temperature will have conductivities of the order of 10^1–10^2 Ω^{-1} cm^{-1}, very possibly with a metallike temperature dependence. Since the M(oep)I_x compounds exhibit Curie-like susceptibility, this suggests that these materials may also exhibit the complete decoupling of orbital and spin degrees of freedom predicted by the one-dimensional Hubbard model in the atomic limit.[22, 23] Single-crystal studies would thus be of considerable interest.

The partially oxidized Ni and Cu complexes behave similarly, but the H_2(oep)I_x complexes have lower conductivities at all iodine stoichiometries. Since magnetic resonance measurements indicate that Ni, Cu, and H_2(oep) all undergo ring oxidation upon iodination, it may well be that the symmetry of the Ni(oep) and Cu(oep) cation state differs from that of H_2(oep) (Figure 36) and that the difference in conductivity reflects different intermol-

TABLE 16. Properties of Small-Ring Porphyrinic Systems

Compound	M	x	Magnetic properties	Conductivity (Ω^{-1} cm^{-1})	References
$M(oep)(I_5)_x$	"H_2"	0.22–1.16	P: $g = 2.0023$ (0.66) $\Gamma = 3.85$ (0.66)	P: $\sigma = 1.0 \times 10^{-8}$ (0.22)– 3.0×10^{-7} (1.16) (sem) $\Delta = 0.51$ eV (0.66)	134
	Ni	0.8–1.14	P: $g = 2.0016$ (0.66) 2.0072 (0.24) $\Gamma = 9.90$ (1.14)– 18.60 (0.34)	P: $\sigma = 1.5 \times 10^{-4}$ (0.08)– 2.8×10^{-2} (1.14, 0.34) (sem) $\Delta = 0.18$ (0.08)– 0.08 eV (1.14)	
	Cu	0.1–0.7	P: $g_{\parallel} = 2.151$ $g_{\perp} = 2.040$ $\Gamma_{\perp} = 36$–49 G CL	P: $\sigma = 2.0 \times 10^{-7}$ (0.1)– 6.0×10^{-3} (0.7) (sem) $\Delta = 0.16$ (0.7) eV	
$M(tmp)(I_3)x$	Ni	0.33	SC: $g_{\parallel} = 2.0067$ $g_{\perp} = 2.0024$ $\Gamma_{\parallel} = 4.65$ G $\Gamma_{\perp} = 6.00$ G $n_{eff} = 0.33$ spins/macr. (EPR, static) MT at $T = 28$ K	SC: $\sigma = 70$–280 (type A) $T_m = 115$ K	135

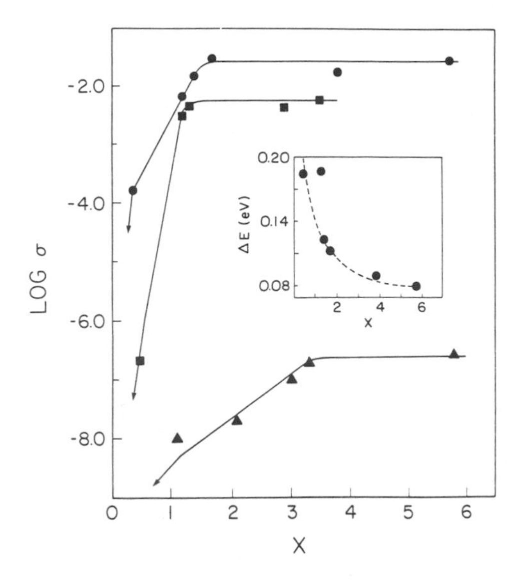

FIGURE 45. Log conductivity as a function of the iodine concentration (x) for pressed pellets of $H_2(oep)I_x$ (▲), $Cu(oep)I_x$ (■), and $Ni(oep)I_x$ (●). Insert: activation energy, ΔE (eV), for conduction as a function of x for $Ni(oep)I_x$ (●). Lines are drawn as guides and have no theoretical significance (Reference 134).

ecular overlaps. Alternatively, the porphine core of H_2(oep) is only twofold symmetric, and disorder along a conducting stack might suppress the conductivity.

4.3.4.2. Ni(tmp). Single crystals of Ni(tmp)I have been prepared by iodine oxidation of the parent complex.[128, 135] An x-ray diffraction study shows metal-over-metal stacks, with surrounding iodine chains, as is seen with the large-ring systems (Figures 2a and 25). The flexibility of the porphyrin ring and the consequent futility in trying to anticipate structural details is evident in the Ni(tmp)I system. Despite the fact that both the unoxidized Ni(tmp) molecule[143] and the Ni(tmp) moiety in the charge-transfer complex [Ni(tmp)][TCNQ] [44] are planar, the Ni(tmp) molecule in Ni(tmp)I has crystallographically imposed $\bar{4}$ symmetry and is definitely puckered. It is tempting to try to attribute this change in conformation to the oxidation of the porphyrin ring. However, since Ni(oep) exists in both a planar and a ruffled form depending on the solvent used for crystallization,[194, 195] and the Ni(pc) [2] and Ni(tbp) [132] systems both retain their planarity upon iodine oxidation, it is most reasonable to attribute the nonplanarity of the porphyrin in Ni(tmp)I to crystal packing forces. It is for such reasons that we have declined to speculate on the crystal packing of the $M(oep)I_x$ compounds.

Bond parameters for Ni(oep) and Ni(tmp) systems are included in Table 14. As was true for the ruffled Ni(oep) system, the pyrrole rings in Ni(tmp)I retain their planarity but are tilted with respect to the mean molecular plane. The ruffling of the porphyrin results in a decrease in the Ni–N bond distance from 1.959(10) and 1.950(4) Å in Ni(tmp) [143] and

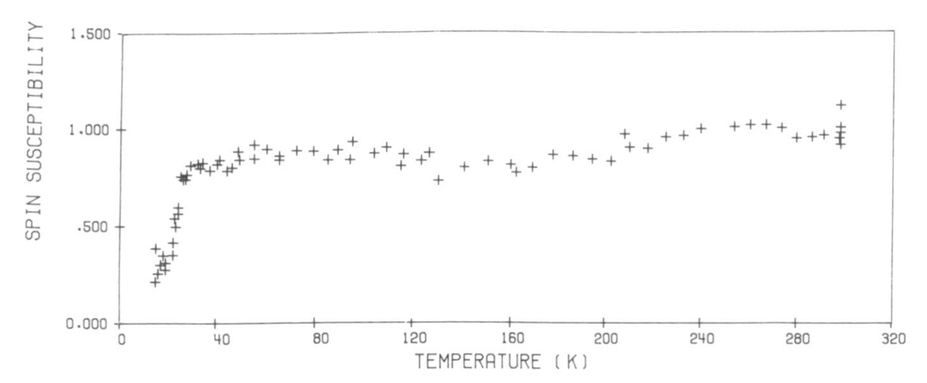

FIGURE 46. Normalized temperature-dependent carrier spin susceptibility (EPR intensity) of Ni(tmp)I (Reference 135).

[Ni(tmp)][TCNQ],[44] respectively, to 1.938(3) Å in Ni(tmp)I. The ruffling of the porphyrin ring has an effect on the crystal packing that is similar to that observed for Ni(omtbp)$I_{1.08}$, where adjacent molecules in the stack cannot approach each other as closely as they might if planarity were maintained. In Ni(tmp)I, the Ni–Ni separation of 3.466(3) Å is 0.3 Å shorter than that found in Ni(omtbp)$I_{1.08}$. The iodine superlattice in Ni(tmp)I is incommensurate with the Bragg lattice ($c' = 1.39c$), and as was the case with Ni(omtbp)$I_{1.08}$, the disorder is so severe that a statistical distribution of electron density was needed to model it.[135]

The physical properties of Ni(tmp)I are given in Table 16. The conductivity of the compound at room temperature is high ($\sigma \sim 100\ \Omega^{-1}\ \text{cm}^{-1}$) and metallike, increasing until it reaches a rounded maximum at $T \sim 115$ K, then decreasing in an activated manner. The spin susceptibility is temperature independent down to a transition at 28 K, well below the conductivity maximum. As the temperature is lowered further, the susceptibility decreases in an activated fashion, with $\Delta/k \approx 60$ K (Figure 46). Similar observations with organic conductors $(TMTTF)_2Y$, $Y = ClO_4^-$, BF_4^-, have been attributed to a Peierls transition,[212] but in the present instance the mechanism has yet to be elucidated.

4.4. *Other Metallomacrocycles*

Some details on other systems are given in Table 17.† Of the many $M(\text{dithiolate})_2^{-q}$ crystals discussed in Sections 2 and 3, none contained a partially oxidized complex or showed a high conductivity. However, the

† Although the temperature and pressure dependence of the conductivity of M(taab)Y complexes, M = Pd and Pt, Y = BF_4^- and I_8^{-2}, have been measured,[179, 180] these data have not been included in Table 17 as they were originally interpreted under the mistaken assumption[181] that the compounds were partially oxidized (Section 4.1).

TABLE 17. Structures and Properties of Other Partially Oxidized Systems

Compound	M	x	Structure	Magnetism	Conductivity (Ω^{-1} cm^{-1})	References
$Li_{1-x}[M(mnt)_2] \cdot 2H_2O$	Pt	0.3	d(Pt–Pt) = 3.62	No EPR signal	SC: $\sigma = 100$ (type A)	120, 121
$M(opd)_2I_x$	Ni	0.97, 2.57		P: $g = 2.0023$	P: $\sigma = 7 \times 10^{-7}$ (0.97) (sem)	173
		5.79		$\Gamma = 25$ G	4×10^{-5} (2.57)	
					1×10^{-7} (5.79)	
$M(dbtaa)I_x$	Ni	$0.8 \leq$		P: $g = 2.0023$	P: $\sigma = 4.5 \times 10^{-1}$ (1.0) (sem)	172, 174
		$x \leq 7.0$		$\Gamma = 30$ G	$\Delta = 0.051$	
					$\sigma = 4.4 \times 10^{-3}$ (1.57) (sem)	
	Ni	1.8			SC: $\sigma = 1$–50	172
	Pd	0.8, 2.0	I_3^-		P: $\sigma = 0.13$ (0.8) (sem)	175, 172
					$\Delta = 0.080$ eV	
					$\sigma = 0.4$ (2.0)	
					$\Delta = 0.04$–0.06 eV	
	Pt	1.35, 1.5	I_3^-		P: $\sigma = 0.12$ (1.35) (sem)	175
					$\Delta = 0.06$–0.12 eV	
					$\sigma = 0.03$ (1.5) (sem)	
	"H_2"	1.4	I_3^-		P: $\sigma = 2.4 \times 10^{-5}$ (sem)	172
	Cu	1.8	I_3^-		P: $\sigma = 1.1 \times 10^{-3}$ (sem)	172
	Co	1.9	I_3^-		P: $\sigma = 1.1 \times 10^{-2}$ (sem)	172
$M(dbtaa)Br_x$	Ni	2.43			P: $\sigma = 1.3 \times 10^{-5}$ (sem)	172
$M(tmdbtaa)I_x$	Ni	$1.76 \leq$			P: $\sigma = 1.4 \times 10^{-2}$ (1.7) (sem)	172
		$x \leq 2.9$			3.8×10^{-2} (2.9) (sem)	
					$\Delta = 0.079$ eV	
	Ni	2.44			SC: $\sigma = 1$–20 (2.44) type A	172
					$\sigma \propto T^{-2.5}$	
					$\Delta = 0.090$ eV	

partially oxidized compound $Li_{1-x}Pt(mnt)_2 \cdot 2H_2O$, where $x = 0.25$, has recently been prepared and found to exhibit a metallike conductivity with a room temperature value of about 200 Ω^{-1} cm^{-1}, which is reduced by dehydration or aging.[120] The temperature dependence of the conductivity is metallike near room temperature (Figure 47). The preliminary observation of high conductivity in the Ni analogue led to the suggestion that the conductivity is at least in part ligand-based, and this is supported by the observation of a large metal–metal distance, 3.65 Å, in the Pt compound.[121] The absence of an observable EPR signal is reminiscent of the integrally oxidized, stacked $M(mnt)_2^-$ systems (Section 3.3).

The successful preparation of porphyrinic molecular conductors has recently led to initial studies of analogous metallomacrocycles with the $M(N_4)$ metal–ligand core, the dihydrodibenzotetraaza[14]annulenes, and related complexes.[172–175] These metal–organic compounds can be viewed as variants on the $M(opd)_2$ compounds, in which *cis*-nitrogen atoms are linked by unsaturated carbon chains, but can also be viewed as variants on

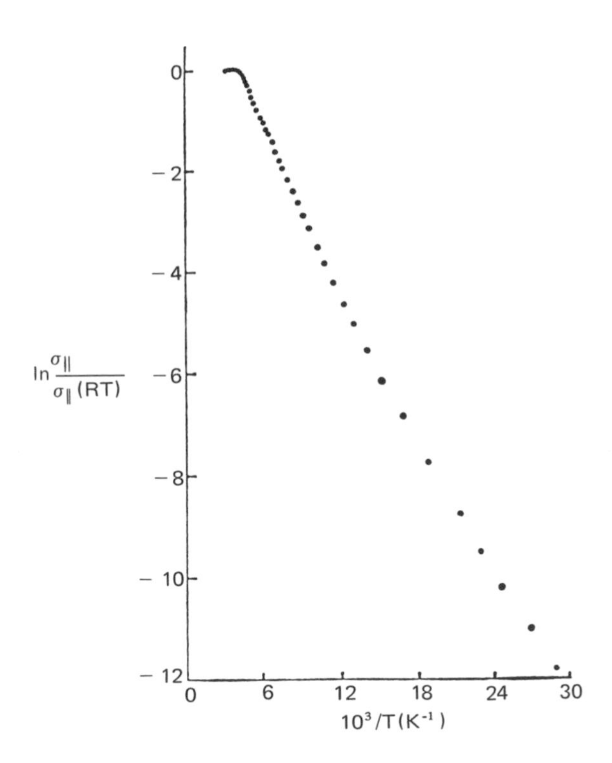

FIGURE 47. Variations of conductivity, $\ln[\sigma/\sigma(RT)]$ with inverse temperature for a single crystal of $Li_{1-x}[Pt(mnt)_2] \cdot 2H_2O$ ($x = \sim 0.25$) (Reference 120).

the M(tatma) compounds in which benzene rings, rather than methyl groups, have been fused on the tetraaza[14]annulene core. An x-ray structural study of the unoxidized complex, Ni(dbtaa) shows it to be essentially planar and packed in slipped stacks (Figure 2b).[175] The Ni–Ni distance in adjacent molecules is 5.228 Å, but the interplanar separations between crystallographically inequivalent molecules are 3.235 and 3.295 Å. There are two crystallographic forms of Pd(dbtaa). One, designated the α form, has the same unit cell parameters as Ni(dbtaa) and is assumed to be isostructural. A second form, β-Pd(dbtaa) exhibits distinctly different packing, with the molecules arranged in pairs having a Pd–Pd distance of 4.328(1) Å.

Iodine oxidation of $Ni(opd)_2$ itself is reported[173] to give rise to $Ni(opd)_2I_x$. Chemical analyses combined with powder x-ray diffraction studies indicate the existence of stoichiometries, $x=0.97$, 2.57, and 5.97. Unfortunately the nature of the iodine species, and therefore the ionicity of the metallomacrocycle in these materials, has not been determined. Powder conductivities were low, the highest being only $4 \times 10^{-5}\ \Omega^{-1}\ cm^{-1}$ for $x=2.57$.

In contrast to the $Ni(opd)_2$ system, halogen oxidation of the closed ring system Ni(dbtaa) and its ring-substituted variants, yields materials with good conductivities.[172, 174, 175] Powder samples of $Ni(dbtaa)Br_{2.43}$ have been prepared, as have $Ni(dbtaa)I_x$ with several stoichiometries in the range $0.8 < x < 7$, although the nature of the iodine-rich materials is open to question. Powders of $Ni(tmdbtaa)I_x$ with $x=1.7$ and 2.9, and powders of $M(dbtaa)I_x$ and $M(tmdbtaa)I_x$ or both have also been reported for $M = H_2$, Co, Cu, Pd, and Pt. In addition, single crystals of $Ni(dbtaa)I_{1.8}$ and $Ni(tmdbtaa)I_{2.44}$ have also been isolated.[172] Resonance Raman spectra indicate that $M(L)I_x$ compounds with $x \sim 3$ contain I_3^-, while for $x > 3$, I_5^-, is the predominant anion. Preliminary x-ray diffraction studies indicate the presence of the structural pattern of Figures 2a and 25, with M(L) stacks surrounded by iodine chains. Preliminary EPR studies suggest that the Ni(L) oxidation is ligand based.

These materials show powder conductivities as high as $\sigma \sim 0.5\ \Omega^{-1}\ cm^{-1}$ [$Ni(dbtaa)I_{1.10}$] and the room temperature conductivities for single crystals of $Ni(dbtaa)I_{1.8}$ and $Ni(tmdbtaa)I_{2.44}$ are $\sim 10\ \Omega^{-1}\ cm^{-1}$ and metallike over a narrow region.[172] It was suggested that these values are limited by crystal perfection. However, since this is also true for most of the porphyrinic $M(L)I_x$ complexes as well, it appears probable that these analogues exhibit conductivities somewhat lower ($\sim$tenfold) than those of the planar porphyrinic conductors, but comparable with those of $Ni(omtbp)I_x$. Thorough characterization of these analogues should permit instructive comparisons between them and the porphyrinic $M(L)I_x$ complexes.

5. *Future Directions*

We have reviewed past and current studies of DA crystals prepared from planar transition-metal complexes. What directions will such research take in the near future? We hazard some guesses.

The use of planar transition-metal complexes to prepare integrated stack DA crystals is shown in Section 2 to lead to a richer array of structural arrangements than that exhibited by the corresponding π-organic DA crystals.[65] However, the physical properties of these new materials, with some notable exceptions, have been much less thoroughly explored. In particular, these crystals are typically described as being either neutral or ionic, whereas the organic DA crystals in fact exhibit intermediate ionicities[15] and can undergo phase transitions from "largely ionic" to "largely neutral."[213] Future studies may uncover an analogous array of physically interesting properties of integrated stack DA crystals containing metal complexes.

The segregated stack, integral oxidation state crystals discussed in Section 3 in general do not display the interesting physical properties observed in many of the other compounds described in this review. These systems will be of little further interest unless new structural motifs are discovered.

It is very clear that the major efforts of the near future will focus on the preparation and characterization of new segregated stack, nonintegral oxidation state systems, such as those discussed in Section 4. Far less effort has gone into the study of these systems than of organic crystals, yet we already have learned that the metal-containing systems exhibit most of the features of importance and interest observed in the organic counterparts. Moreover, the exploitation of the additional degree of freedom offered by the metal systems—the ability to fine-tune the relative energy of metal and ligand orbitals—has only begun. Work to date shows that the promise offered by the chemical flexibility of these systems is not illusory. We expect a rapid growth in the preparation, characterization, and utilization of this class of materials.

ACKNOWLEDGMENTS

This work has been supported under the NSF–MRL program through the Materials Research Center of Northwestern University (Grant No. DMR79-23573) and by National Science Foundation Grant Nos. DMR77-26409 to BMH and CHE80-09671 to JAI. LJP acknowledges receipt of an NSF Graduate Fellowship. We thank Ms. Jan Goranson and Ms. Carol Lewis for their excellent typing of this review.

References

1. Interrante, L. V., Bray, J. W., Hart, H. R., Jr., Jacobs, I. S., Kasper, J. S., Piacente, P. A., and Bonner, J. C., *Quasi One-Dimensional Conductors II*, Lecture Notes in Physics, No. 96, Springer, New York (1979), pp. 55–68.
2. Schramm, C. J., Scaringe, R. P., Stojakovic, D. R., Hoffman, B. M., Ibers, J. A., and Marks, T. J., *J. Am. Chem. Soc.* **102**, 6702 (1980).
3. Schramm, C. J., Stojakovic, D. R., Hoffman, B. M., and Marks, T. J., *Science* **200**, 47 (1978).
4. Cowie, M., Gleizes, A., Grynkewich, G. W., Kalina, D. W., McClure, M. S., Scaringe, R. P., Teitelbaum, R. C., Ruby, S. L., Ibers, J. A., Kannewurf, C. R., and Marks, T. J., *J. Am. Chem. Soc.* **101**, 2921 (1979).
5. Miller, J. S., and Epstein, A. J., *Prog. Inorg. Chem.* **20**, 1 (1976).
6. Williams, J. M., Schultz, A., Underhill, A. E., and Carnerio, K., in *Extended Linear Chain Compounds,* Vol. 1, J. Miller, Ed., Plenum Press, New York (1981), p. 73; Underhill, A. E., Watkins, D. M., Williams, J. M., and Carnerio, K., in *Extended Linear Chain Compounds,* Vol. 1, J. Miller, Ed., Plenum Press, New York (1981), p. 119.
7. Keller, H. J., Ed., *Low-Dimensional Cooperative Phenomena,* Plenum Press, New York (1975).
8. Miller, J. S., and Epstein, A. J., Eds., *Synthesis and Properties of Low-Dimensional Materials, Ann. N.Y. Acad. Sci.* **313** (1978).
9. Keller, H. J., Ed., *Chemistry and Physics of One-Dimensional Metals,* Plenum Press, New York (1977).
10. Hatfield, W. E., Ed., *Molecular Metals,* Plenum Press, New York (1979).
11. Devreese, J. T., Evrard, R. P., and Van Doren, V. E., Eds., *Highly Conducting One-Dimensional Solids,* Plenum Press, New York (1979).
12. Berlinsky, A. J., *Contemp. Phys.* **17**, 331 (1976).
13. Barišić, S., Bjeliš, A., Cooper, J. R., and Leontić, B., Eds., *Quasi One-Dimensional Conductors I*, Lecture Notes in Physics, No. 95, Springer, New York (1979).
14. Barišić, S., Bjeliš, A., Cooper, J. R., and Leontic, B., Eds., *Quasi One-Dimensional Conductors II*, Lecture Notes in Physics, No. 96, Springer, New York (1979).
15. Soos, Z. G., *Ann. Rev. Phys. Chem.* **25**, 121 (1974).
16. Soos, Z. G., and Klein, D. J., in *Molecular Association,* Vol. 1, R. Foster, Ed., Academic Press, New York (1979), pp. 1–119.
17. Williams, R. M., and Wallwork, S. C., *Acta Crystallogr.* **23**, 448 (1967).
18. Torrance, J. B., in *Synthesis and Properties of Low-Dimensional Materials, Ann. N.Y. Acad. Sci.* **313** (1978), pp. 210–233.
19. Torrance, J. B., *Acc. Chem. Res.* **12**, 79 (1979).
20. Gutman, F., and Lyons, L. E., *Organic Semiconductors,* John Wiley & Sons, New York (1967).
21. Mott, N. F., and Davis, E. A., *Electronic Processes in Non-Crystalline Materials,* Clarendon Press, Oxford (1971).
22. Beni, G., Holstein, T., and Dincus, P., *Phys. Rev. B* **8**, 312 (1973).
23. Klein, D. J., *Phys. Rev. B* **8**, 3452 (1973).
24. Bailey, A. S., Williams, R. J. P., and Wright, J. D., *J. Chem. Soc.,* 2579 (1965).
25. Koizumi, S., and Iida, Y., *Bull. Chem. Soc. Jpn.* **46**, 629 (1973).
26. Prout, C. K., and Powell, H. M., *J. Chem. Soc.,* 4882 (1965).
27. Koizumi, S., and Iida, Y., *Bull. Chem. Soc. Jpn.* **44**, 1436 (1970).
28. Iida, Y., *Bull. Chem. Soc. Jpn.* **44**, 2564 (1971).
29. Kamenar, B., Prout, C. K., and Wright, J. D., *J. Chem. Soc. A* 661 (1966).
30. Kamenar, B., Prout, C. K., and Wright, J. D., *J. Chem. Soc.,* 4851 (1965).

31. Bailey, A. S., and Prout, C. K., *J. Chem. Soc.*, 4867 (1965).
32. Murray-Rust, P., and Wright, J. D., *J. Chem. Soc. A* 247 (1968).
33. Schmitt, R. D., Wing, R. M., and Maki, A. H., *J. Am. Chem. Soc.* **91**, 4394 (1969).
34. Wheland, R. C., and Gillson, J. L., *J. Am. Chem. Soc.* **98**, 3916 (1976).
35. Keller, H. J., Leichert, I., Mégnamisi-Bélombé, M., Nöthe, D., and Weiss, J., *Z. Anorg. Allg. Chem.* **429**, 231 (1977).
36. Keller, H. J., Leichert, I., Mégnamisi-Bélombé, M., Moroni, W., Nöthe, D., and Weiss, J., *Mol. Cryst. Liq. Cryst.* **32**, 155 (1976).
37. Mayerle, J. J., *Inorg. Chem.* **16**, 916 (1977).
38. King, A. W., Swann, D. A., and Waters, T. N., *J. Chem. Soc. Dalton Trans.*, 1819 (1973).
39. Castellano, E. E., Hodder, O. J. R., Prout, C. K., and Sadler, P. J., *J. Chem. Soc. A*, 2620 (1971).
40. Cassoux, P., and Gleizes, A, *Inorg. Chem.* **19**, 665 (1980).
41. Matsumoto, N., Nonaka, Y., Kida, S., Kawano, S., and Ueda, I., *Inorg. Chim. Acta* **37**, 27 (1979).
42. Grigg, R., Trocha-Grimshaw, J., and King, T. J., *J. Chem. Soc. Chem. Commun.*, 571 (1978).
43. Piovesana, O., Bellitto, C., Flamini, A., and Zanazzi, P. F., *Inorg. Chem.* **18**, 2258 (1979).
44. Pace, L. J., Ulman, A., and Ibers, J. A., *Inorg. Chem.*, **21**, 199 (1982).
45. Jacobs, I. S., Hart, H. R., Jr., Interrante, L. V., Bray, J. W., Kasper, J. S., Watkins, G. D., Prober, D. E., Wolf, W. P., and Bonner, J. C. *Physica B + C* **86–88**, 655 (1977).
46. Bray, J. W., Hart, H. R., Jr., Interrante, L. V., Jacobs, I. S., Kasper, J. S., Watkins, G. D., Wee, S. H., and Bonner, J. C., *Phys. Rev. Lett.* **35**, 744 (1975).
47. Kasper, J. S., and Interrante, L. V., *Acta Crystallogr.* **B32**, 2914 (1976).
48. Jacobs, I. S., Bray, J. W., Hart, H. R., Jr., Interrante, L. V., Kasper, J. S., Watkins, G. D., Prober, D. E., and Bonner, J. C., *Phys. Rev. B* **14**, 3036 (1976).
49. Delker, G., Ph.D. thesis, University of Illinois, 1979.
50. Bonner, J. C., Wei, T. S., Hart, H. R., Jr., Interrante, L. V., Jacobs, I. S., Kasper, J. S., Watkins, G. D., and Blôte, H. W. J., *J. Appl. Phys.* **49**, 1321 (1978).
51. Kasper, J. S., and Moncton, D. E., *Phys. Rev. B* **20**, 2341 (1979).
52. Moncton, D. E., Birgeneau, R. J., Interrante, L. V., and Wudl, F., *Phys. Rev. Lett.* **39**, 507 (1977).
53. Ehrenfreund, E., and Smith, L. S., *Phys. Rev. B* **16**, 1870 (1977).
54. Smith, L. S., Ehrenfreund, E., Heeger, A. J., Interrante, L. V., Bray, J. W., Hart, H. R., Jr., and Jacobs, I. S., *Solid State Commun.* **19**, 377 (1976).
55. Wei, T. S., Heeger, A. J., Salamon, M. B., and Delker, G. E., *Solid State Commun.* **21**, 595 (1977).
56. Geiger, W. E., Jr., and Maki, A. H., *J. Phys. Chem.* **75**, 2387 (1971).
57. Singhabhandhu, A., Robinson, P. D., Fang, J. H., and Geiger, W. E., Jr., *Inorg. Chem.* **14**, 318 (1975).
58. Vu Dong, Endres, H., Keller, H. J., Moroni, W., and Nöthe, D., *Acta Crystallogr.* **B33**, 2428 (1977).
59. Endres, H., Keller, H. J.; Moroni, W., and Nöthe, D., *Acta Crystallogr.* **B35**, 353 (1979).
60. Rundle, R. E., *J. Am. Chem. Soc.*, **76**, 3101 (1954).
61. Yamada, S., and Tsuchida, R., *Bull. Chem. Soc. Jpn.* **30**, 715 (1957).
62. Hara, Y., Shirotani, I., and Onodera, A., *Solid State Commun.* **19**, 171 (1976).
63. Wing, R. M., and Schlupp, R. L., *Inorg. Chem.* **9**, 471 (1970).
64. Jacobs, I. S., Interrante, L. V., and Hart, H. R., Jr. *A.I.P. Conf. Proc.* **24**, 355 (1975).
65. Herbstein, F. H., *Perspect. Struct. Chem.* **4**, 166 (1971).
66. Chappell, J. S., Bloch, A. N., Bryden, W. A., Maxfield, M., Poehler, T. O., and Cowan, D. O., *J. Am. Chem. Soc.* **103**, 2442 (1981).

67. Reis, A. H., Jr., Preston, L. D., Williams, J. M., Peterson, S. W., Candela, G. A., Swartzendruber, L. J., and Miller, J. S., *J. Am. Chem. Soc.* **101**, 2756 (1979).
68. Mayoh, B., and Prout, C. K., *J. Chem. Soc. Faraday Trans.* **2**, 1072 (1972).
69. Mazumdar, S., and Soos, Z. G., *Synth. Met.* **1**, 77 (1979/80).
70. Soos, Z. G., and Mazumdar, S., *Phys. Rev. B* **18**, 1991 (1978).
71. Bonner, J. C., and Fisher, M. E., *Phys. Rev.* **135**, A640 (1964).
72. McConnell, H. M., and Lynden-Bell, R., *J. Chem. Phys.* **36**, 2393 (1962).
73. Chestnut, D. B., *J. Chem. Phys.* **45**, 4677 (1966).
74. Pincus, P., *Solid State Commun.* **9**, 1971 (1971).
75. Beni, G., and Pincus, P., *J. Chem. Phys.*, **57**, 3531 (1972).
76. Beni, G., *J. Chem. Phys.* **58**, 3200 (1973).
77. Dubois, J. Y., and Carton, J. P., *J. Phys.* (*Paris*) **35**, 371 (1974).
78. Luther, A., and Peschel, I., *Phys. Rev. B* **12**, 3908 (1975).
79. Pytte, E., *Phys. Rev. B* **10**, 4637 (1974).
80. Yamada, S., *Bull. Chem. Soc. Jpn.* **24**, 125 (1951).
81. Eisenberg, R., *Prog. Inorg. Chem.* **12**, 295 (1970).
82. Gray, H. B., *Trans. Met. Chem.* **1**, 239 (1965).
83. Livingstone, S. E., *Q. Rev. Chem. Soc.* **19**, 386 (1965).
84. McCleverty, J. A., *Prog. Inorg. Chem.* **10**, 49 (1968).
85. Hove, M. J., Hoffman, B. M., and Ibers, J. A., *J. Chem. Phys.* **56**, 3490 (1972).
86. Endres, H., Keller, H. J., Moroni, W., Nöthe, D., and Vu Dong, *Acta Crystallogr.* **B34**, 1823 (1978).
87. Kasper, J. S., Interrante, L. V., and Secaur, C. A., *J. Am. Chem. Soc.* **97**, 890 (1975).
88. Interrante, L. V., Browall, K. W., Hart, H. R., Jr., Jacobs, I. S., Watkins, G. D., and Wee, S. H., *J. Am. Chem. Soc.* **97**, 889 (1975).
89. Interrante, L. V., Bray, J. W., Hart, H. R., Jr., Kasper, J. S., Piacente, P. A., and Watkins, G. D., *J. Am. Chem. Soc.* **99**, 3523 (1977).
90. Interrante, L. V., Bray, J. W., Hart, H. R., Jr., Kasper, J. S., Piacente, P. A., and Watkins, G. D., in *Synthesis and Properties of Low-Dimensional Materials, Ann. N.Y. Acad. Sci.* **313** (1978), pp. 407–416.
91. Bray, J. W., Hart, H. R., Jr., Interrante, L. V., Jacobs, I. S., Kasper, J. S., Piacente, P. A., and Watkins, G. D., *Phys. Rev. B* **16**, 1359 (1977).
92. Forrester, J. D., Zalkin, A., and Templeton, D. H., *Inorg. Chem.* **3**, 1507 (1964).
93. Davison, A., Edelstein, N., Holm, R. H., and Maki, A. H., *Inorg. Chem.* **2**, 1227 (1963).
94. Maki, A. H., Edelstein, N., Davison, A., and Holm, R. H., *J. Am. Chem. Soc.* **86**, 4580 (1964).
95. Schmitt, R. D., and Maki, A. H., *J. A. Chem. Soc.* **90**, 2288 (1968).
96. Hamilton, W. C., and Bernal, I., *Inorg. Chem.* **6**, 2003 (1967).
97. Weiher, J. F., Melby, L. R., and Benson, R. E., *J. Am. Chem. Soc.* **86**, 4329 (1964).
98. Williams, R., Billig, E., Waters, J. H., and Gray, H. B., *J. Am. Chem. Soc.* **88**, 43 (1966).
99. Kobayashi, A., and Yukiyoshi, S., *Bull. Chem. Soc. Jpn.* **50**, 2650 (1977).
100. Perez-Albuerne, N. A., Isett, L. C., and Haller, R. K., *J. Chem. Soc. Chem. Commun.*, 417 (1977).
101. Davison, A., Edelstein, N., Holm, R. H., and Maki, A. H., *J. Am. Chem. Soc.* **85**, 2029 (1963).
102. Fritchie, C. J., Jr., *Acta Crystallogr.* **20**, 107 (1966).
103. Baker-Hawkes, M. J., Billig., E., and Gray, H. B., *J. Am. Chem. Soc.* **88**, 4870 (1966).
104. Baker-Hawkes, M. J., Dori, Z., Eisenberg, R., and Gray, H. B., *J. Am. Chem. Soc.* **90**, 4253 (1968).
105. Enemark, J. H., and Ibers, J. A., *Inorg. Chem.* **7**, 2636 (1968).
106. Lauher, J. W., and Ibers, J. A., *Inorg. Chem.* **14**, 640 (1975).

107. Cassoux, P., Interrante, L., and Kasper, J., *Mol. Cryst. Liq. Cryst.* **81**, 293 (1982).
108. Bekaroglu, Ö., El Sharif, M., Endres, H., and Keller, H. J., *Acta Crystallogr.* **B32**, 2983 (1976).
109. Beurskens, P. T., Blaauw, H. J. A., Cras, J. A., and Steggerda, J. J., *Inorg. Chem.* **7**, 805 (1968).
110. Beurskens, P. T., Cras, J. A., and van der Linden, J. G. M., *Inorg. Chem.* **9**, 475 (1970).
111. Alcácer, L., and Maki, A. H., *J. Phys. Chem.* **78**, 215 (1974).
112. Alcácer, L., and Maki, A. H., *J. Phys. Chem.* **80**, 1912 (1976).
113. Alcácer, L., Novais, H., and Pedroso, F., in *Molecular Metals,* W. E. Hatfield, Ed., Plenum Press, New York (1979), pp. 415–418.
114. Plumlee, K. W., Hoffman, B. M., Ibers, J. A., and Soos, Z. G., *J. Chem. Phys.* **63**, 1926 (1975).
115. Plumlee, K. W., Ph.D. thesis, Northwestern University, 1975.
116. Plumlee, K. W., Hoffman, B. M., Ratajack, M. T., and Kannewurf, C. R., *Solid State Commun.* **15**, 1651 (1974).
117. Davison, A., Howe, D. V., and Shawl, E. T., *Inorg. Chem.* **6**, 458 (1967).
118. Hoffman, B. M., unpublished results.
119. Senftleber, F. C., and Geiger, W. E., Jr., *Inorg. Chem.* **17**, 3615 (1978).
120. Underhill, A. E., and Ahmad, M. M., *J. Chem. Soc. Chem. Commun.*, 67 (1981).
121. Underhill, A. E., and Ahmad, M. M., *Mol. Cryst. Liq. Cryst.* **81**, 223 (1982).
122. Petersen, J. L., Schramm, C. J., Stojakovic, D. R., Hoffman, B. M., and Marks, T. J., *J. Am. Chem. Soc.* **99**, 286 (1977).
123. Hoffman, B. M., Phillips, T. E., Schramm, C. J., and Wright, S. K., in *Molecular Metals,* W. E. Hatfield, Ed., Plenum Press, New York (1979), pp. 393–398.
124. Phillips, T. E., and Hoffman, B. M., *J. Am. Chem. Soc.* **99**, 7734 (1977).
125. Phillips, T. E., Scaringe, R. P., Hoffman, B. M., and Ibers, J. A., *J. Am. Chem. Soc.* **102**, 3435 (1980).
126. Hoffman, B. M., Phillips, T. E., and Soos, Z. G. *Solid State Commun.* **33**, 51 (1980).
127. Schramm, C. J., Ph.D. thesis, Northwestern University, 1979.
128. Euler, W. B., Martinsen, J., Pace, L. J., Hoffman, B. M., and Ibers, J. A., *Mol. Cryst. Liq. Cryst.* **81**, 231 (1982).
129. Nohr, R. S., Kuznesof, P. M., Wynne, K. J., Kenney, M. E., and Siebenman, P. G., *J. Am. Chem. Soc.* **103**, 4371 (1981).
130. Wynne, K., and Nohr, R. S., *Mol. Cryst. Liq. Cryst.* **81**, 243 (1982).
131. Martinsen, J., Schramm, C. J., Euler, W. B., Pace, L. J., Hoffman, B. M., and Ibers, J. A., manuscript in preparation.
132. Martinsen, J., Pace, L. J., Phillips, T. E., Hoffman, B. M., and Ibers, J. A., *J. Am. Chem. Soc.* **104**, 83 (1982).
133. Euler, W. B., Martinsen, J., Pace, L. J., Hoffman, B. M., and Ibers, J. A., manuscript in preparation.
134. Wright, S. K., Schramm, C. J., Phillips, T. E., Scholler, D. M., and Hoffman, B. M., *Synth. Met.* **1**, 43 (1979/80).
135. Pace, L. J., Martinsen, J., Ulman, A., Hoffman, B. M., and Ibers, J. A., manuscript in preparation.
136. Schoch, K. F., Jr., Kundalkar, B. R., and Marks, T. J., *J. Am. Chem. Soc.* **101**, 7071 (1979).
137. Schneider, O., Metz, J., and Hanack, M., *Mol. Cryst. Liq. Cryst.* **81**, 273 (1982).
138. Hanack, M., Mitulla, K., Pawlowski, G., and Subramanian, L. R., *J. Organomet. Chem.* **204**, 315 (1981).
139. Smith, K. M., Ed., *Porphyrins and Metalloporphyrins,* American Elsevier, New York (1975).
140. Dolphin, D., Ed., *The Porphyrins,* Academic Press, New York (1978).

141. Lever, A. B. P., *Adv. Inorg. Chem. Radiochem.* **7**, 27 (1965).
142. Jackson, A. H., *The Porphyrins,* Vol. I, Academic Press, New York (1978), pp. 365–386.
143. Ulman, A., Gallucci, J., Fisher, D., and Ibers, J. A., *J. Am. Chem. Soc.* **102**, 6852 (1980).
144. Schramm, C. J., and Hoffman, B. M., *Inorg. Chem.* **19**, 383 (1980).
145. Calleri, M., Ferraris, G., and Viterbo, D., *Inorg. Chim. Acta* **1**, 297 (1967).
146. Calleri, M., Ferraris, G., and Viterbo, D., *Acta Crystallogr.* **22**, 468 (1967).
147. Murmann, R. K., and Schlemper, E. O., *Acta Crystallogr.* **23**, 667 (1967).
148. Ferraris, G., and Viterbo, D., *Acta Crystallogr.* **B25**, 2066 (1969).
149. Godycki, L. E., and Rundle, R. E., *Acta Crystallogr.* **6**, 487 (1953).
150. Williams, D. E., Wohlauer, G., and Rundle, R. E., *J. Am. Chem. Soc.* **81**, 755 (1959).
151. Gomm, P. S., Thomas, T. W., and Underhill, A. E., *J. Chem. Soc. A,* 2154 (1971).
152. Bowers, R. H., Banks, C. V., and Jacobson, R. A., *Acta Crystallogr.* **B28**, 2318 (1972).
153. Frasson E., and Panattoni, C., *Acta Crystallogr.* **13**, 893 (1960).
154. Frasson, E., Panattoni, C., and Zannetti, R., *Acta Crystallogr.* **12**, 1027 (1959).
155. Endres, H., Keller, H. J., van de Sand, H., and Vu Dong, *Z. Naturforsch.* **33b**, 843 (1978).
156. Leichert, I., and Weiss, J., *Acta Crystallogr.* **B31**, 2877 (1975).
157. Brill, J. W., Mégnamisi-Bélombé, M., and Novotny, M., *J. Chem. Phys.* **68**, 585 (1978).
158. Leichert, I., and Weiss, J., *Acta Crystallogr.* **B31**, 2709 (1975).
159. Endres, H., Mégnamisi-Bélombé, M., Little, W. A., and Wolfe, C. R., *Acta Crystallogr.* **B35**, 169 (1979).
160. Mégnamisi-Bélombé, M., *J. Solid State Chem.* **27**, 389 (1979).
161. Mégnamisi-Bélombé, M., *J. Solid State Chem.* **22**, 151 (1977).
162. Brown, L. D., Kalina, D. W., McClure, M. S., Schultz, S., Ruby, S. L., Ibers, J. A., Kannewurf, C. R., and Marks, T. J., *J. Am. Chem. Soc.* **101**, 2937 (1979).
163. Endres, H., Keller, H. J., Moroni, W., and Weiss, J., *Acta Crystallogr.* **B31**, 2357 (1975).
164. Endres, H., Keller, H. J., Mégnamisi-Bélombé, M., Moroni, W., Pritzkow, H., Weiss, J., and Comès, R., *Acta Crystallogr.* **A32**, 954 (1976).
165. Endres, H., Mégnamisi-Bélombé, M., Keller, H. J., and Weiss, J., *Acta Crystallogr.* **B32**, 457 (1976).
166. Pritzkow, H., *Z. Naturforsch.* **31b**, 401 (1976).
167. Foust, A. S., and Soderberg, R. H., *J. Am. Chem. Soc.* **89**, 5507 (1967).
168. Endres, H., Keller, H. J., Lehmann, R., and Weiss, J., *Acta Crystallogr.* **B32**, 627 (1976).
169. Kalina, D. W., Lyding, J. W., Ratajack, M. T., Kannewurf, C. R., and Marks, T. J., *J. Am. Chem. Soc.* **102**, 7854 (1980).
170. Underhill, A. E., Watkins, D. M., and Pethig, R., *Inorg. Nucl. Chem. Lett.* **9**, 1269 (1973).
171. Edelman, L. E., *J. Am. Chem. Soc.* **72**, 5765 (1950).
172. Lin, L.-S., Marks, T. J., Kannewurf, C. R., Lyding, J. W., McClure, M. S., Ratajack, M. T., and Whang, T.-C., *J. Chem. Soc. Chem. Commun.,* 954 (1980).
173. Wuu, Y.-M., and Peng, S.-M., *J. Inorg. Nucl. Chem.* **42**, 205 (1980).
174. Wuu, Y.-M., Peng, S.-M., and Chang, H., *J. Inorg. Nucl. Chem.* **42**, 839 (1980).
175. Hatfield, W. E., private communication.
176. Scaringe, R. P., and Ibers, J. A., *Acta Crystallogr.* **A35**, 803 (1979).
177. Teitelbaum, R. C., Ruby, S. L., and Marks, T. J., *J. Am. Chem. Soc.* **100**, 3215 (1978).
178. Teitelbaum, R. C., Ruby, S. L., and Marks, T. J., *J. Am. Chem. Soc.* **102**, 3322 (1980).
179. Mertes, K. B., and Ferraro, J. R., *J. Chem. Phys.* **70**, 646 (1979).
180. Jircitano, A. J., Timken, M. D., Mertes, K. B., and Ferraro, J. P., *J. Am. Chem. Soc.* **101**, 7661 (1979).
181. Jircitano, A. J., Colton, M. C., and Mertes, K. B., *Inorg. Chem.* **20**, 890 (1981).
182. Wolberg, A., and Manassen, J., *J. Am. Chem. Soc.* **92**, 2982 (1970).
183. Lever, A. B. P., and Wiltshire, J. P., *Can. J. Chem.* **54**, 2514 (1976).
184. Dolphin, D., Ed., *The Porphyrins*, Vol. V, Academic Press, New York (1978).

185. Bobrovskii, A. P., and Sidorov, A. N., *J. Struct. Chem.* (Engl. Transl.) **17**, 50 (1976).
186. Tomkiewicz Y., Taranko, A. R., and Torrance, J. B., *Phys. Rev. Lett.* **36**, 751 (1976).
187. Mégnamisi-Bélombé, M., *Z. Anorg. Allg. Chem.* **473**, 196 (1981).
188. Kubel, F., and Strähle, J., *Z. Naturforsch.* **36b**, 441 (1981).
189. Endres, H., Keller, H. J., Lehmann, R., Poveda, A., Rupp, H. H., and van de Sand, H., *Z. Naturforsch.* **32b**, 516 (1977).
190. Egneus, B., *Anal. Chim. Acta* **48**, 291 (1969).
191. Mehne, L. F., and Wayland, B. B., *Inorg. Chem.* **14**, 881 (1975).
192. Reddy, J. W., Knox, K., and Robin, M. B., *J. Chem. Phys.* **40**, 1082 (1964).
193. Miller, J. S., *Inorg. Chem.* **16**, 957 (1977).
194. Cullen, D. L., and Meyer, E. F., Jr., *J. Am. Chem. Soc.*, **96**, 2095 (1974).
195. Meyer, E. F., Jr., *Acta Crystallogr.* **B28**, 2162 (1972).
196. Robertson, J. M., and Woodward, I., *J. Chem. Soc.*, 219 (1937).
197. Mason, R., Williams, G. A., and Fielding, P. E., *J. Chem. Soc. Dalton Trans.*, 676 (1979).
198. Robertson, J. M., *J. Chem. Soc.*, 1195 (1936).
199. Scheidt, W. R., and Dow, W., *J. Am. Chem. Soc.* **99**, 1101 (1977).
200. Brown, C. J., *J. Chem. Soc. A*, 2488 (1968).
201. Robertson, J. M., and Woodward, I., *J. Chem. Soc.*, 36 (1940).
202. Brown, C. J., *J. Chem. Soc. A*, 2494 (1968).
203. Ukei, K., *Acta Crystallogr.* **B29**, 2290 (1973).
204. Gouterman, M., in *The Porphyrins*, Vol. III, D. Dolphin, Ed., Academic Press, New York (1978), pp. 1–165.
205. Euler, W. B., and Hoffman, B. M., manuscript in preparation.
206. Hoard, J. L., in *Porphyrins and Metalloporphyrins*, K. M. Smith, Ed., American Elsevier, New York (1978), pp. 321–327.
207. Hilti, B., and Mayer, C. W., *Helv. Chim. Acta* **61**, 501 (1978).
208. Bloch, A. N., Cowan, D. O., Beckgaard, K., Pyle, R. E., Banks, R. H., and Poehler, T. O., *Phys. Rev. Lett.* **34**, 1561 (1975).
209. Thomas, G. A., Schafer, D. E., Wudl, F., Horn, P. M., Rimai, D., Cook, J. W., Glocker, D. A., Skove, M. J., Chu, C. W., Groff, R. P., Gillson, J. L., Wheland, R. C., Melby, L. R., Salamon, M. B., Craven, R. A., De Pasquali, G., Bloch, A. N., Cowan, D. O., Walatka, V. V., Pyle, R. E., Gemmer, R., Poehler, T. O., Johnson, G. R., Miles, M. G., Wilson, J. D., Ferraris, J. P., Finnegan, T. F., Warmack, R. J., Raaen, V. F., and Jerome, D., *Phys. Rev. B* **13**, 5105 (1976).
210. Esposito, J. N., Sutton, L. E., and Kenney, M. E., *Inorg. Chem.* **6**, 1116 (1967).
211. Meyer, G., Hartmann, M., Wöhrle, D., *Makromol. Chem.*, **176**, 1919 (1975).
212. Delhaes, P., Coulon, C., Amiell, J., Flandrois, S., Toreilles, E., Fabre, J. M., and Giral, L., *Mol. Cryst. Liq. Cryst.* **50**, 43 (1979).
213. Batail, P., La Placa, S. J., Mayerle, J. J., and Torrance, J. B., *J. Am. Chem. Soc.* **103**, 951 (1981).

Index